H.-P. Blume, H. Eger, E. Fleischhauer,
A. Hebel, C. Reij, K. G. Steiner (Editors)

Towards Sustainable Land Use

Furthering Cooperation Between People And Institutions

VOLUME I

ADVANCES IN GEOECOLOGY 31

 A Cooperating Series of the International Society of Soil Science (ISSS)

ISBN 3-923381-42-5

Die Deutsche Bibliothek - CIP Einheitsaufnahme

Towards sustainable land use : furthering cooperation between people and institutions / H.-P. Blume ... (ed.).
- Reiskirchen : Catena-Verl..
(Selected papers of the ... conference of the International Soil Conservation Organisation (ISCO) ; 9)
(Advances in geoecology ; 31)
ISBN 3-923381-42-5
International Soil Conservation Organization:
Selected papers of the ... conference of the International Soil Conservation Organisation
(ISCO). - Reiskirchen : Catena-Verl.
(Advances in geoecology ; ...)
Vol. 1. - (1998)
Vol. 2. - (1998)

Managing Editor "Advances in GeoEcology":
Margot Rohdenburg

The English language editing of this volume "Advances in GeoEcology 31"
has been carried out by:
Simon Berkowicz, Jerusalem

Sponsors
German Federal Ministry for the Environment, Nature Conservation and Nuclear Safety
German Federal Ministry for Economic Cooperation and Development
Deutsche Gesellschaft für Technische Zusammenarbeit (GTZ) GmbH
Deutsche Bundesstiftung Umwelt
Technical Centre for Agriculture and Rural Co-operation (CTA)
German Foundation for International Development (DSE)
Commission of the European Union
Swiss Development Cooperation
German Scientific Foundation
Food and Agriculture Organization of the United Nations
Kunst- und Ausstellungshalle der Bundesrepublik Deutschland
Stadtwerke Bonn
Verwaltung des Deutschen Bundestages
Rheinbraun AG
Misereor
Brot für die Welt
Deutsche Lufthansa AG

© Copyright 1998 by CATENA VERLAG GMBH, 35447 Reiskirchen, Germany
All rights are reserved. No part of this publication may be reproduced, stored in a retrieval system or transmitted in any form or by any means, electronic, mechanical, photocopying, recording or otherwise, without prior permission of the publisher.
This publication has been registered with the Copyright Clearance Center, Inc.
Submission of an article for publication implies the transfer of the copyright from the author(s) to the publisher.

ISBN 3-923381-42-5

Contents Volume I

F. Holzwarth & H.-P. Blume
Preface

Policies for sustainable land use

R. Süßmuth
Protecting the soil and preserving it for future generations — I

A. Merkel
Striving for sustainable land use - A challenge for all countries - — III

C.-D. Spranger
Food security and poverty alleviation
Essentials for sustainable land use and development - — VIII

E. Dowdeswell
Extent and impacts of soil degradation on a world-wide scale — XI

Chief Leonard Little Finger
Cecilia Violet Makota-Mahlangeni
Two voices from the grass roots — XVI

E. Fleischhauer & H. Eger
Can sustainable land use be achieved?
- An introductory view on scientific and political issues - — XIX

Part I: Soil degradation – diagnosis, appraisal and reversing measures — 1

A. Hebel
Introduction — 1

Chapter 1: Soil functions and land quality indicators — 3

C.R. De Kimpe & B.P. Warkentin
Soil functions and the future of natural resources — 3

T.F. Shaxson
Concepts and indicators for assessment of sustainable land use — 11

Z. K. Filip
Soil quality assessment: An ecological attempt using microbiological and
biochemical procedures — 21

O. Dilly & H.-P. Blume
Indicators to assess sustainable land use with reference to soil microbiology — 29

J. B. Aune & A. Massawe
Effects on soil management on economic return and indicators of soil degradation
in Tanzania: A modelling approach — 37

F. De Graaff & J.W. Nibbering
Difficulties in monitoring and evaluating watershed development –
a case Study from East Java — 45

O. Tietje, R.W. Scholz, A. Heitzer & O. Weber
Mathematical evaluation criteria — 53

Chapter 2: Land resources information systems for decision-making 63

S.A. El-Swaify
Multiple objective decision making for land, water and environmental management 63
H. Diestel, M. Hape & J.-M. Hecker
A conceptual framework for planning and implementing water and
soil conservation projects 73
K. Terytze & J. Müller
Harmonization of soil investigation methods within the framework of
cooperation with countries in central and eastern Europe 79
T. Oweis, A. Oberle & D. Prinz
Determination of potential sites and methods for water harvesting in central Syria 83
V.I. Lyalko, L.D. Voulfson, A.L. Kotlyar, V.N. Shevchenko, A.D. Ryabokonenko,
K.-H. Marek & S. Oppitz
Application of remote sensing to assess soil moisture capacity and depths of
ground-water table in the Chernobyl disaster area 89
B. Hornetz
Monitoring vegetation and soil degradation dynamics in the semi-arid areas
of northern Kenya 97
A. Castrignanò, M. Mazzoncini & L. Giugliarini
Spatial characterization of soil properties 105
P. Leinweber, C. Preu & C. Janku
Classification and properties of soils under different land uses in
south-west Sri Lanka 113
T. Scholten & P. Felix-Henningsen
Site properties and suitability of eroded Saprolites for reclamation and
agricultural use 121
R. Dilkova, G. Kerchev & M. Kercheva
Evaluating and grouping of soils according to their susceptibility to
anthropogenic degradation 125
F. Delgado & R. López
Evaluation of soil degradation impact on the productivity of Venezuelan soils 133
V. Stolbovoi & G. Fischer
A new digital georeferenced database of soil degradation in Russia 143
P. Zdruli, R. Almaraz & H. Eswaran
Developing land resource information for sustainable land use in Albania 153
D. Schaub & V. Prasuhn
A map on soil erosion on arable land as a planning tool for sustainable
land use in Switzerland 161
H. Vogel, J. Utermann, W. Eckelmann & F. Krone
The soil information system "FISBo BGR" for soil protection in technical cooperation 169

Chapter 3: Soil alteration by climate change 175

J.M. Kimble, R. Lal & R.B. Grossman
Alteration of soil properties caused by climate change 175
R. Lal & J.M. Kimble
Soil conservation for mitigating the greenhouse effect 185

H.-W. Scharpenseel & E.M. Pfeiffer
Impacts of possible climate change upon soils; some regional consequences 193
Victor R. Squires
Dryland soils: Their potential as a sink for carbon and as an agent in
mitigating climate change 209
W. Amelung, K. W. Flach, Xudong Zhang & W. Zech
Climatic effects on C pools of native and cultivated prairie 217
Xudong Zhang, W. Amelung & W. Zech
Can amino sugars serve as indicators of climate variations and land use
on soil organic matter? 225
R.M. Bajracharya, R. Lal & J.M. Kimble
Soil organic carbon dynamics under simulated rainfall as related to
erosion and management in central Ohio 231
H.R. Khan, U. Pfisterer & H.-P. Blume
Nitrous oxide, nitrate and ammonium dynamics as influenced by
selected environmental factors 239

Chapter 4: Water and wind erosion 247

C. W. Rose
Water and wind erosion – Causes, impacts and reversing measures 247

Chapter 4.1: Water erosion 257

C. Huang, D.S. Gabbard, L.D. Norton & J.M. Laflen
Effects of hillslope hydrology and surface condition on soil erosion 257
J.S. Samra
Assessment and concepts for the improvement of soils affected by water erosion in India 263
P. Böhm
Scale-effects in the assessment of water erosion 271
K. Klima
Effect of mountain crop rotation on surface runoff, subsurface flow and soil loss 279
Shi Xuezheng, Yu Dongsheng, Xing Tingyan & J. Breburda
Field plot measurement of erodibility factor K for soils in subtropical China 285

Chapter 4.2: Wind erosion, landslides and modeling 291

D.W. Fryrear
Mechanics, measurement and modeling of wind erosion 291
G.P. Glazunov
Potential hazard of wind erosion in regions affected by Chernobyl's fallout 301
Y. Shao, R.K. Munro, L.M. Leslie & W.F. Lyons
A wind erosion model and its application to broad scale wind erosion
pattern assessment 307
L.-O. Westerberg & C. Christiansson
Landslides in East African highlands: Slope instability and its interrelations
with landscape characteristics and land use 317

D.K. McCool, J.A. Montgomery, A.J. Busacca & B.E. Frazier
Soil degradation by tillage movement   327
A. Merzouk & H. Dhman
Shifting land use and its implication on sediment yield in the Rif Mountains (Morocco)   333
L. Laajili Ghezal, T. Aloui, M.A. Beji & S. Zekri
Optimisation of soil and water conservation techniques in a watershed of the
Tunisian semi-arid region   341

Chapter 4.3: Impacts of soil erosion on crop productivity   355

A. Tengberg, M. Stocking & S.C.F. Dechen
Soil erosion and crop productivity research in South America   355
A. Moyo
The effect of soil erosion on soil productivity as influenced by tillage:
With special reference to clay and organic matter losses   363
A.A.C. Salviano, S.R. Vieira & G. Sparovek
Erosion intensity and *Crotalaria Juncea* yield on a Southeast Brazilian Ultisol   369
F.B.S. Kaihura, I.K. Kullaya, M. Kilasara, J.B. Aune, B.R. Singh & R. Lal
Impact of soil erosion on soil productivity and crop yield in Tanzania   375

Chapter 4.4: Measures to halt erosion   383

Chia-Chun Wu & A-Bih Wang
Soil loss and soil conservation measures on steep sloping orchards   383
B. Jankauskas & G. Jankauskiene
The extent of soil degradation on hilly relief of western Lithuania
and antierosion agrophytocenoses   389
Z. Boli Baboule & E. Roose
Degradation of a sandy Alfisol and restoration of its productivity under
cotton/maize intensive cropping rotation in the wet savannah of Northern Cameroon   395
U. Bosshart
Catchment discharge and suspended sediment transport as indicators of the
performance of physical soil and water conservation in the Ethiopian Highlands   403

Chapter 5: Humus degradation and nutrient depletion   415

H. Tiessen, E.Cuevas & I.H. Salcedo
Organic matter stability and nutrient availability under temperate and
tropical conditions   415
M. Körschens
Soil organic matter and sustainable land use   423
G. Sparovek
Influence of organic matter and soil fauna on crop productivity and soil restoration
after simulated erosion   431
J.O. Agbenin
Quantifying soil fertility changes and degradation induced by cultivation techniques
in the Nigerian savanna   435

E.S. Costa, R.C. Luizão & F.J. Luizão
Soil microbial biomass and organic carbon in reforested sites degraded by
Bauxite mining in the Amazon 443
L. Reintam
Impact of intensive agriculture on changes in humus status and chemical properties
of arable Luvisols 451

Chapter 6: Combating desertification and salinisation 457

A. Ayoub
Degradation of dryland ecosystems: Assessments and suggested actions to combat it 457
M. Akhtar
The impact of desertification on Sahelian ecosystems - A case study from the
Republic of Sudan 465
I. Szabolcs[†]
Concepts, assessment and control of soils affected by salinization 469
J. Barros & P. Driessen
Quantifying soil salinity in a dynamic simulation of crop growth and production 477
G. Xie, J. Breburda, A. Kollender-Szych & A. Battenfeld
Relations between soil salinity and water quality as well as water balance in
Yinbei Plain, PR China 485
M. Nitsch, R. Hoffmann, J. Utermann & L. Portillo
Soil salinization in the central Chaco of Paraguay: A consequence of logging 495
M. Badraoui, B. Soudi, A. Merzouk, A. Farhat & A. M'Hamdi
Changes of soil qualities under pivot irrigation in the Bahira region of Morocco:
Salinization 503
M. M. Moukhtar, M. H. el Hakim & A.I.N. Abdel Aal
Drainage for soil conservation 509

Part II: Growing impacts of industrialised agriculture and urbanisation on soils 515

H.-P. Blume
Introduction 515

Chapter 7: Compaction and surface sealing 517

B.D. Soane & C. van Ouwerkerk[†]
Soil compaction: A global threat to sustainable land use 517
R. Horn
Assessment, prevention and rehabilitation of soil degradation caused by compaction
and surface sealing 527
C.J. Chartres & G.W. Greeves
The impact of soil seals and crusts on soil water balance and runoff and their
relationship with land management 539
A.M. Mouazen & M. Neményi
A finite element model of soil loosening by a subsoiler with respect to soil conservation 549

A.F. Ferrero & G. Nigrelli
Influence of establishment and utilisation techniques of hillside pastures on
soil physical characteristics 557
H. Schulte-Karring, D. Schröder & H.-Chr. Von Wedemeyer
Subsoil amelioration 565
J. Baumann, G. Werner & W. Moll[†]
Reclamation of hardened volcanish ash soils in the central Mexican highlands –
their productivity and erodibility 573
D. Norton & K. Dontsova
Use of soil amendments to prevent soil surface sealing and control erosion 581
K. Helming, M.J.M. Römkens & S.N. Prasad
Roughness and sealing effect on soil loss and infiltration on a low slope 589
N.M. El-Mowelhi, M.S.M. Abo Soliman, S.A. Abd El-Hafez & S.A. Hassanin
Effect of different land preparation practices on crop production and soil compaction 597

Chapter 8: The use of agrochemicals, manure and organic wastes 607

S.E.A.T.M. van der Zee & F.A.M. de Haan
Monitoring, control and remediation of soil degradation by agrochemicals,
sewage sludge and composted municipial wastes 607
B. Kranz, W.D. Fugger, J. Kroschel & J. Sauerborn
The influence of organic manure on *Striga hermonthica* (Del.) Benth. infestation in
Northern Ghana 615
T.K. Haraldsen, V. Jansons, A. Spricis, R. Sudars & N. Vagstad
Influence of long-term heavy application of pig slurry on soil and water quality in Latvia 621
M. Hämmann, S.K. Gupta & J. Zihler
Protection of soils from contamination in Swiss legislation 629
A. Freibauer, J. Lilienfein, M. Ayarza, J.E. da Silva & W. Zech
Fertilizer P transformations in loamy and clayey Oxisols of central Brazil 637
A.K. Patra
Nitrogen losses from soils in the Indian semi-arid tropics 645
K. Voplakal
Changes in plant available phosphorus and its soil indices following the introduction
of moderate fertilization 653

Chapter 9: Effects of polluted air and water 661

E. Matzner & N. Diese
Acidic deposition on forest soils: A threat to the goal of sustainable forestry 661
J. Csillag, A. Lukács, K. Bujtás & T. Németh
Release of Cd, Cr, Ni, Pb and Zn to the soil solution as a consequence of
soil contamination and acidification 673
A. Badora & T. Filipek
An assessment of rehabilitation of strongly acidic sandy soils 681
W. Wilcke, J. Kobza & W. Zech
Small-scale distribution of airborne heavy metals and polycyclic aromatic hydrocarbons
in a contaminated Slovak soil toposequence 689

A. Kollender-Szych, J. Breburda, P. Felix-Henningsen & H. Trott
Heavy metal pollution of irrigated soils in Ningxia, China — 697
A. Karczewska, L. Szerszeń & C. Kabała
Forms of selected heavy metals and their transformation in soils polluted by the emissions from copper smelters — 705
M. Gebski, S. Mercik & K. Sommer
Elution of Zn, Pb and Cd in soil under field and laboratory conditions — 713
S. Kretzschmar, M. Bundt, G. Saborió, W. Wilcke & W. Zech
Heavy metals in soils of Costa Rican coffee plantations — 721
A. Prüeß
Action values for mobile (NH_4NO_3-extractable) trace elements in soils based on the German National Standard DIN 19730 — 727
E. Podlešáková & J. Němeček
Criteria for soil contamination of organic and inorganic pollutants in the Czech Republic — 735
K. Wenger, T. Hari, S.K. Gupta, R. Krebs, R. Rammelt & C.D. Leumann
Possible approaches for in situ restoration of soils contaminated by zinc — 745

Chapter 10: Effects of mining industries and disposal sites — 755

W.E.H. Blum
Soil degradation caused by industrialization and urbanization — 755
M.J. Haigh
Promoting better land husbandry in the reclamation of surface coal-mined land — 767
M.J. Haigh
Towards soil quality standards for reclaimed surface coal-mined lands — 775
Qinglan Wu, H.-P. Blume, L. Rexilius, S. Abend & U. Schleuß
Sorption of organic chemicals in Urbic Anthrosols — 781
J. Ammosova & M. Golev
Monitoring of soil degradation caused by oil contamination — 791
Th. Baumgartl, B. Kirsch & M. Short
Soil physical processes in sealing systems of waste dumps affecting the environment — 797
A. Sidorchuk & V. Grigorév
Soil erosion on the Yamal Peninsula (Russian arctic) due to gas field exploitation — 805
F.G. Kupriyanova-Ashina, G.A. Krinari, A.L. Kolpakov & I.B. Leschinskaya
Degradation of silicate minerals by *Bacillus mucilaginosus* using *Bacillus intermedius* RNase — 813

Contents Volume II

Part III: From soil and water conservation to sustainable land management   819

C. Reij, H. Eger & K. Steiner
From soil and water conservation to sustainable land management:
Some challenges for the next decade   819

Chapter 11: Sustainable land use management: policies, strategies and economics   827

Chapter 11.1: Policies and strategies   827

H. Hurni
A multi-level stakeholder approach to sustainable land management   827
J. Pretty
Furthering cooperation between people and institutions   837
A. Steer
Making development sustainable   851
M. Stocking
Conditions for enhanced cooperation between people and institutions   857
J.-C. Griesbach & D. Sanders
Soil and water conservation strategies at regional, sub-regional and national levels   867

Chapter 11.2: Economics   879

R. Clark, H. Manthrithilake, R. White & M. Stocking
Economic valuation of soil erosion and conservation - A case study of Perawella, Sri Lanka   879
P.C. Huszar
Including economics in the sustainability equation: Upland soil conservation
in Indonesia   889
Ram Babu & B.L. Dhyani
Economic evaluation of watershed management programmes in India   897
E.H. Bucher, P.C. Huszar & C.S. Toledo
Sustainable land use management in the South American Gran Chaco   905
M. Zlatić, N. Ranković & G. Vukelić
Improvement of soil management for sustainability in the hilly Rakovica Community   911

Chapter 11.3: Case studies   919

A. Arnalds
Strategies for soil conservation in Iceland   919
J. Hraško
Influence of socio-economic changes on soil productivity in Slovakia   927
F. Dosch & S. Losch
Spatial planning tasks for sustainable land use and soil conservation in Germany   933

I. Hannam
An ecological perspective of soil conservation policy and law in Australia — 945

S. Ewing
Australia's community 'Landcare' movement: A tale of great expectations — 953

J.S. Gawander
The rise and fall of the vetiver hedge in the Fijian sugar industry — 959

A.S. Kruger
Closing the gap between farmers and support organisations in Namibia — 965

Chapter 12: Institutional and planning aspects of soil and water conservation — 971

M. Maarleveld
Improving participation and cooperation at the local level: Lessons from Eeonomics and psychology — 971

L. van Veldhuizen
Principles and strategies of participation and cooperation challenges for the coming decade — 979

A. Oomen
Political and institutional conditions for sustainable land management — 985

H. Zweifel
Sustainable land use - A participatory process — 991

H. Eger
Participatory land use planning: The case of West Africa — 1001

T. Bekele & W. Zike
Village level approach to resource management: A project in a marginal environment in Ethiopia — 1009

Lixian Wang
Mountain watershed management in China — 1017

M. Schneichel & P. Asmussen
Dry forest management - Putting campesinos in charge — 1023

P.S. Cornish
A farmer-researcher-adviser partnership designed to support changes in farm management needed to meet catchment goals — 1029

Chapter 13: Approaches to soil and water conservation and watershed development — 1037

Chapter 13.1: Case studies from Africa — 1037

H. Liniger, D.B. Thomas & H. Hurni
WOCAT - World Overview of Conservation Approaches and Technologies - Preliminary results from Eastern and Southern Africa — 1037

G.O. Haro, E.I. Lentoror & A. von Lossau
Participatory approaches in promotion of sustainable natural resources management experiences from South-West Marsabit, Northern Kenya — 1047

M.J. Kamar
Soil conservation implementation approaches in Kenya — 1057

K.C.H. Mndeme
Participatory land use planning approach in North Pare Mountains, Mwanga, Kilimanjaro Region, Tanzania — 1065

S. Minae, W.T. Bunderson, G. Kanyama-Phiri & A.-M. Izac
Integrating agroforestry technologies as a natural resource management tool for smallholder farmers — 1073

T. Defoer, S. Kantè & Th. Hilhorst
A participatory action research process to improve soil fertility management — 1083

I. Yosko & G. Bartels
Nomads and sustainable land use: An approach to strenghten the role of traditional knowledge and experiences in the management of the grazing resources of Eastern Chad — 1093

H. Paschen, D. Gomer, L. Kouri, H. Vogt, T. Vogt, M. Ouaar & W.E.H. Blum
Management of watersheds with soils on marls in the Atlas Mountains of Algeria - A proposal for a non-conventional watershed development scheme — 1099

Chapter 13.2: Case studies from Asia — 1107

B. Adolph & T. G. Kelly
What makes watershed management projects work? Experiences with farmer's participation in India — 1107

J. Mascarenhas
The participatory watershed development process
Some practical tips drawn from outreach in South India — 1117

R. Chennamaneni
Watershed management and sustainable land use in semi-arid tropics of India: Impact of the farming community — 1125

K. Mukherjee
People's participation in watershed development schemes in Karnataka - Changing perspectives — 1135

J.S. Samra & A.K. Sikka
Participatory watershed management in India — 1145

Sameer Karki & S.R. Chalise
Improving people's participation in soil conservation and sustainable land use through community forestry in Nepal — 1151

C. Setiani & A. Hermawan
Conservation farming land use on critical upland watersheds
- Social and economic evaluation study - — 1161

Chapter 14: Technologies for soil and water conservation and sustainable land management — 1167

Chapter 14.1: The role of tillage practices and ground cover — 1167

H. Liniger & D.B. Thomas
GRASS
<u>G</u>round Cover for <u>R</u>estoration of <u>A</u>rid and <u>S</u>emi-arid <u>S</u>oils — 1167

R. Derpsch & K. Moriya
Implications of no-tillage versus soil preparation on sustainability of agricultural production — 1179

E. Chuma & J. Hagmann
Development of conservation tillage techniques through combined on-station and participatory on-farm research ... 1187
Th. Rishirumuhirwa & E. Roose
The contribution of banana farming systems to sustainable land use in Burundi ... 1197
A. Calegari, M.R. Darolt & M. Ferro
Towards sustainable agriculture with a no-tillage system ... 1205
H. Morrás & G. Piccolo
Biological recuperation of degraded Ultisols in the province of Misiones, Argentina ... 1211
D.K. Malinda, R.G. Fawcett, D.Little, K. Bligh & W. Darling
The effect of grazing, surface cover and tillage on erosion and nutrient depletion ... 1217
Lin Kai Wang, Xiao Tian, Li Yili, Su Shuijin & Xie Fuguang
Preliminary results of the grass-tree system for rehabilitation of severely eroded red soils ... 1225

Chapter 14.2: Nutrient management practices ... 1233

R.D. Lentz, R.E. Sojka & C.W. Robbins
Reducing soil and nutrient losses from furrow irrigated fields with polymer applications ... 1233
A. Calegari & I. Alexander
The effects of tillage and cover crops on some chemical properties of an Oxisol and summer crop yields in southwestern Paraná, Brazil ... 1239
T.N. Mwambazi, B. Mwakalombe, J.B. Aune & T.A. Breland
Turnover of green manure and effects on bean yield in Northern Zambia ... 1247
J. Lehmann, F. v. Willert, S. Wulf & W. Zech
Influence of runoff irrigation on nitrogen dynamics of *Sorghum bicolor* (L.) in Northern Kenya ... 1255
M.C.S. Wopereis, C. Donovan, B. Nébié, D. Guindo, M.K. Ndiaye & S. Häfele
Nitrogen management, soil nitrogen supply and farmers' yields in Sahelian rice based irrigation systems ... 1261

Chapter 14.3: Mechanical control of soil erosion ... 1267

Edi Purwanto & L.A. Bruijnzeel
Soil conservation on rainfed bench terraces in upland West Java, Indonesia: Towards a new paradigm ... 1267
N. Sinukaban, H. Pawitan, S. Arsyad & J. Armstrong
Impact of soil and water conservation practices on stream flows in Citere catchment, West Java, Indonesia ... 1275
R.N. Adhikari, M.S. Rama Mohan Rao, S. Chittaranjan, A.K. Srivastava, M. Padmalah, A. Raizada & B.S. Thippannavar
Response to conservation measures in a red soil watershed in a semi arid region of South India ... 1281
K. Michels, C. Bielders, B. Mühlig-Versen & F. Mahler
Rehabilitation of degraded land in the Sahel: An example from Niger ... 1287

W.P. Spaan & K.J. van Dijk
Evaluation of the effectiveness of soil and water conservation measures in a
closed sylvo-pastoral area in Burkina Faso — 1295
F.A. Mayer & E. Stelz
Gully reclamation in Mafeteng District, Lesotho — 1303
R.G. Barber
Linking the production and use of dry-season fodder to improved
soil conservation practices in El Salvador — 1311

Chapter 15: Adoption of soil and water conservation practices — 1319

Chapter 15.1: Demographic, socio-economic and water conservation practices — 1319

B. Rerkasem & K. Rerkasem
Influence of demographic, socio-economic and cultural factors on
sustainable land use — 1319
M. Tiffen
Demographic growth and sustainable land use — 1333
W. Östberg & Ch. Reij
Culture and local knowledge - Their roles in soil and water conservation — 1349
Yohannes Gebre Michael
Indigenous soil and water conservation practices in Ethiopia:
New avenues for sustainable land use — 1359
H.K. Murwira & B.B. Mukamuri
Traditional views of soils and soil fertility in Zimbabwe — 1367
A.S. Langyintou & N. Karbo
Socio-economic constraints to the use of organic manure for soil fertility
improvement in the Guinea savanna zone of Ghana — 1375
J. Hellin & S. Larrea
Ecological and socio-economic reasons for adoption and adaptation of live barriers
in Güinope, Honduras — 1383
J. Currle
Farmers and their perception of soil conservation methods — 1389
R.J. Unwin
Farmer perception of soil protection issues in England and Wales — 1399

Chapter 15.2: The role of incentives, research, training and extension — 1405

M. Giger
Using incentives and subsidies for sustainable management of agricultural soils -
A challenge for projects and policy-makers — 1405
Ch. Reij
How to increase the adoption of improved land management practices by farmers? — 1413
D. Sanders, S. Theerawong & S. Sombatpanit
Soil conservation extension: From concepts to adoption — 1421
K. Herweg
Contributions of research on soil and water conservation in developing countries — 1429

M. Douglas
Training for better land husbandry — 1435
J. Hagmann, E. Chuma & K. Murwira
Strengthening peoples capacities in soil and water conservation in Southern Zimbabwe — 1447

Chapter 15.3: Women and soil and water conservation — 1463

L.M.A. Omoro
Women's participation in soil conservation: Constraints and opportunities.
The Kenyan experience — 1463
D. Kunze, H. Waibel & A. Runge-Metzger
Sustainable land use by women as agricultural producers?
The case of Northern Burkina Faso — 1469
Swarn Lata Arya, J.S. Samra & S.P. Mittal
Rural women and conservation of natural resources: Traps and opportunities — 1479

Chapter 16: Land tenure and soil and water conservation — 1485

M. Kirk
Land tenure and land management: Lessons learnt from the past,
challenges to be met in the future? — 1485
H.W.O. Okoth-Ogendo
Tenure regimes and land use systems in Africa: The challenges of sustainability — 1493
V. Stamm
Are there land tenure constraints to the conservation of soil fertility?
A critical discussion of empirical evidence from West Africa — 1499
D. Effler
National land policy and its implications for local level land use:
The case of Mozambique — 1505
H.M. Mushala & G. Peter
Socio-economic aspects of soil conservation in developing countries:
The Swaziland case — 1511
M. Chasi
Impact of land use and tenure systems on sustainable use of resources in Zimbabwe — 1519
K. Goshu
Assessing the potential and acceptability of biological soil conservation techniques
for Maybar Area, Ethiopia — 1523
B.J. Rao, R. Chennamaneni & E. Revathi
Land tenure systems and sustainable land use in Andhra Pradesh:
Locating the influencing factors of confrontation — 1531
H. Cotler
Effects of land tenure and farming systems on soil erosion in Northwestern Peru — 1539

Conclusions and Recommendations of the 9th ISCO Conference — 1545

Personal Records Editors — 1557

Authors' Index — 1559

Organisers of the Conference

The Conference was held under the auspices of
ISCO, the International Soil Conservation Organisation
German Society of Soil Science (DBG)
German Federal Ministry for the Environment, Nature Conservation and Nuclear Safety
German Federal Ministry for Economic Cooperation and Development
Deutsche Gesellschaft für Technische Zusammenarbeit (GTZ) GmbH
German Federal Environmental Agency
in collaboration with
World Association of Soil and Water Conservation (WASWC)
International Society of Soil Science (ISSS)
European Society for Soil Conservation (ESSC)

Acknowledgements

The editors and the publisher sincerely thank the reviewers of the articles:
K. Auerswald, Freising-Weihenstephan, H.H. Becher, Freising-Weihenstephan, L. Beyer, Kiel, H.-P. Blume, Kiel, J. Breburda, Gießen, J. Breuer, Stuttgart-Hohenheim, G.W. Brümmer, Bonn, R. Bunch, Tegucicalpa/Honduras, K. Bunzl, Oberschleißheim, B. Bussian, Berlin, C. de Castro Filho, Londrina/Brasil, W. Critchley, Sovenga/South Africa, S. Dabbert, Stuttgart-Hohenheim, O. Dilly, Kiel, L. Dinkloh, Bonn, H. Eger, Eschborn, W. Ehlers, Göttingen, S. El-Swaify, Honolulu/USA, P. Felix-Henningsen, Gießen, E. Fleischhauer, Bonn, H.-G. Frede, Gießen, M. Frielinghaus, Müncheberg, J. Hagmann, Gundelfingen-Wildtal, A. Hebel, Bonn, K. Herweg, Bern/Switzerland, R. Horn, Kiel, H. Hunzinger, Eschborn, R. Jahn, Halle/S., K. Janz, Berlin, I. Kögel-Knabner, Freising-Weihenstephan, J. Kimble, Lincoln/USA, M. Kirk, Marburg, B. Knerr, Witzenhausen, G. Krumnöhler, Bonn, H.-J. Krüger, Göttingen, M. Laman, Amsterdam/Netherlands, H. Liniger, Bern/Switzerland, M. Maarleveld, Wageningen/Netherlands, E. Matzner, Bayreuth, G. Michlich, Hamburg, D. Millar, Tamale/Ghana, W. Musgrave, Sydney/Australia, D. Nill, Reichelsheim, R. Oldeman, Wageningen/Netherlands, S. Pätzold, Bonn, C. Reij, Amsterdam/Netherlands, G. Richter, Trier, C. Rose, Brisbane/Australia, D.W. Sanders, Bristol/Great Britain, D. Sauerbeck, Braunschweig, H.-W. Scharpenseel, Hamburg, U. Schwertmann, München, T. Selige, Oberschleißheim, F. Shaxson, Dorset/Great Britain, K. Stahr, Stuttgart-Hohenheim, K.G. Steiner, Eschborn, H. Sticher, Zürich/Switzerland, L. Stroosnijder, Wageningen/Netherlands, S. Thiele, Bonn, C. Toulmin, Edinburgh/Great Britain, A. Trux, Paris/France, P.L.G. Vlek, Göttingen, H. Vogel, Hannover, Von Bargen, Bonn, M. von Oppen, Stuttgart-Hohenheim, H. Waibel, Hannover, H.-R. Wegener, Gießen, G. Welp, Bonn, W. Werner, Bonn, H. Wiechmann, Hamburg, W. Zech, Bayreuth

Special thanks are due to W. Nonnenmacher and H.-U. Walde, both Federal Ministry for the Environment, Nature Conservation and Nuclear Safety in Bonn, P. Hugenroth, German Soil Science Society, P. Bakker, Center for Development Cooperation Services in Amsterdam/Netherlands, S. Witt, Kiel, C. Berger and A. Feise from Bonn who gave invaluable support to the editors' work.

The editors sincerely thank the CATENA VERLAG, by name M. Rohdenburg and M. Kaiser, and S. Berkowicz as language editor.

Preface

The world's great interest in the 9th Conference of the International Soil Conservation Organisation held in Bonn, August 1996, which met under the theme "Towards Sustainable Land Use", shows us that developing sustainable forms of soil use and combating soil degradation are seen as urgent international tasks on the way towards sustainable development in line with Agenda 21. Combating soil degradation and active steps to conserve soils for future generations are part of important international environment policy challenges. When foodstuffs and self-regenerating raw materials are being produced, the natural fertility of the soil must be maintained in the long-term, taking into account water pollution control, protection of the earth's atmosphere and nature conservation, as well as conservation of biological diversity. At the World Food Summit held in Rome, November 1996, the need for global protection of soil resources and their sustainable manage-ment was once again emphasized as one of the central preconditions for securing food for future generations. We need a worldwide environmental partnership to achieve these goals.

Adverse ecological and economic developments that are, at least in part, a result of soil degradation can be seen in many regions of the world. According to UNEP and the ISRIC, 24% of the world's soils on settled land are already greatly impaired as a result of anthropogenic degradation. Inappropriate soil use, the input of pollutants, and soil sealing arising from infrastructure preparation and the building of housing estates damage, or even destroy, soils. This trend is worrisome. Growing populations and increasing industrialisation have gone hand in hand with a marked rise in soil pollution - both in industrialised and in developing countries. Pollution causes special social and economic problems which need specific actions. Rural communities cannot shoulder the burden for society as a whole of off-site impacts of land degradation. The careful design of incentives such as cost-sharing of pollution control and other legal frameworks is needed in order to target action responses in the appropriate sections of society.

A number of contributions to this publication contain a wide variety of successful, practical and often innovative solutions for sustainable land use. Examples include improvements to soil quality in sites where the soil has already been degraded as well as precautions for soil protection. They should strengthen the international community in its commitment to further support programmes to improve land quality. National and sub-national governments should support financial and economic baseline studies on the implications of continued degradation, and use the results to design appropriate policy responses in, for example, support to extension, research marketing facilities and related programmes in healthcare, education, etc. Transboundary eco-regional land conservation and basin-wide watershed development can be facilitated under the auspices of international conventions, action programmes or regional frameworks. "Prevention" should be the leading principle in soil conservation policy, since "rehabilitation" is extremely difficult and often not possible at all, once symptoms of soil degradation are detected.

From a global point of view, the balance of the successes achieved in combating soil degradation in comparison to the deteriorations in soil quality occurring at other sites has been negative for quite some time. Therefore, on the one hand, efforts to implement sustainable forms of land management appropriate to a particular site and for the development and trials of new and innovative approaches must continue to be intensified. On the other hand, the dissemination of knowledge about responsible land management to land users and to broad sections of the population must be further improved.

The experience discussed at the Conference proves that integrated approaches, in which soil conservation plays a central role, are very promising for the solution of the multi-faceted problems of land use. These approaches include the fact that all stakeholders are sufficiently incorporated in land-use planning and that specific socio-economic framework conditions are taken into account. An integrated planning, development and management approach to land resources, as called for in Chapter 10 of Agenda 21, is what is required. It must systematically record and consistently take account of the needs of man and the environment. It must be based on sound, timely and reliable information which includes available information about the properties of the land, the actual and potential land uses and their influence on the resilience of ecosystems. It is necessary to identify and reconcile the different sets of objectives of land users' communities, governments, short-term needs such as food production, and long-term requirements such as the preservation of soil productive capacity and biological diversity. This integrated approach to land resources planning and management requires full-scale inter-institutional cooperation and development of goodwill of all stakeholders concerned through the functioning of platforms for conflict resolution and participatory decision-making. A planning system oriented to the goal of land use appropriate to a given site in question contributes both to the protection of the entire ecosystem as well as to soil conservation.

We consider it essential that soil conservation, as a cross-sectoral task, be assigned a status comparable to climate protection in global environmental politics in the medium term in the efforts to prevent negative global environmental changes. This is the only way to ensure that the key functions of the soil remain preserved for achieving the quality targets as laid down in the international conventions for climate protection, the conservation of biological diversity and the combating of desertification as well as in the Forests Declaration. An exact updated global review, assuming certain preconditions, of the currently predicted extent of destruction, damage and loss of soils is urgently required. In this way the risks to humans and the environment posed by soil destruction could be assessed in more detail than previously, with regional differentiations, for which resultant steps for internationally coordinated action can be derived.

In December 1996 the International Convention to Combat Desertification (CSD) entered into force. It is to be expected that the implementation of this Convention will lead to an intensification of the efforts for soil conservation, so far regionally limited to the semi-arid tropics and parts of the sub-humid tropics. The ratification of the Desertification Convention is without doubt an important step on the way to sustainable land use.

In a political document prepared by the CSD for the special session of the United Nations General Assembly (UNGASS), held in New York in June 1997, the Chapter entitled "Implementation in areas requiring urgent action" points out the urgent need to combat or reverse the worldwide trend of accelerating soil degradation. It goes on to say that the international community has recognised the need to address this problem with an integrated approach for the protection and sustainable management of land and soil resources in which all stakeholders have to be considered. For this reason the issue of "Integrated planning and management of land resources" has been earmarked as a focus of the deliberations for the year 2000 in the provisional CSD Work Programme.

We anticipate that these volumes of conference documents will be widely distributed and thus contribute not only to a broader discussion on soils but also to provide the impetus for definite measures. We would like to thank the Conference delegates for their active participation and also express our warm thanks to everyone who helped make the Conference a success.

Dr. Fritz Holzwarth
Chairman of the 9th ISCO Conference
Federal Ministry for the Environment, Nature Conservation and Nuclear Safety

Prof. Dr. Hans-Peter Blume
President of the German Soil Science Society

Policy for Sustainable Land Use

Protecting the Soil and Preserving it for Future Generations

Welcoming statement by the President of the German Bundestag Rita Süssmuth to the 9th Conference of the International Soil Conservation Organisation (ISCO) held at the "Waterworks", Bonn, August 26-30, 1996

I take great pleasure in welcoming you today to the 9th Conference of the International Soil Conservation Organisation on the subject "Towards Sustainable Land Use - Furthering Cooperation between People and Institutions" being held here at the Bonn "Waterworks", a former meeting place of the German Parliament.

The fact that this conference is taking place in Germany, on the premises of the German Bundestag, reflects, on the one hand, the importance that is attributed to environmental and resource protection in our country. Indeed it was written in our constitution (Article 2oa) in 1994. On the other hand, it shows how important we feel it is to engage in a process of information exchange and technology transfer relating to the worldwide struggle against soil degradation and the development of sustainable and environmentally appropriate forms of soil use.

Soil is the foundation of our ecosystem and, along with air and water, an important and sensitive interface for the interchange of energy and matter. Here the transition takes place between life and death, the development and decomposition of organisms. It is the medium for a large-scale symbiosis of plants and animals which, together with water, generates new growth. Protecting the soil and preserving it for future generations is a key international task alongside climate protection. Sustainable, environmentally appropriate soil protection means protection of life; it means the preservation, so very necessary today, of the foundations of the human food supply as well as of human prosperity.

This year's 9th Conference of the International Soil Conservation Organisation is meeting under the banner of the "Earth Summit" of Rio de Janeiro 1992. The programme of action approved there, "Agenda 21", calls upon us to pursue a course towards sustainable development. The two central messages of Agenda 21 are more important than ever today:

1. Economic productivity, social responsibility and environmental protection belong together as an unseparable whole, with a view to guaranteeing equitable development opportunities for all countries and preserving natural resources for coming generations.

2. Unilateralism in the area of environmental and development policy must be given up in favour of joint solutions. We are standing here on the threshold of a new global partnership formed to address the challenges of the 21th century.

In view of current demographic and ecological trends, sustainable development in the areas of soil use and soil protection is one of the most pressing problems of our time.

ISBN 3-923381-42-5
© 1998 by CATENA VERLAG, 35447 Reiskirchen

II

The preservation of soils and their fertility as well as a stable balance of natural processes within them are essential prerequisites for the food security of a growing population.

We still have more questions than answers; we are still not certain as to the path we will need to take to achieve sustainable development in these areas. One thing would be certain, however: Without a change in people's attitude and actions and without a change in the priorities set by national governments, we will not succeed in effectively implementing the "spirit of Rio".

In our country, too, soil abuse can be seen in connection with such matters as old industrial pollution, waste disposal practices, farming methods, urban sprawl and the paving over of land surfaces. Soil damage accumulates gradually, but once the damage has been done it is impossible to repair, or only at considerable expense, due to the non-expandability of available land resources as well as the very slow and limited capacity of soil for self-renewal. To deal with this situation we need viable policies and a change in public awareness. As such, comprehensive public debate and the involvement of all the major interest groups in society must be at the centre of any measures and processes undertaken with a view to achieving sustainable soil use. This dialogue between government and society must also have the appropriate international forums.

So, let us be aware that what is being dealt with at this ISCO conference is an issue that will extend far into the future. Around the world people are in need of responses to trends that are threatening the foundations of our existence. This is a reference in particular to the politicians who are faced with the responsibility of making decisions. But simple answers are not enough here. Complex issues associated with global change that relate to sustainable soil use first and foremost require scientific analysis. This will not only place significant demands on the scientific community, it will make it necessary to focus research at universities more strongly on solutions, to focus applied research more strongly on very specific questions. Political practice and scientific analysis must be brought together on a high level. This will constitute an important foundation for policy improvement.

An equally important social prerequisite for sustainable soil use is the need to create in the public at large a stronger environmental awareness in this area. This means it will be necessary to promote public recognition of the problem to a much greater extent than has been the case in the past and to initiate educational measures with a view to enhancing environmental awareness. Attitudes and modes of behaviour that are either good or bad for the environment are learned - and possibly forgotten again - from an early age on. Family, friends, schools, universities and employers are all involved in this lifelong learning and socialisation process. With this in mind, there is a need to promote the development of environmental education as a subject of research and as an environmental policy strategy alongside concrete measures taken for the purpose of achieving soil protection and ecologically appropriate soil use.

In my view Germany has taken an important step here in the right direction. Since 1992 government policy makers have received support from the Scientific Advisory Committee on Environmental Global Change (WBGU) and the Council of Experts on Environmental Questions (SRU) in dealing with core problems of global change such as soil pollution. The German Bundestag Commission of Enquiry of "Protection of Man and the Environment" is currently formulating objectives and general conditions for sustainable development. This applies in particular with regard to land use, erosion and soil compaction, pollution and ground water protection. Last but not least I would like to mention the draft soil protection bill submitted by the German Cabinet and which will be placed before the Bundestag for deliberation in the days to come. These are all steps on the road to sustainable development in the Federal Republic of Germany, something we have felt more committed to than ever since Rio, and there is a need for more such steps to be taken. We are also called upon more than ever to undertake the appropriate scientific studies as well as the appropriate political action.

With this in mind, I would like to wish the 9th ISCO Conference and all the participants interesting and informative discussions.

Striving for Sustainable Land Use
- A Challenge for All Countries -

Angela Merkel
Federal Minister for the Environment, Nature Conservation and Nuclear Safety

The title of the 9th Conference of the International Soil Conservation Organisation - Towards Sustainable Land Use - presents us with a great challenge. Sustainable land use means that the natural fertility of soils is to be preserved permanently in the production of foodstuffs and regenerative raw materials. At the same time, the term includes the concept that human beings pay sufficient attention to the protective functions of the soil for the other areas of the environment, waters, atmosphere and biological diversity whenever they use or transform terrestrial ecosystems.

However, what does it actually mean to use soil under the various conditions in such a way that its biological health is permanently guaranteed? Or so that the spatial distribution of the various soil uses, which also include areas for settlement, industry and transport, take into account the buffer and compensation functions of the soils for the ecobalance? Since you are involved in research on these problems or are seeking practical solutions for the various problems in your countries, you have the experience to answer these questions.

In my speech I would also like to elaborate on the link between the subject matter of the conference and the political task of creating the conditions for sustainable land use and thus also for sustainable development. Over 25 years ago the Club of Rome report triggered an intensive discussion on the ecological opportunities and limits of development on Earth. Some things developed differently than was anticipated at the time by the Group following Meadows. However, scientists and politicians had to recognize that a continuation of the exponential population and economic growth would exceed the carrying capacity of the Earth in the not too distant future. As well as warning against the destruction of the environment and economic depression, the report contained two further conclusions:

Firstly, it was still possible to change trends and sustainably maintain a humane environment and, secondly, the chance of achieving ecological and economically satisfactory growth was greater the quicker we acted.

In many countries these two optimistic conclusions were seen as major challenges which could, however, be met. There are encouraging signs that an increasing number of people are guided by the idea of sustainable development in economic and political decisions. At the same time, the fall of the "Iron Curtain" showed what enormous damage can be done in a relatively short historical period by an economic system that ignores the environment. In many developing countries, too, the protection of resources has not yet acquired the necessary political standing. In this context we must not forget that the developed countries are definitely partly responsible for the insufficient compliance with the conservation requirements in these countries.

The United Nations Conference for Environment and Development in Rio de Janeiro in June 1992 supplied with the Agenda 21 Programme of Action the approaches to sustainable development in the various regions of the world. The fight against soil degradation and the promotion of sustainable agriculture and rural development are central tasks within this. The situation with regard to the changes of soil resources and their possible uses is probably more dramatic than it appears in the documents of the UNCED Conference. At the beginning of this year Lester Brown from the World Watch Institute entitled an essay "In the Face of Hunger". He outlined that the acquisition of sufficient food will become the greatest challenge facing humanity. The reasons he gives for this are indeed pessimistic. Grain production in the world is stagnant: areas of land given over to the cultivation of grain has fallen throughout the world per capita by about fifty per cent since 1950 due to population growth, soil degradation and economic reasons. An increasing lack of water resources and the apparent reduced economic effectiveness of chemical fertilisers are unfavourable conditions for a new "green revolution".

What must we do in this situation to nevertheless find ways towards sustainable land use? To put it briefly we have to increase productivity in the production of foodstuffs and regenerative raw materials on current arable land and additional land areas that may be available for cultivation in the future, without affecting the stability of the ecobalance. What scientific knowledge do we need in order to be able to perform this task? How can parliaments and governments, national and international financing and advisory institutions be supported in drawing up and carrying out programmes to allocate scarce finance reserves not only effectively but also in an environmentally sound way?

As in other global areas - climate protection and conservation of biological diversity - the answers to these questions can only be obtained by targeted international coooporation between scientists, experts and administrators in the various institutions and governments. From their studies and activities they know the biological and physical-chemical conditions of sustainable agriculture, animal husbandry and forestry, which help to return soil fertility in all climate zones. Methods of soil cultivation used over long periods of time, were, as far as I am aware, characterized by the fact that the amount of organic substances in the soil could be maintained at a certain level. However, sustainable land use cannot simply be defined by biological and physical-chemical soil parameters. Rather it is yields and income that provide society with a sound foundation - and that also means land rights. Demographic pressure, distorted product prices, unclear land tenure rights and political instability are major sources of negative developments which ultimately lead to rural poverty or sustain poverty. Increasing soil degradation is usually the result of this process. To effectively help people in developing countries to help themselves, it is important to recognize the links between causes and effects. In this context we must also ask how developed countries can further improve the institutional conditions for fair trade. Measures to combat soil degradation can only be successful if people are confident that their work and their savings will actually secure their future and that of their children.

This Conference can help to identify those factors which determine both the structure and direction of favourable or unfavourable soil cultivation and land use from the point of view of sustainability. We need a clear idea of priorities and their interdependences, i.e. ideas that can also be implemented politically. I believe it is an extremely difficult task to draw up concepts for an increase in soil productivity that are not to the detriment of the ecobalance and that keep land use options open. It is therefore important that the presentations and discussions at this Conference outline key processes for sustainable land use and development under different cultural and socio-economic conditions. I am sure the representatives from self-help groups and non-government organisations will be able to contribute additional experience, such as how new scientific approaches to economic and technical progress can be combined with traditional ecological knowledge.

If we want to conserve the environment we must overcome poverty. Knowledge of more productive and more sustainable forms of land management can only be passed on if, at the same time, the other tasks of combating poverty are tackled simultanously. These include the promotion of education and vocational training for the young generation, easier access of the rural population to the information relevant to it, improved communications as well as overcoming land right structures that hinder development. I would particularly like to emphasise the responsibility of governments for measures to strengthen the economic and social rights of women. I thus agree with Carol Bellamy, the Director of UNICEF, who considers this the prerequisite for the conservation and further development of a humane environment in developing countries. In an essay she quotes the old saying: "women hold up half the sky" and goes on to say that, in the long term, the environment cannot be protected without the active and knowledgable participation of those who are best suited to this task.

We all hope that the international community will be successful in increasing food production, improving living conditions and hygienic standards and in extending opportunities for cultural exchange, which also includes the concept of higher mobility. Economic growth is a precondition for this. This will lead to an increase in energy and substance transformation processes, i.e. an increase in entropy. We have to ask ourselves: Are ecosystems, and in this Conference this means terrestrial ecosystems, resilient enough against an increasing substance and energy flow to ensure that the newly formed dynamic equilibriums of the ecosystems will be conducive to people. In history soil erosion and salinization have been the consequence of overutilisation of soils and thus the main causes of environmental destruction. The transition towards industrialisation, urbanisation and mobility, upon which the developing countries are about to embark, brings to light other fundamental problems which they will have to solve with our support.

At this point I would like to discuss two questions that are of key significance to the soil conservation debate in Germany and in other developed countries. These questions deal with the ecological assessment of persistent substances in soils. This pollution has various causes, including deposits from air pollution and dirty irrigation water which are particularly frequent in developed countries. The question also is to what extent can changes in land use be tolerated. In other words, to what extent can free areas of land be used if buffer and compensation functions for the ecobalance and biological diversity are largely lost. One of the key aspects for sustainable land use is the principle outlined by the Federal Government's council of experts on environmental issues as one of the guidelines for sustainable, environmentally sound development: "The release of substances must not be greater than the absorption capacity of the environmental media". In areas with high industry and population densities we find concentrations of certain heavy metals and persistent organic substances. Some concentrations are considerably higher than natural substance concentrations from the interaction of rock weathering and vegetation. They may even be completely of anthropogenic origin.

One example is the development of dioxine concentrations in the soils. In regions with a long industrial history, high population density, a large number of heating systems and high levels of motorized transport, the low degradability of many dioxine compounds has led to a concentration level in the soil whose changes must be monitored although they do not pose any threats to health. The guideline I just mentioned refers to the environment's absorption capacity for released substances. We must note that environmental research, i.e. in particular ecosystem research, cannot yet specify the pollution limits of soils in larger areas for dioxine compounds and other persistent substances. The same is true for the resilience of ecosystems to discharges and substances that cause soil acidification. In forest soils in Germany the inputs of substances from air pollution have partially led to a marked reduction in alkali saturation sometimes in deeper layers. The particular danger of these processes is that the chemical properties of the groundwater have changed due to an increase of toxic substances resulting from soil acidification in certain areas. If this groundwater is used to provide drinking water, special water treatment measures are necessary which cause considerable costs. By this I mean to say that important functions of other areas of the environment depend on soil quality. The term "sustainable land use" therefore also includes the classification of soil functions and land use as a sink for the substance transformation processes associated with economic growth. In the developed industrialized countries we increasingly recognize that we need integrated environmental and economic accounting for political and economic decisions to take care of this task.

I am pleased that air pollution control policy in Germany has succeeded in clearly reducing substance inputs into soils. Environmental policy is about to face a new challenge: the recovery of organic wastes, which is legally bound into the Closed Substance Cycle and Waste Management Act in Germany. In the future considerable amounts of sewage sludge and probably even more compost will be applied to agricultural land and used for the recultivation of devastated landscapes. This means that a large proportion of the nutrients removed from the soil by crops can be reintroduced and that soils can be renaturalized. At the same time the concentration of organic substances in soils will increase and the soil structure will be better suited to agricultural and horticultural use or even the formation of biomass. In this respect the trend is towards greater sustainablility of land use. However, for the time being it is unavoidable that with sewage sludges and composts soils also absorb certain loads of heavy metals and persistant organic substances which lead to the concentrations mentioned. If we want to define the conditions for sustainable land use the following questions arise: Must appropriate measures be taken to ensure that soil properties deviate as little as possible from those properties that result from soil formation with natural vegetation? Or can we accept greater deviations without having to fear that the dynamic equilibrium resulting from other substance concentrations will no longer enable sustainable land use in the long-term? And where are the concentration limits for persistent substances or changes in filter and buffer properties beyond which long-term and large-scale risks arise for the biological soil processes, water bodies, the atmosphere and human health?

Answering these questions places great demands on scientists. They have to describe in detail soil quality objectives, i.e. optimum soil properties and functions for certain usages, and soil functions for the ecobalance when soil is used in these ways. But environment politicians and administrators must also develop new ideas and instruments. To ensure that soils as sinks are not overloaded with pollutants, monitoring and documentation of environmental changes must become a major instrument of environmental policy. Planning elements will have to be incorporated in environmental law to a greater extent. All economic players must be allocated greater responsibility for the long-term environmental effects of substance releases. We thus need an intensive dialogue with all groups in society and the general public about the acceptance of risks and conduct in the event of uncertainty.

Just like pollution from the insufficient monitoring of substance fluxes, the consumption of free areas of land for settlements, industrial activities and the expansion of infrastructures is looming on the horizon in many developing countries. The social and cultural functions that soils have acquired with the uses just mentioned are part of human ecology. However, land use cannot be expanded infinitely, otherwise the ecobalance would severely restrict the development options for future generations. People in densely populated countries and large cities should be made more aware that they are living on credit from an ecological point of view. The biological functions of soils and ecosystems, which no longer prevail due to building and sealing, must be compensated for by the remaining open spaces. These ecological costs, too, must appear in an integrated environmental and economic accounting system, which is to express the changes in the wealth of a society more appropriately. With these examples I want to show that we, too, in developed countries are obliged to future generations. We, too, have to think about the conditions of sustainable land use as one of the preconditions for sustainable development.

The scientific body advising the Federal Government on technical and political requirements resulting from global environmental changes has recommended that the Federal Government support more comprehensive agreements to protect soils above and beyond the current Desertification Convention. It is undoubtedly correct that the international community must make similar efforts to maintain soil resources and socio-economic stabilization in threatened mountain regions, similar to the efforts provided for in the Desertification Convention. Above and beyond this, greater international cooperation is required to effectively combat the risks of overloading soils with pollutants. Nevertheless, I believe that it is important to first of all gather experience with the new instrument, the Desertification Convention, and to then decide whether we need further-reaching international agreements to conserve soil resources, to develop sustainable land uses and decide on the form they should take. To solve these complex tasks both financial means and an institutional frame must be provided. Bilateral cooperation to protect the environment and to promote sustainable development in developing countries will continue to be significant. Successful examples make it easier to increase the awareness of the population in developed countries and draw their attention to the need to "help people to help themselves". The young generation in particular must be made aware of how close the link is between the economical use of natural resources and a life worth living in the future, and that this link is the same all over the world. It is the difficult task of environmental education to spread knowledge about living conditions in developing countries that leads to empathy with people in other parts of the world and to release these forces for actions of solidarity.

I would like to conclude with a few comments on ethical values which must determine our relationship with the environment. It must be our objective to strengthen an environmental conscience based on respect for nature and creation and the obligation to pass on our natural heritage to future generations. I am of the opinion that independent ways will have to be found in every cultural sphere for achieving sustainable development. However, it is absolutely essential that the principles of human dignity and human rights are observed. In Germany we can be guided by an ethical tradition which goes back to the idealistic philosophy of the 18th century in Europe. The Königsberg philosopher Immanuel Kant defined the guiding principle for moral action then: "Act so that you can will that the maxim of your action be made the principle of a universal law". The philosopher Hans Jonas, who emigrated from Germany in 1933, adjusted this principle for our times in his book published in 1979 "The Imperative of Responsibility - In Search of an Ethics for the Technological Age": "Act so that the effects of your action are compatible with the permanence of genuine human life". He also expresses this principle negatively: "Act so that the effects of your action are not destructive of the future possibility of such life".

Food Security and Poverty Alleviation
- Essentials for Sustainable Land Use and Development -

Carl-Dietrich Spranger
Federal Minister for Economic Cooperation

The conference is dedicated to the conservation of soil, the most important basis of production for food security. The decreased yield from soil which is cultivated and exploited using methods which are not site-appropriate and which cannot therefore be sustained, is an enormous threat to food security for the human race. The methods used are often the result not of a lack of knowledge on the part of the land-users about the need to protect and conserve the soil, but rather the direct result of poverty and population growth, unregulated land-use rights and unsound agricultural policies. Smallholders often have no access to credit or good soil, the necessary means of production. They are therefore forced to cultivate land which is not suitable for farming. The resultant damage is a foregone conclusion.

The rapidly escalating deterioration of the soil in many countries is a danger to the vital natural resource base of 900 million people. According to World Bank estimates, about two-fifths of productive land in Africa, a third in Asia and a fifth in Latin America, is at risk from desertification. This causes floods of refugees whose numbers keep on growing. Ten million people, who were unable to remain in their homes because of deteriorating soil quality and the resultant impoverishment, have already become so-called environmental refugees. To these numbers must be added the refugees fleeing as a result of wars and civil wars, at the root of which the scarcity of cultivable land is often to be found. In Rwanda and Burundi, the current arenas of major human disasters, experts say that this is in fact the main cause of the crisis and the ethnic conflict. At present there are more than 40 conflicts being fought in Africa, Asia, Latin America and Europe too. The result is 30 million refugees and almost as many internally displaced persons. When many thousands of people are obliged to crowd together in refugee camps or other temporary accommodation, often with little more than the clothes on their backs, the inevitable result is over-exploitation and destruction of natural resources such as land, water and trees.

The disastrous cycle goes still further: over-exploitation and environmental destruction result in bio-diversity losses of shocking dimensions. The rainforests are particularly at risk: 16 million hectares - an area almost half the size of the present-day Federal Republic of Germany - are cleared each year to provide new pasture and arable land, or living space, or just for the timber - without being replanted. Since this land is for the most part barely suitable or even unsuitable for planting or pasture, it is often deserted after a very short time, since a few harvests suffice to use up the last reserves of humus and nutrients in the soil.

I do not wish to conjure up horror scenarios, but all these developments and prognoses make one thing clear: the situation is serious. Developing and industrialised countries alike are facing a major challenge, a challenge they must tackle purposefully.

Awareness of the global problems which affect all of us has grown. The international community of nations has recognised the importance of protecting our vital natural resource base, including the soil. The World Population Conference in Cairo, the Earth Summit in Rio, the Social Summit in Copenhagen, the Climate Conference in Berlin and also the World Conference on Women in Beijing, have drawn attention to the unabated growth of the world's population and the challenges this brings with it for securing our food supplies and preserving natural resources, and have called for appropriate action. The World Food Summit which the FAO is organizing this November is intended to renew at the highest political level the obligations of the community of nations to reduce hunger and malnutrition and achieve food security for all the peoples of the world. The summit should not just be limited to joint declarations of intent and good will, rather an action plan is also to be drawn up and passed.

Development co-operation is the focus of great expectations. The prime goal of all development efforts is to help improve the economic and social situation of humanity, especially the poor sections of the population in the developing countries. To do this we should concentrate our funds on key areas where the need for action is the most urgent: poverty alleviation, food security and rural development, environmental and resource protection, the promotion of basic education and vocational training and, last but not least, women in development. These are the focal areas of German development co-operation. The principles behind our co-operation are popular participation both in the political process and in the co-operation projects, help towards self-help and orientation towards the capabilities of our target groups and partners. Thanks to the close interaction and links between the individual areas and the knowledge we have gained that one-dimensional solutions are rarely successful, cross-sectoral co-operation is becoming more and more important. This is no longer a mere question of financial transfers, rather it is a matter of overall policies which are co-ordinated and pursued in joint responsibility between industrialised and developing countries, with the goal of global human security.

Alleviating poverty is a cross-sectoral task of our development co-operation. Every promotional measure is examined to decide to what extent it contributes to poverty alleviation. Effective poverty alleviation is in its turn a precondition for responsible stewardship of the environment: not until there are long-term perspectives for the population in our partner countries and the daily struggle to survive is no longer at the forefront of their minds, will it be possible for people to give a thought to responsible stewardship of their environment.

Poverty alleviation cannot be separated from food security and the development of productive potential in the developing countries: only when all people have access at all times to the food necessary for an active and healthy life, has food security been achieved. A key to food security for the poor in the towns and in the countryside is the development and enhancement of their purchasing power and income earned by their own labours.

Our concept for sustainable development demands that the vital natural resource base also be guaranteed for the generations to come. Sufficient food production for both present and future generations, and protection and preservation of the vital natural resource base are not mutually exclusive. Both goals, food security and the preservation of natural resources, must be achieved simultaneously. Only then will we begin to approach our supreme goal, namely improving the lives of the people in this one world.

Our partner country's own efforts at creating a suitable political and economic climate for productive and ecologically balanced agriculture and rural development are a fundamental prerequisite for the success and sustainability of all development policy measures. Removing the obstacles which prevent people from developing their creative talents is a task for our partners. Our development co-operation supports these efforts by our partners - but we have neither the ability nor the wish to act as substitutes!

What is also necessary is closer co-operation and improved co-ordination at the international level. This affects both political and development co-operation and also co-operation in the field of research and development.

Unless we include all those affected and gain their active participation in planning and land use, we will not make the necessary progress we desire. Access to land and soil, particularly for the poor stakeholders and the landless population, must be secured through the creation of legal certainty and the removal of injustice with regard to the possession of land. This is generally a very sensitive area of political discussion. We use exchanges of opinions on development policy and our negotiations to remind the politicians and parliamentarians in our partner countries of their responsibilities. Improving agrarian structures and land laws is at the top of the political agenda in many developing countries, but also in the countries under transition.

What we need is decentralized public administration structures, the reallocation of efforts to the level at which the problem is being experienced, with a simultaneous improvement in crucial services such as rural extension services and information systems, and rural financial systems. Only then will the users of the land be in a position to fulfil their tasks as "protectors of the land" as well.

International and national agricultural research needs to be directed more towards the problems of sustainable soil protection. Only then can planners and land-users be given suitable and tested suggestions for targeted measures to protect resources. Without corresponding investigations, there will be no chance of achieving sustainable success.

At the same time it is worth recalling tried and tested traditional conservation systems, which are at risk of being lost if they are not identified and further developed in co-operation with the farmers. These traditional systems are generally characterized by a high degree of social and ecological compatibility, require little in the way of technology and are therefore easy to realize and acceptable to broad sections of the population.

What is also needed is the creation of an internationally binding framework for soil protection, a necessary task which needs to be realised as quickly as possible. Soil conservation is a matter of global interest. Like most major issues facing development policy, these problems can no longer be solved in national isolation. It is the job of every country to create the legal framework needed as a foundation for sustainable and ecologically sound exploitation of our vital natural resource base. Including women, who are the most affected by the problems in developing countries, in the planning and implementation of soil conservation measures is a necessary prerequisite for success. We need courage and determination, if we are to design within the framework of our political possibilities the necessary suitable protective measures in such a way that they are both acceptable and affordable. Because of the scale of the task facing us, the development of site-appropriate soil conservation systems is not just the duty of those immediately affected, but rather a task for every level of society, to which all concerned need to make their respective contributions.

The motto of this ISCO conference is: "Towards Sustainable Land Use: Furthering Co-operation Between People and Institutions". It's all about us, people who are the perpetrators, the victims, and the shapers of our own destiny, and the question of how we should deal with the problem of sustainable land use.

Extent and Impacts of Soil Degradation on a World-wide Scale

Elizabeth Dowdeswell
Executive Director
United Nations Environment Programme

I am indeed honoured and pleased by this opportunity to be with you in Bonn today and to address this important Ninth Conference of the International Soil Conservation Organization. Once again, Germany is assuming a leadership role in caring for this Earth. This meeting is particularly well timed as the world community reconvenes next week to continue its work on the desertification convention.

In the ancient nature cults the earth was seen as the source of fertility, the site of germination and regeneration, indeed the womb of life. The earth was held sacred as the embodiment of a great spirit, the creative power of the universe. The earth spirit was believed to give shape to the features of the landscape and to regulate the seasons, the cycles of fertility, and the lives of animals and humans. Rocks, trees, mountains, springs, and caves were recognized as receptacles for this spirit. And even today modern cultural figures like Michael Jackson are singing Earth songs - admonishing that we have turned the "Kingdom to dust".

Perhaps our most precious and vital resource, both physical and spiritual, is that most common matter underfoot which we scarcely even notice and sometimes call **dirt**, but which is in fact the mother of terrestrial life and the medium wherein productivity is regenerated.

As we meet today, the human treatment of the environment has grown worse and ... in our generation, it has brought us to a point of crisis.

Salinization, erosion, denuding of watersheds, silting of valleys and estuaries, degradation of arid lands, depletion and pollution of water resources, abuse of wetlands - all are now occurring more intensively and on an ever-larger scale. Added to these problems, are the problems of pesticide and fertilizer residues, domestic and industrial wastes, the poisoning of groundwater, air pollution and acid rain, the mass extinction of species and finally the threat of global climate change.

Today, there is clear and urgent reason for us to be concerned over the adequacy of land resources to satisfy the demands of our profligate civilization. Our concern is not merely for the availability of these resources but for their quality as well. In the global endeavour to save the environment, a far deadlier toll and perhaps an even greater threat to future human welfare than that of pollution of our air and water is that exacted by the undermining of the productivity of the land itself through accelerated soil erosion, increased flooding and declining soil fertility.

There are encouraging developments too. We know much more about the natural and man-induced processes at work. We can and do anticipate some of their consequences. Degradation and pollution are not inevitable. They can be controlled. We can avoid the major abuses and devise better modes of environmental management.

It is in the interest of promoting and disseminating such an understanding that we are meeting here in Bonn today.

By any definition - we do not practice sustainable land use throughout the world.

Degradation of the world's soils has been called a silent emergency. The soil slips away in such small increments that its loss is hardly recognized, except at times of major flooding or dust storms.

In 1990, the area of arable land available on a per capita basis was 0.3 hectare but this will be reduced to a quarter-hectare by the end of this decade. If the current trends of population growth and soil degradation continue, this will drop to 0.15 hectare by the year 2050.

Current studies reveal that three-quarters of the world's soil degradation occurs in the tropical areas. The reasons for this are obvious. Tropical soils are not only ancient and highly weathered, needing frequent fertilizer use but also suffer repeatedly from droughts and floods. Farmers thus have also to cope with a vast array of pests and diseases. Erosion of soils is also exacerbated by deforestation, overgrazing or inappropriate agricultural practices on inherently poor soils.

Estimates are that well over 100 million people are facing total loss of productivity from their land. Moreover, as the demand for food mounts, land-hungry farmers are pushing their farms higher onto steep slopes and also into semi-arid regions where ploughed land is vulnerable to water and wind erosion.

Intensification of agriculture in many parts of the world during the last three centuries, coupled with over-population and over-stocking, has led to accelerated human-induced soil degradation. Since the beginning of agriculture, more than half the earth's stock of productive land has been irreversibly lost.

The rate at which this arable land is being lost is increasing and is currently 30-35 times the historical rate. The loss of potential productivity due to soil erosion world wide is estimated to be equivalent to some 20 million tonnes of grain per year or 1% of global production. Water erosion annually carries about 20 billion tonnes of topsoil towards the oceans. This loss is equivalent to between 5 million and 7 million hectares of arable land per year. Out of the 20 billion tonnes of topsoil lost in the oceans, 45% comes from South and Southeast Asia, and 1.5 billion tonnes from the Amazon Basin.

The United States was reported in 1993 to have lost equivalent of $18 billion in fertilizers and nutrients to soil erosion. Soil erosion rates in sub-Saharan Africa have increased 20-fold in the last 30 years. Erosion rates as high as 200-300 tonnes per hectare per year are commonplace in certain parts of the continent. A study in 1991 by ODA reported a five-fold increase in silt concentration in the Blue Nile originating from the Ethiopian highlands compared to 1930 levels. Only a century ago 40% of Ethiopia was covered with forest, now only 3% of the forest cover remains. As a result, the annual Ethiopian highland topsoil loss is estimated at 1 to 3 billion tonnes, a loss equivalent to about 1.5 million tonnes of grain a year.

Major problems of soil salinity are now appearing in Asia too. About 487 million hectares are currently affected by soil salinity in that region and 23% of China's irrigated land and 21% of Pakistan's are affected by salinization. Thus, money spent on irrigation is contributing directly to land degradation.

The World Map of the Status of Human-induced Soil Degradation (GLASOD) produced by UNEP and the International Soil Reference and Information Centre (ISRIC) in the Netherlands, shows that out of the world total land area of 13.4 billion hectares, about 2.0 billion hectares is degraded to varying degrees. Soil degradation in Africa and Asia taken together itself accounts for a total of 1.24 billion hectares or 62% of global soil degradation.

Thus unsustainable practices cost us money, undermine productivity leading to food insecurity and ultimately poverty and, cause ecological damage, in ways we have yet to fully understand.

But - you are the experts - you know these statistics well. My role is to put this issue in context - to paint the large picture.

Broad based rural development through agriculture, it is increasingly recognized, can ameliorate simultaneously many of the world's most acute problems.

Not only will more food be grown under the auspices of such programs, but it will be grown by those who most need more nourishment and income. By providing greater income and employment in rural areas, properly designed agricultural development can help stem the flow of migrants to cities. And with improved social conditions, the rural poor may well follow the historical path toward smaller families.

Yet somehow, in many regions of many countries, programmes do not seem to be working. All the money, all the research, all the experts have done little for those on the bottom. What has gone wrong?

Analysts find many culprits. Depending on personal prejudices, the economist may see a failure to generate adequate capital for raising productivity and technology that impede efficiency. The sociologist may see traditional cultures incapable of assimilating the requirements of modernization or socio-economic structures that compel the poor to live recklessly. The political scientist may stress the absence of the administrative capacity to implement social change, or outline the power relationships that prevent the poor from taking control of their own destiny.

All these explanations contain truths, but an additional perspective that in many ways subsumes them all, and is almost always overlooked, is an ecological view. A common factor linking virtually every region of acute poverty, virtually every rural homeland abandoned by destitute urban squatters, is a deteriorating natural environment.

Ecological degradation is to a great extent the result of the economic, social, and political inadequacies. It is also, and with growing force, a principal cause of poverty. If the environmental balance is disturbed, and the ecosystem's capacity to meet human needs is crippled, the plight of those living directly off the land worsens, and recovery and development efforts become all the more difficult.

The glaring disregard of the ecological requisites of progress is at least partially attributable to the rigid compartmentalization of professions, both in the academic world and in government policies.

When reading the analyses of economists, foresters, engineers, agronomists, and ecologists it is sometimes hard to believe that all are attempting to describe the same country. The actions of experts frequently show the same lack of mutual understanding and integration. Engineers build one dam after another, paying only modest heed to the farming practices and deforestation that will, by influencing river silt loads, determine the dams' lifespan. Agricultural economists project regional food production far into the future using elaborate, computerized models, but without taking into account the deteriorating soil quality or the mounting frequency of floods that will undercut it. Foresters who must plant and protect trees among the livestock and firewood gatherers of the rural peasantry receive excellent training in botany and silviculture, but none in rural sociology.

So ignoring the environment and all of its interconnections is clearly one aspect of the problem.

A second lies in our seeming inability or reluctance to listen to the people. The phrase "listening to the people" is an important aspect of the new paradigm of sustainable development in the post-UNCED period. It imbues the concept of development with a meaning which goes far beyond the conventional one. Listening to the people is about building development around the priorities, needs and objectives of the people it seeks to benefit. It is about empowering people to gain control over their lives through active participation in their own development. It is about recognizing the value of indigenous knowledge in sustainable development. It is about the centrality of people in development projects rather than technology.

In our relationship with nature, we have been poor students of history. Despite the all too visible panorama of ruined landscapes and abandoned civilizations, mistakes of the past seem time and again to serve as models instead of usable lessons.

But perhaps, the greatest obstacle in achieving soil conservation is the general inability to protect the environment as a whole. This stems, in part, from inadequate legislation, deficient enforcement of existing laws, poor organization, shortage of trained personnel, information and technology and, lack of financial resources. Desertification is not simply a scientific and technical problem - it is a socio-economic and political one.

Over the years UNEP has tried to play its role. For example, in 1987, UNEP conducted a survey on national soils and land use policies and the extent to which these policies were operational and discovered more than half of the countries surveyed had no operational national soils and land-use policies. African countries ranked the lowest in this list. We also discovered commendable initiatives.

The Dust Bowl experience was a pivotal event in the formation of dryland management practices in the United States. Efforts to hold back desertification in China have been carried out on a massive scale with some success. Actions taken towards sustainable food production in India, particularly at the community level are highly commendable. Australia's "Land Care" programme, is another innovative and determined approach to tackle this problem.

In 1982, the UNEP Governing Council adopted a World Soils Policy, an important element of which was the development of methodologies to monitor global soil and land resources.

A landmark event was the commencement in July 1987 of GLASOD (Global Assessment of Soil Degradation), which led to the publication in July 1990 of the World Map of the Status of Human-induced Soil Degradation (GLASOD). The publication of this map was intended to create awareness among policy and decision-makers of the seriousness of the problem of global and continental soil degradation.

For agricultural planning at national and sub-national levels, a more detailed and quantitative information system has been developed by UNEP, ISRIC and FAO. This is called the World Soils and Terrain Digital Database (SOTER). The SOTER methodology for storing, analyzing and retrieving soils and terrain information at various scales has been illustrated by four UNEP-supported regional pilot projects in Africa, Eastern Europe, Latin America and the Middle East. Systematic interpretations of land degradation assessment, risks of soil erosion, and food producing capacities of certain localities have been successfully done.

Another project of great interest to UNEP is the World Overview of Conservation Approaches and Technologies (WOCAT). WOCAT is an initiative of the World Association of Soil and Water Conservation, the implementation of which is coordinated by the Centre for Development and Environment of the University of Berne. The overall goal of WOCAT is to contribute to sustainable utilization of soil and water resources on cropland, grazing land and forest and mixed land use, and to reduce soil degradation world-wide.

You will hear more about these efforts during the week in workshop sessions. UNEP hopes to continue its efforts in establishing dynamic land resources information systems, with updating and purging capabilities, and the publication of derived digital maps and tabulated data on aspects such as soil degradation status and risk, potential productivity, irrigation suitability and landcover changes.

UNEP hopes to continue to support long term monitoring of soil degradation and other types of land degradation. We understand the importance of social systems and their relationship to soil degradation as also the importance of action-oriented data collection and appraisal besides GIS and remote sensing.

In the future, reinforcement of the interlinkages within the different areas of land management - soils and agriculture, biodiversity and biotechnology, desertification and dryland ecosystems, and freshwater resources - as well as with activities such as environmental economics, assessment and monitoring and, legislation. UNEP will continue calling upon countries to address the problems of soil degradation by formulating and implementing their national soil policies as integral parts of their national developmental and environmental action plans. Experience has shown that soil degradation problems cannot be overcome only through isolated short-term programmes and projects.

UNEP will enhance its efforts in clarifying the relationship between climate change and global warming and management of soils. Agricultural practices have important impacts on global climatic regimes. Recent studies suggest burning of savannas for agriculture contributes significant amounts of carbon dioxide to the atmosphere. Methane gas emissions, mainly from agricultural activities are estimated at 500 million tonnes per year. Developing state-of-the art synthesis of the current knowledge and research needs on the science and practice of soil carbon management and identifying land use management policies that optimize both sustainable production and carbon sequestration will be supported by UNEP.

Let me conclude with a special word about the negotiation of the Desertification Convention. Desertification needs a comprehensive, integrated approach at the local level within a world-wide framework of favourable or at least neutral trade and economic policies, as well as supportive national leadership and administration.

Certainly since Rio the international community has been calling for sustainable development and improved environmental management. Integrated environment management is made up of many focussed concrete actions: fixing potholes, training teachers, increasing access to credit, enabling women, harvesting water, etc. Concrete, diverse actions working together within a coherent framework to contribute to a common goal within a common time frame.

The biggest challenge is to show how this international convention can persuade a whole ladder of bureaucrats to support something that will be done in the field; to create the enabling environment within which local initiative and effort can achieve the necessary results. The issue is not so much WHAT needs to be done in the field but HOW to get it done. Above all the convention calls for LEADERSHIP and COMMITMENT to making improved environment management happen in the field - but from a bottom-up approach.

All terrestrial life ultimately depends on soil and water. So commonplace and seemingly abundant are these elements that we tend to treat them contemptuously. But in denigrating and degrading these precious resources, we do ourselves and our descendants great and perhaps irreparable harm. The once prosperous cities of Mesopotamia are now, mute time capsules in which the material remnants of a civilization that lived and died there are entombed. This happened because of their neglect of their precious natural resource - their soil.

Meetings such as this one give me hope that we have learned from the lessons of the past and are not doomed to repeat our mistakes.

Two Voices from the Grass Roots

I.
Chief Leonard Little Finger
Lacota Garden Association

I want to thank the ISCO Organizing Committee for giving me the opportunity to come before you. I am pleased to answer the question: Was the conference a success and did we receive anything from this? Of course, the answer could be very easily "Yes" or "No", because ... or "maybe", but I would like to answer this by telling you who I am and where I come from. To many I perhaps am the first native American from the United States that you have met. I come from a long line of people. I come from a land which is very ancient. The black hills of South Dakota are the oldest mountains, far older than the Alps in the European countries. I come from a people that I can trace my lineage back to 19 generations. I come from a land, a land that was in all essence the purity of life. I come from a land where buffalos once roamed in range of 30 million. The area that I live in, the Upper Plains of America or the United States is very fragile and yet it supported 30 million buffalos. I come from a people that at one time ranged near 30 million throughout the United States. Today we are less than a million. Today there is less than a million buffalos, more like maybe 3,000 to 4,000. Today our land is fragile and it has had its share of natural calamities, of natural abuse, misuse to the extent that huge droughts occurred in dust storms that swept across our country and not only the native Americans, but also the farmers and ranchers suffered equally. It is a land that has a small piece of grass which is known as Buffalo Grass. This Buffalo Grass only goes down to maybe 5 to 7 centimetres. And once that is destroyed then there is nothing to hold the structure of the soil and the soil can be depleted of its nutrient values. My people today live in the poorest county in the United States, where the unemployment rate is 85% which means that 8 out of 10 people do not have employment. And yet at one time we had 219 million acres of land which was self-sustaining and self-supporting. I myself am a survivor as we all are perhaps are survivors. I come from a chief whose name was Chief Big Foot and he and his people were massacred at Wounded Knee Creek in 1890. My grandfather who was a boy escaped at that time. During these past days not only have I had a chance to join the Conference and talk with people, I have gone to the museums and have looked at 15,000 years of history of mankind here from the Neanderthal man to the man stepping on the moon. High technology, many, many things that have occurred. And so we are all survivors and so I felt that I cannot say that the money that was spent here at this congress can be used somewhere else. The most important thing I see is the ability to communicate. Fifty years ago I would not be able to communicate with you because I did not know my English language. My father did not know this language, my grandfather did not know this language. And yet we had years and years and centuries and centuries of experience to draw from. It is this that now comes from the people that are in this room, from the Africa countries to the Baltic countries, to our country and all inbetween. The people are, as they say, grass roots.

And as I have been such a grass root I do not like that term grass roots simply because I am a person. Perhaps at the beginning when Euro-American people came to my country they thought of us as children. We raised children appropriately. Children need to be taught the right way. But what is the right way without giving credit to the native people? I come from a very small community. I come from a people that is unheard of to travel anywhere, less to come to this place. And yet I have the opportunity to meet people, many different people. And that in itself I think is a beginning of what we as a people in the various countries need to know. Perhaps I am of a lost tribe, I do not know. Perhaps I have brothers in India that are closely related to me. Perhaps we are all of one kind. I think technicians and administration people from our and your country can share thoughts with others about things which should be done.

I will close by sharing some things with you. In our society of Lakota people we have what is known as a sacred circle. Each one of us as an individual has the secret circle. It is where we put our feet and as we turn 360 degrees all the things that are seen, heard, felt, with all the senses that we have. When we turn in this secret circle, one must realize that we are no better than all those things that we see, hear, feel and touch. With that comes respect. The abuses of people, of the soils, of the plants which went to extinction, and of animals such as the buffalo. Had there been respect these things perhaps would still be. We talk in our society about the 7th generation. I represent the 5th generation of grandfathers. We have some grandmothers yet and young women and we must recognize them. My son becomes the 6th generation. I would like to explain to you that the 7th generation is the generation that will not change. So it is up to us as a people to prepare for that 7th generation because we will not have another chance. The 7th generation becomes eternity. Those things are the responsibilities for each and everyone of us that are here at this conference and when we return to our homes and help by listening. So my underscore is to look back. It is easy to say this. Everyone must examine themselves. What I say is: "I know who I am, because I know where I came from - spiritually, physically, mentally, economically - and I know what I am. I am a creation of our creator, I come from a lineage of people that I am proud of. And knowing these things of who I am and who I came from, I know where I am going spiritually, physically, mentally and economically. Economically only so that I have adequate food, shelter, water and clothing".

I thank you with a word in my language: *Mitakuye oyas'in*. All my relatives thank you.

II.
Cecilia Violet Makota-Mahlangeni
Zambian Woman in Agriculture

Good morning to you all participants, scientists, researchers and administrators. I come from Zambia which has a population of about 12 million people. I am here because I am the initiator of an organization for women called Zambian Women in Agriculture. This organization was initiated by us rural women farmers, because we realized that there was too much hunger, too much starvation, and too many deaths from malnutrition in my country. So we elderly women came together and put our experiences together. The objectives of this organization are the alleviation of poverty and the elimination of starvation and hunger by using soil conservation and striving for sustainable land use. What we had in our hand was the soil which is so very precious to us because it is the womb of life.

But how should we use the soil when it was already overused. There were no high yields that would have helped to get out of poverty. Because the time was so short, we had to ask the government to help us, to give us chemical fertilizers. We now realize it was not good for our and for our children's health, because the chemical properties have side-effects. So we are deciding now to use compost manure, cow dung and mix it with soil and conserve the soil for sustainable land use. The justification for only targeting women was that in Zambia 80% of the food is produced by women. They till the land with their hands, they do the weeding with their hands and they harvest with their hands. We cannot afford equipment and machinery, we are unable to maintain the machinery. So we thought to go back to use our own knowledge, our known implements. I was very intimidated when I read this small conference book "Precious Earth" I was given by the Zambian National Farmers' Union when I was asked to represent the rural farming woman in my country. I read that there would be perhaps some hundred soil scientists and researchers. When I came I felt encouraged because I saw although they were talking on high technical levels, they were mixed with down-to-earth people. I could understand what they meant when they said "You must stop salinity of the soil" and I could say we know how we do it. We put ashes in the soil and we mix it with droppings from animals. And we use grass and we also use the crop residue from the soja beans from the mais and we bury it underground in order to fertilize the soil. While they were speaking in scientific terms, I remembered talking my mother and I knew why we used ashes and why we were putting the grass and the plant residue together and making compost. My organization, which is called Women in Agriculture, aims to facilitate the development of affordable and sustainable agriculture production. Affordable which means that we should use the things which we can afford doing. And sustainable that means it should be continuous, it should not end somewhere, because we are living now for our children and our grandchildren so that they too can sustain the fertility of the soil for years to come. With the results from this conference in mind, I would ask the scientists and the researchers which work in their laboratories that when they have found results they should come to us as paritas. We are the people who do the work and we the women in agriculture are going to prove whether what they have discovered in their laboratories is a failure or is a success.

To end my speech I would like to thank the organizing committee of ISCO for bringing me here to rub shoulders with scientists and people from research and administration. I will give you this copper plaque from Zambia. It is an ending resource in my country. I am going back now and I believe we can achieve sustainable land use and fertile soils will maintain.

Can Sustainable Land Use be Achieved?
- An Introductory View on Scientific and Political Issues -

E. Fleischhauer & H. Eger

Summary
This introduction tries to show that, in the decades ahead, the development of new sustainable forms of land use will be among the greatest scientific and political challenges not inferior to, and highly interdependent with, the global tasks to ensure climate and biodiversity. Weight is given particularly to the question of how the opportunities to feed people, to enhance the sources of energy, to invent and produce new products and to improve mobility and communication can be applied concurrently so that the flows of energy and substances in ecosystems, their resilience and the diversity of nature are maintained in order to secure the welfare of an increasing world population. The answer must remain incomplete and provisional due to the boundaries of knowledge concerning coming developments and the complexity of the ecoregional, economic and social characters of land use. But together with the many profound results of research in these proceedings of the 9th ISCO Conference these thoughts might be helpful to strengthen the momentum of the scientific and political discussion about approaches which may lead to land uses which first of all have to be ecologically sustainable before they can provide economic and social security for people.

1 Concepts of sustainable land use

When discussing sustainable land use, different concepts are often meant. The term sustainable soil use is common in German-speaking countries. There is no equivalent for it in English terminology. Sustainable soil use is a form of land management which retains the natural fertility of the soil and allows for the production of high quality foodstuffs and renewable natural resources on a long-term basis. Sustainable soil use implies that in the case of agricultural and forestry land management, cycles and energy fluxes between the soils and other spheres of the natural environment (bodies of water, the atmosphere, nature reserves), are considered.

The term sustainable land use is more comprehensive. "Land" is used commonly to mean a section of the earth's surface with all the physical, chemical and biological features which influence the use of the land. The term "land" thus refers to soils, forms of terrain, climate, hydrology, vegetation and fauna and also includes improvements in land management such as drainage schemes, terraces, etc (Sombroek and Sims, 1995). In FAO terminology "quality of land" is defined to cover complex factors which influence the suitability of land for specific purposes; the term thus spans a wide range of meanings. Land quality can differ with respect to the availability of water and nutrients for agricultural and forestry purposes, the resistance of the soils and soil use with respect to erosion, the natural productivity of pastureland and forests, and terrain attributes affecting mechanization (FAO, 1976, 1993). The following characterisation of principles of sustainable land management refers primarily to these uses: Environmental Sustainable Land Management (ESLM) "combines technologies, policies and activities aimed at integrating socio-economic principles with environmental concerns so as to simultaneously:
- maintain or enhance production services (productivity)
- reduce the level of production risk (security)

- protect the potential of natural resources and prevent degradation of soil and water quality (protection)
- be economically viable (viability)
- be socially acceptable (acceptability)

These five objectives of productivity, security, protection, viability and acceptability are called the pillars of sustainable land management, and they must be achieved simultaneously if true sustainability is to be predicted. Attainment of only one or several of the objectives, but not all, will result in partial or conditional sustainability" (Smyth and Dumanski, 1993).

This definition of sustainable land use can be applied in its broader sense. The term land use encompasses not only land used for agricultural and forestry purposes but also land uses for settlements, industrial sites, roads, etc and, in soil surveys on land use and in land use planning also land whose economic use is renounced in preference of preserving the habitats of wild plants and animals. Land use including these uses can only be termed sustainable if a spatial distribution, i.e. land quality, is achieved in which the soil functions are guaranteed for the eco-balance and biodiversity, at least within a broader landscape framework. In other words, land use which limits the interaction of soils, bodies of water and the atmosphere and degrades the habitat criteria vital to the biological diversity of flora and fauna may only occupy a certain part of an area. In this sense an assessment and weighing of land quality criteria of different land uses is carried out in landscape and regional planning processes in Germany. In order to apply categories such as land quality and landscape value for international comparisons or planning towards sustainable land use, the scientific basis for a framework of concepts and indicators needs to be considerably improved.

A number of basic questions can now be approached concerning the conditions of sustainable development with respect to ecology, economy and the social needs of mankind as stated by the Brundtland Commission: "Sustainable development is development that meets the needs of the present without compromising the ability of future generations to meet their own needs." (World Commission on Environment and Development, 1987).

The World Bank has defined Environmentally Sustainable Development (ESD) as an interdependent triangle framework (Figure 1).The triangle demonstrates the close ecological, economic and social interactions as prerequisites for sustainable development. It is not irrelevant to pose the question as to the priority of maintaining the dynamic equilibria of substances and energy in the natural environment and of biological diversity and the role they play in the concepts of sustainable land use and sustainable development. To date only a small number of the ecological prerequisites have been formulated precisely enough for them to be used as an operational tool for implementing sustainable development. Ecosystem integrity, biodiversity and carrying capacity are abstract terms for extremely complex relationships and processes. Evaluating natural resources consistently raises questions as to methodology when changes in national prosperity are determined and economic programmes are being planned. Irrespective of this, progress towards sustainable development is only possible if the changes in the state of the environment in terms of quality and quantity, such as the amount of natural resources, biological diversity and the resilience of eco-systems vis-à-vis anthropogenic interventions, are seen as key yardsticks. As history shows, former civilisations were able to overcome their economic and cultural decline as long as the ecosystems which made up their environment remained intact and free of interference. Avoiding detrimental changes in the material cycles and energy fluxes and preventing the loss of biological diversity in man's natural environment are of utmost priority among the prerequisites of sustainable development. In our times where man-made impacts are spreading to nearly every part of the world the ecological key processes within the ecological web need to be adequately understood to develop the necessary means to strive for economic and social progress. Only when this has been achieved will it be possible to evaluate changes in material cycles and energy flows in ecosystems and in biodiversity. This is a task all countries are faced with given the present state of population density. Finding a solution calls for international cooperation. An important start has been made with the conventions on climate change, biodiversity and desertification and with further agreements which either directly or indirectly enhance the protection of the natural environment.

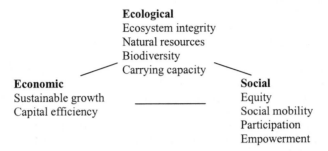

World Bank (1994), modified corresponding to the text

Figure 1: The framework of sustainable development

To achieve a better understanding of the prerequisites of a sustainable use of natural resources in the various climatic regions, then those challenges that ISCO conferences have always called attention to need to be faced: How can irreparable environmental changes through land uses as we have defined them in this context be avoided or limited?

Irreversible changes are in particular:
- soil erosion caused by wind and water as a result of specific forms of land use,
- nutrient loss and eutrophication in entire landscapes,
- transposition and enrichment of persistent chemical and radioactive substances in soils, bodies of water and food chains,
- soil acidification and salinization,
- compaction and soil sealing,
- impairment of the biological diversity of soil organisms, flora and fauna as brought forth by evolution caused by overtaxing ecosystems and accumulation of toxic substances in food chains.

As a rule, the negative consequences of changes in the ecosystem end up being paid for by future generations. Spreading the word on improvements towards man's well-being - e.g. advances in medicine, harnessing energy and increasing energy efficiency, mobility and communications - takes place at an ever greater speed and reaches more people. This applies only to a limited extent to knowledge and experience about maintaining the resistance and diversity of different ecosystems. Furthering this knowledge and handing down traditional practices used by the rural community have become imperative if sustainable development is to take shape.

Frequently, long-term and far-reaching effects of changes in the natural environment cannot be adequately assessed because the processes can either not be reliably predicted by ecosystem researchers or the processes are completely uncertain due to their complexity and dependency on time. So-called residual risks (i.e. unpredictable risks) which mankind now regards as politically and socially acceptable, determine the development for future generations. Responsible action should therefore be guided by refraining interference with the environment as far as possible if irreparable damage to the resilience of ecosystems cannot be excluded.

This action based on an ethic with the "principle of responsibility" would consider the warning:

"But the future is not represented, it is not a force that can throw its weight into the scales. The nonexistent has no lobby, and the unborn are powerless. Thus accountability to them has no political reality behind it in present decision-making, and when they can make their complaint, then we, the culprits, will no longer be there." (Jonas, 1984).

2   Can land resources and intensive agriculture meet future food needs?

The tasks policymakers have to solve in facing the demographic, economic and ecological changes expected in the next decades are unprecedented if they feel committed to this "principle of responsibility". Extensive parts of the terrestrial ecosystems have been degraded by human impacts because soils have lost their fertility to a greater or lesser extent, material cycles and energy fluxes are interrupted and the habitats of wild flora and fauna have been damaged. The oppurtunity for natural regeneration exists only to a limited extent. Population density does not, as a rule, permit a withdrawal from degraded landscapes. In the developed countries technology based on the use of fossil fuels and nuclear energy, and the extent to which they are transferred to developing countries, enables forms of economic development which allows degradation of the soils and ecosystems to be neglected for the time being. Without a doubt there are different levels of degradation. It is safe to assume that in most of the inhabited climatic zones, resilience of the soils and ecosystems is still sufficient to support the evolution of the soil ecosystem and wild plants and animals in narrower landscape frameworks. The preconditions for evolution in sizeable landscapes, which for example fulfil the habitat preferences of large mammals can only be found in a limited number of regions which are either unsuitable for human settlements or which have already been designated nature reserves.

Today's world is the starting point for the future development of land use in the next century. The world population is expected to continue to grow until the middle of the 21st century, requiring a dramatic increase in the production of food. Though food supplies in the developing countries improved between 1960 and 1990 in spite of exponential population growth, one should not feel too optimistic about this task being mastered. To be sure, considerable progress was made in the battle against hunger during this period. At the beginning of the 1960s, 2.14 billion of the world population of 3.13 billion were living in these countries (75% of these 2.14 billion people was surviving on less than 2,100 cal/day). The average available food supply of 1,835 cal/day was only slightly above a daily calorie diet in the chronic undernourishment category. Thirty years later the situation has improved somewhat as to average food intake, although the population in the developing countries had doubled and risen to 4 billion. Whereas in the developing countries in Asia, food intake per capita had crossed the 2,300 cal/day threshold, the situation in other developing countries had, on average, barely improved at all. As a matter of fact, in sub-Saharan Africa the situation had worsened (FAO, 1995a; 1996).

Growth in the production of foodstuffs resulted mainly from an increase in cultivated land (in particular irrigated land) and higher yields of grains (wheat, rice, maize, millet). In the developing countries grain production was increased by 75% from the end of 1960 to the beginning of 1990. Mainly responsible for this increase in yields and production, propagated as the "green revolution", were improved farming practices, particularly the use of high-yielding varieties of seeds, commercial fertilisers and plant protection agents. More than 50% of the increase in the production of foodstuffs was due to an increase in yields, and 20% to an increase in cultivated land. This improvement in the food situation in the developing countries, which is just as impressive as the increase in crop production in the developed countries, enabling the expansion of food aid, has no doubt been achieved at the risk of depleting the ecological resource base. In many regions, and this also applies to the developed countries, the functions of the soils and vegetation in the environment have been weakened and biological diversity is threatened. The Global Assessment of Soil

Degradation (GLASOD Study), based on information from 1989-1990, states that the extent of soil degradation due to human activities amounts to 1,965 million ha of worlds 8,735 million ha arable land, grassland, woodland and forests. About 300 million ha face serious degradation from erosion, loss of nutrients and salinization (Oldemann et al., 1990; German Advisory Council on Global Change, 1996). As long-term studies on the changes in soils are not available, one can only assume the extent of soil degradation, which ran parallel to the increase in crop yields and productivity between 1960 and 1990. Numerous regional studies, however, show that soil fertility has severely declined within this period of one generation. An increase in productivity in the agricultural sector was certainly not the only cause. At the same time overexploitation of forests and urbanisation gathered momentum in the developing countries and were major forces in soil degradation.

For the preparation of the World Food Conference in November 1997 the FAO tried to cover as many details as possible in reply to all questions pertaining to the forthcoming demand for foodstuffs, food production and land use with respect to an increase in the world population from 6 billion in the year 2000 to 8 billion at the end of the first quarter of the next century. Average food supply per capita has to be increased to 2,800 cal/day, which would enable large parts of the population to avoid malnutrition. In the developed countries no essential change to current food consumption (3,200 to 3,500 cal/day per capita) is expected; this is one quarter higher than average daily energy requirements (FAO, 1995b). The studies do not go into further detail on the effects this development might have on the material cycles and energy flow of the natural environment, the resilience of the ecosystems and on biological evolution in the future. In the years to come an anticipatory policy is called on to efficiently direct the scientific, economic and political discourse on these pressing problems. It is relatively safe to say that the measures needed to achieve a 175% increase in the food supply by the middle of the next century (prognosis with the most probable assumptions) will be designed primarily to intensify use on land currently farmed. In developing countries, large areas of land are still potentially available for food production. From a total of 2,537 million ha of land for which sufficient precipitation allows farming, 721 million ha were being used at the beginning of the 1990s. Extending land used for agricultural purposes is, however, restricted by various factors such as an uneven distribution of this land in a few geographical regions - land reserves can be found in sub-Saharan Africa and in South America in particular - unfavourable terrain and soil properties in various agroecological areas, forest cover and areas set aside as nature reserves. The FAO estimates that by 2010 only an additional 90 million ha will be cultivated together with the 721 million ha already under cultivation. Based on this estimation, an extension of arable land for the production of food and plant natural resources is possible in the longer term. It will, however, be necessary in the future to set areas aside that are now under cultivation as reserves which are left to themselves to enable threatened eco-systems to restore their resilience. In the developed countries land reserves have come about in recent years due to the trend in the market and price policies of the USA and the European Community. These new reserves are comparatively small and will not all be used again for the production of food, not even with rising prices in the agricultural sector. The effects of climate changes on soil fertility in the various agroecosystems and how this will influence the availability of land resources for the production of food for a world population reaching 9 to 10 billion people by 2050 is one variable for the prognosis whose impacts cannot be foreseen . It will be a number of years before we will be able to estimate the extent to which climate changes alter agroecosystems.

3 Curtailing risks in the process of intensifying land uses

To give a positive view, this ISCO conference has demonstrated again that in practically all agroecological zones there is an enormous potential for an increase in productivity. The main measures are:

- improvement in the management of water and nutrient resources not only in regions with heavy precipitation but also in arid regions,
- enhancement of the diversity of agricultural and forestry systems, included animal husbandry,
- the selection and application of high-yielding strains and varieties of plants and
- widespread use of integrated pest management.

Raising the level of education of the rural population, strengthening the economic and social standing of women, participative forms of communication and consultation, reliable land use rights, efficient and non-confrontational governmental and administrative policies provide the foundation for dynamic rural development. Traditional soil and water conservation techniques and indigenous knowledge in farming will be the basis for farm families to survive and prosper in all developing countries for a long time. Social networks on a local institutional level are helping to do so successfully. A major challenge in all these countries is to create the macro-economic and administrative preconditions based on responsible policy which induces farmers to invest and to improve their resource management. Then the rural population will also be open to more innovative approaches and eventually will try to bring the productivity of their farming closer to the results achieved in agricultural research and training centres (Reij et al., 1996). Not least, the marketing and price incentives for agricultural products as well as the terms of trade for the means of production - in particular fertilizers - are decisive for the intensity of agricultural production. It will also in the coming years not be easy to improve the nutrient balance of soils in the developing countries, especially in regions where small farms are predominant.

The importance of new findings from agricultural research, especially in the field of biotechnology, gains a different perspective when seen in the light of the socio-economic framework within which the productivity and intensity of farms are to be enhanced. One has to accept that only in the longer run and when progress in other sectors has been made will farming families be able to apply the experience acquired by research centres because they cannot do without the benefits brought by higher yields and crop security. Examples for prospective gains include:
- genetically engineered rice strains which grow on soils with a high salt content as well as maize and millet strains better adapted to acid soils and drought, respectively,
- biotechnologies to cultivate and spread more rapidly growing trees and shrubs whose roots can utilize nutrient deposits in deeper soil layers which help to counter the salinisation of irrigated land in arid regions,
- genetically engineered varieties of sweet potatoes highly resistant to disease which, if cultivated on a larger extent, could improve the food supply in various African countries,
- disease resistant varieties of bananas, the seedlings of which can be grown in large numbers from plant tissue culture (Wambugu, 1997).

A debatable example is what has been reported about treating agricultural land by transgene nitrogen-fixing bacteria to enhance yields and thus reduce fertiliser use. The release of transgene bacteria in soils would actually be unacceptable under the European regulations for biological safety. This underlines the nessecity to agree on commitments internationally to apply appropriate security standards to certain realms of biotechnology as soon as possible, particularily to the release of genetically modified organisms into the environment (Casper, 1995).

Doubtlessly, an intensification in agricultural production will profoundly affect the material cycles, energy fluxes and biodiversity of ecosystems. Changes need to be accurately monitored and their long-term effects and ecological risks estimated. However, the disadvantages for the natural balance and evolution will, in all probability, be even greater if use of the land reserves mentioned above have to be drastically curtailed to provide the resource base for food production. The extent to which biotechnological interventions in the natural environment can be tolerated must be studied by means of ecosystem research using the methods of environmental impact assessment, in particular with regard to the resilience of ecosystems and the sustainability of increases in biomass in the various agroecological zones (Fränzle et al., 1992). Agricultural research and ecosystem research have to cooperate more closely, if a second "green revolution" is to serve the cause of agriculture in the battle against hunger with a minimum of ecological risks. However, it is not only scientists who are called on to learn and understand the tolerance levels of the environment, it is also the duty of governments and institutions with influence on economic development and social development. The time-period during which the food supply has to be expanded on a wide scale and measurable progress towards sustainable development has to be made covers the present generation of children or those to be born in the coming years. In the foreseeable future, scientists will most likely be in a position to interpret the prerequisites of soil fertility to set up sustainable land management schemes which will lead to a considerable increase in food production and promote regenerative natural resources. It will be a greater challenge to ensure that changes in material cycles and energy flows and in biodiversity which occur when, along with the growth in urban population, cities expand, new cities and industrial regions are founded and traffic infrastructure increase their demand for land, are kept within environmentally sound limits.

By the year 2000 approximately half of the world's population will live in cities. If the trend towards concentration of population growth in cities is extrapolated, it seems probable that approximately 5 billion people will be living in urban areas and 3 billion in a rural environment at the end of the first quarter of the next century. Urban conglomeration will be accelerated in the developing countries as a result of the high rate in population growth. More than 200 cities have already topped the one million mark. In about 20 regions the population and local authorities are fighting the problems of food and water supply, of providing housing and heating fuel and the disposal of waste and faeces which, given the 10-15 million inhabitants and a population density of more than 3,000 persons/km^2, apparently can only be solved with time lag. Widespread poverty and a high percentage of the population living in slums are typical of these conurbations and aspirations of implementing sustainable land use and development dwindle to a faint glow on the horizon. Past experience in the developed countries has shown that economic development and urbanisation have profound effects to the detriment of soils used for agricultural purposes, forests and unused ecosystems, not only in the urban regions themselves, but also in peripheral areas. In the period from 1960 to 1990 a linear correlation could be drawn between economic growth and the conversion of open land into areas for settlements and traffic roads in West Germany. During this time, each increase in the gross national product by DM 1 billion based on constant prices was coupled with an increase of 1,000 ha required for these land uses (Hoffmann-Kroll and Wirthmann, 1993). An inevitable consequence of this is that processes of the soil and plant life in natures balance are severely hampered. Similar trends can be expected for the developing countries. Experience now gained in regional and landscape planning should in the future enable economic development with lower land consumption and associated loss of ecological functions. However, it will be possible only to a limited extent to avert the conversion of agricultural land, forests and unused areas into land for settlements and business, for the extraction of raw materials, for the construction of roads and other investments in the infrastucture of a region and the associated stress on biological systems and biodiversity, particularly in those countries facing a great increase in population in the decades to come. This conclusion again underlines the assessment predicted by the FAO that most of all the management of soil and water resources has to be intensified to meet the demand for food.

4 The importance of non-agrarian development

The improvements in living conditions needed in the decades ahead, particularly for people in the developing countries who are caught in a vicious circle of poverty and population growth cannot be brought about solely by an increase in agricultural productivity. Developing countries are characterized by a population distribution of more than 50% living in rural areas and with daily life organized according to the dictates of farming as a livelihood. In the developing countries, where that part of the population is lower than 10%, history has taught us that the expansion of means of earning a living outside the agricultural sector with a simultaneous intensification of agricultural production and growth in inter-regional trade were the foundation stones for economic and social development. These factors can also be shown as mutually dependent elements of a triangular system:

Prerequisites for economic development

Increase in land and labour productivity by the use of sound technology in agriculture and forestry to ensure food security and the production of renewable resources.

Enhancement of means of earning opportunities and sources of income outside the agricultural sector due to scientific and technological innovations.

Expansion of interregional trade and use of comparative cost advantages through higher mobility and communications.

If a dramatic increase in poverty with international repercussions is to be avoided, changes in employment in the agricultural and commercial sectors, in the exploitation of natural resources and in industry as well as in the service sector, which developed countries have gone through in the last two centuries, will have to take place in the developing countries in the next 50-75 years. This applies to the many institutions of society which are important for the civilisation process. Only a small number of prerequisites for positive development in the above-mentioned areas can be mentioned here: improvements in education, in vocational training, in research, in the legal, administrative and social security systems have already been referred to. Recent advances in the fields of communications and mobility have opened the horizons to new forms of cooperation transcending local borders to an extent that only a decade ago would have seemed impossible. An efficient banking system is the precondition for rapid changes in economic structures. Fundmental requirements for economic and social development are the prevention of war and a reduction in expenditure on weapons. Since the middle of this century war has been avoided in many parts of the world, unfortunately, though, it could not be kept out of many developing countries.

In the coming decades completely new tasks will arise as a result of the ecological challenges due to population growth and changes in the economic structure. It is the responsibility of scientists and politicians to formulate the nature of these tasks now so that practical and applicable proposals and measures to promote sustainable development can be worked out. It seems in principle that man has a chance to prevent detrimental changes in material cycles and energy flows in rural and urban regions and to preserve the biological diversity of flora and fauna in landscapes large enough to enable evolutionary processes to continue thus keeping options for development open for future generations. But there is no guarantee of success, given man's present understanding of ecological processes and current economic trends and demands for social change. A number of questions pertaining to the ecology of agroecosystems have been discussed above. Those problems arising from a growing concentration of population in an increasing number of big cities and conurbations will be addressed here too.

Cities are our common future (Töpfer, 1996). Are they also a chance that two-thirds of a world population of 9 billion in the middle of the next century may achieve a sustainable livelihood? The second United Nations Conference on Human Settlements (Habitat II) in 1997 tried to translate the goal of 'sustainable settlements development' into political action. Criteria for city planning based on steps towards sustainable land use in highly populated areas and towards a reduction in the negative impacts exercised on the ecosystems of other regions by urban areas need to be worked out (Barton et al., 1994; Dilks, 1996).

Important criteria for urban areas include:
- the proportion of solar energy directly employed including wind power and geothermy in overall energy consumption,
- the proportion of recycled resources in overall resource consumption; the amount of waste burnt in incinerators, stored in landfills and the size of landfills,
- the proportion of treated water in overall water consumption; discharging untreated and treated waste water into water bodies; the amount of sewage sludges and their use or discharge,
- totals of air pollutants and greenhouse gas emissions as well as persistent toxic substances such as heavy metals and dioxins emitted and their dispersion areas,
- the proportion of foodstuffs produced in urban areas themselves in overall food consumption; the quantities of recycled nutrients and water in this urban agricultural sector.

To permit comparison, criteria should be based on population figures. Further criteria required for assessing economic development and social change include:
- overall population density and data specified for sections of conurbations, and
- the distribution of income in conurbations (e.g. measured with the Gini coefficient) and specific data on poverty (World Bank, 1995).

Energy and substance turnover in conurbations influence soil, water and air quality not only there, they also affect adjacent regions and those which are important for the supply of food, water, fuel and raw materials and for the discharges of waste water and solid wastes of these conurbations. The 'footprints of this giants' have to be drastically mitigated if urbanization shall shape the future of mankind (Giradet, 1997). It goes without saying that sustainable development is destined to fail both in the cities and in the regions affected by them if no interactions can be created by means of which substance and energy losses are considerably reduced. Important ecological targets in this connection include waste avoidance to the greatest extent possible, the treatment of waste water and utilisation of compost and sewage sludge. All of these measures require, however, that the emission of toxic substances from the technical processes can be effectively limited, in particular at industrial sites, at power plants and in transport. More and more cities in developing countries have taken to producing vegetables and meat within the city limits. Numerous cities in Africa and Latin America are already largely self-sufficient with regard to these products (Smit, 1995). Agriculture within city limits can greatly help to mitigate the food problems of the poor inhabitants, if land use rights for these families and land use planning ease the way. The task of systematically studying the locational issues concerning agricultural production within the urban environment, as impressively shown at the beginning of the last century by von Thünen with the methods of marginal cost and opportunity cost analysis, will expand considerably if the substance and energy flows between cities and rural areas with their physical and economic values are incorporated into the study (v. Thünen, 1826).

5 Land use planning - a tool to achieve sustainable land use

Influences of the natural environment and society on the land use system are never constant but they change over time and can also be influenced by human impact (positive and negative). Models of sustainable land use claiming to be close to reality are therefore always complex. As a consequence, theories and models to promote sustainable land use would be inevitably incomplete. Recommendations deduced from those models should therefore not be implemented categorically or follow predetermined procedures. They should only be taken as an assistance for a gradual learning process.

This also means that strategies have to be multi-sectoral and interdisciplinary and have to be constantly negotiated. The elaboration of these strategies is in many cases supported by bilateral, international or NGO cooperation projects. Sustainable land use is considered as a central topic in and for projects in development cooperation worldwide. With the tasks which can be derived from the Sustainable Land Use Triangle in mind, we have to elaborate new means of land use planning to meet the following objectives for different geographic areas.

Ecological targets in land use planning under the aspect of rising population density can be described as follows:

Land use objectives in peripheral areas (global land use and non-use of ecosystems):
- large scale preservation of material cycles and energy fluxes in the environment and of biological diversity which restricts human intervention and allows for the evolution of soil organisms and wild flora and fauna.
- protection of sufficiently large areas in which anthropogenic substance inputs and other interventions are only slightly or not in the least detrimental to dynamic equilibria and habitats are not fragmented.

Land use objectives in rural areas:
- preservation of soil quality for biomass production appropriate to the site.
- preservation of water infiltration and retention capacity as well as biological diversity and activity of soils.
- enhancement of the sequestration of carbon in soils and vegetation and the prevention of greenhouse gases.
- preservation of habitats of wild plants and animals also in agricultural landscapes.
- maintaining of the nutrient balance and regeneration of depleted soils through fertilizing.
- avoidance of accumulation of persistent toxic substances in soils and in food chains and the eutrophication of areas worthy of nature protection.

Land use objectives in urban areas:
- minimization of sealed land and preservation of the infiltration capacity of rainfall, and the biological activity and the gas exchange of soils.
- preservation of habitats for wild plants and animals and of soils for the production of foodstuffs in urban landscapes.
- cycling of nutrient flows from supply for and discharge from the urban population as far as possible in the same region.
- minimization of depositions of toxic substances in soils, bodies of water and the air and avoidance of enrichment of persistent substances in soils and sediments.

The inadequate utilization of natural resources in developing countries is usually a consequence of the rural population's lack of access to land, capital, means of production, information, education, markets and political power, a situation which is aggravated by the high rate of population growth. The fundamental political development objectives are therefore: poverty orientation, sustainability, target group orientation, participation, multi-sectoral approach and regional orientation.

'Poverty Orientation' is an attempt to combat mass poverty through directly addressing development measures to the poor rural population to enable them to satisfy their basic needs. The objective of 'Sustainability' tries to secure the livelihood of the affected population groups so that the benefits of the project work continue, at least in the medium term, once the external support has been withdrawn. It is therefore important to enable the people to develop their ability to secure their own livelihoods through a continuous process of adjustment. Sustainability, in this context, also includes conserving the natural environment. 'Target Group Orientation' means to take the needs of different population groups into account using target group analysis for specific development measures. 'Participation' stands for the involvement of local groups in the project work in the form of a dialogue instead of undermining their ability to solve problems for themselves. The 'Multi-sectoral Approach' is beyond sectoral boundaries and tries to solve problems caused by a wide range of determining factors which arise in several interrelated fields/sectors. 'Regional Orientation' combines spatial units in order to improve the living standard within a broad area with common natural and economic characteristics and shared socio-cultural conditions. One of the main objectives of development cooperation is to identify locally compatible solutions. The role of rural development projects is also to develop and disseminate these innovations (GTZ, 1993; Reiche and Carls, 1996).

In this context, two more strategies are important. The 'Institutional Strategy' which means the integration of disadvantaged groups into the development process and 'Promoting self-help organisations without overtaxing them'. Personal responsibility and the transfer of functions from the central State to decentralized entities and the affected parties themselves should be promoted as well as the organization of the rural population and the improvement of their capacities to represent their own interests (GTZ, 1995).

Other important strategy elements for sustainable land use projects which have been developed during the past decade are:
- orientation on participative action,
- promotion of motivation to participate and self-help potential,
- process orientation,
- combination of effective short and medium term potential,
- process orientation,
- combination of effective short and medium term measures,
- development of local organizations and institutions as well as
- orientation on negotiation and conflict management.

The task/purpose of land use planning is "not [to] be a sectoral effort, executed unilaterally by government institutions. Instead it should develop into an interdisciplinary, holistic approach that gives attention to all functions of the land and that actively involves all stakeholders through participatory process of negotiation platforms, be it on national, province or village level. ... The aim of land use planning is to create the preconditions to achieve a sustainable environmentally sound, socially desirable, and economically appropriate form of land use. Such preconditions are best met by a decentralised approach" (Sombroek and Eger, 1996).

Instruments and methods to achieve sustainable land use include:
- Participative Land Use Planning,
- Conflict Management,
- Participatory Rural Appraisal,
- Remote Sensing (interpretation of area photographs and satellite images),
- Regional Oriented Rural Planning,
- Geographical Information Systems and
- Participatory Development of Sound Soil Management.

6 Partnership - the key for cooperation on all levels

From now on scientists will have to work closer together at an international level and take an interdisciplinary approach if the complex ecological, economic and social problems are to be solved as progress towards more sustainable land use is made. It will be necessary for scientists in collaboration with experienced experts in practical professions to identify the structures of the problems to be solved in the next two decades. Governments and international institutions will have to provide the necessary political and financial support. Soil conservation and the setting up of a strategy defining measurable steps towards sustainable land use have to be more firmly fixed in the agenda of international cooperation.

At the instigation of UNCED, a Convention to Combat Desertification (CCD) was ratified by 50 countries in December 1996. For the first time, a legally binding international framework for action was established. Among the most important aspects of this Convention are the following principles: the participation of affected population groups, partnership between governmental and non-governmental entities, consultation among donors and their involvement in the implementation of the Convention, learning from past experiences, and the establishment of effective control mechanisms in countries afflicted with desertification. Furthermore, it calls for harmonization of existing resource management programmes and plans and for decentralized decision-making structures. The Convention is therefore an important tool for achieving sustainable resource management in the arid and semi-arid regions of the world. It even goes beyond the environmental aspect and could be considered a 'Convention for Development'.

The experience that will be gained within the framework of this Convention in the coming years can surely be transferred to many other regions not directly included in the Convention. This is one reason why FAO and UNEP activities dealing with land use, the protection of water bodies and the climate and the conservation of biological diversity, all of which have common ground, need more efficient coordination. Chapter 10 of Agenda 21 is a sound basis for setting up a statutory planning process for land use and development planning which takes a holistic approach in solving the problems in rural and urban regions discussed above. The objectives and measures of an integrated concept for the planning and use of soil resouces have been described in detail there. An international agreement should be reached as soon as possible on the organization and implementation of a reliable monitoring system of land resources and land use. Data continuously elaborated on the required scale would provide the background for ecological landscape and land use planning to realize concepts for sustainable land use. This interest can be assumed among the international donor institutions for financial and technical support of development cooperation and also among the organizations that carry out the practical work.

Today we can not predict how the trends and developments traced here will merge into new patterns of land use in the lifetime of our children and grandchildren. We only can hope that the present and coming stakeholders in this process will be more and more conscious that finding the way to sustainable land use in the broad sense described here will also be the only way for mankind to survive physically and culturally. What we can do now is help to increase cooperation for responsible action: between people, between people and institutions and insofar as the opportunity exists, between institutions involved in land use planning on a national and on an international level (Hurni, 1996).

References

Barton, C., Bernstein, J., Leitmann, J., Eigen, J. (1994): Toward environmental strategies for cities. The World Bank, Washington, D.C.

Bundesministerium für Wirtschaftliche Zusammenarbeit und Entwicklung / Gesellschaft für Technische Zusammenarbeit (1995): Promoting Sustainable Soil Management in Developing Cooperation. Eschborn.

Casper, R. (1995): Beitrag der Gentechnik zur nachhaltigen Landwirtschaft in den Entwicklungsländern. In: Franzen, H., Begemann, F., Wolpers, K.H., v. Urff, W. (Hrsg.): Auswirkungen biotechnologischer Innovationen auf die ökonomische und soziale Situation in den Entwicklungsländern. Bonn

Dilks, D. (1996): Measuring urban sustainability: Canadian indicators workshop. Environment Canada, Ottawa Toronto

Food and Agriculture Organisation (1976): A framework for land evaluation. FAO Soil Bulletin 32 FAO, Rome.

Food and Agriculture Organisation (1993): Guidelines for land-use planning. FAO, Rome

Food and Agriculture Organisation (1995a): Food, agriculture and food security - Historical development, present situation, future prospects. FAO, Rome

Food and Agriculture Organisation (1995b): Lessons from the green revolution - Towards a new green revolution. FAO, Rome

Food and Agriculture Organisation (1996): World Food Summit - Technical background documents. FAO, Rome

Fränzle, O., Jensen-Huss, K., Daschkeit, A., Hertling, Th., Lüschow, R., Schröder, W. (1992): Grundlagen zur Bewertung der Belastung und Belastbarkeit von Böder als Teilen von Ökosystemen - Forschungsbericht im Auftrag des Umweltbundesamtes. Geographisches Institut der Universität Kiel

German Advisory Council on Global Change (1995): World in transition: The threat to Soils. Bonn

Gesellschaft für Technische Zusammenarbeit (1993): Regional rural development (RRD) - Elements of a strategy for implementing the RRD in a changed operationa context. Eschborn

Gesellschaft für Technische Zusammenarbeit (1995): Landnutzungsplanung - Strategien, Instrumente, Methoden. Eschborn

Giradet, H. (1996): Giant Footprints. Our Planet, Vol 8 No 1

Hoffmann-Kroll, R., Wirthmann, A. (1993): Wandel der Bodennutzung und Bodenbedeckung. Wirtschaft und Statistik, H. 10

Hurni, H., with the assistance of an international group of contributors (1996): Precious Earth: From soil and water conservation to sustainable land management. International Soil Conservation Organisation (ISCO) and Centre for Development and Environment (CDE), Bern

Jonas, H. (1984): The imperative of responsibility - In search of an ethics for the technological age. Chicago & London

Oldeman, L.R. Hakkeling, R.T.A., Sombroek, W.G. (1990): World map of the status of human induced soil degradation - An explanatory note. International Reference and Information Center, Wageningen; United Nations Environment Programme, Nairobi

Pieri, Ch., Dumanski, J., Hamblin, A., Young, A. (1995): Land quality indicators. 315 World Bank Discussion Papers, The World Bank, Washington, D.C.

Rat von Sachverständigen für Umweltfragen (1996): Konzepte einer dauerhaft-umweltgerechten Nutzung ländlicher Räume. Wiesbaden

Reiche, R., Carls, J. (1996): Modelos para el desarollo de una agricultura sostenible - Las ventanas de sostenibilidad como alternativa. In: Development and Cooperation, 4

Reij, Ch., Scoones, I., Toulmin, C. (1996): Sustaining the soil - Indigenous soil and water conservation in Africa. International Institute for Environment and Development, London

Serageldin, I. (1996): Sustainability and the wealth of nations - First steps in an ongoing journey. Environmentally Sustainable Development Studies and Monographs Series No. 5, The World Bank, Washington D.C.

Serageldin, J., Steer, S. (eds.) (1994): Making development sustainable: From concepts to action. The World Bank, Washington, D.C.

Smit, J. (1995): Urban agriculture prospects in Africa, Latin America and Asia. In: Vegetable production in periurban areas in the tropics and subtropics - Food, income and quality of life. German Foundation for International Development, Council for Tropical and Subtropical Research, Feldafing and Zschortau

Smyth, A., Dumanski, J. (1993): FESLM: An international framework for evaluating sustainable land management. World Soil Resources Report 73 FAO, Rome

Sombroek, W.G., Eger, H. (1996): What do we understand by land use planning - a state-of-the-art report. In: Entwicklung und Ländlicher Raum, 30, 2
Sombroek, W.G., Sims, D. (1995): Planning for sustainable use of land resources: Towards a new approach. FAO, Rome
Thünen, J.H. v. (1826): Der isolierte Staat in Beziehung auf Landwirtschaft und Nationalökonomie. Jena
Töpfer, K. (1996): Cities: our common future. In: Our Planet, Vol.8 No.1
Wambugu, F. (1997): Potential impact of biotechnology in sustainable agricultural Development in Africa. In: Agriculture between environmental concerns and growing food needs. AIDA Forum, Council for Tropical and Subtropical Research, Feldafing and Zschortau
World Commission on Environment and Development (1987): Our Common Future. New York.
World Bank (1995): Monitoring environmental progress - A report on work in progress. Washington, D.C.

Addresses of authors:
Eckehard Fleischhauer
Muffendorfer Hauptstraße 46
D-53177 Bonn, Germany
Helmut Eger
Deutsche Gesellschaft für Technische Zusammenarbeit
(GTZ) - IP
Postfach 5180
D-65726 Eschborn, Germany

Part I: Soil Degradation – Diagnosis, Appraisal and Reversing Measures

Introduction

A. Hebel

The finite resource *soil* with its various functions is of direct and indirect benefit to man and the environment. While soils have increasingly been used by man mostly for economic reasons, the degradation of soils has become an alarming global environmental problem threatening the world's sustainable development. Through the GLASOD (= global assessment of soil degradation) study, carried out by UNEP and ISRIC in the beginning of the 1990's, it was shown that already 24 % of the inhabited land area exhibits human-induced soil degradation. Causative factors are deforestation and removal of the natural vegetation, overgrazing, agriculture, overexploitation of vegetation and industrial activities.

In chapter 1 ecological aspects of the soil and the role of indicators to assess the soil and land quality are pointed out and discussed. How can shifts in soil and land quality be measured in a practical, cost-effective way? Which are relevant indicators? To answer these questions it is necessary to realize that the data on land quality should be sufficiently robust to enable a proper assessment of the sustainability of the various land use systems - for present use and potential future uses. Currently there are many attempts underway to derive useful, practical and "close-to-reality" soil quality indicators with high expressiveness - particularly in the field of soil biology.

Once it is decided which soil data are to be collected, the data should be electronically stored in an information system (see chapter 2). Modern technologies nowadays enable the efficient storage and retrieval of soil and land resources data. Data collection is increasingly being made easier by the use of remote sensing. The results of the appraisal of land resources information for specific purposes are the product of a reproduceable step-by-step land resources evaluation procedure and serve as the sound expert knowledge based advice to decision-makers. Multiple objective decision-making will become a powerful tool in the future providing integrated information to decision-makers on land resources and on socio-economic policy factors that influence the sustainability of land use systems.

How will the predicted climate change affect soil properties, and in how far may soil properties have an impact on climate change? Does the change lead to soil degradation or to beneficial developments in soils as related to fertility? Where will anticipated changes occur in soil properties and to what extent? And can soils significantly contribute to a mitigation of global climate change by their capability to sequestrate carbon? The current state-of-the-art of the still limited knowledge in all these issues is presented by renowned experts in chapter 3.

Water and wind erosion are the predominant soil degradation types in the world (see chapter 4). The importance of this subject is reflected by the high quantity and quality of papers in chapter 4. Fundamental soil erosion research has become increasingly advanced and successful, mainly due to new possibilities for modelling techniques. The factors being responsible for the relationship between soil management practices and soil erosion are more and more understood. Thus, in many cases scientifically sound recommendations for optimal soil management with minimum soil erosion in the various agro-ecological regions can be made. Humus degradation and nutrient depletion as subtypes of soil chemical deterioration also contribute a substantial share to soil degradation. Among others, the very interesting issues of soil organic matter transformations under tropical as compared to temperate conditions as well as the effect of soil organic matter application on soil restoration are discussed in chapter 5.

Soil and land degradation often imply drastic negative consequences for man's well-being such as income loss and/or a reduction in food for living. This holds particularly true, if land users - from an economic point of view - are highly dependent on agricultural, soil based land use, such as in many developing countries in the world's drylands. In such regions this problem had reached a level such that the international community felt the need to take action and therefore agreed upon the International Convention to Combat Desertification (CCD) in the world's drylands (December 1996). This mechanism includes action programmes, particularly aimed at the most affected regions. The issue of degradation of dryland ecosystems in general (desertification) on the one hand and salinization as an example for soil degradation in the drylands on the other hand are dealt with in chapter 6.

Soil degradation occurs all over the world and is not restricted to the agroclimatic zone "drylands". Water and wind erosion, humus degradation and nutrient depletion occur all over the world. So does the degradation caused by industrialisation and urbanisation which, of course, is more widespread in the industrialised than in the developing regions of the world. These growing impacts of industrialised agriculture and urbanisation on soils, sometimes referred to as "new forms of soil degradation", are dealt with in chapters 7 to 10 and in the introductory text by Hans-Peter Blume.

Experience with a large number of soil and water conservation projects has shown that the tackling of degradation problems (either in already degraded areas or areas at risk) can only be successful, if the technologies/approaches used are adapted to the local conditions. In addition to land resources information the background conditions such as the local and the socio-economic context are to be considered as equally important for the decision-making process. New experiences and promising approaches in soil and water conservation and sustainable land management are presented and discussed in much more detail in chapters 11 to 16 in Volume 2 of this book.

Address of author:
Axel Hebel
German Federal Ministry for the Environment,
Nature Protection and Nuclear Safety
Ahrstr. 20, 53175 Bonn

Soil Functions and the Future of Natural Resources

C.R. de Kimpe & B.P. Warkentin

Summary

Soils perform five functions (biochemical and geochemical cycling, partitioning of water, storage and release, buffering, energy partitioning) that are essential for the support of ecosystems. When shifting from natural/semi-natural systems to managed systems, the balance among these functions can be extensively changed. This is especially true in the case of agroecosystems. Food supply is one of the most important questions in the domain of interactions between humans and their environment. Land available to agriculture is limited for topographical and climatic reasons, and half of that land is under production. Increasing food production to meet the future needs of the world's growing population will require management practices that make optimal use of available land and water resources, while maintaining the balance among the soil functions. A large imbalance results in land degradation. Increasing productivity on the best arable lands will require additional research towards establishing the tolerable intensity of soil use for sustainable agriculture. A better understanding of soil ecology will help in developing more environmentally-friendly sustainable farming systems.

Keywords: Soil functions, agroecosystems, food production, sustainable farming systems, soil quality.

1 Introduction

Recent developments in making agriculture more sustainable have forced soil scientists to reconsider the role of soils, and hence the priorities for soil science studies. In addition to the role of soils in agricultural production, their role in ecosystem functions is being more widely appreciated.

Population pressures on land resources increase every day. The ravages of soil degradation and the need for costly soil restoration are more and more recognized, and have prompted a conservation ethic amongst the urban population: social concerns and requests for gentler use of natural resources are becoming more cogent and are being heard more frequently. In the fortunate parts of the world where food availability is not the most urgent problem, land, especially soil, and water resources are being used to provide a wider range of environmental amenities such as recreation and ecotourism, leading to an increased quality of life. In those countries, the percentage of the population actively engaged in direct food production is often less than 5%.

Advances in technology, e.g. plant breeding for better response to fertilizer and water inputs, have made it reasonable to suggest that highly productive soils should be used more intensively (FAO, 1993). This is one topic in the debate on sustainable development. At the same time, we are becoming more aware of the important linkages between soils and society. At the time where technical advances in food production were a major concern in soil science, soil studies concentrated on technical aspects such as soil physics and soil fertility. There is presently a need for a broader concept of soils, and the integration of soils into a broader context of societal studies. For example, land degradation is a human concern that includes the elements of poverty, lack of choice in the use of natural resources, lack of information on resource use, and community structures that are not really functional.

In recognition of these trends, the International Society of Soil Science has created a new Commission on "Soils and the Environment", which considers soils in ecosystems. Soil is a component of the natural habitats that includes humans, and therefore also the important considerations of humans as the dominant species managing habitats. A look at the role of soils in ecosystems, both natural and agroecosystems, provides a basis for evaluating land use and soil conservation, planning for water quality and more general aspects of quality of life for humans dependent upon the soil. The long-term sustainability of soils depends upon maintaining soil functions. In the process of degradation, these functions are diminished or even lost, and restoring soil quality depends upon restoring the soil functions.

2 The soil in ecosystems

As we learn more about our environment and ecosystems, and when we try answering questions about sustainability of food production, a major issue is the soil considered in the ecosystem context. Environmental health encompasses the maintenance and quality of the natural resources, soil, water, air, and biota. As emphasized by the Brundtland Commission (WCED, 1987), conserving and enhancing the land resource base is essential for our common future.

An ecosystem is the complex of a community and its environment functioning as an ecological unit in nature. Ecozones are large units characterized by distinct sets of non-living (abiotic) and living (biotic) resources that are ecologically related. Ecological land classification is the process of defining areas with common landform, soil, water, vegetation, climate, wildlife and human factors. Ecological systems can be altered by a succession of natural events (some being sometimes catastrophic, e.g. volcanic activity), or by continuous large-scale human activity. An ecosystem approach to the study of soils requires a balanced and integrated approach between the physical, chemical, mineralogical, and biological components of the soils.

Each site-specific ecosystem has characteristics above and below the soil surface that must be taken into account in order to understand how the ecosystem is functioning. These characteristics are essentially related to soil functions. The soil provides:
- habitat diversity for biota in soils
- habitat stability, including the buffering potential against rapid changes, and
- storage, transformation and transport in soils (partitioning of water, cycling of elements, accumulation, dispersion of pollutants and wastes)

This is performed through the soil functions in natural/semi-natural ecosystems or agroecosystems:
- *biochemical and geochemical cycling*: weathering of minerals and release of nutrients; recycling of nutrients and carbon to produce new biomass; decomposition of toxic materials by soil organisms,
- *partitioning of water*: infiltration of water for storage in the rhizosphere, and to aquifers; transport of water across the soil surface,
- *storage and release*: store nutrients against leaching; store water against drainage; release nutrients and water to roots and biota; immobilization of residual toxic materials,
- *buffering*: dampen temperature and water content variations; moderate solution composition changes; purification of water flowing through,

- *energy partitioning*: determines temperature of biotic environment; global air mass circulation.

Humans, in order to satisfy their needs, often have a "predatory" approach to the ecosystems (source of food and fiber, and of raw materials, support for buildings and roads ...), that neglects the basic functioning of soils. They consider the soils and ecosystems as a natural sink to dispose of the wastes generated by human activities. Soils are also considered a renewable resource which is not to be used exclusively for human benefit. On the other hand, it is increasingly recognized and accepted that soils provide the physical support for maintaining civilizations (memory of archaeological information and cultural heritage), and thus play a major role in ensuring the sustainability of human beings in their ecosystem.

3 Population growth and food production - a historical perspective

Food supply is the most important question in the domain of the interactions between humans and their environment. As of 1995, the world as a whole was consuming directly, or indirectly through animal products, an average of just over 300 kilograms of grain per person per year (Postel, 1996). According to the United Nations Population Division (UNPD, 1992), the world population was estimated at 5.48 billion in 1992, and is expected to reach 8.3 billion by the year 2025. On a worldwide basis, over the period 1990-95, the average growth rate was 1.7%. The greatest challenge for the agrifood sector in the years to come will be to provide enough safe and nutritious food to nourish the planet under expected conditions of increasing pressures for the use of natural resources. Agriculture is an important issue in land-use decisions, and agricultural production is also at the origin of environmental problems.

Mankind has a history of alternating successes and failures in providing food, and it records major periods of starvation in several parts of the world. In the early stages, there were few or moderate pressures on natural resources, thus no overexploitation. Collecting food was obviously a major activity. The hunting/picking/gathering way of nomadic living was probably sustainable, but provided limited returns: food gathering permitted the collection of about 15 calories of food per calorie of energy invested (Lee, 1968). In a food search experiment conducted in Turkey (Harlan, 1965), hand-harvesting of a mixed stand of wild relatives of barley, wheat and oats resulted in the collection of about one kilogram of clean grain per hour. Human energy invested per hour of labor, rather than unit of land, was used to measure yield; there was no conscious effort to increase food production.

With the onset of a more organized agriculture some 10,000 years ago, the approach became very different: people began to till the soil, and intentionally selected the most desirable plants found in the fields. Yield in terms of land unit was substituted for energy input as the measurement of productivity. Cereal yields of one-half to three quarters of a tonne per ha were probably common. Under irrigation, wheat yields of about 2 tonnes per ha and sometimes more were recorded in Mesopotamia around 2400 B.C. (Jacobsen and Adams, 1958). The higher yields were not maintained because an increase of the salt content in irrigation water led to salt accumulation in the soil. At the beginning of the present era, yields of more than 3.5 tonnes of wheat per ha were obtained in Galilea (Feliks, 1963), and were sustained as long as soil fertility was maintained with animal and green manures, and the irrigation water was of adequate quantity and quality.

Much closer in time, between 1950 and 1985, the success of the Green Revolution in reducing food shortage was linked primarily to new seed varieties designed to maximize yields, facilitate multiple cropping, and resist diseases. In order to achieve the full potential yield of these improved varieties and to protect them against diseases and pest infestation, it was also essential to provide the crops with increasing amounts of chemical nutrients, the consumption of which rose more than nine-fold, and to use more pesticides and similar chemicals, the use of which increased 32-fold.

However, application of fertilizers and pesticides to the crops was likely to create environmental problems, not only at the farm level but for the entire ecosystem.

4 Land availability for food production

On the basis of the Soil Map of the World (FAO, 1991), out of the estimated 13,340 million ha of the world's surface free of permanent ice, 3,030 million ha are potentially arable. The larger part of the soil cover around the world is either too cold, too dry, too wet, too steep or too shallow to support profitable cultivation. The present cultivated area in the world - arable land and permanent crops - is estimated at about 1,475 million ha (FAO, 1989a); 3,200 million ha are under permanent pasture, and 4,050 million ha are forests and woodlands. Other lands account for 4,615 million ha, of which 200 million ha are estimated to be covered by towns and permanent infrastructure.

The world average of 0.28 ha cropland per capita in 1990-91 would decline to 0.17 ha by the year 2025 if population projections are accurate (World Resources Institute, 1990), whereas in Asia, cropland per capita would decline to 0.09 ha. For much of the reserves of potential cropland in Latin America and sub-Saharan Africa, the land is only moderately or marginally suitable for crop production: rainfall is unreliable, or it is suitable only for perennial tree crops (FAO, 1989b).

Land suitability for agriculture is determined by soil quality and climate (temperature and precipitation) (King, 1990), and land availability for crop production is limited by landscape, physiography, and the need to protect forest and wild areas, a reservoir for genes and wild relatives essential to improve cultivated varieties and livestock.

How will future attempts to meet food needs of the expanding population affect the soil resource? Either new land must be used for food production, or the level of production must be increased on the most suitable land being presently used. The latter course is preferable, because of potential for both increases of food production and increased land degradation in using new lands less suitable for crop production. Food production has to rely on management practices that make optimal use of available land, and on continuing scientific outputs such as those that made the Green Revolution a success. This will maximize yield on the best agricultural lands and maintain soils with lesser potential for agriculture available for wildlife and biodiversity (FAO, 1993; Avery, 1995).

It has been proposed that increased technical inputs, e.g. fertilizers and irrigation, should be used on land suitable for agriculture in order to prevent further conversion of forest land for food production. This strategy would be successful only if it were clearly established that no degradation would result from the intensified use of these best arable lands. Degradation could result from loss of organic matter, breakdown of soil structure, or change in biological components of the soil. More global effects on the ecosystem could result from losses of fertilizers and agricultural chemicals to the off-farm environment, and from changes in biological functions.

The productivity and stability of soils as a medium for plant growth depend on a balance between their living and non-living components. Under natural conditions, energy from the sun and nutrients essential for growth are stored as chemical energy in plant tissue. These resources are then recycled for use through decomposition by soil micro- and macroorganisms. Sustaining this productivity requires the replacement of whatever is exported. Examples of changes occurring upon converting natural systems to agroecosystems are:

* removal of the natural soil cover, leading to a decrease in organic matter content, either by more rapid turnover and oxidation under tillage, or more simply from harvesting the biomass,
* loss of structure from tillage operations, compaction by machinery, pore discontinuity, plowing under of the humus layer,
* salt accumulation, flood irrigation preventing soil aeration.

5 Land degradation

Deforestation, improper grazing and inappropriate cropping systems are examples of human activities that lead to soil degradation; land degradation reduces the capacity to produce food, which then leads to further degradation, e.g. by expanding crop production to areas of only marginal suitability for agricultural production.

Land degradation is a decrease in the optimum functioning of soils in ecosystems. It results from mismatches in the interaction between natural and human systems. Land degradation, economic growth, and poverty are intractably linked: people living in the lower part of the poverty spiral are in a difficult position to provide the necessary land stewardship. Worldwide, some 305 million ha of land are seriously degraded, and about 910 million ha are moderately degraded. Up to 140 million ha are expected to become seriously degraded over the next 20 years (Oldeman et al. 1990). For instance, the natural fertility of the large agricultural plains in the temperate regions around the world is decreasing and much affected by erosion and flooding (Wicherek, 1994, 1995).

The source of the land degradation process is an imbalance among the soil functions. For example, a loss of organic matter resulting from soil erosion will generate compaction, therefore less gas exchange with the atmosphere (energy partitioning), less possibility for water infiltration and more surface runoff (partitioning of water), less sorptive capacity for nutrients and toxic materials (storage and release), less biological activity, and less potential for biological detoxification (biochemical cycling). Soil erosion is a socio-economic problem as much as a biophysical one: the obvious result of decreased soil productivity and capability for food production will also result in a decrease of socio-economical benefits.

Under conditions of severe pollution of the environment, the conjugation of various processes, including soil functions, is disturbed in the biosphere, and the natural biogeochemical cycles of many elements are transformed. This leads to an alteration of human adaptability due to the closing of biogeochemical trophical chains (for example as a consequence of the contamination of food by pesticides and heavy metals). Assessment of human population health in an ecozone needs to consider, in addition to the traditional indices of mortality and morbidity, preventive criteria such as the degree of adaptability to living media. It is thus necessary to research biogeochemical models of regulation of physiological functions of people independent to anthropogenic geochemical media (Evstafjeva et al, 1994).

6 Outlook for the future

Concepts of soil functions in healthy ecosystems are available: sustainability refers to keeping soil functions operating, and resilience refers to how easily a soil can recover lost functions or restore the balance among the functions (Warkentin, 1995). Concepts of health of soils evolve as the concepts of ecosystem health change with better understanding and with articulation of the changing demands of human society. Questions remain about the level of functions, about how agroecosystems affect the levels, about differences among ecosystems, and about the levels of functions required to ensure sustainability.

* Soils in ecosystems need concise characterization and understanding because ecosystem functioning cannot be reduced to a few simple terms or parameters. This is true especially for soil ecology, being one of the last major „black boxes" in agriculture-related science. Scientists excelled at characterizing the physical and chemical properties of soils, for which much detailed knowledge has been collected. Less is known in the biological area. Soil organism populations are dynamic in nature and are easily affected by the physical disturbance caused by tillage and by the chemical one resulting from the application of fertilizers and pesticides.

Determining biotic activity and species diversity in agroecosystems and managing these populations is a challenge: an exhaustive analysis of the soil biological components in soils is probably impossible, yet not really essential for the understanding of soil functioning. Modern technology is available to improve our knowledge on the importance of the functions filled by soil organisms, and this approach will assist in a better understanding of the role of the biological soil components.

Habitat is a general term that fits better the purpose of future analyses than concepts such as soil area or soil volume; functions have to be reviewed and interpreted at that scale.

* The degree to which a soil can supply nutrients in the absence of external inputs is the inherent fertility. It is a function of the chemical and mineralogical composition of the soil, of the climate under which the soil occurs and has developed, of the length of time the soil-forming processes were active, and of the vegetation that has evolved in response to soil properties and climate (King, 1990). To what extent, and in what ways, can the inherent fertility be augmented to produce more food in agroecosystems without degrading the soil? This question, previously raised in relation to maximum production, must be raised again in relation to sustained soil functions. Where human activities lead to non-sustainability, research is needed to develop alternative acceptable practices.

* Environment and agriculture are closely linked. However, agriculture aims at some controlled and selected biomass production that is largely removed from the production site. Biomass collection for food or forage alters the physical and biological integrity and the health of the ecosystems; increasing the price of food to protect the health of the ecosystems would have a major impact on the poorest 20% of the world population. Research and development must provide the tools to make agriculture more environment-friendly, especially through an increase of soil ecological studies. Future research will have to include crop rotation systems leading to greater biodiversity, multiple cropping systems, crop adaptation to land quality, multi-purpose crops, including food and non-food use, and crop production systems ensuring a more permanent soil cover.

In order to integrate agriculture in the ecosystems, the complexity and stability of the ecosystems must be understood and taken into account. For example, the factors that affect the health of tropical rain forests, temperate zone hardwood forests, estuarine wetlands, freshwater marshes may differ, but some common approaches to their investigation may be developed (Yuill, 1994).

* In the same way that a doctor makes medical diagnosis through tests and symptoms analysis, and then proposes a treatment, ecosystem health will have to be assessed from the point of view of soil functions and their balance, so that the appropriate restoration measures can be proposed.

7 Conclusions

Soil functions and their interactions are the key to determining the suitability of soils for a variety of uses in both natural and agroecosystems. An adequate balance has to be maintained to ensure the sustainability and the health of the land resource. Scaling up this information at the ecosystem and ecozone levels will assist planners and policy makers to develop acceptable systems that will protect the natural resources for our benefit and that of the future generations.

What does this approach to soils based on functions of soils in ecosystems, taking also into account the soil-human interactions in ecosystems, tell us about management of natural resources in the future? What activities need to be undertaken to allow the soils to meet in a sustainable way the human needs in both natural and agroecosystems? What do we still need to know as we plan for sustainable use of the soils and of the environment of which humans are a part? Integration of, and balance among, soil functions are different in natural and disturbed ecosystems. Decline in land productivity is related to some degree to poor stewardship. Health measures developed for natural ecosystems are not fully adequate for managed ecosystems: elements of agroecosystems with an obvious human value will be protected and nurtured, whereas those considered detrimental to humans may be systematically destroyed. Elements with both positive and negative human attributes will be managed in order to achieve the greatest human good at the least human cost (Ikerd, 1994).

Soil quality is an integrative concept in soil science, closely related to human needs. Soil quality links directly to human health. It tries to combine all soil functions under one single idea, rather than considering chemistry, or physics, or biology separately. The concept is easier to grasp than to define. A high quality soil performs all soil functions in an optimum (i.e. a balanced) way. A measure of soil quality is how well soil functions are performed. Some parameters have been used for years from an economic point of view, for example when classifying land suitability for crop production and other human activities: support for biomass production (how much energy and inputs are required to produce the same amount of food), economic structures (land suitability for roads, buildings, railroads...), and waste disposal facilities (landfill sites, wastewater treatment plant). Other parameters are more difficult to define from the societal and economical points of view, for example the importance of biological habitats or wetlands. Their societal relevance has to be defined and integrated in a social context.

Sustainable farming systems are characterized ecologically by high biological diversity, internal recycling of nutrients, cropping systems that rely on more perennial vegetation, and pest management based on biological and cultural control (Stinner and Stinner, 1994). Shifting from increasing productivity to enhancing sustainability will emphasize the fact that agriculture is a major component of environmental management. Reliance only on organic inputs for crop production has not been shown to provide the level of food production needed to meet population needs.

Research is needed toward a better definition of the tolerable intensity of soil use to maintain its resilience. Restoration of degraded land can be very expensive, and often such land is simply abandoned. A concept is required of how a soil responds to increased inputs for increased production. Land use must match land quality. If this is the preferred method to meet increased food demands, a rating of soils based on an interpretation of the balance among soil functions in the context of sustainability will be needed to predict this response. Otherwise increased degradation of the best lands could result. Such research activities would no longer be conducted at the plot or field level, but at the ecodistrict or ecozone level. National policies must be implemented to ensure that land use is not out of balance with ecosystem potential. Land users must have control of, and commitment to, maintaining quality of the soil resource. Government policies can ensure that this is feasible.

References

Avery, D.T. (1995): Saving the Planet with Pesticides and Plastic. The Hudson Institute, Indianapolis, Indiana.

Evstafjeva, E.V., Orlinskiy, D.B. and Bashkin, V.N. (1994): Biogeochemical structure of the biosphere - anthropogenic pollution - adaptation of human populations. pp.30-31, Abstr. 1st Int. Symp. on Ecosystem Health and Medicine, June 19-23, 1994, Ottawa, Ont., Canada.

FAO (1989a): FAO Production Yearbook. World Soil Resources Report No 43. Food and Agriculture Organization of the United Nations. Rome.

FAO (1989b): Food Outlook, No 12. Food and Agriculture Organization of the United Nations. Rome.

FAO (1991): World Soil Resources. World Soil Resources Report No 66. Food and Agriculture Organization of the United Nations. Rome.

FAO (1993): Agriculture: towards 2010. C93/24. Food and Agriculture Organization of the United Nations. Rome.

Feliks, J. (1963): Agriculture in Palestine in the period of the Mishna and Talmud. Magnes Press, Jerusalem.
Harlan, J.R. (1965): A wild wheat harvest in Turkey. Archaeology **20**: 197-201.
Ikerd, J.E. (1994): Assessing the health of agroecosystems from a socioeconomic perspective. pp.48-49, Abstr. 1st Int. Symp. on Ecosystem Health and Medicine, June 19--23, 1994, Ottawa, Ont., Canada.
Jacobsen, T. and Adams, R.M. (1958): Salt and silt in ancient Mesopotamian agriculture. Science, **128**: 1251-1258.
King, L.D. (1990): Sustainable soil fertility practices. Chap. 5, In: C.A. Francis, C.B. Flora, and L.D. King, (eds.), Sustainable Agriculture in Temperate Zones. John Wiley and Sons, New-York.
Lee, R.B. (1968): What hunters do for living, or how to make out on scarce resources. In: R.B. Lee and I. de Vore (eds.), Man the Hunter. Aldine Publish. Co., New-York.
Oldeman, L.R., Hakkeling, R.T.A. and Sombroek, W.G. (1990): World map of the status of human induced soil degradation: An explanatory note. Rev. 2nd Ed. International Soil Reference and Information Centre. Wageningen, The Netherlands. 14 pp.
Postel, S. (1996): Forging a sustainable water strategy. The Worldwatch Institute, W.W. Norton (ed.), New-York.
Stinner, B.R. and Stinner, D.H. (1994): Ecosystem approaches to agriculture: towards a process of sustainability. Abstr. 1st Int. Symp. on Ecosystem Health and Medicine, June 19-23, 1994, Ottawa, Ont., Canada, 95-97.
United Nations Population Division (1992): Long-Range World Population Projections: Two Centuries of World Population Growth, 1950-2150. United Nations, New-York. 22p.
Yuill, T.M. (1994): Ecosystem health: what? For whom? Abstr. 1st Int. Symp. on Ecosystem Health and Medicine, June 19--23, 1994, Ottawa, Ont., Canada, 112-113.
Warkentin, B.P. (1995): The changing concept of soil quality. J. Soil Water Cons. **50**: 226-228.
Wicherek, S. (1994): L'érosion des grandes plaines agricoles. La Recherche, **25**: 880-888.
Wicherek, S. (1995): Inondations: histoire d'eau à fleur de peau. La Recherche, **26**: 692-693.
The World Commission on Environment and Development (1987): Our Common Future. Oxford University Press, Oxford.
The World Resources Institute (1990): World Resources 1990-91. Chap. 6 Food and Agriculture. Oxford University Press.

Addresses of authors:
Christian de Kimpe
Agriculture and Agri-Food Canada
Sir John Carling Building
930 Carling Avenue
Ottawa K1A OC5, Canada
B.P. Warkentin
Department of Crop and Soil Science
Oregon State University
Corvallis, OR 97331, USA

Concepts and Indicators for Assessment of Sustainable Land Use

T.F. Shaxson

Summary

'Sustainable land use' has both agro-ecologic and socio-economic overtones within the time dimension. Soils' organic materials and processes are vital for recuperation of soil architecture after damage, it being a key feature in the quality of the rooting environment. Assessing trends in sustainability of land use requires not only regular observation of indicators but also a trade-off between detail and the time available for processing, interpretation of cumulative results, and timely feedback to decision-makers. Rural people's perceptions of what is changing, how, and how quickly, and the indicators by which they determine this, make a pragmatic start-position at which to decide what other qualitative indicators might be important and whether (and if so, what) quantitative measurements are needed to complement them. For clear understanding of whether sustainability is being achieved, three types of indicators should be monitored at the same time. They should cover: (a) the condition of the soil as a rooting environment, (b) people's attitudes and capacities to continue living on the land, and (c) peripheral factors - such as weather, politics and markets - which indirectly affect farmers' decision-making about how they will look after the land.

Keywords: Sustainability, husbandry, soil, people, land-use, indicators.

1 Introduction

Agricultural populations are rising in the tropics, and areas of land which are safely-usable for sufficient biomass production to satisfy needs are diminishing. This requires that productivity and stability be maintained or increased on areas already in use, and that newly-opened land must also be husbanded in conservation-effective ways. It is necessary to check regularly whether this is the case and, if not, to provoke people to make appropriate adjustments in their husbandry of the land.

2 Concepts

2.1 Sustainable land use

The idea of 'land use' is a human construct, implying man's allocation and management of a particular area for chosen purposes. Unless maintained in relatively undisturbed condition, use of land involves altering the ecological succession and trying to maintain a modified end-point condition - ranging in general from silvo-pastoral, through tillage agriculture, to urban/industrial landscapes.

'Sustainable land use' refers to maintaining the three ecological functions of soils: (a) bio-mass production; (b) filtration, buffering and transformation of incoming materials and water; (c) habitats for organisms, including people. It also refers to maintaining its suitability for three spatial attributes: (a) space for housing, industry, infrastructure, etc.; (b) space for extraction of mineral and other non-renewing resources; (c) areas of cultural heritage. In many places, two or more of these functions may often be in competition with each other (Blum & Santelises, 1994, 535-6), as for example where biomass production and provision of housing vie for the same land, or where surface-mining of minerals destroys species' habitats. 'Sustainable land use' can have two meanings: (a) with ecological overtones, maintaining all the land in a particular area in a condition of usefulness (as opposed to dereliction), long-term productivity and ecological stability; or (b) with socio-economic overtones, maintaining a particular use on a particular piece of land, for example food security and income generation. Farmers may also have their own perceptions and definitions of what they consider to be sustainable land use. Rural people's decisions and actions play a significant part in determining whether present land use is sustainable or unsustainable.

2.2 Period of sustainability

The concept of sustainability of a given land use is qualified by time. Over timescales of more than a century, land-use conditions have seldom remained constant, due to significant changes in the combinations of bio-physical factors (such as climate) and socio-cultural factors. Changes in economic factors, alone or in combination with others, can have very drastic effects on land uses and on people's livelihoods in the span of only a few years. An example of this was the rapid displacement of relatively permanent coffee plantations by annual plantings of soya-beans in southern Brazil in the 1980s. This was a result of frosts killing the coffee on the one hand and world soya prices rising on the other. This altered forms of soil management and marginalised large numbers of rural workers, from being permanent employees to becoming casual labourers or 'boias frias' ('cold lunches'). A corollary is that, at least over shorter timescales, keeping a particular land-use sustainable requires attention to balancing ecologic, socio-cultural, and economic factors in the face of their (independent) tendencies to change. Over medium time-scales of a decade or more, land uses may oscillate over a sequence of years in response to cyclical changes in weather or economic conditions; what may appear to be an unsustainable use in a particular year may in fact be part of a land-use system which is sustainable over a sequence of years, as in the case of use of natural vegetation for animal production on the edges of the Sahel (see e.g. Earthscan Publications, 1996).

2.3 Good land husbandry

Good husbandry of land implies managing it for chosen purposes in ways which will favour the sustaining of its stability, productivity and usefulness for chosen purposes into the future (Shaxson, 1993). This description is as applicable to urban/industrial situations as to rural conditions.

Different areas of land have varying degrees of assorted constraints, particularly for rural production. These include moisture stress, excessive wetness, soil acidity, limited or excessive supplies of plant nutrients, compact horizons, low organic-matter content, difficult topography, shallow depths, etc. Our actions in managing the land for chosen uses should not exacerbate any of these but rather should ameliorate them to optimise and stabilise the land's condition for productivity and continued usefulness.

2.4 Land-use 'flexibility'

Land in undisturbed condition at a given place is most 'flexible' with respect to keeping maximum options open for other land uses in future. Similarly, rural land uses as a whole, being predominantly related to biomass production, are more flexible than urban/industrial uses - in the latter case it is very costly or difficult to return them to plant-production uses, thereby effectively foreclosing other future options. This implies that agro-ecological land uses in rural areas will become unsustainable and almost irreversibly damaged when overtaken by creeping urbanisation or industrialisation.

2.5 Land resilience

Resilience of land when under stresses due to inadequate management is central to sustainability. An appropriate definition of resilience in agro-ecosystems is:
"The ability of a disturbed system to return after new disturbance to a new dynamic equilibrium". (Blum & Santelises, 1994, 540).
Central to the concept of resilience in agricultural systems is the soil architecture and its recovery after damage. The spaces between the particles and structural units regulate movement and retention of water as well as fluxes of oxygen and carbon dioxide in the root zone, affect root growth and function, and house the mass and species-diversity of soil-inhabiting micro-, meso- and macro-organisms. The re-formation of relatively stable soil architecture - after it has been damaged due to collapse, compaction, interstitial sealing, or pulverisation - is achieved primarily through the activity of organisms acting on organic materials produced *in situ* and/or brought in from elsewhere. In the absence of such soil-biologic activity, organic materials (if present) are not transformed into the humic gums, fungal hyphae etc. which cause aggregation of soil particles and which also contribute markedly to the Cation Exchange Capacity of the soil.

2.6 Maintaining soil vitality

Maintaining ecological capabilities of land for plant production in the face of apparently entropic forces (notably the ageing and erosion of soils) over relatively long periods is possible by making use of organic self-recuperative capacities inherent in soils and perennial vegetation. Other contributing factors are the release of mineral plant-nutrients by rock weathering processes, and the ingenuity and skills of people in adapting to altering conditions, both ecologic, social and economic. Thus living as well as non-living factors at micro- and macro-level are involved in sustainability of land and its chosen uses. If the soil fails, for whatever reason, in its biological, physical, hydrologic and/or chemical functions as a rooting medium for living crops, pasture species and trees, then socio-cultural and economic considerations about sustainability are over-ridden (see e.g. Hudson, 1981, 216)

2.7 The time-dimension

In addition to the time-scale for considering sustainability (above), the processes of self-recuperation of soil and perennial vegetation, as in bush-fallow cycles, also require time for full restoration. For instance, even in the hot humid conditions of Malaysia where rates of recovery might be expected to be more rapid than elsewhere, fallow periods of ten years or more are considered necessary for 'an adequate crop' on subsistence farms where additional fertilisers are not applied (Siong, 1990, 20).

2.8 The people-dimension

Both individually and in groups, rural people decide about opening, maintaining or abandoning particular areas and/or uses of land. Their decisions, and what factors affect these, are therefore of prime importance in determining whether a land use or area will be sustainable - if they abandon it, the use will not be sustainable. Erosional exposure of low-fertility subsoils, loss of soil architecture, or pollution, may cause communities to abandon damaged lands because they are no longer able to sustain preferred land use systems which formerly were capable of providing their subsistence. Yields of favoured crops may fall so low in such areas that it becomes uneconomic to continue to grow them under market conditions which elsewhere might be remunerative. An example is the negative effect of increasing soil compaction on the yields and profitability of grain-growing in New Zealand. As compaction increased, gross margins were squeezed on the one side by falling income because of declining yields and on the other by increased costs (Shepherd, 1992).

Because both rural and urban populations may have strong affiliations with place and be unwilling to move, they may be keenly conscious of, and be able to describe, adverse changes occurring in their environment. For instance, a small farmer in Northern India described how the wide torrent-bed (in these days subject to ephemeral floods) in which he was standing used to be the line of a narrow placid perennial stream in his childhood. Whereas his father's land on each bank had been stable through the years, presently the flood-flows of the annual torrents scour away more of the land from each bank every year. In rural areas they may make considerable efforts to cope with and ameliorate problems of water shortage, soil loss, and falling productivity, to the best of their existing knowledge and abilities, and in the light of new knowledge and skills which they may acquire. Even though their observations may be qualitative, and their knowledge incomplete about the causes and effects of such problems, their perceptions and statements can be of great value if brought to light.

2.9 Rural/urban comparisons

Urbanisation and industrialisation have been increasingly impacting on rural land uses and diminishing their sustainability by (a) increased areas being taken out of relatively flexible uses; and (b) causing nearby and more distant rural areas to be affected by pollutive emissions. In agricultural areas leaching, erosion and crop export deplete nutrient reserves and may 'starve' the soil micro-organisms and plants. In and around industrialised areas the rates of accretion of pollutants as wastes - from traffic, industry, settlements, sewage, animal feedlots etc. - may be so high as to 'suffocate' the chemical and biological capacities of soils and waters to filter, buffer and transform them. This can be true even in the case of compounds such as nitrates and ammonia which are potentially usable by plants. An example: it has been calculated that in cattle-rearing parts of NW Europe the net annual input of nitrogen to the soil exceeds the reduction of the nutrient in the production of vegetal biomass by more than 100 kg/ha/year (Fleischhauer, 1996).

In situations of undesirable losses (e.g. soil erosion) and of undesirable gains (pollutants), the aim must be to ensure that rates of biological and chemical transformations in soils and waters, which can return them to their productive conditions, exceed rates of exposure or deposition (or both) of plant-unfavourable materials. Both rates need to be monitored to determine the trends, relative to 'tolerable levels' and 'saturation levels' according to current knowledge.

3 Indicators

3.1 Purpose

To be effective for assessing whether or not land-use sustainability is increasing or declining over time, indicators (as pointers) need to be used within the framework of a formal system of monitoring. Selected indicator-parameters must be observed qualitatively and/or quantitatively on a regular intermittent basis, and the results compared with (a) previous readings; (b) baseline conditions; and (c) desired conditions and threshold values, thus providing a picture of trends towards or away from resilience, stability, productivity, and continued usefulness of the land and use for the chosen purpose. For example, a change in soil pH would alter, differentially, the availabilities of different elements with respect to plant nutrition (FAO, 1985, 150)

The data from monitoring of chosen indicators need to be collated, compared, analysed and interpreted in order to show what may need attention, particularly where a trend is in undesir-able direction, and/or at an unacceptable rate (e.g. Carson, 1965).

3.2 Feedback and action

Observation, comparison, analysis and interpretation are of little practical value if, when undesirable trends appear, no timely feedback takes place to land-users/farmers, and/or administrators, and/or policy-makers. If there is no feedback, then no appropriate remedial decisions and actions to adjust the situation are likely to be made as a result of the monitoring (see Casley & Kumar, 1987 and 1988 for guidelines to monitoring and evaluation in Project situations) (Figure 1).

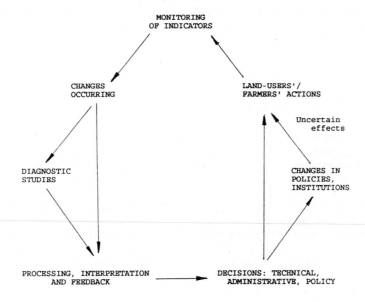

Fig. 1: Monitoring, feedback and policy-uncertainty

Farmers use indicators, feedback and adjustments every day in deciding what to do, where and when. Farmers' experience with local conditions allows them to 'fine-tune' their operations speedily and pragmatically. For example, on vertisols in northern India, small farmers judged to within a few minutes when soil conditions were suitable and unsuitable for tillage after rainfall (Author's personal observations, 1977). Non-farm agriculturists (from government agencies, etc.) seldom achieve such rapid or pragmatic feedback or turn it into action. Monitoring in most project situations is chiefly confined to measuring physical and financial progress towards pre-determined targets and objectives, rather than monitoring what changes in land and livelihoods may be occurring, and why.

3.3 The Pressure:State:Response Framework

Interpretation of what has been observed on the ground, in terms of present conditions and change, may require an understanding of the nature and changes of peripheral factors and observations. The Pressure:State:Response (PSR) approach aims to address this need by using indicators beyond those relevant to an immediate in-field situation. 'Pressures' are driving forces; 'States' are present conditions and current changes; 'Responses' are results of feedback and effects of decisions (Pieri et al., 1995). In retrospect, these can be illustrated by changes in agricultural methods during the 1980s and 1990s in the southern Brazilian State of Santa Catarina. Under 'Pressures' would be included the listed product prices for pig-meat, inadequate spread of effective extension advice about how to tackle severe erosion, low crop yields and concerns about polluted water supplies in wells and streams. 'State' would include much soil exposed to erosive rainfall, ephemeral streamflows, low levels of organic matter and activity in soils, and poor yields. 'Response' would include the development of no-till farming with extensive use of high-bulk leguminous catch- and cover-crops, better spread of advice on avoiding erosion and pollution by organic and inorganic materials such as pig slurry and pesticides, crop diversification, and group actions for rainwater infiltration and erosion minimisation on a catchment basis (De Freitas, 1997).

3.4 Alert versus detail

At regional/country/global scales there have been many attempts to influence policy-makers by using broad 'State' indicators to build 'doomsday scenarios' about environmental failure and possible collapse. While this may raise awareness that there are widespread problems across the globe - e.g. soil erosion, global warming, acid rain, etc. - they are too abstract and uninformative to guide decision-makers in what must be done in practical terms to alleviate the problems.

At the other extreme, the greater the number of variables and of locations monitored as indicators, the more knowledgeable we will be concerning all aspects of land condition. But this can generate huge data sets, many of which will probably lie un-processed and un-interpreted for weeks or months, till long after they might have been useful for relevant and rapid decision-making about needed adjustments. An effective sequence of monitoring, feedback and decision-making involves a trade-off between these two extremes in the choosing of effective and appropriate indicators. These have to be sufficiently informative to provoke action yet not so profuse as to hinder rapid feedback and timely adjustments (e.g. Becerra, 1993).

Concepts & indicators, assessment of sustainable land use 17

3.5 Quantitative and qualitative indicators

Although there may be much pressure always to provide quantified data for characterising land qualities, conditions and changes, excessive reliance on them alone can lead to unnecessary delays and costs, and to provision of data with spurious assumptions of accuracy. An example is estimates of soil erosion losses using the Universal Soil Loss Equation which are quoted on a per-ha basis to one decimal place, even though the validity of the USLE had not been checked where conditions of its factors markedly differed from those in which it was derived (Wischmeier, 1976). Quantitative measurement of chosen indicators may be essential in particular situations, but again a trade-off is needed between accuracy of information required and the limited time and other resources available to collect the data and make use of it.

Qualitative information from discussions and question/answer meetings with key informants, for example, can provide insights and in-depth understandings of people's perspectives, perceptions, attitudes, motivations and behaviour-patterns which may be crucial to understanding a given situation. Such understandings may be quite different in either content or emphasis from what non-farm staff understand of a particular situation, unless truly participatory methods were used in achieving them. In addition, qualitative data derived from time-series of ordinary snap-shot photos of particular items, places, conditions etc., can provide quick, cheap, information-rich records of complex or fleeting situations whose quantification would not be physically or economically practicable.

3.6 Proxy indicators

Proxy indicators provide information indirectly, via interpretation and inference, about important but hard-to-characterise factors for sustainability. Four key proxy indicators are:
* Stability of plant production
* Signs of land degradation
* Keenness and adaptability of farmers' groups
* Land-users' comments about changes.

An example of an in-field proxy indicator is using the height of maize plants as a proxy for the percentage ground cover by the leaves, once a quantified relationship between the two measures has been determined.

3.7 Key indicators

Alongside quantifiable factors, the most appropriate and pragmatic starting-points may be comments, observations etc. of land-users themselves, which provide an integrative view of ecologic, economic, social, cultural and other pressures and constraints. Their comments and observations can point to variables that could be quantified, or at least recorded qualitatively on a regular basis. For example, a small farmer in Costa Rica said, of the effects of more organic matter and less tillage: "The soil has become more fluffy". This suggests the possibilities for characterising this by quantifying of bulk density, porosity, structural stability (Shaxson 1997, 175).

In any situation to assess sustainability of land use, key technical indicators must include factors relating to the biological vitality and health of the soil, as affecting biomass production and biodiversity. These would include (a) vegetative cover close to the soil; (b) soil organic matter plus biological activity and soil biodiversity; (c) soil structure and porosity (architecture); (d) structural stability; (e) available water capacity; (f) plant-available nutrients; (g) effective cation exchange capacity; (h) soil acidity; (i) soil salinity; (j) depth of rooting and crop vigour. All these might be monitored with relative ease in the field, and backed with any necessary laboratory analyses (Baker, 1996: pers. comm.), but in reality in the field, practical problems may make even these difficult to monitor effectively.

3.8 Land users' own indicators

Land-users' observations and perceptions will mostly relate to what they can see and perceive unaided. On the one hand these items may or may not all be relevant to sustainability; on the other hand they may not cover small and unrecognised objects (such as hidden eggs of pest insects, or very small seeds of weeds such as *Striga spp.*) or national economic or political factors of which they are not directly aware.

Choosing to use people's own indicators, as well as other technical indicators, is a useful way of engaging their interest and concern, and developing partnerships in more formal monitoring activities in their area. Benites et al. (1997, 64-69) show how this approach has been used in Costa Rica. Which indicators technical staff may use to complement land users' indicators will depend on their analyses of what farmers and others report, and on their assessments of which other factors additionally need monitoring.

4 Conclusion

To obtain a view of trends towards or away from sustainability of land use, it is necessary to choose indicators from each of three different series of indicators, sufficient to be able to provide insights into what changes are occurring (or not), why they are occurring, and which might be the most appropriate actions to take if adjustments are needed. One series would be indicators of conditions, dynamics and changes of the soil's physical, chemical, hydrologic and biologic condition as a rooting environment. A second series would be indicators of people's own attitudes, willingness and abilities to continue sustaining their land and the chosen uses. A third series would be indicators of peripheral factors affecting land-users' decision-making about how the land will be used, and hence their management-effects on soil conditions. All three types of indicators need to be observed concurrently if we are to be able to determine if land use in a given area is likely to be sustainable.

References

Becerra E.H. (1993): Monitoreo y Evaluacion de Logros en Proyectos de Ordenacion de Cuencas Hidrograficas. Rome: FAO Guia FAO Conservacion **24**. 159pp.

Benites J.R., Shaxson F., Vieira M. (1997): Land condition change indicators for sustainable land resource management. In: Rome: FAO Land & Water Bulletin 5. Land Quality Indicators and their Use in Sustainable Agriculture and Rural Development. 212 pp.

Blum W.E.H., Santelises A.A. (1994): A concept of sustainability and resilience based on soil functions: the role of ISSS in promoting sustainable land use. In: Greenland & Szabolcs (Eds), Soil Resilience and Sustainable Land Use. Wallingford: CAB International, Chap. 30, 535-542.

Carson R. (1965): Silent Spring. London: Penguin Books.

Casley D.J., Kumar, K. (1987): Project Monitoring and Evaluation in Agriculture. World Bank/Johns Hopkins Univ. Press (USA). 159pp. Together with:

Casley D.J., Kumar, K. (1988): The Collection, Analysis and Use of Monitoring and Evaluation Data. World Bank/Johns Hopkins Univ. Press (USA). 174 pp.

De Freitas V.H. (1997): Transformations in the micro-catchments of Santa Catarina, Brazil. In: F Hinchcliffe et al., (Eds) The Economic, Social and Environmental Impacts of Participatory Watershed Development. London: International Institute for Environment and Development. (in press).

Earthscan Publications (1996): Sustaining the Soil. C. Reij, I. Scones & C. Toulmin (eds.), London: Earthscan Publicns. Ltd., 260 pp.

FAO (1985): Guidelines: Land Evaluation for Irrigated Agriculture. Soils Bull. no. 55. Rome: FAO. 231pp.

Fleischhauer E. (1996): Background paper for Ch. 2 of ISCO-9 Preparatory Brochure. Draft, Jan. 1996.

Hudson N.W. (1981): The Place of Rotation. Section 11.3 in Soil Conservation (2nd. edn). London: Batsford., 214.

Pieri C., Dumanski J., Hamblin A., Young A. (1995): Land Quality Indicators. World Bank Discussion Paper 315. Washington: World Bank. 63pp.

Shaxson T.F. (1997): Land quality indicators: ideas stimulated by work in work in Costa Rica, North India and Central Ecuador. In: FAO, 1997. FAO Land & Water Bulletin 5: Land Quality Indicators and Their Use in Sustainable Agriculture and Rural Development. Rome: FAO, 165-181.

Shaxson T.F. (1993): Conservation-effectiveness of farmers' actions: a criterion of good land husbandry. In: E Baum, P Wolff, M A Zobisch (Eds): Acceptance of Soil and Water Conservation. Vol.3 of: Topics in Applied Resource Management in the Tropics. Witzenhausen (Ger.) DITSL, 103-128.

Shepherd T.G. (1992): Sustainable soil-crop management and its economic implications for grain growers. In: P R Henriques (Editor): Proc. International Conference on Sustainable Land Management, 17-23 Nov. 1991, Napier, Hawkes Bay, New Zealand, 17-23.

Siong T.C. (1990): Indigenous conservation farming practices: Malaysia Country Review. In: Indigenous Conservation Farming Practices. ASOCON Report no. 7. Rept. of Jt. ASOCON + Commonwealth Workshop in Goroka, Papua-New Guinea, 3-7 Dec. 1990. London: Common-wealth Secretariat/Jakarta: ASOCON.

Wischmeier W.H. (1976): Use and misuse of the Universal Soil Loss Equation. J. Soil & Water Cons. **31**, 5-9.

Address of author:
T.F. Shaxson
Greensbridge
Sackville Street
Winterbourne Kingston
Dorset DT11 9BJ, U.K.

Soil Quality Assessment:
An Ecological Attempt Using
Microbiological and Biochemical Procedures

Z. K. Filip

I always avoid prophesying beforehand,
because it is a much better policy
to prophesy after the event has already taken place.
(Winston Churchill)

Summary

In an international project which is presently in progress, scientists from five European countries are investigating biological and biochemical characteristics of different natural and anthropogenically affected soils. The aim is to select and harmonize methods capable of indicating soil quality by respecting the nutrient turnover processes of ecological importance. From the data obtained in the first year, it is suggested that nitrogen-fixing bacteria, dehydrogenase activity and CO_2 release determined in soil samples may represent suitable indicators of soil quality.

Keywords: Soil quality, biological parameters

1 Introduction

In the last few years, soil quality became a matter of awareness throughout the world. This is not only a reflection of a negative development in the field of environmental protection. Rather, its roots go back some decades ago, with the growth in anthropogenic exploitation of soil resources in agriculture, forestry, industry and engineering. An individual "prophecy" made in the seventies (e.g., Filip, 1973; Kovda, 1975), which pointed to different risky developments in soil quality and their possible consequences for environment and mankind, was not heared. The reason was the widely spread belief in an almost endless resilience and self-cleaning capacity of soil. In between "the event" (see in motto) has taken place, and at many different locations we have to compete with problems associated with deteriorated soils. A deficiency in reliable indicators of soil quality also creates problems, both for environmentalists and law makers. An urgent need exists to define the present degree of soil deterioration and to adequately evaluate different remedial measures. On the other hand, there is also a need for early warning indicators which should help to predict further developments in soil quality and possibly prevent soil degradation in years to come.

Recently, Howard (1993) summarized the current soil assessment and soil protection approaches which exist in the European Union. Fleischhauer (1993) and Holzwarth (1993) summarized the respective developments and regulatory efforts in Germany. Hunsaker (1993) elucidated new concepts in environmental monitoring as developed in the USA. In our present effort to evaluate soil quality, both the principle of forecasting in environmental policy and the requirement for remediation which may appear in places should be respected. The aim is to develop reliable, ecologically based and internationally harmonized criteria of soil quality.

2 Conceptual approach

The current concept of soil quality includes different attributes such as (i) soil productivity, (ii) food quality and safety, (iii) human and animal health, and (iv) environmental quality (Parr et al., 1992). Until recently, efforts to evaluate soil quality have focused mainly on soil chemical and physical properties which can be easily characterized by using relatively simple, fully instrumentalized and partly standardized procedures. In our opinion this approach is not adequate. Undoubtedly, chemical and physical measurements may provide much spatial and temporal information. However, only a static type of data can be obtained in this way. Such data represent endpoint values and have a retrospective character as to the quality of the mineral and/or organic matter pool in soil.

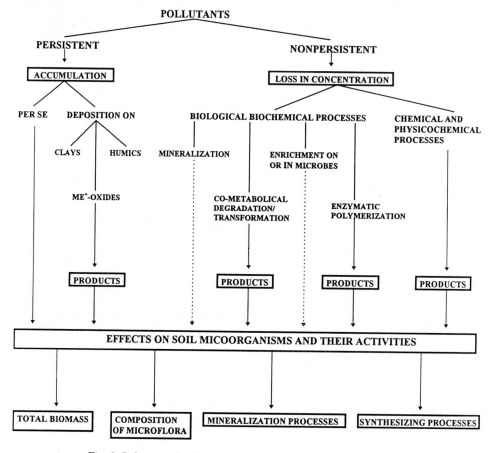

Fig. 1: Behavior of pollutants in soil and their interactions with microbes.

Soil quality, however, does not depend primarily on pool sizes but on transfer of matter and elements between the individual pools in soil and between soil and other constituents of environment. The processes included in this transfer make soil a dynamic part of the biosphere with soil microorganisms and invertebrates vitally involved. For this particular reason we seek biologically and biochemically, i.e., process linked, indicators which can sensitively respond to anthropogenic and/or environmental stresses on soil as a dynamic system. We believe that indicators of this type will be also predictive as to changes of soil quality. In order to select reliable parameters we had to evaluate theoretically the behavior of anthropogenic stresses, e.g., pollutants in soil and to consider their possible effects on soil biota (Fig. 1). Either directly or via different degradation or transformation products, pollutants can exert effects on (i) total soil microbial bio-mass, (ii) composition of soil microflora, (iii) mineralization processes, and (iv) synthesizing processes.

In a comprehensive review, Smith and Paul (1990) characterized microbial biomass as a source and sink for C, N, P and S in soil. Fluctuation in microbial biomass can cause significant increases or decreases in nutrient pool size and turnover, which control the fate of plant nutrients in the soil system.

In association with total biomass, the estimation of diversity in a microbial population represents a powerful tool in evaluating soil as a part of the ecosystem. Thus, the composition of soil microflora is also to be followed. An essential link in the cycle of life on earth represents the recycling in soil of C, N and other elements in different residues of plants, animals and microbes. Therefore, selected processes indicating the intensity of mineralization of organic matter should not fail when evaluating soil quality. Furthermore, the change in quantity and quality of soil organic matter, i.e., humic substances, can be indicative for aggradative and/or degradative developments in a soil ecosystem.

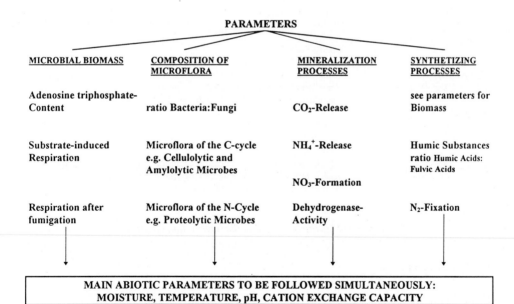

Fig. 2: Selected parameters to evaluate ecologically important soil characteristics.

In Fig. 2 different methodical approaches are shown which can be used to investigate the parameters of soil quality mentioned above. The respective investigation should be accompanied by a simultaneous determination of the basic physicochemical soil characteristics such as pH, moisture content, cation exchange capacity, and temperature. These abiotic chracteristics can allow a better understanding of the biological and biochemical data obtained and support the final evaluation of soil quality.

To enhance the efficiency of our effort and to underline the necessity of international cooperation in this field, we included scientists from five European countries in the proposal: Czech Republic (M. Tesarova, J. Kubat), Hungary (T. Szili-Kovacz), Russia (D.G. Zvyagintsev), Slovakia (P. Bielek) and Germany (Z. Filip, Co-ordinator).

Country	Soils
Czech Republic	
	Brown Soil
"Clean" Sites	Natural and renovated grassland with or without NPK
	Arable land with or without organic or mineral fertilizers
Polluted Sites	Natural grassland near to a motorway (Nox)
	Natural grassland in an urban area (org. pollutants)
	Arable land (airborne pollutants, e.g. SO_2,)
	Mine spoil (containing ash from a power plant using a low quality coal)
Hungary	
	Chernozem
"Clean" Sites	Arable land on loess (Univ. Exp. Station)
	Arable land on tuff and loess
Polluted Sites	Arable land near to a lignite using power plant (Nox, heavy metals)
	Arable land near to an oil refinery (hydrocarbons, heavy metals)
Russia	
	Soddy-Podzolic Soil
"Clean" Sites	Arable land weakly cultivated (Univ. Exp. Station)
Polluted Sites	Same land contaminated with 500 ppm Pb
	Same land contaminated with 2000 ppm Pb
	Same land contaminated with 500 ppm Pb, Cd, Zn
Slovakia	
	Pseudogley/Chernozem
"Clean" Sites	Arable land (Pseudogley)
	Arable land (Calcaro-Haplic-Chernozem)
Polluted Sites	Arable land (Pseudogley) near to metallurgy plant (Pb, Cd, As, Cu)
	Arable land (Pseudogley) near to metallurgy plant (MgO)
	Arable land (Chernozem) polluted by mineral oil

Table 1: Soil sites included in the international proposal

3 Materials and methods

The soil sites included in the internationally based proposal on soil quality assesment are listed in Table 1. They represent typical soils of the individual regions and also various types of long-term anthropogenic effects. During 1995 sampling was made from plough layers (0-20 cm) of the individual soils up to six times. To avoid any discrepancies both in this initial step of the soil assessment and in the analytical methods, a manual of methods has been compiled by the author and distributed to the participating laboratories. It consists of detailed descriptions for the selected microbiological and biochemical parameters as summarized in Fig. 2. In general, the individual methods were based on recommendations made by the International Organisation for Standardisation (ISO) or the American Soil Science Society (ASSS). Those methods which will finally prove to deliver significantly indicative data will be described in detail after the proposal will be finished.

4 Results and discussion

In the past years several researchers demonstrated the usefulness of microbiological and biochemical methods in determining side-effects of heavy metals and agrochemicals in soil (Domsch et al., 1983; Malkomes and Wöhler, 1983; Kandeler et al., 1994; Filip, 1995). However, a critical evaluation of the methods as indicators of soil quality still remains an open question. In Table 2 a relative value of the individual parameters is presented which reflects a preliminary experience made in our proposal. Nitrogen fixing bacteria, the enzymatic activity (dehydrogenase), and the CO_2 release (respiration activity) sensitively indicated the enhanced concentration of lead in a russian soddy-podzolic soil, e.g. (Fig. 3).

Parameter	Relative sensitivity
Total Biomass	*/**
Composition of Microflora	
Total Bacteria	**
Pseudomonas Bacteria	-/*
Actinomycetes	**
Fungi	**
Microbes participating in N cycling	
N_2 fixing bacteria (e.g. Azotobacter sp.)	****
Proteolytic sporeforming bacteria	-
Microbes participating in C cycling	
Cellulose decomposer	*
Oligotrophic bacteria	*/**
Biochemical processes/activities	
Respiration (CO_2 release)	***
Ammonification	**
Nitrification	***
Dehydrogenase activity	***/****

Table 2: Prelimary evaluation of the selected microbiological and biochemical soil parameters as indicators of soil quality.
(* for minimum, and **** for maximum depression in relation to a control soil)

Results obtained in Hungary showed that seasonal variation plays an important role even for the most sensitive N_2 fixing bacteria. Their indicative value was clearly highest in the June '95 soil sampling and was minimum in the samplings of March '95 and '96, respectively. In Slovakian chernozem, the soil respiration was not diminished but enhanced by a mineral oil related soil pollution (Fig. 4). On the other hand, slightly but complex industrially polluted Hungarian arable soils showed a decrease in CO_2 releases during the whole vegetation period in samples enriched with lucerne meal.

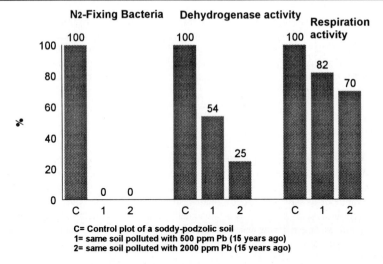

Fig. 3: Relative sensitivity of selected microbiological and biochemical indicators to soil pollution. C = Control plot of a soddy-podzolic soil; 1 = same soil polluted with 500 ppm Pb (15 years ago); 2 = same soil polluted with 2000 ppm Pb (15 years ago).

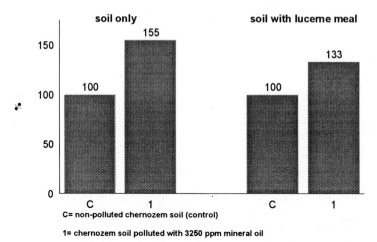

Fig. 4: Relative CO_2 release from non-polluted and mineral oil polluted soil. (C = non-polluted chernozem soil (control); 1 = chernozem soil polluted with 3250 ppm mineral oil.

These are typical examples of the sensitivity of some biological and biochemical procedures as indicators of anthropogenically affected soil quality. Since the international project is still in progress, final result will be presented later. These shall also include observations of the potential indicative power of some soil invertebrates if an apparent relation to ecologically important soil processes will be established. It is our final goal to replace any prophesying of soil quality and to contribute to the development of scientific prediction in this important field of environmental policy.

References

Domsch, K.H., Jagnow, G. and Anderson, T.H. (1983): An ecological concept for the assesment of side-effects of agrochemicals in soil microorganims, Residue Rev. **86**, 65-105.

Filip, Z. (1973): A healthy soil - the basis of a healthy environment, Vesmir **52**, 291-293 (in Czech.).

Filip, Z. (1995): Einfluß chemischer Kontaminanten (insbesondere Schwermetalle) auf die Bodenmikroorganismen und ihre ökologisch bedeutenden Aktivitäten, UWSF – Z. Umweltchem. Ökotox. **7**, 92-102.

Fleischhauer, E. (1993): Welche Aufgaben stellt der Bodenschutz? in BMU Hrsg., Untersuchungsmethoden, Bewertungsmaßstäbe und staatliche Regelungen für den Bodenschutz,Tagungsbericht, Schmallenberg 1.-4. Dez. 1993, 1-47.

Holzwarth, F. (1993): Vorwort. In: BMU (Hrsg.), Untersuchungsmethoden, Bewertungsmaßstäbe und staatliche Regelungen für den Bodenschutz, Tagungsbericht, Schmallenberg 1.-4. Dez. 1993, 001-003.

Howard, P.J.A. (1993): Soil protection and soil quality assessment in the EC, Sci. Tot. Environ. **129**, 219-239.

Hunsaker, C.T. (1993): New concepts in environmental monitoring: the question of indicators, Sci. Tot. Environ, Supplement 1993, 77-95.

Kandeler, E., Pennensdorfer, Ch., Bauer, E., Braun, R. (1994): Mikrobiologische Beurteilung biologischer Bodendekontaminationsverfahren im Modellversuch, Z Pflanzenernähr Bodenk **157**, 345-359.

Kovda, V.A. (1975): Biogeochemical cycles in nature and their disturbances through man, Nauka (Moscow), 72 pp., (in Russian)

Malkomes, H.-P., Wöhler, B. (1983): Testing and evaluating some methods to investigate side effects of environmental chemicals on soil microorganisms, Ecotox. Environ. Safety **7**, 284-294.

Parr, J.F., Papendick, R.I., Hornick, S.B., Meyer, R.E. (1992): Soil quality: Attributes and relationship to alternative and sustainable agriculture, Am. J. Alter. Agric. **7**, 5-11.

Smith, J.L., Paul, E.A. (1990): The significance of soil microbial biomass estimations. In: J.M. Bollag, G. Stotzky (eds.), Soil biochemistry, Vol. 6, New York: Marcel Dekker, 357-396.

Address of author:
Zdenek K. Filip
Federal Environmental Agency
Institute for Water, Soil and Air Hygiene
P.O. Box 1468
D-63204 Langen
Germany

Indicators to Assess Sustainable Land Use with Reference to Soil Microbiology

O. Dilly & H.-P. Blume

Summaryt
Microorganisms play a key role for soil functioning. Therefore, the task of soil classification regarding sustainability should essentially include microbiological features. Indicators for the assessment of sustainable land use in space and time should refer to significant ecological soil functions. Difficulties arise with the selection of integrative and sensitive microbiological features covering heterogeneity, complexity and diversity of the microbial communities and activities in soil. Approaches were developed to classify or qualify the soil microbiology in sustainability-related aspects of soil fertility and effects of pollution or 'stress'. The soil microbiological index, the fertility and the productivity index, and indicators for 'stress' were suggested for holistic estimates. A combination and improvement of these concepts may provide useful tools to obtain an integrated signal of functional ecophysiology of the microbiota. The microbial biomass content as well as general and specific microbiological activities and their interdependence are discussed.

Keywords: Indicator, microbial activities, microbial biomass content, microbial communities, soil functions, 'stress'

1 Functions of microorganisms in soil

The microbiota control striking properties for soil functioning: The microbial biomass is a labile pool of nutrients for plants. Furthermore, microorganisms are food for soil fauna. The high and specific activities of heterotrophic microorganisms are essential for rapid mineralisation of organic matter. With respect to ecotoxicology, microorganisms are able to degrade toxic organic compounds and thus help to decontaminate the environment. Microorganisms as symbionts support the nutrient supply of plants. In addition to biochemical aspects, the microbiota play a key role for soil structure. Soil aggregate stability will be increased by the microbial excretion of polysaccharides. Despite the positive role of microorganisms for system functioning, their activities also have negative effects such as the liberation of greenhouse gases, e.g. CO_2, CH_4, N_2O and the production of easily leachable inorganic compounds, e.g. NO_3^-.

2 Complexity and integrity of microbial life in soil

Microbial communities in soil comprise many genera of organisms that are able to live and grow under different properties of their environment. For different soil types under various vegetations from locations around the world, the biomass content of the microbiota ranges according to Smith & Paul (1992) between 110 and 2240 kg microbial C ha^{-1}. The major components of the microflora in terrestrial soil are mostly bacteria and fungi. It is well known that the most probable number (colony forming units) of microorganisms varies between about 1 and 1000 million within one gram of soil. However, these culturables usually represent only 1 to 10% of the microbial communities (Lorch et al. 1995). Thousands of different species seem to contribute to the microbial biomass in soil varying in distribution in microscale, e.g. microaggregate. In addition, microorganisms differ in size, on average between 0.2 and 20 µm in diameter, enzymatic potential and specific activity, which highlights the difficulty concerning the classification and the evaluation of soil microbiota particularly with reference to sustainability. The greatest diversity of the microbiota at small scales appears to live in the soil habitat (Tiedje et al. 1997).

Microbial communities adjust to ecological conditions in soil. Yet, microorganisms also control properties of this compartment, e.g. concentration of elements and compounds. Due to their activity, for example respiration and nitrification, microorganisms additionally control the level of oxygen concentration, the redox potential and the pH value. Thus, they indirectly influence nutrient availability particularly in soils that are saturated with water at some period of or throughout the year. Microbial communities and the habitat are obviously fine-tuned and build up an integrated unit. There is certainly still a gap in understanding microbial processes in soil. For the assessment of sustainable land use, both the microbiota and the habitat need to be studied simultaneously (Munch 1995).

3 Indices and indicators in soil microbiology

Several approaches have been tried to evaluate the activity of the microbiota in soil. In the 1950s, a number of investigators hoped that correlated information on extracellular enzymes in soil would provide a tool for determining the total biological activity in soil and, consequently, a 'fertility index' (FI) of soils usable for practical purposes in agriculture (Skujins 1978). In the 1970s, obtaining a 'fertility index' by the use of soil enzyme activity values seemed unlikely (Skujins 1978). However, Beck (1984) proposed a superordinate index of microbiological soil features with the development of the "Bodenmikrobiologische Kennzahl". This enzyme activity number (EAN) characterises the intensity of microbial transformations in agricultural soils (Beck 1984) combining values of microbial biomass content (on the basis of O_2-consumption), extracellular hydrolases and overall, cellbounded reductases, in all six microbiological features.

Typically, values of microbiological features, e.g. enzymatic activities, soil respiration and microbial numbers, do not correlate although there are some exceptions (Skujins 1978). Evidently enzymes are substrate-specific and individual enzymatic measurements cannot reflect the total fertility status of the soil. Soil microbial biomass content, microbial activity or individual soil enzyme measurements, however, might answer questions about specific decomposition processes in soil or questions about nutrient cycling. Therefore, the simultaneous measurements of the activity of a range of enzymes might provide a more valid estimation of the metabolic response of soil microbial communities to management, environmental 'stress' and climatic conditions than the determination of the activity of a single enzyme (Nannipieri et al. 1990, Nannipieri 1994).

Several combinations of microbial features were applied for the evaluation of environmental conditions: The metabolic quotient (qCO_2; the ratio between microbial respiration rate and microbial biomass content) is elevated under 'stress' as for instance high proton concentration (Anderson & Domsch 1993), monocultural land use (Anderson & Domsch 1990) and high heavy metal activity (Valsecchi et al. 1995). Doelman et al. (1994) presented a sensitivity-resistance index that reflects potential degradation of aromatic compounds and indicates the effect of heavy metals on microbial diversity in soil. The potential use of soil enzymes as an indicator of sustainability was considered by Nannipieri (1994). In addition to the above-mentioned approaches of FI and EAN, the ecological dose 50% concept was included in his overview. Welp et al. (1991) additionally mentioned the 10 % inhibitory effect, the ED_{10}/ED_{50}-ratio and the evaluation of dose response for the calculation of the confidence intervals of effective doses. Nannipieri (1994) summarised that conclusions of enzyme activity measurements still represent a problem, and future research is needed to devise methods to distinguish 'intracellular' (related to living microbial cells) from 'extracellular' enzyme activity in soils.

The so called 'microbial activity in soil' encompasses a broad spectrum of activities. Therefore, it can only be determined if general (e.g. CO_2 evolution, heat output, rates of nucleic acid synthesis) and specific (e.g. enzyme activities) criteria are measured (Nannipieri et al. 1990, Nannipieri 1994). In conclusion, an integrative approach for the assessment of sustainability does not seem to be available.

4 Concepts in Ecosystem Research

4.1 General introduction

The interdisciplinary project "Ecosystem Research in the Bornhöved Lake District" aims to analyse and model structures, dynamics and functions of terrestrial (agricultural and forest) and limnique ecosystems. In the scope ecosystem theory, emergent properties that may change in the course of development are assigned to evaluate systems, e.g. unit energy flow related to biomass, supported and expressed in metabolic quotient qCO_2 (Odum 1969). Furthermore, the "orientor" approach allows one to evaluate trends. Orientors that are essential for the viability of a system were integrated in a star plot. Values of the functions of evaluation were inserted in the plot for a selected system. If all orientors reach or go beyond the window of viability the system is suggested to be viable (Bossel 1992). This approach resembles the *law of Liebig*, which states that the growth of plants (viable systems) is controlled by the availability of essential nutrients (orientors). A restricted supply of essential nutrients causes deficiency or reduces harvest. Similarly, the orientor approach (Bossel 1977; 1992) determines the viability or nonviability of a system. If one essential factor (orientor) is restricted, the system may suffer without being able to compensate. The microbiota in soil is embedded in and interrelated with this complex system. As a consequence of the above-mentioned complexity of microbial communities and physiology, the theoretical concept of 'total' systems seem reasonably to be focussed on soil microbiology. Whole soils, and their constituent parts, reflect general properties of ecosystems (Elliott 1994).

4.2 Application on microbial communities and activity in soil

The qCO_2 is apparently enhanced under 'stress' conditions and in juvenile in contrast to the mature stage. Comparing the topsoils under different land uses within the Bornhöved Lake district, qCO_2 values significantly decreased in the order, wet site of the alder forest ('wet alder forest') > maize monoculture > dry site of the alder forest ('dry alder forest') > dry grassland, crop rotation > wet grassland > beech forest (Tab. 1; data are taken from Dilly 1994). Low efficiency in utilisation of organic carbon compounds and high intensity of carbon mineralisation related to the size of microbial biomass are obvious for the eutric soil under alder forest and for the soil under monocultural land use.

Star plots, also called sun rays, sun ray plots or amoeba, were already applied for the comparison of data in soil microbiology (Baumgarten & Kinzel 1990, Bachmann & Kinzel 1992). In the event that the ray reflects an essential microbiological feature, it can be called an orientor. Bachmann & Kinzel (1992) suggested that the area of the stars in the plots indicate vitality, and

Site	Depth [cm]	pH [H_2O]	C_{org} [$mg \cdot g^{-1}$ d. w.]	$q_{FE}CO_2$ [$mg\ CO_2$-$C \cdot g^{-1}\ C_{mic}\ h^{-1}$]	Soil unit
Crop rotation	0-20	6.4	32	3.1d	Eutri-cambic Arenosol
Maize monoculture	0-20	5.4	26	4.7b	Dystri-cambic Arenosol
Dry grassland	0-10	6.1	34	3.0d	Cumuli-aric Anthrosol
Wet grassland	0-20	5.8	177	2.6e	Eutri-terric Histosol
Beech forest	0-5	4.2	59	2.4f	Dystri-cambic Arenosol
Dry alder forest	0-20	4.1	623	4.0c	Dystri-fibric Histosol
Wet alder forest	0-20	6.2	500	8.9a	Eutri-terric Histosol

Table 1: pH value, organic C content and metabolic quotient of the microbial biomass (qCO_2; microbial biomass content estimated by fumigation-extraction method, FE) in topsoils (without considering the O-horizon) and soil units of the examined sites in the Bornhöved Lake district, Northern Germany (Jan 1992 to Oct 1993; n = 22; different letters indicate significant differences applied the Student-Newman-Keuls Method, p < 0.05).

that the variability of the star shapes may reflect the diversity or abundance of the involved microorganisms. In Fig. 1, estimates for microbial biomass content, general and specific microbial activities were selected as rays and sorted according to these groups (data according to Dilly 1994; here, substrate-induced respiration with 1 ml CO_2 h^{-1} corresponding to 30 mg The scaling of every ray was adapted to the range of all values. The area of the sun ray for C_{mic}). the topsoils showed highest values for wet grassland and dry grassland soil, and lowest values for maize monoculture. High vitality may be associated with high sustainability. Consequently, data suggested high sustainability for the grassland soils and low for those under arable monoculture. For soil under maize monoculture, the obviously serrated shape suggests a disrupted transformation pattern associated with, or as a consequence of, low biodiversity and poorly fine-turned microbial communities. The lower the serration of the star (as in case of wet grassland topsoil), the higher the association between the microbial features and the link between microbial processes. Looking at the topsoil of wet alder forest, general and specific microbial activities are high in contrast to the size of microbial biomass, demonstrated by the decentralised star. This reveals intensive transformations of the microbiota.

5 General discussion and conclusions

During the last few years, symposia on sustainability were held at different locations around the world, e.g. in Hungary (Brussaard & Greenland 1994) and Australia (Pankhurst et al. 1994). If erosion is not greater that the rate of soil formation, the major factor that regulates soil quality is its biological features (Elliott et al. 1996). Sustainability, which is a more or less unclear term, particularly emphasises biodiversity and functions in elemental cycling (Mathes & Breckling 1995). These features as well as the above-mentioned functions should be taken into consideration for the evaluation of the microbial life in soil.

Several approaches already exist for an integrative evaluation of microbial functions in soils. Despite the fact that combining data within an index may mask relevant microbiological features,

Indicators to assess sustainable land use, reference to soil microbiology 33

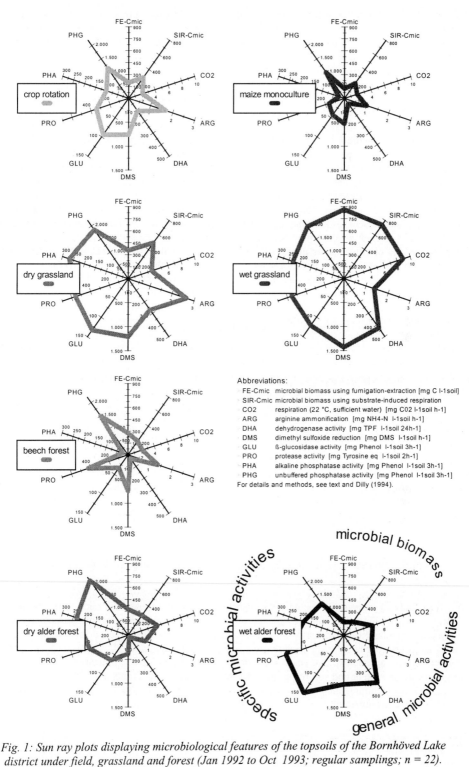

Fig. 1: Sun ray plots displaying microbiological features of the topsoils of the Bornhöved Lake district under field, grassland and forest (Jan 1992 to Oct 1993; regular samplings; n = 22).

the assessment of indicators, e.g. the qCO_2 and an improvement of the orientor approach seem to be adequate to achieve more holistic information about the system. Area and shape of the star may provide an integrated signal for the assessment of sustainable land use regarding soil microbiology. Difficulties arise in the selection of the microbiological features. Baumgarten & Kinzel (1990) and Bachmann & Kinzel (1992) selected 6 microbiological features (activities of phosphatase, dehydrogenase, saccharase, protease, urease) and 2 abiotic soil properties (C_{org} as well as pH value and sugar as well as amino acid content, respectively). The sun rays in Fig. 1 cover 10 orientors of microbial biomass content (fumigation-extraction method, substrate-induced respiration), general microbial activities (rates of basal respiration, dehydrogenase activity, arginine ammonifi-cation, DMSO reduction) and specific microbial activities (activities of glucosidase, protease, alkaline and unbuffered phosphatase). Orientors that reflect essential microbiological features, e. g. relevant activities and estimates for the structure of the microbial biomass, needed to be included in an orientor star (Fig. 2): (i) Greenhouse gas emission and (ii) diversity of the microbial communities, e.g. reflected in the proportion of r- and K-selected organisms. The choice of adequate features, that estimate microbial potentials and *in situ* microbial activities without methodological restrictions, remains a task of research. In addition, the position, the angle and the scale of each orientor may control the shape of the orientor star and, thus, a reasonable structure of the orientor star is necessary. For an holistic evaluation of total (soil) systems with reference to sustainability, orientors for yield, water quality, threat of erosion, temperature, soil water content and soil fauna also need to be considered. Nevertheless, the star shape permits a quick overview and is therefore recommended for a visual comparison of significant features.

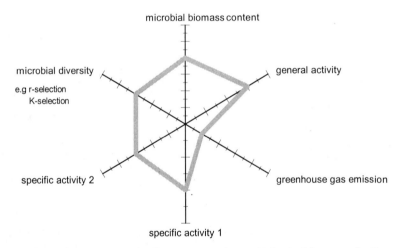

Fig. 2. *General orientor star displaying essential microbiological features of soils.*

Sustainability necessarily includes a dimension in time and trends that need to be monitored. Uncertainties of long-term changes in microbiological features are inevitable particularly with respect to uncertainties of scenarios for global climatic changes. Responses of microbiota to simulated temperature change may be less significant (Sarathchandra et al. 1989). Yet, ecosystem modelling over a period of 100 years with altered temperature of 2.7°C indicated a decline in soil organic carbon content and soil respiration (Kutsch & Kappen 1997) and an increase in N mineralisation at the beginning and a decrease later due to water restriction (Reiche et al. 1995). Environmental conditions apparently control the structure, the content and the activity of the microbiota.

Limitations of the indicators (orientors) in soil microbiology may arise due to the methodology (Ohtonen 1994). In addition, microbiological features were usually estimated without special regard to specific soil conditions, e.g. temperature, oxygen and water content. It might be misleading to apply laboratory results using sieved and homogenised soil to a field situation (Hunt & Parton 1986). Thus, conclusions concerning site-specific microbiological processes necessarily need to take these factors into consideration: Similar values of microbial biomass content in temperate and tropical climate cause different intensities of transformation; soil properties as pH value and organic C content are of great importance. More progress in the scope of soil microbiology is required if the significance of the results of different estimates will be enhanced. A point of issue is the irreversibility, reversibility, resilience and, in addition, the scale, the hierarchy and the boundaries of the structure of microbiota and microbial activity in soil. A system is defined as an arbitrary window of attention that is convenient for synthesising relationships (Odum 1995). In order to assess sustainable land use with reference to soil microbiology, the structure, the content and activities of the microbiota must be taken into consideration. The complex interference of these components restricts the evaluation of changes and sustainability. Homeostatic controls of ecosystems sometimes involve outside factors (organism x environment interaction), but in many instances self-regulating processes (organism x organism interaction) are responsible (Richards 1987).

These studies were supported by the German Ministry of Education, Science, Research and Technology (BMBF), project no. 0339077E, and the state of Schleswig-Holstein.

References

Anderson, T.-H. and Domsch, K.-H. (1990): Application of eco-physiological quotients (qCO_2 and qD) on microbial biomasses from soils of different cropping histories. Soil Biol Biochem **22**, 251-255.

Anderson, T.-H. and Domsch, K.-H. (1993): The metabolic quotient for CO_2 (qCO_2) a specific activity parameter to assess the effects of environmental conditions, such as pH, on the microbial biomass of forest soils. Soil Biol Biochem **25**, 393-395.

Bachmann, G. and Kinzel, H. (1992): Physiological and ecological aspects of the interaction between plant roots and rhizosphere soil. Soil Biol Biochem **24**, 543-552.

Baumgarten, A. and Kinzel, H. (1990): Mikrozonen im Stammfußbereich von Buchen: Untersuchungen der bodenbiologischen Aktivität. In: R. Albert, K. Burian and H. Kinzel (eds.), Zustandserhebung Wienerwald, Wien, Verlag der Österreichischen Akademie der Wissenschaften.

Beck, T. (1984): Mikrobiologische und biochemische Charakterisierung landwirtschaftlich genutzter Böden. I. Die Ermittlung der Bodenmikrobiologischen Kennzahl. Z Pflanzenernaehr Bodenk **147**, 456-466.

Bossel, H. (1977): Orientors of nonroutine behaviour. In: H. Bossel (ed.), Concepts and tools of computer-assisted policy analysis, Basel: Birkenhäuser Verlag, 227-265.

Bossel, H. (1992): Real-structure process description as the basis of understanding ecosystems and their development. Ecol Modelling **63**, 261-276.

Brussaard, L. and Greenland, D.J. (1994): Soil resilience and sustainable land use. Proceedings of a symposium held in Budapest, 28 September to 2 October 1992, including the Second Workshop on the Ecological Foundations of Sustainable Agriculture (WEFSA II). CAB International, Wallingford, UK

Dilly, O. (1994): Mikrobielle Prozesse in Acker-, Grünland- Waldböden einer norddeutschen Moränenlandschaft. EcoSys Suppl **8**, 1-127.

Doelman, P., Jansen, E., Michels, M. and van Til, M. (1994): Effects of heavy metals in soil on microbial diversity and activity as shown by the sensitivity-resistance index, an ecologically relevant parameter. Biol Fertil Soils **17**, 177-184.

Elliott E.T. (1994): The potential use of soil biotic activity as an indicator of productivity, sustainability and pollution. In: C.E. Pankhurst, B.M. Doube, V.V.S.R. Gupta and P.R. Grace (eds.), Soil Biota. Management in sustainable farming systems. Australia: CSIRO, 250-256.

Elliott, L.F., Lynch, J.M. and Papendick, R.I. (1996): The microbial component of soil quality. In: G. Stotzky and J.-M. Bollag (eds.), Soil biochemistry. Vol 9., Marcel Dekker Inc., New York, 1-21.

Hunt, H.W. and Parton, W.J. (1986): The role of modeling in research on microfloral and faunal interactions in natural and agroecosystems. In: M.J. Mitchell and J.P. Nakas (eds.), Microflora and faunal interactions in natural and agro-ecosystems. Nijhoff M. / Dr. Junk W. Publishers, Dordrecht, 443-494.

Kutsch, W.L. and Kappen, L. (1997): Aspects of carbon and nitrogen cycling in soils of the Bornhöved Lake district. II. Modelling the influence of climate changes on soil respiration and soil organic carbon content for arable soils under different management. Biogeochemistry **39**, 207-224.

Lorch, H.-J., Benckieser, G. and Ottow, J.C.G. (1995): Basic methods for counting microorganisms in soil and water. In: K. Alef and P. Nannipieri (eds.), Methods in applied soil microbiology and biochemistry, Academic Press, London, 146-161.

Mathes, K. and Breckling, B. (1995): Nachhaltige Entwicklung: Aufgabenfelder für die ökologische Forschung. Ecosys **3**, 71-73.

Munch, J.-C. (1995): Bodenökolgie aus Sicht eines Mikrobiologen. Mitteilgn Dtsch Bodenkundl Gesellsch **78**, 121-122.

Nannipieri, P. (1994): The potential use of soil enzymes as indicators of productivity, sustainability and pollution. In: C.E. Pankhurst, B.M. Doube, V.V.S.R. Gupta and P.R. Grace (eds.), Soil Biota: Management in sustainable farming systems, Australia: CSIRO, 238-244.

Nannipieri, P., Grego, S. and Ceccanti, B. (1990): Ecological significance of the biological activity in soil. In: J.M. Bollag and G. Stotzky (eds.), Soil biochemistry. Vol 6., Marcel Dekker Inc., New York, 293-355.

Odum, E.P. (1969): The strategy of ecosystem development. Science **164**, 262-270.

Odum, H.T. (1995): Energy systems concepts and self-organiszation: a rebuttal. Oecologia **104**, 518-522.

Ohtonen, R. (1994): Accumulation of organic matter along a pollution gradient: Application of Odums theory of ecosystem energetics. Microb Ecol **27**, 43-55.

Pankhurst, C.E., Doube, B.M., Gupta, V.V.S.R. and Grace, P.R. (1994): Soil biota: Management in sustainable farming systems, CSIRO, Victoria, 3002, Australia.

Reiche, E.-W., Schimming, C. and Branding, A. (1995): Auswirkungen auf Stickstoffhaushalt und Bodenversauerung. In: Leitungsgremium (ed.), Auswirkungen einer Temperaturerhöhung auf die Ökosysteme der Bornhöveder Seenkette. EcoSys **2**, 69-90.

Richards, B.N. (1987): The microbiology of terrestrial ecosystems. Longman Scientific and Technical, Essex.

Sarathchandra, S.U., Perrott, K.W. and Littler, R.A. (1989): Soil microbial biomass: Influence of simulated temperature changes on size, activity and nutrient-content 2. Soil Biol Biochem **21**, 987-994.

Skujins, J. (1978): History of abiontic soils enzyme research. In: Burns R.G. (ed.), Soil enzymes. Academic Press, London, 1-49.

Smith, J.L. and Paul, E.A. (1992): he significance of soil microbial biomass estimations. In: G. Stotzky and J.-M. Bollag (eds.), Soil biochemistry, Marcel Dekker, New York, 357-396.

Tiedje, J.M., Nuesslein, K. and Zhou, J.Z. (1997): Microbial diversity in soil. BIOspectrum Sonderausgabe, 28

Valsecchi, G., Gigliotti, C. and Farini, A. (1995): Microbial biomass, activity, and organic matter accumulation in soils contaminated with heavy metals. Biol Fertil Soils **20**, 253-259.

Welp, G., Brümmer, G.W. and Rave, G. (1991): Dosis-Wirkungsbeziehungen zur Erfassung von Chemikalienwirkungen auf die mikrobielle Aktivität von Böden: I. Kurvenverläufe und Auswertemöglichkeiten. Z Pflanzenernaehr Bodenk **154**, 159-168.

Address of authors:
Oliver Dilly
Hans-Peter Blume
Ökologiezentrum
Universität Kiel
Schauenburgstraße 112
24118 Kiel, Germany

Effects on Soil Management on Economic Return and Indicators of Soil Degradation in Tanzania: a Modelling Approach

J. B. Aune & A. Massawe

Summary

A model has been developed to predict long term changes in yield, economic return and environmental indicators such as erosion, pH and soil organic nitrogen. The data required to run the model are low as only standard soil chemical and physical characteristics are needed. Return to investments is calculated based on the Net Present Value (NPV) method. The model has a optimisation package which makes it possible to determine the optimal level of investment in soil conservation practices and liming in the initial year and how much should be allocated to running costs (fertiliser and mulch) in order to optimise NPV over a ten-year period.

Based on data from the Kilimanjaro region, the model predicted that NPV could be doubled with relatively low investments in terracing and by using fertilisers. Such moderate investments had considerable effects on indicators of soil degradation. Soil erosion was reduced from 29 Mg soil/ha and year without investment to 8 Mg/ha with a moderate investment in terracing. For soil organic nitrogen, it was found that terracing and use of fertilisers slightly increased the level compared to the scenario without any use of inputs.

Keywords: Model, maize, soil conservation, soil properties

1 Introduction

Indicators can provide (qualitative and quantitative information, simplify complex phenomena that can be readily understood by decision-makers, and can best capture improvement or deterioration in environment and land resource quality((WRI, 1995). The World Bank and others have made use of a pressure-state-response framework to analyse relationships between decision-making and environmental degradation (Pieri et al., 1995). Pressure indicators are indicators of pressures exerted upon land resources by human activities, state indicators are indicators of the state of land resources, and response indicators are the response by societies to pressures on, and changes in the state of land quality. The model concept presented here closely resembles a pressure-state-response framework, except that pressure and response indicators are combined. The reason for this is that it has been found difficult to differentiate between pressure and response indicators. Pressure/response indicators will refere here to decisions in relation to soil management and state indicators will refer to yield, level of soil erosion, pH and soil organic nitrogen (Figure 1).

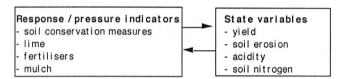

Figure 1. Simplified pressure-state-response framework for the relationship between soil management and soil degradation indicators.

The specific objective of this study was to develop a resource management and planning tool with the following characteristics:
- to predict how yield, economic return and environmental indicators such as soil erosion, pH and soil organic matter develop over years as a function of initial soil properties, input use and soil conservation practices
- to identify resource management practices which minimise soil degradation and optimise economic return (best practices),

The model was specifically developed for Tanzania, but the data used to develop the model originate from a geographically larger area (Aune and Lal, 1995).

2 Model description

Calculation of crop yield is based on the Mitcherlich and Baule principle (Black, 1993). Maize yield is calculated as follows:

$$Y_t = \text{Potential yield} * N_t * A_t * C * k \qquad \text{Eq. 1}$$

Potential yield is the climatically determined yield level in Mg/ha (the highest possible yield determined under non-limiting soil conditions). N_t, A_t, are time dependent indices (1 year time steps) ranging from 0 to 1 for nitrogen supply and soil acidity (pH), respectively. If the index for a specific factor is 1, the factor is not constraining productivity (Aune and Lal, 1995). C is the effect of soil conservation practices on yield, and this effect is not time dependent. A calibrating factor k is used to adjust the yield level in the first year.

The model is programmed in the Microsoft ® Excel spreadsheet. This makes the model user-friendly and replacement of functions to adjust to local conditions is easier.

2.1 Nitrogen (Nt)

The nitrogen index was developed as a relationship between available nitrogen and relative yield. The relationship was based on data from fertiliser experiments in Tanzania (Nyaki et al., 1993; KILIMO/FAO, 1992), and was described as follows:

$$N_t = 1 - 1.067 * e^{(-0.0095 * N_{min(t)})} \qquad \text{Eq. 2}$$

Available nitrogen $N_{min(t)}$ is calculated as the sum of nitrogen mineralised from soil organic matter, residues and additions from fertiliser (Equation 3).

$$N_{min(t)} = N_{F(t)} + \frac{N_{rm(t)} + N_{SON(t)} + N_{A(t)}}{2} \qquad \text{Eq. 3}$$

$N_{min(t)}$ is the amount of nitrogen available in the mineral form in year t, $N_{F(t)}$ is kg fertiliser N/ha, $N_{rm(t)}$ is the amount of N mineralised from residues, $N_{SON(t)}$ is the amount of nitrogen mineralised from soil organic matter and $N_{A(t)}$ is addition through atmospheric deposition. Availability of organic sources of nitrogen is tentatively set to half that of fertiliser because the organic sources of nitrogen have been found to have a lower recovery ratio than that of fertiliser N (Myers et al., 1994).

Release of nitrogen from soil organic matter $N_{SON(t)}$ is described by first order kinetics (Eq. 4) with a mineralization constant of 4 % (Nye and Greenland 1960; Young, 1989)

$$N_{SON(t)} = N_{s(t-1)} (1-e^{-r}) \qquad \text{Eq. 4}$$

where $N_{s(t-1)}$ is the total amount of organic nitrogen per hectare and r is the mineralisation constant.

Soil organic matter changes over the years as a function of decomposition, addition through residues and removal by soil erosion. This is described as follows:

$$N_{s(t)} = N_{s(t-1)} e^{-rt} + N_{r(t-1)} \beta - N_{e(t-1)} \qquad \text{Eq. 5}$$

where $N_{r(t-1)}$ is the amount of nitrogen added by residues, β is the proportion of residue N entering into soil organic matter and N_e is the annual loss of nitrogen through soil erosion (Aune, 1995). The loss of nitrogen through soil erosion is calculated according to the FAO version of the Universal Soil Loss Equation (USLE) (FAO, 1979). The effect of stone terraces and mulching on soil erosion is based on data from Kassam et al. 1992. The annual loss of nitrogen through soil erosion has been made time dependent as the cover factor in the USLE equation changes over the years according to yield level (Aune, 1995).

2.2 Acidity (At)

The relationship between soil pH and relative yield is based on results from different parts of the tropics (Aune and Lal, 1995), and is described as follows:

$$At = 1 - 82.5 * e^{(-1.21 * pHt)} \qquad \text{Eq. 6}$$

This relation implies that if pH is above 5, yields are not much affected while if pH drops below 4, hardly any yield is obtained. Changes in pH are described as a function of added fertiliser, type of fertiliser, years of cultivation and of liming (Aune, 1995):

$$pH_t = pH_{(t-1)} - 0.00091 N_{(t-1)} \Omega - 0.018 + 0.27 \, CaCO_3 \, Mg/ha \qquad \text{Eq. 7}$$

where pH_t is pH in year t, pH_{t-1} is pH year t-1, N is kg N/ha in fertiliser applied in the previous year, Ω is the acidifying effect of different types of fertiliser and 0.018 is the annual change in pH as a result of cultivation. This equation was developed based on analysing 6 long-term fertiliser experiments in the Southern Highlands of Tanzania (Uyole Agricultural Centre, 1991).

2.3 Soil conservation (C)

Effect of stone terraces on yield is difficult to quantify. In a survey of 50 farms in Kenya, yields on terraced land were 42 % higher than yields on non- terraced land (Figueiredo, 1986). However, the effect of terraces was less pronounced in the more humid coffee growing zone. Under these conditions, terraces increased yield by only 5 %. In this model, a 20 % increase of yield on terraced land is assumed. This corresponds to a C factor of 1.2 in Eq. 1. This effect of conservation has a direct effect on yield through improved water conservation. However, soil conservation measures will also affect yield indirectly by reducing nitrogen loss through soil erosion.

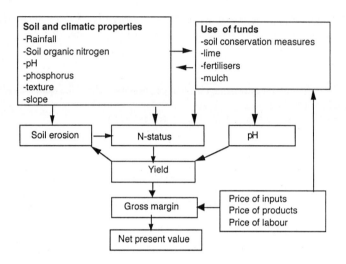

Fig. 2: Model flow chart

3 Economic analyses and resource optimisation

Yield is calculated based on initial soil properties and on soil management. The yield obtained and the expenditures to obtain that yield are used further in economic calculation as shown in Fig. 2. The method used to calculate return to investment over years is the Net Present Value (NPV) (Eq. 8), which discounts future net income to present:

$$\text{NPV} = \sum B_t(1+i)^{-t} - \sum C_t(1+i)^{-t} \qquad \text{Eq. 8}$$

where B_t is the yield in the individual year (calculated according to Eq. 1) multiplied by the price of maize, C_t is expected cost in year t, i is interest rate and t is years which run from t=0 to t=T (the last year) (Hanley and Spash 1994). If the NPV is above zero, the undertaking is economically viable. The costs in the first year are related to investments in terracing and lime, and to the running costs of fertiliser and mulch. In the following years the costs considered are expenditures related to fertilisers and mulch. The cost to construct the stone terraces was calculated according to data from Kassam et al. (1992) . The NPV calculation is based here on a 10 year period and on a 10 % interest rate.

In order to determine how funds should be allocated in order to maximise net present value, the model uses Excel's optimisation package; the Solver. The Solver identifies the use of funds which maximise NPV. Constraints to be taken into consideration when an optimal solution is sought may be introduced into the Solver. Those used in the following examples are related to liquidity. This is because the choice of management practice in tropical farming systems is frequently limited by liquidity.

4 Model test

A model has to be tested before using it for practical purposes. This proved to be difficult as there was a lack of data that could be used for this purpose. However, an experiment which could be used for testing was a long-term fertiliser experiment in northern Zambia (Woode, 1983). The model was able to describe 75 % of the changes in yield over time across three treatments.

The model's ability to quantify the effect of soil erosion on crop productivity was tested by comparing model predictions with data from a field study where the effect of top soil depth on productivity was studied in Tanzania (Kaihura et al., 1996). Maize grain yield decreased on average by 64 kg/cm of eroded soil on 7 plots with natural soil erosion, while the model predicted a reduction in maize yield of 92 kg/cm eroded soil.

5 Economic return and change in environmental indicators as affected by use of funds

The model was run using agroecological data from the Kilimanjaro region and based on prices in the 1995 season (labour = 600 shillings/day, 1 kg N = 600 shillings, 1 kg CaCO3 = 60 shillings, 1 kg maize = 150 shillings). Data for initial soil properties were collected from soil surveys in the Kilimanjaro region (Massawe, 1996). Initial soil organic carbon was 1.6 %, soil organic nitrogen was 0.16 %, pH 6.43 and bulk density was 1.16 Mg m^{-3}. The slope was set at 5 %.
Three different scenarios were tested:
1. Without use of inputs
2. Optimisation of NPV with budget restriction, 30, 000 Tanzanian shillings available for investments in terracing and lime and 20, 000 T. shillings available for fertilisers and mulch. Terracing and lime are investments in the first year.
3. Optimisation of NPV without budget restrictions.

Optimal resource use under budget restriction almost doubled net present value (NPV) as compared with no use of inputs (Table 1). All available funds were used for fertiliser and terracing under this scenario. The fertiliser use in this scenario corresponds to 33 kg N/ha. NPV increased from 2,300,000 T. shillings in the scenario with budget restrictions to 2,800,000 T. shillings when budget restrictions were removed. Expenditures on fertilisers increased to 69,000 T. shillings which is equivalent to an application of 114 kg N/ha. The model did not allocate more than 30,000 shillings to stone terraces even without budget restriction. The reason for this was that the cost of construction of terraces for one hectare of land with a slope of 5 % is 30,000 T. shillings.

Indicators of soil degradation such as soil erosion, pH and soil organic nitrogen were also strongly influenced by resource allocation (Table 1). Soil erosion was reduced from 29 Mg/ha to 8 Mg/ha when stone terraces were constructed. Without budget restrictions soil erosion was 7 Mg/ha. This reduction in soil erosion from 8 Mg/ha to 7 Mg/ha was related to increased yields, which improved soil cover. pH was also strongly influenced by management practices. In the scenario without budget restrictions, pH was 5. 2 in the final year, while it was 6.2 when no fertiliser was used. This severe change in pH was due to the fact that ammonium sulphate was used and this nitrogen source has a strongly acidifying effect. Soil organic nitrogen (SON) was also to some extent influenced by resource management. SON (%) was higher in the two scenarios with input use than in the scenario without the use of inputs. This was due to higher yields when fertiliser was used which increased root growth and thereby contributed to increase SON (%).

	Without investments	Optimal resource allocation- budget restriction	Optimal resource allocation - no budget restriction
Resource allocation			
Stone terraces (Tanzanian shillings)	0	30, 000	30, 000
Liming (T. shi.)	0	0	0
Fertilisers (T. shi.)	0	20, 000	69, 000
Mulch (T. shi.)	0	0	0
Economic return			
Net present value (T. shi)	1, 300, 000	2, 250, 000	2, 810, 000
Average yield Mg/ha	1.42	2.56	3.49
Change in environmental variables			
Soil erosion Mg/ha and year	29	8	7
pH in final year	6.2	5.9	5.2
SON % in final year	0.087	0.103	0.104

Tab. 1: Effect of resource use on economic performance and change in environmental indicators.

The model can also be used to examine the investments required to reach a specific target with regard to environmental quality. The model was set up to examine how resources should be spent if soil erosion was to be reduced to a level of 3 Mg/ha without budget restrictions. In this scenario the model allocated 38,000 shillings to fertiliser and 5,000 shillings to mulch. However, this reduced NPV to 2,500,000 T. shillings as compared to 2,800,000 T. shillings without this target of soil erosion.

The model was also tested under the same agroecological condition as in the Kilimanjaro region except that initial soil pH was set as low as 4.3. Optimisation without budget restriction gave an NPV of only 1,000,000 T. shillings. The model allocated the maximum amount to terracing, but in additions it proposed 25,000 T. shillings to liming. This amount of liming is equivalent to 0.5 t $CaCO_3$ /ha. Allocation to fertilisers was 6,000 shillings and 32,000 to mulch. It is worth noting that under these acidic condition the model proposed mulch as N-source rather than fertiliser. The reason is that mulch has a much less acidifying effect than ammonium sulphate. The amount allocated to lime was not sufficient to significantly raise pH. When a lower lime price was tested, the model allocated much more funds to lime.

The model presented here has weaknesses related to the effect of soil conservation practices on yield, to the effect of soil conservation on the level of soil erosion and to costs in relation to soil conservation. Field testing of the model is therefore required before the model can be used for planning purposes. However, the model represents a tool to examine the complex interactions between soil management, soil degradation and economic return.

References

Aune, J.B. (1995): Predicting soil degradation in Tanzania- a system analysis approach. Norwegian Journal of Agricultural Research., Supplement **21**, 47-60.

Aune, J.B. and Lal, R. (1995): The tropical soil productivity calculator- a model for assessing effects of soil management on productivity, in R. Lal and B. Stewart eds., Soil Management. Experimental Basis for Sustainability and Environmental Quality, Lewis Publishers, 499-520.

Black C.A. (1993): Soil Fertility Evaluation and Control., Lewis Publishers, Boca Raton.
FAO (1979): A Provisional Methodology for Soil Degradation Assessment, FAO, Rome.
Figueiredo, P. (1986): The yield of food crops on terraced and non-terraced land. A field study of Kenya. Working Paper 35. Swedish University of Agricultural Sciences, Uppsala.
Hanley N. and Spash, C.L. (1994): Cost Benefit Analyses and the Environment. Edvard Elgar, Vermont.
Kaihura, F.B.S., Kullaya, I.K., Kilasara, M., Aune, J.B., Singh, B.R. and Lal, R. (1996): Impact of soil erosion on soil productivity and crop yield in Tanzania. In: Structural Adjustment Policies and Environmental Degradation in Tanzania. Paper presented at Seminar in Dar-es-Salaam 10 April 1996, Agricultural University of Norway.
KILIMO/FAO (1992): Yield increase through the use of fertilisers and related inputs. Results of fertiliser trials and demonstration. Rukwa, Iringa, Mbeya and Kilimanjaro regions, United Republic of Tanzania.
Kassam A.H., Velthuizen, H.T., Mitchell, A.J.B., Fischer, G.W. and Shah, M.M. (1992): Agroecological land resource assessment for agricultural development planning. A case study of Kenya. Resources data base and land productivity. Technical Annex 2. Soil erosion and productivity, FAO.
Massawe, A.P. (1996): Assessment of soil erosion and soil fertility in Tanzania and their effects on crop productivity- with special reference to selected agro-ecological zones. M.Sc. thesis, Agricultural University of Norway.
Myers, R.K.J., Palm, C.A., Cuevas, E., Guntatilleke, I.U.N. and Brussard, M. (1994): The synchrony of nutrient mineralisation and plant nutrient demand. In: P.L. Voomer and M.J. Swift (eds), The Biological Management of Tropical Soil Fertility, John Wiley & Sons, Chichester, 81-116.
Nyaki, A.S., Jasperse, S.A. and Kullaya, I.K. (1993): Maize - northern zone. In: J.G. Mowo, J.P. Magoggo, A.E.T. Marandu, E.G. Kaitaba and J. Floor (eds.), Review of fertiliser recommendations in Tanzania. Part 1. NSS Soil Fertility Report F9, 12-29.
Nye, P.H. and Greenland, D.J. (1960): Soils under Shifting Cultivation.. Technical communications No 51, Commonwealth Bureau of Soils, Harpenden, UK.
Pieri, C., Dumanski, J., Hamblin, A. and Young, A. (1995): Land quality indicators., Technical Paper, World Bank, Washington.
WRI (1995): Environmental Indicators: A Systematic Approach to Measuring and Reporting on Environmental Policy Performance in the Context of Sustainable Development. By A. Hammond, A. Adriaanse, E. Rodenburg, D. Bryant & R. Woodward. DC: World Resources Institute.
Woode P.R. (1983): Long-term fertiliser experiments for Kasama, Zambia, M.Sc. thesis. University of Aberdeen.
Uyole Agricultural Centre (1991): Annual Research Report 1988/89. Part 1, Mbeya Tanzania, 41-44.
Young, A. (1989): Agroforestry for Soil Conservation,, CAB, Exeter.

Addresses of authors:
Jens B. Aune
Centre for International Environment and Development Studies
Noragric
Agricultural University of Norway
P.O. Box 5001
N-1432 Ås, Norway
Apia Massawe
Rodent Control Centre
P.O. Box 3047
Morogoro, Tanzania

Difficulties in Monitoring and Evaluating Watershed Development
A Case Study from East Java

J. de Graaff & J. W. Nibbering

Summary

Soil conservation and watershed development activities are hard to evaluate since neither their effects nor their beneficiaries can be easily detected. In monitoring and evaluating the effects of these activities, indicators play a crucial role.

The Konto River Project in East Java, Indonesia, paid much attention to monitoring. Hydrological and erosion research was undertaken to assess effects of land use changes on erosion and streamflow. Physical monitoring focused on tree survival rates, the state of terraces, etc, and a socio-economic monitoring programme looked at farmers' participation, costs and benefits. For this monitoring, various physical and socio-economic indicators were used, part of which were reviewed during a brief post-evaluation mission.

It proved hard to find effective indicators, which should be unambiguously defined, consistent, specific, sensitive to small changes and easy in data collection (Casley and Kumar, 1987). The land use changes covered about 2,000 ha, or almost 10% of the area, but effects on erosion and hydrological parameters were small. Suitable and easy (proxy) indicators were tree survival rates, combined with tree height and diameter for remaining trees, for reforestation results, and participation rates for the involvement of the population. Despite the use of many proxy indicators for changes in productivity and socio-economic status, the results of the socio-economic monitoring activities were much affected by autonomous developments. A plea is made to improve monitoring systems.

Keywords: Monitoring, evaluation, watershed development, indicators, soil erosion.

1 Introduction

Watershed development or rural development in hilly or mountainous areas, is usually aimed at the two often conflicting objectives of increasing production and income while reducing harmful effects of soil erosion. These objectives are seldom expressed in quantitative terms and are interpreted in different ways. Even though soil erosion is apparent, farmers and government agencies are often reluctant to invest in soil conservation and watershed development activities. Investment costs are high, the effects are hard to predict, benefits may only come after many years, and it is not clear to whom these benefits will accrue. Therefore much attention should be paid to monitoring of the activities, which could then be evaluated at different stages of the project cycle. This could be done by selecting for each activity one or more indicators, defined as specific and objectively verifiable measures of changes or results brought about by an activity (UN ACC, 1984).

In this paper basic principles of monitoring and evaluation and the use of indicators are discussed first, followed by a presentation of the Konto River case. Thereafter, some issues in monitoring and evaluation which have emerged from this case study are discussed.

1.1 Principles of monitoring and evaluation

Monitoring has been defined by FAO (1988) as the periodic or continuous surveillance of project activities by participants, management or donors. It is mainly used by project management to follow different activities and to check whether they are heading towards the desired results. Evaluation is a process for determining systematically and objectively the relevance, effectiveness, efficiency and wider impact of activities in the light of their objectives (UN ACC, 1984). Activities can be considered effective when results correspond with the aims, and efficient when overall benefits exceed total costs. Impact stands for the eventual outcome of activities, with their net effect on economic, social and ecological status. Evaluation can take place prior to, during, or after the expiration of a project. The most important evaluation methods are cost effectiveness analysis, cost-benefit analysis (CBA) and multi-criteria analysis (MCA).

Monitoring and evaluation of project activities is greatly facilitated when objectives are well-defined and quantified. In project planning objectives can be specified as concrete targets or goals, which can be translated in either inputs required (activities) or output (effects) to be obtained. The criteria applied in evaluations pertain to the degree and manner in which objectives are reached. For each criterion several attributes can be derived from the objectives. Attributes represent targets, output or inputs. Indicators can be chosen to monitor the value of these attributes. Effective indicators should be unambiguously defined, consistent, specific, sensitive to small changes and easy in data collection (Casley and Kumar, 1987). The more concrete the attribute the more specific indicators can be applied.

Criteria and attributes	Direct or proxy indicators
Efficiency	
Direct, on-site effects on production and income	
Marketable production (local or export)	Production; tree numbers
Indirect, downstream effects on production and income	
Productivity losses due to sedimentation	Prod. losses; reservoir capacity
Equity	
Intragenerational equity	
Income to different social groups	Gini coefficient; state of welfare
Intergenerational equity	
Income to present and future generations	Rate of savings; state of resources
Conservation	
Conserving functions of land	
Erosion control	Soil depth; annual soil loss
Conserving functions of water	
Water supply	Streamflow; Q_{min}/Q_{max} ratio
Conserving functions of vegetation	
Wood supply	Annual yield; remaining stand

Table 1. Examples of attributes on criteria and possible indicators in watershed development

1.2 Monitoring watershed development

Soil conservation and watershed development projects are usually initiated to reduce the degradation of land and water resources, but when these resources are managed by large numbers of poor people, projects are also aimed at rural development and poverty alleviation.

Evaluation criteria for soil conservation and watershed development projects can be grouped under three headings: efficiency, equity and conservation (adapted from van Pelt, 1993). Indicators should relate to attributes of these criteria. The most common attributes of the efficiency criteria are, on the benefit side, marketable and non-marketable goods, and, on the cost side, costs of labour and of man-made and natural resources. Although requiring much data and calculations, effects of erosion, sedimentation and changes in streamflow could be expressed in monetary terms as costs and benefits accruing to various groups.

A distinction can be made between intragenerational and intergenerational equity, which both can play an important role in watershed development projects.

Whereas it is already difficult to express some attributes on the efficiency and equity criteria in monetary terms, this is even more so for attributes on the conservation criterion, for which the functions of land, water and vegetation are possible attributes (Table 1).

2 The Konto River Project: its setting and monitoring and evaluation

The 279 km^2 Upper Konto River watershed area is situated in Malang district of East Java, and consists of an upland plateau surrounded by steeply sloping volcanic mountains. Altitude ranges from 620 to 2,868 m above sea level. Two thirds of the area is state forest land, while the remainder consists of 'village lands', containing 23 villages and all agricultural land. The area coincides with the sub-districts Pujon and Ngantang. On the lowest side of the area is the multi-purpose Selorejo reservoir. Average annual rainfall is about 2400 mm. The steep slopes, fine-textured soils and high rainfall intensities would make the area susceptible to erosion when closed vegetation is removed and the land cultivated. The high permeability of the deep, young volcanic soils has some neutralizing effect.

Total population in the area was about 100,000 in the late 1980s. Population density on village lands is high (1,200 persons/km^2). Farming activities are labour-intensive.

Although the large area still under forest cover made it a relatively privileged area compared with other parts of Java, the situation in the watershed in 1979 was changing rapidly, with a fast decrease of the natural forest area (taken over by shrub) and mounting pressure on the forest land to provide firewood and fodder.

Soil erosion was, on the basis of the USLE formula, estimated at about 1.2 mm per year for the whole catchment. The main sources of erosion were assumed to be the rainfed annual crop fields (with 55% of total erosion) and the built-up areas (Murdiono and Beerens, 1991). As a result the Selorejo reservoir has a high sediment yield, which threatens to reduce its economic life.

2.1 Objectives and activities

After an inventory phase (1979-1983) and a planning phase (1984-1985) the implementation of various watershed development activities was undertaken during the third phase of the project (1986-1990). The ultimate objectives for the implementation were formulated as follows:
1. To reduce the rate of erosion to acceptable levels and simultaneously to improve the hydrological balance in the area as a consequence of better infiltration in both forest land and village land, in order to improve the water supply for irrigation and other purposes (downstream);

2. To increase the productivity of forest and agricultural land and to raise income levels for the local population;
3. To achieve an involvement of the local population in soil conservation and watershed development.

The project's main activities were: 1. reforestation; 2. perennial cropping; 3. soil conservation; 4. livestock husbandry and grass planting; 5. infrastructural support activities.

2.2 Monitoring of project activities

During implementation it was realised that project objectives were vague and not quantified. In the first two phases of the project no hydrological and erosion research was undertaken, and all data about erosion were 'guesstimates' obtained with the USLE formula. To make up for this, in the third phase much emphasis was given to hydrological and erosion research (Rijsdijk and Bruijnzeel, 1990). Rainfall, erosion, run-off, streamflow and sedimentation data were collected for three sub-watersheds during three years (1987-1989).

On the other hand, much data about forest area, farming systems and socio-economic conditions was collected before the implementation phase (Nibbering, 1986). These base data and the information from the third phase monitoring programme, made it possible to monitor and evaluate the effects of development activities on the second and third project objective. The monitoring programme included physical monitoring to assess tree survival rates, state of terraces and gully plugs, and socio-economic monitoring to assess impact on welfare.

2.3 Evaluation at completion of the project

At project completion a preliminary cost-benefit analysis was undertaken to assess the effectiveness, efficiency and preliminary impact of the respective activities (de Graaff and Dwiwarsito, 1990). It was not possible then to assess quantitatively whether erosion had declined to acceptable levels and whether the hydrological balance had improved. For that the hydrological and erosion programme had started too late. For all major components the efficiency or cost-benefit relationship was assessed. In the reforestation programme the 'taungya' system, whereby small farmers undertake the planting and are allowed to interplant annual crops for two years, had become more attractive through inclusion of firewood and tree crops. The other system, with paid labour on steeper slopes, remained costly and had been less successful. Coffee growing and planting of fodder seemed promising, whereas terracing of rainfed land seemed to be not very efficient.

Because of the long gestation period of most implementation activities it was considered too early to make a thorough assessment of the impact of the various activities.

2.4 *Ex post* evaluation

For the *ex post* evaluation three main activities were undertaken, focusing on the main project objectives: erosion control, forest and agricultural production, and involvement of local population. To assess the effects of land use and land management changes on soil erosion, sedimentation and streamflow, a multiple spreadsheet watershed model was developed (van Loon et al., 1995). This allowed comparison of actual land use development with three other scenarios. It was used to evaluate the project activities (de Graaff, 1996), and to put hydrological and erosion data in a historical perspective (Nibbering and de Graaff, 1995). During the *ex post* evaluation, field visits were paid to the sites of a sample of the activities on forest and on village land. Discussions were held with officials in four villages and extensive interviews were held with 25 farmers in these villages. Visits were also paid to government services that had been involved in the project activities. The *ex post* evaluation made it clear that some autonomous developments and changes of farm prices had greatly influenced the development in the area after the termination of project activities.

3 The use of indicators in monitoring and evaluation

In the *ex post* evaluation, use was made of several indicators to assess the effectiveness (or success rate), efficiency (cost-benefit relationship) and impact (increased welfare) of the activities. For long-term effects of the activities on the first project objective (i.e. erosion control), erosion rates (in mm/ha/yr) and the ratio between the dry and wet season streamflow (Qmin/Qmax ratio) were calculated with the spreadsheet model. The base data were calibrated with the results of the hydrological studies. For the assessment of the effects of the activities in relation to the other two objectives two examples are presented hereunder.

3.1 Land-based indicator for the performance of reforestation

The most important activity to reach the second objective of increasing production on forest and agricultural land was the reforestation programme with its agroforestry components. It was also assumed to contribute indirectly to the first project objective. From 1986-1990 a total of 1,802 ha (140 blocks) of shrub land was reforested, 1,010 ha with the 'taungya' method and 792 ha with wage labour. Day-to-day monitoring in the establishment phase was undertaken by officers from the State Forestry Corporation. The project had assisted in defining new reforestation systems and had established several pilot plantations. To evaluate these new reforestation systems a survey was held in 1989 to assess the results of the 1986 and 1987 plantings. For this purpose 15 'taungya' and 10 'wage labour' blocks were studied. Number, height and diameter of remaining trees were recorded on 25 m by 25 m sample plots. This 1989 reforestation survey showed that about 80% of tree seedlings in the 'taungya' system and 70% of those planted with wage labour were still alive and had attained an average height of 80 cm. To reassess eventual results during the *ex post* evaluation in 1994 about half (13) of these plots were revisited. This reassessment showed that reforestation was no longer as promising as it was in 1989. The small sample indicated that survival rates of main trees had dropped to 50-60% (Table 2). However, the results of the reforestation varied considerably. One plantation established with wage labour could not be found back in the shrub area, while one 'taungya' plot had earned much fame as a demonstration plot.

District / reforestation type	1989			1994		
	No of plots	Survival rate (%)	Height (m)	No of plots	Survival rate (%)	Height (m)
Pujon, wage labour	7	74	0.5	3	39	2.7
Pujon, taungya	8	87	0.7	5	50	5.4
Ngantang, wage labour	3	71	0.7	2	55	4.0
Ngantang, taungya	7	90	1.2	3	72	6.0
Total	25	81	0.8	13	53	4.7

Sources: de Graaff and Dwiwarsito, 1990; *Ex post* evaluation survey.

Table 2. Survival rates and height of timber trees in sample plots in reforested areas.

3.2 People-based indicators for the impact on socio-economic conditions

While it was relatively easy to assess the involvement of the local population in the soil conservation and watershed development activities (about 10,000 households participated in one or more of the activities), it was much harder to assess the eventual benefits accruing to these people. The 'beneficiary monitoring' surveys (among 270 households) yielded information about crop and livestock activities and off-farm employment and made it possible to estimate household income and changes therein resulting from project and autonomous initiatives. In the period 1987-1990 income levels had increased with 30% at current and with 10% at constant prices. To be able to get an im-pression about the eventual impact on welfare, several indicators of socio-economic conditions were applied. During the *ex post* evaluation a small stratified sample survey was held among 25 house-holds in 4 villages, whereby for the sake of comparison the same indicators were used (Table 3).

Indicators (Sample of households)	Monitoring survey 1988 (270)			Evaluation survey 1994 (25)		
	Pujon	Ngantang	Total	Pujon	Ngantang	Total
Farm land (ha)	0.52	0.67	0.60	0.60	0.61	0.60
Number of dairy cattle	1.8	1.2	1.5	2.8	0.5	1.7
Number of goats/sheep	0.5	1.2	0.9	0.2	3.2	1.6
Households with (%):						
kerosene stove	18	19	19	46	8	28
motorcycle	5	3	4	38	8	24
bicycle	4	7	5	8	22	16
electricity	40	21	33	92	75	84
television	5	5	5	23	25	24
house of stone	11	21	16	62	58	60
house of bamboo	80	72	77	15	33	24
Firewood (t/yr)	3.8	4.1	4.0	3.0	4.2	3.6
Fodder (t/yr)	18.2	15.1	16.5	24.5	11.6	17.5

Sources: de Graaff and Zaeni, 1989; *Ex post* evaluation survey

Table 3. Comparison between socio-economic welfare indicators in 1988 and in 1994.

Although the small sample size in the evaluation survey does not permit a statistically justified comparison, it is clear that important changes took place in the period 1988-1994. The most conspicuous change in the villages was the fast improvement of housing conditions and electrification. In Pujon more use was made of kerosene (stoves), while the use of firewood had decreased somewhat. Cattle numbers and fodder consumption had increased.

4 Lessons for monitoring and evaluation

The modest scope of the *ex post* evaluation study, the lack of recent aerial photographs and the fact that hydrological and erosion research as well as socio-economic monitoring had ended abruptly in 1990, have made it extremely difficult to get a clear idea about changes and the degree to which they have been brought about by the project. However, several lessons could be learned from this evaluation study, some of which are discussed below.

4.1 Scope for and limitations of indicators

Indicators only refer to one or a few qualities of the object concerned. Tree survival rates, for instance, do not tell anything about the condition and eventual production value of the living trees. To avoid erroneous interpretations it is therefore better to use a few indicators. Pieri *et al.* (1995) suggest that indicators be representative of a group of features and convey the most significant information of these features in summary form. Tree survival rates, crude though they appear, could be taken to indicate overall conditions (e.g. forest production, hydrological conditions, care by people). The extent to which tree survival contributes to these objectives requires additional research, however. When circumstances change, (proxy) indicators at some stage no longer convey relevant information about a change or trend in physical or socio-economic conditions. Kerosene stoves cease to be an analytically sound indicator of socio-economic welfare, when wood becomes scarce and is no longer collected.

4.2 Autonomous developments

Watershed development projects may create much employment and income in an area, but there are always autonomous developments as well. To make it easier to separate changes due to the project from those resulting from autonomous developments, base-line surveys and monitoring should also consider data not directly related to the project. Some autonomous developments in the Konto River watershed were part of trends going on for some time (Nibbering, 1993). Some created favourable conditions for carrying out project activities.

4.3 Land-based and people-based indicators

In watershed development activities it can be difficult to disentangle the role of physical factors and human factors with respect to effects on soil and water conservation. Tree survival rates may relate to climatic and biological conditions, and also to management, more protection and better care by the local population. Reduced sedimentation may be due to soil conservation measures but also to less destructive rainfall intensities. Clearly, monitoring and evaluation should comprise the entire range of relevant factors that bear on the performance and results of project activities and well selected indicators could play a role here.

References

Casley, D.J. and Kumar, K. (1987): Project monitoring and evaluation in agriculture. The John Hopkins University Press, Baltimore and London.
FAO (1988): Participatory monitoring and evaluation. FAO Regional Office, Bangkok.
Graaff, J. de (1996): The price of soil erosion; an economic evaluation of soil conservation and watershed development. Doctoral Thesis, Agricultural University, Wageningen.
Graaff, J. de, and K. Dwiwarsito, 1990. Economic monitoring and evaluation of Konto River Project implementation activities. Project Com-munication no. 15. KRP, Malang.
Graaff, J. de, and Zaeni, W.A. (1989): Socio-economic monitoring of Konto River Project implementation. Project Communication no. 13. Konto River Project, Malang.

Loon, E.E. van, de Graaff, J. and Stroosnijder, L. (1995): A method to estimate effects of land use change on plant production, water and sediment flows at the watershed scale, with reference to the Konto watershed, Java, Indonesia. Department of Irrigation and Soil and Water Conservation, Agricultural University Wageningen, Internal paper.

Murdiono, B. and Beerens, S. (1991): Soil conservation in an East Java watershed: socio-economic and institutional aspects. In: Arsyad, S., Amien, I., Sheng, T. and Moldenhauer, W. (eds.), Conservation policies for sustainable hillslope farming, pp. 298-309. Soil and Water Conservation Society, Ankeny.

Nibbering. J.W. (1986): Socio-economic conditions and developments in the Kali Konto Project area, East Java. Kali Konto Project, Malang.

Nibbering, J.W. (1993): Agricultural diversification in the Upper Konto Area. In: Dick, H., Mackie, J. and Fox, J. (eds.), Balanced Development; East Java in the new order. Oxford University Press, Oxford.

Nibbering, J.W. and de Graaff, J. (1995): Using historical data in a hydrological model: an example from the upper Konto watershed area in Java. In: Indonesian environmental history newsletter, No. 5, Leiden.

Pelt, M.J.F. van (1993): Sustainability-oriented project appraisal for developing countries. Doctoral thesis, Agricultural University Wageningen.

Pieri, C., Dumanski, J., Hamblin, A. and Young, A. (1995): Land quality indicators. World Bank Discussion Papers No. 315. The World Bank, Washington, D.C.

Rijsdijk A. and Bruijnzeel, L.A. (1990): Erosion, sediment yield and land use in the Upper Konto Watershed. Konto River Project, Malang.

UN ACC (1984): Monitoring and evaluation; guiding principles. IFAD, Rome.

Addresses of authors:
Jan de Graaff
Department of Environmetnal Sciences
Erosion and Soil & Water Conservation Group
Agricultural University
Nieuwe Kanaal 11
6709 PA Wageningen, The Netherlands
Jan Willem Nibbering
Proyecto Reserva de Biósfera Alto Orinoco-Casiquiare
Residencia SADA - Amazonas
Ví Alto Carinagua
Puerto Ayacucho
Estado Amazonas, Venezuela

Mathematical Evaluation Criteria

O. Tietje, R.W. Scholz, A. Heitzer & O. Weber

Summary
One of the most important steps towards the evaluation of sustainability is to define appropriate criteria. Two promising methods for such an evaluation are risk analysis and life-cycle assessment. Both these approaches need appropriate strategies for the identification of such criteria. This paper focuses on quantitative evaluations and gives examples of mathematical evaluation functions as used in life-cycle and risk assessments of a heavy-metal polluted soil. The mathematical evaluation models are presented together with their corresponding word model (i.e. their intended meaning expressed in words). Hence with limited effort mathematical evaluation functions are available which are simple, intuitive, and easy to understand and which at the same time may be used to consistently aggregate soil variables.

Keywords: Risk analysis, life-cycle assessment, multi-attributive utility theory, multi-criteria decision making, mathematical evaluation.

1 Introduction

Evaluation is one of the key tasks when investigating environmental impacts. Sustainability is a generally accepted objective, but how to reach it and which criteria are to be applied is a matter of discussion. The main objective of this paper is to show how mathematical models may be used to prepare generally acceptable evaluations.

An integrative evaluation of remediation technologies is in progress using mathematical evaluation modelling within the Swiss Priority Program 'Environment'. Several ongoing investigations assess variants of phytoremediation using different plants and different applications.

Many investigations try to avoid individual evaluation which is referred to as *subjective*. They describe environmental systems using *objective* information and try to deduce evaluation criteria from a normative perspective. Recently there has been much effort directed towards evaluation specially within the frameworks of life-cycle assessment (Goedkopp, 1996; Hofstetter 1996), risk analysis (Scholz et al., 1996; Slovic, 1987; Luce and Weber, 1986), and welfare analysis (system modeling combined with a utility based evaluation). The application of these frameworks is mostly divided into several 'steps', one (preferably the last one) of which is *evaluation*. Within each step a certain method is applied (e.g. inventory analysis as part of life-cycle assessments). The method for evaluation is mostly a kind of multi-criteria analysis.

This paper assumes that valuation consists of two parts, a procedural and a structural one, building an integral whole and depending on each other. In order to make evaluation more comprehensible we show which structures are relevant during the evaluation process and how they combine.

We present here an evaluation concept and show its applicability to life-cycle assessments and risk analysis. The advantage of this scheme is that it facilitates questions about practical evaluations (e.g. the integral evaluation of gentle remediation techniques which includes technical, economic, ecological, and social criteria) and leads to quality criteria for evaluations (i.e. criteria which indicate whether an evaluation will be acceptable).

2 Material

In this paper we use an example of a heavy metal contaminated area in Dornach (Switzerland). The contamination is due to a metal factory with large emissions during the last century. Due to environmental protection measures the emissions are now reduced to a very low level.

The soil is contaminated (e.g. soil Cadmium concentrations of about 4.5 ppm and Copper at about 1.5 g/kg; Geiger and Schulin, 1992). The questions which arise for evaluation are:
- which parts of the area (of about 15 km^2) need remediation?
- what kind of remediation should be applied?
- how to integrate the evaluations of different stakeholders?

With respect to the remediation options the following evaluations are to be made:
1. life-cycle analysis (evaluation of the ecological impacts)
2. risk analysis (evaluation of human and ecotoxicological risks)
3. cost benefit analysis (economic evaluation)
4. error analysis (accuracy, variability, reliability of remediations, technical evaluation)
5. multi-criteria analysis (integral evaluation including ecological, economic, technical and social perspectives).

Special questions relate to administrative policies, e.g. whether environmental standards are appropriate for the development of remediation strategies. Alternative simple assessment strategies are the development of trigger and remediation standards (Hämmann et al., 1996), the application of fuzzy set theory (Pohl et al., 1996) or using probability characteristics (see below).

As an illustrative example, only the concentration of cadmium in soil is shown. Let us consider hypothetically that at the end of these investigations there may be two results (as shown in Fig. 1):
1. Case: the mean Cd concentration after phytoremediation is reduced to 0.8. Due to a rather small variance in this case there is a spatially distributed but small risk.
2. Case: the mean Cd totals are reduced to 0.5 ppm, which is lower than 0.8, but there is a large variance in the measurements and the risk of 'hot spots' remains.

Thus, the following questions have to be addressed:
1. Which system variables (e.g. bioavaliability, costs etc.) must be included in the evaluation?
2. How to measure the *value* of the cadmium contamination: Take the mean? Take the probability of exceeding 0.8 ppm? This adresses the consideration of elementary evaluation functions.
3. What if lead exhibits the opposite results (changed cases): This adresses the composition rule which has to express the preference of cadmium contamination over lead. This preference is generally influenced by toxicological knowledge about the specific consequences of different contaminations.
4. Is it possible to decide whether one alternative is to be preferred over another? This is one of the most applied assumptions. If not, one of the preference axioms is violated and this limits the application of possible evaluation models.

For the evaluation, the first step must be a systems analysis, followed by the definition of the alternatives under consideration (Scholz et al., 1996). Then the ensemble of evaluation variables has to be determined. Defining an elementary evaluation function means to assign values with an intended meaning. These may be utility, gain, damage, welfare, probability or other. For the whole range of the evaluation variable (e.g. for all possible Cd concentrations) one has to specify the value corresponding to the intended meaning (e.g. the probability of measuring such a value, cf. Fig. 1). The probability density function for the Cd concentrations (measured at the site, which is used here for illustration, called 'Mattenweg') is an example for such an elementary evaluation function.

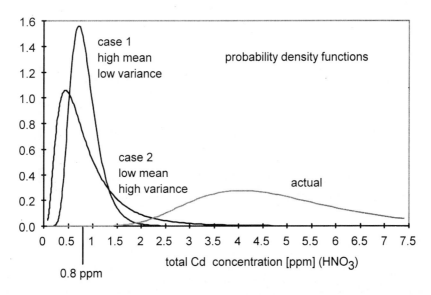

Fig. 1: Measured Cd concentration before remediation (Geiger and Schulin, 1992) and two hypothetical distributions after remediation.

3 Method: Mathematical Evaluation Modeling

3.1 Process and structure of evaluation models

The process of evaluation consists of four parts:
- **decomposition** of alternatives into the evaluation variables used as criteria (e.g. cadmium concentration in soil)
- assigning a **value** to the single criteria (e.g. giving a damage measure for cadmium)
- **composition and aggregation** of the single values (e.g. weighing cadmium vs. lead, and damage vs. probability)
- **comparison** of the alternatives.
  This leads to the structural integration of the evaluation (Fig. 2).

The intended meaning of the evaluation should be consistent with the structures in each part of the evaluation process. This may be viewed as a general scheme which is especially applied to multi-criteria decision making as, e.g. the utility approaches of MAUD (multi-attribute utility decomposition, Berkeley and Humphreys, 1991). For illustration let us consider that different heavy metals (e.g. Cd, Pb, Cu) must be assessed whether additional measurements have to be made due to measurement uncertainty.

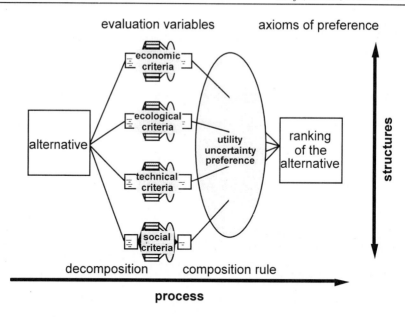

Fig. 2: Process and structures of evaluation models.

1. The decomposition part results in the **evaluation structure 1**: The ensemble of the evaluation variables, which are considered to be indicative for the differences between the alternatives under consideration (e.g. the concentrations of heavy metals (among others) for the intended meaning *uncertainty of soil pollution with heavy metals*).
2. The evaluation of these variables is referred to as **evaluation function 1**: elementary evaluation functions. Each variable is mapped onto the value space defined by the intended meaning (e.g. the *probability density function* shown in Fig. 1). If special functions (e.g. probability density, utility or fuzzy) are used here, please note that they do not automatically represent an intended meaning. Rather they are a tool for this representation and the intended meaning of them has to be defined separately.
3. The **evaluation function 2** consists of applying a composition rule which transforms multiple single values into a single composite value. This is different for different intended meanings as, e.g. in utility, uncertainty or outranking concepts (cf. Berkeley and Humphreys, 1991; Walley, 1991; Vincke, 1992). In the case of heavy-metal polluted soil, a first approach may be the sum of the *coefficients of variation* of all evaluation variables.
4. The **evaluation structure 2** is the preference structure for the comparison of the aggregated values of the alternatives. This structure consists of the axioms that apply to the comparison of the alternatives. Examples are completeness, weak ordering, the sure thing principle, independence, continuity, and transitivity (cf. Fishburn, 1988; Savage, 1972; Vincke, 1992). To assess *uncertainty of soil pollution with heavy metals* using the *coefficient of variation* yields a weak ordering of the alternatives satisfying most of the axioms mentioned above.

Please note that this example shows only a small part of an integrated evaluation. The final decision for certain additional measurements may include additional economic criteria (e.g. the costs of the measurements), ecological criteria (e.g. damage to plant growth), technical criteria (e.g. spatial variability of contamination), and social criteria (e.g. creating worry about risks for the population).

If mathematical operations are included, the four structures mentioned above are the main mathematical elements involved in evaluation modelling. The functions (evaluation function 1 and 2) are mathematical operations constrained to certain axioms and rules and hence are constituents of the mathematical model structure as well.

3.2 Families of evaluation dimensions

From a systematic point of view it is necessary to find elements of mathematical evaluation modelling. Such elements are - besides the intended meaning, the process and the structures described above - the elementary families of evaluations dimensions. The *evaluation function 1* consists of a mapping of each (single, one-dimensional) evaluation variable into the value space which of course may be multi-dimensional (evaluation dimensions may be, e.g., damage and probability). There may be very many such evaluation dimensions corresponding to the intended meaning. Many of these evaluation dimensions belong to the utility family, or to the uncertainty family.

Please note that there are more such families, but from a pragmatic point of view it is sufficient to describe these families. It is necessary to describe the intended meaning of a evaluation with a value space consisting of elements of one or more families of intended meanings. Within technical applications, risk is often assumed to consist of damage and probability.

The **utility family** contains evaluation dimensions which result in a 'better-worse-definition'. This family consists of a large amount of evaluation dimensions. We include utility, damage, costs, or the greenhouse effect (as an environmental impact). Word models include: „Cd concentrations lower than 0.8 ppm are 'good', exceeding 0.8 ppm are 'bad'". Another word model for an elementary evaluation dimension may be: „the larger the Cd concentration, the lower is the quality of the soil".

The **uncertainty family** contains evaluation dimensions which describe the variation associated with other evaluation dimensions. This may be done using probability, possibility, width of the interval of measured values, or other scatter measures. An example is the probability density function presented in Fig. 1.

Table 1 summarises characteristics for the evaluation structures within the evaluation process (cf. 3.1) for different families of elementary evaluation dimensions (cf. 3.2).

Family of evaluation dimension	Evaluation structure 1: Ensemble of evaluation variables	Evaluation function 1: elementary evaluation functions	Evaluation function 2: composition / aggregation	Evaluation structure 2: axioms of preference
Utility	system variables which are different for the alternatives	if MAUT: 1 for best alternative, 0 for worst, interpolate linearly	if MAUT: weighed sum	if MAUT: complete order, linear, sure thing principle, full compensation, ...
Un-certainty	system variables which are not exact	uncertainty measures: probability, possibility (fuzzy), quartiles, scatter	e.g. operations on random variables, fuzzy inference, interval arithmetic	generally no complete order, but in many practical cases

Table 1: Structures within the evaluation process that characterise three families of intended meanings of elementary evaluation dimensions (MAUT = multi-attribute utility theory).

3.3 Aggregation and composition

It is important to note that most of the evaluations are multi-dimensional. Two kinds of multi-dimensionality - and their handling - have to be examined: aggregation and composition. Both of them yield - or try to yield - a single resulting value.

If an intended meaning cannot be described using a single evaluation dimension, more dimensions have to be included. While the probability of a certain event is a one dimensional evaluation, risk for example is not: most technicians perceive risk as a combination of damage (or utility) and probability (as a model for uncertainty). Hence for each evaluation variable these two dimensions have to be specified and later aggregated.

Moreover different sources of risk - as e.g. due to Cd or Pb concentrations in soil - often are to be included. Then for each evaluation dimension (e.g. for damage and utility) the contributions of every evaluation variable has to be composed. From a mathematical point of view the sequence of aggregation and composition is not important. For risk and for life-cycle assessments the composition may precede the aggregation. But it is important to distinguish between these two essential subprocesses of evaluation (Fig. 3). From a psychological point of view the sequence may be very important because the empirical investigation of evaluations is very much influenced by different framings, e.g. due to the logical sequence of questions (cf. Tversky and Kahnemann, 1981).

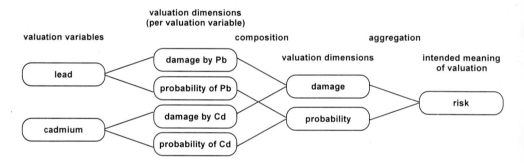

Fig. 3: Aggregation and composition of evaluation dimensions.

4 Application and Results

This scheme is capable of describing a wide range of environmental evaluations. As examples we apply this scheme to a risk assessment and to a life-cycle assessment.

4.1 Risk

For the heavy metal contaminated area in Dornach (Section 2), several studies are available which include qualitative assessments. In order to show the difficulties of quantification let us consider a risk assessment for the hypothetical distributions shown in Fig. 1. Because it is impossible here to capture all aspects of risk, the evaluation is drastically simplified to illustrate the structures within the evaluation process, using the above-mentioned 'Mattenweg' site in Dornach:
- Evaluation structure 1 (ensemble of evaluation variables): Cadmium, copper, and lead concentrations.

- Evaluation function 1 (elementary evaluation functions): frequency distributions derived from the measured concentrations.
- Evaluation function 2:
  a) Composition: cadmium and copper are expressed in units of lead equivalents (Goedkopp 1996), where the cadmium and copper concentrations are considered to exhibit 3 and 0.005 times (respectively) the ecological impacts of lead. The composed concentration is then calculated as the sum of Cd, Cu, and Pb concentrations expressed in lead equivalents, the probabilities are composed neglecting statistical dependence. This is similar to the approach used in error analysis (ISO, 1993).
  b) Aggregation: an acceptability function is derived from the environmental standard for Pb assuming an accuracy of factor 2. The areas under the curves where the composed concentration exceeds the acceptability function quantifies the unacceptable risk.
- Evaluation structure 2 (preference axioms): The definition where exceeding the acceptability function will be allowed yields the indication whether the risk is acceptable or not, or whether further action is needed. A crucial preference axiom is whether or not to allow compensation (e.g. whether a large risk due to low concentrations may be compensated by a small risk due to high concentrations).

The acceptability functions may be expressed by the relative frequency of given concentrations (using a probability density function) or equivalently by the probability of exceeding given concentrations (using decreasing probability functions). They may be derived from probabilistic risk estimates given by Suter (1994) for ecological risk assessments and the conjoint expected risk model (CER model, Luce and Weber,1986) mostly applied to assess financial risks (Holgrave and Weber, 1993).

Intended meaning	Evaluation structure 1: ensemble of evaluation variables	Evaluation function 1: elementary evaluation functions	Evaluation function 2: composition/ aggregation	Evaluation structure 2: axioms of preference
Environ-mental impacts	impacts: CFC Pb, Cd, .. PAH, VOC Dust, SO_2, CO_2, ...	effect scores: ozone layer depletion heavy metals carcinogenics greenhouse effect acidification eutrophication ...	calculation of the indicator value using stepwise aggregation to fatalities, health and ecosystem impairment	complete order linearity, additivity, sure thing principle, full compensation

Table 3: *Structures within the evaluation process of the eco-indicator 95 life-cycle assessment.*

4.2 Life-cycle assessment

In a life-cycle assessment the objective is to add up all environmental impacts of a product, including all relevant processes necessary for production and waste disposal. For each of these processes all environmentally relevant impacts (e.g. emissions and energy use) must be considered. Hence the eco-indicator method (Goedkopp, 1996) defines the intended meaning as 'environmental effects that damage ecosystems or human health on a European scale'. Table 3 describes the evaluation structures for this life-cycle assessment method.

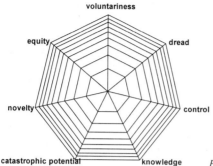

Fig. 4: Dimensions of risk as investigated by Slovic (1987).

5 Discussion and Conclusions

5.1 Perception

Aggregated evaluation includes different components (sometimes referred to as dimensions). Risk is often measured accounting for the damage and its probability. For the reason of describing the risk perception more adequately, more components (dimensions) have been included. For health risks, Slovic (1987) showed seven dimensions to be relevant for individual risk perception (Fig. 4). Not only different dimensions may establish a risk measure. Luce and Weber (1986) found different parts of a probability distribution defining financial risks:

$$R_{CER}(f) = B_1 \int_{-0}^{+0} dF(x) + B_2 \int_{-\infty}^{-0} dF(x) + B_3 \int_{+0}^{\infty} dF(x) +$$
$$+ A_1 \int_{-\infty}^{-0} |x|^\beta dF(x) + A_2 \int_{+0}^{\infty} x^\theta dF(x)$$

(where F(x) is the distribution of the gain (or loss), the vector (B_1, B_2, B_3, A_1, A_2) defines individual risk perception leading to the risk function R_{CER} of the conjoint expected risk model (CER model, Luce and Weber, 1986, cit. Brachinger and Weber, 1996).

Hence different risk measures are constructed in order to model the intended meaning of risk which depends on the specific situation (Tietje and Scholz, 1996). While risk research is well aware of the importance of risk perception, similar investigations are missing which describe the perception of the intended meaning of life-cycle assessments.

5.2 Quality criteria for the modelling of an intended meaning

Using the presented approach quality criteria for evaluations can be derived. These quality criteria divide into three groups:
- quality criteria for the measured data, which assess the uncertainty in the data: sufficiency, representativity, reliability, and indicativity (Nothbaum et al., 1994),
- quality criteria for the model, because the preference axioms constitute a evaluation model for the intended meaning (sufficiency, representativity, uniqueness, efficiency, objectivity, accuracy, reliability, and indicativity), and
- quality criteria for the evaluation, especially for the composition and aggregation, which are mainly comprehensability, equity, consistency, acceptability and relevance.

The presented formal concept of evaluation (which is similar to approaches used in multi-criteria analysis) shows different structures which are referred to during the evaluation process. For practical evaluations, possibilities of specific evaluation designs are presented. Quality criteria of the evaluation, e.g. the consistency of the intended meaning and the comprehensibility, may only be assessed if the perception of the evaluation is considered.

References

Berkeley, D. and Humphreys, P. (1991): Modelling and supporting the process of choice between alternatives: the focus of ASTRIDA. In: H. G. Sol und J. Vecsenyi (Hrsg.): Environments for supporting decision processes. Elsevier, 59-81.

Brachinger, H. W. and Weber, M. (1996): Risk as a primitive: A survey of measures of perceived risk. OR Spektrum **18 (4).**

Fishburn, P.C. (1988): Nonlinear preference and utility theory. Johns Hopkins Series in the Mathematical Sciences, Baltimore, Johns Hopkins, 259 p.

Geiger, G. and Schulin, R. (1992): Risikoanalyse, Sanierungs- und Überwachungsvorschläge für das schwermetallbelastete Gebiet von Dornach. Solothurn, Amt für Umweltschutz, Berichte Nr. 2.

Goedkopp, M. (1996): The eco-indicator 95 (final report). Bilthoven, The Netherlands, Nat. Inst. of Publ. Health and Environ. Protect. (RIVM).

Hämmann, M., Gupta, S.K., Zihler, J. and Hari, T. (1996): Protection of soils from contamination in Switzerland. In: Towards sustainable land use (9th Conference of the International Soil Conservation Organisation), Bonn.

Hofstetter, P. (1996): Towards a structured aggregation procedure. In: A. Braunschweig, R. Förster, P. Hofstetter and R. Müller-Wenk (eds.), Developments in LCA valuation, St. Gallen, IWOe-HSG, 123-211.

Holgrave, D. R. and Weber, E.U. (1993): Dimensions of risk perception for financial and health risks. Risk Analysis **13**, 553-558.

ISO (International Organization for Standardization) (1993): Guide to the expression of uncertainty in measurement. Genf, Bureau International des Poids et Mesures. 101 p.

Luce, R. D. and Weber, E.U. (1986): An axiomatic theory of conjoint, expected risk. Journal of Mathematical Psychology **30**, 188-205.

Nothbaum, N., Scholz, R.W. and May, T.W. (1994): Probenplanung und Datenanalyse bei kontaminierten Böden. Berlin, Erich Schmidt, 164 p.

Pohl, C., Ros, M., Waldeck, B. and Dinkel, F. (1996): Imprecision and uncertainty in life-cycle-assessment. In: Schaltegger, S. (Ed.), LCA - Quo Vadis? Birkhäuser, Basel, Boston

Savage, L. J. (1972): The foundations of statistics (2 Hrsg.). New York, Dover Publications, 310 p.

Scholz, R. W., Heitzer, A., May, T., Nothbaum, N. and Tietje, O. (1996): Datenqualität und Risikoabschätzung - zum Umgang mit Unsicherheiten bei Probenplanung und Datenanalyse bei kontaminierten Böden. In: Symposium CONLIMES 94 'Altlasten - Gefährdungsabschätzung: Datenanalyse und Gefahrenbewertung'. ecomed-Verlag. Neuherberg (BRD)

Slovic, P. (1987): Perception of risk. Science **236**, 280-285.

Suter, G. W. (1993): Ecological risk assessment. Lewis Publishers, Chelsea, 538 p.

Tietje, O. and Scholz, R.W. (1996): Concepts of probability and environmental systems (in German). In: Gheorghe, A. and H. Seiler (Eds.), Was ist Wahrscheinlichkeit? Die Bedeutung der Wahrscheinlichkeit beim Umgang mit technischen Risiken. vdf. Zürich, 31-49.

Tversky, A. and Kahnemann, D. (1981): The framing of decisions and the psychology of choice. Science **211**, 453-458.

Vincke, P. (1992): Multicriteria decision-aid. Wiley, Chichester.

Walley, P. (1991): Statistical reasoning with imprecise probabilities. Chapman and Hall, London.

Address of authors:
Olaf Tietje
R.W. Scholz
A. Heitzer
O. Weber
Environmental Sciences: Natural and Social Science Interface
ETH Zentrum HCS, Hochstraße 48, CH-8092 Zürich, Schweiz

Multiple Objective Decision Making for Land, Water, and Environmental Management

S.A. El-Swaify

Summary

Implementing such paradigms as "sustainable development" and "conservation-effectiveness" has exposed potential conflicts between "conventional" and "environmentally-sensitive" objectives of natural resource management. Often-encountered conflicts include: (a) Enhancing food production, commonly by using intensive inputs, while also controlling soil degradation and water pollution, and (b) Selecting land uses and management practices to simultaneously counter several forms of degradation; e.g. reducing runoff and erosion by maximizing water infiltration while also preventing groundwater contamination by leached agri-chemicals. Such conflicts need not always exist, but when they do, they need to be resolved. Tools that allow such resolution are referred to as "multiple objective decision support systems" (MODSS). These are specialized forms of "decision support systems (DSS)" that use "multi-criteria" to define choices, predict and evaluate tradeoffs, and achieve optimal, integrated, and harmonious decisions on land use and management. To achieve appropriate balance, MODSS attempt to integrate the physical, biological, management, and socio-economic-policy factors that influence the productivity and sustainability of agroecosystems.

The components, ingredients, tools, integrative aspects, and appropriate applications of multiple objective decision support systems are presented in this paper. It is concluded that available natural resource models and databases, as well as computer-based expert systems can be combined into powerful tools that allow quantitative, *ex ante* evaluation of land use and formulation of specific management practices for productive, sustainable, and environmentally sound land use.

Keywords: Natural resource management, sustainable development, modeling, decision support systems, multicriteria decison making, knowledge integration and synthesis

1 Introduction

Assessing and controlling soil degradation, planning and evaluating management options for sustainable land use systems, and managing such systems effectively and profitably must be based on the best information available. The intensifying worldwide "revolution" in innovative and efficient technologies for information retrieval, management and dissemination now allows and facilitates "well-informed" decision making. Decision makers must rely on sound information from scientists who must, in turn, target their efforts to addressing critical knowledge gaps (El-Swaify, 1994).

This paper addresses the complexity and means of facilitating decision making when one is confronted with several competing goals in natural resource management. Making rational decisions in such situations requires quantitative decision aids (often called decision support systems or DSS) that are capable of meeting multiple objectives (termed here as multiple objective decision support systems or MODSS). These definitions of DSS and MODSS, as specific forms of decision aids, are used throughout the paper. It builds on the accomplishments of a recent international conference that addressed the need for MODSS, and the concepts, approaches, and tools required for designing and applying them for natural resource use planning and management at the local, national, regional, and global levels (El-Swaify and Yakowitz, 1998).

2 Rationale and need for multiple objective decision making

People continually exercise the selection of choices and engage in making compromises for conducting many aspects of their lives, e.g. when determining food preferences or when making investments for secure capital or its rapid growth. In land use, decision makers cover a wide spectrum of actors including first and foremost the land users and also extension advisors, planners, administrators, policy makers, and scientists. Technological innovations in the field of agriculture have concentrated on enhanced production and succeeded in assuring impressive increases in global food supplies. Even historically, many viable and long lasting agriculturally-based civilizations were quite sustainable and some contributed positive contributions to the quality of land and soil (aggradation). Some occurred in rather harsh environments. As an example, Hughes (1992) provided an elaborate analysis of how "sustainable development", including agriculture was realized by the successive societies of ancient Egypt before the onset of perennial irrigation. Unfortunately, such successes are now rare due to the decreasing availability of new productive lands and usable water resources, increasing dependence on unsustainable intensive inputs by agriculture, the alarming rates of ongoing human-induced land, soil, and water resource degradation, and the increasing population encroachment on and over-stressing of marginal, fragile lands (Hurni et. al., 1996).

"Sustainable land use", "conservation-effectiveness", "land husbandry" are among the relatively recent concepts or paradigm shifts that emerged in response to these concerns. All aim to integrate "conventional, production-driven" and "protective, environmentally-sound" uses and management of land, water, and other natural resources. Realizing such ideals is often hampered by conflicts among competing development or management objectives. El-Swaify (1998), contends that such conflicts are not inevitable and provided some reasons why they do arise. These reasons include the facts that the impacts of human actions within ecosystems may not be fully foreseen or predictable, may require different time horizons to be manifested, may be simultaneously enhancing of certain ecosystem attributes but degrading of others, or may result from inadequate "accounting" and ignoring the full economic "cost" of environmental degradation vis-a-vis the economic "benefits" of environmental enhancement.

Examples of conflicting objectives in land use and management include the effect of enhancing food production by intensifying inputs on the quality of soil and water; effect of reducing runoff, erosion, and surface water pollution on leaching and groundwater contamination by agricultural chemicals; consequences of strip-cropping, hedge-row agroforestry, and other live vegetative methods for erosion control may be accompanied by reduced crop yields particularly on limited cropping areas; maximizing the economy of water use on irrigated lands may result in insufficient salt leaching and subsequent soil salinization; and meeting the needs of certain clientele, e.g. land users, may run counter to the needs of others, such as land use planners or regulatory agencies. Maintaining a holistic ecosystem-based perspective on the multitude of impacts of land use and management will often reveal many more examples of conflict and also of harmony. Proliferation of national environmental laws and international treaties presents another potential source of conflict in land use planning. Diverse stakeholders and clients also have different needs and preferences; identifying optimal choices and reconciling competing objectives can be a complex task. Such needs and preferences must be addressed not with single options but by a spectrum of choices; prescribing optimal choices underpin the need for multiple objective decision making.

3 Steps required for reconciling competing objectives

All the likely impacts of land use and management and potential for management conflicts such as those discussed above need to be addressed in order to exercise optimum decisions. The following elements are necessary to facilitate planning and to achieve sustainable land use systems:
1. Careful holistic articulation of all (multiple) land use objectives,
2. Recognizing potential problems from all relevant perspectives,
3. Defining the "multiple-criteria" that correspond to the articulated multiple objectives and that must be addressed to evaluate the ecosystem's responses to alternative uses and selected management alternatives,
4. Selecting the appropriate databases and/or models for quantifying ecosystem responses and tradeoffs between competing objectives,
5. Setting a framework and threshold values for judging changes in ecosystem attributes as a result of certain actions, and
6. Formulating recommendations for optimal actions

4 State-of-the-art in multiple objective decision support

Significant progress has been made in understanding the influence of physical, biological, management, and socio-economic-policy factors on the productivity, quality, and sustainability of agro-ecosystems. Many natural resource models, multiple resource databases, and computer-based expert systems have been developed or compiled. These are powerful tools which, when used appropriately, allow quantitative, *ex ante* evaluation of complex land use alternatives and specific management practices. Decision support systems (DSS) or decision aids integrate these tools to allow formulating recommendations for "multiple objective decision making". Increased adoption of this and other operations research (OR) techniques is envisaged in the coming years for aiding decision making in natural resources and environmental management. Emerging emphases are summarized, by themes, in the next section. Applications in the following areas were addressed to demonstrate the utility of MODSS in addressing land use issues (El-Swaify and Yakowitz, 1998):
1. Water supply, management, quality, and conservation,
2. Environmentally compatible agriculture, including crop selection, tillage systems, pesticide and nutrient management,
3. Conservation planning for combating soil physical, chemical, and biological degradation,
4. Assessing market demand and enhancing farm profitability,
5. Land use planning and rehabilitation, economic, social and environmental focus points,
6. Waste management and remediation of contaminated sites with concern for human, wildlife, water, and vegetation health,
7. Forest management with concern for wild-life, erosion control, recreation value, and economic potential,
8. Aquaculture, with concern for food supply, competition, and water quality, and
9. Livestock management, with concern for economics, food supply, and waste cycling

Option/Alternative	Scores for Selected Management Objectives / Sustainability Indicators (Open-ended ---->)						Total Score
	Enhance crop production (& profits)	Minimize erosion & runoff	Minimize pest incidence	Minimize surface water pollution	Minimize groundwater pollution	Improve soil fertility	Improve soil structure
(Open-Ended) ---\->							
Current land use system							
New system or crop			Insert Scores here (Response curves?)				
Modified Management Minimum tillage Zonal tillage Residue cycling Hedge-row farming Alley cropping			Weighting for relative importance				
Ground cover use Legume-ley farming Bench terracing Integrated Pest Management (IPM)			Information gaps?				

Tab. 1: A simplified multiple objective matrix for evaluating land use

Table 1 provides a specific simplified example for areas 1, 2, and 3 above. Land use and management options for a specific location (including the system already in place) are compared in view of seven (multiple) criteria or indicators. More management alternatives and indicators can be added as dictated by specific site needs. The impact of each intended land use or management practice is judged (subjectively or objectively using available databases or models), weighted as to its importance, and assigned a score that is inserted in the table. The scores for each land use or practice are added to obtain a total score (last column in Table 1). The scores provide the basis for comparing the contemplated options with view of the stated (multi) criteria.

Such a quantitative comparison provides a basis for recommending changes in land use or management practices. The exercise also allows identifying information gaps and research priorities where data bases are unavailable or models are inapplicable for conducting required evaluations and assigning scores.

5 Conclusions and recommendations

A number of general conclusions were formulated from the material presented and published in Jones et al. (1998). These are summarized by "Theme" in the following synthesis.

5.1 Guidelines for making MODSS useful to clients

To be effective in meeting the needs of concerned stakeholders, developers of decision support systems (DSS) must be mindful of the following:

a. Building an MODSS that users need and can use: The MODSS must function in a working environment and direct users of the tool must be identified early in the development stage. Users should be informed of the data, model, and DSS limitations and assumptions should be made explicit so that the model is more believable. Developers of MODSS should aim for a range of complexities to provide the decision maker with a choice based on one's capability and preference. As discussed above, Table 1 is an example of a matrix device which allows multiple objective decision making in a simplified manner.

b. Involving the users: Users should be part of the MODSS development process to ensure that relevant issues are addressed and are easily understood. Also, users are more likely to use the MODSS if they feel a sense of ownership. Involvement of users in DSS development also creates a unique learning environment that will result in a more meaningful DSS. In industry, this is called "total engineering".

c. Overcoming institutional and disciplinary barriers: Communication across disciplines, ethnic groups, cultures, and socio-economic groups is often difficult. When human energy is expended in breaking down real or perceived barriers, the result is often not a product, but instead a process. Disciplinary, vague, and complex jargon also detracts from effective communication, even with intelligent, but non-technical, users.

d. Enhancing the reward system: Bureaucracies tend to reward outputs not outcomes. Reward systems rarely extend special recognition to researchers for effective communications with and involving the users.

e. Choosing an appropriate approach: If decision makers are many, knowledge is largely insufficient, and the problem is value-laden, then the process of making decisions needs to be an iterative rather than a linear one. Multiple objective problems are quite difficult to sort out and solve. This is especially true when many decision makers are involved. MODSS can facilitate the decision making process by iteratively helping decision makers come to a consensus decision to which all choose be committed.

The importance of addressing user needs as part of the overall success of a MODSS was illustrated at the conference by case studies from several countries. These included investigations on biomass energy in the United states and the "Landcare" program in New Zealand which uses a community-based approach to seek land use solutions. Sharing an understanding of how different groups see the world and what they do in it creates a learning environment that can promote constructive and voluntary behavioral change.

5.2 MODSS methodologies, components and integration

Keeping in mind the discussion on the "State-of-the-Art", above, the following observations are added here:

a. MODSS appear to be developing from a truly multidisciplinary team perspective. Teams include social, physical and biological scientists, engineers, model developers, decision support systems specialists, stakeholders and the ultimate user or client. Such joint involvement strengthens communication and provides the team with a better understanding and appreciation of each other's perspective and priorities.

b. There is an abundance of MODSS models and computer programs available and being developed. Nevertheless, important issues are being raised that impact further development and refinement of MODSS. Data is often gathered from many different research sources. As a result, databases for MODSS may be incomplete, inconsistent, and inaccessible at the micro or macro levels. Data quality and availability, model accessibility and level of sophistication, and multiplicity of MODSS and stakeholder objectives are among the priorities that need to be increasingly addressed by MODSS developers.

c. MODSS tools may be advancing more rapidly than MODSS methodologies. While MODSS is complex, it may be that the tools being developed are too complex for the end user. Alternative MODSS and tools of varying complexity are needed to provide choices that are suitable for different clientele. The validity, appropriateness, and complexity of evolving MODSS need to appeal to decision makers for use in making sound resource management decision.

d. MODSS developers need to be more holistic in their approaches. Current interests and documented history tell us about a select group of land use and management alternatives. However, other (innovative) options, that are comprehensive in scope and cognizant of spatial and temporal contexts, should also be part of the complement of alternatives using MODSS as a vehicle for change.

Examples of developing MODSS included work in Egypt on the use of integrated DSS consisting of three components, namely crop management expert systems, databases, and multimedia capabilities, is underway. While the purpose of the first two is obvious, the latter maximizes user-friendliness by acquiring and displaying major symptoms of disorders that may inflict growing crops. This substitution for narrative descriptions also minimizes the errors in making diagnostic characterization of crop disorders. In the U.S., other approaches to natural resource management resulted in MODSS that include such features and components as a Windows user interface, a simulation model, a decision model, simple databases and a report generator. A multi-step approach is used to order decision variables and rank management alternatives using utility functions. Applications of this approach include the effects of dryland agricultural management systems on water quality, evaluation of irrigation management systems on water quality, and evaluation of trench cap design systems for mixed hazardous waste burial sites.

5.3 Socio-cultural-economic, policy and sustainability issues

Decision making is a human-based sociological process. Quantitative and qualitative parameters as well as values and other intangibles impact this process. Thus, sociology, economics, policy, culture and concerns with risk, uncertainty, and sustainability considerations must be incorporated into the design of DSS, when and where appropriate. Several factors that determine success in dealing with these issues include accessibility to DSS, dealing with DSS in a human context, effective communications, and the nature of leadership. For example:

(a) Differences in culture and standard of living among countries determine our ability, at the global level, to be effective stewards of land, water, and air resources. Individual countries generally survive within a limited geographical niche and so differ in access to wealth. It is likely that wealth, to a large extent, will strongly influence what we choose to do as stewards of natural and renewable resources, both on a short and long-term basis. Monetary wealth gives countries ready access to pertinent knowledge, facilitates the acquisition of quantitative information and the development of tools such as DSS, and allows enhancing their population's intellectual capacity.

(b) Models are our best "interpretation" of the way we understand the real world and our best translation of it. Most DSS emphasize the incorporation of quantitative information which then provide different outcomes that reflect risk or uncertainty. Even if a quantitative measure of risk is not directly incorporated in a model or DSS, decision makers can use it to manage risk by simply deriving multiple outcomes and interpreting the associated consequences as part of the decision making process.

(c) Decision support systems are intended to provide information that will aid in decision making; decision making itself is a subsequent, very value-laden process. A DSS, especially one that is computer-based, should not be used as a tool to provide the one perfect solution to any given problem. Rather, it should be used as a tool to help support and incorporate as much information as possible that is likely to impact a decision. This represents a unique opportunity for scientists to be an integral part of shaping not only a decision making process but also long term policy.

(d) Contrasts between short term and long term considerations also pose challenges to MODSS modelers. Compared to most physical and biological processes, social, economic and political conditions can change quite rapidly over time. Thus, to be effective, models and DSS should describe current conditions and also allow projecting future conditions. Producing dynamic outcomes broadens the views of decision makers.

(e) DSS facilitates the decision making process by helping people see the world in a different light and providing them with options that are different from those that might otherwise be chosen. A beneficial MODSS will inspire decision makers to view the problem under consideration in a way that is different from their past views of the same problem. Thus DSS allows improved decision making processes and also can be effective tools for promoting behavioral change.

(f) In addition to the external information incorporated into the DSS, each decision maker brings a unique set of valuable internal information to the process. This internal information may be gained from experiences, intuition, values, age, culture and other sources. The synergy between both external and internal information can help decision makers to better formulate instructional problems, acquire deep insight into the issues at hand, and ask the right questions.

(g) DSS modelers are human. If modelers do not consider all of the factual information available or misrepresent that information, a bias will result that will likely influence the alter-native solutions presented by the DSS. Interpretation of data brings with it some value judgment so scientific objectivity is essential. Scientific credibility and ethics will be among the priority issues by which external groups and individuals judge the scientific community and our ability to provide unbiased information for their decision making process. In any case, it is unlikely

that the even the best DSS will capture the entire range of alternative decisions, so decision makers should be flexible in their thinking and look beyond outcomes addressed strictly by the DSS.
(h) Communications and working relationships between stakeholders, scientists/engineers and decision makers must be strong. A historical perspective indicates that, in too many instances, the lack of stakeholder and decision maker input in the DSS development stages produces irrelevant, misleading, or unrealistic products from an economic, social, political and scientific point of view.
(i) The research community has been labeled by some as "a land of excellence and knowledge surrounded by a sea of non-communication". Many of today's scientists and engineers were educated in a university setting where the importance of communication skills, sociological considerations and economics may have been diminished. Within the DSS context, communica-tion must be improved with stakeholders and decision makers, as well as among the scientists and engineers themselves.
(j) Achieving sustainability will reflect the role of land managers and decision makers as leaders of their community. Production, economics, environmental concerns and societal desires must be in balance and be allowed to evolve over time. What is considered good and correct, i.e. sustainable, today may change tomorrow because of our changing values, knowledge base, or socio-economic-policy factors. Sustainability is not static but, rather, a dynamic and broad goal whose complexity is often misunderstood or ignored.

Case studies included the high erosion rates, runoff, sedimentation and mass wasting that are among the indirect consequences of misuse of natural and renewable resources in India. Past efforts to mitigate such problems have been unsuccessful because they were evaluated singularly and in isolation, planned and administered in a top-down manner, and carried out with little coordination. Integrated watershed management approaches, used since the mid-1980's, have proved successful in accomplishing diverse and competing goals for farmers and the nation. Participatory, bottom up and client first approaches yielded desirable results as reduced runoff, erosion, drought and flood hazards. Dealing with non-market values which do not lend themselves to a quantitative measurement but should be of primary consideration in natural resource management and policy formulation is a challenge that lies ahead. Approaches such as conjoint analysis and game theory are being tested and appear to be promising for addressing these issues.

5.4 Global and regional issues

Among the issues raised and contrasting views exposed in dealing with natural resource use and environmental issues are the following:
(a) Sustainability may be the luxury of the rich. A subsistence standard of living is dominated by the day to day concerns of producing enough food for survival. Sustainability, by its very definition, considers production in relation to other priorities for profitability, environmental quality and social justice. Without adequate food supplies, these other priorities do not even enter into the production equation. Some authors noted, however, that sustainability is a also necessity of the poor, including those who have little or no purchasing power (Jones et al., 1998). Their very survival depends upon the ability to utilize the resource base in a way that sustains produc-tivity. At the national level, a short-term or narrow focus of natural resource management will not suffice as ecosystem productivity and quality are so strongly entwined that, given enough time, the decline of one is a serious threat to the other.
(b) Global environmental awareness will continue to increase. Environmental treaties may result in imposing trade barriers that require continual efforts to improve agricultural, forest and other land uses so that they are more ecologically sound and environmentally safe. This will

provide many opportunities for MODSS tools to help evaluate current and future land use and management practices from an environmental perspective. International action programs such as Agenda 21 and conventions will continue to address international environmental issues. An International Environmental Court of Law could very well be established, thus providing additio-nal opportunities and challenges for MODSS on a global scale.

(c) Knowledge will be increasingly globalized. Computer technology and electronic communications all serve to break down traditional barriers to the exchange of knowledge. As knowledge becomes more portable, more people will be able to make better decisions. Also, there is a likeli-hood that people trying to solve similar problems may converge in their approach to problem solving; thus, using the best approaches from all different sectors of the planet.

(d) Restoration, rehabilitation, and recycling of natural and renewable resources will be growth industries. DSS will clearly have an important role here to provide acceptable options based on well-defined sustainability criteria.

(e) Many DSS lend support to the current power structures and decision making processes. In a few instances MODSS may support farmers but these tools do not appear to be supporting non-government organizations, indigenous people or other organizations within society. Concur-rently, all too often, scientists may choose not to be involved in shaping policy or may be excluded from the power structure in terms of making meaningful contributions to decision making on environmental and sustainability issues. If more such involvement takes place, MODSS can serve as mechanisms by which scientists gain some power and be able to influence decisions and shape policy.

(f) DSS are largely designed, developed, and tested by people who are rather well educated, have an above average standard of living, and not necessarily representative of mainstream stakeholders. On the other hand, a review of the global literature clearly portrays that women are the major decision makers at the farm level. Very few decision support tools attempt to address the issues raised by different genders or indigenous people.

(g) Many evolving DSS reinforce individual rights rather than collective rights. To date, there has been an emphasis to make individual decisions based exclusively on farm level information with little regard for the rights or needs of the larger community. Many DSS actually reinforce litigation rather than conflict resolution. Now, more emphasis is being place on problem solving at the watershed level. However, the watershed approach tends to group individual farms within geographic boundaries rather that to consider the watershed as a community of individual farms. Collective decision making, collective involvement, and collective rights constitute the model of how indigenous people successfully solve environmental problems.

(h) Most DSS are used for tactical or operational decision making rather than strategic decision making. Strategic thinking and decision making has been out of favor for several decades. Most existing DSS and decision makers focus on "how to get the job done" instead of "which job is to be done". Fortunately, the visioning process appears to be returning and will likely mark the end of one millennium and the beginning of another in terms of natural resource and environ-mental management.

These issues will be interpreted differently by diverse cultures because of their differential access to wealth, education, political land use policies and global knowledge of resource management. On Oahu, Hawaii, U.S.A., water may be presently abundant on the whole, but is not well distributed geographically. This situation was highlighted in the context of rapidly changing land use, including increased urbanization. Issues include water rights, geographic and use-based water allocation, impacts on the quantity and quality of groundwater recharge and sustainable water yield from aquifers, and alternative patterns of tourist industry and residential development. Among the socio-economic factors affecting the resolution of these issues are the impact of preserving open space on tourism and the need to replace diminishing plantation agriculture with diversified agriculture. Addressing these issues and factors represents a classic need for applying

MODSS. Interactive multi-criterion optimization allows the identification of most desirable land use options. Some of the additional considerations for guiding further development of MODSS in support of sound management of natural resources include prioritizing the global issues where DSS can help people make better decisions, upscaling individuals and organizations with the skills needed to work with public multi-stakeholder groups, and providing sufficient funds that stipulate direct involvement of the users to assure developing relevant MODSS and relevant research information.

We advocate continued efforts to hold follow-up conferences periodically in order to communicate further state-of-the-art developments, identify global issues of mutual concern, seek collaborative problem solving opportunities, promote MODSS transferability and adoption, and deploy appropriate efforts towards shaping sound environmental policy at the national, regional, and global levels.

References

El-Swaify, S. A. (1994): Issues critical to furthering the cause of soil and water conservation. In: Soil and Water Conservation: Challenges and Opportunities, Special Publication for Inaugural Addresses, 8th ISCO Conference, Indian Association for Soil and Water Conservationists, Dehra Dun, India.

El-Swaify, S.A. (1998): Land use planning for agriculture - are environmental conflicts inevitable?. In: Multiple Objective Decision Making for Land, Water, and Environmental Management. St. Lucie Press, Delray Beach, FL, USA.

El-Swaify, S. A. and Yakowitz, D.S. (Eds) (1998): Multiple Objective Decision Making for Land, Water, and Environmental Management. St. Lucie Press, Delray Beach, FL, USA.

Hughes, Donald J. (1992): Sustainable agriculture in ancient Egypt. Agricultural History **66**,12-22.

Hurni, H. and an International Group of Contributors (1996): Precious Earth: From Soil and Water Conservation to Sustainable Land Management. International Soil Conservation Organization ISCO) and Centre for Development and Environment (CDE), Berne. 89 pp.

Jones, A. C., El-Swaify, S. A., Graham, R., Stonehouse, D. P. and Whitehouse, I. (1998): A Synthesis of the state-of-the-art on multiple objective decision making for land, water and environmental management. In: Multiple Objective Decision Making for Land, Water, and Environmental Management. St. Lucie Press, Delray Beach, FL, USA.

Address of author:
S.A. El-Swaify
Department of Agronomy and Soil Science
1910 East-West Road
University of Hawaii at Manoa
Honolulu, Hawaii 96822
USA

A Conceptual Framework for Planning and Implementing Water and Soil Conservation Projects

H. Diestel, M. Hape & J.-M. Hecker

Summary

The water cycle plays an important role in planning and implementation programs. In the approach to land classification and subsequent implementation of environmental management measures presented here, a scheme of action is proposed for defining the planning aims, establishing a tailormade classification of local factors determining the limits of landscape units and classification of hydrologic landscape suitablities. The conceptual framework offered is applicable under a wide range of conditions and includes the definition of a hydrotope as a landscape unit with a specified hydrologic suitability for a defined planning aim, the choice of a district as a preferred regional unit for planning and implementation, as well as a proposal for a basic structure of an environmental management procedure.

Keywords: Hydrotope, water and soil conservation planning, management tool.

1 Introduction

When a land classification is carried out, relatively static site properties are considered. When landscape units have to be defined with regard to their hydrologic suitability for a specific planning aim, a dynamic system with a higher degree of unpredictability has to be forced into classes with specified limits. Assumptions have to be made regarding the expected hydrologic dynamics in the planning zone. Additional uncertainties stem from the planning and implementation process itself as well as from the reactions of the affected population.

There are many planning and implementation cases in which the direct and/or indirect consequences of the deviations of the water cycle from an "average" behavior become important for the evaluation of landscapes. The aim of this paper is to present a conceptual framework applicable under a wide range of conditions for evaluating the influence of hydrologic dynamics on the suitability of an area for a specific objective.

2 Landscape planning zones

2.1 The regionalization of land properties vs. the regionalization of properties of the hydrologic cycle

The "suitability" definition which forms the basis for most land classification work is strongly influenced by economics. Even when mapping is carried out for soil conservation measures, the potential of the "pedotopes" to serve as a basis of production and of the site to serve as a medium to convert water into a crop is classified, directly or indirectly. The locality of the classified soil resource is fixed and can be assigned a monetary value. In contrast, water is not as localized, and the monetarization of its value is problematic. Most of the indicators which have been used in the past for the classification of "land suitability" (texture, etc.) are properties of an area which can be directly mapped. When attributing them to their locality, few assumptions must be made. Even the mapping of parameters such as depth to groundwater can still be carried out with quite high confidence limits. Whenever components of the hydrologic cycle such as precipitation, evapotranspiration or groundwater recharge (expressed in various units of volume or volume per time) are allocated to an area, more or less simple arithmetic operations – which are based on assumptions – have to be carried out.

Sombroek (1995) gives a holistic definition of land, in which "land" encompasses attributes such as the near surface climate, the surface hydrology and the geohydrological reserve. Such a definition makes it possible and feasible to delineate on maps land properties and land suitability even if dynamic attributes are included into the evaluation. This approach should be applied exhaustively even to hydrologic objectives, making use of the established and available instruments and concepts of land evaluation. However, the stronger the influence of hydrologic variability (e.g. of irregularities in rainfall distribution or precipitation intensities) on land suitability for a specific objective of use, the more it becomes difficult to produce maps which depict the regionalized suitabilities in mapping units in the classical fashion. In such cases, the definition of hydrotopes and of hydrotope properties will prove useful.

2.2 Size of area of planning and / or implementation

2.2.1 General considerations

Becker (1992) defines scale ranges for hydrological work in which different methods should be applied in hydrologic regionalization. Sombroek (1995) analyzes landscape units from the point of view of land characterization and development. Ongley (1987) discusses the influence of spatial scale on basin assessment. Kiemstedt et al. (1990) tabulate and examine, for Germany, planning scales between 1:500 and 1: 4 million in relation to planning activity, administrative unit subjected to the planning activity and the legal planning frame.

Depending on the objective, it may be of interest to quantify causes and effects on a daily, weekly, monthly, seasonal or yearly basis or even for cycles of years. The contaminant loads from extended rainfalls, for instance, can be quantified with some reliability on a monthly basis while prognosis of sediment-born contamination may be possible on the basis of yearly loads (Walther, 1995). The time scale also has an influence on the size of the area taken into consideration. It is not possible to fix a rigid frame in which the sizes of areas to be dealt with are linked to methodologies of work. A large desert area or a small biotope can represent – depending on the objective – one hydrologic unit. Catchments frequently contain zones with very different hydrologic behavior. The administrative entity of a district (see below) or any other political zones in which landscape planning or conservation measures are carried out by administrative bodies may be larger or smaller than a "natural" unit.

2.2.2 The district level

This framework has been conceived for the lowest administrative level at which governmental bodies execute, commission or supervise planning and implementation work. In many cases, the areal extension of districts corresponds to the size of regions in which work can be executed with a rather homogeneous set of methodologies. Frequently, the most appropriate level for the achievement of realistic compromises between the interests of the target groups and the intentions of politicians and/or development agencies will be the district (or a comparable entity). In many industrialized countries, legal and administrative provisions, including mechanisms for the internalization of damages to the environment, exist which facilitate the execution of measures of soil and water conservation. Decisions can be taken or fulfilled by district authorities (or comparative units). The statements made here certainly are less valid in those developing countries in which the political power at the district level (if an organizational body at that level exists at all) is low. The recommendation to apply the proposed conceptual framework, if possible and feasible, at the district level, has a guideline character based on methodological considerations. Although the framework is conceived for that level, its potential applicability is not restricted to it.

2.3 Hydrologic indicators

There are objectives which can only be achieved if the planning or implementation measures take into account the dynamics of the hydrologic cycle. Examples are the preservation of nomadic pastoralism in the Sahel or water harvesting in semi-arid areas, salinity control (Diestel, 1993), changes in the intensity of agricultural production, revitalization of streams, biotope protection, groundwater protection, reduction of diffuse contamination of surface waters, recultivation of mining areas and extension of forested land in temperate zones, and mangrove conservation or sewage disposal from animal husbandry units in tropical zones. In order to make a realistic prognosis of the areas in which a project aim can be achieved with different probabilities of success, an appropriate, aim-oriented compilation of hydrologic indicators must be defined and regionalized in addition to the classical land suitability indicators. Depending on the specific project aim, more or less detail is required regarding the acquisition and processing of hydrologic data. Such hydrologic indicators can be precipitation (e.g. monthly data or data on intensive storms), precipitation profiles over several years (Akhtar-Schuster, 1995), evapotranspiration rates in different degrees of accuracy, groundwater recharge rates, stream discharges, various precipitation/runoff relationships or tide influence. Processed hydrologic data may be required such as the total yearly sediment or contaminant load. Often, such data are required for "wet" or "dry" years which have to be defined. In a very rough analogy between classical land classification and hydrologic land suitability evaluation, one could say that, comparable to the soil map, a thematic hydrologic map – containig additional information inputs – must be produced which shows the regionalization of different hydrologic suitability levels of the landscape with regard to a specific planning aim.

3 The proposed conceptual framework

3.1 Definition of the hydrotope

A pragmatic approach to the problem was chosen which is based on the concept of the "Hydrotope". A hydrotope – as defined for these purposes – is a landscape unit with a **specified** hydrologic suitability for a **defined** planning aim (Diestel, 1994, Diestel et al., 1994). The water balance of such "hydrotopes" may be largely uninfluenced by man or (like in irrigation schemes) may be highly anthropogenic. The time range in question may be long or short. The limits of hydrotopes, their definitions and the numerical ranges assigned to them depend on the planning aim. A single map showing the water balance-related landscape potential for a given objective may have a too complex legend. A high degree of abstraction may make it useless for planning purposes. Thus, it may not be possible to represent the required hydrotopes on one single map.

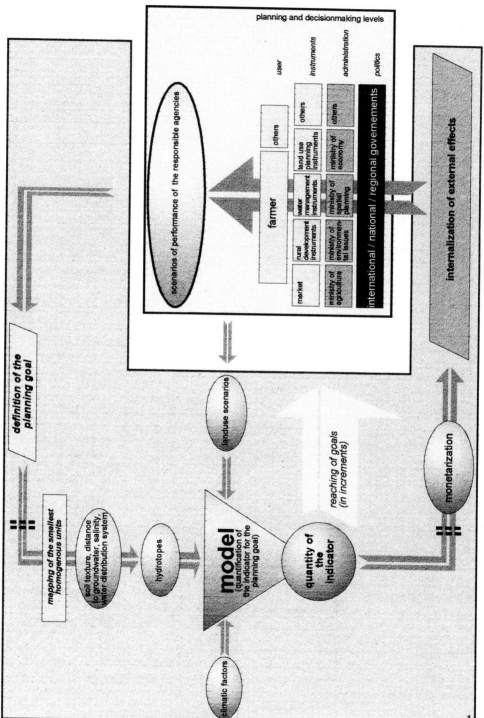

Fig. 1: Basic structure of the environmental management tool: Principle sketch.

Ideally, the data base to elaborate hydrotopes is obtained by using lysimeters, representative test slopes, sample catchments and by executing hydrologic model calculations for "what if"-scenarios (Diestel et al., 1994). Such calculations can be carried out for different planning aims with various assumptions, with variables such as climate, soil, topography, vegetation, wet and dry years, etc.

A land classification for "upland crops" can be used successfully for many objectives, because one set of static soil properties signifies a certain potential for a wide range of uses. But the variance in the types of water cycle dynamics is high. Furthermore, the same hydrologic dynamics (e.g. a long term precipitation pattern) may be favorable for one objective and detrimental to the other. It is unfeasible and very expensive to elaborate hydrotope quantifications and regionalizations for the wide ranges of objectives which are related to the various types of water cycle dynamics. Thus, an indispensable prerequisite for the elaboration of hydrotopes is that, from the very beginning of planning activities, the decision-making entity or the donor agency starts with the work on quite specific definitions of the planning aims. This specification can be modified progressively during the work in accordance with the emerging results.

3.2 The basic structure of an environmental management tool

Figure 1 shows the overall conceptual framework proposed as a tool to execute the complex work of quantification and regionalization of the water-balance related landscape potential. A detailed discussion cannot be given here. The approximately horseshoe-shaped field in this diagram is a rather idealized representation of an optimal situation regarding landscape potential assessment processes. In addition, it shows where the elaboration and specification of hydrotopes should be located in the suitability appraisal processes, with some examples of indicators. The broken double lines indicate that a satisfactory definition and verification of planning goals as well as a true monetarization and internalization usually do not take place.

The vertical column at the left can also be seen as a representation of the standard hydrological working steps, with the supplemental hydrotope element and the illustration that – in good hydrologic work – scenarios with different planning variants are simulated. The rectangular element inserted at the right is very frequently a "black box". The processes taking place in the implementing agency (ideally, a district authority) and the mechanisms which rule them, the flow of information as well as the interactions between this unit, the beneficiaries and the higher level institutions, are usually not included into the planning procedure. But the capacity of the implementing agency to operate according to the objectives which have been defined based on engineering considerations is an essential component of the hydrologic suitability of a landscape unit for the defined aims. It is an indispensable prerequisite of success to analyze the performance of these administrative structures. Thus, coupled to the physical "what-if" scenarios are scenarios of how the responsible agency will perform in different situations.

In its entirety, Figure 1 is a representation of an ideal environmental assessment and landscape planning situation. It is not at all a vital precondition that GIS-techniques, sophisticated hydrologic models, extensive data bases or high-tech monitoring installations are used to cope successfully with the problems discussed above. Essential information can be made available and processed in a conventional manner. The important precondition is the sequence and the networking of the elements shown. Figure 1 can serve as a guide for flow charts which can be used to plan, manage and control project work.

4 Concluding remarks

In routine project work and in most of the relevant literature, one finds that the limits of land classification are given by accepted or specifically defined ranges of parameters such as soil and land characteristics, water quality and / or quantity, land tenure, various socioeconomic factors, habits of nomadism, dominant climatic characteristics, distance to water sources or vegetation. Essentially, maximization of production is the basis for the specification of these ranges. Land classification for projects with the explicit aim of water and soil conservation may, however, require the establishment of specifically adapted classification frames. In the approach presented here, we attempted to impose in each individual planning case – via the scheme of action presented in Figure 1 – the definition of planning aims, the establishment of a tailormade classification of those local factors which will determine the limits of landscape units which will have to be managed differently in a district and the aim-oriented regionalization of levels of hydrologic landscape suitability.

It remains to be proven whether in future, the use of this planning procedure in landscape assessments for soil and water conservation or for other applications will lead to a higher success rate. The elaboration of this concept has been prompted by project failure, inefficiency of planning and the heterogeneity of planning processes and results. An essential next step must be the execution of tests of the validity of the framework and of its elements in a project network. Only some individual components have been elaborated by the authors up to a modest level of development. The definition of the hydrotope can also be applied to purposes other than those described here. The same is true for management tools such as the one that has been sketched. A more or less standardized procedure in which the technological state of the art and the reality of planning and implementation processes are merged might lead to improvements in cooperation between beneficiaries, engineers and institutions as well as a sharper identification of information deficits in the relevant field of engineering, economics and sociology.

References

Akhtar-Schuster, M. (1995): Degradationsprozesse und Desertifikation im semiariden randtropischen Gebiet der Butana / Rep. Sudan. Göttinger Beiträge zur Land- und Forstwirtschaft in den Tropen und Subtropen, Heft 105.

Becker, A. (1992): Methodische Aspekte der Regionalisierung. In: Regionalisierung in der Hydrologie. Deutsche Forschungsgemeinschaft, Mitteilung XI der Senatskommission für Wasserforschung:16-44.

Diestel, H. (1993): Reactions of water management to the salinity of soil and water. DVWK (Deutscher Verband für Wasserwirtschaft und Kulturbau) Bulletin 19. Ecologically Sound Resources Management in Irrigation. Verlag Paul Parey. Hamburg/Berlin, 187-198.

Diestel, H. (1994): Abschätzung von Niederschlag-Abfluß-Beziehungen als Grundlage für Schutz- und Entwicklungskonzepte. Mitteilungen aus der Norddeutschen Naturschutzakademie 4: 33 - 38.

Diestel, H., Bekurts, V., Bismuth, C., Markwart, N., Schmidt, M., Hape, M., Hasch, E. and Wolf, R. (1994): Hydrotop-Kennzeichnung als Grundlage für die Planung von Maßnahmen zum Gewässerschutz auf landwirtschaftlich genutzten Flächen. Report (unpublished) to the Ministerium für Umwelt und Naturschutz und Raumordnung, Brandenburg.

Kiemstedt, H, Wirz, S. and Ahlswede, H. (1990): Gutachten "Effektivierung der Landschaftsplanung". Umweltbundesamt, Texte 11/90.

Ongley, E.D. (1987): Scale effects in fluvial sediment -associated chemical data. Hydrological Processes 1: 171 - 179.

Sombroek, W.G. (1995): Development of a framework for holistic land characterization and development at different scales. In: Land and Water Integration and River Basins Management. FAO Land and Water Bulletin 1: 15 - 24.

Walther, W. (1995): Über den Stoffhaushalt der Landschaft und über die diffuse Stoffbelastung von Böden, Fließgewässern und Grundwasser, dargestellt an ausgewählten Standorten. Habilitationsschrift, Techn. Univ. Braunschweig, FB Bauingenieur- und Vermessungswesen.

Address of authors:
Heiko Diestel, Martina Hape, Jens-Martin Hecker
Fachgebiet Wasserhaushalt und Kulturtechnik, Technical University of Berlin
Albrecht-ThaerWeg 2, D-14195 Berlin, Germany

Harmonization of Soil Investigation Methods Within the Framework of Cooperation with Countries in Central and Eastern Europe

K. Terytze & J. Müller

Summary

For assessing soil contaminants it is essential to use standard methods to produce comparable analytical data. Inorganic (heavy metals) and organic substances (PAH, PCB, chlorinated pesticides) of ecotoxicological relevance were analysed in topsoils of contaminated and uncontaminated areas in countries in central and eastern Europe. The comparability of the data produced was assured by using the same analytical procedures. The results serve as input for the international harmonization and standardization of methods used in the area of soil protection.

Keywords: Soil protection, harmonization, standardization, analytical methods

1 Importance of harmonizing soil evaluation methods

In assessing of soil contaminants it is essential that analytical data are generated using standard methods. Recently, soil protection has become a major concern for environmental protection. However, only a few methods suited to large-scale investigations of contaminants in soil have been developed and generally accepted in terms of standards. The large diversity of methods at the national level is even more pronounced at the international scale (Terytze et al., 1996). To ensure the generation of intercomparable data harmonized methods must be used in any investigation (harmonization of methods). The provision of detailed procedural rules should ensure that the results obtained for a given method are reproducible (standardization of methods).

2 Harmonization and standardization efforts in Central and Eastern Europe

The study required a comprehensive determination of the pollutant concentrations in soils in central and eastern european countries. Therefore topsoils of uncontaminated and contaminated areas in the Federal Republic of Russia (Fränzle et al., 1995 a) and the Republic of Bulgaria (Terytze et al, 1997) were analyzed.

Before start of the projects each institution selected the areas to be investigated. The areas had to be as representative as possible of the respective countries. Fourteen inorganic substances (heavy metals) were selected as contaminants of ecotoxicological relevance; organic substances selected included the non-volatile polycyclic aromatic hydrocarbons (PAHs), polychlorinated biphenyls (PCBs) as well as a number of chlorinated pesticides (Table 1). Total concentrations and, where appropriate, mobile fractions (heavy metals) were determined.

The same scheme was used in various areas (Fränzle et al., 1995 b) and the results were subsequently compared. Based on this comparison the methods used were assessed and recommendations derived for future work to investigate soil contamination over large areas. The results of this exchange of experiences served as an input for the international harmonization and standardization of methods used in the field of soil protection.

Heavy Metals:	* lead * chromium * nickel * mercury * arsenic * wolfram * cobalt	* cadmium * copper * zinc * thallium * antimony * vanadium * tin
Polychlorinated Biphenyls:	* PCB 28 (2,4,4`-trichlorobiphenyl) * PCB 52 (2,2`-5,5`-tetrachlorobiphenyl) * PCB 101 (2,2`-4,5,5`-pentachlorobiphenyl) * PCB 138 (2,2`-3,4,4`,5`-hexachlorobiphenyl) * PCB 153 (2,2`4,4`,5,5`-hexachlorobiphenyl) * PCB 180 (2,2`,3,4,4`,5,5`-heptachlorobiphenyl)	
Polycyclic Aromatic Hydrocarbons (PAHs):	* naphthalene (NAP) * acenaphtylene (ACY) * acenaphthene (ACE) * fluorene (FLU) * phenanthrene (PHE) * anthracene (ANT) * fluoranthene (FLA) * pyrene (PYR) * benz[a]anthracene (BaA) * chrysene (CHR) + triphenylene (TRP) * benzo[b]fluoranthene (BbF) * benzo[k]fluoranthene (BkF) + benzo[j]fluoranthene (BjF) * benzo[a]pyrene (BaP) * indeno[1,2,3-cd]pyrene (IND) * dibenz[a,h]anthracene (DBahA) + dibenz[a,c]anthracene (DBacA) * benzo[ghi]perylene (BghiP)	
Chlorinated Pesticides:	* p,p`- and p,p`-DDT * o,p`- and p,p`-DDE * o,p`- and p,p`-DDD * α-, β- and γ-hexachlorocyclohexane (HCH) (γ-HCH = lindane) * hexachlorobenzene (HCB)	

Table 1: Soil contaminants analyzed

3 Results

To better understand the descriptive statistical data generated, so-called 'Box-and-Whisker' plots were prepared. These allow a multitude of distribution parameters of a given collective of data to be represented in graphic form (e.g. minimum, 25 percentile, median, 75 percentile, maximum, outliers).

Figure 1 shows an example of the total content of lead in soil and litter from different types of forest. For the PAHs, PCBs, the HCH isomers as well as DDT and its metabolites, plots show the sum of the individual substances in soil. To ensure that the data obtained on the contaminants could be compared in spite of differences in bulk density at the various sampling sites (which sometimes were substantial), the results were depicted not only in relation to mass (μg/kg) but also to volume (μg/l) (Fig. 2).

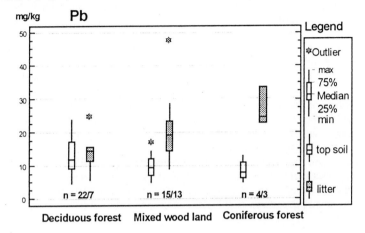

Fig. 1: Total content of lead in soil and litter from different types of forest (Federal Republic of Russia, Ostashkov Region).

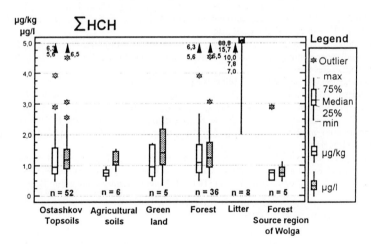

Fig. 2: Sum of HCH isomers in soils of different land uses (Federal Republic of Russia, Ostashkov region).

The results of the evaluation of soil contamination projects can be summarized as follows:
– Extraction with aqua regia has proved to be a practicable method to determine the 'total content' of heavy metals at the national and international level and provides sound and comparable results.
– Extraction with ammonium nitrate for the determination of the mobile fraction of heavy metals is more vulnerable to departures from the method of extraction and analysis. This is because the

concentration level is very low particularly in the case of soils from 'uncontaminated' areas. The method is better suited for use in investigating soils from 'contaminated' areas.

The reproducibility of the results obtained by various laboratories using differing analytical methods was found to be adequate to the determination of organic soil contaminants. Nevertheless, reproducibility could be further improved by increasing the harmonization of the analytical methods.

The main topics to be dealt with in further cooperation with countries in Central and Eastern Europe in the field of soil protection are:

- Harmonization of methods for sampling, sample treatment, sample analysis and data evaluation.
- Generation of data on the current status of soil contamination as a function of the magnitude of contamination, the type of land use and the parent material involved in soil formation.
- Improving knowledge of the contents and dynamics of substances in soils from background territories of different geographical regions.
- Tracing trends in soil contamination by environmental chemicals and monitoring the effectiveness of initiated measurements by dating terrestrial and subhydric (aquatic) soils.
- Aquisition of knowledge of the effect and behaviour of substances under different conditions (type of ecosystem, input patterns, pedological, geographical and climatic situation).

4 References

Fränzle, O., Krinitz, J., Schmotz, W., Delschen, Th., Leisner-Saaber, J. (1995 a): Harmonisierung der Untersuchungsverfahren und Bewertungsmaßstäbe für den Bodenschutz mit der Russischen Föderation, UBA-Texte 60 / 95, Umweltbundesamt Berlin.

Fränzle, O., Schmotz, W., Krinitz, J., Hertling, Th., Florinski, M., Permitin, V., Pochlebkina, L., Müller, J. (1995 b): Handlungsanleitung für Schadstoffuntersuchungen in Böden, Teil I und II, UBA-Texte 26 / 95, Umweltbundesamt Berlin.

Terytze, K., Fleischhauer, E., Schwerz, A., Müller-Wegener, U., Filip, Z., Müller, J., Gzyl, J., Kucharski, R., Sas-Nowosielska, A., Piesak, Z., Wcislo, E., Curlik, J., Fulajtar, E., Fiala, K., Fulajtar, E. jun., Matuskova, L., Dzatko, M., Kuklik, M., Marek, V., Sanka, M., Staoa, J., Szabo, P., Bartalos, T., Busas, M., Csenke, Z., Horvath, A., Koncz, I., Kosa, S., Nemestothy, B., Stefanovits, P. (1996): Methods in soil protection used in Poland, Slovakia, the Czech Republic and Hungary, UBA-Texte 77 / 96, Umweltbundesamt Berlin.

Terytze, K., Müller, J., Böhmer, W., Kölln, A., Knoche, H., Berg, C., Ricking, M., Peitchev, I. (1997): Schulung bezüglich Probenahme und Bestimmung von Schadstoffen in Böden der Republik Bulgarien, Bericht Teil 1 (II 3.1-91005-26/0) und Teil 2 (310 01 004) an das Umweltbundesamt Berlin

Addresses of authors:
Konstantin Terytze
Umweltbundesamt und Freie Universität Berlin
Postfach 33 00 22
D-14191 Berlin
Germany
Josef Müller
Fraunhofer-Institut für Umweltchemie und Ökotoxikologie
Auf dem Aberg 1
D-57392 Schmallenberg
Germany

Determination of Potential Sites and Methods for Water Harvesting in Central Syria

T. Oweis, A. Oberle & D. Prinz

Summary

Water harvesting is an efficient means of proper managing conditions of low annual and erratic rainfall to provide sufficient water for sustainable agricultural production. Determination of appropriate sites and suitable methods of water harvesting on a large scale present a great challenge.

This work tried to develop a methodology for the application of the remotely-sensed data and geographical information system for identifying appropriate sites and methods of water harvesting in the dry areas of West Asia and North Africa (WANA).

The interpretation of the Landsat TM scenes of a pilot project area in central Syria, taken at different periods before, during, and after the rainy season helped to identify areas of high potential for water harvesting. The classified image was incorporated in a geographical information system comprising digital data sets on topography, soil, vegetation, hydrology and meteorology. This permitted the classification of the study area according to suitability for a given water harvesting method.

Keywords: Water harvesting, remote sensing, hydrology

1 Introduction

The dry areas of WANA receive low annual rainfall with poor spatial and temporal distribution. Renewable water resources in the dry areas of WANA are about 1,250 m^3 per capita, compared to about 7,420 m^3 for the world and 15,000 m^3 per capita for Europe (World Resources Institute, 1994/1995). Over 75% of the available water in the dry areas is used for agriculture. To guarantee more food production with less water, the efficiency of use of the available resources must be substantially increased.

The project area lies mainly in the steppe region of central Syria receiving between about 100 and 300 mm of annual rainfall. This area is characteristized as follows:

1) Total annual rainfall amounts are inadequate for economic crop production. Agricultural use without full irrigation by non-renewable resources is limited. Most of the land is considered as rangeland and used for grazing.
2) Rainfall distribution is erratic. The rainy season lasts from November to April. Most of the rain comes in sporadic storms which are unpredictable, leaving frequent drought periods during the growing season. If rain falls on the typical crusty soil, mostly Aridisols, characterized by a low infiltration rate, uncontrolled runoff occurs depriving cultivated land its share of rainfall.

3) Natural vegetation cover consisting mainly of annual herbs and shrubs is too low to allow sufficient water storage. In addition, overgrazing causes severe degradation of the vegetation, which leads to soil degradation, since the soil looses its stability and becomes vulnerable to water erosion caused by runoff following a heavy rainstorm.

An important option for enhancing the effectiveness of rainfall in the area and for improving its management for sustainable agricultural production is water harvesting (Critchley and Siegert, 1991), defined as the process of concentration of rainfall into smaller areas, through runoff, and storing it for beneficial use (Oweis and Taimeh, 1996). According to the ratio between collecting and receiving area, two major types can be classified: micro- and macro-catchment systems (Prinz, 1995). Micro-catchments include e.g semi-circular bunds, contour ridges, and other small basins suitable to collect runoff water of an area betwenn 20 m^2 to 200 m^2 and store rain water directly in the root zone of trees, shrubs and field crops planted in the lowest portion of the infiltration basin. Macro-catchments collect runoff from large catchments of area as small as a few hundred square meters to several hundred square kilometers. Runoff water is directed to and stored in various constructions - small earth dams, water spreading structures, or ground-water recharge systems.

Many of the modern projects for water harvesting in dry areas have had little success (Reij et al., 1989) because of deficiencies in planning, design, and implementation, and negligence of local socio-economic aspects. Selection of the appropriate sites and the suitable methods under the prevailing conditions are among the most important prerequisites for successful water harvesting systems. Planning water harvesting on a large scale through field visits is often tedious and costly especially in areas like central Syria with low infrastructure. The basic information needed for proper design of water harvesing systems should include properties: topography, vegetative cover, drainage systems, rainfall patterns, land use, soil and land tenure. Recent studies (Tauer and Humborg, 1992) in the Sahel zone show that remote sensing and geographical information systems are economical and reliable methods for the selection of the appropriate sites and methods suitable for water harvesting for large areas with deficient infrastructure and databases.

2 Objectives

The general objective of this study is improved sustainable agricultural production in the steppe zone of WANA by more efficient utilization of rainfall through proper water harvesting planning.
Specific objectives include:
1. Identification of potential sites and methods of water harvesting at a pilot site in central Syria.
2. A methodology for the application of remotely-sensed data and geographical information systems to determine the potential sites for water harvesting in the dry areas.

3 Methodology

A Landsat Thematic Mapper scene of Path 179 and row 36, covering an area of 33,000 km^2 (Fig. 1) between latitudes 33.7^0 N and 35.5^0 N and between longitudes of 36.8^0 E and 38.8^0 E in central Syria has been used. As this sensor uses the spectral bands of the near and mid infrared (0.76 - 0.90 µm, 1.55 - 1.75 µm, 2.08 - 2.35 µm), such data are helpful in the assessment and the evaluation of land vegetative cover and geological features. In order to compare different vegetation periods, scenes taken in October 1993 and July 1994 corresponding to the periods before the rainy seasons of 1993/94 and 1994/95, respectively, were selected.

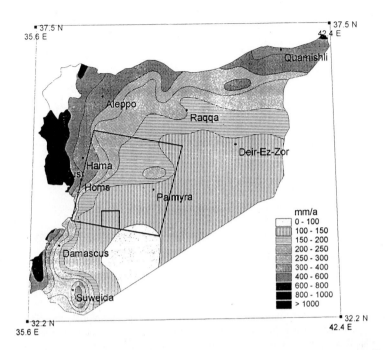

Figure 1: Location of project site, Landsat TM scene in central Syria and isohyetes of average annual rainfall.

The image processing scheme is handled by ERDAS Imagine, software on DEC Workstation. The image is rectified to the coordinate system of the Universal Transverse Mercator projection to allow implementation of the satellite image into a geographical information system using ground control points identified with the Global Positioning System. False color images (bands 4,3,2) after filtering are used for the classification according to the maximum likelihood procedure and checked by ground truthing.

The classified image is incorporated into a geographical information system where different data such as topography, soil maps, etc, are superimposed. The topography is represented by a digital elevation model computed by the ILWIS program according to topographic maps at a scale of 1 : 100,000 and digitized through ArcInfo. A digital soil map produced by the Soil Department in Damascus provide information about soil types according to the taxonomy of the U.S. Department of Agriculture. Other datalevels comprise soil depth, rainfall data statistically evaluated and hydrological data. The last ones will be incorporated in a later stage of the project.

Criteria were developed matching each of the water harvesting methods and techniques to a set of ground, rainfall, crops and socio-economic conditions. These criteria were based on water harvesting experience in the area and from similar areas. The major requirements of each method concerning slope, minimum rainfall, vegetative cover, minimum soil depth and crop are included in the criteria, which was assessed through the geographical information system. According to certain criteria such as geomorphological settings of the study area, a first evaluation to assess suitability of the area for different water harvesting methods can be done (Fig. 2). The geomorphological setting of a hilly region with slopes between 1 to 20 % can permit the use of semi-circular bunds or contour bunds, if parameters like soil cover characterized by high runoff coefficient and fertile soils are considered.

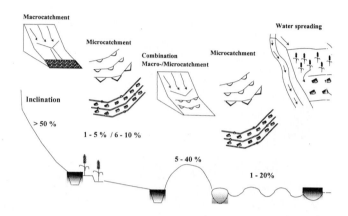

Figure 2: Water harvesting systems suitable to various conditions in the area

Finally, the evaluation of various digital datasets via geographical information procedures results in a classified image illustrating the suitability of the region for various water harvesting systems. The developed methodology is expected to be suitable for use in other arid regions with similar conditions.

4 Preliminary Results

The first classification of the image according to vegetative and soil cover led to the differentiation of eight classes: rock and scree slopes, bare ground, bare ground covered with flintstones, water bodies, ploughed ground and three different levels of vegetation that is very sparse (widely scattered bushes), sparse (grasses and green bushes) and dense vegetation (in flooded depressions or tree orchards in nearby villages). This classified image is overlaid by the slope image, computed by the digital elevation model. The slope parameter was used to classify the area according to suitability for different water harvesting systems of macro- or microcatchments. Analysis of these datalevels resulted in an image showing that about 12% of the whole area is unsuitable for any type of water harvesting. Most of the area (about 68%), however, is generally suitable. Out of this 24% are favourable for the application of a macro-catchment water harvesting system (Fig. 3). About 4% of the area is suitable or favourable for micro-catchment water harvesting systems, whereas the remaining 3% and 12% are suitable and favourable, respectively, for using both methods.

The classification indicates that most of the region can be used for the application one or more types of water harvesting methods. An improved classification of the image can be generated by incorporating more data delivered by a multi-temporal approach. In addition, the implementation of further data levels like meteorological data will allow a more precise differentiation of the area according to differing suitability for each water harvesting method. Nevertheless, these preliminary results demonstrate the applicability of the presented methodology and illustrate that the application of remotely-sensed data and geographical information systems is technically and economically possible, and feasible, for identifying suitable areas for water harvesting in the dry areas.

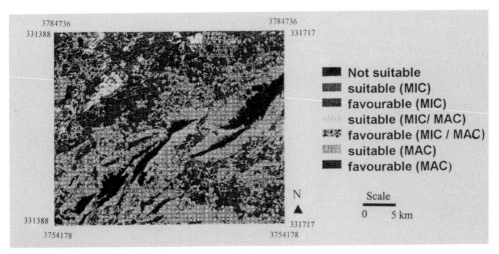

Figure 3: Result image illustrating suitability of the area to various water harvesting systems (MIC = Micro-catchment, MAC = Macro-catchment) ; Projection: Universal Mercator Projection, Zone 37.

Acknowledgment

The project is funded by the German Ministry for Economical Cooperation (BMZ).

References

ARC INFO, PC version 3.4.2 (1994): Environmental Systems Research Institute, Redlands, California, USA.
Critchley, W. and Siegert, C. (1991): Water Harvesting FAO paper no. AGL/MISC/17/91, Rome.
ERDAS IMAGINE, vrsion 8.2 (1995): Erdas Incorporation, Atlanta, Georgia, USA.
Integrated Land and Water Information System, ILWIS, version 1.4 (1992): ITC, Enschede, Netherlands.
Lillesand, T. and Kiefer, R. (1994): Remote Sensing and Image Interpretation. Third ed. John Wiley and Sons, Inc. New York.
Oweis ,T. and Taimeh, A. (1996):Evaluation of a Small Basin Water Harvesting System in the Arid Region of Jordan Water Resources Management **10**: 21-34.
Prinz, D. (1995): Water Harvesting in the Mediterranean Environment - Its Past Role and Future Prospects. In: Tsourtic, N. X. (Ed.) Water Resources Management under Drought or Water Shortage Conditions: Proceeedings of EWRA 1995 Symp. 14-18 March 1995, Nicosia, Cyprus, Balkema, Rotterdam, pp 135-144.
Reij C., Mulder, P. and Begemann, L. (1989): Water Harvesting for Plant Production. World Bank technical paper number 91. The World Bank, Washington, D.C.
Tauer W, and Humborg, G. (1992): Runoff Irrigation in the Sahel Zone CTA (Technical Centre for Agricultural and Rural Cooperation)
World Resources Institute, Water Resources Report (1994/1995): New York, Oxford University Press.

Addresses of authors:
T.Oweis
Annette Oberle
International Center for Agricultural Research in the Dry Areas ICARDA)
P.O. Box 5466
Aleppo, Syria
Dieter Prinz
University of Karlsruhe
D-76131 Karlsruhe, Germany

Application of Remote Sensing to Assess Soil Moisture Capacity and Depths of Ground-Water Table in the Chernobyl Disaster Area

V.I. Lyalko, L.D. Voulfson, A.L. Kotlyar, V.N. Shevchenko, A.D. Ryabokonenko,
K.-H. Marek & S. Oppitz

Summary
For studying problems associated with environmental contamination caused by the Chernobyl nuclear accident, a method has been developed for remote sensing estimation of soil moisture (W) and depths of ground-water table (H) using nadir airborne radar operating at a wavelength of $\lambda = 70$ cm. This method has been designed to use data acquired by both airborne and spaceborne (Space Shuttle L-Band SAR - Synthetic Aperture Radar operating at a wavelength of $\lambda = 24$ cm) surveys. The purpose is to find functions for homogeneous landscape sites in order to relate the SAR-reflection values to the response signal and therefore to W and H. The homogeneous landscape sites have been separated by SPOT satellite data examination using the image processing system ERDAS IMAGINE.

Keywords: Soil moisture, depth of ground-water table, remote sensing, SAR data, Chernobyl disaster

1 Introduction

Scientists from the Center of Aerospace Research of the Earth, Kiev, have developed an ecological monitoring system for soil and ground water using remotely sensed data. This has been linked to a project sponsored by the German Space Agency (DARA) to study environmental problems in the Ukraine after the Chernobyl nuclear accident.

The goal of the remote sensing procedure is a fast and effective detection of soil moisture (W) and depth of ground-water table (H) for vast areas in order to solve a number of problems associated with geology, natural resources, and environmental protection.

There are important reasons for determining W and H by remote sensing methods in regions which have been exposed to radioactive contamination as a result of the Chernobyl accident. First, ground samples may affect the personnel and, secondly, ground-water contamination would occur if wells were drilled.

The studies have been carried out within the southern part of the 30-km alienated zone around the Chernobyl nuclear power plant and cover an area of 1,750 km^2.

2 Models and Methods

The Chernobyl environmental problems have stimulated the development of remote sensing methods for soil moisture capacity and depth of ground-water table estimations by nadir airborne radar at a wavelength of λ = 70 cm.

A well-known possibility for making such estimations consists of the determination of various radiowaves reflection of soils with different W (Brekhovsky 1973, Finkelstein et al. 1977, Ji et al. 1994). The calibration function of W = W(E) and H = H(E) connecting response E, which is taken by remote sensing, with W and H for various types of soil and vertical profiles of moisture W = W(Z), respectively, may be obtained by taking ground samples, well-drilling at the test sites, and by computation of *a priori* data which are very important for the work to be done in the contaminated areas.

Taking into account that the response signal E is a function of the soil dielectric constant, the semi-empirical model developed by Ulaby for the 0.3-1.3 GHz range is used to obtain the values of the real ε' and imaginary ε'' part of the soil dielectric constant (Peplinski et al. 1995, Ulaby et al. 1986). According to this model, ε' and ε'' are functions of the following soil physical properties: volumetric soil moisture W, bulk density ρ_b, specific density of solid soil particles ρ_s, mass fractions of sand S and clay C, respectively, the static dielectric constant for water ε_{w0}, the high frequency limit of the real part of the dielectric constant for water $\varepsilon_{w\infty}$, the relaxation time for water τ_w, the temperature t(^{0}C) and the frequency f in hertz :

$$\varepsilon' = \varepsilon'(W, \rho_b, \rho_s, S, C, \varepsilon_{w0}, \varepsilon_{w\infty}, \tau_w, t, f),$$

$$\varepsilon'' = \varepsilon''(W, \rho_b, \rho_s, S, C, \varepsilon_{w0}, \varepsilon_{w\infty}, \tau_w, t, f).$$

Figures 1 and 2 show the relations W = W(E) and H = H(E) obtained, following these expressions using *a priori* data on grain-size distribution for sandy, sandy-loam, and loamy soil which are typical soil categories for the 30-km alienated zone of the Chernobyl nuclear power plant (Table 1), as well as experimental measurements of W, H, E for the test sites.

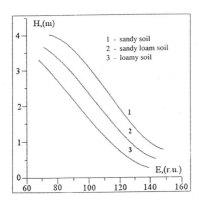

Fig. 1: Calibration dependencies between remote signal E in relative units and groundwater depths H in m

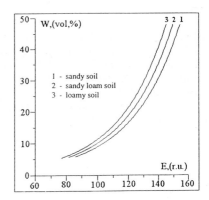

Fig. 2: Calibration dependencies between remote signal E in relative units and volumetric moisture W in %

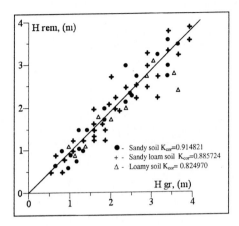
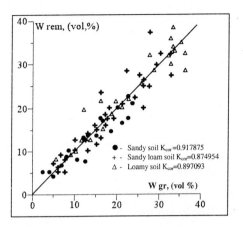

Fig. 3: Results of comparison between remote (rem) and ground-based (gr) measurements of groundwater depth H W in %

Fig. 4: Results of comparison between remote (rem) and ground-based (gr) measurements of volumetric moisture

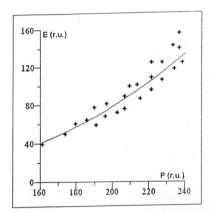
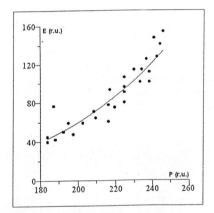

Fig. 5: Results of comparison between airborne radar E and Shuttle SAR data P (in relative units) for meadows, $K_{cor} = 0.925018$

Fig. 6: Results of comparison between airborne radar E and Shuttle SAR data P (in relative units) for fallow lands, $K_{cor} = 0,857466$

	2 mm<d≤0.05mm	0.05mm<d≤0.003mm	d<0.003mm
Sandy	49	47	4
Sandy-Loam	45	47	8
Loamy	40	40	20

Table 1: Grain-size distribution for the main soil types of the study area.

Figures 3 and 4 show results of remote sensing estimations for W and H and ground measurements of W and H for typical sites covered with grass canopies for the three soil types shown in Table 1. The correlation coefficient between remotely-sensed and ground data for W is at least 0.87 and the mean square deviation for the determined parameter does not exceed 15%. As for H, the correlation coefficient is no less than 0.82 and the mean square deviation is no more than 20%.

In order to improve the efficiency of remote sensing studies, a procedure of combined application of airborne radar data taken in the nadir position and SAR-reflection values acquired from the Shuttle spacecraft has also been developed. Essentially, this procedure includes the relations between E and phototone density (P) from the radar image negative within the limits of homogeneous parts with regard to surface roughness of soil and vegetation types. Examples of this relation are shown in Figures 5 and 6. Experimental data verification has proved that the mean square deviation for the signal restoration E from the phototone density is no more than 14% for the test sites with long-fallow land and meadows.

3 Results and Discussion

The results of airborne surveys aimed at obtaining W, H and γ-radiation distribution shown in Figures 7, 8 and 9 can be considered as outline maps of the above parameters.

From remotely sensed data regions with higher soil moisture and high levels of ground-water table have been detected. As a rule such places are located along the Prypiat, Uzh, Veresnia valleys and the tributaries of these rivers. Here the H level mark varies from 0 to 2 m. The presence of a valley bottom relief, a deterioration in the functioning of drainage systems, the embankment of flood plains to prevent radionuclides being washed out into the Pripyat river favours the natural and artificial formation of swamps and flooding here. This process may also be called the regeneration of natural conditions typical of the Polissia region.

The inclined parts and watershed terrain are characterized by H of 3 - 4 m and more. At the same time many upper swamps which are formed on the local clay confining layers have been marked. The overall area of submerged lands amounts to 28% for the study area, i.e. 10% more than for the pre-accident period. Regarding the contaminated area, 35% has γ-exposure exceeding

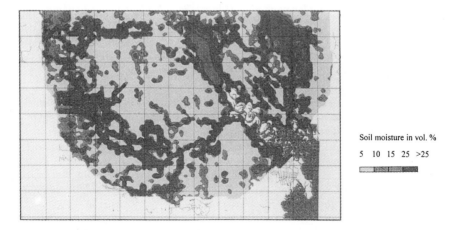

Fig. 7: Schematic map of soil moisture within the alienated zone of the Chernobyl nuclear power plant (from airborne microwave survey data acquired by ZAKIZ in 1994)

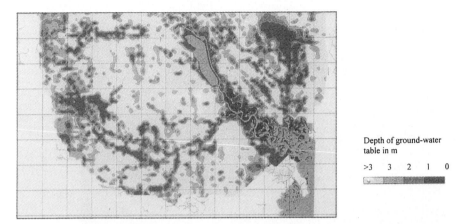

Fig. 8: Schematic map of ground-water table depth witihin the alienated zone of the chernobyl nuclear power plant (from airborne microwave survey data acquired by ZAKIZ in 1994)

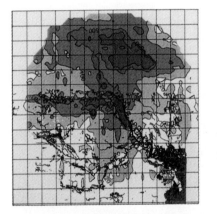

Fig. 9: Schematic map of exposure dose of γ -radiation within the alienated zone of the Chernobyl nuclear power plant (from airborne γ -survey data acquired by ZAKIZ in 1994)

20 µR/h. Here, radioactivity are determined by two principle radioisotopes, ^{90}Sr and ^{137}Cs. In this case 20.5% of the contaminated territory is submerged, with 60% of this figure belonging to sites with H varying from 0 to 1 m, and less than 40% of this figure with H varying from 1 to 2 m.

These sites are known to have outcrops of the capillary fringe to the surface and periodical elevation of the ground-water level to the ground surface during floods. This leads to a higher contamination probability for ground water not only with the most mobile forms of radioisotopes in solution, but also with less mobile ones which are absorbed on soil grains as a result of gravitation redistribution (^{137}Cs).

Figure 9 presents the results of landscape classification from multispectral scanner data in the visible and NIR part of the spectrum. Six land cover types could be marked out: fallow land, grassland, agricultural crops, pine, oak-hornbeam, aspen-birch forests.

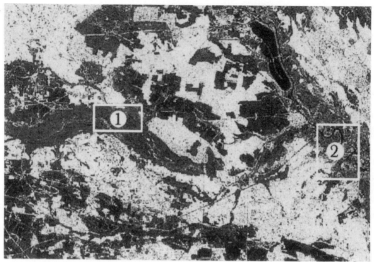

Fig. 10: Fragment of Chernobyl nuclear power plant zone Space Shuttle SAR imagery, 24 cm band
Date: 1 October 1994

Test site

Fig. 11: Distribution of soil moisture at test site 1

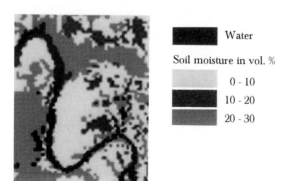

Fig. 12: Distribution of soil moisture at test site 2

To reveal the moisture distribution from the Shuttle SAR-image (Figure 10), two test sites within the Pripyat flood plain (Site 1) and the Uzh valley (Site 2) were selected. In both sites there are two landscape types such as grassland and fallow land where the above-mentioned procedure was used to determine E as $E = E(P)$ (see Figures 5 and 6) and W values by means of the $W = W(E)$ function. The results of these computations are presented in Figures 11 and 12. These Figures show that the following main gradations of W are detected for each test site: 0-10%, 10-20%, 20-30%. At the same time the structure of W-fields in these sites is different. The first site has three segments with various moisture values, whereas the main part of the area has moisture values more than 20%. The second site embraces dry and medium humidified soils where the W-distribution has a more differentiated spotted nature. Both distributions correlate quite well with the airborne survey results.

4 Conclusions

1. Results of remote sensing on correlations between airborne radar survey data, moisture content and depth of ground-water table measurements using airborne and spaceborne imagery in the visible and NIR part of the spectrum can be used for further study.
2. Soil moisture estimation based on the correspondence between airborne and spaceborne radar imagery as well as ground measurements carried out on test sites is cost efficient and sufficiently accurate.
3. Remote sensing was able to outline submerged parts of the territory, especially those areas which grew by 10% after the accident. More than 20% of lands contaminated by radionuclides are submerged now. These areas have the highest probability of radioisotopes penetration into ground water.

Symbols and Abbreviations

ρ_b	bulk density
ρ_s	specific density of solid soil particles,
τ_w	relaxation time for water
$\varepsilon_{w\infty}$,	high frequency limit of the real part of the dielectric constant for water
ε_{w0}	static dielectric constant for water
ε', ε''	real and imaginary part of the soil dielectric constant, respectively
E	nadir airborne radar reflection signal
f	frequency in hertz
H	depth of ground-water table
L-band	24 cm band of the spectrum
P	phototone density of a Space Shuttle SAR
S, C	mass fractions of sand and clay, respectively
SAR	synthetic aperture radar
t	temperature in ^{0}C
W	soil moisture
λ	wave length

References

Brekhovsky, L.M. (1973): Waves in the layered media, Nauka Press, Moscow, 344 pp. (in Russian).

Finkelshtein, M.I. , Mendelson, V., Kushev, V.A. (1977): Radiolocation for layered land covers, Sov. Radio Press, Moscow, 176 pp. (in Russian).

Peplinski, N., Ulaby, T., Dobson, M. (1995): Dielectric properties of soil in the 0.3-1.3 GHz range, IEEE Trans. Geosci. Remote Sensing. Vol. 33, N 3, 803--807, 3 May.

Ulaby, F.T., Moore, R. K., Fung, A.K. (1986): Microwave remote sensing. Vol. 3. Dedham, MA: Artech House, Appendix E.

Ji, J., Skriver, H., Gudmandsen, P. (1994): Estimation of soil moisture from the MAESTRO-1 SAR data of Flevoland, in Proceed. 14th EARSeL Symposium, Goteborg, Sweden, 6-8 June 1994, 103--109.

Addresses of authors:
V.I. Lyalko
L.D. Voulfson
A.L. Kotlyar
V.N. Shevchenko
A.D. Ryabobokenko
ZAKIZ, Centre of Aerospace Research of the Earth
Olesia Gonchara St. 55
252930 Kyiv 30
Ukraine
K.-H. Marek
S. Oppitz
UVE Remote Sensing Centre Potsdam
Berliner Str. 50
D-14467 Potsdam
Germany

Monitoring Vegetation and Soil Degradation Dynamics in the Semi-Arid Areas of Northern Kenya

B. Hornetz

Summary

On-station and on-farm experiments are being conducted on climate-plant-soil-water-relationships and on soil fertility and landuse practices in close cooperation with the Kenya Agricultural Research Institute (KARI) and the Marsabit Development Programme (MDP-GTZ). It was found that the phenomenon of 'dwarf shrub encroachment' in the ecoclimatic zones 5 and 6 of Jätzold (1988) was originally caused by high grazing pressure several decades ago. Further degradation of the soils favours the establishment of unpalatable herbs (e.g. *Solanum* spp.), shrubs, dwarf shrubs (e.g. *Indigofera* spp.) and annual grasses from neighbouring arid areas. Ecophysiological records and botanical observations on differently degraded plots of *Duosperma eremophilum* revealed a high correlation between soil quality/level of soil degradation and a *Duosperma-Indigofera*-Index, which can be used for biomonitoring purposes.

Keywords: Bio-indicators, dwarf shrub encroachment, *Duosperma-Indigofera*-Index, Geographical Information System (GIS), soil erosion hazards, semi-arid areas of northern Kenya

1 Introduction

Food shortages are occuring more frequently within the last years among the pastoralistic and agropastoralistic communities in the semi-arid areas of northern Kenya. This can be mainly attributed to prolonged droughts, but also reduced migration facilities (e.g. due to political instability in neighbouring countries) and the natural growth of the population causing an increase of human settlements, a concentration of herds and vegetation and soil degradation due to overgrazing. Meanwhile population density around rural centres, where relief food is distributed, has increased dramatically as was confirmed by Schiff (1994) at Ngurunit/Marsabit-Samburu District in April-May 1993 (about 77 persons/km^2). Many of the pastoralists have lost a great number of their animals during the drought forcing them to engage in non-pastoralistic activities such as small-scale farming without any prior agricultural advice from relevant authorities.

Since 1990 on-station and on-farm experiments have been conducted on climate-plant-soil-water-relationships as well as on soil fertility and land use practices (e.g. agroforestry, matuta) in cooperation with Kenya Agricultural Research Institute (KARI) and the Marsabit Development Programme (MDP-GTZ). A particular objective of this study was to develop new approaches to monitoring vegetation and soil degradation dynamics on a temporal-spatial level by means of bio-indicators and GIS.

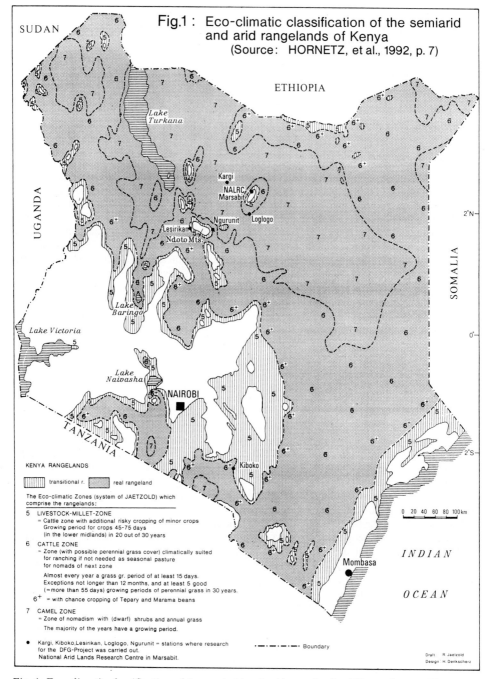

Fig. 1: Eco-climatic classification of the semiarid and arid rangelands of Kenya (source: Hornetz et al., 1992, p.7)

2. Materials and methods

The field experiments on vegetation and soil degradation at the KARI National Arid Lands Research Centre (NALRC) Ngurunit, Marsabit-Samburu District/N-Kenya (Fig. 1) deal with:
- biomonitoring and vegetation succession (ecophysiological experiments, field interviews);
- effectiveness of soil and water conservation techniques (e.g. matuta, agroforestry, reseeding and N_2-fixation);
- acceptance of soil and water conservation techniques (field interviews, participatory approaches);
- unoff and soil erosion;
- monitoring of soil physical and chemical properties.

A detailed description of the study areas' socioeconomic and landuse dynamics is contained in Hornetz (1997). The study area belongs to the agroecological (ecoclimatic) zones LM/L 5 and 6 of Jätzold (1988) and is characterised by a bimodal median rainfall of between 500-600mm and savanna vegetation dominated by *Acacia tortilis* and *Duosperma eremophilum*.

The biomonitoring experiments were carried out on lime-enriched Cambisols of the footslopes/ pediments of the Ndoto Mountains. Under normal conditions they contain very favourable physical and chemical properties (e.g. high permeability and infiltration rates, high plant available water, pH-values of 6.5-7.5, high phosphate and potassium contents, but climatically-induced low nitrogen). Overgrazing in some areas has resulted in severe gully and surface erosion, initiated by heavy torrential rains. Using botanical methods (soil-cover and composition, physiognomy) and pedological parameters (height of the exposed plant roots, depth and width of gullies, soil-type and colour, surface sealing), different intensity classes of soil degradation were delineated and then used to map the Ngurunit area. Degradation dynamics were recorded on randomised protected plots in the area. The different degradation classes were evaluated using soil chemistry analysis (N_{min}, P_2O_5, K_2O, CEC, pH, C/N, organic carbon) as well as plant physiology, botanic and biometric measurements and observations of *Duosperma eremophilum* and *Indigofera* spp. as the key indicator plants of degradation during the long rains 1993 (Schiff, 1994). Ecophysiological and biometric records included: Leaf Area Index (LAI) (LAI 2000; LICOR); rate of shoot growth; leaf water potential (C-52; WESCOR); transpiration and diffusive resistance (LI 1600; LICOR); actual soil moisture (gypsum electrodes, LF 90; SOILMOISTURE, WTW); climatological parameters; soil cover and vegetation composition.

Further objectives of this research were to establish the vegetation and nutrient dynamics from climax vegetation to aridification conditions and to calibrate a *Duosperma-Indigofera*-Index, which can be used to evaluate different degradation levels and suggest the measures for stabilising soil fertility (resource conservation). The above calculations were preliminarily calibrated for Ngurunit with the help of runoff and soil loss records depending on the type of landuse (Plankermann, 1995).

The problem with all field experiment results is that they are site specific and one cannot easily apply such results elsewhere. With the availability of GIS software, it is now possible to digitally superimpose such data on climate, soil, vegetation and landuse maps.

3 Results and discussion

From field interviews of elders (April-May 1993) conducted by Schiff (1994) it became evident that the undisturbed vegetation of the Ndoto pediments, which is presently dominated by *Duosperma eremophilum* in the undergrowth, is not the original climax vegetation of this ecosystem. The settlement dynamics in this area offer the best explanation of the present situation (Schiff, 1994). This was also confirmed in a vegetation succession experiment on a protected plot at NALRC Ngurunit station. After clearing the plots, the *Duosperma* vegetation was overshadowed by dense grass vegetation consisting of perennial grasses such as *Eragrostis superba* and *Chloris roxburghiana*. The *Duosperma* plants could not compete effectively against the grasses (see Fig. 2).

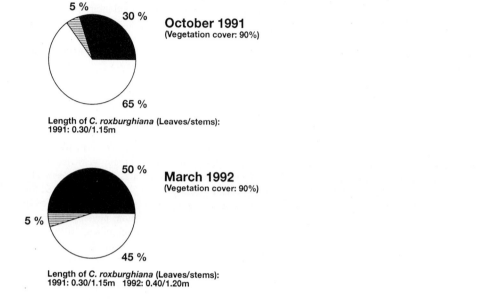

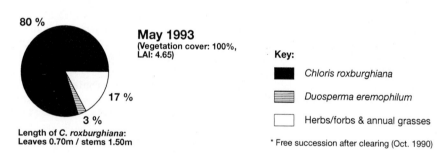

Fig. 2: *Vegetation succession on Cambisols at Ngurunit/N-Kenya**

Using the earlier defined botanic and pedological criteria, 6 degradation classes were found to occur in the study area of northern Kenya. By means of this classification the site-specific ecophysiological conditions and growth performance of the indicator plants *Duosperma eremophilum* and *Indigofera spinosa/cliffordiana* were examined and compared on differently degraded areas. The physiological measurements clearly showed that with increased degradation and soil loss, water balance is affected particularly through surface sealing which leads to a reduction of the infiltration capacity. Records of transpiration, diffusive resistance, leaf water potential, LAI and growth rate of *Duosperma eremophilum* demonstrated a strong dependency on the soil water balance. A reduction of soil moisture below 40% field capacity led to a significant reduction in plant physiological activities (Schiff, 1994). Rangeland plants from the more arid neighbouring Hedad Plains such as *Indigofera* spp., however, can find optimum growth conditions on already degraded sites resulting in a process of encroachment into the *Duosperma* and grass communities.

As a result of soil loss in the upper layer, the first to be affected are lightly mobile nutrients (e.g. NO_3-, P_2O_5) as well as humic substances which are important for the nutrient exchange. There is a decrease in phosphate, potassium and nitrate with increasing degradation class (Fig. 3). The most serious degradation combines an enrichment of nitrate and phosphate in the lower soil layers, which could be explained in terms of reduced extraction or inflow from neighbouring vegetation covered areas. The dynamics of soil degradation as related to runoff and sedimentation rates can be shown clearly by the reduction of nutrients within one rainy season in an experiment on soil and water conservation. Reduced soil protection led to a rapid decrease of phosphate on a freshly cleared area during the short rains (Nov. 1990 - Feb. 1991), in the upper soil layer by as much as about 25% and organic carbon even by as much as 50% (Hornetz, 1997). Water and soil conservation measures such as the simple 'matuta' system were effective in controlling such degradation. For example, stabilization of the potassium content in the soil could be observed during the same rainy season.

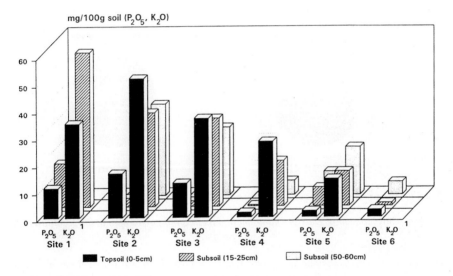

* KARI/NALRC Ngurunit, April 1993;
Analysis by D. Zühlke, Uni. Trier

1 no records for K_2O on site 6 and 1 (subsoil 50-60cm)

Fig. 3: Content of phosphate and potassium on differently degraded Cambisols in a semi-arid area of northern Kenya*

Unfavourable soil moisture and nutrient conditions caused a reduction in growth dynamics and temporal shifts in the phenology of more susceptible rangeland plants such as *Duosperma eremophilum*. It was found that the plants require a longer period to produce generative plant organs, have few flowers and fruits and start maturing early during periods of water stress. On the other hand the drought adapted *Indigofera* spp. were able to complete their vegetation cycles about 15 days earlier than *Duosperma* and are well adapted to arid sites due to favourable stress patterns such as low water requirements, small LAI, parahelionastic properties and osmotic adjustment (Hornetz et al., 1992). The dominance of *Indigofera* at the expense of *Duosperma* plants turned out to be a reverse indication of soil quality. For example, on less degraded sites, there is no presence of *Indigofera* (the *Duosperma-Indigofera*-Index is 1.0; see Fig. 4); on seriously degraded areas (degradation class 5) it is common to find less than 10% *Duosperma* and more than 90% *Indigofera* (the index is 0.1-0.0).

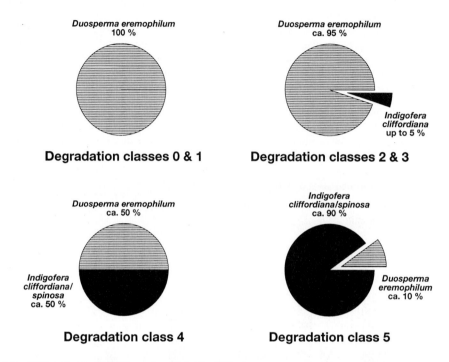

* May 1993; pediment plains of the eastern Ndoto Mtns/N-Kenya

Fig. 4: Development of Duosperma eremophilum and Indigofera spp. at different levels of land degradation*

By means of the *Duosperma-Indigofera*-Index it is possible to evaluate qualitatively vegetation and soil degradation. GIS software, however, allows the quantification of soil erosion hazards using the Universal Soil Loss Equation (USLE). Therefore digitizing was carried out of the relief (1:50,000), rainfall and soil maps (1:500,000). This basic ecological information was combined with other parameters such as rainfall erosivity, soil erodibility, soil gradients (acc. to calculations with the Structured Elevation Model, SEM) and slope length factors in order to delineate soil erosion hazard classes on a spatial level (1:50,000). The first results of the runoff and soil erosion experiments under different landuse treatments at NALRC Ngurunit, during the short rains of 1993 (Plankermann, 1995), confirm the suitability of the GIS calculations of the sedimentation rates.

4 Conclusions

The biomonitoring method developed by Schiff (1994) for the semi-arid pediments of the Ndoto mountains in northern Kenya provides a simple, quick and cost effective approach to evaluating soil fertility in degraded semi-arid areas. This allows for quick control measures to be implemented within the shortest time possible. The greatest advantage of this method is that it can easily be applied in other degraded semi-arid ecosystems provided that bio-indicators with their respective ecophysiological properties for the area can be determined.

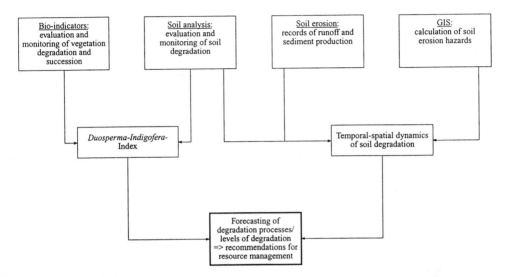

Fig. 5: Approach for monitoring and forecasting vegetation and land degradation dynamics.

By means of GIS, it is possible for site specific measurements and simulation results to be evaluated and their spatial representativeness tested. These results can then eventually be presented in the form of maps. The advantage here is that complex eco-processes like soil degradation through erosion can be spatially interpolated and finally configurated in maps. From such maps, it is possible to predict which areas are likely to be seriously affected by erosion hazards.

The combination of approaches allows the estimation and forecasting of temporal-spatial dynamics of vegetation and soil degradation (by defining the levels of degradation) as well as recommendations for further resource management (Fig. 5).

References

Hornetz, B., Jätzold, R., Litschko, T. and Opp, D. (1992): Beziehungen zwischen Klima, Weideverhältnissen und Anbaumöglichkeiten in marginalen semiariden und ariden Tropen mit Beispielen aus Nord- und Ostkenya. (=Materialien zur Ostafrika-Forschung, 9), Geographische Gesellschaft Trier, Trier (with an english summary).

Hornetz, B. (1997): Ressourcenschutz und Ernährungssicherung in den semiariden Gebieten Kenyas - Ein holistischer methodischer Ansatz unter Verwendung von agrarökologischen Experimenten, Computersimulationsmodellen und Geographischen Informationssystemen (GIS). Dietrich Reimer Verlag, Berlin (with an english summary).

Jätzold, R. (1988): Range management handbook of Kenya, Vol.II,1. Republic of Kenya, Ministry of Livestock Development (MOLD), Nairobi, Map 1-11.

Plankermann, P. (1995): Untersuchungen zu Bodenerosion und Oberflächenabfluß bei verschiedenen Flächennutzungsarten auf einer semiariden Pedimentfläche im Trockengrenzbereich des Anbaues Nordkenyas (unter besonderer Berücksichtigung bodenphysikalischer und -chemischer Parameter). Report (with an english summary) for GTZ/TÖB, Trier/Eschborn.

Schiff, C. (1994): Erfassung und Bewertung von Bioindikatoren im Hinblick auf ein Biomonitoring zur Früherkennung von Erosions- und Degradationserscheinungen in N-Kenya. Report (with an english summary) for GTZ/TÖB, Trier/Eschborn.

Address of author:
Berthold Hornetz
FB VI / Geography-Geosciences
University of Trier
D-54286 Trier
Germany

Spatial Characterization of Soil Properties

A. Castrignanò, M. Mazzoncini & L. Giugliarini

Summary

Soil samples were collected over an area of 4 hectares at the Interdepartmental Agro-Environmental Study Centre "E. Avanzi" of the University of Pisa (Italy) to evaluate spatial variability of soil characteristics (particle-size classes, cation exchange capacity, pH, organic matter, total N, available P, available K, exchangeable Na).

Geostatistical procedures and punctual Kriging were then used to generate maps at the isolines of some soil parameters. The maps indicated high spatial variability for each Principal Component and allowed to divide the study area into homogeneous zones. Such an approach might be applied knowing within-field variation of properties such as soil water retention curve, nutrient levels or soil salinity in order to optimize agronomic tecniques such as irrigation and fertilization.

Keywords: Spatial dependency, geostatistics, Kriging, map at the isolines, homogeneous areas.

1 Introduction

A knowledge of the spatial characterization of soil properties is necessary in order to locate homogeneous areas to be carefully managed for agricoltural "sustainable development" (Stelluti et al., 1996). The characteristics of the soil are continuous variables whose values at various points differ according to changes in direction and distance from nearby samples (Burgess and Webster, 1980). Most variations show that the properties of the soil have a spatial dependency within a given area. It is useful to follow a relatively new methodology, called Geostatistics, that allows the spatial dependency of soils properties to be considered directly in the interpolation.

2 Materials and methods

The experiment began in 1993 in an area of 4 hectares at the Interdepartmental Agro-Environmental Study Centre "E. Avanzi" of the University of Pisa (Italy). The soil was classified as "Typic Xerofluvent" according to Soil Survey Staff, 1992: Keys to Soil Taxonomy, Us Soil Conservation Service. Sampling took place at the nodes of a semi-regular grid (11 x 22 m) and was studied to explore the profile of the ploughed ground (25-30 cm deep). For each point, two depths were sampled: a surface-level layer of 0-10cm and a deeper layer of 10-30cm. Subsequently, the (disturbed) samples of earth were heater-dried at 40 °C, sieved at a diameter of less than 2 mm, then analyzed in the laboratory for ten variables: three particle-size classes [sand, silt and clay in %, Soil Science Society (ISSS) classification], cation exchange capacity (%, $BaCl_2$ exchange method), pH in H_2O (potentiometric method), organic matter (%, Walkley Black method), N total (%, Kjeldahl method), available P (ppm, Olsen method), available K (ppm, International method, NH_4OA_c extraction), exchangeable Na content (ppm, International method, NH_4OA_c extraction).

The procedure used to process the data was the following: a) Principal Component Analysis (PCA); b) Structural analysis of the Principal Components; c) Linear interpolation using the punctual Kriging; d) Mapping. PCA was carried out in order to simplify the description of the set of intercorrelated variables and to study their correlations. The amount of information contained in each component represents its variance; this allowed us to analyze a smaller number of uncorrelated Principal Components (PC), which account for most variance and to which a physical interpretation can be given. Before the PC analysis the variables were standardized to mean value = 0 and variance = 1 as they were not homogeneous in nature and measurement units. The Principal Components were rotated using the Varimax procedure to better highlight their structure. Structural analysis has been applied to the PC following geostatical procedures.

Structure of the spatial distribution of each of these components was analyzed through the study of their variograms, i.e., curves describing changes in the semi-variance between pairs of points sampled at a certain distance as a function of such distance. An estimation of the semi-variogram is half the expected value of the squared difference between the values separated by a given distance or lag, h (Trangmar et. al., 1985). A traditional estimator of the semivariogram at lag h is the following:

$$\gamma(h) = \left[\frac{1}{2n(h)}\right] \sum_{i=1}^{n(h)} [z(x_i) - z(x_i + h)]^2$$

where z is a regionalized variable, $z(x_i)$ and $z(x_i+h)$ are samples measured at the points x_i and $x_i + h$ and n(h) is the number of pairs separated by distance or lag h. The variogram has three characteristics: nugget, sill and range (Burgess and Webster, 1980). The value $\gamma(0)$ is zero in theory.

When, for h→0, $\gamma(h)$ tends towards a value different from 0, this is called the Nugget effect (Co). The Nugget effect can be caused by sampling errors or by variability at a smaller scale than the sampling one. If the semivariogram looks limited, sill represents the semivariance value where it reaches a plateau. The lag value at which the variogram reaches sill is called range (a) and represents the distance of sample influence, i.e., the greatest mean distance at which two samples can be considered correlated. One of the authorized mathematical models has been adapted to the experimental variogram using a weighted least squares procedure. In our case, a spherical model was chosen and validated according to the Jackknife method (Isaak and Srivastava, 1989). In order to evaluate the correctness of the adaptation, two particular statistics were chosen, the reduced mean and reduced variances and, as validity criterium, the approximation of mean to zero and of variance to one (Russo,1982).

The Punctual Kriging interpolation method has been applied to the Principal Components in order to intensify the number of spatial points at which they are estimated; this is to improve the representation of spatial variability and, therefore, the subsequent production of maps at the isovalues. The analysis of the principal components and the production of maps were carried out using the SAS\STAT and SAS\GRAPH software packages (SAS Institute Inc., 1994, Release 6.08). The geostatistical analysis was carried out using the GEOPACK freeware program from the United States Environmental Protection Agency (A User's Manual for the GEOPACK Version 1.0 Geostatistical Software System).

3 Discussion

The application of PCA to the data collected at the two depths shows that no substantial difference can be seen between the surface-and the deeper layer; hereafter data refer to the 0-10 cm layer only. Four Principal Components (PC) were outlined which accounted for about 81% of total variance for each depth. Table 1 presents the structure of the first four components, from which it is possible to assign physical meaning. The first component was positively influenced by the finer components of the texture and by the CEC, whereas it was negatively influenced by the coarser component. We can define this component as being a characteristic of the texture of the soil. The second component was connected to soil fertility and is significantly influenced by total N and organic matter content and by soil pH. The third component is influenced positively by available P and less by available K, which also has the same load on the first component. Therefore, we can define the third component as characteristic of phosphopotassic or essentially phosphatic fertility of the soil. The fourth component is influenced positively and almost exclusively by exchangeable Na, which also reveals the lack of correlation between this variable and other soil parameters. This component may be defined as relating to the content of exchangeable Na in the soil.

	C 1	C 2	C 3	C 4
SILT	94 *	6	9	1
CLAY	91 *	1	6	24
CEC	82 *	26	- 5	12
SAND	- 97 *	- 5	- 7	- 14
PH	1	82 *	2	- 18
N tot.	- 2	77 *	- 15	42
O. MAT.	28	72 *	25	1
P	- 7	19	91 *	- 6
K	50 *	- 24	57 *	7
Na	29	0	- 1	91 *

Table. 1. Structure of the Component, depth 0-10cm Communality: 80.9 %

The experimental semivariograms, relating to the first four principal components (Fig. 1), look all limited, indicating a sufficient degree of local stationarity of the parameters. Significant and interesting differences between the various components can be seen: the first two are well-structured with a relatively small Nugget effect (20% Sill); the other two components, on the other hand, show a much greater Nugget effect with the casual component of variability (60% Sill) exceeding the structured component (40%), this means that the variability associated to phosphorus and sodium content is essentially random. Table 2 contains the results of fitting of a spherical-type model to the experimental semivariograms, can be seen that the component linked to the texture has a larger-scale variability probably deriving from the genetic development of the soil. On the other hand, the remaining three components, linked to the chemical characteristics of the soil and to man's management, have a shorter-range variability. The model's level of adaptation was always quite good for all four components as the reduced mean and the reduced variance were close to 0 and 1, respectively (Table 3). However, the third and fourth components had the reduced mean farther from 0.

The data measured were then interpolated with Punctual Kriging, with interpolation steps of 2 m and giving 584 interpolated points allowing us to obtain isoline maps regarding the first 4 Principal Components with satisfactory accuracy. The maps allowed us to better highlight the structural differences in spatial variability of each component.

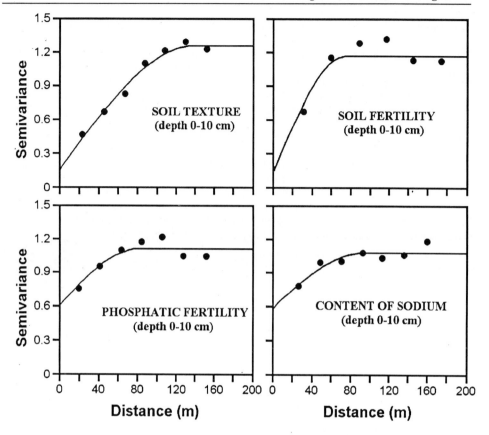

Fig. 1: Experimental semivariograms depth (0-10cm) describing spatial variability of the four Principal Components.

COMPONENT	MODEL	NUGGET	SILL-NUGGET	RANGE (m)
C 1	Spherical	0.1650	1.086	139
C 2	Spherical	0.2085	0.8747	55
C 3	Spherical	0.6355	0.4165	85
C 4	Spherical	0.6252	0.3930	88

Table 2: Results of fitting depth 0-10 cm.

COMPONENT	SUM OF SQUARES	REDUCED MEAN	REDUCED VARIANCE
C 1	0.0015	0.0082	1.046
C 2	0.0100	0.0044	1.061
C 3	0.0028	-0.0131	1.086
C 4	0.0016	0.0228	0.971

Table 3. Results of jackknifing semivariogram models at 0-10 cm depth

Regarding the first Principal Component (Fig. 2), the field can be divided into three zones: left and right sides with essentially horizontal flow lines and a central transition zone in which the flow lines are tendentially vertical. The right half has a larger percentage of coarse component than the left side and has a higher variability with many individual areas of different texture. For the second Principal Component (Fig. 3), it can be seen how the direction of the isolines changes radically to horizontal; this is probably due to its previous management. In fact, in the preceding years, the tillage operations were made perpendicularly to the present field partition.

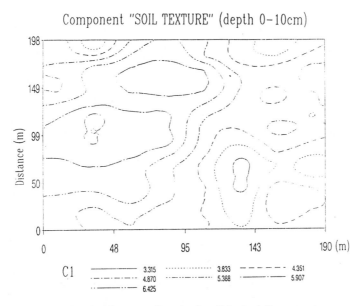

Fig. 2: Map regarding the first Principal Component.

Once again, the field can be divided in three parallel zones: one lower, sufficiently large and homogeneous characterized by high nitrogen fertility levels; one central with a decreasing degree of nitrogen fertility and one upper in which the fertility increases together with the variability. Furthermore, the upper right half has a lower fertility degree in comparison with the left half. It corresponds to a zone with higher texture in which the organic substance is "burned" more rapidly. On the other hand, the left zone has a higher percentage of fine components in which the mineralization processes are slower. The map that refers to the third component (Fig. 4) is somewhat similar to that of fertility with essentially horizontal flow-lines, however it has a higher non-structured variability. The tight link between the nitrate and phosphatic fertility components can be explained in that the organic matter raises the cationic absorption and exchange of the soil; thus the phosphate anions can unite with the humic acids creating phosphohumates that impede the retrogradation of phosphorus. At the deeper layer, the phosphatic fertility seems more structured and homogeneous. Sodium, as already seen in the structural analysis, has a high random component in its variability which increases at deeper depths therefore the map is not included.

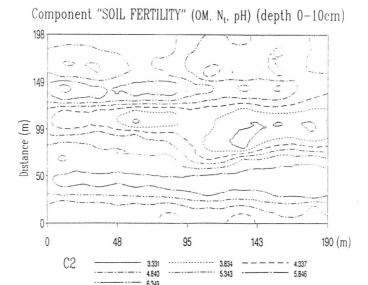

Fig. 3: Map regarding the second Principal Component.

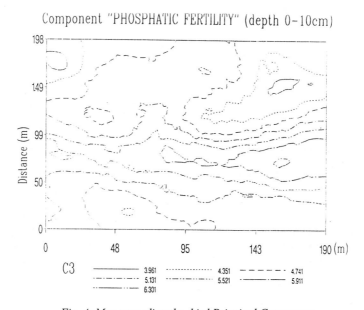

Fig. 4: Map regarding the third Principal Component.

4 Conclusions

Analysis of the spatial variability of soil, based on a small number of independent variables (4 Principal Components), and the successive interpolation with the Punctual Kriging method, has been demostrated to be a highly effective method to share the area in homogeneous zones in order to optimize their management. Such an approach might be applied knowing within-field variation of properties such as soil water retention curve, nutrient levels or soil salinity in order to optimize agronomic techniques (irrigation, fertilization, or soil improvement). Identification of a spatially dependent component of "random" error allows us to reduce the residual error in the analysis of variance, giving more effectiveness to tests ordinarily used in agricultural trials. A knowledge of the spatial dependence may also help in identifying optimal experimental plot size and sampling schemes (Vieira et al. 1981).

References

Burgess, T. M. and Webster, R. (1980): Optimal interpolation and isarithmic mapping of soil properties. The semivariogram and punctual Kriging. J. Soil Sci. **31**, 315-331.
Isaaks, E. H. and Srivastava, R.M. (1989): An Introduction to Applied Geostatistics, Oxford University Press, New York.
Journel, A. G. and Huijbregts, C.H. (1978): Mining geostatistics, Accademic Press, New York.
Russo, D., and E. Bresler, (1982), Soil idraulic properties as stochastic processes: II Errors of estimates in a heterogeneous field, Soil Sci. Soc. Am. J. **46**, 20-26
Soil Survey Staff (1992): Keys to Soil Taxonomy USDA. Soil Conservation Service, Washington.
Stelluti, M., Castrignanò, A. and Lopez, G. (1996): Uso della Geostatistica e del GIS ai fini della razionalizzazione della concimazione. Rivista di Agronomia, **1**, 9-16.
Trangmar, B. B., Yost, R.S. and Uehara, G. (1985): Application of geostatistics to spatial studies of soil properties, Adv. Agron. **38**, 45-94.
Vieira, S. R., Nielsen, D. R. and J. W. Biggar (1981): Spatial variability of field measured infiltration rate, Soil Sci. Soc. Am. J. **45**, 1040-1048.

Addresses of authors:
Annamaria Castrignanò
Istituto Sperimentale Agronomico
Via C. Ulpiani, 5
70125 Bari
Italy
M. Mazzoncini
Dipartimento di Agronomia e Gestione dell'Agro-Ecosistema dell'Università degli Studi di Pisa
Via San Michele Degli Scalzi, 2
56124. Pisa
Italy
L. Giugliarini
Centro Interdipartimentale di Ricerche Agro-Ambientali dell'Università degli Studi Pisa
Via Vecchia Di Marina, 6
56010 Sampiero a Grado (Pisa)
Italy

Classification and Properties of Soils Under Different Land Uses in South-West Sri Lanka

P. Leinweber, C. Preu & C. Janku

Summary

The classification of 30 soil profiles, arranged in 9 catenas, showed the predominance of Ferralsols in the wet climatic zone of south-west Sri Lanka. They are characterised by (1) 18-67 % clay but very low cation exchange capacities based either on whole soils (2-13 $cmol_c$ kg^{-1}) or on clay fractions (0.4-31 $cmol_c$ kg^{-1} $clay^{-1}$), and (2) small total reserves of bases (1-11 $cmol_c$ kg^{-1}). The soil properties which determined the units at the second and third classification level, as well as important soil fertility characteristics, depended on the actual land use and on the topographical position of the soil profiles. Land uses with dense vegetation and canopy cover, for example homesteads and rubber plantations, had greater soil organic matter contents, cation exchange capacities, base saturation and contents of available P.

1 Introduction

Soil classification and mapping are essential for land use planning and sustainable management strategies. The most recent systematic survey of the soils of Sri Lanka is based on a morphological classification system in which 15 Great Soil Groups were distinguished (Moormann and Panabokke, 1961). Many of these categories (e.g. "Red-Yellow Podzolic Soils") are not used in modern classification systems such as the new "World Reference Base for Soil Resources" (Spaargaren, 1994). This disagreement between national and international soil classifications limits scientific communication and comparability of data. Large areas in the wet climatic zone of SW-Sri Lanka were classified as "Red-Yellow Podzolic Soils" (Survey Department Sri Lanka, 1975). The assignment of this category to the actual soil units of the "World Reference Base for Soil Resources" is not known. Furthermore, basic fertility properties of the different soil units and land uses have to be determined for planning sustainable management strategies.

The objectives of the present study were
(1) to classify representative soil profiles in the wet climatic zone of Sri Lanka according to the "World Reference Base for Soil Resources",
(2) to investigate basic physical and chemical properties determining the fertility of these soils, in relation to topographic position and the most important land uses in this area.

2 Study area in south-west Sri Lanka and analytical methods

- Study area: 05°57'N - 06°02'N and 80°25'E and 80°29'E, located in the lower catchment of the Polwatta Ganga that flows into Weligama Bay on the south-west coast of Sri Lanka.
- Basic rock: highly metamorphosed charnockites of Precambrian and Cambrian age, Quaternary fluvial sediments in the valleys, Holocene beach rock overlaid by fossil beach sediments near the coast.
- Climate: mean annual precipitation 2000-2500 mm, mean annual temperature $\approx$ 27°C, generally high humidity around 80 %.
- Land use: intensive agricultural use since the beginning of the colonial period, natural lowland forests have been completely removed, coconut plantations and gardens (homesteads), paddy cultivation on the fluvial sediments, tea-, rubber- and cinnamon plantations, small forests or scrub.
- Fieldwork: 30 soil profiles, arranged in 9 catenas, representing topographic positions and major land uses (excluding paddy soils). Figure 1 shows the location of the study area in south-west Sri Lanka, the distribution of the major soil units and the arrangement of the investigated 9 catenas of 30 soil profiles in the Polwatta Ganga catchment.
- Analyses: particle-size distribution, pH, organic C (C_{org}), total nitrogen (N_t), cation exchange capacity (CEC) and exchangeable cations, total reserves of bases (TRB), H_2O-, $NaHCO_3$-, NH_4F/HCl-extractable phosphorus.

3 Important soil properties

The morphology of soil profiles was characterised by very low moisture contents, except under dense, closed vegetation (rubber plantations) and near paddy fields. The soil horizons were distinguished according to colours, texture, occurrence of plinthite or petroplinthite and redoximorphic features. The very dry soils in scrub-land and tea plantations always had a firm consistence. In contrast, the profiles in homesteads and rubber plantations had stable microaggregates or, at high C_{org} contents, a crumb structure. The texture was characterised by (1) variable contents of gravel and stones within the soil profiles and along catenas, (2) moderate to high clay (mean: 35 %) and low silt (mean: 11 %) contents, and (3) often increased clay contents in subsoils (argic horizons). The colours of most soil samples ranged from "bright yellowish brown" (10YR 6/8) to "reddish brown" (5YR 4/8).

The pH-values ranged from 2.3 to 8.7 (mean 5.2 for pH (H_2O) and 4.7 for pH ($CaCl_2$)). Often the Ah-horizons had the highest pH-values in the profiles because of the circulation of basic cations with growth and decomposition of the vegetation. The concentrations of $\mathbf{C_{org}}$ were in the range <0.1 to 68.6 g kg^{-1}, and N_t ranged from 0.1 to 4.4 g kg^{-1}. The corresponding means for the topsoil horizons are 17.8 g kg^{-1} C_{org} and 1.4 g kg^{-1} N_t. As expected, the C_{org} concentrations often decreased with profile depth. The CEC (1.5-25.5 cmol$_c$ kg^{-1}) were significantly correlated to C_{org} concentrations (CEC [cmol$_c$ kg^{-1}] = 0.31 × C_{org} [g kg^{-1}] + 1.4; r = 0.934; n = 108). The CEC of clay fractions (<1-31 cmol$_c$ kg^{-1}; mean 6 cmol$_c$ kg^{-1}) and effective (e)CEC (<1-19 cmol$_c$ kg^{-1}; mean: 3 cmol$_c$ kg^{-1}) are characteristic of strongly weathered soils and a major diagnostic criterion of ferralic horizons (CEC <16 cmol$_c$ kg^{-1}, eCEC <12 cmol$_c$ kg^{-1}). This was confirmed by very low TRB (mean: 4-5 cmol kg^{-1}, maximum: 11 cmol kg^{-1}). Hence, most of the soil profiles were classified as Ferralsols. Within this major soil group, the categories "Humic", "Eutric", "Plinthic", "Gleyic" and "Haplic" were distinguished at the second, and the categories "Posi"-, "Lateri"-, "Xanthi"-, "Dystri"-, and "Eutri"- at the third classification level.

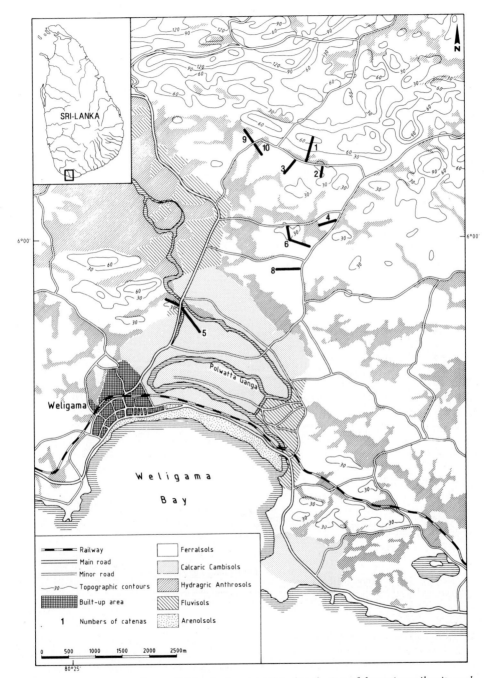

Fig. 1: Study area in south-west Sri Lanka (upper right), distribution of the major soil units and arrangement of the investigated soil profiles in 9 catenas.

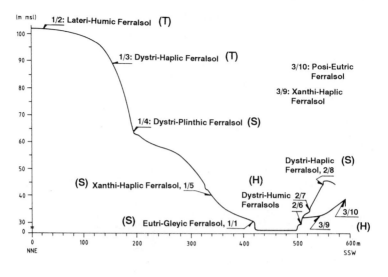
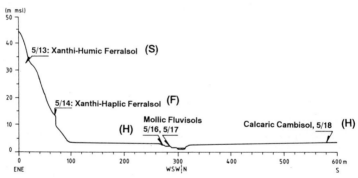
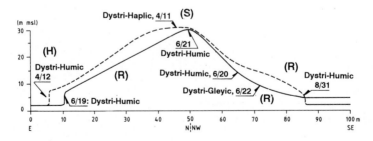

Fig. 2: Topographic positions of soil profiles along the surveyed catenas (catena/number of soil profile; for location of the catenas in the study area see Figure 1) and forms of land use (in parentheses: T = tea plantation, S = scrubland, R = rubber plantation, C = cinnamon plantation, H = homestead, F = native forest). All soil profiles in catenas 4, 6, 8, 9 and 10 were classified as Ferralsols.

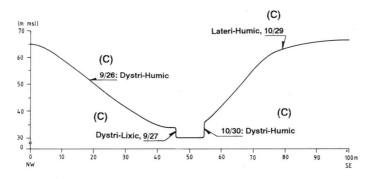

Fig. 2: Continuation

4 Soil units in catenas

Figure 2 shows the distribution of soil profiles along the investigated catenas. The classification of these soils at the second and third level depended largely on topographic position and land use. The profiles 2 and 29 in the top positions of catenas 1 and 10 were classified as Lateri-Humic Ferralsols (petroplinthe in 40-50 cm depth). These positions probably represent old land surfaces from which the horizons overlying the plinthite had been eroded (Driessen and Dudal, 1991). Plinthite was also found in mid-slope positions under sparse vegetation cover such as old tea or scrub, and on steep slopes (1/4: Dystri-Plinthic Ferralsol). Sediment accumulation and soil development in downslope positions resulted in Xanthi-Haplic Ferralsols (1/5, 3/9, 5/14). They were characterised by colour hues of 7.5 YR or yellower, pH (H_2O)-values in the range 4.6-5.1 and BS of about 8-34 %. Further downslope, the accumulation of base-richer sediment from top positions (1/1), stronger influence of freshly weathered rock (3/10) and additional bases input with plant residues, animal manure or human waste from homesteads resulted in base saturations >35 % (Posi-Eutric- [3/10] and Eutri-Gleyic Ferralsols [1/1]). Gleyic Ferralsols generally occurred only in downslope positions close to the adjacent paddy fields (1/1, 6/22).

Most soil profiles were either Haplic and Humic Ferralsols, again dependent on vegetation cover and relief position. In top or upper slope positions with sparse vegetation cover (4/11: homestead; 1/3, 2/8: scrub), limited production of primary organic matter and soil loss through erosion did not allow SOM enrichments >10 g C_{org} kg^{-1} in the upper 50 cm (Haplic Ferralsols). Humic Ferralsols occurred in homesteads, tea plantations in deep positions, under permanent forest (5/13), often in rubber plantations (6/19-22) and on mid-slope positions of cinnamon plantations (9/26). Table 1 shows average Ferralsol profiles and means of physical and chemical characteristics that were important for the classifications of Humic and Haplic Ferralsols. The combination of moderate to high clay contents and low CEC confirms the ferralic character of the horizons. Humic Ferralsols had higher C_{org} concentrations, in particular in subsoils, and slightly higher CEC. Most of these Ferralsols were further assigned to the Dystri-subgroup because of their low BS (5-29 %).

The few other soil groups in the study area included Mollic Fluvisols from the Polwatta Ganga sediments (5/16, 5/17) and a Calcaric Cambisol on beachrock and dune sand (5/18). Their occurrence is restricted to either fluvial or marine sediments, which predominate in the Polwatta Ganga valley and the fossil lagoon, respectively (Figures 1 and 2).

Soil units	Horizons	Thickness (cm)	Clay (%)	Silt (%)	C_{org} (g kg^{-1})	CEC (cmol$_c$kg^{-1} / kg$_{clay}^{-1}$)	CEC$_{clay}$	BS (%)
10 Humic Ferralsols	Ah	28 (21)	38 (13)	10 (3)	18 (6)	7 (2)	5 (3)	16 (11)
	BwsAh	38 (21)	43 (13)	10 (2)	14 (5)	5 (2)	4 (2)	7 (2)
	AhBws	64 (34)	44 (13)	12 (4)	10 (8)	4 (2)	3 (2)	14 (9)
6 Haplic Ferralsols	Ah	13 (10)	36 (7)	10 (3)	15 (10)	6 (3)	5 (2)	23 (7)
	B(t)ws1	36 (20)	41 (6)	11 (2)	8 (5)	4 (2)	5 (2)	10 (7)
	B(t)ws2	75 (49)	37 (16)	15 (8)	4 (3)	3 (1)	7 (4)	13 (4)

Table 1: *Average soil profiles and means of analytical data from Humic and Haplic Ferralsols in south-west Sri Lanka (standard deviations are given in parentheses).*

5 Influence of land use

The influence of land use was investigated in 25 Ferralsol profiles (0-70 cm depth) and 20 separate samples from adjacent Ah-horizons. The medians in Table 2 show that:
- soils in rubber plantations had the highest C_{org} and N_t concentrations, CEC and concentrations of H_2O-P and $NaHCO_3$-P (resulting from the accumulation of SOM and not related to differences in the mineral soil materials),
- the homestead soils had the highest BS and concentrations of exchangeable Ca, Mg and K, probably because of nutrient input with fertiliser, manure or waste from households,
- soils in cinnamon and tea plantations and scrubland had the lowest CEC,
- $NaHCO_3$ was the most efficient extractant for "available P" in these soils, and
- the three different P-fractions did not equally reflect possible influences of land use.

Land use	clay (%)	pH (CaCl$_2$)	C_{org}	N_t (g kg^{-1})	CEC (cmol$_c$kg^{-1} / kg$_{clay}^{-1}$)	CEC$_{clay}$	BS (%)
Cinnamon	28	4.3	12	1.0	4.3	3.8	15.1
Homestead	36	4.4	13	1.1	5.7	7.4	27.5*
Rubber	46	4.3	23***^{a}	1.9***	7.4**	2.9	9.0
Scrub	30	4.2	11	0.9	4.0	5.6	13.5
Tea	42	4.2	12	1.1	4.6	4.8	12.7

Land use	Ca	Mg	K	Na	H_2O-P	$NaHCO_3$-P	NH_4F-P
			(cmol$_c$kg^{-1})			(mg kg^{-1})	
Cinnamon	0.28	0.15	0.06	0.07	0.6	7.6	4.3
Homestead	0.82	0.46	0.13	0.07	0.4	13.7	2.2
Rubber	0.10	0.31	0.10	0.16	1.9	11.4	2.3
Scrub	0.18	0.19	0.12	0.07	0.8*	9.7*	1.8
Tea	0.18	0.18	0.11	0.07	0.7	9.7	1.6

^{a} Significance levels of differences between bold printed medians: * $P = 0,20$, ** $P = 0,05$, *** $P = 0,01$ (U-Test)

Table 2: *Medians of analytical characteristics of Ferralsols under different land use.*

6 Conclusions

- A simple, unchecked translation of the term "Red-Yellow Podzolic Soils" into recent units of the "World Reference Base for Soil Resources" would result in Acrisols, Alisols or Lixisols

(Spaargaren, 1994). Detailed investigation of profiles in the field and by physical and chemical analyses clearly showed that Ferralsols were the most important soil group in south-west Sri Lanka.
- Topographic position and land use determined their further classification in the second and third level categories. This was not reflected by the older classification as "Red-Yellow Podzolic Soils" (Moormann and Panabokke, 1961; Survey Department Sri Lanka, 1975) and shows the suitability of the proposed "World Reference Base for Soil Resources".
- Scrubland resulting from shifting cultivation and the traditional tea and cinnamon cropping in plantations leads to reduced soil fertility probably due to mineralization of SOM and erosion of fertile topsoils.
- Most soils in rubber plantations were classified as Humic Ferralsols according to relatively high SOM contents. Because they occurred uniformly among the catenas, soil translocation by erosion and sedimentation dynamics has been strongly reduced.
- As the investigated rubber plantations were established some 20 years ago on degraded tea plantations, land use with dense vegetation cover and high input of organic matter appears necessary for soil conservation and improvement of soil fertility in this region.

References

Driessen, P.M. and Dudal, R. (Eds.) (1991): The major soils of the world. Lecture notes on their geography, formation, properties and use. Agricultural University Wageningen, The Netherlands, and The Catholieke Universiteit Leuven, Belgium, 145-187.

Moormann, F.R. and Panabokke, P.C. (1961): Soils of Ceylon. Tropical Agriculturist **67**, 1-71.

Spaargaren, O.C. (Ed.) (1994): World Reference Base for Soil Resources. Wageningen, NL, International Soil Reference and Information Centre, 161 pp.

Survey Department Sri Lanka (1975): Sri Lanka - Approximate Distribution of Great Soil Groups (compiled by C.R. Panabokke).

Addresses of authors:
P. Leinweber
Institute of Soil Science
University of Rostock
D-18051, Rostock, Germany
C. Preu
C. Janku
Institute for Spatial Analysis and Planning
In Areas of Intensive Agriculture (ISPA)
P.O. Box 1553
D-49364 Vechta, Germany

Site Properties and Suitability of Eroded Saprolites for Reclamation and Agricultural Use

T. Scholten & P. Felix-Henningsen

Summary

Investigations of hydrological, physical and chemical site properties were undertaken on saprolites in the Middleveld of Swaziland, Southern Africa, to determine the suitability of saprolites for reclamation following erosion and for agricultural use. The results show that their hydrological properties are of great importance for plant growth after erosional phases and drought periods, leading to a stabilization of the land cover. Gullies deeply incised into the saprolite can be reclaimed by planting grass, shrubs, and trees due to favourable hydrological properties and the sufficient nutrient reserve of the saprolite. In spite of low nitrogen and organic carbon reserves, the suitability for agricultural use of eroded soils underlain by saprolites is good, especially in regions with a climatic water deficiency. However, because of the low structure stabilities of the saprolites, they are prone to erosion. Therefore, effective erosion control by local farmers is a precondition for a sustainable rehabilitation of eroded areas on saprolite and their further agricultural use.

Keywords: Saprolite, Swaziland, gully erosion, reclamation, site properties

1 Introduction

The term 'saprolite' was first used by Becker (1894). He described saprolite as 'a general name for thoroughly decomposed, earthy, but untransported rock'. According to Fölster (1971), Eswaran & Bin (1978), Calvert et al. (1980), Felix-Henningsen (1990), and others, saprolites developed in situ for millions of years exclusively through isovolumetric chemical weathering of rocks under warm and humid climatic conditions in the tropics and subtropics. They are relicts of ancient warm and humid periods (e.g. the Mesozoic-Cenozoic) and, therefore, are found widely on the earth's surface.

Investigations in Swaziland have shown that saprolites underlying the soil cover with a thickness of up to 20 m form the largest portion of the weathering mantle and are wide-spread in the study area (Scholten et al., 1995). In areas with changes in vegetation following a more seasonal climatic regime and, more recently, human clearance of the plant cover for cropping and grazing, runoff and erosion have increased and the topsoil has been removed (cf. Rohdenburg, 1989). Hence, the underlying saprolites are affected by gully, sheet and rill erosion and, moreover, form the rooting zone and essentially determine the site properties of eroded areas. Therefore, knowledge of the site properties of saprolites is important when considering erosion control, gully reclamation and the suitability of land for agriculture and other uses.

2 Material and methods

The Kingdom of Swaziland is located in South-East Africa and bordered by the Republic of South Africa and Mozambique. Climatically, it is a subtropical region with summer rain. Three study sites in the Middleveld (31° 07' E, 26° 35' S, 500-1,200 m a.s.l, 600-1,200 mm annual rainfall, rainy season from October to March, distinct dry season) were chosen for comprehensive investigations. At deep incised erosion gullies and road embankments, ten representative soil-saprolite profiles from three different bedrocks were described and sampled. Soil classification and designation of horizons are according to FAO (1989). The saprolites were divided into zones as described in Scholten et al. (1996). Standardized methods of physical, chemical, and mineralogical soil analysis were used. Detailed descriptions of these procedures and study sites are given in Mushala et al. (1994) and Scholten et al. (1995).

3 Results and discussion

The saprolites of the magmatic and metamorphic rocks of Swaziland's Middleveld developed under a tropical humid climate during the Cretaceous and Early Tertiary (Fränzle, 1984). The greatest thickness of the saprolite cover in the Swaziland Highveld and Middleveld was formed at midslope positions with inclinations < 30 % (Scholten & Felix-Henningsen, 1993). The saprolites are missing in the Highveld at crest positions and steep backslope positions. In the Middleveld the saprolite cover is often preserved at the ridges and the following backslope positions. Due to actual erosional processes, the saprolites are subjected to erosion as well as the soils. Furthermore, they are the initial material for soil formation in erosion landscapes after stabilization (Scholten & Felix-Henningsen, 1993).

The saprolitisation process can be considered as polygenetic subdividing saprolites into a near-surface oxidation zone and an underlying reduction zone due to different oxygen supply during weathering (Scholten et al., 1996). Isovolumetric weathering of saprolites led to an increase in pore volume of more than 50 vol.-%. High silt and low clay content led to predominate medium pores amounting to 70 % of total pore space. Compared to the soils, the available water capacity of the saprolites is two to four times higher and the saturated hydraulic conductivity is two times higher (Table 1). According to their high pore volume and large water-storage capacity, the saprolites can serve as water reservoir during high-intensity storms thereby lowering the risk of mass movement.

The chemical paleo-weathering of the saprolites led to a decrease in contact surfaces between minerals which resulted in a reduced structure stability. Together with low clay contents, absence of organic matter and distinctly lower values for cementing pedogenic oxides compared to the soils, this results in a weak structure. Accordingly, the shear strength of the saprolites with values of 2.3-4.4 kPa was distinctly lower than that of the soil horizons (Table 1). The high K-factors of the saprolites of about 0.5-0.8 (t·h)/(N·ha) point to the low cohesion of the material and, therefore, its vulnerability against erosion. This explains the concentration of gully erosion in areas with soil-saprolite complexes, which is in accordance with the findings of WMS (1990) in Swaziland, Downing (1968), Fränzle (1984), and Skakesby & Whitlow (1991) in Southern Africa, and for other parts of the world, e.g. observations of Ireland et al. (1939), Brenon (1952), de Meis and de Moura (1984), and Wells et al. (1991). Accordingly, saprolites are one essential precondition for the rapid development of deep incised gullies especially at midslope and lower slope positions at the Middleveld of Swaziland.

The pH-values of the saprolite differ between slightly acid and neutral. The CEC is medium in the upper zone dominated by kaolinite and medium to high in the middle and lower zone due to the high smectite content in the clay fraction (Scholten et al., 1995). The CEC of the saprolite

	Saprolite	Soil
Site potential		
C [wt.-%]	0	0.4 – 3.2
N [wt.-%]	0	0.03 – 0.25
P [mg/kg]	0.1 – 2.1	0.1 – 0.5
K [kg/m^2]	6.6 – 12.0	3.5 – 5.7
Ca [kg/m^2]	0.8 – 18.0	0.4 – 0.8
Mg [kg/m^2]	4.3 – 9.0	1.8 – 2.7
pH [H$_2$O]	5.1 – 6.5	4.5 – 5.4
CEC [cmol$_c$/kg]	8.1 – 25.1	5.3 – 14.5
AWC [m^3/m^3]	0.25 – 0.36	0.07 – 0.24
Erosion risk		
K-factor [(t· h)/(N·ha)]	0.5 – 0.8	0.1 – 0.3
Shear strength [kPa]	2.3 – 4.4	6.5 – 12.8
Infiltrability [mm/h]	20 – 75	9 – 33
K$_S$ [10^{-7} m/s]	4.2 – 11.8	0.4 – 5.7

Table 1. Site potential and erosion risk of saprolite versus soil (C: organic carbon; N: total nitrogen; P: available phosphorus from Bray-I-method; K, Ca, Mg: nutrient reserve calculated for 1 m depth; CEC: cation exchange capacity at pH 7; AWC: available water capacity; K$_s$: saturated hydraulic conductivity).

(Table 1) is not correlated to the low clay content of about 10 wt.-%. Clay mineralogical and micromorphological studies show that the silt fraction of the saprolite is dominated by aggregates of kaolinite (Felix-Henningsen et al., 1993; Scholten et al., 1995). It can be assumed that these mineral aggegates contribute considerably to the exchange capacity of the saprolite. Except for nitrogen and carbon, the total amounts of nutrients are higher in the saprolite than in the soil (Table 1) due to the higher amount of primary minerals in the saprolite. The exchangeable cations are dominated by Ca and Mg. These cations have not been completely leached out from the saprolite, probably due to more arid climatic conditions in Swaziland (cf. Bruhn, 1990).

4 Conclusions

For areas affected by sheet erosion, most of the soil cover is denudated and the underlying saprolites essentially determine the site properties of the rooting zone. Therefore, the hydrological properties of the saprolites are of great importance for plant growth in areas affected by soil erosion especially during drought periods, leading to a stabilization of the land cover.

Gullies deeply incised into the saprolite can be reclaimed by planting grass, shrubs, and trees due to favourable hydrological properties and sufficient nutrient reserves of the saprolites.

Except for low nitrogen and organic carbon reserves, eroded soils underlain by saprolites are suitable for agricultural use, especially in regions with a climatic water deficiency.

The low structure stabilities of the saprolites create a high erosion risk and must be taken into account. Therefore, effective erosion control by local farmers is a precondition for a sustainable rehabilitation of eroded areas on saprolite and their further agricultural use.

Acknowledgements

This research is part of the STD3-project 'Soil erosion and river sedimentation in Swaziland' (Contract-No. TS*CT90-0324), which was carried out in collaboration with the University of Swaziland, Department of Geography, Environmental Science and Planning (H.M. Mushala) and

Cranfield Institute of Technology, Silsoe College, Great Britain (R.P.C. Morgan and R.J. Rickson). We would like to thank the EU for funding.

References

Becker, G.F. (1894): Reconnaissance of the gold fields of the southern Appalachians, U.S. Geological Survey, Annual Report **16**, 251-331.

Brenon, P. (1952): Géomorphologie de l'Antsihanaka et de l'Antano-simboangy, in: Erosion des sols à Madagascar, Antananartvo, 6-22.

Bruhn, N. (1990): Substratgenese-Rumpfflächendynamik, Bodenbildung und Tiefenverwitterung in saprolitisch zersetzten granitischen Gneisen aus Südindien, Kieler Geogr. Schriften **74**, Selbstverlag, Kiel.

Calvert, C.S., Buol, S.W. and Weed, S.B. (1980): Mineralogical characteristics and transformations of a vertical rock-saprolite-soil sequence in the North Carolina piedmont: I. Profile morphology, chemical composition, and mineralogy, Soil Sci. Soc. Am. J. **44**, 1096-1103.

De Meis, M.R.M. and De Moura, J.R.S. (1984): Upper Quaternary sedimentation and hillslope evolution - southeastern Brazilian plateau, Am. J. Sci. **284**, 241-254.

Downing, B.H. (1968): Subsurface erosion as a geomorphological agent in Natal, Trans. of the Geological Society of South Africa **81**, 131-134.

Eswaran, H. and Bin, W.C. (1978): A study of a deep weathering profile on granite in peninsular Malaysia. I: Physico-chemical and micromorphological properties, Soil Sci. Soc. Am. J. **42**, 144-149.

FAO (1989): Soil map of the world - revised legend, ISRIC, Technical Paper 20, Wageningen.

Felix-Henningsen, P. (1990): Die mesozoisch-tertiäre Verwitterungsdecke (MTV) im Rheinischen Schiefergebirge - Aufbau, Genese und quartäre Überprägung, Relief, Boden, Paläoklima 6, Borntraeger, Stuttgart.

Felix-Henningsen, P., Schotte, M. and Scholten, T. (1993): Mineralogische Eigenschaften von Boden-Saprolit-Komplexen auf Kristallingesteinen in Swaziland (südliches Afrika), Mitteilgn. Dtsch. Bodenkundl. Gesellsch. **72**, 1293-1296.

Fränzle, O. (1984): Bodenkunde - Südafrika. Beiheft zu Blatt 4, Serie S, Afrika-Kartenwerk, Borntraeger, Stuttgart.

Fölster, H. (1971): Ferrallitische Böden aus sauren metamorphen Gesteinen in den feuchten und wechselfeuchten Tropen Afrikas, Göttinger Bodenkd. Ber. 20, Selbstverlag, Göttingen.

Ireland, H.A., Sharp, C.F.S. and Eargle, O.H. (1939): Principles of gully erosion in the Piedmont of South Carolina, USDA Tech. Bulletin No. 633, Washington/DC.

Mushala, H.M., Scholten, T., Felix-Henningsen, P., Morgan, R.P.C. and Rickson, R.J. (1994): Soil erosion and river sedimentation in Swaziland, Final Report to the EU, Contract number TS2-CT90-0324, Brussells.

Rohdenburg, H. (1989): Landschaftsökologie - Geomorphologie, CATENA Cremlingen.

Scholten, T. and Felix-Henningsen, P. (1993): Gully-Erosion in Boden-Saprolit-Komplexen auf Kristallingesteinen in Swaziland (südliches Afrika), Mitteilgn. Dtsch. Bodenkundl. Gesellsch. **72**, 1247-1250.

Scholten, T., Felix-Henningsen, P. and Mushala, H.M. (1995): Morphogenesis and erodibility of soil-saprolite complexes from magmatic rocks in Swaziland, Z. Pflanzenernähr. Bodenk. **158**, 169-176.

Scholten, T., Schotte, M. and Felix-Henningsen, P. (1996): Hydrologische Eigenschaften von Saproliten aus Kristallingesteinen in Swaziland (Südliches Afrika), Zbl. Geol. Paläont., Teil 1 **3/4**, 507-520.

Shakesby, R.A. and Whitlow, R. (1991): Perspectives on prehistoric and recent gullying in central Zimbabwe, GeoJournal **23(1)**, 49-58.

Wells, N.A., Andriamihaja, B. and Rakotovololona, H.F.S. (1991): Patterns of development of lavaka, Madagascar's unusual gullies, Earth Surface Processes and Landforms **16**, 189-206.

WMS (1990): Investigation of the causes and hydrological implications of gully erosion in Swaziland, WMS Associates Ltd., Fredericton, New Brunswick Canada.

Address of authors:
T. Scholten
P. Felix-Henningsen
Justus-Liebig-University Giessen, Institute of Soil Science and Soil Conservation, Wiesenstraße 3-5
D-35390 Giessen, Germany

Evaluating and Grouping of Soils According to Their Susceptibility to Anthropogenic Degradation

R. Dilkova, G. Kerchev & M. Kercheva

Summary

The present status of the arable and virgin soil layers of the main Bulgarian soil units was assessed using well known classifications in literature and data for soil particle size distribution, humus content, pH, bulk density, soil aggregate stability, pore size distribution, and water peptizable clay. Discriminant analysis of soil properties in the A-horizons of virgin lands and their relative changes in arable layers was applied to distinguish different groups according to soil susceptibility to anthropogenic impacts.

The main physical degradation process was the destruction of the soil structure. The multiple linear regression equations between soil aggregate stability and humus and clay content obtained for the virgin A-horizons were used to calculate the aggregate stability of the arable layers. Only 20% of the structure deterioration of the Kastanozems, Chernozems and Haplic Luvisols was due to changes in humus and clay content; the remaining 80% were due to other factors, such as mechanical impact of heavy machinery and/or changes in humus quality. In contrast, in the Vertisols the influence of the other factors is only 26%. The differences between calculated and measured aggregate stability of the Chromic Luvisols and Planosols vary greatly, which does not allow to point out the leading risk factor.

Keywords: soil structure, soil physical degradation, soil physical classification

1 Introduction

Bulgaria is a relatively small country (111,000 km^2) but has a great variety of soil resources. About 61% of the land is cultivated and is subject to anthropogenic activities which often lead to soil acidification, dehumification, disaggregation and other changes that are hard to control and which are detrimental to soil fertility. Accordingly special attention is required to protect the limited cultivated land from degradation and to improve its productive potential.

According to the GLASOD map (Oldeman et al., 1991), one of the most extensive soil degradation processes in Europe is that of physical deterioration. In their explanatory notes, they stress that there is a need for more detailed local information on quantitative soil attributes, and the type, degree and causative factors of soil degradation.

The object of the present study was to analyse the data available about the current state of soil properties in the arable layers under cultivation and to group them according to the degree of anthropogenic changes.

ISBN 3-923381-42-5
© 1998 by CATENA VERLAG, 35447 Reiskirchen

2 Materials and methods

The arable layers of 30 soil profiles of Haplic Kastanozems, Haplic and Luvic Chernozems, Eutric Vertisols, Haplic and Chromic Luvisols, Eutric and Mollic Planosols, Humic Cambisols and Lithic Kastanozems, and the respective layers of genetically analogous soils from virgin lands, have been used in this study. The following characteristics have been analysed: soil particle size distribution (Katchinski, 1958), organic matter content (Tjurin, 1965), pH in KCl, bulk density at field capacity or close to that (100 cm^3 ring samples), soil aggregate stability, expressed by the ratio (MWDR) of the meanweight diameters of the aggregate after and before wet sieving (Vershinin and Revut, 1957), pore size distribution (suction plate method) and water peptizable clay (Gorbunov, 1977). The coefficients of variation of these properties are presented in Table 1.

In a previous study (Dilkova et al., 1995), the discriminant analysis of these properties in the A and B horizons of soils from virgin land confirmed the division of the above-mentioned soils into 8 groups. The present study employed the same methods of analysis but only for the upper 0-20/30 cm layers. The soil susceptibility to physical degradation has been classified as low, average, high and very high, taking into account the present state of the arable layers with respect to the virgin layers. The classifications in Dregne and Boyadgiev (1983), Hall et al. (1977), and Kramer (1969) have been used for qualifying the current status of the physical properties.

Soil groups		n	Clay	Humus	BD	MWDR	AWC	pH$_{KCl}$
1. Haplic	virgin	3	6	34	4	31	1	10
Kastanozems	arable	3	9	26	10	42	18	8
2. Haplic	virgin	4	10	17	6[a]	40	10[a]	12
and Luvic Chernozems	arable	4	18	24	2[a]	37	8[a]	8
3. Haplic	virgin	3	22	24	4[a]	35	17[a]	18
Luvisols	arable	3	22	21	9[a]	14	24[a]	19
4. Chromic	virgin	2	23	40	1	1	3	29
Luvisols	arable	2	25	4	4	5	3	23
5. Mollic	virgin	5	11	25	7	25	19	13
Planosols	arable	5	18	52	14	34	36	17
6. Eutric	virgin	8	58	51	13	46	25	17
Planosols	arable	7	52	31	14	27	27	22
7. Vertisols and	virgin	3	25	12	3	31	15	5
Vertic Chernozems	arable	3	8	16	7	52	15	6
8. Humic	virgin	2	9	82	48	12	56	47
Cambisols	arable			usually not cultivated				

[a] n=2

Table 1: Soil physical property's coefficients of variation (%) of the upper 0-20/30 cm of virgin and arable soils.

3 Results and discussion

The upper humic horizons of the virgin soils are characterised by high and average organic matter contents, very good and good aggregate stabilities, as well as optimal proportions between the aeration pores and the pores containing available water (Dilkova et al., 1995). In most of the soils the silt particle size fraction dominates.

Under the impact of prolonged, mechanised, and intensive land use of these soils, serious changes have occurred in the humus content and the aggregate stability of the arable layers. The clay (particles smaller than 1μm) contents in the arable layers of the soils with high profile differentiation, such as Luvisols and Planosols, are greater than those from corresponding virgin soils (Table 2). That is the result of mixing soil from the illuvial B horizon into the ploughing layers.

The present humus status of most of the arable layers could be described as very deteriorated (Table 2). Most of the soils (86%) had a very low (1-2%) or low (2-3%) organic matter content. Under virgin conditions, 36% of them were high (4-5%) and average (3-4%) humic soils. The low humic virgin soils (1-2%) stayed the same (O.M.<1.7%). Only in a few (7%) cases did the organic matter increase in the arable layer.

Soil groups		Clay %	Humus %	BD g/cm^3	MWDR	AWC %	PH$_{KCl}$
1. Haplic	virgin	27	3.5	1.22	0.68	24.6	6.6
Kastanozems	arable	26	2.9	1.25	0.12	24.3	6.6
	suscept.		average	low	Very high	low	
2. Haplic	virgin	30	3.0	1.22	0.43	17.8	4.8
and Luvic	arable	30	2.0	1.47	0.19	14.5	4.8
Chernozems	suscept.		average	high	High	average	
3. Haplic	virgin	24	3.4	1.21	0.44	24.2	4.4
Luvisols	arable	28	2.1	1.49	0.14	17.1	4.7
	suscept.		average	high	High	average	
4. Chromic	virgin	21	1.3	1.42	0.52	18.3	4.7
Luvisols	arable	27	1.6	1.33	0.31	21.2	5.4
	suscept.		low	low	Average	average	
5. Mollic	virgin	17	2.6	1.25	0.59	23.9	4.3
Planosols	arable	20	2.2	1.30	0.27	20.6	4.4
	suscept.		low	low	Average	low	
6. Eutric	virgin	13	1.9	1.38	0.39	20.5	4.6
Planosols	arable	21	1.5	1.39	0.19	18.0	4.2
	suscept.		low	low	Average	average	
7. Vertisols and	virgin	44	4.3	1.18	0.44	25.9	5.6
Vertic	arable	47	2.9	1.20	0.17	18.8	5.9
Chernozems	suscept.		average	low	high	average	
8. Humic	virgin	24	10.5	0.89	0.86	22.3	5.3
Cambisols	arable		usually not cultivated				
^{a}	**very good**		**good**	**average**		**poor**	

Table 2: Present physical status^a of the upper 0-20/30 cm of virgin and arable soils and the assessments of the soil susceptibility to anthropogenic changes

The reduction of the organic matter content in the arable soils as a result of mineralization predominating over humification could be classified (according to Dregne and Boyadgiev, 1983) as low in Chromic Luvisols and Planosols and average in Kastanozems, Haplic and Luvic Chernozems, Vertisols and Haplic Luvisols (Table 2).

Under virgin conditions, about 60 % of the investigated soils show a very good to good water stability of the aggregates (MWDR>0.4), 37 % have a mean water stability of the aggregates (MWDR=0.4-0.2) and only 3% have a poor aggregate stability (MWDR≤0.2). Due to the long-term agricultural use of the soils their aggregate stability has deteriorated (fig. 1 and table 2) and has to be classified as average and more often as poor.

In order to improve the productive potential of the arable soils it is necessary to determine the reasons causing the strong deterioration of the soil structure. The multiple linear regression equations between soil aggregate stability (MWDR) and organic matter and clay content (Table 3) obtained for the virgin A-horizons were used to calculate the aggregate stability (MWDRc) of the arable layers.

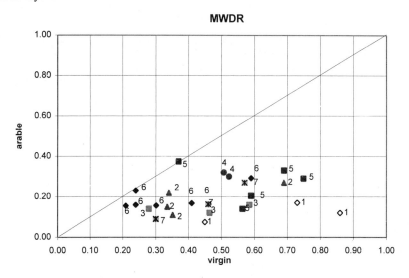

Fig. 1: *The water stability of the aggregates, expressed as the ratio (MWDR) of the meanweight diameters of the aggregates after and before wet sieving in the upper 0-20/30 cm of the arable and the corresponding virgin layers. 1 – 7 soil groups (see Table 1 and 2).*

Soil groups	Equations	R^2	SE
1, 5, 7, 8	MWDR=0.48+0.18*OM-0.008*OM^2-0.015*clay	88	0.09
2, 3, 4, 6	MWDR=0.38+0.076*OM-0.037*clay+0.001*clay^2	91	0.03

Table 3: *Multiple regression equations between aggregate stability (MWDR) and clay and organic matter (OM) content of the virgin soils.*

The values of the calculated (MWDRc) and the measured (MWDRm) aggregate stability of the arable soils can greatly differ. Only 20% of the structure deterioration of the Kastanozems,

Chernozems and Haplic Luvisols was due to changes in organic matter and clay content; the remaining 80% were due to other factors, such as mechanical impact of heavy machinery and/or changes in humus quality. In contrast, in the Vertisols the influence of the other factors is only 26%. The soils formed on Loess and those with silt domination, Kastanozems, Chernozems and Haplic Luvisols, are very vulnerable to mechanical destruction (Antipov-Karataev et al., 1960). The differences between calculated and measured aggregate stability of the Chromic Luvisols and Planosols vary greatly, which does not allow us to point out the leading risk factor. This could be due to the great variability of the parent materials, the domination of the coarse soil particles, and could also be due to the unfavourable changes in the organic matter characteristics.

The soil aggregate stability is one factor determining the soil density. The data in Fig.2 show a tendency for an increasing content of the water peptizable clay in almost all arable layers accompanied by an increasing bulk density (Table 2). In addition to the decreasing organic matter content and deterioration of the soil aggregation, the available water content in the soils also diminishes (Table 2).

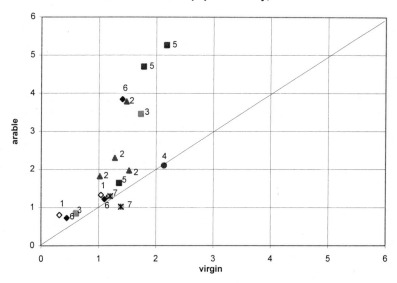

Fig. 2: Water peptizable clay in the upper 0-20/30 cm of the arable land and the corresponding virgin layers. 1 – 7 soil groups.

In order to compensate for the effects of the physical properties' deterioration of the arable layers, some types of ameliorations (liming, organic and mineral fertilization) have been applied. This could be seen from the changes in pH values. The data show that 31% of the soils became less acidic while 26% of the soils became more acidic. Some of the Vertisols (with slight acid and neutral reaction), and the Kastanozems (with alkaline soil reaction) with chemical conditions naturally favourable for plants, show no changes.

Discriminant analysis of the present status of clay content, humus, pH, bulk density, and MWDR illustrates the overlap between the soil groups when data of virgin A-horizons were used (Fig. 3). Discriminant analysis of the soil properties in the A horizons of virgin lands and their relative changes in arable layers (Fig. 4) allowed us to distinguish 4 soil groups according to soil susceptibility to anthropogenic impacts:

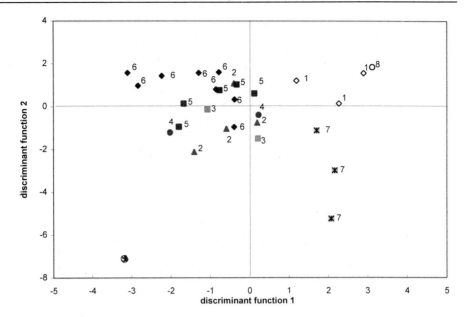

Fig. 3: Soil grouping according to the first and second discriminant function of the clay content, humus, pH, bulk density and MWDR of the upper 0-20/30 cm virgin layers. 1 – 8 soil groups.

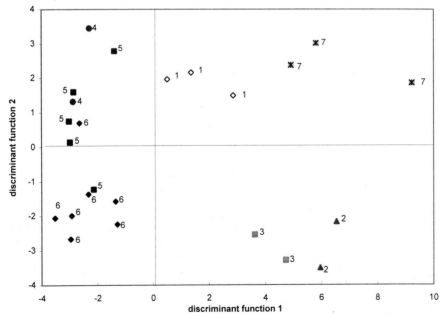

Fig. 4: Soil grouping according to the first and second discriminant function of the clay content, humus, pH, bulk density and MWDR of the virgin A-horizons and their relative changes in the arable layers. 1 – 7 soil groups.

Group I - Low susceptibility to compaction and changes in clay content, low and average susceptibility to changes in water available content, average susceptibility to dehumification, and high susceptibility to aggregate stability deterioration (Vertisols and Haplic Kastanozems);

Group II - High susceptibility to compaction and aggregate stability deterioration, average susceptibility to dehumification and changes in the available water content (Haplic and Luvic Chernozems, Haplic Luvisols);

Group III - Low susceptibility to compaction, dehumification and changes in the water available content, average aggregate stability deterioration (Mollic Planosols);

Group IV - Low susceptibility to compaction and dehumification, average susceptibility to aggregate stability deterioration and changes in clay content and in the available water content (Eutric Planosols).

4 Conclusion

Bases on an assessment of the present status of the arable and virgin soil layers, we were able to classify human-induced changes in soil physical properties of the main Bulgarian soil units. The main physical degradation process was the destruction of the soil structure which resulted not only due to the reduction in the humus content. Discriminant analysis of soil properties in the A horizons of virgin lands and their relative changes in arable layers permitted us to distinguish 4 soil groups according to soil susceptibility to anthropogenic impacts.

References

Antipov-Karataev, I.N., Galeva, V., Gerasimov, I.P., Enikov, K., Tanov, E. and Tjurin, I.V. (eds.) (1960): The Soils in Bulgaria, Zemizdat, Sofia.

Dilkova, R., Kerchev, G. and Kercheva, M. (1995): Physical Properties Classification of Soils In Bulgaria, Summary papers of 3rd International Meeting on Red Mediterranean Soils. May 21-26, 1995. Halkidiki, Greece, 257-260.

Dregne H. and Boyadgiev, T.G. (eds.) (1983): Provisional methodology for assessment and mapping of desertification, FAO and UNEP, Rome, Italy.

Gorbunov, N.I. and Orlov, D.S. (1977): Nature and strength of the bonding of organic substances with soil minerals, Pochvovedenie **7**, 89-100.

Hall, D.G.M, Reeve, M.R., Thomasson, A.J. and Wright, V.F. (1977): Water retention porosity and density of field soils, Soil Survey Technical Monograph **9**, Harpenden.

Katchinski, N.A. (1958): Soil texture and soil microaggregate contents, methods for their study, Academy of Sciences of the USSR, Moscow.

Kramer, P.I. (1969): Plant and Soil Water Relationships, McGraw Hill Book.

Oldeman, L.R., Hakkeling, R.T.A. and Sombroek, W.G. (eds.) (1991): World Map of the Status of Human-Induced Soil Degradation. An Explanatory Note. Second revised edition, ISRIC, Wageningen, UNEP, Nairobi.

Tjurin, I.V. (1965): Agrochemical methods of soil analysis, Nauka, Moscow.

Vershinin, P.V. and Revut, I.B. (1957): Methods of soil structure analysis, Bulletin for scientific-technical information of agrophysics **3**, Agrophysical Institute, Leningrad.

Addresses of authors:
Raina Dilkova
G. Kerchev
M. Kercheva
N. Poushkarov Institute of Soil Science and Agroecology
P.O. Box 1369, 7 Shosse Bankya str.
1080 Sofia
Bulgaria

Evaluation of Soil Degradation Impact on the Productivity of Venezuelan Soils

F. Delgado & R. López

Summary

A Productivity Index (PI) model was adapted and validated to establish soil productivity values as a function of several edaphic variables. It was based on the premise that under similar conditions of climate, crops and management, productivity can be directly related to edaphic conditions which guarantee an adequate medium for root growth. The Productivity Index and respective crop yields have been obtained for seven different soils in the Andean mountains, piedmont and the western plains of Venezuela, using corn (*Zea mays* L.) and black bean (*Phaseolus vulgaris* L.) as crop indicators. Another important application of the model is to predict erosion effects on soil productivity. The results show significant differences for the productivity values among the various soils with important implications related to the crop yield and root growth parameters. The development of a method to estimate the soil loss tolerance is an important achievement of this research. The model can be used for preliminary estimation of relative soil productivity for inferring the impact of the effects of the degradation process on soil productivity and for producing useful information in the evaluation and planning of land use.

Keywords: Soil productivity, soil erosion, soil properties, soil productivity index, soil prediction techniques, Venezuelan soils.

1 Introduction

The productivity of agricultural lands is the result of complex interactions between soil, climate, crops and management. However, where climatic conditions are relatively homogeneous over extensive areas, with specific crops and under given management systems, variations in productivity are closely associated to soil conditions.

The productive capacity of soils can be evaluated directly or indirectly. Direct evaluations are carried out in the field or in greenhouses by means of agronomical experiments which allow the properties of the soil to be related to the yield of the specific crop under given climatic and management conditions. Indirect evaluations consist basically in developing and applying models of varying complexity, thereby attempting to estimate productivity based on soil and crop conditions under defined environmental and management situations.

The magnitude, complexity and great number of variables required by many of the models have created the need for simple models, which allow the productivity of the soil to be estimated as related to edaphic variables which are easy to obtain and to interpret. Notable among such models are the so-called Productivity Indexes (Kiniry et al, 1983), which are based on the premise that under similar soil, climatic and management conditions, productivity is directly related to the edaphic properties which guarantee an optimum medium for root growth. Accordingly, the objectives of this study were to adapt and validate a Productivity Index for soils in the western region of Venezuela, and to establish relations between this index and soil degradation processes, principally erosion, for the purpose of evaluating the impact of such processes on the productivity of these soils.

2 Materials and Methods

2.1 Development of the Model

The model used consists fundamentally of a modification to the Productivity Index (PI) initially developed by Neill (1979), later modified by Pierce et al. (1983) and adapted by Delgado and López (1995) and Delgado (1995).

The model assumes that, under given climatic, crops and management conditions, yield depends on edaphic conditions which provide an adequate environment for good root system growth and in general is as follows:

$$PI = \sum_{i=1}^{n} (A_i . B_i . C_i . K_i)$$

In this model, PI is the Productivity Index of the soil and has a range value of between 0 and 1; A_i is the parameter which evaluates the water-air relation of horizon i starting with the most limiting condition, comparing between plant-available water-holding capacity and aeration of the soil (clay content as related to soil structure); B_i is the parameter which evaluates the conditions that favor the radical exploration of the crop in horizon i starting with the most limiting condition, considering compaction of the soil (bulk density) and content of coarse fragments (volumetric %); C_i is the parameter which evaluates the potential fertility conditions of horizon i starting with the most limiting condition, expressed by soil reaction (*pH*) and/or exchangeable aluminum (saturation %); and K_i is the parameter which evaluates the relative importance of horizon i in the profile (weighting factor of horizon i). All of these parameters are evaluated in each one of the horizons of the soil studied, up to a depth of 100 cm and on a scale of 0 to 1, with 1 corresponding to the conditions of the parameter which most favor the radical growth of the crop. As can be seen, each of parameters A_i , B_i , and C_i is composed in turn of two sub-parameters. To calculate PI, **only the most limiting** respective sub-parameter is taken into consideration, that is, the parameter acquires the value of the sub-parameter having the least value on the scale 0-1. Each sub-parameter is calculated from the following equations:

2.2 Parameter *A*:

Sub-Parameter A_1 : Plant-available water-holding capacity (water retained at a suction of between -33 and -1500 KPa).

$$A_1 = 0.05 \, W; \quad 0 \le W \le 20$$
$$A_1 = 1.00 ; \quad \quad W > 20$$

where: A_1 = value of the sub-parameter available water capacity
W = gravimetric content (%) of available water.

Sub-Parameter A_2 : Clay content as related to soil structure
1. For soils with a weak structure:
    If clay ≤ 20% $A_2 = 1.0 - 0.01 \, (cl)$
    If clay > 20% $A_2 = 1.2 - 0.02 \, (cl)$
2. For soils with a moderate structure:
    If clay ≤ 30% $A_2 = 1.0 - 0.0066 \, (cl)$
    If clay > 30% $A_2 = 1.3 - 0.016 \, (cl)$
3. For soils with a strong structure:
    If clay ≤ 40% $A_2 = 1.0 - 0.005 \, (cl)$
    If clay > 40% $A_2 = 1.3 - 0.0133 \, (cl)$

where: A_2 = value of the sub-parameter clay content as related to soil strcture
cl = clay content (%)

2.3 Parameter B:

Sub-Parameter B_1 : Bulk density of the soil as related to texture type.
1. For fine texture (clayey, fine silty)
    $B_1 = 3.6 - 2 \, (BD)$ $1.30 \leq BD \leq 1.40$
    $B_1 = 9.6 - 6 \, (BD)$ $1.40 < BD \leq 1.60$
2. For medium textures (loamy, coarse silty)
    $B_1 = 1.87 - 0.67 \, (BD)$ $1.30 \leq BD \leq 1.55$
    $B_1 = 6.00 - 3.33 \, (BD)$ $1.55 < BD \leq 1.80$
3. For coarse textures (coarse loamy, sandy):
    $B_1 = 1.52 - 0.4 \, (BD)$ $1.30 \leq BD \leq 1.80$
    $B_1 = 8.00 - 4.0 \, (BD)$ $1.80 < BD \leq 2.00$

where: B_1 = Value of the sub-parameter bulk density
BD = Bulk density of the soil ($Mg.m^{-3}$)

Sub-Parameter B_2 : Volumetric content of coarse fragments (fragments with an equivalent diameter greater than 2 mm).

$$B_2 = (1 - f)^r$$

where: B_2 = Value of the sub-parameter content of coarse fragments
f = Fraction of coarse fragments (volume)
r = Coefficient of root exploratory capacity
The values of the coefficient r appear in Table 1.

r	Exploratory Capacity	Types of crops
1.20	low	tubers, leaf and root vegetables
0.80	moderate	cereals, legumes, oil plants
0.60	high	feed grasses
0.40	very high	trees

Table 1: Values of the coefficient r for the calculation of sub-parameter B_2

2.4 Parameter C:

Sub-Parameter C_1: pH of the soil

$C_1 = 0$	if	pH < 2.8
$C_1 = 0.50$ (pH) - 1.35	if	$2.8 \leq$ pH ≤ 4.5
$C_1 = 0.45 + 0.1$ (pH)	if	$4.5 <$ pH ≤ 5.5
$C_1 = 1$	if	$5.5 <$ pH ≤ 7.0
$C_1 = 1.905 - 0.13$ (pH)	if	$7.0 <$ pH ≤ 8.5
$C_1 = 4.2 - 0.4$ (pH)	if	$8.5 <$ pH ≤ 10.5
$C_1 = 0$	if	pH > 10.5

where: C_1 = Value of sub-parameter pH of the soil
pH = pH in the soil-water extract, relation 1:1

Sub-Parameter C_2: percentage of aluminum saturation
1. When organic matter is greater than or equal to 4%:
    If AS $\leq$ 40% $C_2 = 1$
    If AS > 40% $C_2 = 1.666 - 0.01666$ (AS)
2. When organic matter is between 2.5 and 4%:
    If AS $\leq$ 35% $C_2 = 1.0033 - 0.000666$ (AS)
    If AS > 35% $C_2 = 1.55 - 0.01625$ (AS)
3. When organic matter is between 1.0 and 2.5%:
    If AS $\leq$ 30% $C_2 = 1 - 0.001$ (AS)
    If AS > 30% $C_2 = 1.429 - 0.0157$ (AS)
4. When organic matter is less than 1.0%:
    If AS $\leq$ 25% $C_2 = 1 - 0.002$ (AS)
    If AS > 25% $C_2 = 1.3 - 0.015$ (AS)

where: C_2 = Value of the sub-parameter aluminum saturation as related to organic matter
AS = Aluminum saturation (%)

When the values of exchangeable aluminum saturation are unknown and the pH of the soil is less than 5.5, sub-parameter C_2 can be evaluated from organic matter values and clay content of the respective horizon. In this case, sub-parameter C_2' can be calculated using the following equation:

$$C_2' = 1 - e^{-ax}$$

where: C_2' = Value of the sub-parameter organic matter content
x = Organic matter content (%)
a = Clay coefficient
The value of coefficient a appears in Table 2.

A	clay (%)
1.0	> 20
0.8	15-20
0.6	10-14
0.4	5-9
0.2	< 5

Table 2: Values of coefficient a for the calculation of sub-parameter C_2'

2.5 Parameter K: weighting factor for the horizon i

Evaluates the relative importance of horizon i of soil to the crop.

$$K_{cum} = 0.024 \times 0.82$$

where: K_{cum} = Accumulative weighting factor up to horizon i
X = Maximun depth of horizon i (cm)
For the horizon i under consideration:

$$Ki = K_{cum}(i) - K_{cum}(i-1)$$

2.6 Validation

Validation of the model was carried out on representative soils from three sub-regions in western Venezuela. The soil evaluated and the respective productivity indicative crops, representative of production systems in the sub-region under study, appear in Table 3.

Sub-region	Soil series	Soil taxonomy	Productivity Indicative crop
Western Plains	Barinas	Kandic Paleustalfs, coarse loamy	
	Torunos	Typic Haplustolls, coarse silty	
	Fanfurria	Fluvaquentic Ustropepts, fine loamy	Zea mays L.
	Guanare	Typic Ustropepts, coarse loamy	
	Mosquitero	Fluvaquentic Ustropepts, fine silty	
Andean Piedmont	Boconoito	Typic Haplustalfs, coarse loamy	Zea mays L.
Andean Mountains	————	Typic Humitropepts, coarse loamy	Phaseolus vulgaris L.

Table 3: Classification of soils studied and productivity indicative crops used in each sub-region.

In the case of the Western Plains and Andean Piedmont sub-regions, the yield measurements were taken during two climatically different years: one dry year and one wet year, under relatively homogeneous crop and agronomic management conditions. Moreover, a methodology to measure root growth and an index to quantify their degree of development were proposed. A variation analysis was made, as well as a comparison of averages using the Duncan multiple range method, and statistical correlations were established between Productivity Index, crop yield and degree of root development.

The method for evaluation of roots is termed here "bounding box" and the resulting index DRD. This consists in digging a hole or trench of 1.0 m x 0.8 m and exposing the roots of the crop on a profile transverse to the planting row, located at a distance of 15 cm from the stalk of a randomly selected plant. Later a sheet of transparent plastic is placed over the profile and on this, a metal screen of 50 cm x 60 cm having 1 cm x 1 cm cells (holes) demarcating the rectangle or "bounding box". It contains all the exposed roots, vertically and horizontally. Each one of the cells in the screen in which at least one exposed segment of root appears is marked with an "X" on the plastic sheet, no matter how small. This procedure produces a graph, temed a **radical profile**, as illustrated in the example in Fig. 1.

The DRD is an index that incorporates both the **depth** (D) reached by the roots in the soil profile and the **profusion** (Z) of their growth, the latter being expressed as a relation between the quantity of cells occupied by the roots and the total number of cells in the area explored, or "bounding box". Both parameters define the DRD as a number on a scale of 1 to 9 (Table 4), derived from the analysis of the extreme values encountered for the radical growth of corn plants in the Piedmont and Western Plains sub-regions of Venezuela.

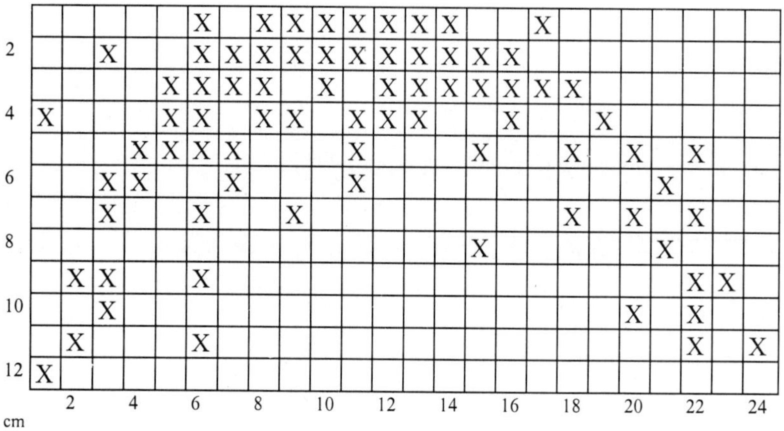

Figure 1: Example of radical profile showing the "bounding box" (12 cm x 24 cm) and the cells occupied by at least one segment of roots (X).

A high degree of radical development corresponds to conditions combining good depth (D) and high radical profusion (Z). The other extreme would be the case of roots with a low degree of radical development, while different combinations of D and Z produce a gamut of intermediate values. In the example in Figure 1, the radical depth (D) is 12 cm and profusion (Z) is 27%, which results from dividing the total of 288 cells contained in the "bounding box" by the 78 cells occupied by roots. This gives a DRD of 3 (Table 4).

		Radical depth (D) (cm)					
		< 10	11-20	21-30	31-40	41-50	> 50
Radical Profusion (Z)	0-10	1	2	3	4	5	6
	11-30	2	3	4	5	6	7
(%)	31-50	3	4	5	6	7	8
	>50	4	5	6	7	8	9

Table 4: Quantification of the DRD as function of D and Z

In the case of the Andean Mountain sub-region, where erosion is the principal soil degradation process, the measurements of crop yield were taken at different **levels of artificial removal** of the superficial soil, to simulate the effect of erosion on productivity (López 1994). To validate the model in the soil representative of this sub-region, three levels of removal of the superficial horizon (0, 50 and 100%) were taken into consideration, each one with two levels of management: the sample treatment, which was cultivated without applying fertilizers (management 1) and treatment with fertilizer, applying to each level of removal formula 15-15-15 fertilizer at a dose of 300 kg.ha^{-1} (management 2).

The validation of the model was determined by correlating the relative PI values **simulated** by the model versus the yield values of the indicative crop **observed** in experimental field conditions. Similar experiences, validating the PI by means of the use of data obtained from artificial erosion, are reported in specialized literature (Rijsberman and Wolman, 1985). Finally, a modified method based on methodology suggested by Pierce et al. (1984) was applied to estimate tolerance to soil loss from erosion (T) starting with the **vulnerability** curve of the soil. The latter relates the amount of soil lost due to erosion versus the respective PI values. Vulnerability is defined as the rate of change in productivity, measured by the changes in the PI value, by unit of soil removed (Pierce et al., 1984).

The modified method to estimate tolerance consists in applying the following equation:

$$PI_f = PI_0 (1-\delta)$$

where: PI_f = Final productivity index, after soil removal
PI_0 = Initial soil productivity index
δ = Permissible productivity loss (decimal-fraction)

With the value of PI_f on the respective vulnerability curve, the corresponding amount of soil loss (cm) is obtained, which, when divided by a previously selected planned horizon value (H, years) allows the calculation of tolerance (T, cm.year^{-1}). Knowing the values for bulk density (Mg.m^{-3}), tolerance can be expressed in Mg.ha^{-1}. The values δ and H are assumed as related to the needs and premises adopted by the planner of land use and soil conservation. Normally, δ varies between 0.05-0.10, and H is generally assumed to be 100 years.

3 Results

In the Western Plains and Piedmont sub-regions, the analysis of variance for PI showed highly significant differences ($p<0.01$) between the soil series for the dry year and significant differences ($p<0.05$) for the wet year, but no significant differences between the two years. This is explained by the fact that PI is a value which depends on soil characteristics but not on climate, which is the changing element between one year and another. The Duncan multiple range test confirmed this. As concerns the DRD variable, the analysis of variance showed highly significant differences ($p < 0.01$) between the dry year and the wet year, indicating that in fact, different moisture conditions between one year and another give different root growth patterns. Likewise, the analysis shows highly significant differences ($p<0.01$) between soil series, and significant differences ($p < 0.05$) for the interaction year with soil series, which suggests that DRD is closely related to the type of soil but with different reactions as a result of humidity conditions in the respective year. The Duncan multiple range test confirmed that DRD averages were significantly different for the two years, emphasizing the fact that in the dry year radical development was significantly greater than in the wet year, a circumstance attributable to an ecophysiological reaction of the crop, which mobilizes reserves for the growth of roots in search of the best soil humidity conditions. Moreover, semi-logarithmic regression curves with highly significant determination coefficients ($p < 0.01$) were adjusted between PI values of the soil and DRD of the crop, between PI values of the soil and corn yield (Y) and between DRD values and corn yield (Y), which shows a close relation among these three variables (Table 5).

Variables	Year	Regression Equation	r^2	n
PI vs DRD	Dry	GDR=8.04 + 3.39 Ln (PI)	0.886	15
	Wet	GDR=6.26 + 2.27 Ln (PI)	0.704	9
PI vs Y	Dry	Y = 4950.8 + 1268.4 Ln (PI)	0.799	15
	Wet	Y = 6027.9 + 1985.6 Ln (PI)	0.798	9
DRD vs Y	Dry	Y = 1890.3 + 1307.6 Ln (DRD)	0.809	15
	Wet	Y = 347.16 + 2727.1 Ln (DRD)	0.798	9

DRD: degree of radical development
PI: productivity index
Y: yield (kg.ha^{-1})

Table 5: Regression equations for variables related with soil productivity and root development. Andean Piedmont and Western Plains soils. Crop: corn (Zea mays L.)

In the Andean Mountains sub-region, the soil studied showed important variations in the rooting area, as the result of the removal of superficial matter. The effect of such an alteration is seen in a significant reduction in the yield of the indicative crop as superficial soil is removed (Table 6). The application of the model to this case indicates, that the initial productivity of this soil is high and that, as superficial soil is removed, productivity values decrease even though they remain between high and moderate (Table 6). This indicates that the soil has low vulnerability.

Level of Management	Sample replica n	Level of Removal (%)	Productivity Index	Yield (kg·ha^{-1})	Tolerance*(T) (Mg·ha^{-1}·year^{-1})
1	1	0	0.427	1373	13.87
	1	50	0.398	736	14.94
	1	100	0.396	625	30.16
	2	0	0.421	1206	14.54
	2	50	0.399	1086	15.62
	3	0	0.423	1722	12.60
	3	50	0.410	1201	16.77
	3	100	0.415	941	29.39
2	1	0	0.397	1549	9.98
	1	50	0.372	782	15.15
	1	100	0.362	560	34.36
	2	0	0.398	736	11.32
	2	50	0.386	910	17.33
	2	100	0.376	681	32.58
	3	0	0.437	1571	14.12
	3	50	0.417	1562	18.80
	3	100	0.356	435	35.22

* For $\delta = 0.05$ and $H = 100$ years

Table 6: Productivity indexes, yields and tolerances for different levels of management and superficial soil removal. Andean Mountain soils. Crop: black beans (Phaseolus vulgaris L.)

The regression analysis shows that there is an acceptable correlation (r) between the variables PI and bean crop yield (Y) for the adjusted regression models. In both cases: level of management 1 and level of management 2 (Table 7), the regression models explain more than 60% (management 1) and 70% (management 2) of the variations observed (r^2).

Level of Management	Regression equation	r^2	n
1	Y= 9397.28 + 9318.01 Ln (PI)	0.62	8
	Y= -22926 + 94343.7 (PI) - 87196.8 (PI)^2	0.62	8
2	Y= 6656.7 + 6003.67 Ln (PI)	0.76	9
	Y=-14736.4+64986.7 (PI) - 62969.5 (PI)^2	0.77	9

Table 7: Regression equations for the variables Productivity Index (PI) vs Yield (Y) (kg.ha^{-1}), for two levels of management. Crop: black beans (*Phaseolus vulgaris* L.)

The previously adjusted logarithmic and quadratic regression models generally explain the relations established inasmuch as yield increases as the soil offers better condictions (high PI) for the development of the crop. However, they tend towards an asymptote due to the other factors that control yield such as the genetic potential of the crop and the climate.

4 Conclusions

The results obtained allow us to conclude that the *Productivity Index*, structured as related to the variables and equations contained herein, can be used to evaluate the relative productivity of soils in the regions under study. The index is defined as related to the characteristics and dynamic properties of the soil, subject to changes, modifications and alterations originated fundamentally by the management of the resource. For this reason, the model allows the analysis of the effects such changes have on productivity, eventually producing information on the impact of soil degradation processes on productivity. Even though the PI depends essentially on the characteristics of the soil, the relations between this value and yield vary as the result of climate, management and the genetic characteristics of the crop. Variations in these parameters can noticeable change the expected results.

In relation to the DRD, the results obtained show that it can be used as a good indicator of the degree of root growth and development, under given environmental conditions. The proposed measurement method, a "bounding box", is practical and relatively simple, offering conditions favorable to *in situ* plant root system study, as well as simultaneous sampling of soils and roots.

With reference to the evaluation of the erosion-productivity relations, the results obtained confirm the generally accepted fact that the relative productivity of soils and their rate of change due to loss of matter in the superficial horizon, depends on the presence of characteristics favorable to the development of the root system of the crop in the soil profile. This is supported by the fact that it was observed in these experiments that: a) productivity decreases with the loss of soil, even though management practices are applied to counteract this; b) there is a good correlation between edaphic variables and soil productivity, a relation that varies according to climate, crop and management.

The application of the PI model proposed in this study gave satisfactory results for the evaluation of the erosion-productivity relations, even though a more extensive validation of this model is required. Given the variety of edaphoclimatic conditions and agro-ecological systems present in tropical mountain zones at considerable hazard of erosion, the results obtained encourage the use of the PI model for preliminary evaluation of the impact that soil loss could have on the productivity of these Andean Mountain soils when sufficient information is not available for the evaluation of erosion-productivity relations.

References

Delgado, F. (1995): Un índice de productividad para la evaluación de suelos agrícolas en la región de Piedemonte y Llanos Occidentales de Venezuela. CIDIAT. Mérida, Venezuela.122 p.

Delgado, F. and López, R. (1995): Validación de un modelo erosión-productividad en suelos de los Andes Venezolanos. XIII Congreso Venezolano de la Ciencia del Suelo. Maracay, Venezuela.

Kiniry, L., Scrivner, C. and Keener, M. (1983): A soil productivity index based upon predicted water depletion and root growth. Research Bulletin 1051. University of Missouri, Columbia, USA. 26 p.

López, R. (1994): Factores y efectos de la erosión hídrica en suelos de los Andes Venezolanos. CIDIAT. Mérida, Venezuela.124 p.

Neill, L. (1979): An evaluation of soil productivity based on root growth and water depletion. M.S. Thesis. University of Missouri, Columbia, USA.

Pierce,F., Larson, W., Dowdy, R. and Graham, W. (1983): Productivity of soils: assessing long-term changes due to erosion. Journal of Soil and Water Conservation **38(1)**: 39-44.

Pierce, F. J., Larson, W.E., Dowdy, R.H. and Graham, W.E. (1984): Soil productivity in the corn belt: an assessment of erosion s long-term effects. Journal of Soil and Water Conservation, **39 (2)**: 131-136.

Rijsberman, F., and Wolman, M. (1985): Effect of erosion on soil productivity: an international comparison. Journal of Soil and Water Conservation, **40**:349-354.

Addresses of authors:
Fernando Delgado
Roberto López
CIDIAT
Parque la Isla
Apartado 219
5101 Merida
Venezuela

A New Digital Georeferenced Database of Soil Degradation in Russia

V. Stolbovoi & G. Fischer

Summary

Information on human-induced soil degradation in Russia has now been compiled in a new digital georeferenced database. It comprises the latest data on the status of soil degradation in Russia, including soil deterioration in non-agricultural regions. The information has been linked to a digital soil database, which has recently been prepared for the FAO by the Dokuchaev Soil Institute. Soil degradation attributes were derived from unpublished maps compiled for the State Committee on Land Resources and Land-Use Planning of Russia. The analysis shows that more than 14.5% (243 million ha) of the Russian territory is affected by soil degradation caused by a variety of reasons, including socio-economic changes, and improper management and technology. The assessment reveals that the rate of soil degradation and loss of soil productivity in Russia has been fairly rapid.

Keywords: Soil degradation, land use, soil database, GLASOD

1 Introduction

Expanding populations and economic development have generated a growing demand for various land-based products, leading to increasing pressure on soils, water resources, and plants. In developing and developed countries, this pressure can exceed critical thresholds and requires land managers to face problems of deteriorating land resources, declining productivity and consequently reduced income. Maintenance of the productive potential of land resources, and checking of land degradation, is a fundamental element of sustainable land use (Pieri, et al., 1995).

The first attempt to combine soil degradation data collected by different ministries and institutes of Russia was undertaken by Dokuchaev Soil Institute in 1988-89 in the frame of the project on Global Assessment of Soil Degradation (GLASOD) (Oldeman, et al., 1991) Since then[1] numerous publications concerning negative human impacts on soil have appeared in scientific and public journals describing types of degradation, their nature, severity, rate of change, extent, consequences, etc. The basic data were collected and published in Government (national) reports on the status and use of land in Russia (Government Report, 1993).

The GLASOD assessment for the Russian territory was based on data of varying quality, ranging from well-documented sources (i.e., on soil erosion) to assessments based on expert opinion (i.e., acidification). Also, the project was limited to degradation of agricultural lands. Thus, several other widespread forms of soil deterioration taking place in Russian forests and permafrost areas were not considered. Another disadvantage resulted from the fact that GLASOD aimed to compile a degradation map "manually". This led to many cartographic restrictions, generalization and loss of collected information presented in tabular and paper formats. There was an enormous discrepancy between the amount of soil degradation data collected and their acceptability and practical application.

---

[1] soil degradation was not widely discussed before, as officially at that time the Former USSR did not have widespread ecological problems.

The project *Modeling Land-Use and Land-Cover Changes in Europe and Northern Asia (LUC)*, established by IIASA in 1995, recognizes the importance of soil degradation as one of the driving factors in land resources alteration. The present georeferenced database on soil degradation for Russia is the result of collaborative efforts by the LUC project, the Dokuchaev Soil Institute, and the State Committee of Russian Federation on Land Resources and Land-Use Planning.

2 Objectives

The objectives of this paper are twofold:
- to introduce a new georeferenced database on soil degradation in Russia;
- to overview the soil degradation status in Russia.

2.1 The georeferenced database on soil degradation in Russia

In the georeferenced database on soil degradation in Russia, the spatial information is represented by the mapping units of the updated FAO soil map of Russia (Stolbovoi et al., 1995). The updating was contracted by the FAO and the map was compiled by the Dokuchaev Soil Institute in Moscow on the basis of the latest *Soil Map of the Russian Soviet Federative Socialist Republic* at scale 1:2.5 M (Soil Map, 1988). In total, 1295 mapping units were created.

The attribute database consists of two parts (Fig. 1). The first set of attributes contains general soil information derived from the soil database and included soils, soil phase, slope and texture classes, according to FAO's Revised Legend of the Soil Map of the World (FAO-Unesco, 1988).

The second set of attributes relates to soil degradation. It was created by coding information from several unpublished paper maps into digital format (Fig. 2). These maps had been contracted by the State Committee of Russian Federation on Land Resources and Land-Use Planning, and compiled by different authoritative organizations for the Government (national) report on the *Status and Use of Land in the Russian Federation* (Government Report, 1993). The following source maps were used:
- *Map of Soil Water and Wind Erosion in Russia*, at a scale of 1:4 M, compiled by the Dokuchaev Soil Institute, 1992;
- *Map of Recent Land Status of Forest Fund of Russia*, at a scale of 1:4 M, compiled by the All Russian Research Institute of Forest Resources, 1993;
- *Map of Natural Grassland Degradation in Russia*, at a scale of 1:4 M, compiled by the All Russian Research Institute of Fodder, 1992;
- *Map of Soil Salinization in Russia,* at a scale of 1:4 M, compiled by the Dokuchaev Soil Institute, 1992.

Soil degradation attributes were compiled following the *Guidelines for the assessment of the status of human-induced soil degradation in South and Southeast Asia (ASSOD)* (Lynden, (ed.), 1995). The georeferencing of degradation attributes was accomplished by overlaying the polygons from the soil map with each of the maps listed above.

The database contains soil degradation attributes describing type and extent of degradation, and indicating the severity of impacts on productivity, rate of change, causative factor, and rehabilitation or protection measures. Degradation types include water and wind erosion, secondary salinization, desertification, underfloods and compaction. For forest areas, two additional types of soil degradation were distinguished: disturbances of organic horizons caused by industrial cutting, and disturbances caused by fires. To account for the specific forms of degradation in permafrost areas, the distribution of thermokarst and surface corrosion were shown. These two types of soil degradation are mainly caused by overgrazing of deer pastures and by industrial activities such as oil drilling.

Digital georeferenced database, soil degradation, Russia

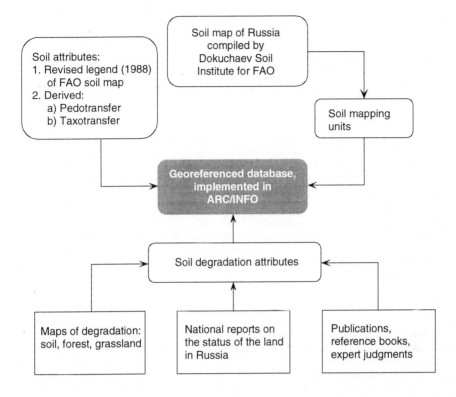

Figure 1. Principal elements and basic sources of the database

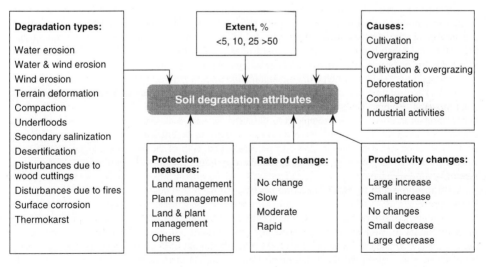

Figure 2. Composition of soil degradation attributes.

2.2 Soil degradation in Russia

When starting with the analysis of soil degradation of a region it is important to take into account the general features of land use (Table 1). This gives an idea of the variety of human impacts on soils, and can help identify and understand the main causative factors.

The cultivated areas are located in the densely populated west European region of Russia and the southern part of the west Siberian plain. Forest land is located both in the European (30% of total forest area) and Asian (70%) parts of Russia.

The total extent of land in Russia affected by soil degradation is estimated at 243 million ha, i.e., 14.4% of the area covered by soil (Table 2). Most soils stable under natural conditions (Table 2) have been formed under undisturbed vegetation (1266 million ha), primarily forests, forest-tundra and tundra. There are some stable soils (30.6 million ha) that have developed under poor vegetation in desert areas, high mountain zones, etc. The third group includes soils that are stable due to human influence (141.8 million ha). The latter comprises soils that are stable due to regulating human influence (reserves, protected areas).

Cultivated land in Russia occupies some 130.7 million ha (Table 1). Almost half of it (63.3 million ha) is affected by various degradation processes caused by cultivation. Soil compaction is the most widespread type of soil degradation influencing agricultural land. Compaction refers to soil conditions with increased bulk density exceeding undisturbed levels by more than 1.2 times. It has been assessed to occur on more than one fourth of the cultivated land (33.2 million ha), mainly in the European part of Russia (Fig. 3). For the first time soil compaction has been widely identified in the northern tundra deer pastures (Fig. 3), caused by overgrazing.

Water and wind erosion are the second most wide-spread types of soil degradation occurring on cultivated land. The total extent suffering from erosion is estimated at 25.8 million ha (Table 2), i.e., approximately one fifth of cultivated land. It occurs in the agricultural regions of both the European and Asian parts of Russia (Fig. 4). The geographical distribution of different types of soil erosion varies with climate aridity. The database indicates (Table 3) that protection measures, comprising land and plant management practices, have been implemented on practically the entire area affected by erosion. However, the effectiveness of these measures appears to be rather low, as can be concluded from the considerable rate of erosion.

Irrigated soils are often influenced by secondary salinization. This type of soil degradation refers to salt accumulation in the upper part of the soil profile resulting from evaporation of irrigation groundwater in the capillary fringe. The estimated extent of secondary saline soils is 3.5 million ha. Protection and rehabilitation measures are used on 3.2 million ha, i.e., some 0.3 million ha of soil affected by secondary salinization are not covered by any protection or rehabilitation activity.

Desertification is the main type of soil degradation affecting moisture deficit zones in steppe, dry steppe, and semi-desert regions. This refers to expansion of desert areas as a result of natural and anthropogenic factors. Desertification extends over 35.7 million ha. It includes a mixture of degradation processes such as compaction, deflation, loss of soil structure, decline of soil water holding capacity, etc. The primary human causative factor of desertification is overgrazing. The rate of desertification has been mostly slow and moderate. One fifth of the area is adjudged a rapid rate, caused by a high degree of human intervention and fragile natural conditions (soil texture and moisture, wind speed, etc.).

Overgrazing is also the main cause of other types of rangeland degradation. In tundra areas it can trigger processes of surface corrosion and thermokarst in gelic soils (solifluction, landslides, etc.). These two types of soil degradation are found (Fig. 5) in the north of the European and West-Siberian parts of Russia as well as in East Siberia and the northern Far East. The total area covered by permafrost is estimated to be more than 1100 million ha (about 65% of the entire Russian territory). Surface corrosion was recorded to occur on 60.2 million ha and thermokarst on 31.2 million ha (Table 2).

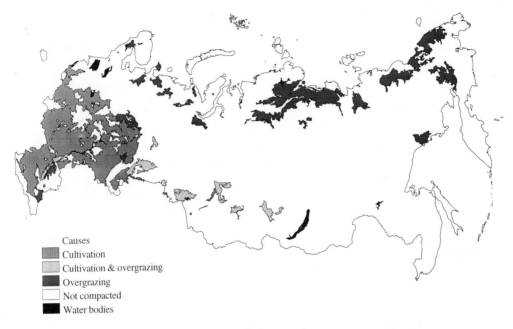

Fig. 3: Soil compaction

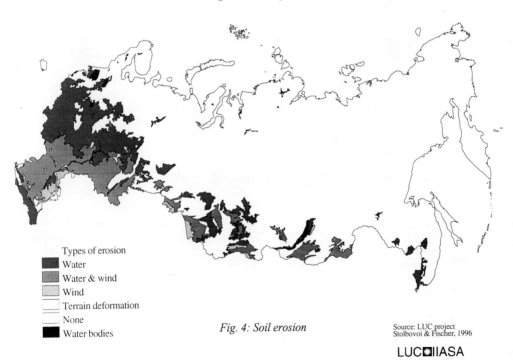

Fig. 4: Soil erosion

Source: LUC project
Stolbovoi & Fischer, 1996

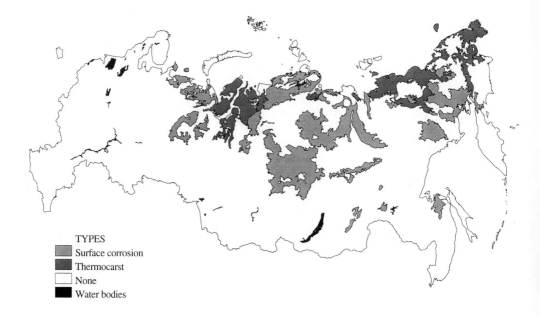

Fig. 5: Degradation of gelic soils

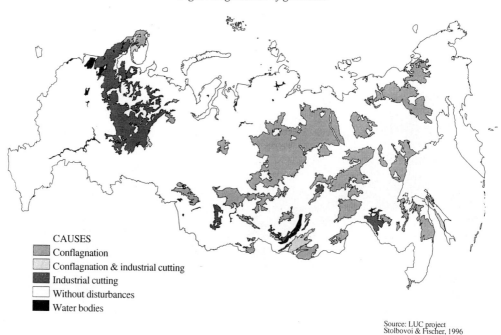

Fig. 6: Disturbances of organic horizons of forest soil

The forested area in Russia is 784.7 million ha (Table 1). In these areas, disturbances of the soil organic horizon caused by fires and industrial wood cutting are the main degradation types identified in the database, indicating insufficient forest management and application of inadequate technology.

The extent of disturbances of the soil organic horizon caused by fires during a 10-year period is estimated at 15.4 million ha, or about 2% of the total forested area. The primary cause of forest fires (90%) is human carelessness (see: Land of Russia, 1995). Fires occur mainly in the Asian part of the country, where 40% of the forest is not protected against fire.

Disturbances of the soil organic horizon caused by industrial wood cutting result from the application of heavy machinery, mechanized skidding, etc. It refers to loss of organic and mineral topsoil. The total extent of such disturbances in Russia is estimated at 10 million ha (1.3% of total forested area), widely spread in the European part of Russia (Fig. 6), where clear-cutting is practiced over the whole year.

3 Conclusions

1. A georeferenced soil degradation database covering the entire territory of Russia now allows one to assess the status of human-induced soil degradation country-wide. The database is available for application to scientific analyses, practical recommendations, and decision- and policy-making. One application is the analysis of actual productivity and management of soils. It contributes to the understanding of the spatial distribution, causes, and rates of soil degradation, and thus provides information relevant for optimizing the use of soil resources and for sustaining agricultural production.
2. The total extent of soil degradation in Russia is estimated at 243 million ha, or 14.5% of the country. Almost 48% of the cultivated land is affected by various degradation processes. This means that soil degradation is widely spread in Russia.
3. The rate of soil degradation processes over the past 5-10 years has been rather rapid. Apparently the effectiveness of soil protection measures has been low. Thus, it can be expected that the recent trend towards further soil degradation and loss of soil productivity in Russia may continue. This undesirable development calls for proper policy-making.

References

FAO-Unesco. Soil Map of the World. Revised Legend (1988): World Resources Report No. 60, FAO, Rome.
Fridland, V.M. (ed.) (1988): Soil Map of the Russian Soviet Federative Socialist Republic at scale 2.5 M All Union Academy of Agricultural Science (GUGK), USSR, 16 sheets.
Government (National) Report on the Status and Use of Land in the Russian Federation (1993): Moscow.
Land of Russia (1995): Problems, Figures, Commentaries, 1996. Moscow.
Lynden, G.W.J. van (Ed.) (1995): Guidelines for the Assessment of the Status of Human-Induced Soil Degradation in South and Southeast Asia (ASSOD). ISRIC, Wageningen, The Netherlands.
Oldeman, L.R., Hakkeling, R.T.A. and Sombroek, W.G. (1991): World Map of the Status of Human-Induced Soil Degradation. An Explanatory Note, second revised version. Global Assessment of Soil Degradation.
Pieri, C., Dumanski, J., Hamblin, A. and Young, A. (1995): Land Quality Indicators. World Bank Discussion Paper, **315**, 63.
Stolbovoi, V. and Sheremet, B. (1995): New Soil Map of Russia, compiled in FAO System. Pochvovedenie No. 2, 149-158. (in Russian).

Addresses of authors:
Vladimir Stolbovoi
Günther Fischer
International Institute for Applied Systems Analysis (IIASA), Schlossplatz 1
A-2361 Laxenburg, Austria

APPENDIX

Land categories	Land use by owners (million ha)							
	Agricultural enterprises	Forest enterprises	Urban	Industrial	Land of water fund	Protected areas	Reserves	Total	% of total
Cultivated	122.9	0.2	5.9	0.2	-	-	1.5	130.7	7.6
Other agricultural	63.7	3.6	18.5	1.0	0.0	0.4	3.9	91.1	5.3
Forest and shrubs	134.4	620.5	5.0	3.8	0.1	13.0	7.9	784.7	45.9
Deer & horse rangeland	253.3	60.1	0.0	0.1	-	1.7	12.8	328.0	19.2
Infrastructure	3.6	1.7	5.0	3.0	0.0	0.1	0.0	13.4	0.8
Other land	54.4	62.5	1.7	8.3	0.4	9.1	45.5	181.9	10.6
Swamps	15.8	77.1	1.6	0.4	0.7	1.6	11.0	108.2	6.3
Water bodies	19.6	12.9	0.9	0.8	18.2	1.4	18.0	71.8	4.2
Total	667.7	838.6	38.6	17.6	19.4	27.3	100.6	1709.8	
% of total	39.1	49.0	2.3	1.0	1.1	1.6	5.9		100.0

Table 1. The structure of land use in Russia, (from Land of Russia, 1995)

Degradation type	Extent[1]		Causative factors (million ha)					
	Million ha	% of soil area	Cultivation	Overgrazing	Cultivation & overgrazing	Deforestation	Conflagration	Industrial activities
Water erosion	12.8	0.76	12.8					
Water & wind erosion	8.4	0.50	8.4					
Wind erosion	4.6	0.27	4.6					
Terrain deformation	1.7	0.10		1.7				
Compaction	58.3	3.50	33.8	19.3	5.2			
Underfloods	0.9	0.05						0.9
Secondary salinisation	3.5	0.21	3.5					
Desertification	35.7	2.14	0.2	35.4				0.2
Disturbances of soil organic horizon due to cuttings	10.0	0.60				10.0		
Disturbances of soil organic horizon due to fires	15.4	0.92					15.4	
Surface corrosion[2]	60.2	3.61		60.2				
Thermokarst	31.2	1.87						31.2
Total area affected by degradation	242.7	14.44	63.3	116.6	5.2	10.0	15.4	32.3
Stable soil								
Stable due to human influence	141.8	8.43						
Naturally stable by vegetation	1265.5	75.30						
Naturally stable without vegetation	30.6	1.82						
Total stable	1437.9	85.56						

[1] the discrepancy in total extents of land between Table 1 and Table 2 is mainly due to some differences between the statistics and extents of mapped inland water bodies
[2] natural and human-induced degradation are combined

Table 2. Extent and causes of soil degradation and stable soils in Russia

Degradation type	Protection measures				Productivity decrease				Rate of degradation change			
	Land management	Plant & land management	Plant management	Others	Small	Moderate	Large	No change	Slow	Moderate	Rapid
Water erosion	12.8				0.1	12.7			2.8	8.5	1.5
Water and wind erosion		7.9			0.6	7.8			0.8	7.5	0.1
Wind erosion		4.6			0.2	1.9	2.4		0.2	1.7	2.8
Terrain deformation							1.7				1.7
Compaction					43.0	15.3			10.8	25.0	22.6
Under floods						0.9			0.9		
Secondary salinisation				3.2	2.9	0.6		1.0	1.6	0.4	0.4
Desertification					23.6	12.2		1.0	13.4	14.1	7.2
Disturbances of soil organic horizon due to cuttings					10.0					10.0	
Disturbances of soil organic horizon due to fires					n.a.	n.a.	n.a.				15.4
Surface corrosion	0.5	0.6	2.3				60.2¹			60.2	
Thermokarst							31.3				31.2
Stable due to human influence	2.2		139.5								
Total	15.5	13.1	141.8	3.2	80.4	51.4	95.6	1.7	30.5	127.4	82.9

¹ natural and human induced degradation are combined

Table 3. Protection measures, productivity decrease and rate of soil degradation in Russia (million ha)

Developing Land Resource Information For Sustainable Land Use in Albania

P. Zdruli, R. Almaraz & H. Eswaran

Summary
Soil erosion, deforestation and overgrazing are perhaps the most serious environmental problems in Albania today. Estimates for soil erosion are from 32 to 185 tons per hectare per year. About 200,000 ha (out of 700,000 ha of total agricultural land), located in the flat zones are affected by nutrient mining, seriously reducing soil fertility. Saline areas are increasing in the coastal plains and flooding is frequently damaging urban areas and putting public health at risk. Site specific stresses, such as heavy metal contamination, urbanization, random disposal of urban waste and loss of biodiversity are yet to be quantified, however, evidence is readily seen. Overall, roughly 85 percent of the land resource stresses are of Anthropogenic nature, which implies that the trend for the future may be directed either towards aggravation of the situation or improvement.

This preliminary work is intended to show the most critical factors affecting the sustainability of agriculture and the status of natural resources.

Keywords: Soil resources, land degradation, erosion, deforestation, resalinization, Albania

1 Introduction

With more than 65 percent of the population living in rural areas, agriculture largely remains the "back bone" of Albania's economy. However, its potential has not been reached and, moreover its long-term stability is yet to be assured. Land resources are limited and agricultural land is finite. The only chance for increasing production must come from existing arable land. This requires good stewardship for the natural resources and strengthening of the support services needed to maintain a sustainable agriculture.

Following privatization, the farming community was left to manage on its own. The work of research and extension institutions, which are responsible for ensuring sustainability of agriculture, has been significantly reduced due to lack of funds. Albania must maintain an adequate incentive framework, reform the financial sector and effectively manage natural resources in response to rapid deterioration of land, water and marine resources. If action is not taken now, the consequences will be continued and increasing reliance on imported agricultural products, little or no earnings from agricultural exports, an increasing water quantity and quality problem, reduced environmental quality with negative effects on biodiversity, a permanent threat to food security, and an increasing gap between the rich and the poor.

Until the early 90's, the only soil classification system used in Albania was a locally-developed method adopted from the Russian system. Modern systems such as the Legend of the Soil Map of the World (FAO-Unesco, 1990) and Soil Taxonomy (Soil Survey Staff, 1996) were unknown. Since 1992 the demand for soil resource information has increased, particularly by foreign organizations working in the country.

During a three year period (1994-1996), the International Fertilizer Development Center (IFDC) of Muscle Shoals, Alabama, the US Department of Agriculture, Natural Resources Conservation Service (USDA-NRCS) and the Soil Science Institute (SSI) of Tirana in Albania, were engaged in a field study initiative in the country. Funding for this activity was provided by the US Agency for International Development (USAID). A total of 30 profiles were sampled and described in detail and hundreds of auger observations were done throughout the country. Soils were analyzed at the Laboratories of NRCS in Lincoln, Nebraska. The results of the study were reported by Zdruli (1997) in the monograph "Benchmark Soils of Albania", the main purpose of which was to:

- compile all available soil information into one source;
- translate the 1958 soil map of Albania into Soil Taxonomy;
- provide detailed and modern characterization data of representative soils; and
- analyze available information and present future needs and activities to support sustainable agriculture.

Based on land use considerations, each polygon of the soil map was assigned with the most constraining attribute and a new map was generated using a Geographic Information System (GIS). The spatial and tabular information enabled the evaluation of the major land resource stresses for the country as a whole.

2 Land resources

Albania has a wide range of climates that permit a great variety of land uses. The southern and western parts have a Mediterranean climate with wet and mild winters and warm and dry summers, and a temperate continental climate in the eastern parts. In the northern sub-alpine regions and eastern mountain ranges, it is cold and humid with severe winter conditions.

Located in a tectonically active zone, many of the mountains are young with steep slopes and near-vertical fault scarps. Albania is one of the few countries in the world where large areas of ultrabasic rocks, represented by dunites, peridotites and pyroxenites occur. However, soils that are formed on these rocks are hyper-magnesic and apart from having toxic levels of heavy metals, are also deficient in molybdenum, boron, sulfur and other trace elements.

According to Soil Taxonomy (Soil Survey Staff, 1996) six soil orders have been identified in the country. Inceptisols occupy the largest areas accounting for about 35 % of the nation's territory. Alfisols cover the second largest group of soils and almost 60 %, dominantly Ustalfs and Xeralfs, occur in areas with thermic soil temperature regimes (STR). Mollisols predominate on the flat to undulating landscapes and are under increasing pressure from agriculture and grazing. They make up about 7 % of the country's territory and are found in areas with different kinds of STRs. Histosols, Vertisols and Entisols occupy about 8 % of the total land area. Miscellaneous soil unit makes up to 32 % of the land mass and includes a variety of soils, rock and bodies of inland water. This is quite a large area, which could be used for different purposes, including mining, forestry, and grazing or as protected natural resources.

3 Land use

Only 25 % of the total land area of Albania is arable; the rest is mountainous with few opportunities for farming. The majority of agricultural land is found in the western coastal plain and in lowlands. Of the 700,000 ha of arable land, about 580,000 ha were cropped annually up to 1990, about 60,000 ha were under fruit trees, 40,000 ha under olive trees and 20,000 ha of vine-yards. Despite its generally fertile soils, the productivity levels of Albanian agriculture, even in its best years, have been far below potentials. This was due mostly to lack of technology and inputs.

The trends and forecasts for short and medium term growth are that farmers will grow less wheat than in the 1980's with consequences on total national production. If processing capacity is restored, the production of agro-industrial crops such as sugarbeet, sunflower and cotton, could increase significantly. Tobacco, potato, fruit and vegetable production is also likely to increase as well as production from greenhouses, however this will depend on private investments and markets (World Bank-European Community, 1992). Albania has excellent possibilities for increasing production of cash crops in the medium and long terms.

4 Land resource stresses

Degradation of the natural resource base has increased in recent years and the consequences are affecting the sustainable development of the country. In certain cases, the environmental damage has led to a critical situation. (Ministry of Health and Environment, 1995). Zdruli et.al., (1997) have also shown that the natural resources of the country are under stress and subject to mismanagement.

Degradation of natural resources deals with two interlocking complex systems: the natural system and the human-induced system. The interaction between the two systems determines the success or failure of resource management programs. Land degradation is increasingly threatening the quality of natural resources with direct impacts on sustainability of agriculture and eventually, to the quality of life (UNDP, 1995). The degradation processes are:
1. human-induced, including: chemical pollution, salinization, nutrient mining of agricultural land, deforestation, overgrazing, and accelerated soil erosion.
2. natural processes or conditions, including: acidification, flooding, stony and shallow soils, and areas of low temperatures and poor accessibility.

Human-induced degradation results from mismanagement and requires both mitigating technology and also a societal commitment through stewardship and awareness. It is estimated that about 85 % of the country's territory is under human-induced degradation stresses. Figure 1 shows the spatial distribution of all the land resource stresses nationwide and Table 1, gives the areas of each.

Soil erosion, deforestation and overgrazing are perhaps the most serious environmental problems in Albania today. Estimates for soil erosion show values from 32 tons per hectare up to 185 tons per hectare per year (World Bank, 1993), however final scientific data are yet to be published. Another form of degradation are the landslides, found in many areas throughout the country, greatly reducing not only the sustainability of agriculture but also the long term security of rural housing.

Deforestation had started hundreds years ago, when most of the oak forests of the country's southern part were cut to build the merchant ships of old Venice, and later because of the burning of wood for fuel, and the expansion of villages onto hillsides. The country's high dependency on fuel wood for heating - amounting to 100 % of household energy in rural areas and over 80 % in the cities - contributes to the over-exploitation of forests.

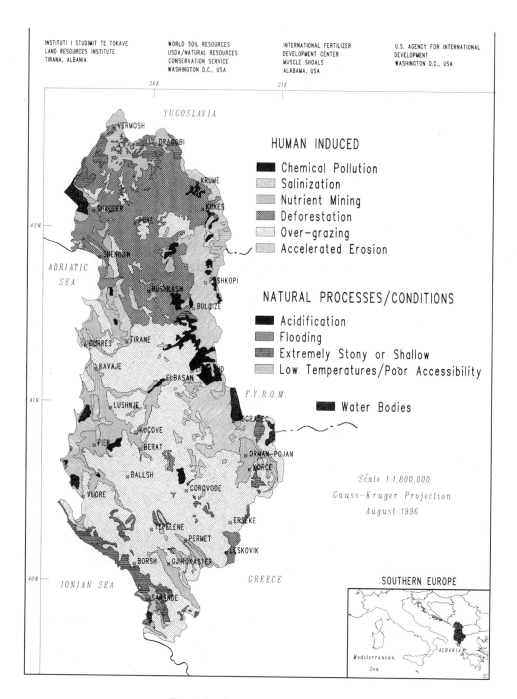

Fig. 1: Land resource stresses in Albania

Land resource stresses	Area in km²	%
Human induced		
Chemical pollution	165	0.6
Salinization	654	2.3
Nutrient mining	2,936	10.2
Deforestation	5,005	17.4
Overgrazing	12,202	42.7
Accelerated erosion	3,585	12.4
Natural processes and conditions		
Acidification	624	2.0
Flooding	461	1.6
Extremely stony or shallow	1,614	5.6
Low temperatures/poor accessibility	1,182	4.1
Water	320	1.1
Total	28,748	100

Tab. 1: Land resource stresses in Albania

FAO has made an assessment of forest change between 1988 and 1993 for over 10 % of the total country's forest cover in two representative areas and found that the area covered by forests had decreased by an average of 15 % over a five year period. (World Bank, 1996). Forest fires are becoming more frequent and add to the damage.

The land reform that followed the chaos of 1991 left the pasture lands under state ownership, therefore damage was far greater than on private lands. Recent data show an increasing number of livestock herds (Henao, 1994), which compete for food, therefore it is expected that the pressure on pastures will increase and *overgrazing* will continue to threaten them.

In the mid 1970's, the government began a program of converting pastures, once used for summer grazing, into cropland production. This resulted in a rapid decrease of their natural fertility and accelerated soil erosion. Those lands are currently no longer used for crops nor for grazing. Both forest and pastures make up to 50 % of the natural resources, representing an extremely large area with high potentials for the development of the country, if managed properly.

Saline soils are located on the western lowland coastal plains and are under a Mediterranean climate. In 1995, evidence of increasing salinization was found in Ishull Shëngjin (Lezhë), Novoselë (Vlorë) and the trend is continuing. The areas represented in Fig. 1 within this context are at the moment, not necessarily salty, but they have high potentials to become saline if good management and maintenance of drainage canals and pumping stations are ignored. Another form of salinity is represented by the excess of magnesium cations, as found typically in the smonitsa (Chromic Haplusterts) soils. They make up about 10,000 ha, mainly located in the north and southeast.

Flooding is increasingly becoming a problem especially in the northwest. In 1996, the area of Lezha was flooded four times through September. A chain reaction due to overgrazing, deforestation and erosion, culminates in flooding, which is also accelerated by the poor maintenance of drainage canals and pumping stations. Waterlogging is reducing yields in low lying areas and the reverse phenomena of swamp and marsh formation is becoming evident. A side effect of erosion is also the siltation of the reservoirs.

Acid soils are mainly located in the northeastern regions and to a limited extent in the southeast of the country. During the old regime some investments were made by subsidizing lime use by collective farms. Today this program is discontinued because of lack of funding and population migration. Since acidification is a natural process influenced by the parent materials and the climate of the area, there is concern that the acid soils will expand, even in those lands that were considered ameliorated a few years ago.

Albania has a relatively short history in fertilizer use and very long tradition in agriculture that dates back more than 2,000 years. Soils have been exploited systematically and have received little inputs from other sources, other than some manure. Even during peak years, the country had used less fertilizers than almost any other nation in Eastern Europe. Nevertheless, in the early 1990's the agricultural sector experienced a drastic fertilizer shortage. It is believed that most of the best agricultural lands in the country today are under *nutrient mining* conditions with decreasing soil fertility. Data from the Soil Science Institute (Qilimi, 1996) show that compared to 20 years ago there is a decrease mainly in organic matter content, nitrogen, and potassium.

Certain areas within the country are affected by several kinds of typical land resource stresses. They can be summarized as follows:
- conversion of high quality farmland to urban land;
- point-source contamination from mining activities, factory wastes and effluents;
- uncontrolled and random disposal of urban waste on agricultural land;
- heavy metal contamination from natural and human activities; and
- loss of biodiversity.

Some estimates (UNDP, 1995) for the urbanization of the surrounding areas of Tirana show that the land lost to unplanned construction is one and a half times larger for the same number of inhabitants, than if the home-construction would have been planned. Agriculture land contaminated by oil and gas industry, factory wastes and effluents is roughly 50,000 ha. Air and water pollution by the processing of copper ore is one of the greatest risks for public health and the quality of the environment in general in the northeastern regions. In some areas of Kukës and Rubik, within a radius of three kilometers the natural vegetation has been destroyed and crop production greatly reduced. Other areas include the vicinity of metallurgical complex in Elbasan and Bulqizë. In the latter the waste disposal site of the chromate factory was placed at a higher elevation of the watershed. Discharge from the site contaminates the whole valley.

5 Concluding remarks

The land resource stresses described above are a first step in making an assessment of the land degradation status in Albania. This study shows clearly that more than 85 % of the nation's territory is estimated to be under pressure of human induced degradation factors.

Albania is blessed with fertile soils, but at the same time, the total amount of arable land is finite and may not be able to support the increases in population expected in the coming years. Now that individuals have ownership of the land, it depends largely on them to provide the necessary stewardship. However, they will need assistance from the government, as for many of them, this is a new experience. To reduce the misuse of land and enhance sustainability in the country, the national decision makers must embark on a program providing:
- detailed land resource assessments;
- monitoring of land and water quality;
- awareness among land users regarding environment;
- research and support services to assist land users;
- implementation of sustainable land management practices; and
- drafting a national natural resources management policy.

Land degradation is a slow process, not easily seen or measured. Rehabilitation of degraded lands is expensive and seldom does the land revert to its original performance. Consequently, if we ignore the land today, future generations will pay the price.

References

FAO-Unesco (1990): Soil Map of the World, Revised Legend. Food and Agriculture Organization of the United Nations, Rome, Italy.

Henao, J. (1994): Agricultural Production in Albania. In: Socioeconomic Survey, 1994-93. Section I - Survey Results. International Fertilizer Development Center and Ministry of Agriculture and Food of Albania. 44 pp.

Ministry of Health and Environment Protection (1995): Environmental Status Report 1993-1994. Committee of Environmental Protection. Publication of the Ministry of Health and Environment, Tirana 1995. 23 pp.

Qilimi, B. (1996): Appraisal of Soil Fertility in Albania. Doctoral thesis. 105 pp. Soil Science Institute, Tirana.

Soil Survey Staff (1996): Keys to Soil Taxonomy. Sixth Edition. USDA Natural Resource Conservation Service, Washington DC. 644 pp.

UNDP (1995): Human Development Report: Albania. United Nations Development Program, New York. 54 pp.

World Bank-European Community (1992): An Agriculture strategy for Albania. 265 pp.

World Bank (1993): Environmental Review and Environmental Strategy Studies. Phase II Framework for Integrated Environmental Management in the Shkumbin River Basin, Washington DC. 150 pp.

World Bank (1996): Albania Forestry Project, Staff Appraisal Report. Report No. 15104-ALB. 77 pp.

Zdruli, P. (1997): Benchmark Soils of Albania. Vols. I and II. Internal Monograph of the World Soil Resources, USDA Natural Resources Conservation Service. 293 pp.

Zdruli, P., Eswaran, H., Almaraz., R. and Reich, P. (1997): Developing the prerequisites for sustainable land use in Albania. Soil Use and Management Journal, CAB International. **13**, pp 48--55.

Addresses of authors:
Pandi Zdruli
JRC-SAI
European Soil Bureau
T.P. 440
21020 Ispra (VA)
Italy
Russel Almaraz
Hari Eswaran
USDA
Natural Resources Conservation Service
P.O. Box 2890
Washington DC 20013
USA

A Map on Soil Erosion on Arable Land as a Planning Tool for Sustainable Land Use in Switzerland

D. Schaub & V. Prasuhn

Summary

In order to provide a tool for identifying areas at high risk of water erosion, an evaluation of soil loss rates at the national scale was made, using the Universal Soil Loss Equation (USLE) as the conceptual guideline and public databases. The areas classified have an average size of 6 km². Due to a good regional adaptation of crop rotations and field sizes to the topographic situation, soil loss rates on arable land in Switzerland are lower than reported from other Central European countries, despite the relatively high rainfall erosivity. Weighting the risk map with the actual percentage of arable land yields a more comprehensive picture of the soil erosion problem. With regard to sustainable land use, soil loss rates in the eastern and western Swiss Plateau as well in the small loess covered part in the north are too high.

Keywords: Erosion susceptibility, national level, USLE, data availability, nonpoint-source pollution

1 Introduction

In the recent governmental report on "Agronomic Policy 2002" (Schweizerischer Bundesrat 1996) pure crop yield numbers are no longer considered to be the only matter that counts in Swiss agriculture. Within this framework, financial compensation shall be directed to sustainable land use systems such as integrated farming or organic agriculture. It is expected that the vast majority of Swiss farmers will have shifted to such programs by the turn of the millennium. The eligibility for these direct payments depends on the observance of several quality aspects, e.g. the avoidance of losses by leaching, surface runoff or soil erosion.

For each farmer, the task of accomplishing quality criteria may greatly depend on given natural conditions. Soil erosion, for example, is certainly more difficult to control in regions with highly erodible soils and steep slopes. The aim of this study was thus to identify areas at high risk of water erosion at the national scale in order to provide a tool for targeting preventative and ameliorative action. Despite the small size of the country (41,285 km², Fig. 1) there is a great variety of types of rocks, soil parent materials and relief forms in Switzerland, entailing an immense wealth of landscape types. Therefore, the results of direct soil erosion measurements in small watersheds (reviewed in Prasuhn 1992) are of local relevance only, and cannot be simply extrapolated to larger areas. Other attempts of macro-scale assessments of soil erosion susceptibility in Central Europe (e.g. Auerswald 1988; Jäger 1994) are based on adaptations of the Universal Soil Loss Equation (USLE) (Wischmeier and Smith 1978). However, the USLE considerably overpredicted soil loss rates in some of the above-mentioned Swiss test areas (Table 1). We thus decided to retain the USLE as the conceptual framework for our evaluation, but to group the single process factors into classes. An ordinal value was assigned to each class, from 1 = very low to 5 = very high.

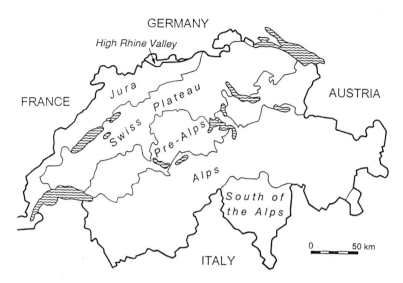

Fig. 1: Switzerland and its five main geographical regions.

2 Method: Creation of the geographical database

Our study refers to soil erosion on arable land. "Arable land" denotes all cropland, including short-term pastures, but without vineyards. The latter was excluded because of the lack of appropriate data on regional differences on sodding and terracing. Due to its high data-demand, a reliable nation-wide soil erosion risk evaluation can only be based on data already available, for example from state agencies. Spatial resolution of the available data thus determines the resulting soil erosion map. For Central Europe, Auerswald (1988) found slope gradient and crop cover (S and C of the USLE) to be the most sensitive erosion factors. Due to its linkage to a 100 m digital elevation model, the Swiss land use statistics (Bundesamt für Statistik 1993) allow the partition of arable land as a function of slope gradient on the level of the municipality. Since the boundary of each municipality is also available in digital form, we used these as elementary spatial units.

The surface area of the 3018 Swiss municipalities varies between 0.32 and 282.3 km^2. Larger municipalities are mainly situated in the Pre-Alpine and Alpine region, where the proportion of arable land is small or even zero. We did not classify the 782 municipalities with an arable area <5% of the total agricultural area, so as not to give too much weight to high rates of "potential" erosion due to the steep slopes and the increasing rainfall depth in the mountainous part of the country. This exclusion also reduced the variability of the spatial units so that the average size of the classified areas is about 6 km^2.

Data compilation and calculations were performed with Microsoft Excel 5.0, for overlay procedures and map drawings the low cost GIS RegioGraph was used. A detailed list of the available databases and the methods of coping with the different levels of resolution is contained in Prasuhn & Schaub (in prep.).

3. Results

3.1 Single process factors

The isoerosivity map for Switzerland by Schaub (1989), based on the correlation of erosivity (R value) and mean rainfall depth in summer by Rogler & Schwertmann (1981), was compared with maps of extreme point rainfalls, thunderstorm and hailstorm frequencies, and the water equivalent of snow cover for an assessment of the effect of thaw water. Only minor modifications of the original isoerosivity map were necessary. In general erosivity increases with higher annual rainfall due to increasing altitudes in the Jura and the Alps. The highest values occur in the region south of the Alps which is influenced by the more aggressive Mediterranean climate zone. In the main agricultural region of the Swiss Plateau R is between 80-100 [N h^{-1}] and thus relatively high for Central European conditions.

For the estimation of erodibility we followed the scheme to derive the K factor of the USLE. Large-scale soil surveys and even evaluations of the K factor are available for some Swiss regions (e.g. Baril & Thöni 1988). For a country-wide determination, however, we had to rely on the 1 : 200,000 map of the Bundesamt für Raumplanung (1980). It includes detailed soil taxonomy data based not only on soil properties but also on pedogenetic attributes. The latter sometimes made the assignment of a K value difficult, so that we had to draw conclusions form analogies. Except for a small loess belt in the north, the main part of cultivated soils in Switzerland has a low to moderate erodibility, corresponding to K values around 0.025 [kg h N m^{-2}].

Due to the small farm acreage (10 ha on average) and the variable topography, the field sizes in Switzerland are usually between 0.5 - 1 ha, corresponding to short erosive slope lengths (50-100 m). We thus considered erosive slope length (L) to be constant, so that the topographic factor depends solely on slope steepness (S). The spatial distribution of arable land can be expressed as a function of freely selectable slope gradient classes (Bundesamt für Statistik 1993). From our slope gradient classes we computed a weighted mean slope steepness for the total arable area of each municipality. Steeper slopes are cultivated in the Jura and the central Pre-Alps, compared to low slope gradients in the Swiss Plateau and the eastern Pre-Alps.

The estimation of the influence of land-use was based on the C-factor of the USLE. Similarly to Auerswald (1988), it was calculated from the spatial distribution of crops at a specific date for each municipality using the census of agricultural enterprises (Bundesamt für Statistik 1992). The slope gradient map and the crop cover map are more or less complementary. Areas with a large percentage of crops with a high erosion risk are predominantly situated in the Swiss Plateau (especially in the eastern part), in the Alpine valley bottoms of the Rhine and Rhône river, and south of the Alps. Crop rotations with a lower erosion risk are typical for the Jura and the Pre-Alps. Thus land use is generally well adapted to the topographical situation.

No erosion prevention factor (P factor of the USLE) was considered, because no specific protection measures are currently used on nearly all of the arable area. Moreover, no appropriate data are available.

3.2 The susceptibility maps

A weighting of the ordinal values of the single process factors produces a soil erosion risk map for arable land only (Fig. 2, Table 1):

$$\text{erosion risk} = \frac{3*C + 3*S + 2*K + R}{9}$$

It is thus of specific agricultural interest. For most areas the different erosion factors cancel themselves out to some extent, thus there are actually no regions with distinct higher erosion risk. Actual soil loss rates can be assigned to the risk classes, using data from direct field measurements (Table 1). According to this, the region with highest measured soil erosion rates falls into the moderate class. Moreover, the comparison of calculated risk classes and measured soil loss rates shows some similarities (e.g. the ranking of the test areas) which support our simplistic modeling approach.

However, even small soil losses on single fields can cause environmental problems in areas with a high proportion of arable land. Despite the relatively low rates, Braun et al. (1994) consider soil erosion as an important cause for nonsource-point pollution of surface waters in Switzerland. Overlaying the risk map with the actual percentage of arable land thus yields a more comprehensive picture of the soil erosion problem (Fig. 3). Many municipalities rise to a higher susceptibility class, so that a more obvious pattern can be observed with the most sensitive regions in the eastern and western Swiss Plateau and the small loess belt in the north. On the other hand, soil erosion is of minor importance in the hilly areas of the Pre-Alps, the Jura and the part south of the Alps, but also in the central Swiss Plateau with a high proportion of short-term pasture in the crop rotation.

Region	Soil parent material; soil texture	Risk value, risk class	Measured erosion rates [t ha^{-1} yr^{-1}]	Duration	Reference
High Rhine Valley	loess; loamy silt	3.10 moderate	4.0	1975-1987	Schaub (1989)
Jura	periglacial till; loamy clay	2.40 low	0.35	1978-1990	Prasuhn (1992)
Central Swiss Plateau	glacial till, molasse; silty loam - loamy sand	2.28 low	0.3-1.0	1987-1989	Mosimann et al. (1991)
Napf (Pre-Alps)	molasse detritus; loamy sand	2.25 low	0.4	1980-1982	Rohrer (1985)
High Rhine Valley	gravel; silty loam	2.10 low	0.3	1975-1987	Schaub (1989)

Tab. 1: *Soil erosion risk classes compared with data already published on measured soil loss rates. Risk classes are based on the weighted mean of the ordinal values for the classes of erosivity, erodibility, crop cover and slope (e.g. low = 1.81--2.60, moderate = 2.61--3.40 etc.).*

4 Discussion and conclusions

With the help of a simple expert system, a soil erosion susceptibility map of Switzerland was made based on officially available data. The purpose of the map is the delineation of problem areas, where soil conservation should be emphasised or, as a further step, restrictions may be applied. With areal units of about 6 km^2, however, the spatial resolution does not allow one to derive site-specific solutions. For identified problem areas a more detailed analysis of the causalities is necessary, for example with the more sophisticated GIS-based soil erosion model by Dräyer & Fröhlich (1994).

Fig. 2: *Soil erosion risk map (for arable land only). Risk classes are based on classes of erosivity, erodibility, crop cover and slope.* →

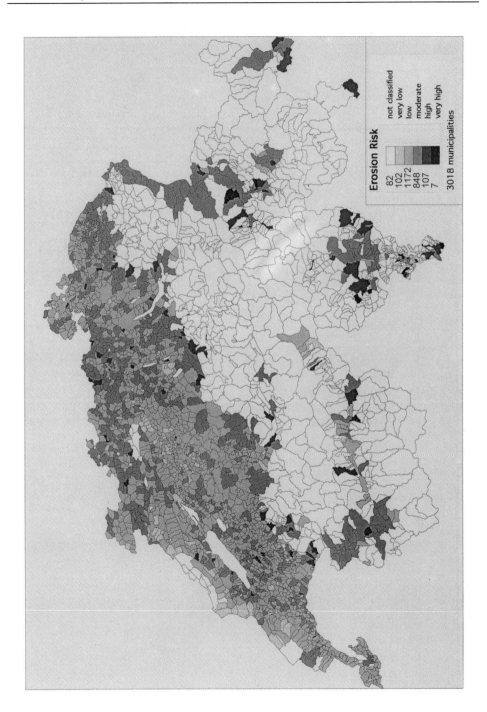

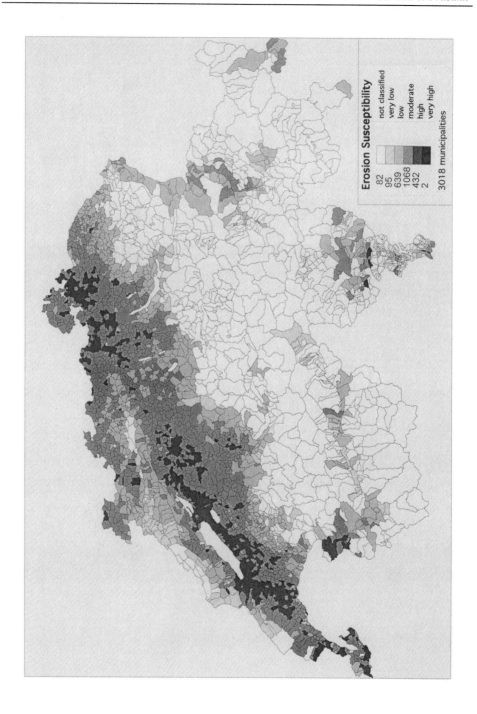

← *Fig. 3: Soil erosion susceptibility map as an overlay of the risk map (Fig. 2) with the actual percentage of arable land for each municipality.*

In Switzerland 10% of the total area or 26% of the productive land area, respectively, are tilled. Due to the natural conditions, arable land is concentrated in the Swiss Plateau. The natural soil erosion risk is rather high, especially rainfall erosivity and slope steepness. However, crop rotations are regionally well adapted to these conditions. The small farm acreage (10 ha on average) and the small field sizes (usually between 0.5-1 ha), corresponding with short erosive slope lengths, and the low degree of mechanisation also reduces the risk. It is thus not surprising that soil losses on arable land in Switzerland are lower than reported from other Central European countries. With regard to sustainable land use including the avoidance of nonpoint-source pollution effects, however, these rates are still considered too high.

A verification of the approach with field data shows that the model produces plausible results within the framework of its specific purpose. With regard to the available databases, soil erodibility was the most difficult factor to estimate. Another source of uncertainty is the C factor, which was determined only on the basis of single crop distribution, not on detailed crop rotations. For a further development of our model we will have to account for this, as well as for an inclusion of vineyards.

References

Auerswald, K. (1988): Erosion Hasard Maps for Bavaria. In: R.P.C. Morgan and R.J. Rickson (eds.), Erosion Assessment and Modelling, Silsoe: CEC, 41-54.

Baril, P. and Thöni, M. (1988): Etude de l'Erodibilité des Sols sur le Plateau Vaudois (Suisse), Proceed. Intern. Symp. on Water Erosion, Varna (Bulgaria), 145-152.

Braun, M., Hurni, P. and Spiess, E. (1994): Phosphorus and Nitrogen SurPluses in Agriculture and Para-Agriculture. Assessments for Switzerland and the Swiss Rhine Hydrological Catchment Area Downstream from Lakes, Schriftenreihe der FAC, No. 18, Liebefeld.

Bundesamt für Raumplanung (1980): Bodeneignungskarte der Schweiz 1 : 200 000, Bern.

Bundesamt für Statistik (1992): Eidg. Landwirtschafts- u. Gartenbauzählung 1990: Kulturland nach Gemeinden, Statistische Resultate, Reihe 7, Land- u. Forstwirtschaft, Bd. 3, Bern.

Bundesamt für Statistik (1993): Die Bodennutzung der Schweiz: Arealstatistik 1979/85. Resultate nach Kantonen und Bezirken, Statistik der Schweiz, Reihe 2: Raum, Landschaft und Umwelt, Bern.

Dräyer, D. and Fröhlich, J. (1994): A GIS-Based Soil Erosion Model for Two Investigation Areas in the High Rhine Valley and the Swiss Jura Plateau (NW Switzerland). Proc. Fifth European Conference and Exhibition on Geographical Information Systems, 29th March - 1st April, Paris (France), 1032-1041.

Jäger, S. (1994): Modelling Regional Soil Erosion Susceptibility Using the Universal Soil Loss Equation and GIS. In: R.J. Rickson (ed.), Conserving Soil Resources, European perspectives, Wallingford: CAB International, 161-177.

Mosimann, T., Maillard, A., Musy, A., Neyroud, J.-A., Rüttimann, M. and Weisskopf, P. (1991): Erosionsbekämpfung in Ackerbaugebieten. Ein Leitfaden für die Bodenerhaltung, Liebefeld.

Prasuhn, V. (1992): A Geoecological Approach to Soil Erosion in Switzerland. In: H. Hurni and K. Tato (eds.), Erosion, Conservation, and Small Scale Farming, Bern: Geographica Bernensia, 27-37.

Prasuhn, V. and Schaub, D. (in prep.), Bodenerosionskarte der Schweiz, Materialien zur Physiogeographie No. 21, Basel.

Rogler, H. and Schwertmann, U. (1981): Erosivität der Niederschläge und Isoerodentkarte Bayerns. Zeitschrift f. Kulturtechnik u. Flurbereinigung 22, 99-112.

Rohrer, J. (1985): Quantitative Bestimmung der Bodenerosion unter Berücksicht-gung des Zusammenhanges Erosion-Nährstoff-Abfluss im oberen Langete-Einzugsge-biet, Physiogeographica, Basler Beiträge zur Physiogeographie, Bd.6, Basel

Schaub, D. (1989): Die Bodenerosion im Lössgebiet des Hochrheintales (Möhliner Feld - Schweiz) als Faktor des Landschaftshaushaltes und der Landwirtschaft, Physiogeographica, Basler Beiträge zur Physiogeographie, Bd. 13, Basel.

Schweizerischer Bundesrat (1996): Botschaft zur Reform der Agrarpolitik: Zweite Etappe (Agrarpolitik 2002). Teil I: Neues Landwirtschaftsgesetz, Bundesblatt, 148 Jahrgang, Band IV, Nr. 40, 08.10.96, Bern.

Wischmeier, W.H. and Smith, D.D. (1978): Predicting rainfall erosion losses - a guide to conservation planning, U.S. Dep. of Agricultur Handbook No. 537, Washington D.C.

Addresses of authors:
D. Schaub
Department of Geography
University of Basel, Spalenring 145
CH-4055 Basel, Switzerland
V. Prasuhn
Swiss Federal Research Sation for Agroecology and Agriculture,
Inst. of Environmental Protection and Agriculture Liebefeld
CH-3003 Bern, Switzerland

The Soil Information System "FISBo BGR" for Soil Protection in Technical Cooperation

H. Vogel, J. Utermann, W. Eckelmann & F. Krone

Summary

The German Federal Institute for Geosciences and Natural Resources (BGR) is tailoring its soil information system ("FISBo BGR") to the needs and requirements of technical cooperation programmes. FISBo BGR is aimed at evaluating the suitability of regionally prevailing soil series for different forms of soil use, and at monitoring and predicting patterns of changes in the soils (e.g. soil contamination) that are relevant to improved soil resource management.

The system is a working tool designed to archive and analyze standardized soil data bases, to link them with spatial data bases, and to map the geo-coded soil characteristics at a variety of scales. It uses a geographic information system (GIS) to maintain the spatial soil resource data.

Keywords: Soil information system, technical cooperation

1 Introduction

To appreciate the human impact on the environment, policy makers and land-use planners require a basic understanding of the nature and properties of soils. Hence, they need to have access to sufficiently accurate and easily readable pedological maps. To help solve problems that arise in everyday planning, such thematic maps need to provide information that is both of a dynamic, is of an analytical nature, and up to date (Ernstrom & Lytle, 1993). Information displayed in static, i.e. inventorial soil survey maps, is usually not the information required by planners (Oldeman & van Engelen, 1993).

2 Soil Information Systems (SIS)

The worldwide demand for soil-related information has risen tremendeously over the last two decades, in particular as a result of increased efforts to protect the natural soil resources that constitute the basic resource for agricultural food production. Decisions and measures aimed at soil protection require a profound knowledge of the spatial distribution of soils, of the quality and quantity of their properties, and of their interactions with other resources (air, water).

In order to organize and handle the resultant complex mass of soil observations, computerised soil information systems (SIS) are required that allow for the storage, computation, retrieval, interpretation and mapping of soil-related data. In a typical case, a SIS is made up of a tool library consisting of (1) a digital database management system maintaining all available soil physical and chemical data sets and related information (e.g. site, sample, and profile descriptions), (2) a set of method moduls (pedotransfer functions and modelling techniques) required for specific data processing (e.g. quantitative erosion risk assessment), and (3) a geographic information system (GIS) to visualize and map soil attributes and derived data via a spatial reference database (Heineke et al., 1995).

3 The Soil Information System (FISBo BGR)

In an effort to better respond to environmental policy-making and planning needs in Germany, the German 'Federal Institute for Geosciences and Natural Resources' [Bundesanstalt für Geowissenschaften und Rohstoffe (BGR)] has developed a soil information system called FISBo BGR (Eckelmann et al., 1995). FISBo BGR is part of a geo-referenced natural resources information system that also embraces geology, geomorphology, hydrology, etc. Together they form an overall geo-information system of both principal and supplemental data and information sets (Vinken, 1992).

FISBo BGR is primarily a computer-based tool library aimed at integrating all soil survey data relevant in the context of soil use and soil protection in Germany, and to subsequently disseminate sound interpretations to policy- and decision-makers all over Germany in the form of thematic maps (Eckelmann & Hartwich, 1996). Apart from helping to advise both central government bodies as well as governmental institutions of the 16 federal states (Länder) in Germany, FISBo BGR is also supporting the research activities of the BGR by acting as a source of system development. For example, algorithms (that is pedotransfer functions) and soil evaluation methods and models have been compiled for FISBo BGR (Hennings, 1994) that need to be optimized and developed further for various soil attributes. Standard pedotransfer functions are essential for the harmonisation of soil information.

4 The Structure of FISBo BGR

FISBo BGR is an integrated soil information system that consists of three main components, namely:
(a) a soil profile and laboratory database that contains both the observations of soil surveys as well as the results of all soil chemical and physical analyses,
(b) a method base that defines the data processing techniques from soil maps and the relevant principal and supplementary data, and
(c) a spatial database that maintains the geometric-topographical data and a number of already existing soil and related maps.

Data integration in FISBo BGR is provided by a relational data management system (i.e. MS Access) that links the 'soil profile and laboratory database' and the 'methods base' for analytical data processing. Since most data manipulations are done (or will be done) in the relational database management system, the geographic information system (GIS) software (i.e. PC ARC/INFO) is mainly used to only maintain the geometric-topographic data ('spatial database'); a crucial advantage in many technical cooperation projects.

5 Contents of the FISBo BGR soil profile and laboratory database

The 'soil profile and laboratory database' stores all soil attribute data extracted from point observations of fully described and analysed reference profiles in sets of digital files for later retrieval. Links between the files, i.e. tables, are maintained through primary keys. Depending on regional or national requirements, the soil database may be set up according to various nomenclatures. In addition to the German soil taxonomy, the FAO soil classification system has been carried out (Fig. 1). The latter has been done in order to specifically cater for technical cooperation programmes and projects. Similarly, a soil database has been developed according to the U.S. Soil Taxonomy (widely used in Asia and the Americas).

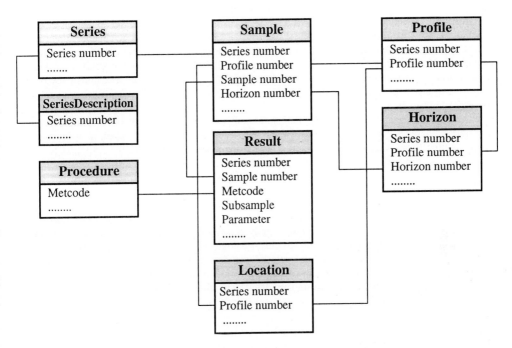

Fig. 1: Data model of soil profile and laboratory database of FISBo BGR (FAO-classification).

In a further step towards harmonisation of global soil information, BGR recently adopted the terminology and components of the Multilingual Soil Database (FAO, ISRIC and CSIC, 1995) for their FAO soil database version. Unlike the FAO-ISRIC-CSIS Soil Database (SDBm), the FISBo BGR soil database allows, however, to be easily tailored according to customer needs and wishes. Based on standard software (i.e. MS Access), soil and other components (e.g. vegetation) can readily be added to or removed from the database by individual users or customers.

6 Contents of the FISBo BGR method base

The 'methods base' holds the available methods for the derivation of pedological potentials and hazards (such as soils' filter capacity for heavy metals, their nitrate retention capacity, their susceptibility to erosion, etc.) from soil maps (Hennings, 1994). In general, the methods base of a SIS may feature three types of interpretation techniques, namely:
- pedotransfer functions, that is algorithms to derive simple soil properties such as rooting depth or soil water characteristics,

- deterministic models based on simple empirical relationships to estimate complex relationships such as the water erosion risk or the filter capacity of heavy metals, and
- analytical and numerical simulation models to predict, for example, the effects of different management practices and land use systems on the environment.

In FISBo BGR, the methods, so far, mostly consist of pedotransfer functions in modular form. BGR is also testing the feasibility of programming (some) pedotransfer functions in the relational database management system (*i.e.* Access). The advantage of this would be greater flexibility of the system since the need to employ too many different software components is reduced.

In future, considerable attention will have to be given to the optimization of algorithms. In order to avoid repetition and to identify the approaches best suited for a objectives-oriented selection, available algorithms are being tested on the basis of existing soil databases (Tietje & Hennings, 1996). The main objectives of this exercise are (1) to quantify the validity of pedo-transfer functions (for example to estimate soil hydraulic properties), (2) to compare existing approaches (using the same databases), and (3) to obtain a ranking on the basis of the accuracy of the predicted values.

In addition, some more methods, guidelines and instructions need to be standardized, for example with regard to the collection of data (minimum data set), field sampling instructions, mapping procedures, classification diagrams, standardized research programs.

7 Contents of the FISBo BGR spatial database

FISBo BGR allows the storage of any geometric-topographic information that a user requires to produce a certain thematic map (including already existing maps). In most cases, initial delineations are derived from conventional topographic, geologic, etc., base maps through digitizing. In the context of development cooperation, such polygon databases may also be derived from aerial photographs and remote sensing images. Once the spatial database has been set up, the original delineation can easily be re-grouped and up-dated depending on individual needs and requirements.

8 BGR and Technical Cooperation

For over 30 years, the BGR has assisted by the German Federal Ministry for Economic Cooperation and Development (BMZ) with the implementation of the German contribution to bilateral technical cooperation projects in various geoscience disciplines. At present, BGR is involved in more than 40 technical cooperation projects in more than 30 countries worldwide. These involvements are mostly in the fields of :
- exploration and exploitation of groundwater resources,
- exploration and assessment of metallic, non-metallic and energy resources,
- and the
- development of pedological concepts for appropriate site-specific soil use.

With regard to the latter, the number of projects and programmes requesting technical assistance in the development of digital soil information systems is increasing. Most partner institutions are fully aware of the need and usefulness of computer-based tools to systematically store extensive alpha-numerical and graphic data sets from which tailor-made thematic maps can be derived on request (Van den Berg & Tempel, 1995). Therefore, the BGR is adapting system components of the (national) FISBo BGR to meet the needs and requirements of their partner organizations. In a first step, a soil database was developed according to the FAO nomenclature as mentioned above (Fig. 1).

In a follow-up experimental exercise to a previously bilateral soil erosion programme carried out in Cyprus between 1980-86 at a total of 7 different research sites, FISBo BGR was tested by developing a soil erosion hazard map. For this, the Universal Soil Loss Equation (USLE) was employed, albeit in a rudimentary form. While some maps containing the USLE factors could be derived from existing data sets (i.e. K and R factor) and a topographical map (i.e. S factor) respectively, the L factor had to be omitted because of the difficulties to derive slope lengths values from digital terrain models (e.g. the raster-based GRID module of PC ARC/INFO). After crossing the available maps, an erosion risk map was obtained that displays soil erosion hazard in the form of qualitative classes rather than quantitatively.

9 Conclusion

FISBo BGR is in a continuous process of development. The ultimate goal of this developmental activity will be a decentralised, that is a PC-based data and information source needed by policy makers, planners, and researchers alike. Consequently, new features to be included in „FISBo BGR" will depend primarily on the the needs of BGR's clientele and the future development of soil evaluation systems.

References

Eckelmann, W., Adler, G., Behrens, J., Hartwich, R., Hennings, V. & Stolz, W. (1995): The soil information system "FISBo BGR". A digital information system for soil protection in Germany. In: King, D., Jones, R.J.A. & Thomasson, A.J. (Eds.): European land information systems for agro-environmental monotoring. Published by the European Commission, Directorate- General XIII, Luxembourg, 235-243.

Eckelmann, W. & Hartwich, R. (1996): Soil mapping in Germany and the Soil Information System FISBo BGR. In: Le Bas, C. & M. Jamagne (Eds.): Soil databases to support sustainable development. Published by INRA and the Joint Research Centre of the European Commission, Orleans, 49-55.

Ernstrom, D.J. & D. Lytle (1993): Enhanced soils information systems from advances in computer technolgy. Geoderma **60**, 327-341.

FAO, ISRIC and CSIS (1995): Multilingual Soil Database. World Soil Resources Reports, No. 81, Food and Agriculture Organization of the United Nations, Rome, Italy.

Heineke, H.J., Filipinski, M. & Dumke, I. (1995): Vorschlag zum Aufbau des Fachinformationssystems Bodenkunde. Profildatenbank, Flächendatenbank und Labordatenbank, Methodenbank., Geologisches Jahrbuch, F 30, Hannover, Germany.

Hennings, V. (Coordinator), (1994): Methodendokumentation Bodenkunde. Auswertungsmethoden zur Beurteilung der Empfindlichkeit und Belastbarkeit von Böden. Geologisches Jahrbuch, F 31, Hannover, Germany.

Oldeman, L.R. & van Engelen, V.W.P. (1993): A world soils and terrain digital database (SOTER) - An improved assessment of land resources. Geoderma **60**, 309-325.

Tietje, O. & Hennings, V. (1996): Accuracy of the saturated hydraulic conductivity prediction by pedotransfer functions compared to the variability within FAO textural classes. Geoderma **69**, 71-84.

Van den Berg, M. & Tempel, P. (1995): SWEAP. A computer program for water erosion assessment applied to SOTER. Documentation version 1.5. International Soil Reference and Information Centre (ISRIC), Wageningen, The Netherlands.

Vinken, Renier (1992): From Digital Map Series in Geosciences to a Geo-Information System. Geologisches Jahrbuch, A 122: 7-25, Hannover, Germany.

Address of authors:
H. Vogel
J. Utermann
W. Eckelmann
F. Krone
BGR-Bodenkunde
Postfach 51 01 53
D-30631 Hannover
Germany

Alteration of Soil Properties Caused by Climate Change

J.M. Kimble, R. Lal & R.B. Grossman

Summary

We can estimate many different alterations of soil properties caused by climate change. In some areas these effects would be positive and, in others, negative. Loss of Soil Organic Carbon (SOC) can have a major effect on soil aggregation and the overall fertility of the soil. The major problem in predicting changes in soil properties is that they take place very slowly. Soil genesis may take hundreds to thousands of years, so we would not expect major genetic changes to be readily evident. Also, the predicted temperature changes are relatively minor in terms of affecting soils. Changes in SOC could be quite dynamic, and resulting changes in soil structure, soil erodibility, crusting, compaction, infiltration rate, runoff, salinity, and cycling of plant nutrients also could be profound.

Data are needed to predict actual changes. The near surface of soil (uppermost cms) is the part most sensitive to changes in moisture and temperature, but many of its properties are temporal and change throughout the year. Tillage operations, plant growth, rainfall impact, wetting and drying cycles, etc., make it hard to give definitive answers about what alteration of soil properties to expect because of climate change.

Keywords: Climate change, soil properties, erodibility, compaction

1 Introduction

Soils are an important component of the terrestrial ecosystem and play a major role in the regional and the global carbon cycle. They are in a dynamic equilibrium at present and alterations can be expected with climate change. Soils have many different properties which may be altered by climatic changes. Important findings related to soils and global change (Lal et al., 1995a, 1995b; and Bowman, 1990) are being published. We make no attempt to list them here, but offer some suggestions as to soil changes as a result of climate change. The factors of soil formation were listed by Jenny (1941) as climate, organisms, topography, parent material, and time. Of these factors climate is the one which is undergoing anthropogenic change. If there is a climatic shift it may well have a major impact on the overall process of soil formation and the soil properties which result from the interaction of these factors. As many predictions show climatic changes to be quite variable, we would also expect the alteration of soil properties to be quite variable.

If soil physical and chemical properties are impacted we expect changes also to affect overall quality and sustainability of the pedosphere. Many changes would be very slow, but others could be quite fast. A potentially useful index for assessing the climate-induced change in soil physical properties is the Least-Limiting Water Range (LLWR) (Kay, 1996). The LLWR is affected by the interactive changes in the available water capacity, soil structure, and air-filled porosity at different moisture potentials, soil strength in relation to the moisture potential, and oxygen diffusion rate for different moisture regimes. Soil organic carbon and its interaction with clay colloids, both affected by the climate, play an important role in the magnitude of spatial and temporal variations in the LLWR.

If global warming occurs, as scientists, dealing with the global circulation models (GCMs) predict, we will need to know its potential impact on terrestrial ecosystems, especially the pedosphere. Slight increases in the overall average air surface temperature have been documented (Jones et al. 1986; Hanson and Lebedeff, 1987; Boden et al. 1991). Also, the extremes have become more variable (hotter summers and colder winters). This paper addresses the question of what effect, if any, climate changes may have on soil properties. Will the quality of the pedosphere be changed? The immediate answer is yes. However, what are the expected changes, and are they all negative, or would some be positive? Will they be evident in short periods of times (1 0r 2 years), or will it take decades or even longer?

In addition to their effects on the LLWR, climate-induced changes in soil structure (loss of aggregation caused by the mineralization of organic matter) may also have profound effects on susceptibility to soil erosion, crusting and compaction, infiltration rate and runoff, salinity, and cycling of several plant nutrients. The nature and magnitude of the effect will depend on site-specific ecoregional characteristics.

The potential climatic change could have a notable effect on the local, regional, and global water balance. Precipitation is filtered as it moves through the pedosphere, and changes in precipitation and/or temperature may affect many pedosphere processes (leaching, mineralization). Less precipitation could reduce leaching and, combined with higher temperatures and resulting increased evaporation, could lead to an increase in the water soluble salts in the near surface layer. Areas which now receive adequate moisture to prevent the accumulation of water soluble salts at or near the surface could become more saline. Also, the overall moisture retention of the pedosphere may be reduced in many areas with reduced precipitation while in areas of greater precipitation it would increase.

One reason why changes in soil properties are hard to predict is that GCMs show some areas having increased precipitation and temperatures while other areas may experience a reduced amount of precipitation and higher temperatures. Most models suggest that the greatest changes in temperatures will be at the high latitudes (2-3 °C increases). This may have a profound effect on the active layer in areas affected by permafrost. The active layer may deepen, and the large amounts of organic carbon (Kimble et al., 1993; Schlesinger, 1984) may be subject to much higher rates of mineralization. As a result major alterations in the properties of the soils would be expected. Some researchers feel there will be increased plant growth and the amount of carbon sequestered actually may increase at the high latitudes.

Soil aggregation could be affected greatly by higher temperatures, which could increase the rates of oxidation of soil carbon, affecting carbon sequestration and therefore lead to lower soil aggregation. A loss of carbon could reduce soil aggregation and have a negative effect on the overall quality of the soil. Changes in carbon sequestration would reduce soil fertility in many areas.

Many soil alterations will be long-term. However, we could expect some changes to occur very quickly, such as a loss of soil organic carbon (SOC) when areas not under cultivation are cultivated. The effects of land use changes on SOC have been well-documented (Houghton, 1995). Loss of SOC causes rapid decreases in the quality of many soil properties (cation exchange capacity, soil aggregation, soil structure), and these changes can be negative (loss of SOC reduced soil quality, etc.) or positive (increases in SOC). The major changes to soil result from human activities and mismanagement by man. We know very little about the effects of minor changes in soil temperature. Finally, the world wide amount of SOC bears directly on changes in atmospheric C. Modest changes in SOC, whether positive or negative, may have an appreciable effect on the rate of change. The reason is that the yearly change in atmospheric C is a small percentage of the total SOC. Therefore, small changes in SOC due to climate change can have a large effect on the change in atmospheric C.

Two values are illustrative (Table 1). Both pertain only to cultivated land; this makes the computations conservative because land in other uses would also be subject to change in SOC. One value, on the assumption of 10 kg/m² total SOC, is that the increase in atmospheric C is 1.7 % relative to the SOC. Another value pertains to a yearly change equal to the increase in atmospheric carbon for an average tillage zone and is 0.087 % absolute. Numerous studies have shown that such a yearly change for tillage zones is readily obtainable.

1. The yearly increase in atmospheric C is 0.005 kg/m².
   a. 0.4 g/cm² CO_2
   b. 4000g/m² CO_2
   c. 1000 g/m² (approximately) atmospheric C.
   d. Assume 0.5 percent relative yearly increase in atmospheric C concentration.
   e. 1000 g x 0.005 = 0.5 g/m² = 0.0005 kg/m².

2. The yearly increase in atmospheric C relative to total SOC of cultivated soils is 1.7 % relative.
   a. SOC = 10 kg/m²
   b. 3 % of the earth surface is cultivated.
   c. $\dfrac{0.005 \text{ kg/m}^2}{10 \text{ kg/m}^2}$ X $\dfrac{100}{0.03}$ = 1.7 %

3. For the tillage zone of a cultivated soil, the yearly increase of atmospheric C as an absolute percentage of the SOC is 0.087 %.
   a. Assume 0.005 kg/m² atmospheric C increase and that the cultivated area is 3 percent of the land surface of the earth.
   b. SOC change for cultivate area. $\dfrac{0.005}{0.03}$ = 0.17 kg/m²/yr
   c. The absolute percentage that the SOC increases under "b" is of the typical tillage zone.
   SOC (kg/m²) = $\dfrac{L \times Db \times Px}{10}$
   Where L is the thickness assumed to be 15 cm, and Db is the bulk density assumed to be 1.30 Mg/m³, and Px is the absolute change in SOC.
   0.17 kg/m² = 15 cm x 1.30 g/cc x Px
   Px = $\dfrac{0.17 \times 10}{15 \times 1.30}$ = 0.087 %

Table 1. Computations pertaining to the evaluation of the change in atmospheric C relative to changes in SOC.

2 The very near surface

The top few centimeters of soil are sensitive to the moisture and temperature regimes that may be associated with global climate change. However, we do not have data for the changes in surface and near surface properties, because relatively small changes have been proposed in soil temperature due to global warming. We do have many measurements and anecdotal observations about the surface and near surface of soils that differ markedly in temperature. These differences are greater than those expected to occur but should be indicative of the kinds of changes. Differences in existing soils suggest what may happen to the pedosphere with warming. Cropping areas could be expected to shift north or south. These more northern areas, at present, tend to have higher levels of organic carbon because of lower mineralization rate.

By comparing soils with different temperature regimes as defined in Soil Taxonomy (Soil Survey Staff, 1975), we can see how warming may change soils. As an example for this paper we selected grasslands with frigid and thermic soil temperature regimes.

Frigid soils (Soil Survey Staff, 1975) have a mean annual soil temperature < 8° C. Thermic soils have a mean annual soil temperature between 15 and 22° C. Frigid and thermic grassland soils show marked differences in the uppermost 5 cm. In frigid rangeland soils, a thin sod is common and the ground surface is covered nearly completely by vegetal material. Cryptogams are common and add to the vegetal cover. In contrast, in thermic rangeland soils, sod is absent. Non sod-forming grasses occur. The soil is largely bare between grass clumps with the vegetal cover less than 20 percent, compared to the nearly 100 percent for frigid soils.

The greater vegetative cover in the frigid areas promotes infiltration. Both wind and water erosion are less because of the much more complete cover. The vegetation also buffers changes in water state in the top very few centimeters and moderates soil temperature change. The vegetal cover during the early growing season commonly acts to increase the albedo, because the dead vegetal residue has a high color value. The color of soils is expressed using the Munsell system, where the color value is the square root of the reflectivity in the visible range (Soil Survey Staff, 1993). In contrast mineral surface horizons of Mollisols have a dry near surface color value of less than 5 below that of surface residue. In thermic soils, the albedo depends on the color value of the soils and the degree of crusting, which acts to increase the albedo.

Albedo of the land surface could be affected in opposite ways by different changes. The effect assumes exposure of the soil to the atmosphere without vegetal cover. Reductions in snow cover due to increased temperatures at higher elevations would reduce albedo. Areas which become drier would have reduced vegetation and soil cover along with stronger soil crusts and a loss of SOC which would lead to an increase in the albedo. As an example, expected color values of the ground surface of a soil with a mollic epipedon are shown in Table 2.

Condition	Value	Reflectance
Moist	2.5	6
Air dry, no crust	4	16
Crust, dry	5	25
Eroded, crust, dry	5.5	30

Tab. 2: Relationship of the soil conditions to the color value and reflectance of a mollic epipedon.

Other differences in the soils occur. In frigid soils, the uppermost 5 cm commonly have bulk densities below unity. For the 0-2 cm, the bulk density is commonly below 0.50 Mg/m^3. In thermic soils, the bulk density varies, being lower at high clay percentages but generally are higher. The changes with depth over the uppermost 5 cm are commonly the reverse of that of the frigid soils. The uppermost 2 cm have higher bulk densities than the 2-5 cm depth because of compaction by water of the mostly bare soils.

Transient ponded infiltration is from initial ponding to deep wetting. Transient ponded infiltration rate decreases with wetting time of the very near surface. Steady ponded infiltration occurs after prolonged wetting and is more influenced by a limiting zone within the upper 1/2 to 1 meter section. Transient ponded infiltration controls the infiltration rate during much of the growing season when wetting is only to shallow depths. Steady ponded infiltration controls the rate when the soils are completely wetted which is more common before or after the growing season in temperate regimes. For either transient or steady infiltration rates are enhanced by the 0-5 cm zones common to frigid soils if there is no shallow limiting zone beneath.

Under thermic conditions, infiltration rates in general would be lower. The reduction depends on near surface organization, which ranges from a strong raindrop impact crust to granular depending on the soil. The thermic soils generally should have lower transient ponded infiltration rates than the frigid soils. Differences in steady ponded infiltration would depend on water transmission properties within the uppermost 50-100 cm. As a consequence, steady ponded infiltration should vary more within frigid than the thermic soils and show less difference between the two temperature regimes. Hence, the model in general outline would indicate, less of an effect on runoff during periods of prolonged water excess. These periods commonly occur outside the growing season. Runoff that occurs from convectional storms during the growing season would depend more on transient ponded infiltration and hence would differ to a greater extent between frigid and thermic soils.

An increase in runoff during the growing season would cause a loss of soil vegetation with warming. This would lead to a loss of SOC, increases in the bulk density of the near surface, and increased runoff in the growing season. Both wind and water erosion rates would therefore be expected to increase.

2.1 Soil Structure

Soil structure depends on climate because of climate-induced changes in soil organic matter (SOM) content and soil moisture and temperature regimes. Changes in SOM content affects soil structure both directly and indirectly.

Direct effects of SOM content are related to development of aggregates through the formation of organo-mineral complexes, leading to changes in: (i) total aggregation, (ii) aggregate size distribution, (iii) aggregate stability to disruptive forces of wind and water, and (iv) aggregate strength and resistance to comparative forces of farm traffic. Indirect effects of SOM content are those related to soil biodiversity and the activity and species diversity of soil fauna.

Earthworms, termites, and other macro- and microfauna have a drastic influence on soil structure. Activities of soil fauna related to structural properties are: (i) burrowing, (ii) mixing and soil turnover, (iii) biomass decomposition, and (iii) fecal ejecta and body fluids that bind and enrich soil particles. Soil fauna thrives best under moist/humid environments within a temperature range of 20-25°C.

There are two possible scenarios of climate-induced changes in soil structure (Figure 1). Scenario 1 relates to possibilities of a decrease in SOM content due to high temperatures, low moisture, low biomass production, and high rates of mineralization of SOM. These trends would lead to low aggregation, high susceptibility to crusting and compaction, high erodibility and potential erosion risks, and high risks of soil degradation and desertification. This is the shift which was discussed by comparing frigid to thermic soils. These adverse changes may occur in regions with coarse-textured soils and where the climate change is likely to cause decreased precipitation, increased temperatures, and low biomass production.

Scenario 2 relates to possibilities of an increase in SOC content through a potential increase in precipitation (Fig. 1) and CO_2 fertilization. The CO_2 fertilization is a result of the rise in the level of CO_2 in the atmosphere (Schlesinger, 1995). The higher levels of CO_2 may lead to a higher net primary productivity and an increase in SOC content. High biomass production, especially beneath the soil surface as root growth, may enhance aggregation and improve soil structure. Favorable structural attributes can set in motion soil restorative trends, leading to lower runoff, less soil erosion, improved soil biodiversity and nutrient recycling, and improved soil quality. These restorative trends may happen in ecoregions with soils of heavy texture, and where precipi-tation may increase due to the greenhouse effect.

These potential climate-induced changes in soil structure may trigger a chain reaction leading tosoil degradation or restoration because of a strong interdependence between soil quality and soil structure. Soil quality is related to "inherent attributes of soils that are inferred from soil character-istics" (SSSA, 1987) and refers to "the capacity of a soil to function within ecosystem boundaries

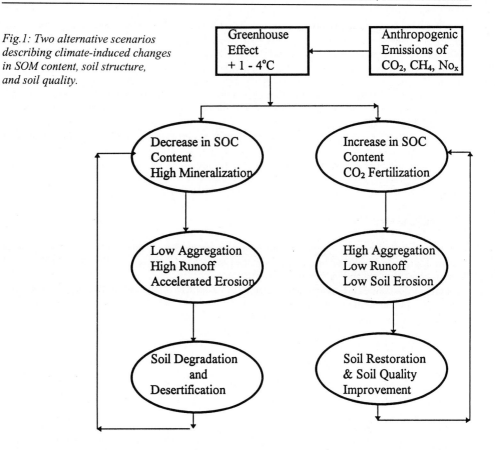

Fig.1: Two alternative scenarios describing climate-induced changes in SOM content, soil structure, and soil quality.

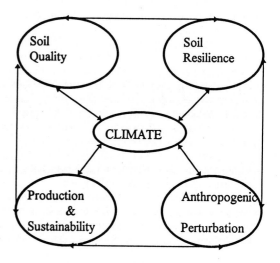

Fig. 2: Interactive effects of human-induced climate change on soil quality, soil resilience, and productivity and sustainability.

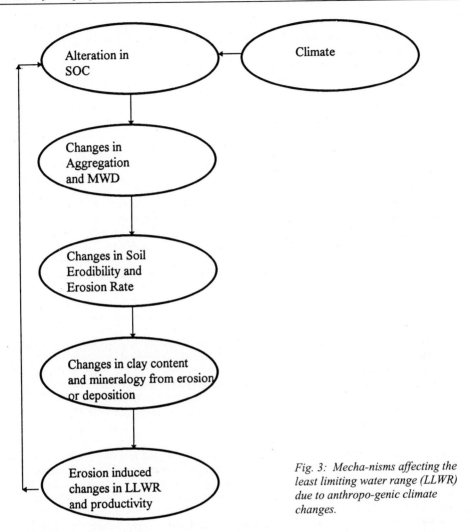

Fig. 3: Mecha-nisms affecting the least limiting water range (LLWR) due to anthropo-genic climate changes.

to sustain biological productivity, maintain environmental quality, and promote plant and animal health" (Doran and Parkin, 1994). Soil structure also relates to soil resilience, the "ability of a soil to restore its productivity and environmental regulatory capacity following a perturbation" (Lal, 1994). Anthropogenic climate change may have a drastic impact on soil quality because of strong interactive effects between climate, soil quality, soil resilience, and agricultural sustainability (Fig. 2). It is important to develop and strengthen our understanding of the processes involved so that soil quality and resilience may be enhanced and the degrading effects of climate change minimized. Least Limiting Water Range

Change in climate may influence soil structure in four principal aspects (Kay, 1996): form, stability, resilience, and vulnerability. The structural form refers to properties that describe the arrangement of void and solid space, e.g., porosity, pore size distribution, and aggregate/peds shape and size. Structural stability refers to a soil's ability to maintain its arrangement of void and solid space against natural or anthropogenic disruptive forces. Structural resilience refers to the ability to recover form and ability following a perturbation, while vulnerability is its inability to recover. Considering the complexity of structural attributes, and considering the fact that measurement ranges from microscopic to field size scale, it is difficult to identify a structural index that is climate-sensitive and universally applicable for all soils, land uses, and ecological environments.

An index that combines the four structural aspects described above and their interaction with soil moisture regime is called the least limiting water range or LLWR (da Silva et al., 1994). It is defined as the water content at which aeration, water potential, and resistance to penetration reach values that are critical to or limiting to plant growth (de Silva et al., 1994). The LLWR is a modification of the original concept of the non-limiting water range proposed by Letey (1985). The upper limit of LLWR is defined by the moisture content at which aeration becomes limiting, and the lower limit by the moisture content at the permanent wetting point at which the soil strength becomes limiting. Basic soil properties affecting the LLWR are SOC content, clay content and mineralogy, soil bulk density, aggregation, and aggregate stability. Pedotransfer functions relating LLWR to soil properties are developed for data extrapolation and for evaluating the effects of management and other perturbations:

These pedotransfer functions are soil-specific and dynamic.

LLWR = F(SOC, WSA, MWD, Cl, BD)

Where WSA is water stable aggregation, MWD is mean weight diameter of aggregates, Cl is clay content, and BD is soil bulk density.

Climate change may affect the LLWR through its effect on key soil properties, e.g., SOC content and aggregation (Figure 3). The principal effect on the properties for which LLWR is sensitive may be due to changes in soil erodibility and erosion rate. The accelerated soil erosion may expose a horizon with different SOC and clay contents, clay minerals, soil bulk density, and soil strength. These changes also may occur where eroded sediments with their different composition have been deposited. Erosion-induced changes in crop yield and biomass production will accentuate differences in SOC content and aggregation and hence in LLWR.

2.2 Chemical Changes

Changes in the chemical properties of soils can be related to soil moisture regimes (Soil Survey Staff, 1975). Increased aridity would increase the accumulation rate of soil carbonates. For the carbonate to be a sink for atmospheric C, the calcium must come from weathering rather than from the atmosphere. Present evidence is that most of the calcium originates from carbonate in dry dust or from calcium dissolved in the precipitation ; hence, the carbonate is not a sink (Grossman et. al., 1995). Areas which may become drier would be expected to have reduced levels of leaching, which may lead to the buildup of more soluble salts in the upper part of the soil profile. Areas which may become wetter would have the reverse effect. The most rapid changes would be in salt buildup in drier areas while the leaching of exchangeable cations would be expected to be much slower (decadal or longer). With warming we also could expect a loss of SOC due to increased mineralization with a resultant loss of CEC, particularly for low activity soils (LAC). Much of the CEC in LAC soils arises from the exchange sites associated with the SOC which is present in the soil and not from sites on the clay particles. A loss of CEC commonly reduces the overall fertility of the soil.

Overall soil fertility would change with climate induced changes. In areas of increased rainfall and temperature, a loss of fertility would occur from increased leaching as well as the previously discussed mineralization of SOC. The reverse would be true in areas of reduced precipitation. Most major chemical changes in soils would be quite slow if not associated with SOC. Very little data is available for predicting expected or possible changes.

3 Conclusion

Soil properties may be altered over time by climate change but the changes are expected to be quite long term. These changes would expect to be manifested in the near surface as this is the part most sensitive to changes in moisture and temperature. These changes would be in SOC and quite dynamic, and could result in changes in soil structure, soil erodibility, crusting, compaction, infiltration rates, runoff, salinity, and the cycling of plant nutrients. If new areas are put into crop land as a result of climate change, we can expect that overall soil quality in these areas to be reduced as it has elsewhere when soils were first cultivated. Loss of fertility, increase erosion (wind and water), increased near surface bulk densities, plow pans, etc. will occur and some of these changes can be reduced through proper soil conservation practices. Very little alteration of soil genesis will take place with climate changes as soil genesis normally takes hundreds to thousands of years so we would not expect major genetic changes to be readily evident..

References

Boden, T., Sepanski, R.J. and Stoss, F.W. (1991): Trends 91: A Compendium of Data on Global Change. Oak Ridge national Laboratory, Oak Ridge, Tennessee.

Bowman, A.F. (ed) (1990): Soils and the Greenhouse Effect. John Wiley and Sons, United Kingdom. 575 p.

da Silva, A.P., Kay, B.D. and Perfect, E. (1994): Characteristics of the least limiting water range of soils. Soil Sci. Soc. Am. J. 58: 1775-1781.

Doran, J.W. and Parkin, T.B. (1994): Defining and assessing soil quality. In: Defining Soil Quality for a Sustainable Environment. SSSA Special Publication 29, Madison, WI: 61-89.

Grossman, R.B., Ahrens, R.J., Gile, L.H., Montoya, C.E. and Chadwick, O.A. (1995): Areal evaluation of organic and carbonate in a desert area. In: R. Lal, J. Kimble, E. Levine, B.A. Stewart (eds). Soils and Global Change. Adv,. Soil Sci. CRC Press, Boca Raton, FL

Hanson, J., and Lebedeff, S. (1987): Global trends of measured surface air temperature. Journal of Geophysical Research **92**, 13345-72.

Houghton, R.A. (1995): Changes in the storage of terrestrial carbon since 1850. In: R. Lal, J. Kimble, E. Levine, B.A. Stewart (eds). Soils and Global Change. Adv,. Soil Sci. CRC Press, Boca Raton, FL

Jenny, H. (1941): Factors of Soil Formation - a System of Quantitative Pedology. McGraw-Hill, New York.

Jones, P.D., Raper, S.C.B.,0 Bradley, R.S., Diaz, H.F., Kelly, P.M. and Wigley, T.M.L. (1986): Northern Hemisphere surface air temperature variations: 1851-1984. Journal of Applied Meteorology **25(2)**, 161-79.

Kay, B.D. (1996): Soil structure and organic carbon: a review. Proc. International Symposium. Carbon Sequestration in Soils, 22-26 July 1996, The Ohio State University, Columbus, OH.

Kimble, J.M., Tarnocai, C, Ping, C.L., Ahrens, R., Smith, C.A.S., Moore, J. and Lynn, W. (1993): Determination of the Amount of Carbon in Highly Cyroturbated Soils. In: D.A. Gilichinsky (ed). Joint Russian-American Seminar on Cryopedology and Global Change: Post Seminar Proceedings. Russian Academy of Sciences, Pushchino, Russia.

Lal, R. (1994): Sustainable land use systems and soil resilience. In D.J. Greenland and Szabolcs (eds) Soil Resilience and Sustainable Land Use, CAB International, Wallingford, U.K.: 41-68.

Lal, R., Kimble, J., Levine, E. and Steward, B.A. (1995a): Soils and Global Change. CRC. Lewis Publishers, Baco Raton. FL

Lal, R., Kimble, J., Levine, E. and Steward, B.A. (1995b): Soil Management and Greenhouse Effect. CRC. Lewis Publishers, Baco Raton. FL

Letey, J. (1985): Relationship between soil physical properties and crop production. Adv. Soil Sci. **1**, 277-294.

Schlesinger, W.H. (1984): Soil organic matter: A source of atmospheric CO_2. In: G. W. Woodwell (ed) The Role of Terrestrial Vegetation in the Global Carbon Cycle: Methods for Appraising Changes. Wiley and Sons, New York.

Schlesinger, W. H. (1995): An overview of the carbon cycle. In: R. Lal, J. Kimble, E. Levine, B.A. Stewart (eds). Soils and Global Change. Adv,. Soil Sci. CRC Press, Boca Raton, FL.

SSSA. 1987 Glossary of soil science terms. Madison, WI.

Soil Survey Staff (1975): Soil Taxonomy: A basic system of soil classification for making and interpreting soil surveys. USDA-SCS Agri. Handb. 436 U.S. Gov. Print. Office. Washington D. C.

Soil Survey Staff (1993): Soil Survey Manual. USDA, Government Printing Office, Washington D.C.

Addresses of authors:
J.M. Kimble
R.B. Grossmann
USDA-NRCS-NSSC
Fed. Bldg. Rm. 152, MS 33
100 Centennial Mall North
Lincoln, Nebraska 68508-3866
USA
R. Lal
The Ohio State University
2021 Coffee Road
School of Natural Resources
Columbus, Ohio 43210
USA

Soil Conservation for Mitigating the Greenhouse Effect

R. Lal & J.M. Kimble

Summary

World soils play a crucial role in the global carbon cycle. Erosion-induced soil degradation affects about 1.0 billion ha of land area by water erosion and 0.5 billion ha by wind erosion. The most severe erosion occurs in the ecologically-sensitive ecoregions of Asia (mean erosion rate 18 Mg/ha/yr), South America (19 Mg/ha/yr) and Oceania (40 Mg/ha/yr). Accelerated soil erosion has severe impacts on both productivity and the environment. A principal off-site environmental impact is due to the disruption of the global carbon cycle, resulting in an erosion-induced efflux of about 1.14 Pg C/yr from soil to the atmosphere. Furthermore, pathways and fate of carbon buried in depressional or depositonal sites, and in lakes and reservoirs are not fully understood. Conservation effective land use and soil mangement systems offer a potential of carbon sequestration in soil and terrestrial ecosystems through: (i) increase in soil aggregation, (ii) improvements in water and nutrient use efficiencies, (iii) decrease in losses of soil organic carbon, and (iv) increase in biomass production. There is a wide range of technological options for soil conservation, e.g. mulch farming and conservation tillage.

Keywords: Greenhouse effect, soil erosion, global carbon cycle, conservation tillage, soil organic carbon

1 Introduction

Soil conservation for carbon sequestration and mitigating the greenhouse effect is a new paradigm, although the importance of the soil's impact and of residue management on agricultural sustainability has long been recognized. In parallel with this broader perspective of soil conservation, the concept of agricultural sustainability also needs to be revisited. Because of the emphasis on both productivity and environmental quality, criteria for agricultural sustainability should include the following: (i) non-negative trends in per capita productivity, (ii) minimal risks of eutrophication of surface waters and contamination of ground waters, (iii) carbon sequestration in soil and terrestrial ecosystems for mitigating the greenhouse effect, and (iv) maintenance or improvements of soil quality. The latter is defined as "inherent attributes of soils that are inferred from soil characteristics or indirect observations" (SSSA, 1987), or "the capacity of a soil to function within ecosystem boundaries to sustain biological productivity, maintain environmental quality, and promote plant and animal health" (Doran and Parkin, 1994). Therefore, there is a close relationship between soil conservation, soil quality, agricultural sustainability, and environmental quality especially the greenhouse effect (Fig. 1).

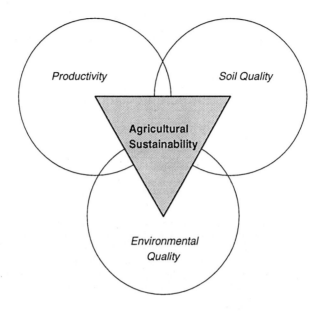

Fig. 1: Soil conservation affects agricultural sustainability though ist interactive effects on productivity, soil quality and environments.

Region	Suspended load	Dissolved load	Total
Africa	530	201	731
Asia	6,433	1,592	8,025
S. America	1,788	603	2,391
Europe	230	425	655
North and Central America	1,462	758	2,220
Oceania/Pacific Islands	3,062	263	3,355
World Total	13,505	3,872	17,377

Table 1: Global estimate of soil erosion (10^6 Mg/yr, Walling, 1987; Lal, 1994).

2 Soil erosion and agricultural sustainability

Reliable estimates of the global land area affected by soil erosion are difficult to make (Lal, 1994). Available statistics are based on two approaches: (i) assessment of sediment transport in the world's rivers (Walling, 1987), and (ii) estimates of land area affected by soil erosion (Oldeman, 1994). The accuracy of the first approach is based on the knowledge of the sediment delivery ratio, and that of the second on the knowledge of the quantitative magnitude of soil erosion, for often subjective and qualitative designation of erosional phases, e.g. slight, moderate and severe. The data in Table 1 show that total sediment load discharged into the ocean is about 17 billion Mg per year. Assuming a mean delivery ratio of 10%, gross soil erosion over the land surface is about 170 billion Mg per year. Global estimates of land area affected by soil erosion shown in Table 2 indicate that the area affected by water erosion is 1.1 billion ha and that by wind erosion is 0.55

billion ha. Combining this information provides estimates of soil erosion rates in different continents (Table 4). Mean soil erosion rates are estimated at 3.2 Mg/ha/yr for Africa, 5.7 Mg/ha/yr for Europe, 18.2 Mg/ha/yr for Asia, 19.4 Mg/ha/yr for South America, 40.4 Mg/ha/yr for Oceania and 15.9 Mg/ha/yr for the world. Low estimates of the mean soil erosion rate for Africa and Asia may be due to the high proportion of land area covered by deserts without water erosion.

Region	Area (10^6 ha) Water erosion	Wind erosion
Africa	227	186
Asia	441	222
South America	123	42
Central America	46	5
North America	60	35
Europe	114	42
Oceania	83	16
World	1094	548

Table 2. Global estimates of land area affected by soil erosion (Oldeman, 1994).

Accelerated soil erosion influences sustainability through on-site and off-site effects. Pimentel et al. (1995) estimated the combined cost of on-site and off-site effects of soil erosion at $44 billion per year or about $100 per hectare of cropland and pasture in USA, and $400 billion per year or $70 per person per year for the world. Dregne (1990; 1992) identified several regions in Africa and Asia where erosion-induced productivity decline is at least 20%. Lal (1995a) estimated that yield losses due to past erosion in Africa range from 2 to 40%, with a mean of 8.2% for the continent and 6.2% for sub-Saharan Africa. If accelerated erosion continues unabated, yield reduction by the year 2020 may be 14.5% for sub-Saharan Africa and 16.5% for the African continent (Lal, 1995a; b). Similar estimates can be made for other continents. These and other statistics have shown that accelerated soil erosion is a major threat to agricultural sustainability at farm field and global scales.

3 Soil erosion and greenhouse effect

Effects of accelerated erosion on soil organic carbon (SOC) dynamics in relation to emissions of radiatively-active gases from soil to the atmosphere are not understood. Possible pathways of SOC displaced by soil erosion are outlined in Figure 2. Two principal components are: (i) dissolved organic carbon (DOC) and particulate organic carbon (POC) transported in water runoff, and (ii) SOC contained in the soil. Some of the carbon displaced by erosion and that contained in runoff (DOC and POC) may be decomposed and released into the atmosphere. Lal (1995b) estimated erosion-induced efflux at 1.14 Pg C/yr. Erosion may also lead to deposition and burial of C in depressional sites over the landscape, lakes, reservoirs, and oceans. While the magnitude of deposition in the ocean floor is estimated at 0.57 Pg C/yr, the amount of C burial in depressions, lakes and reservoirs are not known. There is a need to quantify these effects at field, landscape, watershed, regional, and global scales. A conservation-effective technique of soil erosion management is the judicious use of crop residue mulch (Lal, 1995c). Properly used, crop residue mulch can decrease soil erosion, enhance soil quality, sequester C in soils and reduce the risks of greenhouse effect.

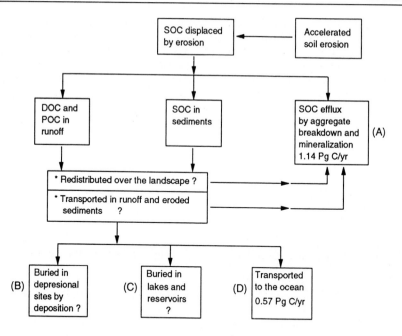

Fig. 2: *Pathways of soil organic carbon as influenced by accelerated soil erosion by water (estimates from Lal 1995b). DOC = dissolved organic carbon, POC = particulate organic carbon.*

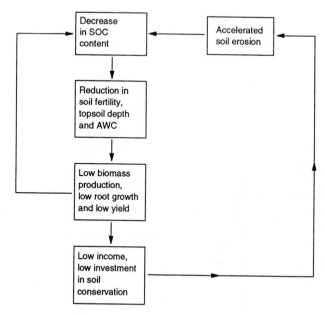

Fig. 3: *On-site effects of accelerated soil erosion on soil organic carbon content (SOC). AWC = available water content.*

Possible mechanisms of on-site effects of accelerated erosion on SOC dynamics are outlined in Fig. 3. Because of preferential removal, as evidenced by high enrichment ratios, erosion has an adverse impact on SOC content of the surface horizon. In contrast, depositional sites may gain in SOC content. Erosion-induced reduction in topsoil depth and its SOC content leads to: (i) decline in soil structure, (ii) depletion of soil fertility, especially N, and (iii) decrease in plant-available water capacity especially in the least-limiting water range (LLWR). Soil degradation and desertification, accentuated by erosion, cause low biomass production, low root growth, low yield, low income, low investment in soil conservation, and higher risks of accelerated soil erosion (Fig. 3). This is a vicious self-perpetuating cycle.

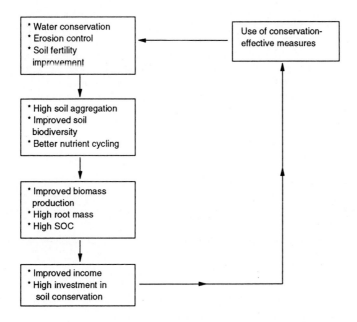

Fig. 4: On-site effects of soil conservation practices on carbon sequestration in soil.

4 Soil conservation and C sequestration in soil

There are at least three mechanisms of C sequestration in soil through soil conservation: (i) high biomass production, (ii) low mineralization rate due to favorable soil moisture regime, especially under semiarid conditions, (iii) low depletion rate of SOC due to erosion control. Although plot and farm scale data on these three mechanisms are available for some soils and farming systems, examples for watershed and regional scales are rare. It is, therefore, difficult to aggregate the data on C dynamics at the global scale. Soil conservation enhances carbon sequestration through improvements in: (i) water conservation, (ii) nutrient use efficiency, (iii) soil aggregation, (iv) soil biodiversity, (v) nutrient cycling, and (vi) biomass production by deep rooted plant species varieties (Fig. 4). High income generated through high production permits investment in soil conservation and adoption of conservation-effective farming systems.

There is a wide range of technological options for soil conservation including those based on: (i) soil management for improving soil structure and decreasing erodibility (e.g. conservation tillage, mulch farming, cover crops), (ii) techniques to minimize runoff by allowing more time for water to infiltrate into the soil (e.g. rough seedbed, contour/tied-ridges), (iii) decreasing runoff velocity by vegetative barriers and agroforestry techniques (e.g. alley cropping, vetiver grass planted on the contour), and (iv) safe disposal of runoff through engineering structures (e.g. terraces, drop structures, chutes).

Two principal scenarios of C sequestration by soil conservation are through: (i) additional biomass produced and crop residues returned to the soil, and (ii) reduction in C efflux caused by mineralization and decomposition as well as by SOC displaced by soil erosion. Lal (1994) estimated total crop residues production at 3462×10^6 Mg/yr containing a total of 1.5 Pg C/yr. If improved soil conservation measures can produce an additional 10% of crop residue, it may lead to additional C sequestration of 0.15 Pg C/yr (Table 3). This is definitely a high estimate scenario which may be justified by the fact that estimates of crop residue do not include the root biomass and that high biomass production decreases erosion-induced losses in SOC. The second mechanism is through reduction in soil erosion. Lal (1995b) estimated that 1.14 Pg C/yr may be emitted from the SOC pool displaced by soil erosion. It is difficult to estimate what fraction of this carbon may be sequestered by soil conservation.

Particular	Crop residue (10^6 Mg/yr)	Carbon content (Pg C/yr)
Amount of crop residue produced	3462	1.5
Additional residue produced with conservation (10% of the total)	346.2	0.15

Table 3. Carbon sequestration in crop residue due to soil conservation practices.

Region	Estimates of[1] gross erosion rate (Mg/ha/yr)	Total SOC lost[2] in eroded sediments (Tg/yr)	C sequestration[3] by soil conservation (Tg/yr)
Africa	3.2	15.9	0.4
Asia	18.2	193.0	4.8
South America	19.4	53.6	1.3
North and CentralAmerica	20.9	43.9	1.1
Europe	5.7	6.9	0.2
Oceania	40.4	91.9	2.3
World total[4]	15.9	405.2	10.1

1 Calculated by dividing figures on total sediment load shown in the last column in Table 1 with land area affected by erosion shown in the first column in Table 2.
2 Computed from suspended load shown in column 1 in Table 1 assuming SOC content of 3%.
3 Assuming erosion control is practiced to sequester 1/10th of the 25% of the C being released into the atmosphere (Lal, 1995b).
4 All figures in Table 4 should be divided by the delivery ratio in fraction e.g. 0.1 if the delivery ratio is assumed to be 10%.

Table 4: Estimates of soil erosion rates in different continents and potential of C sequestration.

Another approach to assess the potential of carbon sequestration by soil conservation is through evaluation of reduction in sediment transport. The data in Table 4 depicts the actual SOC transported to the ocean at 405 Tg/yr. The last column in Table 4 shows that if only 1/10th of the 25% of the C being released is sequestered through conservation, total sequestration is about 10 Tg/yr. These calculations, however, do not involve the delivery ratio factor of 10%. Thus, potential of total C sequestration through soil conservation is about 10 times the 10 Tg or 0.1 Pg/yr. Therefore, the total potential of carbon sequestration through soil conservation at global scale is about 250 Tg C/yr or 0.25 Pg C/yr (Table 5). The potential is much greater if a large proportion of 1.5 Pg C/yr produced in the residue can be sequestered in soil through judicious management.

This potential is achievable through adoption of conservation tillage, agroforestry techniques, vegetative barriers, conservation reserve program, use of deep rooted varieties, strip cropping, terraces and other engineering structures. There exists a much larger scope in restoration of degraded soils.

Activity	Carbon sequestration rate (Tg C/yr)
(i) Erosion control	100.0
(ii) SOC through additional residue return	150.0
Total	250.0

Potential of C sequestration through management of 1.5 Pg C/yr contained in crop residues produced worldwide (Lal, 1995a) is much greater.

Table 5: Potential C sequestration as soil organic carbon by global soil conservation.

5 Conclusion

1. Agricultural sustainability refers to non-negative trends in per capita productivity, minimal risks of contamination of surface and ground waters, and carbon sequestration in soil and terrestrial ecosystems.
2. Soil conservation plays an important role in agricultural sustainability while meeting all three criteria.
3. Potential of C sequestration through erosion control is about 100 Tg C/yr and that through additional crop residue produced is about 150 Tg C/yr, with a total potential of about 0.25 Pg/yr.
4. Much larger potential of C sequestration in soils and terrestrial ecosystems exists through residue management and restoration of degraded soils and ecosystems.
5. Special impact concerning mitigation of greenhouse effect

References

Doran, J.W. and Parkin, T.B. (1994): Defining and assessing soil quality. In: Defining Soil Quality for a Sustainable Environment, SSSA Special Publication No. **35**, 3-21.
Dregne, H.E. (1990): Erosion and soil productivity in Africa. J. Soil Water Conserv. **45**, 431-436.
Dregne, H.E. (1992): Erosion and soil productivity in Asia. J. Soil Water Conserv. **47**, 8-1
Lal, R. (1994): Global overview of soil erosion. In: Soil and Water Science: Key to Understanding Our Global Environment, SSSA Special Publication **41**, 39-51.
Lal, R. (1995a): Erosion-crop productivity relationships for soils of Africa. Soil Sci. Soc. Am. J. **59**, 661-667.
Lal, R. (1995b): Global soil erosion by water and carbon dynamics. In: R. Lal, J.M. Kimble, E. Levine and B.A. Stewart (eds) "Soils and Global Change", CRC/Lewis Publishers, Boca Raton, FL: pp. 131-141.

Lal, R. (1995c): The role of residues management in sustainable agricultural systems. J. Sustainable Agric. **5**, 51-78.

Oldeman, L.R. (1994): The global extent of soil degradation. In: D.J. Greenland and I. Szabolcs (eds) "Soil Resilience and Sustainable Land Use" CAB International, Wallingford, U.K., pp. 99-118.

Pimentel, D., Harvey, C., Resosudarmo, P., Sinclair, K., Kurz, D., McNair, M., Crist, S., Shpritz, L., Fitton, L., Saffouri, R. and Blair, R. (1995): Environmental and economic costs of soil erosion and conservation benefits. Science **267**, 1117-1123.

Soil Science Society of America (SSSA) (1987): Glossary of Soil Science Terms. SSSA, Madison, WI.

Walling, D.E. (1987): Rainfall, runoff and erosion of the land: a global view. In: K.J. Gregory (ed), "Energetics of the Physical Environment", J. Wiley & Sons, U.K., 89-117.

Addresses of authors:
R. Lal
School of Natural Resources
The Ohio State University
210 Kottman Hall
2021 Coffey Road
Columbus, OH 43210, USA
J.M. Kimble
National Soil Survey Center
Federal Building, Room 152
100 Centennial Mall North
Lincoln, NE 68508-3866, USA

Impacts of Possible Climate Change Upon Soils; Some Regional Consequences

H.-W. Scharpenseel & E.M. Pfeiffer

Summary

Impacts of climate change upon soils have not been studied to the same extent as the effects of soil born contributions, such as GH (greenhouse) forcing trace gas emissions to climate change. Quite obvious interrelations exist between continuous functioning of the thermohaline heat transport to N-and W-Europe, that could be interrupted by a temperature rise related glacier melting and termination of the salt rich heavy deep-water conveyor transport to the Caribbean and Antarctic regions. This would greatly impede the Gulf Stream related temperature rise, and eventually soil and land use productivity.

Similarly, the S-Asian soil stability and land use system could suffer from shortening or suppression of the monsoon due to a rise of aerosol concentrations in the atmosphere by vigorous industrial development in S- and SE-Asia.

The ENSO (El Nino Southern Oscillations) related climatic extremes contribute to the more catastrophic soil destruction, and to the mainly anthropogenic forced erosion of presently ca 10 million ha per annum.

This paper tries to identify and to assess the effects of climate change upon soils of different ecosystems, mainly via its influence upon the carbon and nutrient cycles.

Problem areas of major geographic regions of the continents are scanned for expected impacts of climate change upon soils and land use.

Keywords: Climate change, causes, soil fertility features, major regional threats

1 Soils and climate change

The interaction between the processes of soil dynamics/degradation and those of regional climate change impacts on land use and soil emissions on climate change have been studied over the last decade. The interaction of soil dynamics on climate change and climate change on soils are rather synchronous forcing complexes, not easily to isolate as individual timely sequences. In Lovelock's Gaia-Hypothesis the world is reflected as a kind of super organism, capable of warding off external impacts and damaging trends by the resilience of a joint curative action from the part of the diverse biospheric species, especially the microflora (Lovelock,1979, 1993). Thus, Lovelock & Kump (1995) demonstrate that in a cold climate under glacial conditions, feedback of cloud albedo, marine algae and plant-mediated CO_2 are all negative: increasing temperature will cause increased albedo and increased CO_2 consumption rates, geared to reduce global temperature.

Similarly, rising temperature in oceans leads to enhanced planktonic dimethylsulfide production, which by oxidation to methylsulfonic acid and sulfuric acid typically reflects solar input and exerts a cooling effect, just as the SO_2 emission, associated with the combustion of fossil C sources or, derived of volcanic actions, after oxidation by hydroxyl radicals to sulfuric acid counteracts temperature rise due to CO_2 related forcings. Climate plus land use related effects of soil degradation are much less (if at all) restored by Gaia-hypothesis-like interactions.

Recently Pimentel et al. (1995) tried to estimate the costs or inversely the possible conservation benefits, arising from climate (wind and water) related soil erosion or its curtailment. Erosion damage, mainly nutrient losses, amounted to ca 400 billion $ per year worldwide (ca 44 billion $ alone in the USA) and a soil loss of about 10 million ha per year. Erosion control and implicit nutrient saving with an emphasis on recycling and relaxing the use of agrochemicals is recommended by Sanchez (1994).

Possible effects of climate change on soils tacitly imply a veritable rise of temperature as well as a change of precipitation and atmospheric ventilation patterns. Their corresponding effects exist on the C cycle (C-sources and sinks), response in biomass production, C-sequestration, type of clay mineral formation, intensity of erosion, soil conservation, rise of sea level, and potential resilience by diverse ecosystems to profound changes.

The GCM's (global circulation models) as a means for predicting GH (greenhouse) forcing suffer from inadequate inclusion of small area varying of water vapor, cloud formation, aerosols, especially sulfate nuclei (from SO_2 and $(CH_3)_2S$ oxidation); an element of uncertainty exists. Is the GH-forcing with the consequence of climate change real or are we just dealing with historical statistical noise variations (Lovelock & Kump, 1994; Crutzen et al., 1995; Taylor & Penner, 1994; Wigley, 1994; Wallace & Gerlach, 1994; Khalil & Rasmussen, 1994; King & Schnell, 1994; Broecker, 1994; Schnell, 1995; Keeling et al., 1995; Kerr, 1995a; Deming, 1995; Regelado 1995; Kerr, 1995b). Recently, Kerr (1995b), Heimann (1995), Keeling et al. (1995) and others made a rather cautious but firm commitment with regard to the real existence of temperature rise and climate forcing, relying on observation of temperature rise versus cooling aerosol-sulfate emission.

These results come from the GCM's, the Mauna Loa Observatory, Hawaii, and a South Pole Station as well as the MPI (Max Planck Institute) in Hamburg. In the Mauna Loa Observatory over the last 35 years, CO_2 as well as ^{13}C contents in the atmosphere were consistently monitored, changes were recorded against the background of El Nino cycles. The arguments in favor of the real existence of temperature rise beyond the statistical noise level stand against the observations, mitigating the predictions of climate change, mainly based on the GCM's. The pro-climate-forcing arguments seem recently to have gained ground.

The latest IPCC-report of 1995 identifies a "discernable human influence on climate". Yet, recently Crowley and Kim (1996) invoke new findings regarding the influence of the sun cycling effect on climate change at the expense of the GH-forcing trace gas contribution. This dispute remains virulent. A kind of contemporary effect may be the findings of Sellers et al (1996). They feel that the temperature rise due to CO_2 increase and also the transpiration (evapotranspiration) decrease of the soil - vegetation system should also be considered, since it enhances the ambient temperature. In contrast, Berner et al (1996) drew attention to the findings, that large changes in CO_2 concentration throughout geological history reflect no parallel to temperature according to C-isotope measurements in carbonates and fossil marine phytoplankton. The test curve stretches back 950 million years. Water vapor and albedo changes are more important as forcing mechanisms than CO_2 concentration, which was anyway in the course of the present Quarternary, lower than before.

2 The soil related part of the carbon cycle

SOM-C (soil organic matter) of the pedosphere, with ca 1550 Pg C is the largest accessible organic C-pool (Eswaran et al., 1993). If we add the ca 1200 Pg C of the secondary soil carbonates, which

to a certain extent, as the $\delta^{13}C$ measurements allow us to assess is of organic C (respiratory C) in origin, then this is even more pronounced. The economically accessible fossil C of oil, gas and coal comprises only about 1200 Pg C according to more recent estimates (Grassl & Klingholz, 1990). From the total organic C in the global vegetation system ca 75% are sequestered in the SOM, and ca 25% of the living biomass including woods. The organic carbon pools in woodlands are the largest, according to Dixon et al (1994) with 787 Pg C in the SOM and with 359 Pg C in the living biomass. This is followed by those in grassland of ca 300-350 Pg C in the SOM, ca 40-50 Pg C in the living biomass, in the 1.5 billion ha cropland area about 120-150 Pg C, and in the 136 million ha riceland 12-15 Pg C in the SOM. Due to land clearing and grassland conversion into cropland ca 55 Pg C of the autochthonous SOM-C are estimated as having been lost and released to the atmosphere (Cole, 1995). Inversely, ca 18-24 Pg as litter and remnants-C developed as a new sink. The C balance in terrestrial soils was recently assessed to be, mainly due to CO_2-fertilization, slightly positive. Carbon pools in major ecosystems and zones comprise, based on 2700 soil profiles, a SOM reservoir of 1400 Pg C according to Post et al (1982) and Degens (1989). They show:
- 24.5% of total C in wetlands
- 13.7% of total C in tundra
- 12.0% of total C in cropland (agricultural areas)
- 9.5% of total C in wet boreal forests
- 9.6% of total C in tropical woodland and savanna
- 8.6% of total C in cool temperate steppe

(plus ca 375 million ha = 1500 Pg C as peat = 400 kg C/m^{-2})

ABOVE GROUND DRY MATTER NPP ACCORDING TO CLIMATE ZONES	
Humid tropics, without dry season	20,000 kg/ha/yr or more
Humid tropics, short dry season	20,000 kg/ha/yr
Subhumid tropics, moist	10,000 kg/ha/yr
Subhumid tropics dry	5,000 kg/ha/yr
Semi arid zone	2,500 kg/ha/yr

Table 1: Natural dry matter production in climate zones (Young, 1989)

Tropical forests without further human pressure have a potential to sequester up to 2.5 Pg C annually (Brown et al., 1992). An ambiguous complex of C sequestration concerns the ca 2 Pg C of the MCF (Missing Carbon Fraction), that remain from the 7 Pg C of annual anthropogenic emission after subtracting the 3.5 Pg C of atmospheric uptake and the 1.5 to 2.0 Pg C, that disappear in the biota and carbonates of the oceans. The initial controversy regarding the MCF's uptake and transport by the N-Atlantic conveyor circulation to lower latitudes, or preferably sequestration by more biomass production in mainly higher northern latitudes, has been diversified by more recent claims for inputs in more biomass of (sub)tropical ecosystems (Table 1). The direct and indirect impact of the MCF-induced rise of biomass and SOM production on soils is obvious. An increase in SOM relates to higher sorption capacity, especially higher CEC (cation exchange capacity) in tropical LAC (low activity clay) soils, rise of biological activity, water holding capacity and buffering potential. Otherwise, more biomass and SOM production may excessively consume CO_2 with the result of CO_2 related temperature reduction. Additional sequestration of atmospheric CO_2 into SOM-C is anyway one of the major chances for land use measures to mitigate CO_2-, temperature rise and climate change.

3 GH-forcing due to rise of CO2 concentration: ist effect on the soil vegetation system

The major global thermostat is water vapor. It sustains the GH-forcing that lifts the global mean temperature from -17°C to presently +15°C. More than half of the radiative forcing by the trace gases comes from CO_2 and contributes with 356 ppmv some 1.56 W/m^{-2}. Methane with 1.7 ppmv renders about 0.5 W/m^{-2}, N_2O with 310 ppbv 0.1 W/m^{-2}, CFC's contributed 0.3 W/m^{-2}, but the unsaturated compounds (after the Montreal Protocol) are so far only 0.05 W/m^{-2}. Tropospheric ozone adds 0.2-0.6 W.m^{-2}.

Inversely, the depletion of ozone in the stratosphere, mainly due to halogen carbon compounds, requires the subtraction of 0.1 W.m^{-2}, and the negative forcing due to sulfate aerosols cuts 0.25 to 0.9 W/m^{2}. A further 0.05-0.6 W/m^2 is subtracted for aerosols from burning of biomass, thus reducing the ca 3 W/m^2 due to radiative forcing by about 1/3 to ca 2 W.m^2 (Tiger Eye, NERC, Summer 1995).

Changes of CO_2 concentration are considerable; the Eemian 80 ppm rise, the Holocene adjustment to ca 280 ppmv, with recent changes of 280 ppmv in 1880 to 315 ppmv in 1959 and to 356 ppmv in 1993.

The climate change promoting radiative forcing of CO_2 could have a positive side effect due to higher biomass production in consequence of the present annual ca 0.5% CO_2 rise (1.5 ppmv), especially in C3 plants, which comprise about two-thirds of all crop plants and most wood plant species. Decrease in transpiration, i.e. water saving by shrinking stomatal aperture, is a valuable by-product. The loss of the cooling effect by the reduction of transpiration may add mildly however to a temperature rise.

In the dispute regarding extent and compatibility of rising biomass production due to CO_2 fertilization with land use and land degradation related CO_2 emission, Goudrian & Unsworth (1990) ascribe 5-10% of the actual worldwide increase in agricultural productivity to CO_2 fertilization by rising atmospheric CO_2.

Esser & Lieth (1989) and Esser & Lautenschläger (1994) arrive by modeling and stable isotope measurements at a slight superiority of CO_2 fertilization related C sequestration compared with CO_2 emission by land use and wood-clearing. In addition, CO_2 burial as $CaCO_3$ (Olson et al., 1985) plays a role for CO_2-C sequestration in continental waters as well as in dryland soils.

Regarding the levels of present annual cereal production of ca 1970 million t, and the estimated demand of 2450 million t by 2000 and 3970 million t for 2025 (Borlaug & Dowswell 1994), yield increases due to CO_2 fertilization receive enormous relevance.

Absolutely optimistic expectations for considerable grain yield increases can be derived from the intensive field experiments in controlled elevated CO_2 atmosphere by Idso (1990), Kimball et al. (1993), Rogers et al. (1994), Culotta (1995) with about a 30-33% rise of biomass production by doubling of ambient CO_2. For forest ecosystems, comprehensive field experiments assessing the NEP (net ecosystem production) under elevated CO_2 conditions, are still required (Post et al., 1992).

The 60 PgC/y of NPP, balanced by CO_2 release from biomass decomposition, represents about 10 times the CO_2 released from burning of fossil fuel and about 1/10 of the living biomass-C. The C-flux components are still under study. Post et al (1992) cite Armentano (1980) with 0.14 Pg C/y accumulation in organic soils, Schlesinger (1985) with 0.01 Pg C/y, adding to further production of soil carbonates, Meybeck (1981) with about 0.45 Pg C/y, transferred into the oceans by fluvial transport. Post et al. (1992) estimated a net terrestrial flux of 1 to 3 Pg C/y; Blanis et al. (1995) using their TCCM (terrestrial carbon cycle model) for 1990 calculated a biosink of 0.9 Pg C.

The different potential for natural dry matter production in various climate zones is listed by Young (1989) (Table 1). An important critical survey of the consistency of rising biomass productivity under elevated CO_2 concentration was undertaken by Bazzaz & Fajer (1992). They identified at least 7 critical preconditions, such as soil resources, N-supply, light and temperature, plant ontogeny, non-linear effects, changes in species composition, or evolutionary changes in ecosystems. Possible genetic effects of continually rising atmospheric CO_2 must also be considered. In consequence, there remains some ambiguity regarding the compound effect of rising CO_2 concentration with other growth factors, those that can vary (temperature, rainfall, wind), those that are temporarily constant (SOM, humus, clay, texture, structure of soils) or those, that decrease in concentration such as plant available nutrients. The impact of the CO_2 as a major radiative forcing trace gas component, plus related rise of biomass production and eventually temperature upon soils of different regions and ecosystems, cannot presently be conclusively quantified.

Process	Methodology used	Author, result
N-Atlantic carries 1.2 ± 0.2 10 PW of heat northwards		
Water volume of conveyor ca. 17 Sv (1 Sverdrup = 10 mil $m^3.s^{-1}$)		
Slowing or shut down of conveyor by release of ca. 0.06 Sv glacial melting water	Modeling	Rahmsdorf,1995 Chance for conveyor to slow down or stop
May be weaker flow of thermohaline, but gradual till total recovery	Modeling	Manabe & Stouffer 95 Relative stability
Dating in water, sediments, especially iceberg invasion related Heinrich sediment layers	U/Th and ^{14}C dating	Broecker, 1996 Stable even during Dryas
Measurements in Atlantic and surface sediments of thermohaline path	$^{231}PA/^{230}Th$	Yu et al.1995 Relative stability
Events in tropical Atlantic consistent with N-Atlantic region, forcing mechanisms of a more global nature	Relative Reflectivity of fresh surfaces, AMS-^{14}C-dating	Hughen et al.,1996 Relative stability

Table 2: Climate change and conveyor stability

4 Effects of climate change upon soils

Non-sustainable agricultural land use is defined as agricultural mining. But Newton (1995) concludes that "we simply do not have a sufficient physical, biological and chemical understanding of the integrated Earth System to do more than balancing a desire for minimal perturbations against short term economics". The last few decades and the immediate future are in focus in most isotope measuring and modeling efforts, and evidence is mounting, that accomodation of the "missing carbon sink" is the most relevant immediate problem. These 1.5 to 2.5 Pg C of the annual anthropogenic C input, unaccountable as such by estimates of the atmospheric, oceanic and terrestrial C sinks, are due to improved methodological sensitivity for gas and isotope ratio analysis lately predominately identified and assigned to the terrestrial biosphere, mainly in the northern hemisphere and the middle/higher latitudes. Since the atmospheric and oceanic pattern of functioning are better understood than the terrestrial biosphere, uncertainties remain regarding the robustness of the large terrestrial sink for anthropogenic C. The possibly limited pool size for continued C-acceptance comes into the picture. Also the background of CO_2 fertilization, of saturation problems, of global impacts on soil vegetation-ecosystems and, not the least, of demands for higher cropland productivity, requires urgent elaboration.

With rising CO_2 concentration and biomass formation the different N supply pattern of the life zones/ecosystems poses major constraints, evident by the widening C/N ratio in some cases of the rising biomass output. Post et al (1985) observed characteristic C/N ranges in the system comprising climate-related biotic processes, such as productivity of vegetation and inversely organic matter decomposition.

The total N-pool in the thin pedosphere (0-100 cm) amounts to ca 95 Pg of N (versus 1550 Pg of C). N storage ranges from 0.2 kg/m^3 in warmer deserts, 1.6 kg/m^3 in subtropical wet rain forests to 2 kg/m^3 in rain tundra soils. Correspondingly, the C/N varied from <10 in tropical deserts to >20 in cool wet forests or rain forests. In cool life zones the C/N ratio ranged from 15-20, and in warm life zones from 10-15. This indicates that in tropical climate zones there are large amounts of decaying humic substances with low C/N, and in temperate regions more variability depending on the C/N of the predominant vegetation. In wet tundra regions with slow biomass turnover, the C/N is high.

Thus, rising CO_2 levels causing climate change, come into conflict with limits of plant available N in their regional impact on soils via SOM-enrichment, complexing with clay or oxides, SOM-solubility and its removal by chelating percolation. Temperature change generally exerts an influence on the functional SOM compartments, for example those according to Jenkinson & Rayner (1977) or those according to Parton et al. (1987) (see also Woomer et al. (1994)), in the Century Model. The effect is mainly to accelerate the turnover of the lower C-residence time compartments.

Human-induced versions of soil degradation according to GLASOD are estimated to be 55.6% caused by water (1093.7 million ha), 27.9% by wind (454.2 million ha), 12.2% chemical related (239.1 million ha) and 4.2% physical related (83.1 million ha), for a total of 1964.4 mil ha.

Commonly one estimates changes due to anthropogenic measures as being more pronounced and destructive than those arising from climate change (Brinkman, 1990). The most striking processes of land degradation with potential impacts by changing climate include:

- Sufficiently moist soil, exposed to rising solar radiation looses SOM (LAC-SOM-complexes >HAC-SOM-complexes in loss of nutrient storage capacity), further loss of porosity and structural aggregate stability.
- Land pressure by rising population, by desertification, climate/weather extremes, by extended drought periods as well as ENSO-related floods.
- Erosion of exposed soils or nutrient rich sediments due to wind-, rain-, and sun-related weather/climate extremes: reducing depth of solum, water infiltration, depth of racination, of the nutrient capital.
- Flood risk increase for more unpredictable precipitation regimes, especially in low-lying land, concomittant with silting-up of dams and river beds in higher areas.
- Irreversible, man and climate-extreme related land destruction by degraded patches forming extended areas.
-
- Efforts of resilience:
- Alternatives to slash and burn as well as to subsistence agriculture.
- Restoration of degraded land by reforestation, agroforestry rotations, pasture improvement with deep rooting varieties.
- Mulch, compost, greenmanure, fertilizer.

Sombroek (1990) states, that "in the tropics the Soil Classification related soil properties will, at not more than 2°C temperature rise not show great changes within 50 to 100 years". The effects of climate change upon soils manifest themselves mainly in the soil temperature regime and hydrological characteristics.

Land degradation as caused by climate change is discussed in IGBP No 34 (1995) and IGBP No 36 (1996). The landscape transects of the GCTE program will hopefully contribute effectively to assessment of soil destruction and resilience.

The growing gap between population increase and arable land availability vs N-fertilizer use was estimated by Cole et al. (1993). While until 2025 in temperate plus (sub)tropical areas the expansion of arable land will be about 36% and of N-fertilizer use 49%, the population increase will amount to 63%.

Soil degradation by climate related actions associated with physical transport, i.e. erosion, or with SOM losses by respiration as well as leaching (Scharpenseel & Becker-Heidmann, 1994). Some measures to enhance resilience and mitigation of soil degradation by SOM loss include:
- Mitigate SOM loss due to biotic or abiotic (protolytic in acidic soils, photochemical in dryland soils with high light intensity) turnover.
- Reducing SOM leaching and respiration in all but Aridisols, reduce turnover under submergence in paddy soils (water temperature > 28°C), further prevent soil fossilization due to burial.
- Limit decline of SOM quality with widening C/N-ratio (eutrophic < mesotrophic < oligotrophic < dystrophic), being enhanced by lack of soil protection and increasing acidification.
- Manage intelligently changes of plant culture and shift of woodland or grassland into cropland, avoiding SOM to become a carbon source.

(Classic C sinks being Entisols and riceland/wetland <28°C water temperature, as well as Andisols during the approximate 10,000 year phase of Allophane-SOM-complex formation)

5 Impact of climate on soils of different regions and ecosystems

The results of the different intensity of biomass production as well as soil degradation is expressed by C storage in soil as well as in plant biomass of the different ecozones; lowest in savanna, highest in tropical forests. Grassland-SOM obviously commands high SOM-C resilience against decomposition (Whittaker & Likens,1973). Regional differences of climate variability can be quite effective, particularly rapid alternations in the tropics (Zahn, 1994) (partly reflected from ice sheet measurements of isotope ratios).

Moisture availability determines the degree of C sequestration and of biomass productivity. In temperate and boreal wetlands C sinks decreased since 1850 by ca 50%, i.e. from 0.2 to 0.1 Pg C/y. Coastal and alluvial wetlands C sinks require protection to minimize loss of C by agriculture and rural development (Downing et al., 1993). Lugo & Wisniewski (1992) estimate the carbon storage in coastal wetlands as amounting to ca 0.3-0.6 Pg C/y. By fertilization with Fe, in southern oceans usually the marine minimal nutrient, as well as by rising P and N loads of rivers, the primary production in coastal ocean areas can rise according to Richard (1992) by about 20 %.

In the subtropics/semiarid areas the depletion of rainfall by climate change enhances salinization, leading to decreasing crop revenues (Kempe, 1993).

In boreal and subarctic peatlands ca 455 Pg C accumulated at a rate of ca 96 Tg C/y in the course of the postglacial period (Holocene).

Through long term drainage, an oxidation to CO_2 can be expected at a rate of 8.5 Tg C/y plus an amount of ca 26 Tg C/y by combustion of peat as fuel plus an emission of 46 Tg C/y as CH_4 by metanogenesis.

Global warming effects of permafrost thawing, thermokarst erosion and thaw lake formations are still little investigated, as is the increasing role of fire in the C-cycle of peatlands, which requires urgent action (Gorham, 1991)

The C sinks in riceland with its 12-15 Pg C in the SOM were reduced to about 60 Tg C/y. Rice soils at <28°C paddy temperature seem to be minor sinks, with >28°C paddy temperature slight C-sources (Downing et al., 1993; Scharpenseel & Pfeiffer, 1995; Scharpenseel et al., 1996).

The rise of CO_2, contributing to slow temperature rise, expected to be of the order of 2°C by the end of the next century, will probably increase rice yields by CO_2 fertilization and decrease the rice yields temperature rice related (Neue et al., 1995). Also the increase in UV-B-radiation as a consequence of ozone depletion is liable to damage rice plants and crop yields The global increase in methane emission, partly produced in the 136 million ha of riceland, contributes ca 20% of the anthropogenic forcing of global warming (Neue et al., 1994; Neue & Scharpenseel, 1987).

Very critical for climate and soil stability is the durability of the conveyor and the thermohaline circulation in the 3-4°C warmed-up European - Atlantic space. It is assessed differently by various authors, based on modeling (Rahmsdorf, 1995), on C-14 dating of iceberg invasions and "Heinrich-sediment" layers by Manabe & Stouffer (1995), and on U/Th and C-14 dating (Broecker, 1996), indicating constancy of the conveyor even during the Younger Dryas, or on Pa-231/Th-230 measurements in Atlantic and surface sediments of the thermohaline path (Boyle, 1995; Yu et al., 1996). So far optimism for ongoing stability seems to prevail (Table 2 and 3).

Region or Ecosystem	Change to advantage	Change to disadvantage
BOREAL CLIMATE	Possibly end of next century by 1-3°C rise, deciduous tree zone migrates to higher latitudes	
SUBTROPICS		Drier, less luscious for livestock, probably ENSO-related, extremely dry phases and floods
TROPICS	probably no great change	
EUROPE	N-Central Europe by higher temperature and possibly more rainfall, higher biomass/humus; S-Europe possibly drier	Possibly decrease of warming by Gulf Stream thermohaline higher than gain by GH-forcing
AMERICA, especially CENTRAL		El Nino related droughts and cyclones
S-AFRICA		El Nino related rain floods and droughts
AUSTRALIA		El Nino related drought periods and rain floods
INDIA		Monsoon decrease due to rising aerosol release by industrialization
CHINA	Higher biomass by CO_2-rise yields more nutrients from organic residues; estimate is 35.6% increase by year 2000 (Chang, 1995)	

Table 3: Impact of climate change upon soils and ecosystems

Hughen et al. (1996) report that abrupt climate changes as well as rapid climate oscillations, such as in the Younger Dryas, were synchronous with forcing mechanisms of a more global nature, such as changing of upwelling rate and nutrient supply, caused by thermohaline affected sea surface temperature. Apparently events occurring in the tropical Atlantic region were consistent with homologs in the N-Atlantic region (Charles, 1997).

The effect of climate change on the agricultural and horticultural potential especially in Europe was assessed by Kenny et al. (1993), see also climate factors involved in soil degradation in World in Transition,Econ. Bonn (1995):

1. Level of solar radiation
2. Level of precipitation
3. Level of humidity
4. Level of air temperature
5. Level of wind speeds
6. Hydrological regime (resulting from 1-5)
7. Effect of 1-6 upon species of (micro)flora and (micro)fauna
8. Exerted effects by human migration in consequence of climate change

The resulting variation in impact is linked to the effect of the individual forcing principles.

The topsoil loss of ca 24 billion t/y and the percentage dryland surface in Africa and N-America that is already under degradation (70-75%), are alarming and liable to increase disproportionally in case of global warming and higher aridity (Wissensch. Beirat der Bundesregierung Bonn, 1994; Down to Earth, Desertification, UNEP, 1995).

Predominant stress factors on soils by global change are:
- Balance between production and decomposition of SOM,
- Climate and land use change versus accumulation and release of substances,
- Biomass production losses by change of CO_2, UV-B-radiation, temperature, precipitations,
- Required protection of fragile soils and ecosystems.
- Residues in water and food chain, soil as filter and buffer,conservation of soils and vegetative soil cover as carbon reservoir.

The response of soils to global change has different facets, triggered by climate change, change of atmospheric composition,and changes in global and human ecology. Among the major contributing degrading principles and effects of land degradation, those which are climate change related are but one important forcing complex compounded by anthropogenic forms of land conversion by soil erosion, groundwater related soil drought, and degradation or change of vegetation.

An overview of vulnerable ecosystems and those exposed to higher risk in relation to effects of climate change is given by the World Wildlife Fund (WWF, 1992).

As front-line ecosystems, those sensitive to temperature and sea level changes are coral reefs in relation to warming periods, mangrove woods in relation to sea level, and arctic marine ecosystems in relation to sea water warming (ice melting).

Highly vulnerable ecosystems include mountain systems, coastal wetlands, tundra and permafrost areas of the arctic (soil erosion) and boreal as well as temperate forests with change of habitats and soil genesis due to warming.

High risk ecosystems, certainly in jeopardy but with the type of impact by climate change less clear, appear to be wet and dry savanna exposed to changing rainfall and temperature, and tropical moist forests where temperature, humidity and rainfall changes are highly disturbing to plants and animals with narrow tolerance. Soils are less affected. McDougall (1995) emphasized that expected temperature rise and corresponding effects will in the southern hemisphere be comparable with trends in the northern hemisphere and not much lower as models predicted.

For grassland, global change effects found so far are less than for woodlands. The high C levels in tundra, boreal and temperate forest soils (Whittaker & Likens,1973) make grassland effects appear very conservative.

Hall et al. (1994) estimate GCM's-C losses of 0-14% in the 0-20 cm epipedon. Soil-C losses were in European grasslands higher than in tropical grassland and savannas. Productivity along with altered climate depended mainly on changes in rainfall and N-mineralization rates, while temperature acted primarily on soil (soil organic matter) decomposition.

CO_2 fertilization effects could reduce the impact of climate alterations by net C accumulation in tropical savanna regions. Detection of climate change effects will require, according to Hall et al. (1994), a minimum of 16% change in NPP and of 1% in soil-C.

Management pressure, overgrazing,and frequent burning can impede the C-sequestration even within a rising CO_2 regime.

According to a study of New Zealand tussock grassland, based on bomb ^{14}C enrichment, global warming is expected to decrease current SOM content up to 10% at a temperature rise of 0.03°K/y . An additional C-source of 0.5 Pg C to the atmosphere could develop (Tate,1992) during the next 60 years.

Savanna ecosystems are, according to Alexandrov and Oikawa (1995), probably a sink of C since the last century at a level of ca 0.2 Pg C/y, i.e. ca 10% of the missing carbon fraction pool. A climate change related transition of savanna grassland to savanna woodland (wooded savanna) seems to contribute considerably. Squires & Glenn (1995) estimate the C sequestration potential of rangeland restoration within the 5.2 billion ha of drylands at ca 0.5-1.0 Pg C/y with costs of $US 179-363 billion over 20 years.

The world forests have, according to Meyers (1995) shrunk by about 1/3 to presently ca 4.1 billion ha. Constancy of albedo makes them second after the oceans regarding impact on global climate ranges. This stabilizing effect saves per ha about $US 1-3000 in prevented global warming injuries.

Woodland is the largest SOM-C-pool with 787 Pg of C (Dixon, 1995) and is in many countries the last unoccupied refuge for landless immigrants. Consequently it is even more exposed to profound, mostly deteriorating changes than the other ecosystems. The socioeconomic conditions contribute visibly more than the climate change related ones.

An overview of forest areas, annual losses by deforestation, carbon pools and fluxes by Dixon et al. (1994) shows that the 4.1 billion ha of woodland, loses 15.4 million ha per annum with 1.146 Pg C (SOM + biomass) and a net flux of ca 0.9 Pg C/y to the atmosphere. The ca 1.7 billion ha tropical woodland, ca 36% of the tropical land surface with 655 million ha (38%) of tropical rain forest, 626 million ha moist and dry tropical deciduous forest and 178 million ha mountain forest, annually loses about 0.9%, ca 16.9 million ha (8.3 million ha, 0.9%, in Latin America; 3.6 million ha, 1.2%, in Asia; 5 million ha, 0.8%, in Africa) (FAO-1993).

Turnover of tropical forest sites by slash and burn is wasteful and GH-forcing. The more than 200 million landless people (shifting cultivation) have brought the fallow periods down to 5 years or even less with an alarming nutrient, SOM, and crop yield decrease.

Alternatives, especially in the realm of agroforestry, and special efforts to conserve biological diversity are presently expanding. They need the fullest support by decision makers with regard to land conservation in conjunction with climate change (Sanchez, 1995; Phillips & Gentry, 1994). Alterations of vegetation systems like wood with C3 photosynthesis mechanism (δ ^{13}C of -25%o) and grassland (savanna) with predominately C4 mechanism (δ ^{13}C of -12%o) are often reflected clearly by a thin layer soil profile scan for δ ^{13}C (see Martin et al., 1990).

Carbon in northern ecosystems is often underestimated, but soils in such northern ecosystems contain an estimated 350-455 Pg C in the permafrost ecozone,i.e. 22.5 to 29.4% of the total world SOM-C pool. The arctic tundra alone contains 192 Pg of soil-C or 12.4% of the global soil-C pool. Soils under forest ecosystems are important C-reserves, accounting for 60% of the terrestrial C pool (Lal & Kimble, 1994). The current climate change is expected to be most pronounced in the higher latitudes with these soils and carbon pools.

The Canadian forest and wetland surface is well assessed for phytomass, NPP, and C-storage to allow the recognition of alterations due to climate change, especially temperature rise and/or rainfall changes (Apps & Kurz, 1991; Gorham, 1991). Canadian forests are considered a weak net sink of C, but are feared to likely come under drought stress by climate change (IMPACT, 1993). Canadian forests comprise ca 45% of the Canadian surface (453 million ha), ca 30% of circumpolar boreal forests and about 233 million ha productive forest wood (24.6 billion m^3).

In the wetlands of the N-boreal and subarctic regions is a phytomass of ca 2 Pg C, a NPP of about 0.3 Pg C/y and ca 140 Pg C as stored C in ca 111 million ha peatlands, increasing by ca 28 G C/m^2/y (0.03 Pg C).

In S-America land use problems due to human actions, climate change, land degradations are difficult to discern. McKane et al. (1995), in an estimate of C-sequestration based on a 2400 km transect in the Amazon basin of Brazil, found a strong relation to nutrient availability. Pla Sentis (1994) predicts, that without enhanced conservation measures the rain-fed agriculture in S-America will decrease by 10 to 30% by the year 2000, and 30% of the land area would lose productivity. In the tropical belt only 3-5% of the land area are considered as being without serious limitations for intensive farming. Forest cutting amounts to 5 million ha/y with less than 10% reforestation. This all has to be harmonized, with a required increase in food production of at least 100% over the next 20 years.

The Mediterranean and (semi)arid lands are almost equally vulnerable to human induced land use changes, to climate variability, and to inadequate water and nutrient reserves. The effects of changing climate upon the grassland dominated ecosystems are mostly exacerbated by intense human land use. However, reduced precipitation or temperature increase accelerates land degradation via loss of plant cover, biomass turnover, nutrient cycling, and SOM-storage, accompanied by higher GH-forcing trace gas emission (Ojima et al., 1995). Yaalon (1990) identifies the effects of increased CO_2 and other gaseous emissions on soils in Mediterranean and other subtropical regions, becoming measurable within a span of 50 years in the form of organic matter reduction, changes of carbonate as well as of the salt regime, erosion, crusting, water logging and fire frequency. Undoubtedly, in the Mediterranean and adjacent drylands normal land use itself is already without replenishment of nutrients, thus promoting soil degradation, and eventually adding to the climate change related effects.

Tiessen et al. (1994) estimate the economic span of land use without fertilization in temperate prairie soils as 65 years, and in tropical semi-arid ecosystems only to a limit of 6 years.

In much of Asia the anthropogenic and climate change produced soil/land degradation are scarcely separable.

According to Richards (1995), between 1880 and 1980 the forest and wetland area declined by 47.7 %. Former high biomass land cover was converted to low biomass categories with the consequence of an emission of ca 29 Tg C/y to the atmosphere as CO_2.

The agro-ecological subregions of India are well studied. The 75% share of semi-arid and hot-subhumid (dry) agricultural regions points to land use sensitivity towards effects of climate change. There is a steep rise of all areas in production and biofluxes from the 1950s to the 1980s (Dadhwal et al.,1994). Recently, modeling results (Mudur, 1995) suggested a shrinking of the monsoon period with the higher aerosol concentration due to intensive S- and SE- Asian industrialization. This will have severe consequences for land use and production for India and for China.

In China, studies regarding nutrient releases from organic matter residence time and turnover versus CO_2 rise predict a gain of 36.6% by the year 2000 (Chang, 1995).

Cheung (1992)(in Chenny,1992), assessing the threat of only slight climate changes to survival of marginally tropical soils in China, fears it would exclude regeneration in logged forests and most likely upset the whole ecosystem.

Table 3 lists estimated first-hand climate change related impacts upon soils and ecosystems. Among the climate zones and major continental areas of the world the least pronounced effects are expected for the humid tropics.

When early in this paper it was mentioned that more attention has so far been devoted to "impact of soil reactions and gas emissions upon climate change" (for example, the World Inventory on Soil Emissions - project) than to "impact of climate change upon soils", it seems to be correct because it is more simple to identify and monitor.

The impact of climate change upon soils obviously requires more time to become measurable. Exceptions are more catastrophic short-term events such as volcanic eruptions with lahars, typhoons or hurricanes, compounded by El Nino conditions, with heavy floods and landslides.

Some hope remains for milder steady state scenarios in both directions. May the major catastrophic events brought about under the contribution of soil emissions and denudations as consequences of anthropogenic mismanagement recede under a consequent regime of mitigation efforts. May we have the self-control, farsightedness and readiness to support essential research to alleviate negative and uncontrollable impacts of both smooth shifts or sharp fluctuations in climate on soil and land use stability.

References

Alexandrov, G.A. & Oikawa, T. (1995): Net ecosystem production resulting from CO_2 enrichment: Evaluation of potential response of a savannah ecosystem to global changes in atmospheric composition.Proc. of the Tsukuba Global Carbon Cycle workshop. Tsukuba, Japan, 117-119.
Apps, M.J. & Kurz, W.A. (1991): Assessing the role of Canadian forests and forest sector activities in the global carbon balance. World Resource Review **3**, 333-334.
Armentano, T.V. (1980): The role of organic soils in the world carbon cycle.CONF-7905135, United States Department of Energy, Washington D.C.
Bazzaz, F.A. & Fajer, E.D. (1992): Plant life in a CO_2 rich world. Scientific American, Vol **266**, 18-24.
Berner, U., Ortisle, G. & Streif, H.J. (1996): Water vapor determines the climate history. Zeitschrift für Angewandte Geologie **41**, 69.
Blanis, D., Matsuoka, Y. & Kainuma, M.M. (1995): Carbon fertilization of the terrestrial vegetation. Proc. of the Tsukuba Global Carbon Cycle Workshop,CGER, Tsukuba, Japan, pp 126-133.
Borlaug, N.E. & Dowswell, C.R. (1994): Feeding a human population that increasingly crowds a fragile planet. Transactions of 15th World Congress of Soil Science, Acapulco, Supplement, Keynote Lecture, 1-15.
Boyle, E. (1996): Deep water distillation. Nature **379**, 679-680.
Brinkman, R. (1990): Resilience against climate change? In: Soils on a Warmer Earth. Developments in Soil Science **20**. H.W. Scharpenseel, M. Schomaker and A. Ayoub (eds), Elsevier, Amsterdam, 51-60.
Broecker, W.S. (1994): Massive iceberg discharges as triggers for global climate change. Nature **372**, 421-424.
Brown, S., Lugo, A.E. & Iverson, L.R. (1992): Processes and lands for sequestering carbon in the tropical forest landscape. Water, Air and Soil Pollution **64**, 139-155.
Chang, X.J. (1995): Prospects of plant nutrient supply from organic residues in China. Ife , international fertilizer correspondent,Vol **XXXVI**, 1, p 5.
Charles, Ch. (1997): Cool tropical punch of the ice ages, Nature **385**, 681-682.
Chenny, C. (1992): Can nature survive global warming? In: WWF (Worldwide Fund for Nature), Intern. Discussion Paper, E. Drucker & S. Russel (eds), Gingins, Switzerland, 32.
Cole, C.V., Flach, K., Lee, J., Sauerbeck, D. & Stewart, B. (1993): Agricultural sources and sinks of carbon. Water, Air and Soil Pollution **70**, 111-122.
Cole, V. (1995): IPCC, Working Group II, Second Assessment Report, 23: Agricultural Options for Mitigation of Greenhouse Gas Emisson.
Crowley, T.J. & Kim, K.Y. (1996): A new dawn for sun-climate links? Science **271**, 1360-1361.
Crutzen, P.J., Grooß, J.U., Brühl, C., Müller, R. & Russel III (1995): A reevaluation of the ozone budget with HALOE UARS Data. No evidence for the ozone deficit. Science **268**, 705-708.

Culotta, E. (1995): Will plants profit from high CO_2? Science **268**, 654-656.
Dadhwal, V.K., Shah, A.K. & Vora, A.B. (1994): Global change studies, scientific results from ISRO Geosphere Biosphere Programme. Indian Space Research Organization, Department of Space, Bangalore, 560094, India, 203-226.
Degens, E. (1989): Perspectives on Biogeochemistry, Springer, p 208.
Deming, D. (1995): Climatic warming in N-America: Analysis of borehole temperatures. Science **268**, 1576-577.
Dixon, R.K. (1995): Carbon pools and flux of global forest ecosystems. Proc.of Tsukuba Global Carbon Cycle Workshop CGER, Tsukuba, Japan, February 1995, 117-119.
Dixon, R.K., Brown, S., Houghton, R.A., Solomon, A.M., Trexler, M.C. & Wisniewski, J. (1994): Carbon pools and flux of global forest ecosystems. Science **263**, 185-190.
Downing, J.P., Meybeck, M., Orr, J.C., Twilley, R.R. & Scharpenseel, H.W. (1993): Land and water interface zones. Water, Air and Soil Pollution **70**, 123-137.
Down to Earth, a simplified guide to the convention to combat desertification (1995): Center of our Common Future in collaboration with Interim Secretariat for the Convention to Combat Desertification (UNEP), Nairobi.
Esser, G. & Lieth, H. (1989): Decomposition in tropical rain forests. In: Tropical Rainforest Ecosystems, Elsevier, Amsterdam, Chapter 32, 571 -579,.
Esser, G. & Lautenschlager, M. (1994): Estimating the change of carbon in the terrestrial biosphere from 18 000 BP to present using a carbon cycle model. Environmental Pollution **83**, 45-53.
Eswaran, H., Van Den Berg, E. & Reich, P.F. (1993): Organic carbon in soils of the world. Soil Science Society Am. J. **57**, 192-194.
FAO (1993): FAO-Aktuell (No.3), BML in Cooperation with FAO. Bonn.
Glenn, E, Squires, V., Olsen, M. & Frye (1993): Potential for carbon sequestration in drylands. Water, Air and Soil Pollution **70**, 341-355.
Gorham, E. (1991): Northern peatlands; role in the carbon cycle and probable responses to climate warming. Ecological Applications : (2), pp 182-195 (1991 by the Ecological Society of America).
Goudrian J. & Unsworth, M.H. (1990): Implications of increasing carbon dioxide and climate change for agricultural productivity and water resources. American Soc. of Agron., Crop Sci. Soc. of America and Soil Sci. Soc. of America, 677 S.Segoe Rd.,Madison, USA. Special Publication No.**53**, 111.
Grassl, H. & Klingholz, R. (1990): Wir Klimamacher. Auswege aus dem globalen Treibhaus. S. Fischer, Frankfurt.
Hall, D.O., Scurlock, J.M.O., Ojima, D.S. & Parton, W.J. (1994): Grasslands and the global carbon cycle : Modelling the effects of climate change. In: The Carbon Cycle,1993 OIES Global Institute Proceedings, Cambridge University Press.
Heimann, M. (1995): Dynamics of the carbon cycle. Nature **375**, 629-630.
Hughen, K.A., Overpeck, J.T., Peterson, L.C. & Trumbore, S. (1996): Rapid climate changes in the tropical Atlantic region during the last deglaciation. Nature **380**, 51-54.
Idso, S.B. (1990): The carbon dioxide/trace gas greenhouse effect: greatly overestimated? Amer. Soc. Agron. Crop Sci Soc. & Soil Sci Soc Amer., 677 Segoe Rd.,Madison, Public. No **53**, 19-26.
IGBP (1995): Global Change Report,Report No 34,1995, IGBP Programme,Stockholm.
IGBP (1996): Global Change Report,Report No 36, 1996, IGBP Programme,Stockholm.
IMPACT (1993): Report on Canadian forests and climate change.
Jenkinson, D.S. & Rainer, J.H. (1977): The turnover of organic matter in some of the Rothamsted classical experiments, Soil Sci. **123**, 298-305.
Keeling, C.D., Piper, S.C. & Heimann, M. (1995): In: D.H.Peterson (ed), Aspects of climate variability in the Pacific and in the Western Americas. Geophys. Monograph **55**, Amer. Geophys. Union.Washington D.C., 303-363.
Kempe, St. (1993): Damming the Nile. Transport of carbon and nutrients in lakes and estuaries. Part 6, SCOPE/UNEP Special Volume, Geolog..Paleontolog. Institute, University of Hamburg.
Kenny, G.J., Harrison, P.A. & Parry, M.J. (eds) (1993): Environmental Change Unit, Univ. of Oxford (ECU), Res. Report No 2.
Kerr, R.A. (1995a): US climate tillts towards the greenhouse. Science **268**, 363-364.
Kerr, R.A. (1995b): It's official: First glimmer of greenhouse warming seen (Report on IPCC decisions), Science **270**, 1565-1566.

Khalil, M.A.K. & Rasmussen, R.A. (1994): Global decrease in atmospheric carbon monoxide concentration. Nature **370**, 39-641.

Kimball, B.A., Mauney, J.R., Nakayama, F.S. & Idso, S.B. (1993): Effects of elevated CO_2 and climate variables on plants. Journal of Soil and Water Conservation, pp 9-14.

King, G.M. and Schnell, S. (1994): Effect of increasing atmospheric methane concentration on ammonium inhibition of soil methane consumption. Nature **370**, 282-284.

Lal, R. & Kimble, J.M. (1994): Soil management and the greenhouse effect. In: Soil Resources and the Greenhouse Effect, R.Lal, J.M.Kimble and E.Levine (eds), USDA -SCS Washington. Chapter 1, pp 1-5

Lovelock, J.E. (1979): Gaia, a new look at life on Earth. Oxford University Press, New York, pp 242.

Lovelock, J.E. (1993): The soil as a model for the Earth. Geoderma **57**, 213-215.

Lovelock, J.E. & Kump, L.R. (1995): Weathering and glacial cycles. Nature **373**, 110.

Lugo, A.E. & Wisniewski, J. (1992): Natural sinks of CO_2, conclusions, key findings and research recommendations from the Palmas del Mar Workshop. Water, Air and Soil Pollution **64**, 455-459.

Manabe S. & Stouffer, J. (1995): Simulation of abrupt climate change induced by freshwater input in the North Atlantic Ocean. Nature **378**, 165-167.

Martin, A., Mariotti, A., Balesdent, J., Lavelle, P. & Vuattoux, R. (1990): Estimate of organic matter turnover rate in a savanna soil by delta ^{13}C natural abundance measurements. Soil Biol. Biochem. **22**, 517-523.

Meybeck, M. (1981): Flux of organic carbon by rivers to the oceans. National Technical Information service, Springfield, Virginia, 219-269

McDougall, T. (1995): In: Global Environmental Change Report Vol VII, No 23, p 5. CSIRO, Hobart, Tasmania.

McKane, R.B., Rastetter, E.B., Melillo, J.M., Shaver, G.R., Hopkinson, C.S. & Fernandes, D.N. (1995): Effects of global change on carbon storage in tropical forests of South America. Global Biogeochemical Cycle,Vol 9, No 3, 329-350.

Meyers, N. (1995): The World's forests : Need for a policy appraisal. Science **268**, 823-824.

Mudur, G. (1995): Monsoon shrinks with aerosol models. Science **270**, 1922.

Neue, H.U. & Scharpenseel, H.W. (1987): Decomposition pattern of C14 labelled rice straw in aerobic and submerged rice soils of the Philippines. The Science of the Total Environment **62**, 431-434.

Neue, H.U., Gaunt, J.L., Wang, Z.P., Becker-Heidmann, P. & Quijano, C. (1994): Carbon in tropical wetlands. Transactions 15th World Congress Soil Science, Acapulco, Vol 9, Supplement, pp 201-220.

Neue, H.U., Ziska, L.H., Matthews, R.B. & Dai, Q. (1995): Reducing global warming - the role of rice. Geo Journal, Feeding 4 Billion People **35**, 351-362.

Newton, P. (1995): Perspectives past and present. Nature **378**, 312.

Ojima, D.S., Stafford-Smith, M. & Beardsley, M. (1995). Factors affecting carbon storage in semiarid & arid ecosystems. Proc. "Combating Global Warming by Combating Land Degradation". Nairobi, V.R.Squires, E.P.Glenn and A.T.Ayoub (eds.). UNEP, Nairobi 1997.

Olson, J.S., Garrels, R.M., Berner, R.A., Armentano, T.V., Dyer, M.I. & Yaalon ,D.H. (1985): Atmospheric carbon dioxide and the global carbon cycle.The Natural Carbon Cycle.USDE, Office of Energy Research J.R.Trabalka (ed), pp 175-213.

Parton, W.J., Schimel, D.S., Cole, C.V. & Ojima, D.S.(1987): Analysis of factors controlling soil organic matter levels in Great Plain Grasslands. Soil Sci. Soc.Amer. J. **51**, 1173-1179.

Phillips, O.L. & Gentry, A.H. (1994): Increasing turnover through time in tropical forests. Science **263**, 954-958.

Pimentel, D., Harvey, C., Resosudarmo, P., Sinclair, K., Kurz, D., McNair, M., Crist, S., Shpritz, L., Fitton, L., Saffouri, R. & Blair, R. (1995): Environmental and economic costs of soil erosion and conservation benefits. Science **267**, 1117-1123.

Pla Sentis, I. (1994): Soil degradation and climate incuced risks of crop production in the tropics. Proc. XV. ISSS Conf., Acapulco, Vol 1, 163-188.

Post, W.M., Pastor, J., Zinke, P.J. & Stangenberger, A.G. (1985): Global patterns of soil nitrogen storage. Nature **317**, 613-616.

Post, W.M., Pastor, J., King, A.W. & Emanuel, W.R. (1992): Aspects of the interaction between vegetation and soil under global change. Water, Air and Soil Pollution **64**, pp 345-363.

Rahmsdorf, St. (1995): Bifurcations of the Atlantic thermohaline circulation in response to changes in the hydrological cycle. Nature **378**, 145-165.

Regalado, A. (1995): Listen up! The worlds oceans may be starting to warm. Science **268**, 1436-1437.
Richard, R.L. (1992): Marine algae as a CO_2 sink. In: Wisniewski, J. and Lugio, A.E. (eds). Natural Sinks of CO_2. Water, Air and Soil Pollution **64**, 289-303.
Richards, J. (1995): In: IGBP Newsletter 24, in report "Natural and Anthropogenic Changes: Impacts on Global Biogeochemical Cycles". P 5.
Rogers, H.H., Brett Runion, G. & Krupa, S.V. (1994): Plant responses to atmospheric CO_2 enrichment with emphasis on roots and the rhizosphere. Environmental Pollution **83**, 155-189.
Rose, W.I., Delene, D.J., Schneider, D.J., Bluth, G.J.S., Krueger, A.J., Sprod, I., McKee, C., Davies, H.L. & Ernst, G.G.J. (1995): Ice in the 1994 Rabaul eruption cloud: implications for volcano hazard and atmospheric effects. Nature **375**, 477-479.
Sanchez, P.A. (1994): Tropical soil fertility research: Towards the second paradigm. State of the Art Lecture, XV. International Soil Science Congress, Acapulco, Mexico, Vol 1, pp 65-88.
Sanchez, P.A. (1995): Science in agroforestry. Agroforestry Systems **30**, 5-55.
Scharpenseel, H.W. & Becker-Heidmann, P. (1994): Sustainable land use in the light of resilience /elasticity to soil organic matter fluctuations. CAB International , Soil Resilience and Sustainable Land Use. D.J.Greenland and I. Szabolcs (eds.), 249-264.
Scharpenseel, H.W. & Pfeiffer, E.M. (1995): Carbon cycle in the pedosphere. Proc. Tsukuba Global Carbon Cycle Workshop. CGER-1018-95, pp 146-160.
Scharpenseel, H.W., Pfeiffer, E.M. & Becker-Heidmann, P. (1996): Organic carbon storage in tropical hydromorphic soils. In: Advances in Soil Science. Structure and Organic Matter Storage in Agricultural Soils. M.R. Carter and B.A. Stewart (eds.), CRC Press, Lewis Publ, 361-392.
Schlesinger, W.H. (1985): The formation of caliche in soils of the Mojave desert, California. Geochim. Cosmochim. Acta **49**, 57-66.
Schnell, R.C. (1995): Carbon cycle species at Mauna Loa Observatory in relation to airmass origins. Proc. of the Tsukuba Global Carbon Cycle Workshop, CGER-1018-95, pp 54-61.
Sellers, P.J., Bounoua, L., Collatz, G.J., Randall, D.A., Dazlich, D.A., Los,S .O., Berry, J.A., Fung, I., Tucker, C.J., Field, C.B. & Jensen, T.G. (1996): Comparison of radiative and physiological effects of doubled atmospheric CO_2 Science **271**, 1402-1406.
Sombroek, W.G. (1990): Soils on a warmer earth : the tropical regions. In: Soils on a Warmer Earth, H.W. Scharpenseel, M. Schomaker and A.T. Ayoub (eds), Development in Soil Science **20**, Elsevier, 157-174.
Squires, V.R. & Glenn, E.P. (1995): Combating Global Climate Change by Combating Land Degradation. V.R.Squires, E.P. Glenn and A.T. Ayoub (eds.), Proc. UNEP, Nairobi, 1997.
Tate, K.R. (1992): Assessment, based on the climosequence of soils in tussock grasslands of soil C storage and release in response to global warming. Journal of Soil Science **43**, 697-707.
Taylor, K.E. & Penner, J.E. (1994): Response of the climate system to atmospheric aerosols and greenhouse gases. Nature **369**, 734-737.
Tiessen, H., Cuevas, E. & Chacon, P. (1994): The role of soil organic matter in sustaining soil fertility. Nature **371**, 783-787.
Wallace, P.J. & Gerlach, T.M. (1994): Magmatic vapor source for sulfur dioxide released during volcanic eruptions: Evidence from Mount Pinatubo. Science **265**, 497-499.
Whittaker, R.H. & Likens, G.E. (1973): The primary production of the biosphere. Human Ecology, Vol 1, pp 299-369.
Wigley, T.M.L. (1994): Outlook becoming hazier. Nature **369**, 709-710.
WBGU Wissenschaftl. Beirat der Bundesregierung, Globale Umweltveränderungen (1994): Welt im Wandel, Gefährdung der Böden. Economia Verlag Bonn.
Woomer, P.L., Martin, A., Albrecht, A., Resck, D. V.S. & Scharpenseel, H.W. (1994): The importance and mangement of soil organic matter in the tropics. Chapter 3 in: The Management of Tropical Soil Biology and Fertility. P.L. Woomer and M.J. Swift (eds.), TSBF, Wiley-Sayce Publication.
WWF (1992): Can nature survive global warming? A World Wildlife Fund Discussion Paper.(Febr. 1992), Media Natura Ltd. London.
Yaalon, D.H. (1990): Soils of a warmer earth, projecting the effects of increased CO2 and gaseous emission on soils in Mediterranean and subtropical regions. In: Soils of a Warmer Earth, H.W.Scharpenseel, M.Schomaker and A.T.Ayoub (eds):. Development in Soil Science **20**, Elsevier.
Young, A. (1989): Agroforestry for soil conservation. Science and Practice of Agroforestry, No 4. ICRAF, Nairobi/Kenya and CABI, Wallingford, UK

Yu, E.F., Francis, R., Bacon, M.P. (1996): Similar rates of modern and last glacial ocean thermohaline circulation, infered from radiochemical data. Nature **378**, 689-694.

Zahn, R. (1994): Fast flickers in the tropics. Nature **372**, 621-62.

Address of authors:
H.W. Scharpenseel
E.M. Pfeiffer
Institut für Bodenkunde der Universität Hamburg
Allendeplatz 2
D-20146 Hamburg, Germany

Dryland Soils:
Their Potential as a Sink for Carbon
and as an Agent in Mitigating Climate Change

Victor R. Squires

Summary

Global climate change is occurring and elevated levels of atmospheric carbon dioxide and other trace gases from fossil fuel burning (principally) have been implicated. Drylands have the potential to be a sink for significant amounts of carbon, especially if they are restored to their ecological potential. Dryland restoration would have a major impact on global climates since the non-forested world's drylands (excluding hyper-arid regions) cover 5.2×10^9 ha (about 43% of the earth's surface) and have the potential to sequester more than 1.0 Gt of C per year. This is a significant figure because it represents about 12% of the C emissions from fossil fuel burning. The possibility of enhancing carbon sequestration through management of dryland plants and soils has gained credibility.

The potential effects of management on carbon storage in dryland soils are substantially greater than that of climate change or CO_2 enrichment. The soil store of C in these systems is a very important pool, since it is stabilized for hundreds to thousands of years, and forms the bulk of the dryland C pool. Biotic carbon offsets are a means of mitigating the adverse effects of elevated atmospheric carbon dioxide levels.

1 Introduction

The well-documented increase in atmospheric carbon dioxide has implications for climate change through an increase in mean global temperature (up to 4^0 C within the next 50 years). While atmospheric scientists now believe that such an increase is inevitable (in fact many believe that the first $0.5 - 0.6^0$ rise has already occurred), the implications of this rise on regional climates will be debated still for many years. And the broader issue of global change involves not only climate change but also the way the land is used and the impacts of expanding human populations.

It is important to identify options to reduce the rate of atmospheric carbon increase even before major climate change occurs. There are some actions that may be taken at no cost or at relatively low cost that will reduce the rate of CO_2 emissions but most steps to reduce atmospheric concentrations will be costly and require sacrifice. The Intergovernmental Panel on Climate Change (IPCC) calls on governments around the world to reduce CO_2 emissions or to invoke the Joint Implementation option whereby C emissions in an industrialized nation might be offset by C sequestration elsewhere. There is an obvious economic linkage then between fossil burning, climate change and C sequestration efforts in drylands (Squires and Glenn, 1995).

ISBN 3-923381-42-5
© 1998 by CATENA VERLAG, 35447 Reiskirchen

2 Biotic carbon offsets: their role and significance

Burners of fossil fuel, principally the operators of coal, gas and petroleum-fired power stations, contribute about one-third of all global fossil fuel emissions (Haggin, 1992; Judkins et al. 1993). They are seeking cost-effective and ecologically sound ways to offset the C they emit (Glenn, et al. 1993). Carbon offsets are an attempt to reabsorb atmospheric C to limit or delay the potential deleterious effects, e.g. global warming and climate change. In environmental economics terminology an offset is an action taken "beyond the stack" in order to compensate for emissions considered excessive by regulations, thus avoiding a tax liability or other penalty for excessive emissions.

Biotic carbon offsets in managed drylands offer a way that is cheaper than some forest-based offset programmes (Squires et al. 1997) and much cheaper than efforts to remove gas from the stack by physical or chemical means which costs about $US 300 per ton (Flour, 1991).

The option of absorbing excess CO_2 into biomass through global-scale revegetation programmes seems attractive (Kinsman and Trexler, 1993) despite problems of cost, availability of land and the difficulty of achieving and monitoring long-term C storage (Trexler, 1993). Possible carbon offsets programmes to remove CO_2 via living vegetation or to store it in the soil have been proposed for every major type of ecosystem (Squires et al. 1997). Projected costs of such carbon offsets range from $US 5 to 200 per ton (Trexler, 1993) - considerably lower than the cost of C removal from the stack. Wisely chosen projects have beneficial social and environmental effects in addition to C removal, especially in many parts of the world's dryland where population pressure is high and quality of life is low (Trexler and Meganck, 1993). Action to increase biomass on drylands will protect biodiversity, ameliorate the living conditions of local people as well retain carbon.

2.1 Carbon offsets - some practicalities

Unless adequately regulated the C offsets concept can be misused. An important principle of the offsets idea is that emitted C is "recaptured". There is risk that an area of dryland that benefited from offset funding could be damaged later by such unpredictable factors as wildfire or climate change or even converted to another use under changed tenure arrangements, thereby resulting in no net reduction. The implications of grazing control for communal grazing lands also need to be addressed but the apparent difficulties might be simplified by the adoption of national plans of action to combat desertification and/or climate change.

Other potential practical problems include the following:

- It may be difficult to distinguish a genuine offsets programme from opportunistic schemes that operate in contradiction to the spirit of the offsets concept. For a discussion of "carbon fraud" see Trexler (1991).
- As with all sustainable development issues, the question of equity must be considered, a technically feasible C offset programme may infringe on the rights of some parties (Trexler and Meganck, 1993).
- Although pioneering work is underway to clarify the carbon savings of various dryland restoration schemes, there remain considerable difficulties in estimating the relative contribution of a C offsets component from anti-desertification measures *per se*.
- The length of time the C needs to be stored before being "offset" is also a matter for debate.

The whole question of carbon offsets is an evolving one. The likelihood of tradable (both locally and internationally) permits being established and the formulation of international protocols has been canvassed (Jones and Stuart, 1994; Trexler and Meganck, 1993).

The degree to which C offsets programmes will be taken up in the world's drylands is highly dependent on the ongoing negotiations of the Framework Convention on Climate Change

(FCCC) which was signed at the Rio Earth Summit in 1992 and the Convention to Combat Desertification agreed in Paris in 1994.

3 Drylands as a carbon sink

Glenn et al. (1993) have argued the case for using drylands as a C sink. This has been further elaborated in Squires et al. (1997). In the past there has been some land degradation with a concomitant loss of C into the atmosphere through oxidation of soil organic matter (SOM). A move away from practices that lead to degradation could reverse this process and begin to sequester C via increasing levels of SOM in the upper layers of the soil (Gifford, 1994, Ojima et al., 1993b; Parton et al . 1993).

Preliminary modeling estimates are derived from the CENTURY model (Parton et al., 1993). Soil C flux is driven in part by net primary productivity (NPP) on the land surface. Land degradation results in reductions to NPP. Hence it is theoretically possible to calculate using the CENTURY model and United Nations Environment Programme (UNEP) data the differences in soil flux between scenarios of "acceptable" management versus current levels of "unacceptable" management projected into the future. Simulations of three different land management scenarios showed changes in C levels of soils in dryland regions (Ojima et al.1993a). Differences in flux between "sustainable" (30% biomass removal) and "regressive" management scenarios (50 to 80% biomass removal) were derived (Ojima et al. 1993b).

Projections based on a double CO_2 climate (including climate driven shifts in biome area) by 2040, resulted in net sinks of 5.6 Gt, 26.8 Gt and 27.4 Gt for the three different land cover projections respectively compared with "optimal" dryland management. The increase in soil C storage in this future projection resulted mainly from climate-induced changes in biome boundaries together with net increases in SOM density resulting from net ecosystem response to climate changes and enhanced CO_2 concentrations. CENTURY predicted that over a 50 year period the difference in carbon emissions between the regressive scenario and the sustainable scenario will be 37 Gt (annual difference = 0.7 Gt) over the whole land base under consideration (4.5×10^9 ha). This is a significant rate of sequestration (c. 12% of fossil fuel C emissions). It is important to note that the half life of the 37 Pg (net) stored in soil C is measured in hundreds of years. This is a much longer storage interval than can be achieved in tree plantations, forest preservation or other sequestration scenarios that depend on storing C in above-ground standing crop.

Calculations based on those of Gifford et al. (1990) suggest that the present C pool in dryland soils (estimated at 417 Gt to a depth of 1m) may be a net sink of 0.6 Gt per year. Owensby (1993) suggested that additional C storage in dryland soils could be as high as 25 to 30% with CO_2 doubling over the next 50 to 70 years, i.e. an annual increase of c. 0.4 %. He suggested that as much as 1.7 to 1.8 Pg could be added annually. More recent estimates derived from a meeting of experts convened by UNEP suggest that the annual C sequestration in the world's drylands would be in the range of 1.0 to 1.3 Gt of C per yr (Table 1).

4 Land degradation and its implications for global climate change

At present on severely degraded lands, release of CO_2 due to overutilization of plant production, ensure that drylands are a net source of CO_2. The benefits of turning this around are obvious. Dryland soils are low in C but there is scope to augment this via better management (Tinker and Ineson, 1990). According to Batjes (1996) dryland soils contain about 300 Pg of organic carbon and about 1200 Pg of carbonate carbon.

Option	Area (mill ha)	Rate (tC/ha/y)	Period (yr)	Cost (US$/tC)	Total (MtC/y)
Dryland crop management	450	0.3-1	5-20	1-5	135
Halophytes	130	0.5-5	Indefinite if harvested 5 yrs if not	170 (irrig & harvested) 20 (??) (dryland not harvested)	65
Bush encroachment	150	0.1-0.5	15-50	10-20	37
Energy crops	20 (5% of dryland crop area)	4-8	indefinite	150	80
Domestic biofuel efficiency	not applicable	not applicable	indefinite	2-5	75
Agro-forestry (arid)	50	0.2	30	2-10	10
Agro-forestry (semi-arid)	75	0.5	20	2-10	38
Agro-forestry (subhumid)	150	1.5	15	2-10	225
Improved Pasture (semiarid Asia)	10 (2500 degraded globally)	0.1	30	10	1
Savanna fire control	900 (globally)	0.5	30	1-5	450
Woodland management	400 (globally)	0.5	30	1-5	200

Tab. 1: Potential carbon sequestration in drylands classified by vegetation type and by land use. Source: UNEP (1995).

The SOM stored in the world's soils as fresh organic matter, stable humus or charcoal, is two to three times higher than the carbon stored in the natural vegetation and in standing crops. The soils of grasslands and cropped drylands store up to ten times as much carbon as the plants growing on them. The predicted global warming may result in a somewhat smaller pool of organic matter in the soil, especially at a lower ratio between rainfall and evaporation. Higher concentrations of atmospheric CO_2 will, however stimulate photosynthesis and thereby formation of dry matter, especially in C_3 plants such as woody plants. If nutrients, water and light don't become deficient, a doubling of CO_2 may yield 30% more biomass production. The CO_2 fertilizing effect is a negative or mitigating factor in any scenario for an enhanced greenhouse effect (Sombroek, 1995)

Only part of the C pool in soils, namely the SOM in topsoils, is dynamic or labile and contributes to CO_2 exchange and nutrient cycling in the soil-water-atmosphere system (Ojima et al., 1993a). The SOM status in natural ecosystems is closely related with the type and age of above ground vegetation. The projected change in global climate is thought to be accompanied by higher soil surface temperatures, increased drought conditions and larger erosion hazard (Tinker and Ineson 1990; Neilson and Marks 1994). Whether the anticipated climatic changes will be positive or negative, the sequestration of C in soils as SOM makes sense in its own right because of the benefits already mentioned. One can go considerably beyond the restoration of SOM levels to those under the original vegetation (Sombroek 1995).

In the long term the most important determinants of the rate of decomposition of SOM are the rate of input of litter, soil moisture content and temperature. Raich and Schlesinger (1992) analyzed global ecosystem data and found a tight linkage between the rate of plant productivity (g C m^{-2} yr^{-1}) and soil respiration rate (g C m^{-2} yr^{-1}). The mechanisms behind the linkage work in both directions. High productivity produces high litter fall which fuels soil respiration (i.e decomposition) High soil decomposition produces high N mineralisation rates creating highly available N supplies for plant growth. Soil respiration and NPP are inextricably linked.

Raich and Schlesinger (1992) found that the soil respiration rate of diverse ecosystems (g C m^{-2} yr^{-1}) is positively related to temperature. This implies that NPP is also positively related to temperature. The median Q_{10} for soil respiration across world ecosystems was 2.4 From this they conclude that "increased soil respiration with global warming is likely to provide positive feedback to the greenhouse effect". Gifford (1994) argues that for ecosystems close to equilibrium, "one would expect NPP and soil respiration to track each other during a gradual increase in annual average temperature with some lag to do with the different turnover times of vegetation and soil organic matter".

5 Ecosystem dynamics and their impact on C fluxes

This question of gradual warming has also been discussed by Brinkman and Sombroek (1996) who write of a *transient* scenario of global change and the possibility of adaptation by soils to the influence of higher photosynthetic rates, and water use efficiency under conditions of higher atmospheric CO_2 conditions. Neilson and Marks (1994) argue that global climate change, as currently simulated, could result "in broad-scale redistribution of vegetation across the planet. Vegetation change could occur through drought-induced dieback and fire". It is predicted though that grassland and shrublands (dryland ecosystems) could expand and that the tundra and boreal forests would contract most (Figure 1). A net result is that under the new equilibrium conditions the terrestrial biosphere may store up to 30% more C above ground than it currently does and this would act as a negative feedback to climate change.

Drylands may be among the earliest systems to exhibit the effects of climate changes (OIES 1991). Drylands are clearly vulnerable to climate change. Sensitivity to climatic change may be a reflection of inadequate reserves of water or soil nutrients. Temperature increases will modify ecosystem processes such as evapo-transpiration, decomposition and photosynthesis (see above). The combined effects of temperature and precipitation changes may alter rates of ecosystem processes in such a way that they may offset each other, or alternatively, they may act synergistically to amplify a positive or negative effect on system C storage.

6 Research perspective

A number of research needs have been identified over the past few years. Work needs to be mounted to answer most of these questions. The key question is whether anti-desertification measures (to reverse land degradation) even if fully funded, can be effective in mitigating climate change? Technical solutions to rehabilitating the drylands are available but if population increases put additional pressure on the landscape, land degradation may continue despite external funding for land rehabilitation.

To support the modeling work, experimental work is needed to characterize the effect of higher CO_2 concentrations on the physiological processes including the response of natural ecosystems, both above and below ground, to elevated CO_2. This work needs to be done across broad climate gradients with particular emphasis on changes in relationships among environmental variables such as temperature, moisture and nutrients. Important aspects include how higher CO_2 levels affect partitioning of above- and below-ground biomass; plant respiration; stomatal conductance and water use efficiency.

Further work is needed on various aspects of ecosystem C dynamics. More field data is needed on the amount of C stored in ecosystems; biomass decay rates and biomass accumulation rates following disturbance. A key question is what is the optimum SOM content for each soil type?

7 Summary and conclusions

There is a clear economic linkage between fossil fuel burning, climate change, dryland degradation and carbon flux. Carbon sequestration is the process of carbon stock aggradation and may be viewed as the key to reverse land degradation. As degradation is reversed so carbon sequestration increases and *vice versa*. In forging a link between global climate change, fossil fuel burning and anti-desertification measures it should be remembered that anti-desertification measures benefit C sequestration and that C sequestration *per se* benefits anti-desertification efforts. There is a synergy between global environmental problems and local aspirations. "Bottom-up" efforts are the key to successful implementation of anti-desertification and/or C offset programmes. Incentive-compatible mechanisms can be devised to enhance C storage in even the poorest areas provided that the local people are actively involved.

The projected C sequestration opportunities in drylands over the next 5 to 50 years suggest that if conservation and rehabilitation measures were implemented this would lead to an annual carbon sequestration of more than 1.0 Gt in the world's drylands

Acknowledgments

It is a pleasure to thank the 9th ISCO Organizing Committee for the invitation and the support to allow my participation. Many of the ideas presented in this paper were refined through discussion with numerous colleagues involved in the UNEP-sponsored workshop in Nairobi in September 1995.

References

Batjes, N.H. (1996): Total Carbon and Nitrogen in the Soils of the World. European Journal of Soil Science, **47,** 151-163.

Brinkman, R. and Sombroek, W.G. (1996): The effects of global change on soil conditions in relation to plant growth and food production. In: F. Bazzaz and W.G. Sombroek (eds) Global Climatic Change and Agricultural Production John Wiley & Sons, Chichester,

Flour, D. (1991): Energy and economic evaluation of CO_2 removal from fossil-fuel power plants. Electric Power Research Institute, Palo Alto. CA Report IE-7365

Gifford, R.M. (1994): The global carbon cycle: a viewpoint about the missing carbon. Australian J. Plant Physiol. **21,** 1-15.

Gifford, R.M., Cheney, N.P., Noble, J.C., Russell, J.S., Wellington, A.B. and Zammit, C. (1990): Australian land use, primary production of vegetation and carbon pools in relation to atmospheric carbon dioxide concentration. In: R.M. Gifford, and M. Barson (eds) Australia's Renewable Resources :Sustainability and Global change. IGBP Symposium Bur. of Rural Resources, Canberra. pp. 151-188.

Glenn, E.P., Squires, V.R., Olsen, M. and Frye, R. (1993): Potential for carbon sequestration in the drylands. Water, Air & Soil Pollution **70,** 341-355.

Haggin, J. (1992): Methods to reduce CO_2 emissions appraised. Chemical Engin. News. **70(38)**, 24.
Jones, D.J. and Stuart, M.D. (1994): The evolving politics of carbon offsets. ITTO Tropical Forest Update **4(5)**, 13-15.
Judkins, R.R., Fulkerton, W. and Sanghvi, S.K. (1993): The dilemma of fossil fuel use and global climate change. Energy and Fuels **7(1)**,14-22.
Kinsman, J. D. and Trexler, M.C. (1993): Terrestrial carbon management and electric utilities. Water, Air & Soil Pollution **70**, 545-560.
Neilson, R. P and Marks, D. (1994): A global perspective of regional vegetation and hydrologic sensitivities from climatic change. J of Vegetation Sci. **5**, 715-730.
OIES (1991): Arid and semi-arid regions: response to climate change. Office of Interdisciplinary Science, Boulder, Colorado, USA.
Ojima, D. S., Dirks, B.O.M., Glenn, E.P., Owensby, C.E. and Scurlock, J.M.O. (1993a): Assessment of C budget for grasslands and drylands of the world. Water, Air & Soil Pollution **70**, 95-110.
Ojima, D.S., Parton, W.J., Schimel, D.S., Scurlock, J.M.O. and Kittel, T.G.F. (1993b): Modeling the effects of climatic and CO_2 changes on grassland storage of soil C. Water, Air and Soil Pollution **70**, 643-657
Owensby, C.E. (1993) . Potential impacts of elevated CO_2 on above- and below-ground litter quality of a tall grass prairie. Water, Air and Soil Pollution **70**: 413-424.
Parton, W. J., Scurlock, J.M.O., Ojima, D.S., Gilmanov, T.G., Scholes, R.J., Schimel, D.S., Kirchner, Menaut, C.J., Seastedt, T., Garcia-Moya, E., Apinan Kamnalrut and Kinyamario, J.I. (1993): Observations and modeling of biomass and soil organic matter dynamics for the grassland biome worldwide. Global Biogeo-chemical Cycles **7**,785-809.
Raich, J.W. and Schlesinger, W.H. (1992): The global carbon dioxide flux in soil respiration and its relationship to vegetation and climate. Tellus **44B**, 81-99
Sombroek, W. G. (1995): Aspects of soil organic matter and nutrient cycling in relation to climate change and agricultural sustainability. In: Chr. Hera (ed) Nuclear techniques in soil-plant studies for sustainable agriculture and environmental preservation. Proceed. of a Symp., International Atomic Energy Agency, Vienna
Squires,V.R. and Glenn, E.P. (1995): Creating an economic linkage between fossil fuel burning, climate change and rangeland restoration. Proc. 5th International Rangelands Congr. ,Salt Lake City, USA. pp.531-32
Squires, V.R., Glenn, E.P. and Ayoub, A.T. (eds) (1997): Combating global climate change by combating land degradation. UNEP, Nairobi, 348 p.
Tinker, P.B. and Ineson, P. (1990): Soil organic matter and biology in relation to climate change. In: H.W. Scharpenseel, M. Schomaker and A. Ayoub (eds) Soils on a Warmer Earth. Elsevier, Amsterdam, pp.71-78
Trexler, M.C. (1991): Minding the carbon store: weighing U.S. forestry strategies to slow global warming. World Resources Institute, Washington D.C.
Trexler, M.C. (1993): Manipulating the biotic carbon sources and sinks for climate change mitigation: can science keep up with practice? Water, Air & Soil Pollution **70**, 579-593.
Trexler, M.C. and Meganck, R. (1993): Biotic carbon offset programs: sponsors or impediments to economic development.? Climate Research 3, 129-136.
UNEP (1995): Combating global climate change by combating land degradation. Desertification Control Bulletin No.29 79-82.

Addresses of author:
Victor R. Squires
Dryland Management Consultant
497 Kensington Road
Wattle Park 5066
South Australia
and
Adjunct Professor
Department of Soil, Water and Environmental Sciences
University of Arizona
Tucson, USA

Climatic Effects on C Pools of Native and Cultivated Prairie

W. Amelung, K. W. Flach, Xudong Zhang & W. Zech

Summary
Understanding of the impact of global change on soil organic matter (SOM) dynamics and its response to agricultural management can in part be deduced from changes of SOM pools. This study was conducted to assess how the concentrations of organic C and total N in < 2 µm (clay), 2-20 µm, 20-250 µm, and 250-2000 µm size fractions are influenced by climate and long-term cultivation. Composite samples were taken from the top 10 cm of 21 native grassland and 20 adjacent continuously cropped (> 60 years) sites along temperature and precipitation transects across the North American prairie.

In the native sites, the SOM concentration in the clay fraction exceeds that of the bulk soil by a factor of 1.9 for C and 2.2 for N. This factor increases significantly from ustic to udic moisture regimes and decreases with increasing mean annual temperature of the sites. After long-term cultivation, SOM is lost from all fractions but preferably from the more labile sand sized pools. This results in a additional shift to significant higher enrichment of C and N in the clay fraction, and the magnitude of this shift tends to increase from the cryic to the hyperthermic temperature regimes.

Keywords: soil organic matter, climate, cultivation, particle size fractions, grassland soils

1 Introduction

In grasslands, the amount of soil organic matter (SOM) represents a balance between primary productivity and decomposition and as such is a sensitive and integrated measure of changes in ecosystems at different climates. However, SOM is known to decline rapidly after grassland has been put under cultivation, thereby approaching, perhaps, a lower steady state level. Understanding the processes that control SOM dynamics and their response to management is essential for sustainable use of agricultural land in different climates. Such understanding can in part be derived from the assessment of organic matter in different C-pools such as those in different particle size fractions in native and adjacent cultivated sites in different climatic regions.

Particlesize separates have been widely used to distinguish pools of different SOM-quality and -turnover rates (review: e.g., Christensen, 1992). Tate and Churchman (1978) indicated that increased precipitation may result in an accumulation of sand-sized, particulate organic matter (POM) in tussock grasslands of New Zealand. Since the turnover rates are highest for POM, it is most rapidly lost from a continuously cultivated soil, resulting in a relative enrichment of SOM in finer particle size fractions (Tiessen and Stewart, 1983; Dalal and Mayer, 1986). These studies suggest that climate and cultivation may interactively affect the relative amount of POM in soil.

With increasing soil temperatures decomposition of soil organic matter also increases (Parton et al., 1987). It is likely that decomposition will affect the labile C-pools, such as POM, more than the more stable ones, such as SOM attached to clay or silt (Christensen, 1992). Yet, little work has been done concerning the effect of temperature on POM and related C-pools in soils of the North American prairie.

The objectives of this study were, therefore, to assess i) the concentration and distribution of SOM within particle size fractions from soils of the North American grasslands, ii) the impact of climate on the native soil organic carbon (SOC) and N distributions among the fractions, and iii) the effect of long-term cultivation on SOC and N in the size fractions at different climates.

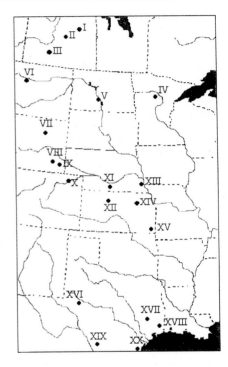

Fig. 1: Locality of the sites under study

2 Material and methods

2.1 Soils

Twenty-one native sites and nearby, 20 paired cultivated ones (same series) were selected for sampling in late spring 1994 along temperature and precipitation transects across the North American Prairie (Fig. 1). The native sites (except site XX) have to our knowledge never been cropped (Table 1). The vegetation composition resembles the "potential" of the respective Land Resource Areas which is assumed to be in equilibrium with climatic elements (USDA-SCS, 1981). Exceptions were brush, such as *Prosopis juliflora* (mesquite), and cacti, e.g., *Opuntia* (prickly pear), which grew up to 2 m at site XIX, and the invasion of sites *Bromus tectorum* L. (downy brome) at sites VII, XI, X, XII, XIV or the invasion of *Sorghum halepense* (L.) Pers. (johnson grass) at site XVIII, probably all because of recent heavy grazing. All U.S. sites had previously been characterised by NRCS (USDA-SCS, 1994), and most sites were under observation of agricultural experiment stations. The cultivated sites have never been irrigated or significantly fertilized with organic manure for at least 60 years. We collected composite samples (0 - 10 cm) from at least 5 sub-sites (200 cm³ core volume at each sub-site) in a radial sampling scheme across every native and cultivated site area (between few ha and some square km).

No.	Series name	classification (Soil Survey Staff, 1996)	MAT °C	MAP mm	Descr.	Texture 0-10 cm	Brief: cultivation history[a]
cryic							
I	Elstow	Fine-loamy, mixed Typic Cryoboroll	1.6	343	dry	l	ca. 60 yr W, the last 5 yr. rape, reduced till
II	Ardill	Fine-loamy, mixed Typic Cryoboroll	3.2	380	dry	l	since 1917 WF, reduced till
III	CutKnife-Naicam	Fine-loamy, mixed Argic Cryoboroll	0.9	456	wet	l	1925-1985 W, then rot, reduced till
frigid							
IV	Svea	Fine-loamy, mixed Pachic Udic Haploboroll	6.1	565	wet	sicl	ca. 75 yr corn-F or W-corn
V	Amor	Fine-loamy, mixed Typic Haploboroll	5.0	419	dry	l	at least since 1947 W, conventional till
VI	Joplin	Fine-loamy, mixed Aridic Argiboroll	6.1	300	dry	l	since 1911 W (plots), conventional till
mesic							
VII	Ulm	Fine, smectitic mesic Ustic Haplargid	7.2	400	dry	sil	ca. 60 yr WF, stubble mulch
VIII	Phiferson	Coarse-loamy, mixed, mesic Aridic Haplustoll	8.9	400	dry	sl	ca.60 yr WF, stubble mulch
IX	Ascalon	Fine-loamy, mixed, mesic Aridic Argiustoll	9.0	400	dry	sl	ca. 60 yr WF, stubble mulch
X	Weld	Fine, smectitic, mesic Aridic Paleustoll	10.8	375	dry	fsl	>60 yr WF, conventional till
XI	Holdrege	Fine-silty, mixed, mesic Typic Argiustoll	11.6	666	dry	sil	ca. 70 yr W-sorgh-corn, last 25 yr W, conventional till (uncertain history)
XII	Armo	Fine-loamy, mixed, mesic Entic Argiudoll	12.2	573	dry	l	after legumes and W at least 40 yr WF, conventional till
XIII	Sharpsburg	Fine, smectitic, mesic Typic Argiudoll	10.9	792	wet	l	since 1870 cult, since 1960 corn-W-sorgh, conventional till
XIV	Reading	Fine-silty, mixed, mesic Pachic Argiudoll	12.4	791	wet	sil	since 1890 W (-F), conventional till
XV	Parsons	Fine, mixed, thermic Mollic Albaqualf	14.2[b]	1000	wet	l	at least since 1960 W-soy, partly sorgh, total > 60 yr, conventional till
thermic							
XVI	Amarillo	Fine-loamy, mixed thermic Aridic Paleustalf	17.1	466	dry	fsl	since 1916 mainly cotton-F, conventional till
XVII	Haletsville	Fine, smectitic, thermic Udertic Paleustoll	20.0	1030	wet	l	ca. 75 yr C, 10 yr ago 7 yr F, conventional till
XVIII	Morey	Fine-silty, mixed, hyperthermic Oxyaquic Argiudoll	20.3[c]	1308	wet	sicl	>50 yr soy.-F, chissel
hyperthermic							
XIX	Moglia	Fine-loamy, mixed, hyperthermic Ustic Haplocalcid	23.4	440	dry	l	none
XX	Racombes	Fine-loamy, mixed, hyperthermic, Pachic Argiustoll	22.2	700	wet	scl	>75 yr W -F-corn-sorgh, conventional till (disk)

[a] W = wheat, F = fallow, sorgh = sorghum, soy = soybean.
[b] taxadjunct, at the mesic-thermic temperature regime.
[c] taxadjunct, at the thermic-hyperthermic temperature regime.

Tab. 1: *Climate, land use history and soil classification of the sites under study.*

2.2 Particle size fractionation

Air dried bulk samples were treated ultrasonically at 60 J mL^{-1} in a soil:water ratio of 1:5 to disperse macroaggregates (> 250 µm). The coarse sand fraction (>250 µm) was isolated by wet sieving. Final ultrasonic dispersion was applied again at 440 J mL^{-1} to the < 250 µm suspension in a soil:water ratio 1:10 to disperse the samples completely. Centrifugation was used to separate the clay fraction (< 2µm). Wet sieving was again performed to separate silt (2-20µm), and fine sand (20-250µm). All fractions were dried at 40 °C (Christensen, 1992) and ground for chemical analysis.

2.3 Soil analysis

Subsamples of all size fractions were analysed for total C, N, and S with a C/N/H/S-analyser (Elementar). Soil inorganic carbon was measured after dry combustion for 5 h at 560 °C, and soil organic carbon (SOC) was calculated by subtracting inorganic-C from total-C.
   Both size fractionation and subsequent analyses were done in duplicate.

3 Results and Discussion

3.1 C, and N in the particle size fractions at the native sites

Large proportions of SOM are attached to clay, comprising on average 43 % of total SOC, and 56 % of total N. The silt fraction contains 20 % of each of the two elements. About one third of SOC, and one fourth of total N is bound in particulate organic matter (POM) of both sand-sized pools. The values for SOC and N intercorrelate significantly in all size separates (r^2 ranges from 0.88*** to 0.94***). Any trend that is, therefore assessed, for the impact of climate or cultivation on SOC is similar for N. Consequently only SOC is discussed in the following figures.
   Usually, SOC pools are characterised by enrichment factors E (Christensen, 1992) which relate the SOC content of a particular size fraction to the average SOC concentration of the corresponding bulk soil, e.g., $E_{SOC, clay}$ = (g SOC kg^{-1} clay)/(g SOC kg^{-1} bulk soil). Preferential enrichment of SOC within a C-pool results in $E_{SOC} > 1$, whereas $E_{SOC} < 1$ indicates preferential SOC depletion. The E_{SOC} factors decrease significantly in the order clay (1.89) > silt (1.05) > fine sand (0.34), but increase again significantly in the coarse sand fraction (3.54). Obviously, particle size fractionation into clay, silt, fine sand and coarse sand yielded four different SOC pools with different E factors.
   With decreasing particle size the average C/N ratio decreases significantly ($p < 0.05$) in the order coarse sand (C/N = 21.8) > fine sand (C/N = 13.8) ≥ silt (C/N = 13) > clay (C/N = 9.5), because the microbial alteration of SOM is higher the finer the particle size (Christensen, 1992; Guggenberger et al., 1994). Apparently, SOM that has been altered by microbes contributes greatly to the SOM stocks in soils of the native prairie.
   Commonly the sand fractions are not subdivided into different C-Pools. We found, however, that C/N of SOM of the fine sand was significantly lower than that of the coarse sand. Apparently SOM of the coarse sand fractions comprised fresh or little altered plant material, whereas decomposed particulate SOM ended up in the fine sand fraction (20-250 µm). Hence we suggest to equate SOM of the coarse sand-sized organic matter pools with "recent particulate organic matter" RPOM (250-2000 µm) and that of the fine-sand-sized pools with "non-recent particulate organic matter" NRPOM (20-250 µm).

3.2 Effects of climate on the enrichment factors for SOC at the native sites

To illustrate the impact of climate on the C-pools, we grouped the sites according to the five temperature regimes cryic, frigid, mesic, thermic and hyperthermic (Soil Survey Staff, 1996). In each temperature regime we additionally defined classes of "moist" and "dry" sites. If the MAP of a site was higher than the mean between the highest and lowest precipitation of the sites in a given temperature regime, we called it "moist". Sites receiving less MAP were named "dry". On the average, the moist sites received 394 mm more precipitation than did the dry sites. Every class consisted of 1-5 sites (Table 1). In the frigid, mesic, and thermic temperature regimes "moist" corresponds to *udic* and "dry" to *ustic moisture regimes* (Soil Survey Staff, 1994).

There were little if any trends assessed for RPOM with climatic elements, suggesting that it was still too young to be affected by climate. Changes of C-Pools as influenced by climate are indicated most clearly by the clay fractions, containing the decomposed SOM (Amelung et al., 1998). Since there is a detailed publication concerning the effect of climate on native C-pools in North American grasslands (Amelung et al., 1998), we concentrate in this paper the discussion of climatic effects on SOC in the clay-sized pools.

The preferential enrichment of organic matter in the clay fraction was significantly less pronounced in the moist sites compared to the dry ones ($p < 0.01$, Fig. 2). The opposite was true for the fine sand-sized pools (Amelung et al., 1998). Since the production of plant biomass relates linearly to the MAP at the sites (Sala et al., 1988), POM contributes more to total SOC at moist than at dry sites, resulting in lower E_{SOC} for clay.

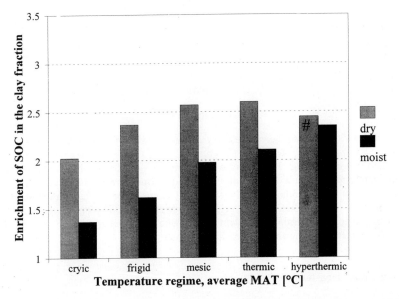

Fig. 2. Enrichment of SOC in the clay fraction as related to mean annual temperature MAT. The "#" indicates that invasion of brush possibly influenced the $E_{SOC,clay}$ at site XIX (redrawn from Amelung et al., 1998).

For both moist and dry sites E_{SOC} of the clay fraction correlates significantly with MAT ($r = 0.89^{**}$ for the dry and $r = 0.99^{***}$ for the moist classes). The higher thus the absolute heat input to the soil, the more SOM seems to be preferentially stabilised by clay.

Bulk SOM decreases exponentially with increasing mean annual temperature and it is lower for the dry classes than for the moist ones ($p < 0.05$). The SOC concentrations in the clay fractions follow the trend for the bulk soil ($r = 0.88^{***}$). This is the inverse of the enrichment factors of the clay fraction (Fig. 3) but similar to the enrichment factors for NRPOM (Amelung et al., 1998). Preferential SOM enrichment in stable, clay-sized pools is therefore not related to higher bulk SOM. Instead the balance between rates of input and losses of SOC (governed by interactions of temperature, precipitation, and other factors) controls both the SOC stock of the bulk soil and the SOC or N enrichment in the clay fraction.

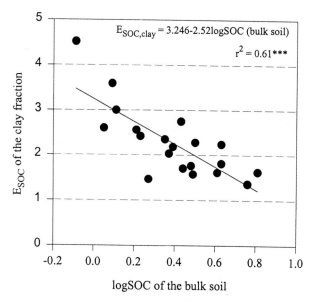

Fig. 3. *Relation between E_{SOC} in the clay fraction and the SOC stocks of the bulk soil*

Since the relationship (Fig. 3.) follows a logarithmic trend, SOC of the clay fraction contributes significantly more to bulk SOC at the mesic and warmer temperature regimes (> 50 % of SOC bound to clay) than at cooler climates, where clay exerts a minor effect on the SOM content of the bulk soil (ca. 40 % of SOC bound to clay in the frigid and cryic temperature region). These results indirectly support McDaniel and Munn (1985) who found better correlations between bulk SOM and clay content in mesic and warmer than in cooler climates.

3.3 Changes after long-term cropping

About sixty years of continuous cropping induces significant losses SOC (47%), and N (44 %). As the clay content of the cultivated sites does not differ significantly from the ones of the native sites, different SOM contents in native and cultivated sites do not reflect any textural differences.

The SOC, and N losses are not restricted to a certain particle size pool. The elements have depleted significantly ($p < 0.001$) in all fractions, yet at different intensities. Concentrations of SOC and N in the clay sized pools, for instance, decreased after long-term cultivation by only 32 and 36 %, in contrast to depletions > 50 % in the sand fractions. The fine fractions show less SOM losses (Tiessen and Stewart, 1983), because they continuously gain metabolic products from decomposition of POM (Christensen, 1992), and because these metabolic products are more resistant to subsequent mineralization than sand-sized SOM (Dalal and Mayer, 1986).

Preferential loss of organic matter from a size fraction affects its enrichment factor. Thus E_{SOC} and E_N of the clay fractions increase significantly after long-term cultivation ($p < 0.01$), whereas those of NRPOM decrease ($p < 0.05$). The general trends for C/N and the enrichment factors, however, were not changed. Apparently long-term cultivation results only in a shift from more recent SOM of sand fractions to more stable, microbially altered SOM in clay (Christensen, 1992)

3.4 Interactive effect of climate and long-term cropping on the enrichment of SOC in clay

In the cultivated fields, the E_{SOC} of the clay fraction is significantly lower in the moist than in the dry sites (p<0.01) and it increases with increasing MAT in both the dry ($r = 0.9*$) and the moist sites ($r = 0.95**$). Obviously, the impact of climate on the distribution of SOM within particle size fractions is not only apparent in the native but also the cultivated sites.

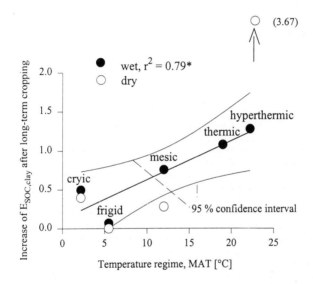

Fig. 4. Increase of E_{SOC} in clay after long-term cropping at the moist sites

Any SOC depletion of the bulk soil due to cultivation effects or increased MAT resulted in a shift to higher E_{SOC} in the clay sized pools. Therefore, the shift to more clay bound organic matter as caused by the agricultural use of the land is more pronounced at sites with thermic or hyperthermic temperature regimes than at sites with frigid and cryic temperature regimes. At moist sites, agriculturally induced changes in E_{SOC} of the clay fraction correlates linearly, positively with MAT (Fig. 4). The warmer therefore the climate, the larger the fractions of SOM attached to clay after long-term farming. It remains unclear whether a similar tendency exists at the dry classes. At the dry site of the thermic temperature region (XVI; Big Spring, TX) E_{SOC} of the clay fraction increases dramatically from 2.64 to 6.0 after cultivation. Since cultivated fields at the dry, hyperthermic regions were deep-ploughed (0 - 80 cm) and therefore not included into this study, a dramatic increase of E_{SOC} after > 60 years cropping remains to be verified at dry hyperthermic or iso-hyperthermic sites.

4 Conclusions

Particle size fractionation into clay (< 2 µm), silt (2-20 µm), fine (20-250 µm) and coarse sand (250-2000 µm) yields four different C-Pools with respect to their SOC-concentration in both native and cultivated soils of the North American prairie. The relative size of these pools can be assessed in terms of enrichment factors E that relate the SOC concentration of a defined size fraction to that of the bulk soil. The corresponding enrichment factors, however, are not constant but change with climate and clay content. If other attributes are held constant, $E_{SOC,clay}$, for instance, increases with increasing MAT, decreasing MAP or after long-term cropping, i.e., it behaves contrary to SOC of the bulk soil.

We conclude that clay plays an important role in ecosystem functioning of the prairie, because it may buffer SOM losses due to cultivation and accelerated decay at high MAT. The higher the total SOM loss, the more of SOC remaining in the soil is concentrated in the clay-sized C-pools. This fraction is thought to be more resistant to further mineralization than the coarser fractions (Dalal and Mayer, 1986; Christensen, 1992).

Acknowledgement
The Deutsche Forschungsgemeinschaft supported the research (Ze 154 / 22-1).

References

Amelung, W., Zech, W., Zhang, X., Follett, R.F., Tiessen, H., Knox, E. and Flach, K.-W. (1998): C,N,S-pools in particle size fractions as influenced by climate, Soil Sci. Soc. Am. J. **62**, 172-181.
Christensen, B. T. (1992): Physical fractionation of soil and organic matter in primary particle size and density separates, Advances in Soil Science **20**, 1-90.
Dalal, R.C. and Mayer, R.J. (1986): Long-term trends in fertility of soils under continuos cultivation and cereal cropping in Southern Queensland. III. Distribution and kinetics of soil organic carbon in particle size fractions, Aust. J. Soil Res. **24**, 293-300.
Guggenberger, G., Christensen, B.T. and Zech, W. (1994): Land use effects on the composition of organic matter in soil particle size separates I. Lignin and carbohydrate signature, Eur. J.Soil Sci. **45**, 449-458.
McDaniel, P. A. and Munn, L.C. (1985): Effect of temperature on organic carbon-texture relationships in mollisols and aridisols, Soil Sci. Soc. Am. J. **49**, 1486-1489.
Parton, W. J., Schimmel, D.S., Cole, C.V. and Ojima, D.S. (1987): Analysis of factors controlling soil organic matter levels in Great Plain grasslands, Soil Sci. Soc. Am. J. **51**, 1173-1179.
Sala, O. E., Parton, W.J., Joyce, L.A. and Lauenroth, W.K. (1988): Primary production of the central grassland region of the United States, Ecology **69**, 40-45.
Soil Survey Staff, USDA-NRCS (1996): Keys to Soil Taxonomy, 7th edition, Virginia
Tate, K. R. and Churchman, G.J. (1978): Organo-mineral fractions of a climosequence of soils in New Zealand tussock grasslands, J. Soil Sci. **29**, 331-339.
Tiessen, H. and Stewart, J. (1983): Particle-size fractions and their use in studies of soil organic matter, II. Cultivation effects on organic matter composition in size fractions, Soil Sci. Soc. Am. J. **47**, 509-514.
USDA-SCS (1981): Land resource regions and major land resource areas of the United States, Agriculture Handbook No. 296, Washington.
USDA-SCS (1994): Soils of the United States 391, CD-ROM, National Soil Survey Laboratory, Lincoln,NE

Addresses of authors:
Wulf Amelung, Xudong Zhang, Wolfgang Zech
Institute of Soil Science and Soil Geography, University of Bayreuth, D-95440 Bayreuth,Germany
Klaus W. Flach, El Macero Drive 4104, Davis, CA 95616, USA

Can Amino Sugars Serve as Indicators of Climate Variations and Land Use on Soil Organic Matter?

Xudong Zhang, W. Amelung & W. Zech

Summary

This study tested the use of amino sugars (glucosamine, mannosamine, galactosamine, and muramic acid) in soils as indicators of climate variations and long-term cropping impact on soil organic matter (SOM) dynamics. Soil samples were taken from surface layers (0-10 cm) of 17 native grassland and 17 adjacent cultivated sites along temperature and precipitation gradients across the North American prairie.

The contents of amino sugars varied markedly among the sites. After long-term cultivation about 53% of amino sugars were lost. The higher the mean annual temperature (MAT) at a site, the more amino sugars were degraded. The ratio of glucosamine to galactosamine increased with increasing MAT as well as with mean annual precipitation (MAP), indicating a preferential contribution of chitin-derived glucosamine to amino sugars in the soils studied. In contrast, bacterial-derived amino sugars declined with increasing MAT and MAP. Obviously, the amino sugar patterns can be used to indicate the effect of climate on SOM dynamics under the sites studied. In addition, long-term cropping resulted in a significant increase ($p<0.01$) of the glucosamine-to-galactosamine and glucosamine-to-muramic acid ratios, suggesting that the two ratios are also sensitive indicators of the influence of cultivation on SOM dynamics.

Keywords: Amino sugars, hexosamine, muramic acid, climate variations, cultivation, indicators

1 Introduction

It is well established that climate affects soil organic matter (SOM) quantitatively and qualitatively. In general, the SOM content increases with increasing mean annual precipitation (MAP) and decreasing mean annual temperature (MAT) (Burke et al. 1989; Honeycutt et al. 1990; Parton et al. 1987). Moreover, climate has an impact on the chemical composition of SOM (Arshad and Schnitzer 1989; Zech et al. 1989; Amelung et al. 1997). Thus characterising SOM in soils along a climate gradient is important to infer the impact of climate change on SOM dynamics.

Many studies have shown that improper management of land can result in rapid depletion of soil fertility. Long-term cultivation can especially result in rapid depletion of soil organic carbon (SOC) and total N. For example, SOC losses of as much as 50% have been documented in the US Central Plains (Burke et al. 1989). Moreover, an increase in the oxidation and decomposition of humic substances (Andriulo et al. 1987; Rosell et al. 1989) was observed when a native site was put under cultivation. Nascimento et al. (1992) also found an increase in the resistance of humic compounds against biodegradation in cultivated Oxisols, Brazil, compared with uncultivated areas. Yet, how long-term cropping affects certain species of organic carbon and nitrogen remains to be clarified.

There is a need to evaluate the impact of climate and land-use practices on SOM dynamics in terms of sustainable land use. Because higher plants do not synthesize amino sugars (Stevenson 1982), the amino sugars in soils can provide a clue to the microbial contribution to soil organic matter turnover (Sowden and Ivarson 1974; Kögel and Bochter 1985). The objective of this study was to test the suitability of amino sugars in soils as an indicator of the impact of climate and long-term cropping on SOM.

2 Materials and methods

Thirty four paired native and cultivated surface-soil samples (0-10 cm) were collected from the North American prairie along a climosequence from Saskatoon, Canada, to Texas, USA, in spring, 1994. Mean annual temperature (MAT) ranged from 0.9 to 22.2 °C and mean annual precipitation (MAP) from 300 to 1308 mm. Soil classification and cultivation history (>60 years) of the sites have been described by Amelung et al. (1998). All samples were air dried and sieved (< 2 mm) prior to analyses.

Four amino sugars, muramic acid, glucosamine, galactosamine, and mannosamine were determined by means of gas liquid chromatography using the method of Zhang and Amelung (1996). Soil total carbon and nitrogen and soil inorganic carbon (lime) after ashing organic matter at 560°C for 5 h were analysed by dry combustion (a C/N/H/S-analyser, Elementar). Soil organic carbon (SOC) was estimated by subtracting inorganic carbon from the total and soil organic matter (SOM) was calculated by multiplying SOC by 1.724. All analyses were conducted in duplicate.

3 Results and discussions

3.1 Amino sugar contents in soils

In native sites the contents of four amino sugars averaged 1873±851 (mg kg^{-1} soil) for glucosamine, 1090±581 (mg kg^{-1} soil) for galactosamine, 160±68 (mg kg^{-1} soil) for muramic acid, and 57±27 (mg kg^{-1} soil) for mannosamine. About 53% of amino sugars, on average, was depleted after long-term cultivation of more than 60 years (p<0.001). Thus in cultivated sites the contents were only 946±479 (mg kg^{-1} soil) for glucosamine, 486±282 (mg kg^{-1} soil) for galactosamine, 58±41 (mg kg^{-1} soil) for muramic acid, and 31±18 (mg kg^{-1} soil) for mannosamine. The depletion of amino sugars was enhanced by MAT (Fig. 1) but not significantly affected by MAP. The higher the MAT, the more amino sugars were obviously degraded.

The amino sugar contents (Y) in both native and cultivated sites could be predicted as a function of MAT, clay, and silt in the soils under study (Fig. 2).

[1] Y_{native} (g kg^{-1} soil) = 0.213 - 0.121MAT (°C) + 0.140Clay (%) + 0.041Silt (%) $R^2 = 0.70$ (P<0.001)

[2] $Y_{cultivated}$ (g kg^{-1} soil) = 0.419 - 0.053MAT (°C) + 0.034Clay (%) + 0.013Silt (%) $R^2 = 0.87$ (P<0.0001)

The different levels of significance in the two equations indicated that the amino sugar dynamics showed better equilibrium with MAT at cultivated sites than that at native ones.

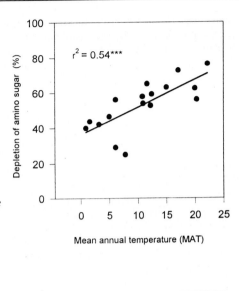

Fig. 1: Relationship between the decrease of amino sugar contents (g kg⁻¹ soil) after long-term cropping relative to those at native sites.

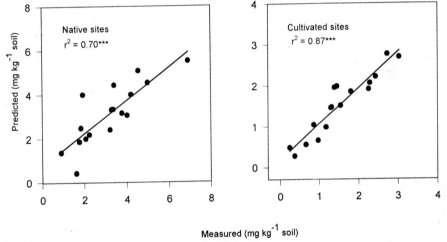

Fig. 2: Relationship between measured amino sugar contents (g kg⁻¹ soil) and predicted ones by Equations 1 and 2.

3.2 Relative accumulations of amino sugars

We used amino sugar-N proportions to soil total N (amino sugar-N g kg⁻¹ total N) to indicate the relative accumulation of amino sugars in soils. The proportions varied greatly (Table 1). After long-term cultivation amino sugar N concentrations (amino sugar-N g kg⁻¹ total N) declined by about 18 %, on average ($p<0.001$, Table 1), indicating that the mineralisation of amino sugars was faster or re-synthesis was slower, or both, compared with other nitrogen-containing compounds. Although the depletion of each amino sugar was pronounced after cultivation, a preferential depletion was found for muramic acid and galactosamine (Table 1). The former uniquely originates from bacteria, the latter is common in bacteria (Wilkinson 1977; Kenne and Lindburg 1983; Parsons 1981). The findings suggested that muramic acid and galactosamine were less stable than glucosamine or galactosamine to microbial decay. In addition, cultivation might selectively inhibit the activity of bacteria and thus the re-synthesis of muramic acid and galactosamine.

Sites	Glucosamine	Mannosamine	Galactosamine	Muramic acid	Total
	Means±SD (g amino-sugar-N kg^{-1} N)[1]				
Native sites	50.7±8.7	1.5±0.4	28.7±6.6	3.2±0.7	84.1±14.4
Cultivated sites	43.6±9.5**[3]	1.4±0.5	21.7±5.7**	1.8±40.9**	68.5±15.2**
Differences (%)[2]	13.4a[4]	3.8b	23.7c	40.9d	18.1a

1): Amino sugar-N proportions to soil total N (averages of 17 sites), 2): Differences between cultivated and native sites in percentages, 3): Significance level at p<0.01 between amino sugar contents in native and cultivated. sites, and 4): Letters are used to compare differences in this row.

Table 1: Amino sugar-N proportions

Parabolic regression curves were suitable to simulate the relationship between MAT and amino sugar-N concentrations in the cultivated sites (r^2=0.67***, Fig. 3) and in the native sites (R^2=0.48**). Obviously, the significance level at the cultivated sites were higher compared with that in the native sites, confirming that the equivalence of amino sugar-N dynamics with MAT became better after long-term cropping. According to the parabolic regression model, the concentration of amino sugars peaked in the mesic temperature regime (Fig. 3), where, apparently, MAT exerted an optimum for the accumulation of amino sugars in SOM.

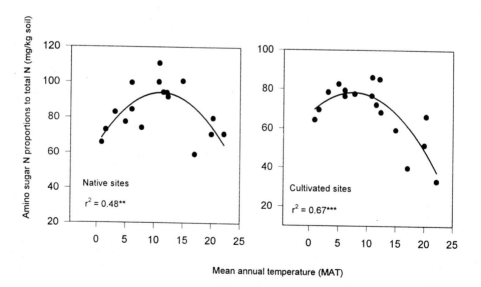

*Fig. 3: Parabolic curves of MAT with total amino sugar-N of total N in soils of the native sites and cultivated sites. **: significance level at p<0.01 and ***: p<0.001*

3.3 Amino sugar ratios

The ratio of glucosamine to galactosamine ranged from 1.45 to 2.61, correlating positively with the logarithm of MAP (r^2 = 0.46**) and MAT (r^2 = 0.45**). Obviously, the contribution of glucosamine, which is common in chitin, to amino sugars in soils was greater at the sites receiving higher temperature and precipitation. As Sowden and Ivarson (1974) demonstrated, little if any galactosamine was synthesised in any of the fungi-incubated materials. The ratio of glucosamine to galactosamine was used for distinguishing the relative contribution between fungi and bacteria to SOM turnover and accumulation (Kögel and Bochter. 1985). If a similar conclusion can be made for soils of the North American prairie, the contribution of fungi to SOM was enhanced relative to that of bacteria as precipitation and temperature increased.

Because of the preferential depletion of the two amino sugars after long-term cultivation, the ratio of glucosamine to muramic acid increased drastically from 12 to 19, on average, (p<0.001) and the glucosamine to galactosamine ratio increased significantly from 1.81 to 2.05 (p<0.01). Because of such significant changes of amino sugar patterns due to cultivation we conclude that both ratios can indicate the impact of land use on SOM dynamics. Apparently, characterising amino sugars is a sensitive tool to indicate climatic and land use effects on SOM dynamics. In how far different soil properties affect amino sugar dynamics remains to be investigated.

Acknowledgements

This project was funded by Deutsche Forschungsgemeinschaft (DFG Ze 154/22-2). The authors wish to acknowledge V.O. Biederbeck, L. Brown, S. Brown, C. Campbell, G.R. Carlson, G. Creinwelge, K.-W. Flach, R.F. Follett, B. Fryrear, E. Knox, R. Molina, E. Montemayor, E.G. Pruessner, C. Richardson, E. Skidmore, D. Sweeney, R. Vredenburk, C. Thompson, H.Tiessen, D Towns, F. Turner, W. Vorhees, and R. Zink for helping with sampling.

References

Amelung, W., Flach, K.W. and Zech, W. (1998): Climatic effects on soil organic matter composition of the Great Plains, Soil Sci. Soc. Am. J. **61**, 115-123.
Amelung, W., Flach, K.W., Zhang, X. and Zech, W. (1998): Climatic effects on C-pools of native and cultivated prairie. Advances in GeoEcology **31** (this issue).
Andriulo, A.E., Rosell, R.A. and Crespo, M.B. (1987): Effects of tillage on organic matter properties of a soil of central Argentina, Sci. Total Environ. **62**, 453-456.
Arshad, M. A., and Schnitzer, M. (1989): Chemical characteristics of humic acids from five soils in Kenya, Z. Pflanzenernähr. Bodenkd. **152**, 11-16.
Burke, I.C., Yonker, C.M., Parton, W.J., Cole, C.V., Flach, K. and Schimel, D.S. (1989): Texture, climate, cultivation effects on soil organic matter content in US grassland soils, Soil Sci. Soc. Am. J. **53**, 800-805.
Honeycutt, C.W., Heil, R.D. and Cole, C.V. (1990): Climate and topographic relations of three Great Plain soils: 2. Carbon, nitrogen, and phosphorus, Soil Sci. Soc. Am. J. **54**, 476-483.
Kenne L.K. and Lindburg B (1983): Bacterial polysaccharides. In . G.O. Aspinall (ed.) The polysaccharides, New York: Academic Press, **2**, p. 287-365.
Kögel I. and Bochter, R. (1985): Amino sugar determination in organic soils by capillary gas chromatography using a nitrogen-selective detector, Z. Pflanzenernähr. Bodenkd. **148**, 260-267.
Nascimento, V.M., Almendros, G. and Fernandes, F.M. (1992): Soil humus characteristics in virgin and cleared areas of the Parana river basin in Brazil, Geoderma **54**, 137-150.
Parsons, J.W. (1981): Chemistry and distribution of amino sugars. In: E.A. Paul and J.N. Ladd (eds.), Soil Biochemistry, New York: Marcel Dekker, Inc. p.197-227.

Parton, W.J., Schimel, D.S., Cole, C.V. and Ojima, D.S. (1987): Analysis of factors controlling soil organic matter levels in the Great Plains grassland, Soil Sci. Soc. Am. J. **51,** 1173-1179.

Rosell, R.A., Andriulo, A.E., Schniter, M., Crespo, M.B. and Miglierina, M.A. (1989): Humic acids properties of an Argiudoll soil under two tillage systems. Sci. Total Environ. **81/82,** 391-400.

Sowden, F.J. and Ivarson, K.C. (1974): Effects of temperature on changes in the nitrogenous constituents of mixed forest litters during decomposition after inoculation with various microbial cultures, Can. J. Soil Sci. **54,** 387-394.

Stevenson, F.J. (1982): Organic form of soil nitrogen. In: F.J. Stevenson (ed.), Nitrogen in agricultural soils, Madison, Wisconsin: American Society of Agronomy, Inc. p. 101-104.

Wilkinson, S.G. (1977): Composition and structure of bacterial lipopolysaccharides. In: I. Sutherland (ed.), Surface carbohydrates of the prokaryotic cell, London: Academic Press. P. 97-174.

Zech, W., Haumaier, L. and Kögel-Knabner, I. (1989): Changes in aromaticity and carbon distribution of soil organic matter due to pedogenesis. Sci. Total Environ. **81/82,** 179-186.

Zhang, X. and Amelung, W. (1996): Gas chromatographic determination of muramic acid, glucosamine, galactosamine, and mannosamine in soils, Soil Biol. Biochem. **28,** 1201-1206.

Addresses of authors:
Xudong Zhang
Wulf Amelung
Wolfgang Zech
Institute of Soil Science and Soil Geography
University of Bayreuth
D-95440 Bayreuth, Germany

Soil Organic Carbon Dynamics under Simulated Rainfall as Related to Erosion and Management in Central Ohio

R.M. Bajracharya, R. Lal & J.M. Kimble

Summary

The movement and transformation of soil organic carbon (SOC) in soil has important implications for agricultural productivity, soil degradation, carbon sequestration, and environmental pollution. Intact monoliths of surface soil excavated from four erosion phases at three sites in central Ohio were subjected to 1 h laboratory rainfall of 30 to 40 mm h-1 intensity. Runoff, percolation water and sediment samples collected at 10 min intervals were analyzed for dissolved organic carbon (DOC), and SOC, respectively. Severely eroded phases had high runoff rates with short times to initiation of flow. Percolation rates were almost always higher than runoff rates initially but diminished to low, steady rates within 40 min of rainfall, indicating surface sealing and reduced infiltration. The erosion phase had no significant effect on runoff or percolation volumes from the soil blocks, though total runoff was greater than percolation. The DOC conentration of both runoff and percolation water peaked early during rain events, diminishing towards the end of the event suggesting rapid removal of accumulated soluble organic D upon re-wetting. Percolation water had significantly higher DOC concentrations that runoff suggesting water picked-up soluble OC as it moved through the surface soil. Total bedload and suspended sediment were higher than percolation sediment for all sites and erosion phases. However, SOC content of percolation sediment was typically higher that those of bedload or suspended sediment indicating that vertical translocation of SOC may be significant.

Keywords: Soil organic carbon, dissolved organic carbon, rainfall simulation, soil erosion phase

1 Introduction

The movement and transformation of soil organic carbon (SOC) within and from the soil profile has major implications for agricultural productivity, soil degradation, carbon sequestration, and environmental impacts (Brown, 1984; Lal et al., 1995). The SOC status of soils may be altered by numerous processes such as, removal by surface runoff water along with eroded soil, deposition of sediment down slope or in depressional areas, translocation within the profile by percolating water, and transport of dissolved organic carbon (DOC) in both runoff and seepage water. Eroded sediment is generally enriched in SOC due in part to selective removal of fine and light soil fractions with higher amounts of associated SOC (Bajracharya, 1990; Olson et al., 1994). Thus, depositional areas typically have higher SOC, particularly in the surface layer, than eroded areas which have had removal of SOC-rich topsoil and exposure of subsoil. Translocation of SOC to lower reaches of the soil profile may, on the other hand, render such SOC less readily decomposed, and hence, contribute to the stable soil C pool which has a long turnover period (Nelson et al., 1994; Tisdall, 1996).

Dissolved or soluble organic carbon comprises a portion of labile SOC (Cook and Allan, 1992) which is readily transported through or over soils. Although DOC typically makes up a small part of total SOC, it is of particular interest from an environmental standpoint because of its role in the transport of nutrients, heavy metals and various other inorganic and insoluble organic compounds to surface water bodies (Dosskey and Bertsch, 1994; Nelson et al., 1993). While the quantity and quality of DOC reaching streams and other water bodies is initially dependent on the nature of the vegetation, litter and organic soil horizons with which the water comes into contact, soil properties and flow pathways can substantially influence DOC quality and quantities in aquatic ecosystems (Boissier and Fontvieille, 1995; Dosskey and Bertsch, 1994; Nelson et al., 1993).

An understanding of the various components and dynamics of SOC, is therefore, necessary for the development of sound land management practices for sustained agricultural production while preventing soil degradation and minimizing adverse environmental effects. Such knowledge is also required to better appreciate the ability of soils to sequester carbon and the implications for global climate change. The objective of this research was to evaluate effects of severity of soil erosion on translocation of SOC and DOC from intact soil blocks obtained from three agricultural fields in central Ohio.

2 Materials and methods

This study was conducted on soil monoliths collected from three farms in central Ohio (hereafter called sites A, B and C). The farms were located in Clark County, near Springfield, Ohio, and contain two predominant soil types: Strawn silty clay loam (fine-loamy, mixed, mesic Typic Hapludalf) at sites A and B; Miamian silty clay loam (fine, mixed, mesic Typic Hapludalf) at site C. Four erosion phases, namely, slight (SLI); moderate (MOD); severe (SEV); and depostition (DEP); were identified at each of the sites using the USDA-SCS (1975) technique. The erosion phases generally corresponded to summit, shoulder, backslope and footslope/toeslope landscape positions, respectively (Ruhe, 1975; Daniels et al., 1985). A summary of soil and site characteristics, as well as management, has been given by Fahnestock (1994). Site A was under a corn (*Zea mays* L.), soybean (*Glycine max* L.), wheat (*Triticum aestivum* L.) rotation with conventional tillage for corn and no tillage performed for soybean and wheat. Site B received disking prior to corn panting with subsequent direct drilling of soybean in corn stubble without tillage. Site C was maintained in a no-till corn, wheat (or clover, *Trifolium repens* L.), soybean rotation.

Three replicates of intact soil blocks were obtained from various erosion phases at each site between September 1995 and May 1996. Galvanized sheet-metal boxes (of 0.2 m width, 0.4 m length and 0.15 m depth) with both ends open and sharpened bottom edges were driven into the ground using a sledge hammer over wooden boards. The soil filled boxes were then excavated by digging around and under them and trimming the bottom end flush with the box edges. The blocks were saturated under tension for several hours and drained for one to two just prior to rainfall simulation. After preparation, the blocks were subjected to laboratory rainfall simulation of 30 to 40 mm h^{-1} intensity for one hour. Runoff and percolation samples were collected at 10 minute intervals and the filtered water samples were analyzed for DOC using a Shimadzu TOC 5000 carbon analyzer. The pH of the water samples were also determined by means of a pH electrode. Suspended soil loss, sediment in percolation and bedload sediment trapped at the bottom of the collection trough were collected, oven dried and analyzed for SOC content by dry combustion at 550 °C (Nelson and Sommers, 1982). This temperature was chosen to measure only organic C and exclude C from secondary carbonates, which combust at 814°C (Dr. W.A. Dick[1]; Dr. T.J. Logan[2]; personal communications).

---

[1] School of Natural Resources, The Ohio State Univ./OARDC, Wooster, OH.
[2] School of Natural resources, The Ohio State Univ., columbus, OH.

3 Results

3.1 Runoff and percolation

Runoff and percolation of simulated rain water did not follow consistent trends across the three experimental sites, reflecting a combination of factors including: differences in management, time of sampling, sampling variability and experimental error. However, some general trends in flow characteristics, which were best expressed for site A monoliths can be noted.

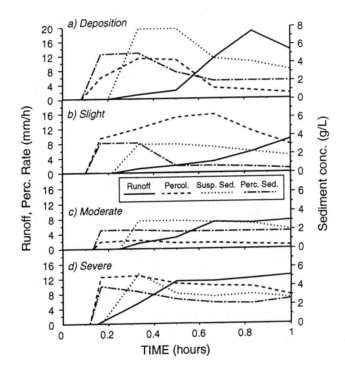

Fig. 1: Plots of typical rainfall simulation paramters for soil blocks from different erosion phases at Clark County site A.

Soil blocks from site A exhibited characteristic differences in runoff rates among the four erosion phases (Fig. 1). Severely eroded phases had typically high runoff rates (10-20 mm h^{-1}) for much of the duration of rain event, as well as, shorter times to initiation of runoff (2-9 min). This was consistent with the observation that SEV areas had substantially exposed subsoil with higher clay content and lower permeability than less eroded areas. Depositional areas generally had low initial runoff rates which eventually reached values comparable to SEV by the end of the rain event. This reflected a gradual sealing of the soil surface with progression of rainfall. Runoff initiation times were, however, considerably longer (5-15 min) for DEP, as well as, SLI and MOD, compared to SEV. The SLI and MOD areas both had low runoff rates (5-10 and 5-12 mm h^{-1}) indicating greater infiltration and water storage capacity than SEV. Water percolation through the soil blocks generally followed the opposite trend of runoff with initially high rates (5-12 mm h^{-1}), which gradually diminished to lower, more steady rates as surface sealing progressed and infiltration rates declined (Bajracharya, 1995).

Erosion phase had no significant effect on total runoff and percolation volumes for all three experimental sites. The lack of statistical significance was due in part to extreme variability of the data and relatively few sample numbers. Average runoff volumes ranged from 277 ml for MOD to 987 ml for SEV and percolation water volumes ranged from 177 ml for DEP to 477 ml for MOD (data not shown). Significant differences were, however, observed due to the effect of sample type (Table 1). Total runoff amounts were usually greater than percolation amounts regardless of erosion phase by nearly a factor of 2 at sites A and C. This suggests relatively poor structure and weak aggregation typical of cultivated and eroded agricultural soils (Bajracharya, 1995; Blevins and Frye, 1994; Lal et al., 1989).

Parameter/ SITE	SAMPLE TYPE		
	Runoff	Percolation	Significance/$LSD_{0.05}$
Water Sample (ml)			
A	648a	310b	*298
B	437	314	Ns
C	725a	386b	**191
DOC Concentration (mg L^{-1})			
A	6.2a	15.0b	**4.3
B	3.6a	6.4b	**1.7
C	2.9a	4.5b	**1.1
Solution pH			
A	6.5	6.6	Ns
B	7.1	7.2	Ns
C	7.5	7.6	Ns

*, **, ns indicate significance at P = 0.05, P = 0.01, and non-significant, respectively.
Across rows, values followed by the same letter are not significantly different.

Table 1. Mean amounts of runoff, percolation, dissolved organic carbon (DOC) and pH according to sample type.

3.2 Dissolved organic carbon and pH of runoff and percolation water

The DOC content in both runoff and percolation water collected from the soil blocks followed similar trends during the course of the rainfall events (Fig. 2). Concentrations of DOC were typically highest (about 10-20 mg L^{-1}) in the early part of the rain event and subsequently diminished to low (5-8 mg L^{-1}) and generally steady values. This likely indicated flushing of most of the DOC in the soil water during the early periods of rainfall events. Early peaks in total and biodegradable DOC upon re-wetting of soils have been reported by Boissier and Fontvieille (1995) who attributed them to development of chemical processes in the soil before any significant biological activity had occurred. Site A had substantially higher DOC concentrations in both runoff and percolation water than sites B and C, which were similar. At site A, the DEP areas had significantly higher mean DOC (15.0 mg L^{-1}) than other phases, although this was not clearly seen at sites B and C (data not shown). Higher DOC concentrations from site A and DEP samples probably reflect the nature and amount of residue, state of decomposition and biological activity in the surface soil (Cook and Allan, 1992; Moore, 1996).

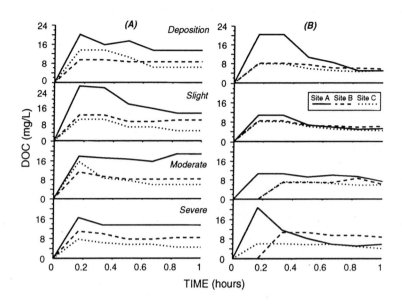

Fig. 2: Dissolved organic carbon (DOC) concentrations in percolation water (A), and in runoff (B) from four soil erosion phases at the Clark County sites.

The effect of sample type, was however, clearer than that of the erosion phase. Percolation water had significantly higher DOC concentrations on average than those of runoff water (Table 1). This was observed for surface soil blocks from all three sites and indicated that greater amounts of DOC were translocated in water percolating through the A horizon than in water which merely ran off the surface. This probably reflects the fact that very little crop residue remained on the soil surface following sampling and transport of the soil blocks. The pH of both runoff and percolation samples were not significantly different due to erosion phase and sample type.

3.3 Sediment loads in runoff and percolation water

No statistically significant erosion phase differences were observed for sediment loads (bedload, suspended load and percolation sediment) apart from weak significance (at P = 0.10) in the case of site A where SEV tended to have the highest amounts of bedload and suspended load sediment removal with an average value for all three of 43.6 g m^{-2} (Table 2). Average amounts of total bedload ranged from 31.0 g m^{-2} for DEP to 76.9 g m^{-2} for SEV; those of total suspended load ranged from 13.0 g m^{-2} for MOD to 46.1 g m^{-2} for SLI; and, total percolation sediment ranged from 3.3 g m^{-2} for DEP to 31.2 g m^{-2} for MOD (data not shown).

No clear trends in concentrations of suspended load and percolation sediment were noted during the coarse of the rain events (Fig. 1). Lack of statistical differences were attributed to high variability of data probably due partly to sampling variability and experimental technique. One noteworthy observation, however, was that mean sediment loss as total bedload and suspended loads were consistently higher (by 25% to more than an order of magnitude greater) than that translocated below the surface 0.15 m layer in percolation water. This was true for all sites and erosion phases as seen from the high statistical significance of the data (Table 3). Thus, much more soil is moved laterally along the surface by overland flow than vertically downward by percolating water (Bajracharya, 1995).

Parameter/ SITE	EROSION PHASE				
	Slight	Moderate	Severe	Deposition	Signif./LSD$_{0.05}$
Sediment load (g m^{-2})					
A	24.1ab	19.5a	43.6b	18.5a	+21.5
B	20.0	33.0	27.4	28.4	ns
C	37.3	31.0	33.0	37.8	ns
SOC Content (% by weight)					
A	1.40	1.62	1.26	1.27	ns
B	1.71a	1.41a	1.73a	2.28b	*0.40
C	1.46a	1.68a	1.53a	2.48b	**0.34

+, *, ** indicate significance at P = 0.1, 0.05 and 0.01 levels, respectively.
Across rows, values followed by the same letter are not significantly different.

Tab. 2: Erosion phase effects on sediment loads and soil organic carbon (SOC) contents.

3.4 Soil organic carbon content of translocated sediment

The effect of erosion phase on SOC content of sediment is shown in Table 2. In all cases sediments were substantiably enriched, by 35 to 100%, in SOC compared to initial soil (Fahnstock et al., 1995; their Table 13). Significant differences were seen for sites B and C with sediment loads from DEP phases having 50 to 70% higher SOC contents than the other phases. Lack of differences for site A samples resulted from some blocks yielding insufficient percolation sediment for SOC analyses, hence causing skewed results. The expectedly high SOC values for DEP were attributed to higher initial SOC status of DEP phases caused by enrichment from eroded areas (Bajracharya, 1990).

Parameter/ SITE	SAMPLE TYPE			
	Bed load	Suspended load	Percolation Sediment	Significance/ LSD$_{0.05}$
Sediment load (g m^{-2})				
A	26.6a	43.9a	8.9b	**18.6
B	25.9a	42.1b	13.6a	**14.7
C	42.9a	46.9a	14.4b	***13.1
SOC Content (% by weight)				
A	1.29ab	1.76a	1.12b	*0.52
B	1.73a	1.41a	2.15b	***0.35
C	1.53a	1.68a	2.11b	***0.29

*, **, *** indicate significance at P = 0.05, 0.01 and 0.001 levels, respectively.
Across rows, values followed by the same letter are not significantly diffrent.

Tab. 3: Mean sediment loads and soil organic carbon (SOC) contents by sample type.

A more notable result was the generally higher SOC contents in percolation sediment as compared to bedload and suspended load sediment (Table 3). This indicates that despite smaller quantities of sediment being translocated vertically, significant amounts of SOC can still be moved to lower depths of the soil profile, which has major implications for C sequestration. The low mean for site A percolation sediment was again due to missing values.

4 Conclusions

The erosion phase had no apparent effect on runoff or percolation volumes from surface soil blocks. Total runoff amounts were greater than percolation regardless of erosion phase, indicating rapid reduction of infiltration and increased runoff rates due to slaking and surface sealing of these cultivated soils. The DOC concentration of both runoff and percolation water showed peaks early during rain events, diminishing towards the end of the event suggesting rapid removal of accumulated soluble organic C upon re-wetting. Erosion phase effects on DOC in runoff and percolation were unclear although DEP areas appeared in general to have higher concentrations than other phases, presumably due to higher initial OC status and possibly soil biological activity (Boissier and Fontvieille, 1995; Adu and Oades, 1978a). Percolation water had significantly higher DOC concentrations than runoff suggesting that water picked-up considerable amounts of soluble OC as it moved through the surface soil as compared to overland flow.

No significant differences in sediment loads were observed among erosion phases, though highly significant differences were noted for sediment types. Total bedload and suspended sediment were higher than percolation sediment for all sites and erosion phases. However, the SOC content of percolation sediment was typically more than those of bedload or suspended sediment indicating that vertical translocation of SOC may be significant. Thus, movement of OC in the form of DOC or sediment-associated SOC vertically to lower depths of the soil profile where it is less accessible/available to biodegradation (Adu and Oades, 1978b; Nelson et al., 1994) could have a major bearing on C sequestration in soil, and hence, global climatic change. Management strategies to improve the OC status of agricultural soils and increase C sequestration are imperative for enhanced productivity and reduced soil and environmental degradation. Further work to elucidated the processes and dynamics of C in the soil profile and across the landscape are needed to develop such strategies.

References

Adu, J.K. and Oades, J.M. (1978a): Utilization of organic materials in soil aggregates by bacteria and fungi. Soil Biol. & Biochem. **10**, 117-122.

Adu, J.K. and Oades, J.M. (1978b): Physical factors influencing decomposition of organic materials in soil aggregates. Soil Biol. & Biochem. **10**, 109-115.

Bajracharya, R.M. (1990): Interrill erodibility of five Ohio soils as influenced by soil properties and seasonal K-variation for a Miamian silt loam. Unpubl. M.S. Thesis. The Ohio State Univ., Columbus, OH.

Bajracharya, R.M. (1995): Crusting and erosion processes on an Alfisol in south-central India. Unpubl. Ph.D. dissertation. The Ohio State Univ., Columbus, OH.

Blevins, R.L. and Frye, W.W. (1993): Conservation tillage: An ecological approach to soil management. In: D.L. Sparks (ed), Adv. Agron. **51**, 34-78. Academic Press. San Diego, CA.

Boissier, J.M. and Fontvieille, D. (1995): Biological characteristics of forest soils and seepage waters during simulated rainfalls of high intensity. Soil Biol. & Biochem. **27(2)**, 139-145.

Brown, L.R. (1984): The global loss of topsoil. J. Soil Water Conser. **39**, 162-165.

Cook, B.D. and Allan, D.L. (1992): Dissolved organic carbon in old field soils: total amounts as a measure of available resources for soil mineralization. Soil Biol. & Biochem. **24(6)**, 585-594.

Daniels, R.B., Gilliam, J.W., Cassel, D.K. and Nelson, L.A. (1985): Soil erosion clases and landscape position in the North Carolina Pedimont. Soil Sci. Soc. J. Am. **49**, 991-995.

Dosskey, M.G. and Bertsch, P.M.. (1994): Forest sources and pathways of organic matter transport to a blackwater stream: a hydrologic approach. Biogeochemistry **24**, 1-19.

Fahnestock, P. (1994): Erosion effects on soil physical properties and crop yield of two central Ohio Alfisols. Unpubl. M.S. Thesis. The Ohio State Univ., Columbus, OH.

Fahnestock, P., Lal, R. and Hall, G.F. (1995): Land use and erosional effects on two Ohio Alfisols: I. Soil properties. J. Sustainable Agric. **7(2/3)**, 63-84.

Lal, R. (1995): Global soil erosion by water and carbon dynamics. In: R., Lal, J. Kimble, E. Levine and B.A. Stewart (eds) Soil Management and Greenhouse Effect. Lewis Publ., Boca Raton, FL. 131-142.

Lal, R., Logan, T.J. and Fausey, N.R. (1989): Long-term tillage and wheel-track effects on a poorly drained mollic ochraqualf in northwest Ohio. 1. Soil physical properties, root distribution and grain yield of corn and soybeans. Soil Tillage Res. **14**, 34-58.

Moore, T.R. (1996): Dissolved organic carbon: Controls on sources, sinks and fluxes and its role in the soil carbon cycle. Proc. Carbon Sequestration in Soil, An International Symposium. July 22-26, 1996. The Ohio State Univ., Columbus, OH. In Press.

Nelson, P.N., Baldock, J.A. and Oades, J.M. (1993): Concentration and composition of dissolved organic carbon in streams in relation to catchment soil properties. Biogeochemistry **19**, 27-50.

Nelson, P.N., Dictor, M.C. and Soulas, G. (1994): Availability of organic carbon in soluble and particle-size fractions from a soil profile. Soil Biol. Biochem. **26**, 1549-1555.

Nelson, D.W. and Sommers, L.E. (1982): Total carbon, organic carbon, and organic matter. In: A.L Page, R.H. Miller, and D.R. Keeney (eds) Methods of Soil Analysis, Pt. II. Chemical and Microbiological Properties, 2nd Ed. ASA Monograph No. 9, ASA-SSSA, Madison, WI. Pp. 539-580.

Olson, D.R., Norton, L.D., Fenton, T.E. and Lal, R. (1994): Quantification of soil loss from eroded soil phases. J. Soil and Water Conser. **49(6)**, 591-596.

Ruhe, R.V. (1975): Geomorphology: geomorphic processes and surficial geology. Houghton Mifflin Co. Boston.

Tisdall, J.M. (1996).. Formation of soil aggregates and accumulation of soil organic matter. In: M.R. Carter and B.A. Stewart (eds) Structure and Soil Organic Matter Storage in Agricultural Soils. Adv. Soil Sci., Lewis Publ., Boca Raton, FL. 57-96.

USDA-SCS. (1993): Soil Survey Manual. Soil Conservation Service, United States Department of Agriculture, Washington, D.C.

Addresses of authors:
R.M. Bajracharya
R. Lal
School of Natural Resources
The Ohio State University
2021 Coffey Road Rm. 210
Columbus, OH 43210, USA
J.M. Kimble
National Soil Survey Center
USDA-NRCS
100 Centennial Mall North Rm. 152
Lincoln, NE 68508-3868, USA

Nitrous Oxide, Nitrate and Ammonium Dynamics as Influenced by Selected Environmental Factors

H.R. Khan, U. Pfisterer & H.-P. Blume

Summary

Influences of temperature, groundwater (Gw), soil condition, microbial activity and incubation time on the dynamics of N_2O, NO_3 and NH_4 in an adjacent natural (never used and covered with natural vegetation) and an arable (under intensive use as cropland) Calcaric Fluvisol were examined using soil monoliths in a laboratory. Soils at different depths and gases from the soil surfaces were collected prior to start of treatment and 1, 10, 45 and 90 days after incubation.

The rates of emission of N_2O and concentrations of NO_3 in the soils increased, whereas the concentrations of NH_4 decreased with time and rising temperature. The emissions of N_2O were more striking during the first 45 days, and drastically decreased after 90 days, regardless of the treatments. The Gw level at a depth of 18 cm significantly ($p=0.01$) increased the rates of emission of N_2O and the concentrations of NO_3, but reduced the concentrations of NH_4 in the soils as compared with the Gw at 12 cm. These effects were even more pronounced for the arable soils. Microbial processes in the soils were significantly influenced by the variations of Gw and soil conditions even at 4°C.

Keywords: Nitrous oxide, denitrification, temperature, groundwater-level, microbial activity

1 Introduction

Knowledge on the dynamics in different forms of N under variable soil and environmental conditions is necessary for their sustainable management. Research on soil N dynamics remains a high priority because of environmental concerns, such as increasing NO_x and NH_4 concentrations in the atmosphere and NO_3 contamination of groundwater. The sources and their strength of N_2O emissions are still fragmentary (Duxbury et al. 1993, Khan et al. 1996). The contribution of agriculture to the present anthropogenic N_2O emissions is estimated to be 70 to 92 % (Mosier, 1993). Developing management practices to minimize NO_3 loss and N_2O emissions from managed ecosystems requires an understanding of the sources and factors controlling soil-N dynamics. Parton et al. (1988) revealed that the variability of N_2O emission rates is related to temperature, moisture and NO_3 and NH_4 concentration in soils. It has also been shown that high concentrations of NO_3 can cause a shift in end products towards N_2O due to denitrification (Firestone et al. 1980). The soil conditions, environmental factors, their interactions and/or biological processes which regulate soil-N dynamics as well as water and atmospheric pollution are not yet fully understood and require extensive studies to increasing our understanding of these processes and interlinkages (Batjes and Bridge 1992, Bouwman 1995). Furthermore, the relationships between soil-N dynamics and environmental factors have been less successfully established in field studies and these relationships can only and more reliably be identified in laboratory studies. Accordingly, a laboratory study was conducted to evaluate the influences of temperature, Gw, soil condition, microbial activity and incubation time on the dynamics of N_2O, NO_3 and NH_4 in adjacent natural (never used and covered with natural vegetation) and arable (under use since 1980s) soils.

2 Materials and methods

A laboratory study was conducted using soil monoliths (0-20 cm depth), which were collected by a PVC core sampler (20 cm height and 15 cm $\varnothing$) during February 1995, i.e. 3 months after harvesting of cabbage from the arable soils. They usually received annual N-fertilisers at rates of up to 300 kg N ha^{-1}. The soil monoliths were put on PE bowls (12 cm height and 20 cm $\varnothing$). The space between the cylinders and bowls were filled with initially-collected basement soils (10-22 cm subsoil) from the sampling site. The soil (intact) monoliths were then subjected to variable temperatures in different climate chambers after autoclaving (by a steam steriliser at 150°C and 3 kg cm^{-2} pressure for 2 hours) of some monoliths (D9 to D16: Natural soils) and adjustment of Gw at 12 and 18 cm beneath the soil surface. The soils used in the basement sections of the autoclaved soil monoliths were also autoclaved. The Gw levels at desired depths of the soil columns were adjusted with tap water for non-autoclaved (D1 to D8: Natural; D17 to D24: Arable) soils and sterilised water (20 mg of Ag[NaCl]$^-$ per L of tap water) for autoclaved soils, and allowed to stand at room temperatures (15-18°C) for 7 days, prior to start of the experiments. Several small (0.5 cm $\varnothing$) holes were made over the surface at the bottom sections of the cylinders to maintain Gw. The Gw levels were kept constant by the addition of tap and sterilised waters to the basement sections of the columns as required. The experiments were set up in a randomised design with four levels of temperature (4, 10, 20 and 30°C), two levels of Gw (12 and 18 cm), two levels of soil microorganisms (M$^+$: with and M$^-$: without), and natural and arable soils having duplicates of each treatment. Soil samples at depths of 0-5, 5-10 and 10-15 cm, and gas samples from the soil surfaces were collected prior to start of the treatment, and 1, 10, 45 and 90 days following establishment with the treatments. Emissions of N_2O were measured following the closed chamber method of Mosier (1990) and analysed by gas chromatography. Some physical and chemical properties of the soils (0-24 cm) were analysed following the methods of soil analysis by Klute (1986). The natural and arable soils contained 14.6, 43.3 % sand; 25.7, 4.3 % clay respectively. They showed silty, loamy texture and had 0.86, 1.34 Mg m^{-3} bulk density, 41.0, 24.9 % field moisture, 7.6, 7.6 pH (1:2.5 water), 2.2, 28.1 mg L^{-1} NO_3-N, 9.9, 6.6 mg L^{-1} NH_4-N, 11.2, 15.0 C/N ratio, 23, 13 g kg^{-1} organic-C and 37, 11 cmolc kg^{-1} CEC. Significance of treatment means was determined by Duncan's New Multiple Range Test.

3 Results and discussion

3.1 Emissions of nitrous oxide

The rates of emissions of N_2O were very low during earlier periods of incubation, even in two treatments (Fig. 1) negative values were obtained from the natural soils at 1 day after treatment, which might be attributed to uptake of nitrogen by plants prior to the (7 days) establishment of the treatments. The pattern of emission of N_2O was characterised by low initial rates, rising during the middle (45 days) and decreasing at the end (90 days) of the experiments. This might be due to the decrease of soil moisture contents with time as well as of substrate availability to microbial turnover. The processes generally declined in course of the incubation. About 80 % of the fluxes fell within the range (1.5 to 3) of the temperature coefficient (Q_{10}) and amounting 1.2 to 2.2 for the

Environmental factors influencing nitrous oxide, nitrate & ammonium dynamics

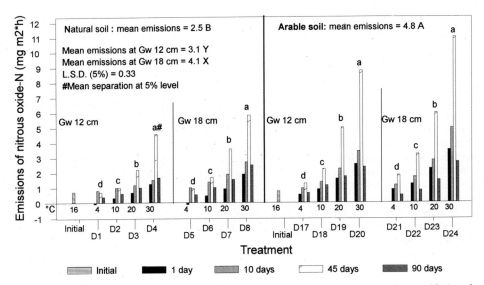

Fig. 1: *Rates of emission of nitrous oxide as influenced by temperature, groundwater (Gw) and soil condition during 90 days of experiment.*

	Emissions of N₂O (mg m⁻² h⁻¹)		NO₃-N (mg L⁻¹)	NH₄-N (mg L⁻¹)
Natural soils: Initial	(0 day) = 0.69		Initial = 2.18	Initial = 9.88
Non-autoclaved	One day	Ten days	Ten days	Ten days
D1 (Gw at 12 cm)	-0.10 d#	0.79 d	1.63 d	11.27 b
D5 (Gw at 18 cm)	-0.14 e	0.97 c	2.10 c	7.45 c
Autoclaved				
D9 (Gw at 12 cm)	0.36 c	0.42 f	148 e.	17.30 a
D13 (Gw at 18 cm)	0.39 c	0.48 e	1.70 d	18.21 a
Arable soils: Initial	(0 day) = 0.76		Initial = 28.1	Initial = 6.57
Non-autoclaved				
D17 (Gw at 12 cm)	0.52 b	0.93 b	25.43 b	5.63 d
D21 (Gw at 18 cm)	0.84 a	1.12 a	30.87 a	5.32 d
L.S.D. (5 %)	0.04	0.05	0.21	0.93

#Mean separation within rows by Duncan's multiple range test at 5%.

Tab. 1: *Emissions of N₂O, concentrations of NO₃ and NH₄ (0-15 cm depth) at 4°C as influenced by soil condition, autoclaving and groundwater (Gw) during 10 days of incubation.*

natural soils and 1.4 to 2.3 for the arable soils. The emissions of N_2O from the arable soils were significantly larger than from the adjacent natural soils. This might be due to the high rates (about 300 kg ha^{-1} y^{-1}) of N-fertilisation, low water holding capacity and higher sand content as well as loamy texture of the arable soils as mentioned in the section of materials and methods. Obviously, low moisture content in the surface soils enhances gas diffusion.

Soil texture may affect rates of emission of N_2O. Arah et al. (1991) reported that the emissions were higher from light-textured soils and discussed the possibility that N_2O reduction is more complete in heavier-textured soils due to restricted diffusion. The fluxes of N_2O measured at different days of incubation showed significant ($p=0.05$) differences in the emissions under variable temperatures, Gw, soil conditions and interactions among these factors (Fig. 1). Moisture contents of the soils showed significant inverse relationships with the N_2O fluxes, while NO_3 concentrations showed significant positive relationships with N_2O emissions, indicating that denitrification dominated the emissions; the findings are in accordance with the results of Ryden (1983). On the other hand, there was no significant relationship between N_2O fluxes and the NH_4 contents in the soils which indicates that nitrification probably did not dominate N_2O emissions, though the results are in contrast with the findings of Parton et al. (1988). The non-autoclaved soils at 4°C showed significant variations between the rates of emission of N_2O in comparison to autoclaved soils (Table 1), indicating that the process of denitrification took place considerably at 4°C. Even at that low temperature, the variations of Gw and soil conditions were proven to have a significant influence on this process.

3.2 Concentrations of nitrate

The concentrations of NO_3 determined during the 90 days of incubation showed significant ($p=0.05$) increments at the higher (20-30°C) temperatures, low Gw (18 cm below) and incubation times (Fig. 2a). The most striking increments were detected at the highest temperature (30°C) during 45 and 90 days after incubation and were induced by the low Gw in both sites. The pattern of temporal variation was similar at both sites, but the effects were more striking at the arable soils and may be due to the higher (about 300 kg ha^{-1} y^{-1}) rates of N-fertilisation. The interactions between temperature, Gw and soil conditions were found to be significant (Fig. 2a). There is (Fig. 2a) no NO_3 concentration bulge in the lower depth (10-15 cm) and these effects were more pronounced with the higher (20-30°C) temperatures during later periods of incubation. Furthermore, the NO_3-enriched water may also have a chance to leach and eventually to contaminate the groundwater.

The concentrations of NO_3 at the higher (20-30°C) temperatures are on a higher level (>45 mg L^{-1}: see Schlichting et al. 1995) during later periods and were adequate to support denitrification activity for all the sampling dates (Fig. 2a). The Q_{10} values of NO_3 ranged from 1.1 to 2.8 for the natural soils and 1.0 to 1.5 for the arable soils during the 90 days and relatively high Q_{10} values in the natural soils were due to the initial high content of organic matter, indicating a depletion of nutrients with time due to humus degradation by the treatments/practices. These results are in agreement with the findings of Evans et al. (1994) and Malhi et al. (1990).

3.3 Concentrations of ammonium

The concentrations of NH_4 decreased significantly ($p=0.05$) with higher (20-30°C) temperatures, low Gw (18 cm below) and incubation times in both sites (Fig. 2b). These trends were more pronounced with the natural soils throughout the incubation. No trace of NH_4 was detected at 0-5 cm of the surface soils with the highest temperature of 30°C at low Gw level after 90 days of

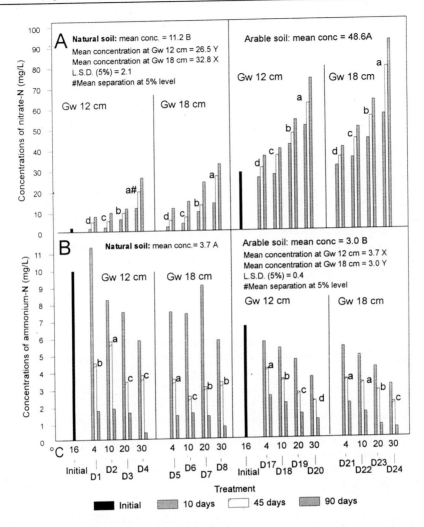

Fig. 2: Concentration of nitrate (A) and ammonium (B) at 0-15 cm depth as influenced by temperature, groundwater (Gw) and soil condition during 90 days of incubation.

incubation (data not shown). The concentrations of NH_4 were higher in the natural soils than in the arable soils, which might be attributed to the high CEC, organic matter and water holding capacity of the natural soils. The Q_{10} values of NH_4 ranged from 0.3 to 1.3 for the natural soils and 0.5 to 0.9 for the arable soils during the 90 days, which remained below the optimum Q_{10} range (1.5 to 3, it was attributed to the use of undisturbed soil monolithe. The NH_4 concentrations, at the time in each incubation period when N_2O and NO_3 production reached a maximum, were very low at all temperatures (Fig. 2b). This might be due to NH_3 volatilisation at higher temperatures and low Gw level, which is supported by the noticeable smelling during the experiment and by the soil pH of 7.6. Natural condition of soils and high Gw levels also showed inverse relationships as compared to those obtained for the N_2O and NO_3 dynamics. On the other hand, NH_4 concentration was found to increase more towards the lower depth (10-15 cm: data not shown), and the increment was higher at 20°C than at 30°C. This indicates a more pronounced release (or ammonification) of organic N at 20°C than at 30°C. Taken together with the data of N_2O emissions, NO_3 and NH_4 concentrations at 4°C indicated that the processes of ammonification, nitrification and denitrification were considerable at that low (4°C) temperature, being significantly influenced by the Gw and land-use condition (Table 1). These results are in accordance with the findings of Dorland and Beauchamp (1991), as well as Green et al. (1994).

4 Conclusions

We hypothesised that variable temperature, Gw, soil condition and their interactions may have considerable influence on soil-N dynamics. The present results derived from cropland and natural soil support this hypothesis. The emissions of N_2O and concentrations of NO_3 were correlated inversely with the soil water contents and positively with the higher temperatures and low Gw level. These trends were more pronounced with the high and intensively fertilised arable soils, suggesting that soil management practices should be adjusted to utilise residual inorganic N, bringing supply and demand into closer balance, which may result in less N_2O entering the atmosphere as well as reduced NO_3 contamination to groundwater.

Acknowledgement

The senior author (H.R. Khan) gratefully acknowledges a postdoctoral fellowship by the "Alexander von Humboldt Foundation" Germany, which enabled him to carry out this research project.

References

Arah, J.R.M., Smith, K.A. and Li, H.S. (1991): Nitrous oxide production and denitrification in Scottish arable soils. J. Soil Sci. **42**, 351-367.

Batjes, N.H. and Bridges, E.M. (eds.) (1992): World inventory of soil emissions. Int. Soil Reference and Information Centre, Wageningen.

Bouwman, A.F. (1995): Compilation of a global inventory of emissions of nitrous oxide. PhD Thesis, Agric. Univ., Wageningen.

Dorland, S. and Beauchamp, E.G. (1991): Denitrification and ammonification at low temperatures. Can J. Soil Sci. **71**, 293-303.

Duxbury, J.M., Harper, L.A. and Mosier, A.R. (1993): Contributions of agroecosystems to global climate change. p. 1-18. In: D.E. Rolston et al. (eds.): Agricultural ecosystem effects on trace gases and global climate change. ASA Sp. Publ. **55**. CSSA and SSSA, Madison, WI.

Evans, S.D., Peterson, G.A., Westfall, D.G. and McGee, E. (1994): Nitrate leaching in dryland agroecosystems as influenced by soil and climate gradients. J. Environ. Qual. **23**, 999-1005.

Firestone, M.K. Firestone, R.B. and Tiedje, J.M. (1980): Nitrous oxide from soil denitrification: Factors controlling its biological production. Science (Washington, DC) **208**, 749-751.

Green, C.J., Blackmer, A.M. and Yang, N.C. (1994): Release of ammonium during nitrification in soils. Soil Sci. Soc. Am. J. **58**, 1411-1415.

Khan, H.R., Pfisterer, U. and Blume, H.-P. (1996): Impact of selected environmental factors on the emissions of trace gases. p. 261-270. In: T.V. Duggan and C.A. Brebbia (eds.): Environmental Engineering Education and Training. Computational Mech. Publ., Southampton, UK.

Klute, A. (ed.) (1986): Methods of soil analysis. Agron Series 9. Am. Soc. Agron. Publ., Madison, WI/USA

Malhi, S. S., McGill, W.B. and Nyborg, M. (1990): Nitrate losses in soils: Effects of temperature, moisture and substrate concentration. Soil Biol. Biochem. **22**, 733-737.

Mosier, A.R. (1990): Gas flux measurement techniques with special reference to techniques suitable for measurements over large ecologically uniform areas. In: A.F. Bouwman (ed.): Soils and the Greenhouse Effect. Vol.2, John Wiley and Sons, 289-301

Mosier, A.R. (1993): State of knowledge about nitrous oxide emissions from agricultural fields. Mitt. Dtsch. Bodenkd. Ges. **69**, 201-208.

Parton, W.J., Mosier, A.R. and Schimel, D.S. (1988): Rates and pathways of nitrous oxide production in a shortgrass steppe. Biogeochemistry **6**, 45-58.

Ryden, J.C. (1983): Denitrification loss from a grassland soil in the field receiving different rates of nitrogen as ammonium and nitrate. Journal of Soil Science **34**, 355-65.

Schlichting, E., Blume, H.-P. and Stahr, K. (eds.) (1995): Bodenkundliches Praktium, Blackwell, Berlin., pp. 248

Addresses of authors:
H.R. Khan
Department of Soil Science
University of Dhaka
1000 Dhaka
Bangladesh
U. Pfisterer
Soil Ecology (FTZ)
University of Kiel
Hafentörn
25761 Büsum
Germany
H.-P. Blume
Institute for Plant Nutrition and Soil Science
University of Kiel
Olshausenstraße 40-60
24118 Kiel
Germany

Water and Wind Erosion – Causes, Impacts and Reversing Measures

C. W. Rose

Summary

The major processes involved in soil erosion by water and by wind are briefly described. The on-site impacts of both forms of erosion in an agricultural context are considered, and the effectiveness of soil conserving practices linked to the process descriptions. Similarities and differences between wind and water erosion processes and their control receive brief mention.

Keywords: Water erosion, wind erosion

1 Introduction

Soil erosion by wind or water erosion, accelerated by human activity, threatens sustainable production and environmental quality. The causes, impacts and requirements for land degradation processes to be reversed, and soil restoration to be achieved, are both biophysical and socio-economic. Socio-economic constraints and marginalisation commonly inhibit the adoption of soil conserving methods of production, even when land degradation is understood by the producer. To change management practices, especially in a bare survival context, the change must be seen as feasible, not increasing risk or the chance of net loss, and with the hope of net gain in the short to medium term. Indeed, in such a context, soil conservation cannot be promoted as an end in itself, but rather as an inadvertent consequence of practices perceived as improving the economic or social well-being of the primary producer.

This ideal description assumes that effective, feasible and acceptable soil conservation technologies exist or can be conceived of in the particular socio-economic and environmental context of concern. This paper is limited to addressing the question of how an understanding of the various processes involved in water and wind erosion can assist in the complex process of matching suitable soil conserving measures to meet the challenge of these erosion processes. This limited objective does not provide the instant answer to soil conservation design and selection. However, it aims to improve the efficiency in judging the likely effectiveness of management alternatives in reducing soil erosion, or in the complex interactive process of assisting the producer in such decisions. Justification of the objective is that, unless a land management option is effective in conserving the soil resource, there is no value in its promotion; furthermore, the most conservation-effective option compatible with all other constraints can be assisted by understanding the process-nature of the erosion threat.

In addition to water and wind, gravity can be a potent cause of soil erosion, as in landslides, slumping, bankfall and some gully-forming and rilling processes. Such processes will not be directly addressed in this paper.

2 The processes of water erosion

Water erosion is chiefly due to stresses induced in the soil, either by impacting raindrops or by flow of water over the soil surface. Whether or not overland flow develops depends locally on whether the rate of rainfall exceeds the rate of infiltration, though the requirement to fill pondage due to irregularities in the soil surface provides a moderating influence. The infiltration rate usually declines with time during a runoff event, is spatially variable, and structural modification of the soil surface through interaction with rainfall commonly reduces infiltration rate, especially in soils prone to sealing or crusting (Sumner, 1995). Whilst soil sealing and crusting are not universally so pronounced as to provide major management problems, they are pervasive, and Hairsine and Hook (1995) have provided a conceptual framework for considering erosion and sealing as complementary surface phenomena.

Erosion results in the formation, by net deposition during the erosion event, of a deposited layer of soil which can quickly cover a large fraction of the soil surface. This deposited layer is initially coarser than the original soil, and eroded sediment correspondingly finer and commonly richer in chemicals than the original undegraded soil. This rapid flush of fine sediment is the major reason why sediment concentration is initially high, followed by a gradual decline, as illustrated in Fig. 1 for a constant erosional situation.

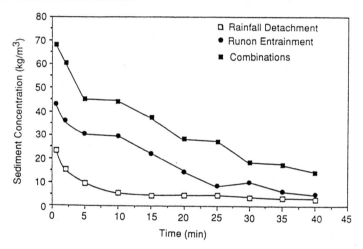

Fig. 1: Comparing sediment concentration as a function of time exposed to simulated constant rainfall of rate 100 mm/h (rainfall detachment), a constant rate of overland flow (giving a streampower of 0.33 W/m^2, denoted runon entrainment), and a combination of both inputs adjusted to give a streampower of 0.33 W/m^2. Soil type a red Podzolic or Ultisol, slope 6% for runon entrainment and combinations, slope length of flume 6 m, bare soil. (From Okwach et al. 1992).

Fig. 1 also illustrates that both rainfall-driven and flow-driven processes, separately or in combination, can add to sediment concentration and soil loss, with the possibility of some interaction between these two types of processes (Proffitt and Rose, 1992). The slow decline in sediment concentration indicates that, even in a constant erosional environment, it takes time for the deposited layer to approach an equilibrium situation assumed to exist in most models of erosion, although dynamic erosion models have recently been developed (Rose et al., 1994).

Rainfall-driven erosion of the original soil is described as detachment, and of the deposited layer as re-detachment. The net outcome of these processes over deposition (which returns sediment to the soil surface) is also termed interrill erosion (Foster, 1982), since these processes can be dominant between proximate rills. Rose and Hairsine (1988) and Hairsine and Rose (1992) have identified the maximum sediment concentration, the transport limit, c_t, as occurring when the mechanically weak deposited layer completely covers the soil surface. Though erosion driven by overland flow can be important, even in the absence of rills (often defined as erosional flow channels obliterated by tillage), the presence of rills is a clear indication of the action of flow-driven erosion. Theory developed by Hairsine and Rose (1992) predicts that the sediment concentration at the transport limit in the presence of rills is given by

$$c_t = \frac{k}{\phi}(\Omega - \Omega_0)\frac{Z}{D}, \tag{1}$$

where k is a known constant, ϕ the depositability of the sediment (defined in Rose, 1993), Ω the streampower (with threshold value Ω_0), Z a geometry factor depending on the shape of the rill (with Z = 1 for sheet flow), and D the depth of water flowing in the rill. Equation (1) helps explain variation, noted by Loch (1996), in the rill sediment concentration with volumetric discharge from experimental plots.

As discussed in Rose (1993) and Ciesiolka et al. (1995), c_t and its average for an erosion event $\bar{c}_t$, can be readily calculated if runoff rate per unit area, Q, is measured and rill geometry noted. These same authors and Marshall et al. (1996) then define the erodibility, β, of a soil which erodes with an average sediment concentration $\bar{c}$ as

$$\beta = \frac{\ln \bar{c}}{\ln \bar{c}_t}, \text{ or } = (\bar{c}_t)^\beta \tag{2}$$

where $\beta \leq 1$, unless erosion processes other than flow-driven erosion add significantly to $\bar{c}$. With $\bar{c}$ measured experimentally, and $\bar{c}_t$ calculated via Equation (1), the erodibility β can be calculated from Equation (2), as illustrated in a multi-country research program partly reported in a special issue of Soil Technology (1995). Fig. 2 illustrates that, partly due to cultivation effects in this case, erodibility β can vary with time.

For simplicity considering a non-rilled situation, Fig. 3 illustrates a relationship between $\bar{c}$ and streampower Ω (assuming $\Omega_0 = 0$), with decreasing erodibility, β, being associated with a reduction in $\bar{c}$ for any given Ω.

Again, for simplicity restricting ourselves to sheet flow, Marshall et al. (1996) show that it follows from Equation (1) that

$$\bar{c}_t = k_1 \frac{\Sigma Q^{1.4}}{\Sigma Q}, \tag{3}$$

where k_1 is also a known constant, and Q the rate of runoff per unit area. It follows from Equations (2) and (3), and the fact that the rate of soil loss per unit area, dM/dt = cQ, that the soil loss per unit area for an event, M, is given by

$$M = k_1^\beta \Sigma Q^{1-\beta} \Sigma Q^{1.4\beta}. \tag{4}$$

It follows from Equation (2) that if $\beta = 1$, $\bar{c} = \bar{c}_t$, and, as is consistent with Equation (3), it follows from Equation (4) that

$$M(\overline{c}=\overline{c}_t) = k_1 \Sigma Q^{1.4} \quad (\beta = 1), \tag{5}$$

which describes the worst scenario of soil loss from bare soil.

Equations (4) or (5) bear the same message, that to reduce soil loss due to flow-driven erosion, it is necessary to reduce the rate of runoff per unit area, Q, or the velocity of runoff flow, which is related to Q. Possible ways in which this may be achieved will be considered in Subsection 4.

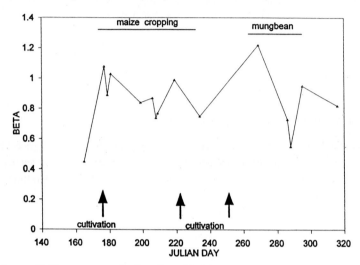

Fig. 2.: The erodibility parameter β plotted as a function of time for a bare soil (Typic Tropudalf) at UPLB Experimental Farm, Laguna, Philippines. The plot was cultivated at times indicated, in phase with cultivations carried out on adjacent plots with maize followed by mungbean. (After Paningbatan et al., 1995).

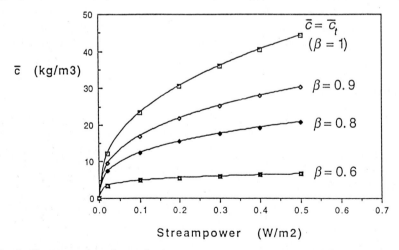

Fig. 3: Illustrating the relationship between streampower and c_t (the transport limit, given by the upper curve with β = 1), and between streampower and c for values of erodibility parameter β shown in the figure.

3 The processes of wind erosion

Wind erosion is dominantly manifested by the near-surface drift in saltation of sand-sized particles (conventionally defined as in the range 60-1000 μm), and the aerial suspension of 'dust' (< 60 μm approximately). Saltation (the hopping motion of sand particles) moves material from metres to kilometres during a wind storm, and is the major cause of ejection of the smaller dust particles which can be dispersed up into the atmosphere by atmosphere turbulence and transported over intercontinental distances by the atmospheric circulation (Knight et al. 1995).

Briefly, the shear stress, τ, exerted by the wind on the land surface, lifts and accelerates sand particles which, on return to the soil surface, dislodge dust and other sand particles in a form of controlled chain reaction (Shao et al., 1993). Dust particles have a settling velocity less than the mean velocity of atmospheric turbulence given by κu_*, where κ (= 0.4) is the von Karman constant, and u_* the friction speed is defined by

$$\tau = \rho u_*^2, \tag{6}$$

where ρ is air density. Most of the shear stress τ is absorbed in providing momentum to the saltating sand particles (the 'saltation stress'), the residual stress being quite small. A minimum threshold friction speed, u_{*t}, is required to initialise particle movement, analogous to the threshold streampower, Ω_0, in Equation (1), but generally of more significance since Ω_0 can be quite small, at least in cultivated soils. In wind erosion, the speed of the wind is the major driving factor, but its erosive effectiveness is modified by soil moisture (which strengthens the soil), by soil characteristics such as structure, texture and stoniness, by cultivation, and, above all, by any vegetation or other cover on the soil surface not removed by the strong wind. The nature of the eroding soil surface can be modified as erosion proceeds. As more erodible components are preferentially removed, the surface becomes less erodible due to limited availability of erodible material, with increased surface roughness and associated sheltering also assisting the decline.

In Subsection 2, the concept of streampower was introduced in describing erosion due to water flow; it is defined as the product of shear stress between water and the soil surface and its velocity of flow. The analogue in wind erosion of streampower is τu_*, which, from Equation (6) is ρu_*^3. This term is a measure of the available wind power, and ρu_*^3 is the core term in equations describing the flux of saltating particles, and thus of dust particles emitted by saltation bombardment.

For example, Shao et al. (1996) use the theory of Owen (1964) to describe the "transport-limited" streamwise flux of saltating particles of uniform size d, denoted $\widetilde{Q}(d)$, as

$$\widetilde{Q}(d) = (c_s \rho\, u_*^3/g)(1 - [\,u_{*t}(d)\_u_*\,]^2), \tag{7}$$

where c_s is a coefficient ($\approx$1), and g is the acceleration due to gravity. The tilde on Q indicates it applies to particles of uniform size d. Leys and Raupach (1991) have extended Equation (7) to multiple-sized soil particles in the transport-limiting situation, interactions between classes being neglected.

Shao et al. (1996) also show that the entrainment or emission rate per unit area of dust particles can be related to $\widetilde{Q}(d)$ for bombarding sand particles. Again assuming limited interaction between the effects of different-sized particles, summing over all the range of sand-sized particles and dust yields an expression for the total vertical flux of dust particles. The United States Department of Agriculture has also provided a comprehensive computer model called WEPS (Hagen, 1991) that predicts wind erosion on a daily time step.

4 Impacts and control of soil erosion

4.1 Impacts and control of water erosion

4.1.1 Impacts

While water erosion can have important off-site impacts on the water quality of lakes, rivers and the near-shore ocean, the focus is restricted here to the on-site effects such as loss of plant nutrients and organic matter and reduced topsoil depth which can reduce the yield of crops and pastures. More soil-sorbed chemicals can be lost in eroded soil than would be expected on the basis of soil loss if the enrichment ratio, E_R, is greater than one, where:

$$E_R = \frac{(\text{kg chemical/kg soil}) \text{ in eroded sediment}}{(\text{kg chemical/kg soil}) \text{ in original soil}}. \tag{8}$$

For fairly well understood reasons, E_R declines with the amount of sediment eroded, and is greater for rainfall-driven than for flow-driven erosion. E_R also depends on soil type, declining somewhat exponentially as clay content increases (Rose and Dalal, 1988). Knowledge of the effect of soil nutrients on crop production can be used to infer the likely effect of soil erosion on the yield of a particular crop (Williams et al. 1983). Whilst in some economic contexts it may be quite feasible to replace the loss of nutrients by erosion through fertilizer application, changes in the soil physical characteristics, lower organic matter levels, and reduced plant-available water storage capacity are effects much more difficult and costly to overcome (e.g. Mbagwu et al., 1984).

Stocking (1984) and Lal (1987) have reviewed literature describing the varied research approaches which have demonstrated a decline in crop productivity associated with soil erosion.

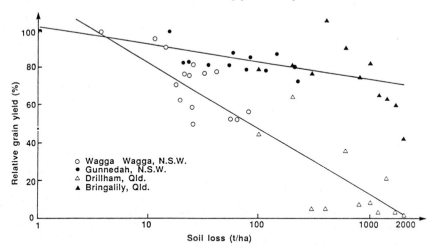

Fig. 4: Relationship between soil loss and relative wheat grain yield under dryland conditions in eastern Australia. Circles represent cumulative soil loss under natural rainfall over 26 years (Gunnedah) and 30 years (Wagga Wagga) followed by two and four wheat crops, respectively (from Aveyard, 1983). Triangles represent soil removal by desurfacing, followed by a wheat crop (P.J. White, unpublished data). Soil profiles at Gunnedah and Bringalily are medium to deep, whereas the others are shallow, with either texture-contrast (Wagga Wagga) or saline subsoil (Drillham). (From Rose and Dalal, 1988).

Whilst decline in productivity can be rapid in tropical steeplands with acid subsoils, decline is often slow. Fig. 4 presents a relation between cumulative soil loss and relative wheat grain yield measured over several decades in a semi-arid environment, the log-linear relationships fitted being reasonably consistent with data obtained using artificial desurfacing of topsoil in an attempt to simulate the effect of greater erosion.

Research designed to improve identification of the various possible yield-reducing effects of soil erosion, and to quantify such reductions, is in an early stage of development, most soil erosion models being unable to predict adequately changes in soil properties due to erosion, let alone the consequences of such change on crop production. Particularly because of the importance of being able to forecast longer term consequences of current or alternative management practices, Pierce and Lal (1994) have emphasised the need for greater research emphasis on predicting the rate and extent of erosion-induced productivity change.

4.1.2 Control

Despite uncertainty in our current ability to predict the long-term productivity consequences, scientific guidelines based on an understanding of water erosion processes have continued to strengthen since the pioneering work summarised in the Universal Soil Loss Equation (or USLE, Wischmeier and Smith, 1978). Fig. 3 can be shown to be consistent with the universal dependence of soil loss on slope and slope length used in the USLE for soils in a cultivated erodible condition (high β); however, Fig. 3 also indicates that the dependence on slope and slope length becomes increasingly muted as β declines, which is generally as soil strength increases, for example by reducing tillage.

Equation (1) shows that c_t (and so c) is reduced if the depositability, Φ, is increased by improving soil structure, by increasing organic matter content for example. However, the control of water erosion is most directly and powerfully achieved by providing the soil surface with cover, such as mulch or previous crop residue, in a form that is so close to the soil surface that it retards overland flow, which is commonly rather shallow. This „surface contact cover" can also bear much of the reduced shear stress that the slowed overland flow exerts. Surface contact cover (or SCC) is thus more effective than above-ground or „aerial" cover which can only assist by protecting the soil surface against raindrop impact, a function also fulfilled by SCC. Surface contact cover can also reduce the likelihood of rill formation and improve infiltration, especially on heavier textured soils.

So effective is SCC in reducing sediment concentration, c, relative to that for bare soil, c_b, that c/c_b typically falls rapidly and somewhat exponentially with increasing SCC, as is illustrated in Fig. 5. A similar plot of c/c_b against aerial cover is much more linear than the relation in Fig. 5.

Since it may not be possible at all times to maintain an adequate level of SCC, protection by grass strips or hedgerows can provide effective longer term protection, though they can break down or be overwhelmed in more extreme events, when soil loss can then be very high.

Restoration of lands degraded by various causes, including erosion, is the topic of a recent volume (Lal and Stewart, 1992). Some of the most extensive challenges lie in economically rehabilitating eroded grazing land, Gichangi et al. (1992) providing an interesting example in Kenya.

4.2 Impacts and control of wind erosion

Commonly, the most effective protection against wind erosion is to reduce the wind velocity at the soil surface by control of stock pressure in grazing lands, or retaining plant residue in a well-anchored form (Fryrear, 1990). Aerial cover is much more effective against wind than water erosion.

Soil management can also play a role in protection, maintaining a cloddy condition, increasing u_{*t} in Equation (7). Developing ridges aligned at right angles to the direction of erosive winds, and the use of porous wind breaks is also helpful (Hagen, 1976).

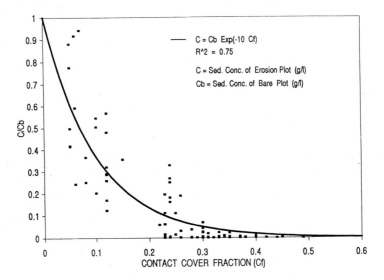

Fig. 5: The ratio of sediment concentration, c, with the level of surface contact cover shown, expressed as a ratio to c_b, the sediment concentration measured from a bare plot of the same size (12 m long by 6 m wide) and slope (≈15%). Location and soil type as for Figure 2. (After Paningbatan et al., 1995).

5 Concluding comments

The outline of the major process of wind and water erosion in Subsections 2 and 3 indicate that there are similarities as well as differences between the erosive processes at work. There is no direct analogy to rainfall detachment in wind erosion, and the sensitivity of wind erosion to soil moisture is lacking in water erosion, though soil strength plays a role in both types of erosion.

Also, there are differences in the effectiveness or appropriateness of soil conservation measures used for water erosion in controlling wind erosion and vice versa. However, dead or living cover, close to the soil surface and anchored to it, is effective against both forms of erosion, which can occur at the same location at different times.

References

Aveyard, J.M. (1983): Soil erosion: Productivity research in New South Wales to 1982, Soil Cons. Serv. NSW Tech. Bull. No. 24.
Ciesiolka, C.A., Coughlan, K.J., Rose, C.W., Escalante, M.C., Hashim, G.M., Paningbatan, E.P. and Sombatpanit, S. (1995): Methodology for a multi-country study of soil erosion management, Soil Technology **8**, 179-192.

Foster, G.R. (1982): Modelling the erosion process, in: C.T. Hann (ed.), Hydrologic Modelling of Small Watersheds, Am. Soc. Agr. Eng. Monograph No. 5, 297-379, St. Joseph, Michigan.

Fryrear, D.W. (1990): Wind erosion: Mechanics, prediction, and control, in: R.P. Singh, J.F. Parr and B.A. Stewart (eds.), Dryland Agriculture: Strategies for Sustainability, Advances in Soil Science, Vol. 13, Springer-Verlag, New York Inc.

Gichangi, E.M., Jones, R.K., Njuarui, D.M., Simpson, J.R.,Mututho, J.M.N. and Kitheka, S.K. (1992): Pitting practices for rehabilitating eroded grazing land in the semi-arid tropics of eastern Kenya: A progress report, pp. 313-327, in: H. Hurni and K. Tato (eds.), Erosion, Conservation, and Small-Scale Farming, Walsworth Publishing Company Inc., 306 N. Kansas Ave., Maceline, Missouri, 64658, USA.

Hagen, L.J. (1976): Windbreak design for optimum wind erosion control, in Shelter-belts on the Great Plains, Ag. Counc. Pub. No. 78, pp. 31-37.

Hagen, L.J. (1991): A wind erosion prediction system to meet user needs, J. Soil and Water Conserv. **46**, 106-111.

Hairsine, P.B. and Hook, R.A. (1995): Relating soil erosion by water to the nature of the soil surface, pp. 77-91, in: H.B. So, G.D. Smith, S.R. Raine, B.M. Schafer and R.J. Loch (eds.), Sealing, Crusting and Hardsetting Soils: Productivity and Conservation, Aust. Soc. Soil Sci., Queensland Branch, Australia.

Hairsine, P.B. and Rose, C.W. (1992): Modelling water erosion due to overland flow using physical principles: I. Uniform flow, Water Resources Res. **28**, 237-243.

Knight, A.W., McTainsh, G.H. and Simpson, R.W. (1995): Sediment loads in an Australian dust storm: Implications for present and past dust processes, CATENA **24**, 195-213.

Lal, R. (1987): Effects of soil erosion on crop production. CRC Critical Reviews in Plant Science **5**, 303-367.

Lal, R. and Stewart, B.A. (1992): Soil restoration, Advances in Soil Science **17**, Springer-Verlag, N.Y., pp. 456.

Leys, J.F. and Raupach, M.R. (1991): Soil flux measurements using a portable wind erosion tunnel, Aust. J. Soil Res. **29**, 533-552.

Loch, R.J. (1996): Using rill/interrill comparisons to infer likely responses of erosion to slope length: Implications for land management, Aust. J. Soil Res. **34**, 489-502.

Marshall, T.J., Holmes, J.W. and Rose, C.W. (1996): Soil Physics, 3rd Edition, Cambridge University Press, Cambridge, CB2 IRP, UK, pp. 453.

Mbagwu, J.S.C., Lal, R. and Scott, T.W. (1984): Effects of desurfacing of alfisols and ultisols in southern Nigeria. 1. Crop performance, Soil Sci. Soc. Am. J. **48**, 828-833.

Okwach, G.E., Palis, R.G. and Rose, C.W. (1992: Sediment concentration and characteristics as affected by surface mulch., land slope and erosion mechanisms, pp. 91-105, in: H. Hurni and K. Tato (eds.), erosion, Conservation and small-Scale Farming, Walsworth Publishing Company Inc., 306 N >kansas Ave., Maceline, Missouri, 64658, USA.

Owen, P.R. (1964): Saltation of uniform grains in air, J. Fluid Mech. **20**, 225-242.

Paningbatan, E.P., Ciesiolka, C.A., Coughlan, K.J. and Rose, C.W. (1995): Alley cropping for managing soil erosion of hilly lands in the Philippines, Soil Technology **8**, 193-204.

Pierce, F.J. and Lal, R. (1994): Monitoring the impact of soil erosion on crop productivity, pp. 235-263, in: R. Lal (ed.), Soil Erosion Research Methods, 2nd Edition, Soil and Water Conservation Society, St. Lucie Press, Beach, Florida, 33483, USA.

Proffitt, A.P.B. and Rose, C.W. (1992): Relative contributions to soil loss by rainfall detachment and runoff entrainment, pp. 75-90, in: H. Hurni and K. Tato (eds.), Erosion, Conservation, and Small-Scale Farming, Walsworth Publishing Company Inc., 306 N. Kansas Ave., Maceline, Missouri, 6458, USA.

Rose, C.W. (1993): Erosion and sedimentation, pp. 301-343, in: M. Bonell, M.M. Hufschmidt and J.S. Gladwell (eds.), Hydrology and Water Management in the Humid Tropics - Hydrological Research Issues and Strategies for Water Management, Cambridge University Press, Cambridge.

Rose, C.W. and Dalal, R.C. (1988): Erosion and runoff of nitrogen, pp. 212-233, in J.R. Wilson (ed.), Advances in nitrogen cycling in agricultural ecosystems, C.A.B International: Wallingford, U.K.

Rose, C.W., Hogarth, W.L., Sander, G., Lisle, I., Hairsine, P.B. and Parlange, J.Y. (1994): Modelling processes of soil erosion by water, Trends in Hydrology **1**, 443-452.

Rose, C.W. and Hairsine, P.B. (1992): Processes of water erosion, pp. 312-316, in: W.L. Steffen and O.T. Denmead (eds.), Flow and Transport in the Natural environment, Springer Verlag, Berlin.

Shao, Y., Raupach, M.R. and Findlater, P.A. (1993): The effect of saltation bombardment on the entrainment of dust by wind, J. Geophys. Res. **98**, 12719-12726.

Shao, Y., Raupach, M.R. and Leys, J.F. (1996): A model for predicting aeolian sand drift and dust entrainment on scales from paddock to region, Aust. J. Soil Res. **34**, 309-342.

Soil Technology (1995): Special Issue: Soil Erosion and Conservation, Vol. 8, No. 3, pp. 258.

Stocking, M.A. (1984): Erosion and soil productivity: A review, Food and Agriculture Organisation Consultants Working Paper No. 1, Soil Conservation Programme, Land and Water Development Division, Food and Agriculture Organisation, Rome, Italy.

Sumner, M.E. (1995): Soil crusting: chemical and physical processes, the view forward from Georgia, 1991, pp. 1-14, in: H.B. So, G.D. Smith, S.R. Raine, B.M. Schafer and R.J. Loch (eds.), Sealing, Crusting and Hardsetting Soils: Productivity and Conservation, Aust. Soc. Soil Sci., Queensland Branch, Australia.

Williams, J.R., Renard, K.G. and Dyke, P.T. (1983): EPIC - A new method for assessing erosion's effect on soil productivity, Journal of Soil and Water Conservation **38**, 381-382.

Wischmeier, W.H. and Smith, D.D. (1978): Predicting rainfall erosion losses - A guide to conservation planning, Agricultural Handbook No. 537, U.S. Department of Agriculture, Washington, D.C.

Address of author:
Calvin W. Rose
Faculty of Environmental Sciences
Griffith University
Nathan Campus
Brisbane, Queensland 4111
Australia

Effects of Hillslope Hydrology and Surface Condition on Soil Erosion

C. Huang, D.S. Gabbard, L.D. Norton & J.M. Laflen

Summary

Seepage, the reemergence of soil water at the surface, is a common occurrence in fields with an impeding soil layer during periods of excessive soil moisture. The seepage zone has been associated with the development of 'seepage steps' and ephemeral gullies. Despite the linkage between seepage and landform development, little data are available for seepage induced surface erosion on gentle slopes when seepage flow alone will not cause any erosion.

Laboratory experiments were designed to recreate the hydrologic conditions of a 5-m segment of the hillslope at different locations on the landscape. Variables considered are: slope steepness, seepage and drainage gradients. Runoff and sediment samples were collected under rainstorms, and rainfall with added inflows to emulate various levels of erosive conditions as the hillslope position is changed.

Results showed that seepage conditions greatly enhanced surface erosion. Sediment delivery was 3 to 4 times higher with seepage flow as compared to a free-drained condition. The near-surface hydraulic gradient has a significant effect on runoff and sediment regime. This is supported by a data set collected during a rainfall run when the seepage condition was suddenly reversed to drainage. The rapid change from exfiltration to infiltration conditions caused a 70% reduction in sediment delivery.

Knowledge of hydrologic effects on soil erodibility and erosion processes may lead to different erosion control strategies and management practices for regions where seepage occurs.

Keywords: Erosion, hillslope hydrology, seepage, surface condition.

1 Introduction

Water movement on the landscape, from both surface and subsurface sources, is the dominant mechanism in shaping the landform and the formation of the soil mantle covering it (Birkeland, 1984). When subsurface flow exfiltrates onto the soil surface, the development of several different landscape features may evolve. Seepage cusps, concavities on shallow slopes serving as an area of deposition and vegetative growth, result from an increase in the downwearing of the backslope in the seepage zones. This increased erosion of the backslopes relative to the summits leads to the development of tors and the preservation of the crests (Bunting, 1961). Despite general knowledge of seepage effects on reducing soil strength and causing slope failure and headward erosion (Dunne, 1990), very few studies have been conducted to measure seepage induced erosion during rainfall events.

A conceptual hillslope process model is illustrated in Fig. 1. It is hypothesized that different hillslope positions will affect the hydrological conditions that will, in turn, affect the erosion processes. For example, drainage conditions are dominant near the summit through upper backlope. The small amount of surface flow contributed from the area upslope leads to interrill-dominated processes in the upper locations of the hillslope. At locations further downslope, the increased water from contributing areas upslope enhances the concentration of surface flow, hence rill erosion processes. Near the toe of the slope, seepage may occur during periods of excessive soil moisture. A simple force diagram to show how the seepage and drainage flow can change the stress on a soil aggregate at the surface is presented in Fig. 2. Under drainage conditions, the additional suction from the downward flow holds down the aggregates in place and thus increases the resistance for soil detachment. Seepage flow enhances the 'quick' condition as interparticle forces are reduced, thus reducing soil strength.

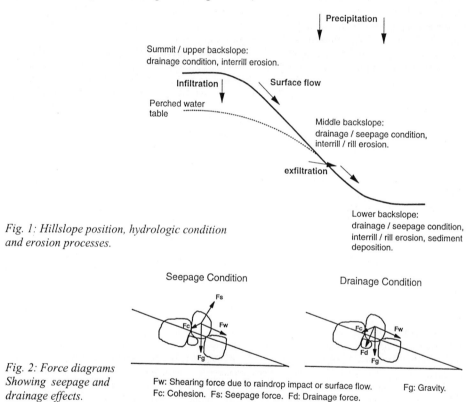

Fig. 1: Hillslope position, hydrologic condition and erosion processes.

Fig. 2: Force diagrams Showing seepage and drainage effects.

Fw: Shearing force due to raindrop impact or surface flow. Fg: Gravity.
Fc: Cohesion. Fs: Seepage force. Fd: Drainage force.

This paper summarizes recent studies on hydrologic conditions, especially seepage and drainage gradients, and erosion processes occurring at a hillslope. A 5 meter long soil box was designed to recreate hydrologic conditions of a segment of the hillslope under various scenarios: subsurface water movement, slope steepness, raindrop impact, and concentrated flow. Experiments were designed to quantify sediment delivery under different combinations of soil erodibilty and flow erosivity representing different hillslope positions. Results of this study will further the understanding of erosion processes and their relationships to the spatially and temporally varying hydrologic conditions on a hillslope.

2 Materials and methods

The soil used in this study was Glynwood clay loam (fine, illitic, mesic Aquic Hapludalf with 22% sand, 49% silt and 29% clay) collected from a farm at Blackford County, Indiana. The soil was collected from the summit and upper part of the shoulder on the hillside. Surface soil, to a depth of 10 cm, was sampled randomly in the field. The soil was air-dried in the laboratory prior to the experiment. The soil was neither sieved nor ground and care was taken to preserve its original aggregation. After drying, the soil had a subangular-blocky structure.

The study was conducted on a 5 m long, 1.2 m wide and 0.3 m deep soil box. The bed slope was adjustable from 0 to 40% in 5% increments. A baffle plate, attached and sealed to the bottom of the soil box, was placed 1.2 m from the downslope end of the box to force the subsurface flow to exfiltrate before the end of the plot. A water circulation system was designed to supply water to the bottom of the soil box with a capability to maintain a constant water level of any position relative to the soil surface. A detailed description of the soil box and water circulation system was given by Huang and Laflen (1996) and Gabbard (1996).

Three rainfall simulation troughs, each with five VeeJet nozzles (Part no. 80100, Spraying Systems Co., Wheaton, IL) at 1.07 m spacing, were placed 2.3 m above the soil bed. These rainfall troughs were programmed to produce 25 to 150 mm hr^{-1} simulated rainfall.

Flow devices were also constructed to introduce added inflows to the soil surface. The flow rate was set by adjusting flow valves according to the reading from an in-line flow meter.

Surface hydrological conditions are controlled by adjusting both the slope steepness and seepage/drainage gradients. The seepage or drainage conditions were set by setting the hydrostatic pressure at the bottom of the soil box relative to the soil surface. Treatments were water tube heights at 20 and 10 cm above the soil surface (20WT and 10WT), at the soil surface (0WT), 10 cm below the surface (-10WT) and no water tubes connected (NWT). Under the three high water tube settings (20, 10 and 0 cm), water seeped out of the soil and flowed onto the surface.

The run procedure consists of an initial preparation rain of 30 minutes at 50 mm hr^{-1} with the soil box at its level position 24 hours prior to the erosion run. After the initial rain, the soil bed is set to the selected slope steepness and the water level was applied. The erosion run started with a 30 minute rainstorm at 50 mm hr^{-1}. After the rainfall-only portion of the run, inflow was added while the rain continued. A total of 5 levels of inflow (7.6, 15.1, 22.7, 30.2, and 37.8 L $min.^{-1}$) were applied, starting with the lowest flow rate. Each level of inflow was applied for 8 minutes with runoff samples taken every two minutes.

3 Results and discussion

3.1 Bed morphology under infiltration conditions

Surface features of the soil bed after the runs with the water tube set at 10 cm below the soil surface were used to illustrate erosion processes under infiltrating conditions. Before the experiment, it was thought that the soil would start rilling under added inflow conditions regardless of the hydrological conditions of the test bed. This was not always the case, especially under infiltrating conditions. Instead of forming rills, knickpoints were formed randomly throughout the bed. In several cases, these knickpoints developed into scallop-shaped depressions and crescents.

At 10 % slope, the crescent-shaped pits eroded headward until they coalesced with the next crescent shaped pit upslope. Over time this process may lead to the development of shallow rills as the flow becomes concentrated in these areas where the crescent-shaped pits coalesce into one another. Insufficient run time may have prevented this from happening under the -10 cm drainage condition but the process becomes evident when the water tube was raised to the surface of the soil. When the water surface was set to the soil surface level, the bed changed from the mechanism of developing crescent shaped pits to the development of headcuts. This occurred at a flow rate of about 22.7 L min^{-1} as the soil eroded into seepage conditions which enhanced detachment and the increase inflow provided extra transport for the development of rills. Rilling processes increased as both seepage gradient and inflow were increased.

3.2 Bed Morphology Under Seepage Conditions

The bed surface after runs at 10% slope and the water tube set at 10 cm above the soil surface was used to show the effects of seepage conditions on processes of erosion. Headcuts started in the general vicinity of the baffle plate installed 1.2 m from the exit end of the bed. The plate was placed so the effect of the seepage would not be influenced by the plot end effects. During the run, intense rilling took place and large aggregates of sediment were carried down the slope. In some instances, extensive collapsing of the sidewalls and rapid headcut advancement showed severe rill erosion in action. The stochastic nature of sidewall sloughing contributed to the high variability in sediment data under high flow rates.

The largest observable difference after the runs with seepage was caused by the slope steepness. After runs at 5% slope, the bed surface had wide, shallow, meandering rills which resembled miniature rivers. There were several areas of cutting and deposition within the meanders after runs at 5% slope. Under 10% slopes, rills were incised much deeper and meandered less. The runs with 10% slope appeared to be dominated by detachment and transport with very little deposition except the immediate area above the plot end.

3.3 Runoff and sediment delivery

The trends of sediment data were presented by plotting sediment delivery as functions runoff. From the conceptual model in Fig. 1, these graphs illustrated the water runoff and soil loss values typical of a 5 m segment on different positions along a hillslope. Hydrologic conditions for summit, shoulder and upper backslope were mostly drainage dominant and represented by the water tubes set at either -10 cm (-10WT) or disconnected (NWT) under the rainfall only portion of the run. The water tube heights of 0 to 20 cm were designed to simulate runoff and erosion that may occur in areas susceptible to seepage. The seepage condition simulated the hydrology on the middle to lower backslopes and footslopes in the landscape (Fig. 1).

Fig. 3 depicts soil loss rates as functions of runoff. One surprising feature from the 5% slope data is that erosion rates were significantly lower under -10WT treatment than under free drainage conditions. Before the experiment, it was perceived that the soil loss rate would decrease as the level of the water supply tube was lowered from 20 cm above the soil surface to the free drainage condition. Therefore, for the two drainage situations, one would also think that the free drainage condition will have less soil loss than the condition with -10 cm water level. One explanation is that capillary rise caused a direct hydraulic connection for the top 10 cm of the surface, which enhanced the drainage flux and, subsequently, produced a higher suction on the soil aggregates.

There appears to be two different sediment trends for 5% slope. Two drainage conditions, -10WT and NWT, seemed to approach sediment detachment limiting regime as the flow transport was increased. However, under seepage conditions, especially under the 20WT treatment, there seems to be a linear association between water runoff and sediment loss, indicating a transport dominated sediment regime. Data also showed an increased scatter under high sediment and runoff conditions. This high variability reflected the inherent nature of the chaotic processes that took place while the soil bed was under severe rilling. Headwall collapse, soil anomalies and variability in inflow all led to higher variability of the results under flow conditions.

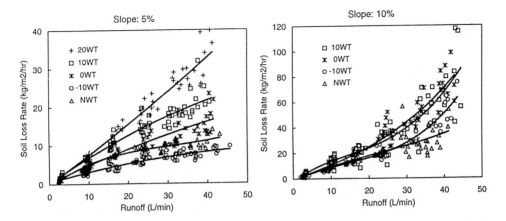

Fig. 3: Soil loss rates as functions of runoff for (a) 5% and (b) 10% slopes.

One different feature of the soil loss at 10% slope is the less significant differentiation as the near-surface hydraulic gradient was shifted from seepage to drainage conditions. This shows that effects of surface condition on soil erodibility demonstrated for the 5% slope condition was masked as the slope gradient is increased. This figure also showed what happened when the soil surface was shifted from drainage to seepage conditions during the run with the -10WT treatment. Soil loss rates were similar for both -10WT and NWT treatments at low runoff rates until the 30.3 L min^{-1} flow rate was introduced. At that point, the trend for the -10WT treatment changed and began to parallel that of the seepage treatment. This indicated that erosion processes must have changed. During the experiment, it was witnessed that this was the approximate time when the bed began to rill.

Under seepage conditions at 10% slope, the difference between the soil loss rates from the 0WT and 10WT treatments is insignificant. This is probably because once rilling was initiated, the bottom of the rill changed to seepage conditions and the high flow transport capacity associated with the higher slopes overshadows the minute difference in the exfiltration gradient. Soil eroded as fast as it could be transported. Intense rilling, sloughing of rill channels, and rapid rill head advancement were witnessed.

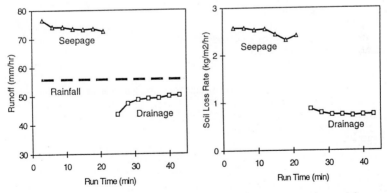

Fig. 4: Changes in runoff and soil loss as the hydraulic gradient was changed from seepage to drainage conditions.

Blume, Eger, Fleischhauer, Hebel, Reij & Steiner (Editors): Towards Sustainable Land Use

To further demonstrate the effects of near-surface hydraulic gradient on soil erosion, we conducted a run in which the hydrologic condition was reversed from 20WT seepage to free drainage during the rainstorm. The runoff and sediment data were plotted in Fig. 4. The reversal from a seepage to a drainage condition caused a reduction of runoff from 75 to 48 mmh^{-1} and sediment delivery from 2.5 to 0.7 kgm^{-2}h^{-1}. This reduction in soil loss can be attributed to two factors: the increased soil strength under drainage conditions and reduced shallow flow transport.

4 Conclusions

This study examined the effects of different hydrological conditions, especially seepage and drainage, on different positions of a hillslope and the dominant erosion processes that occurred throughout. Different slope and hydrological conditions were used to simulate sections of a hillslope. Results showed that, under drainage conditions, rilling is limited and the surface contained scattered crescent-shaped pits after the run. Under seepage conditions, rilling started during rainfall-only and the added inflow accelerated the headward erosion of the rills.

At the landscape scale, erosion from the summit or upper backslope areas was simulated by drainage condition under rainfall- only portion of the run. As the location on the landscape moves down slope, the surface flow is increased and the dominant hydrologic condition may also shift from drainage to seepage situations. If the 5-meter soil box represents the condition of a hillslope segment in the landscape, erosion rates would vary as much as 60 to 70 times as location moves down slope. These values were obtained by comparing two extreme sediment regimes: from a low slope, rainfall-only under drainage condition to a high slope, rainfall plus added inflow under seepage conditions. This indicated that rills and gullies on lower levels of the backslope may be catalyzed by seepage conditions rather than hydraulic shear alone.

Current farming practices need to be evaluated for their impacts on hillslope hydrology, which may affect erosion and sediment production. Further research may lead to different management practices and erosion control strategies in regions where seepage may occur, such as using drainage as an erosion control measure.

References

Birkeland, P.W. (1984): Soils and Geomorphology, Oxford Univ. Press, New York, NY.
Bunting, B.T. (1961): The role of seepage moisture in soil formation, slope development, and stream initiation. Am. J. Sci. **259**, 503-518.
Dunne, T. (1990): Hydrology, mechanics and geomorphic implications of erosion by subsurface flow, In C.G. Higgins and D.R. Coates (ed.), Groundwater Geomorphology, the Role of Subsurface Water in Earth Surface Processes and Landforms. Geol. Soc. of America, Special Paper 252, 1-28.
Gabbard, D.S. (1996): The effects of landscape position and hydrology on erosion, M.S. Thesis, Agronomy Dept., Purdue Univ., W. Lafayette, IN.
Huang, C. and Laflen, J.M. (1996): Seepage and erosion for a clay loam soil, Soil Sci. Soc. Am. J. **60**, 408-416.

Address of authors:
C. Huang
D.S. Gabbard
L.D. Norton
J.M. Laflen
USDA-ARS, National Soil Erosion Research Laboratory, 1196 SOIL Bldg., Purdue University West Lafayette, IN 47907-1196, USA

Assessment and Concepts for the Improvement of Soils Affected by Water Erosion in India

J.S. Samra

Summary

Land providing livelihood gathering opportunities is excessively eroded in India due to ever-increasing demographic pressures. An average erosion rate of 16.35 t/ha/year is ecologically undesirable. Soil conservation technologies for non-arable land, arable land and mass wasting such as landslides, mining, gullies and torrents (flash floods) have been reviewed. Application of bio-physical measures on prioritized catchments since 1962 was effective in preventing excessive sedimentation of natural water bodies. Contour bunding, masonary as well as loose boulder structures and non-monetary input in arable lands were partially successful. Integration of vegetative barriers, structural measures and water harvesting techniques found greater acceptability by farmers. Resource conservation on the basis of naturally occurring geohydrological units of a watershed was taken up on a large scale since 1991. People's participation, greater role of non-government organizations, voluntary agencies, self-help groups and women form the current paradigms of soil conservation. Joint forest management, people's empowerment, transparency, equity, socio-economic factors, local level institution building, sustainability, environmental concerns, income generation for landless and creation of the productive employment are recent policy initiatives. Application of modern tools and procedures, quantification of environmental economics, ground water recharging and diversification of production base are also being discussed for trans-regional integration.

Keywords: Soil conservation, bio-physical, socio-economics, participation, watershed management.

1 Introduction

The importance of soil as a source of biomass production and environmental sink for sustainable development is very well understood by the World Community. For an agrarian economy like India, improvement and maintenance of soil qualities is vital for providing a sustainable life support system and environmental security. Water erosion due to over-exploitation of forest resources, excessive human and livestock pressure, cropping on very steep slopes, shifting cultivation, inadequate agronomic practices, mining, ever-increasing communication network, insufficient policy support and lack of stakeholders participation, account for 123 million ha (37.5% of total geographical area) of land degradation. Ancient Indian literature is replete with scriptures highlighting the importance of resource conservation. Such views have been adopted in a field project taken up for restoration of ravine in North India (Uttar Pradesh) in 1884. The Panjab Land Preservation (Choes) Act was enacted in 1900 followed by the Bombay State Land Improvement

Scheme Act (1942), Damodar Valley Corporation Act (1949) and many other Acts were passed by almost all states of the Union of India. In addition to legislative and policy measures, an extensive research and development infrastructure was created especially through the establishment of a chain of Soil and Water Conservation, Research Demonstration and Training Centres in 1954. Initial efforts generally endeavored for more food production whereas environmental externalities and sustainable restoration on a watershed basis are being focussed in the current strategies.

2 Assessment

Estimation of erosion status through expressive mapping units denotes kind, degree, extent, distribution and overall average degradation. This exercise was initiated in quantitative terms with the establishment of runoff plots at five locations in 1933. The Central Soil and Water Conservation Research and Training Institute, Dehra Dun set up several standard runoff plots, 42 gauged micro-watersheds and 367 gauged plots at nine locations in the country. Apart from that many provincial states set up their own institutions for managing water erosion. Sedimentation data is being recorded (Singh et al. 1990) in almost all river valley projects (29 reservoirs) at 5-10 years interval (Table 1). An overall assessment of erosion computed by pooling all the available information by Dhruva Narayana & Babu (1983) estimated an annual loss of 5334 million tonnes or 16.35 t/ha/year. Out of this, 1572 million tonnes (29%) is permanently lost to the sea, 480 million tonnes (10%) is deposited in the reservoirs and 3282 million tonnes (61%) is redistributed. About 5.37 to 8.4 million tonnes of nutrients are lost due to water erosion. Eutrification of lakes, lowering of water qualities and suspended sediments has reduced hydro-biological productivity of the natural water bodies.

Catchment	Area ('000 km^2)	Annual rate of silting (tonnes/ha)		Ratio of observed to assumed rates
		Assumed	Observed	
Beas	12.51	6.44	35.84	5.6
Mayuraksh	1.87	5.42	30.14	5.6
Ramganga	3.63	6.44	25.95	4.0
Ghod	3.64	5.42	23.26	4.3
Maithon	18.19	2.43	18.22	7.5
Mahi	20.80	1.94	13.49	7.0
Nizamsagar	21.69	0.44	9.51	21.6
Sutlej	18.20	6.44	9.33	1.4
Ukai	62.40	6.71	7.44	1.1
Chambal	26.00	5.42	5.42	1.5
Tawa	5.96	5.4	4.01	0.7
Machkund	5.02	4.79	3.29	0.7

Tab. 1: Sedimentation rate in some representative Indian reservoirs selected out of 27 River Valley Hydro-electric Projects, India

Spatial distribution of soil erosion was further estimated by Universal Soil Loss Equation and Iso-erosion rates in India prepared by Singh et al. (1992). Applications of remote sensing and GIS are being made by several organizations and large-scale maps for micro-watershed planning are finding acceptability with development agencies. A resolution of 5.9 m of the recently launched satellite (IR-1C), with contouring capabilities, have added new dimensions to the assessment strategies. Process based runoff and erosion prediction (numerical solutions) and semi-process based models of gully head extension are at various stages of their development. Cost-benefit analysis and environmental externalities of soil conservation practices have been computed for different agro-ecological regions of the country. Quantitative evaluation of glacier erosion, on-site and off-site effects, i.e., immediate, short-term and long-term consequences of water erosion by remote sensing and GIS based modeling are important concerns of the scientists.

3 Concepts for improvements

Demographic pressure, scarcity of land resources and concerns for the prevention of excessive siltation of reservoirs as well as natural water bodies set in motion a programme of soil conservation in the 1950's. Unfortunately, till very recently, independent resource conservation strategies were developed on two arbitrarily devised non-arable and arable land use systems due to administrative reasons. Through various federal and state sector schemes, India has treated 37.3 million ha of 123 million ha eroded by water with a total investment of Rs. 35,114 millions upto the end of 1996 with indigenously developed practices.

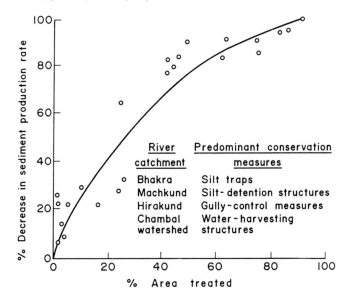

Fig. 1: The effect of percentage of prioritized catchment area treatment on sediment production rate.

4 Non-arable lands

Greater attention was given to the treatment of River Valley and Flood Prone River catchments. For optimizing the use of limited resources, catchment areas were prioritized on the basis of Sediment Yield Index. Treatment of only 25% of the prioritized critically eroded subsets of catchment reduced sediment production by 50% (Fig.1).

5 Bio-engineering measures

Sustainable restoration of highly eroded catchments was realized by initiating the process of in situ conservation of rain water, sediments and other fluxes through structural measures. This process of conservation was supplemented and subsequently taken over by the regenerated vegetation promoted through social fencing of people's participation (Table 2).

Particulars	Pre-treatment (Ave. 1970-79)	Post treatment (Ave. 1979-89)
Average annual		
Monsoon rainfall (mm)	1003	953
Monsoon water yield (mm)	295	63 (4.7)
Runoff (% of monsoon rain)	29	7
Monsoon sediment yield (t/ha)	140	18 (7.8)

(Values in brackets indicate times of reduction)

Tab. 2: *Effect of integrated soil conservation measures and social protection in Sukhna Lake catchment (4207 ha) on sediment and water yield, India*

6 Landslides

The Himalayas are generally affected at a rate of 5-15 landslides per km^2 with a loss of 0.5 to 1.0 ha of area per landslide. Mass wasting due to pore volume pressure, geology, denudation and accelerated developmental activities was stabilized with a series of structural and vegetative treatments (Table 3).

Description	Post-treatment (1978)	Pre-treatment (1960)
Surface runoff, % of rainfall	54	32
Sediment load, tonnes/ha	320	5
Vegetative cover, %	Negligible	95
Lean flow, days	100	220
Channel slope, %		
Upper reach	54	44
Middle reach	23	14
Lower reach	12	7

Tab. 3: *Effect of bio-engineering measures on 60 ha landslides affected watershed (Nalota Nala), India*

Description	Pre-treatment (1984)	Post-treatment (1994)
Surface runoff, % of rainfall	57	36
Sediment load, tonnes/ha/yr	550	8
Vegetative cover, %	10	70
Lean flow, days	75	250
Equivalent slope, %	38	21
Cost of debris clearnce, million Rs.	0.1	.002
Water quality	Not potable	Potable

Tab. 4: *Effect of integrated measures on restoration of 64 ha mine-spoiled watersheds (Sahastradhara), India*

7 Mine spoil restoration

Open mining without adequate environment management plans is a major concern of resource conservation in India. Restoration technologies consisting of slope stabilization structure on steeper slopes, grade moderation structures in the channels and vegetative measures in the entire spoil have been successfully tested at the field scale (Table 4).

8 Eco-development of ravines

Erosion processes like pot hole formation, piping, slope failures, bed and side scouring in differentially layered and denuded alluvial formations is extending gully formation at the rate of 0.24% per annum in 170 districts of India. Ecologically sustainable reclamation practices consist of land shaping, bunding and safe disposal of runoff in the runoff contributing peripheral zone. Shallow and medium gullies are terraced and channels are stabilized with suitably designed gully plugs for the production of fruits and other perennials. The deep gullies are left to the natural permanent vegetation.

9 Shifting cultivation

This is a traditional tribal farming system of slash burning, cultivation for 3-5 years, and temporary fallow of the area for fertility restoration through natural regenerative processes, mainly in the North-Eastern part of India. This shifting (zhooming) cycle has been reduced to 5-6 years from the earlier rest period of 15-20 years. This reduced cycle has accelerated the water erosion processes in nearly 4 million ha. Its improvement is less of a bio-physical problem but more of an ethanic as well as socio-economic issue.

10 Arable land

In the early formative years of the programme, costly structural or engineering measures were relatively highly focussed. Bunding was done quite extensively during the 1960s and also as a famine relief ad hoc activity in the subsequent years. It was later on discovered that nearly half of the rainfed area (Vertisols) were affected by excess of water during the rainy season and in situ moisture conservation was of secondary importance. Drainage type of graded bunds as well as broad bed and furrow system of cultivation were promoted. Regular maintenance of the bunds was not a part of the programme and the local community was not motivated enough to discharge this function. In this way, these practices did not find favour with the masses and unfortunately were replaced completely with the so-called cost effective vegetative technology where local material and skills could be deployed. In the initial phase of this World Bank aided programme, greater emphasis was placed on a single species such as *Vetivera zizanioides* for very diverse agro-ecological conditions throughout India. Later on, options were given for adopting local grasses and materials but it was too late. In this process, practices much desired by farmers such as land shaping, leveling and terracing were lost to the programme of soil conservation and improvement.

Fortunately, research efforts remained unbiased and continued to strive for a proper mix of structural and vegetative measures. Larger issues of sustainability, environment and people's participation were focussed with site-specific technologies. Ground water recharge (Fig. 2) and drought-proofing (Fig. 3) with the combined structural and vegetative measures found favour with the stakeholders.

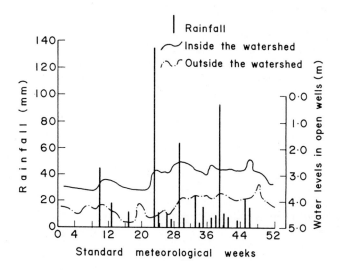

Fig. 2: Effect of watershed management programme on groundwater recharge in the Chinnatekur watershed of Andhra Pradesh (1120 ha)

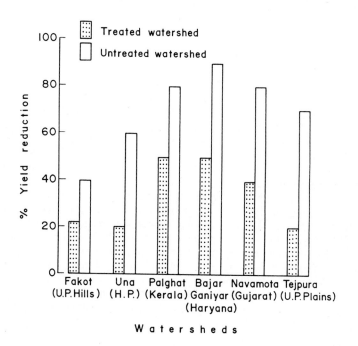

Fig. 3: Effect of watershed management on drought moderation (1987)

11 Integrated watershed management

Resource utilization and conservation in a programme mode at the scale of a naturally defined hydrological unit of a watershed proved to be ultimately sustainable. This aims at optimizing rain water, forests, grazing lands, agriculture, horticulture, livestock, allied activities and human skills in a harmonized manner (Table 5). Water harvesting, storage, use and equitable democratic management by the community has been found to be an important activity (Samra et al. 1996).

Product	Average level of attributes during		
	Pre-project 1974-75	During project 1975-86	After financial Withdrawal 1987-95
A. Food crops (q)	882	4015	5843
B. Fruit (t)	Neg.	62	1962
C. Milk ('000 lit.)	56.6	184.5	237.6
D. Flowers ('000 Rs.)	*	*	120.0 (1994 & 95)
E. Cash crops ('000 Rs.)	6.5	24.8	202.5
F. Animal rearing method	Heavy grazing	Partial grazing	Stall feeding
G. Fodder dependency on forest (%)	60	46	18
H. Runoff (%)	42	18.3	13.7
I. Soil loss (t/ha/annum)	11	4.5	2.0

* Community diversified into floriculture in 1994.

Tab. 5: Impact of integrated watershed management programme at different stages of development, Village Fakot, UP. India (327 ha)

12 Coastal ecosystem

The area lying between the sea coastline and the 50 m contour has very unique issues. In many parts rains generally come in the form of a tornado and cause severe erosion as well as damage to the standing crops. Coastal shelter belts of grasses, screwpines, Casuarina equisetifolia, coconuts and cashew nuts could fix sand and protect crops in the high rainfall zone. Aquifer recharge in the semi-arid Kutch region with the help of check dams, percolation tanks, recharging wells and storage tanks prevented the advance of saline water and lowered salinity of the ground water. In the 3000 mm rainfall coastal zone of Maharashtra, many soil conservation practices were found to be viable and environmentally desirable (Table 6).

Practice	Cost/Benefit ratio
Nala bunding	1.3
Terracing	1.6
Land leveling	2.1
Mango orchards	3.1
Cashew orchards	3.5

Tab. 6: Cost-benefit analysis of different practices and land uses in Kumbhave coastal watershed (Maharashtra), India

13 Participatory concept

This approach realizes that sustainable biomass production through resource conservation strategies is a massive programme and is difficult to achieve with traditional development models using the of "top down" approach. The dynamics of group actions of stakeholders where government functionaries become the facilitators are being popularized. An ideal bottom-up participatory approach encompasses involvement of beneficiaries inmediately at the project initiation stages, and equitable sharing of costs and benefits irrespective of gender or other social stratifications. This concept has been applied in India in a modified form, wherein some basic infrastructure is created with the funds provided by the government and contribution of the community is generally in the form of manual labour or some local materials. Subsequent management and maintenance of the infrastructure as well as equitable sharing of goods and services is left to the institutionalized user or self-help group.

References

Dhruva Narayana, V.V. and Ram Babu (1983): Estimation of soil erosion in India. J. Irrigation and Drainage Engg., Amer. Soc. of Civil Engineers, **109(4)**, 409-434.
Samra, J.S., Sharda, V.N. and Sikka, A.K. (1996): Water harvesting and recycling:Indian Experiences. Central Soil & Water Conservation Research and Training Institute, Dehradun 248195, India.
Singh, S., Pandey, C. and Das, D.C. (1990): Sedimentation in small watersheds. Tech. Series No. 4H85/1990. Soil Cons. Div., Min. of Agriculture, Govt. of India, New Delhi, India.
Singh, G., Ram Babu, Narain, P., Bhushan, L.S. and Abrol, I.P. (1992): Soil erosion rates in India. J. Soil and Water Cons. **47**, 97-99.

Address of author:
J.S. Samra
Central Soil and Water Conservation
Research and Training Institute
218, Kaulagarh Road
Dehradun 248 195
India

Scale-Effects in the Assessment of Water Erosion

P. Böhm

Summary

For the catchment area of the Karabalçik creek (Konya District, Turkey), soil erosion patterns were studied using results of aggregate stability tests, microplot experiments (rainfall simulation), long term plot data, erosion damage mapping and a catchment analysis. Plot measurements contained a high risk of spatial and temporal errors. They should be used for erosion assessment only with additional support by field mapping. Catchment-scale measurements showed no qualitative and quantitative consistency with results of the other scales.

Keywords: Water erosion, aggregate stability, plot measurements, erosion damage mapping, Turkey

1 Introduction

An sound knowlegde of soil erosion dynamics is a requirement for sustainable land use practices. Erosion is a result of combined effects reflecting spatial and temporal variability as well as a dependence on the scale (Vandaele & Poesen 1995). At different scales, different groups of processes tend to become dominant. Therefore, the understanding of erosional systems requires distinct information at various scales. In this paper, results are presented of various measurement scales on soils of a small catchment in Central Anatolia.

2 Study area

The study area is the catchment of the Karabalçik creek (11 km^2), located in South-Central Turkey (Konya District). The climate is subhumid with hot summers and cold winters (9-10°C, 490 mm). The Konya Uplands are an area of dry farming. The dominant soils are Reddish Brown Chestnut Soils and Brownish Calcareous Colluvial Soils (Table 1). Due to the long history of cultivation, beginning with Neolithic farming at about 5000 BP, human activity has severely degraded the soils.

3 Study design

To determine differences between the various methods and measurement scales, aggregate stability tests (microscale) and miniplot experiments were compared with longterm plot data, erosion damage mapping (field scale) and runoff and discharge of solid loads (catchment scale).

Aggregate stability was measured after a 5 minute shower of 120 mm/h (480 J/h, 0.3-0.4 mm drops). The design of the rainfall simulator, the miniplot-experiments and the plot characteristics are reported in Böhm & Gerold (1995). Fifteen individual sites representing typical slope angles (5-15%), soil types and land-use forms were selected. Forty-three individual site and soil characteristics were related by stepwise multiple regression analysis to variables describing infiltration and soil loss. Linear soil erosion damage was mapped in April 1988, 1992 and 1993 for an area of 10 ha arable land plus 3 ha pasture (Böhm 1995). Plot and runoff data were taken from reports of the Agricultural Research Institute Konya. In 1992, runoff, solid and dissolved load discharge of the Karabalçik creek were calculated and the runoff from contributary waterways was estimated. During the main snow melt event in march, field observations and estimations of the snow cover, runoff generation and erosional damage were undertaken.

	Chestnut Soil		Brownish Colluvium	
	topsoil	subsoil	topsoil	subsoil
Clay (%)	40.9	41.1	30.0	31.7
Gravel (vol.-%)	6.9	1.0	11.0	10.4
Humus (%)	0.9	0.2	0.3	0.3
Carbonate (%)	3.2	5.1	31.5	33.0
Bulk density (g/cm^3)	1.5	1.6	1.4	1.4
Shear strength (kg/cm^2)	0.8	1.1	0.5	0.4

Table 1: Characteristics of the soil groups in the Karabalçik catchment

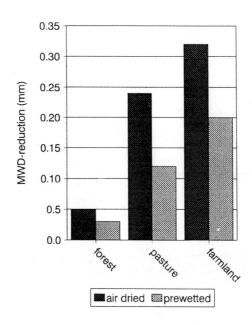

Fig. 1: Aggregate stability (Mean Weight Diameter reduction) of Soil aggregates (3-5 mm)

4 Results

One of the parameters characterizing soil erodibility is *aggregate stability*. With the aggregate stability test the effect of *slaking by air entrapment* and the *splash erosion* are examined. Figure 1 show that air-dried aggregates were less stable than prewetted aggregates. This result confirms the strong effect of the antecedent moisture on the aggregate stability (Bissonnais 1990). Among various soil physical and chemical characteristics only the organic matter content had a significant effect on the aggregate stability (r=.84 for air dried, r=.69 for prewetted aggregates). Both characteristics, water and organic matter content, were the main controlling factors for aggregate stability. The results indicate a seasonal variation of aggregate stability. If wet soil surfaces are more stable than dry surfaces, hence aggregate stability is lowest in the dry summer and highest in the humid winter period. The land use forms were closely interrelated to the organic matter content. Thus aggregate stability depended strongly on the type of land use.

Dependent variable		multiple R	Independent variable
I80	1.	.75	cover
	2.	.83	shear strength
	3.	.86	bulk density
	4.	.90	clay
	5.	.92	gravel
CONC80	1.	.76	cover
	2.	.83	shear strength
SEDALL80	1.	.72	cover
AO5SED	1.	.46	cover

I80 = infiltration rate (mm/h) after 80 mm cumulative rainfall
CONC80 = sediment concentration (g/l) in the runoff after 80 mm cumulative rainfall
SEDALL80 = total sediment loss (g) after 80 mm cumulative rainfall
AO5SED = sediment loss (g) of first 5 mm runoff

Table 2: Infiltration and soil loss on miniplots: stepwise multiple regression analysis

On the *miniplots* the processes of *rainsplash and sheet erosion* were mainly generated. Cover percentage was the main controlling factor for infiltration and erosion (Table 2). This fact is mostly considered in erosion prediction (e.g. Wischmeier & Smith 1978). Plant cover decreased the rainsplash energy as well as runoff generation and therefore reduced soil sealing. Physical resistance (aggregate stability, shear strength) was only of secondary importance.

Soils under forest, pasture and established cereals provided high infiltration rates and therefore good erosion protection (Fig. 2). Nevertheless, explaining the findings only with the effect of cover as a shelter from raindrop energy is inadequate. In this data, cover was an indicator for various soil characteristics which controlled infiltration. Interrelations appeared between the land use and soil characteristics, especially the organic matter (r=.49), macropore (r=.68) and the root content (r=.69), and shear strength (r=.50). The results show that land use (cover percentage) is a good indicator for infiltration *and* erosion, since it integrates the controlling effects of important soil characteristics. High infiltration rates on most sites indicate, that high rain volumes generally will be infiltrated and rain induced runoff, and soil loss will only appear on arable land when rain energy is high. Sensitive to erosion are bare soil surfaces. Because of low rainfall erosivity in winter (4% of the annual rainfall energy occurs between December and March), rain induced soil losses are unlikely for that season.

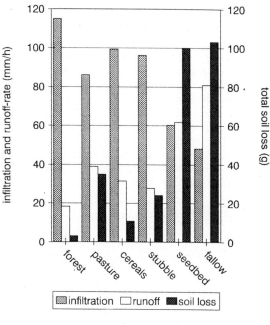

Fig. 2: Infiltration, runoff and total soil loss on miniplots after 80 mm cumulative rainfall (intensity: 120 mm/h)

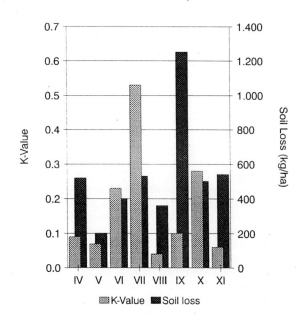

Fig. 3: Plots – Mean Soil Losses and K-values

year: K=0.12; A=1800 kg/ha

data: annual reports of the Agricultural Research Institute Konya; n = 40 events

The combined effects of *rainsplash, sheet* and *rill erosion* were determined by use of *plot measurements*. No relation existed in the data between rainfall energy (R-value) and erosion (r=.14). Soil loss depended on the rainfall volume (r=0.45), hence the infiltration. The premise of the Universal Soil Loss Equation (USLE) - the relation between rainfall energy and soil loss rate - is not valid for that site. On the pasture plot no surface runoff and no erosion occurred in 9 years (Önmez 1991). Soil losses on arable land are reduced by five times when wheat is being sown (C=0.16). The seasonal variation of the soil losses and the K-factor (Fig. 3) indicate a higher erodibility in summer (dry surface, high rain energy). Thirty percent of the annual soil losses occurred in the period between tillage and seedbed preparation. In fact, missing losses and the low erodibility in winter can not be put down to a low erosional susceptibility. Simply the kinetic energy of the winter rain is too low to generate soil losses. Snow melt runoff did not occur on the plots. The determined controlling factors and the seasonal erosion dynamics correspond to the qualitative results of the aggregate stability and miniplot tests. Nevertheless, the R-factor of the Universal Soil Loss Equation did not work adequately in Karabalçik.

With *erosion hazard mapping linear erosion* was calculated. A low mean soil loss rate of 1.6 t/ha/winter was determined (Table 3). Almost no transportation of the material to waterways was observed. Most losses were deposited on the parcels and contributed to the formation of colluvial soils. The main losses occurred on fields with a fine granulated soil surface (seedbed, 2.9 t/ha) and up and down hill culture, on slopes with angles > 8° and on north-west exposed slopes. Controlling factors were the slope - the frequency of erosional events increased with the slope angle, the exposure - high losses on the cooler and more humid NW-exposed slopes, and the type of land use - no or only very low losses appeared on pasture, rough fallow and stubble. Since the snow cover of the north-facing areas remained for a longer time, greater damage occurred in the melting period.

The small scale controlling factor, *land use*, was checked by field mapping. Additional information was obtained about the role of slope and microclimatic effects. In contradiction to the plot results, where no losses were measured in the winter period, the mapping showed that snow melt caused erosion in winter.

Mean of winter 1988, 1992 + 1993		channels (number)	rills (number)	soil loss (t/form)	(t/ha)
slope	2-4°	4	14	0.4	
	5-8°	4	8	0.7	
	8-12°	7	39	0.7	
exposure	SW	0	1	0.1	0.0
	SE	2	23	0.4	2.9
	NW	13	37	0.8	2.0
land use	seedbed	14	52	0.7	2.9
	fallow	0	8	0.4	0.3
	stubble	1	1	0.4	0.2
	pasture	0	0	0.0	0.0
	downslope	14	45	0.7	2.1
	contour	1	16	0.3	0.5
mean/total		15	61	0.6	1.6

channel = width > 25 cm or depth > 15 cm
rill = width < 25 cm and depth < 15 cm

Table 3: Linear soil losses in winter -- investigated with erosion damage mapping

The *catchment* balance integrates all processes taking place in the watershed. Generally, surface runoff of the Karabalçik starts in early winter with a maximum at the end of winter (snow melt). The creek runs dry in early summer (Yilmaz 1991). The solid load discharge (1992) of 12 t/km²/a is of a very low magnitude. Sixty-five percent of the solid load catchment discharge appeared during one week of snow melting. Observations in March/April 1992 showed severe snow melt induced erosion damages on agricultural sites of flat open areas of the higher, northern parts of the catchment. The severe floodings occurred because the meltwater could not infiltrate into the frozen ground. Generally, altitude, exposure and land use controlled the snow cover (distribution, depth and duration) and therefore runoff generation and soil loss in winter. Due to microclimatic effects, forest and areas with brushes were a sink for runoff and sediment. There, the snow melted only slowly and was able to infiltrate into the soil. In contrast, on the warmer central part of the catchment, especially on the south-exposed slopes, the snow melted away before the main melt event started. There, no surface runoff appeared at all. Due to a missing snow cover, no erosion appeared during the melting period on the USLE plots.

		1979-1992	1992
Precipitation	mm	471	454
Total runoff	mm	60	63
Surface runoff	mm	32	38
Runoff peak	l/s	2350	448
Highest daily runoff	l/s	470	396
Solid losses	t/ km²		12
Dissolved losses	t/ km²		22

Precipitation and runoff after KHAE-Konya, Annual Reports 1979-1992

Table 4. Catchment balance of the Karabalçik

5 Conclusions

Aggregate stability and erosion on microplots were well-suited for determining relative differences between sites. A main controlling factor, land use, was determined by all scale studies. However, upscaling led to a modification of the set of controlling factors. Factors of spatial significance became dominant at the catchment scale. Slope had only a modifying effect on microplots but was very important at the field scale. In contrast, on the catchment-scale, variations of the microclimate (snowmelt) became important. The low magnitude of the catchment discharge depended on restricted transport capacities, mainly in summer. Only sediments which reached a channel are calculated with the balance. The catchment discharge did, therefore, not correspond with the plot data and the field mapping (no plot losses in winter, low field damage). The small slope length and width of the plots, the specific exposure and geographic situation (location) of the plots and the mapping area were important methodical restrictions.

The comparison shows, that type and magnitude of the soil loss measured depended on the method used and its specific boundary conditions as well as on the scale of the measurements. The results of methods using different experiment durations to determine different erosional processes are not, or only with great reservation comparable (Table 5). Generally, a qualitative comparison of the results (controlling factors) can be made. However, small scale measurements cannot be extrapolated to larger scales.

method	scale	Duration	erosional processes	land use	compara-bility	Spatial compara-bility	losses (t/ha)
Aggregate stability	mm²	5 min.	splash	various	qualitative	None	8-139
miniplots	cm²	45 min.	splash, sheetwash	various	qualitative	None	0-1.9
USLE plot	m²	Year (1992)	splash, sheetwash, linear	fallow wheat	qualitative quantitative	with reservation	1.5 0.9
hazard mapping	ha	winter (1992)	linear	seedbed (wheat)	qualitative, quantitative	Given	1.3
catchment balance	km²	Year (1992)	all possible processes	various	quantitative, qualitative	Given	0.1

Table 5. Soil losses in the Karabalçik catchment according to various methods and scales

The pattern of erosion in the area cannot be adequately understood using only small scale and plot data. Supplementary field observations must be undertaken. Hence, conclusions arrived at on the basis of plot measurements should at least be qualitatively checked by field-scale studies. Since locally-severe erosion damage occurs, further studies especially concerning the role of extreme events, must be carried out to improve our knowlegde of spatial erosion dynamics.

References

Böhm, P. (1995): Bodenerosion und pedohydrologische Gebietsvarianz im Einzugsgebiet des Karabalçik (Türkei). Göttinger Beiträge zur Land- und Forstwirtschaft in den Tropen und Subtropen **102**, Göttingen.

Böhm, P. & Gerold, G. (1995): Pedo-hydrologic and sediment responses to simulated rainfall on soils of the Konya-Uplands (Turkey). Catena **25**, 63-76.

Bissonnais, Y. le (1990): Experimental study and modelling of soil surface crusting processes. Catena Suppl. **17**, 13-28.

Önmez, O. (1991): Konya - Beysehir Sartlarinda Üniversal Toprak Kaybi Denkleminin R, K, C, ve P Faktörleri (Ara Raporu 1980-1989). Konya Köy Hizmetleri Arast. Enst. Müd. Yay. 147.

Vandaele, K. & Poesen, J. (1995): Spatial and temporal patterns of soil erosion rates in an agricultural catchment, central Belgium. Catena **25**, 213-226.

Wischmeier, W.H. & Smith, D.D. (1978): Predicting rainfall erosion losses - a guide to conservation planning. Agric. Handbook 537.

Yilmaz, A. (1991): Konya - Çiftliközü, Karabalçik Deresi Havzasi Akimlari (Ara Raporu 1979 - 1988). Konya Köy Hizmetleri Arast. Enst. Müd. Yay. 141.

Address of author:
Peter Böhm
UFZ, Centre for Environmental Research Leipzig-Halle
Postfach 2
D-04301 Leipzig, Germany

Effect of Mountain Crop Rotation on Surface Runoff, Subsurface Flow and Soil Loss

K. Klima

Summary

An experiment was conducted between 1992-94 at the Czyrna experimental station situated near Krynica (southern Poland) to determine the influence of mountain slope steepness and quantity of clover and grass mixture in crop rotations on the amount of surface runoff and sheet wash, and subsurface flow. The results of the studies showed that the greatest soil losses occurred in the uppermost steep part of the slope. The highest subsurface flow was found in the lower and middle parts of the slope. The protection of soils from erosion by crop rotation may be enhanced by increasing the quantity of mixed clover and grasses.

Keywords: Anti-erosion crop rotations, mountain slope, surface runoff and sheet wash, subsurface flow.

1 Introduction

Studies on water erosion occurring in croplands have dealt mainly with the issue of soil protection properties of particular cultivated crop plants. These studies concluded that a mix of perennial papilionaceous plants and grasses inhibits water erosion to a greater extent than do cereals or root crops (Starkel, 1979). According to definition by Könnecke (1967), crop rotation consists of alternate planting of leafy plants and corn on the basis of a special yearly schedule.

Our study tried to determine the amount of rainwash and surface sheet runoff, as well as sub-surface flow, within three crop rotations and on meadows in relation to their location on a mountain slope.

2 Method

The results refer to the period 1992-94 when a two-factor experiment was conducted by the method of random blocks at the Czyrna Station near Krynica (Southern Poland). Three mountain slope zones of varying steepness constituted the first factor of our experiment; the upper zone, 570-560 m a. s. l., with an average steepness of 16.6%; the middle zone, 560-552.8 m a. s. l., with an average steepness of 11.6% and the lower zone, 552.8-545 m a. s. l., with an average steepness of 12.4%. The second factor of the experiment was based on 4 crop rotation schemes: "A": fodder beet (*Beta vulgaris L. sp. var. crassa*), oats (*Avena sativa L.*), broad bean (*Vicia faba L. var. minor*), winter Triticale (*Triticosecale Wittmack*); "B": fodder beet, oats complemented by red clover (*Trifolium pratense L.*) as an underplant crop, red clover, winter Triticale; "C": fodder beet, oats complemented by red clover and timothy (*Phleum pratense L.*) as an underplant crop, red clover with timothy; "D": a mountain meadow, as a control.

Each crop rotation and the control were repeated four times in each zone. The soil is a Cambisol (FAO-Unesco, 1990) composed of detrital flysch rock. The rainwash as well as the surface sheet and subsurface flow were surveyed by means of Slupik bag catchers (Slupik and Gil 1974), each 2 m wide. The arrangement of bag catchers within the plot was such that they were placed one after another every 5 m. Thus, the measuring strip was 2 m x 5 m = 10 m². The subsurface flow was checked by a Gerlach (1966) rill inserted down to a depth of 0.5 m. This rill served to catch and measure the quantity of water flowing down through the eluvium layer within that soil cross-section. This layer holds almost the entire root mass of the plants planted for our experiment.

3 Results

In the first year the rate of soilwash increased substantially in the third decade of May 1992, as well as in the first decade of June and September 1992. In the second year sequence (1993), a similar rainfall occurred in the second decade of June, July and August and, in 1994, in the first decade of April, June and August (Fig. 1).

Date	rain (mm) Max. intensity mm min^{-1}	Fodder beet			Broad bean			Mountain meadow		
		A	B	C	A	B	C	A	B	C
Storm rain 28.05.92	41.0 ---- 1.5	4.3	0.002	35.3	1.9	0.1	4.2	0.1	0.03	0.5
Extensive rain 4-6.06.94	90.5 ---- 0.4	12.7	2.8	4.9	8.3	3.2	1.7	10.7	1.9	1.1

A - surface sheet runoff (mm), B - subsurface flow (mm), C - water erosion in kg ha^{-1} (soil loss)

Tab. 1: Values of surface and subsurface sheet and wash-down during various types of rains

To determine the different soil protecting properties of certain plants planted within the framework of our crop rotation program, two variants of rainfall were selected. As a result of thundershowers, the highest surface sheet runoff and soil denudation was discovered in places where beets had been planted, and the lowest values were measured on the mountain meadow (Table 1). During prolonged rain, the quantity of surface sheet runoff for individual plants was similar while the average quantity of washed away soil under beets was twice as high as that for broad beans and for the mountain meadow.

During thundershowers, the mean surface runoff within the broad bean rotation was 44 kg ha^{-1} (soil loss), within the clover rotation 42 kg ha^{-1}, within the two-year clover with timothy 38 kg ha^{-1}, and on the meadow not more than 0.5 kg. The corresponding quantities for extensive rainwash were: "A" 9.6 kg ha^{-1}, "B" 8.4 kg ha^{-1}; "C" 8.9 kg ha^{-1} and "D" 1.1 kg ha^{-1}, respectively. Thundershowers and extensive rainfalls occurring in periods other than April to June did not entail extended soil denudation. The thaw after little and average snowfalls during the winters of 1992-1994 did not generate excessive soil wash on ploughed or sowed plots because the thaw runoff dissipated and never turned into a concentrated runoff. Concentrated runoff could actually cause acute soil losses through rill networks.

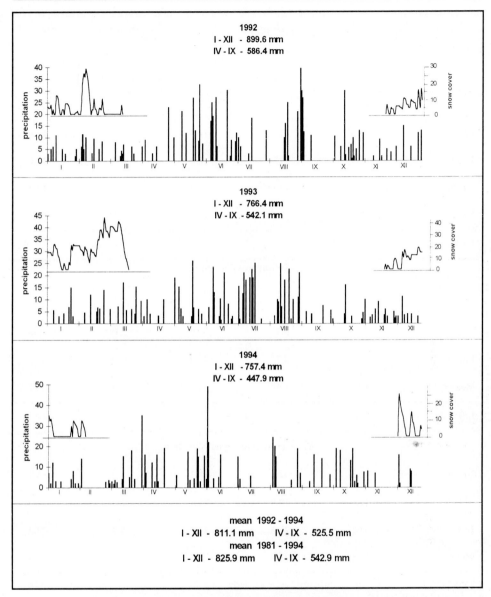

Fig. 1: Precipitation (mm) and snow cover (cm) in the years 1992-94.

The mean quantities of runoff from (soil loss) plots with individual plants were registered during the period of experiments in the years 1992-1994 as follows: fodder beet 82.5 kg ha^{-1} of soil dry mass; oats 23.2 kg ha^{-1}; broad bean 22.3 kg ha^{-1}; winter Triticale 19.5 kg ha^{-1}; red clover 3.4 kg ha^{-1}, and mountain meadow 2.5 kg ha^{-1}. The soil protecting power of individual crop rotated plants increased with increasing quantities of red clover or of added mixed clover and grasses. More extensive soilwash/denudation and runoff took place in the upper part of the mountain slope

Zone	Crop rotation			Mountain meadow	Mean of
	A	B	C	meadow	zone
Surface sheet runoff (mm)					
Upper	14.4	12.9	11.8	8.4	11.9
Middle	12.5	11.7	10.9	7.5	10.6
Lower	12.3	11.9	10.4	7.4	10.5
Mean	13.1	12.1	11.0	7.7	11.0
LSD $(p.=0.05)$	0.59				0.18
Subsurface flow (mm)					
Upper	4.04	3.27	2.87	1.56	2.93
Middle	4.61	3.75	3.35	1.79	3.38
Lower	4.66	3.77	3.29	1.74	3.36
Mean	4.43	3.59	3.17	1.70	3.22
LSD $(p=0.05)$	0.165				0.086
Water erosion in kg ha^{-1} (soil loss)					
Upper	47.5	42.4	38.2	3.2	32.7
Middle	31.1	28.0	25.0	2.3	21.6
Lower	33.2	28.5	24.9	2.2	22.2
Mean	37.3	33.0	29.3	2.5	25.5
LSD $(p=0.05)$	0.92				1.15

Tab. 2: Surface and subsurface flow and water erosion under different crop rotations.

with the steepest slope (Table 2). In this zone the smallest subsurface flow was reported. This was due to better water permeability of the soils as compared to the water permeability of soils in the middle and lower zones. The subsurface flow amounted to an average of 29% of the surface sheet runoff; this quantity in turn amounted to an average of 0.4% of the total quantity of rain. The rainfall water penetrated the soil profile, and also entered tunnels made by soil fauna and rodents, as well as ducts 15 cm long and cracks. Such water draining could develop also along the roots of plants. This has already been documented by Whipkey (1965) working in Kentucky (USA) who found that the subsurface flow constituted 15-62% of total rainfall in the clay soil of plots artificially watered by spraying.

The typical values that depict water erosion in our experiment are not large. The main factors responsible for the number of denuded soils are; the extent and intensity of rainfalls, the type of plants sowed and the downgrade of the mountain slope. Gil and Slupik (1974), who carried out research at the Szymbark Station 50 km from the Czyrna Station, stated that a really intensive rainfall of 20 mm or more might generate runoff whereas a regular extensive rainfall will not cause denudation, especially if it occurs after a longer period of no rain. Under such circumstances, soils are able to absorb the rainwater thanks to their retention capacity. However, results of soil denudation demonstrate essential inconsistencies. Gil (1990) described the following results for plots of 10 m (width) x 60 m (length) in Szymbark; denudation on plots with potatos 21.8 t ha^{-1}, with cereal 2.5 t ha^{-1} and on meadows 0.11 t ha^{-1}.

On the other hand, Kopec and Misztal (1990) reported the following results on small plots at the Jaworki Station 40 km from the Czyrna Station; denudation on plots with potatoes 25.4 - 108 kg ha^{-1}, with winter wheat 15.8 to 73.4 kg ha^{-1}, with spring barley 18.3 - 86.5 kg ha^{-1}, and on meadows 4.5 - 21.8 kg ha^{-1}. It should be stated that Kopec and Misztal (1990) applied a methodology similar to that implemented at the Czyrna Station. The discrepancies in the results of the various experiments dealing with water erosion derive from the differences in climate and soils, the different conditions under which the experiments were carried out as well as methodological differences (plot sizes, mountain slope steepness, type of devices to catch runoff water, selection of plants used, the planting method and planting principles). All the above-named factors have an influence on the results obtained and are responsible for the fact the majority of them cannot be compared with others.

Moreover, it often happens that there may only be one rainfall during a longer measuring period (for example a hydrological half-year) that becomes decisive for water erosion, or even only one thaw period. Thus, surface sheet runoff and denudation magnitude cannot necessarily be evaluated merely from the rainfall quantity; the character of the rainfall must be taken into consideration. The circumstances described here show that results obtained in our experiment could be compared with results of studies carried out under similar climatic and soil conditions with an analogous methodology.

4 Conclusions

1. The most intensive surface sheet runoff and soil wash was noticed in the upper part of the mountain slope exhibiting the steepest slope.
2. Soil protection by crop rotation can be enhanced by increasing the quantity of clover and mixing with grasses and winter plants.
3. The highest subsurface flow occurred in the lower and middle zone of the mountain slope. The average subsurface flow amounted to 29% of surface sheet runoff and to 0.4% of rainfall.

5 References

FAO-Unesco (1990): Soil map of the world. Revised legend., Rome: 1-55

Gil, E. (1990): Racjonalne uzytkowanie ziemi na stokach pod katem ochrony przeciwpowodziowej i przeciwerozyjnej. (Rational land use on mountain slopes from the point of view of anti-flood and anti-erosional protection (in Polish). Problemy Zagospodarowania Ziem Gorskich, **30**, 31-48.

Gerlach, T. (1966): Wspolczesny rozwoj stokow w dorzeczu Gornego Grajcarka (Developpement actuel des

varsants dans le basin du Haut Grajcarek „in Polish). Prace Geogr. IG PAN **52,** 104 pp.
Könnecke, G. (1967): Fruchtfolgen. VEB Deutscher Land., Berlin: 384 pp.
Kopec, S. & Misztal, A.. (1990): Wplyw roznej okrywy roslinnej na ochrone przed erozja gleb uzytkowanych rolniczo w warunkach gorskich. (The effect of varying plant cover on protecting agriculturally utilised soils from erosion in mountain conditions. in Polish). Problemy Zagospodarowania Ziem Gorskich, **30,** 127-138.
Slupik, J., Gil, E. (1974): The influence of intensity and duration of rain on water circulation and the rate of slope-wash in the Carpathian Flysch. Abhand. der Akademie der Wissen. in Göttingen, Dritte Folge, **29,** 386-402.
Starkel L. (1979): On some questions of the contemporary modelling of slopes and valley bottoms in the Carpathians Flysch. Stud. Geomorph. Carpatho.-Balcan., **13**: 191-206.
Whipkey, R.Z. (1965): Subsurface stormflow from forested slopes. IASH Bull.No. **10/2,** 109-118.

Address of author:
Kazimierz Klima
Agricultural University of Cracow
Al. Mickiewicza 21
31-120 Cracow
Poland

Field Plot Measurement of Erodibility Factor K for Soils in Subtropical China

Shi Xuezheng, Yu Dongsheng, Xing Tingyan & J. Breburda

Summary

This paper addresses *in situ* measurement of the erodibility factor K of seven different soil types in subtropical China using field plots without any vegetation cover under natural rainfall. Results show that factor K varied sharply from type to type; the calcaric Regosol on purple shale and the cultivated Cambisol derived from red sandstone being the highest (about 0.44), and eroded Acrisol derived from Quaternary red clay being the lowest (about 0.10). Wischmeier's nomograph was also applied to calculating factor K of soils. A comparison of the measured and calculated factor K values was carried out.

Keywords: Soil erodibility, factor K, nomograph, subtropical China

1 Introduction

Soil degradation is a major problem in the world. In subtropical China, erosion is the most prevalent, serious and widely distributed cause of soil degradation. The amount of soil loss depends not only on external erosivity, but also on the soil's resistance to erosion, which is usually measured as soil erodibility in terms of factor K. K can be used for predicting soil erosion and planning erosion control. In Wischmeier & Smith (1978), they defined the soil erodibility factor K and its calculation. Wischmeier et al. (1971) developed a nomograph to determine the soil erodibility factor K of cultivated lands, but studies have indicated that it is not applicable everywhere in the world.

Because of weak basic research on soil erosion and conservation, China has not yet established the erodibility factor K for its soils (Shi, 1983). Wischmeier's nomograph method is thus used to calculate soil erodibility factor K in China. A few researchers even took K as a constant as 1, thus leading to low accuracy in prediction and lack of a benchmark for observation data from thousands of erosion plots. This paper discusses how to conduct *in situ* measurements to establish comparable tables of erodibility factor K of various types of soils in subtropical China.

1.1 Design of test plots, samples and method

The test plots (No. 9-15) with seven different soil types (see Table 1) were established in Yingtan, Jiangxi Province. The site was formerly a tract of wasteland with sparse masson pines and an average slope of 8%. Prior to the experiment, this wasteland had only one dwarf masson pine for each 30 m², and the soil was an Acrisol derived from Quaternary red clay, with a entirely eroded A horizon.

The soils in plot 9, plot 10 and plot 11 were all derived from Quaternary red clay, but plot 9 was an eroded Acrisol with a plinthitic horizon, exposed on the surface. Plot 10 was a cultivated Acrisol which had been cultivated as upland for over 40 years, and plot 11 was an Acrisol in the wasteland. The soils of plot 12 and 13 were both derived from red sandstone with differences only in land use; plot 12 was wasteland and plot 13 upland. The soils of plot 14 and plot 15 were calcaric Regosols derived from purple shale and Cambisols derived from granite, respectively. Every year at the end of March all the plots were cultivated. Runoff and silt content of eroded sediment were determined after each rainfall. Soil fraction was determined by the pipette method.

2 Results

2.1 Mechanical composition of the soils

As the soils in plot 9, plot 10 and plot 11 were all derived from Quaternary red clay (Table 1), their particle distribution was very regular and their mechanical composition was dominated by clays and coarse silts. With regard to soil texture, apart from the soil in plot 10 which barely falls into the category of clay, the other ones are loamy clays. The soils derived from red sandstone (plot 12 and plot 13) were sandy loam with dominantly fine sands (0.25-0.1mm). The Calcaric Regosol, quite unique, was of silty loams with its silt content reaching more than 70%. The Cambisols were derived from granite and contained mainly fine sands, very fine sands and clays (Shi & Xi 1995).

Plot Soil No.	Depth (cm)	2-1 mm	1-0.5 mm	0.5-0.25 mm	0.25-0.1 mm	0.1-0.05 mm	0.05-0.005 mm	0.005-0.002 mm	<0.002 mm	Texture
9 Eroded Acrisol	Cw 0-20	1	1	2	5	11	26	10	45	Loam clay
10 Cultivated Acrisol	Ap 0-15	1	1	3	8	8	24	4	50	Clay
11 Acrisol	Ah 0-20	0	1	2	7	11	32	5	42	Loam clay
	Bt 20-40	0	1	2	7	11	29	7	44	Loam clay
12 Cambisol	Ah 0-20	1	1	7	38	22	12	2	12	Sandy loam
	Bw 20-40	0	1	5	33	19	4	11	25	Sandy clay loam
13 Cultivated Cambisol	Ap 0-15	0	1.	7	43	21	14	2	12	Sandy loam
	Bw 15-40	0	1	8	37	21	12	4	18	Sandy loam
14 Calcaric Regosol	Cw 0-20	0	1	1	1	12	62	9	14	Silty loam
15 Cambisol	Ah 0-15	4	7	10	20	20	10	7	22	Sandy clay loam
	Bw 15-40	4	6	11	25	12	10	9	22	Sandy clay loam

Table 1: Soil texture (%)

2.2 Precipitation

The precipitation (Table 2), is unevenly distributed throughout the year. The years 1993 and 1994 were basically normal and had their rainfalls concentrated mainly in the period April to June. In

1995, rainfall fell mainly between March to June. June 1995 had the highest, about 593 mm, which is 30% of the annual rainfall.

Year	J	F	M	A	M	J	J	A	S	O	N	D	Total
1993	74	118	127	187	490	372	300	76	86	69	48	38	1909
1994	72	137	156	288	209	546	29	151	58	47	7	230	1930
1995	74	89	207	419	275	593	129	95	91	15	0	0	1987

Table 2: Monthly precipitation distribution, 1993-1995 (mm)

2. 3 Runoff and soil loss

The mean annual runoff coefficient (1993 and 1994) of different soil types differed from plot to plot (Fig 1). The seven plots can be roughly divided into 3 groups. The first group with the highest runoff coefficient (greater than 0.42), included plot 13 (cultivated Cambisol derived from red sandstone), plot 14 (calcaric Regosol) and plot 15 (Cambisol derived from granite). The group with the lowest runoff coefficient (0.16) had only plot 9 of eroded Acrisol from Quaternary red clay. The third group consisted of plot 10, plot 11 and plot 12 with runoff coefficients lying in middle of the other two groups.

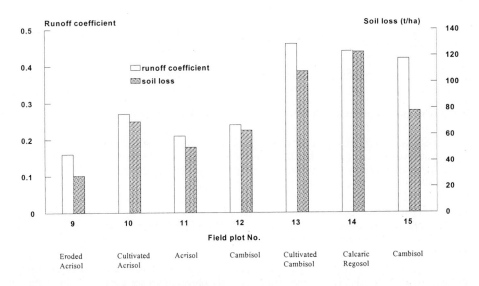

Figure 1: Mean annual runoff coefficient and soil loss for different soil types

Soil loss varied greatly, showing roughly the same trend as for the runoff coefficient. The highest was for plot 14 of calcaric Regosol on purple shale, and plot 13 of cultivated Cambisol derived from red sandstone (Fig 1). Their mean annual soil loss (1993 and 1994) reached 122.6 tons and 108.1 tons per hectare, respectively. The lowest was plot 9, an eroded Acrisol from Quaternary red clay, being only 28.6 tons per hectare, even less than 25% of that from calcaric Regosols on purple shale.

3 Discussion

3.1 Soil erodibility factor K calculated with Wischmeier's nomograph

A number of soil scientists in China have begun to adopt this method to calculate K (Chen & Wang, 1992; Ma, 1989). Further research is required for Chinese soil conservationists to verify whether it is appropriate for conditions in China. The nomograph equation for determining K is as follows:

$$K = 0.0277 M^{1.14}(10^{-4})(12-a) + 0.043(b-2) + 0.033(4-c) \quad [1]$$

where M = (% silt + very fine sand)(100 - %clay), a = % Organic matter (OM), b = Soil structure class, c = Soil permeability class. In Table 3 it is quite clear that the soil with the highest erodibility factor K is Quaternary red clay, ranging between 0.26 and 0.31. The soil with the lowest factor K is found in plot 12 and plot 15, both being 0.22. Due to high silt content of over 70%, the calcaric Regosol does not fit Wischmeier's nomograph.

Plot No.	Mechanical composition 2.0-0.1 mm	0.002-0.05+ 0.05-0.1mm	OM (%)	Soil structure class	Soil permeability class	K $(0.132 T \cdot hr/MJ \cdot mm)$
9	8	47	3	4	5	0.31
10	13	37	12	4	6	0.26
11	10	48	9	4	5	0.31
12	47	36	4	2	3	0.22
13	51	37	8	2	3	0.24
14	3	84	5	2	3	--
15	40	38	34	3	4	0.22

Table 3. Soil erodibility factor K of seven soils calculated with Wischmeier's nomograph

3.2 *In situ* measurement of soil erodibility factor K under natural rainfall

To evaluate the soil erodibility factor K, it is important to first obtain the erosivity factor R. There are a number of methods to calculate R, but the "classic" one is $E \cdot I_{30}$ method, i.e. $R = E \cdot I_{30}/100$, where E is rainfall energy, I_{30} is the maximum 30-min rainfall intensity (Lal, 1990). Based on the natural rainfall curve recorded by the auto-rainfall gauge, using Wishmeier's kinetic energy formula $e_i = 210.35 + 89.04 \lg(I i)$, $E = \Sigma e_i P i$ and $R = E \cdot I_{30}/100$, the rainfall erosivity R of the rainstorm could be calculated, the annual value of R is the sum of them (Lal, 1990). Thus, R (1993) = 459, R (1994) = 297 and R (1995) = 385. In this experiment the cover and management factor, C, = 1, and the supporting practices factor, P, = 1, thus the Universal Soil Loss Equation (USLE) could be simplified as: $A = R \cdot K \cdot LS$, where A is soil loss (t/ha), LS is landform factor, L is slope length and S is slope gradient. Soil erodibility factor, K, could be evaluated in line with the equation: $K = A / (R \cdot LS)$. Under natural rainfall, observations were made for the plots with two treatments, i.e. leaving the crust in 1993 and breaking the crust in 1994 and 1995. The *in situ* measurement of the soil erodibility factor K of the seven different types of soils under natural rainfall (Table 4) shows that in 1993 the cultivated Cambisol derived from red sandstone in plot 13 and the calcaric Regosol in plot 14 had the highest *in situ* measured soil erodibility factor K (0.20). The eroded Acrisol derived from Quaternary red clay in plot 9 was the lowest, only 0.051. In 1994 and 1995 a new treatment raised the K values, about twice as much as those in 1993. The calcaric Regosol reached 0.44 in 1994 and 0.45 in 1995. The eroded Acrisol from Quaternary red clay was the lowest, only 0.10 and 0.11, which is less than a quarter of the calcaric Regosol.

Plot No.	LS	1993 With crusting			1994 Without crusting			1995 Without crusting		
		Soil loss	R	K	Soil loss	R	K	Soil loss	R	K
9	0.47	25	459	0.05	32	297	0.10	43	385	0.11
10	0.45	57	459	0.12	83	297	0.28	93	385	0.24
11	0.39	41	459	0.10	59	297	0.23	85	385	0.25
12	0.44	58	459	0.13	68	297	0.23	84	385	0.22
13	0.43	91	459	0.21	125	297	0.43	148	385	0.39
14	0.49	102	459	0.20	144	297	0.44	188	385	0.45
15	0.43	82	459	0.19	74	297	0.26	93	385	0.25

Unit: R, $17 MJ \cdot mm/ha \cdot hr$; Soil loss, T/ha; K, $0.132 T \cdot hr/MJ \cdot mm$

Table 4. Soil erodibility factor K measured in field plots under natural rainfall

3.3 Comparison of calculated and *in situ* measured soil erodibility factor

The K measured *in situ* with broken crust under natural rainfall should be taken as a benchmark. Therefore, in order to evaluate the applicability of Wischmeier's nomograph to the soils of subtropical China, a comparison was carried out of the factor K. In Table 5 the Cambisol in plot 12 and the cultivated Acrisol in plot 10 had lower calculated K values than the *in situ* measured ones, respectively, with the relative difference less than 5%. Consequently, Wischmeier's nomograph can be applied to the two types of soils with a relative error less than 5%. A greater difference was found for the Cambisol derived from granite in plot 15. Its error was 15.4%. The eroded Acrisol in plot 9, the Acrisol in plot 11 and the cultivated Cambisol in plot 13, exhibited significant differences in their measured and calculated K values. The calcaric Regosol in plot 14, due to a high silt content of over 70%, also does not fit Wischmeier's nomograph. It is suggest that the factor K for soils in subtropical China, which were calculated with Wischmeier's nomograph, should be adjusted taking the coefficient C into account.

Plot No.	9	10	11	12	13	14	15
Soil type	Eroded Acrisol	Cultivated Acrisol	Acrisol	Cambisol	Cultivated Cambisol	Calcaric Regosol	Cambisol
$K1_{(natural\ rainfall)}$	0.11	0.26	0.25	0.23	0.42	0.45	0.26
$K2_{(nomograph)}$	0.31	0.26	0.31	0.22	0.24	--	0.22
$\|K1-K2\|$	0.20	0.00	0.06	0.01	0.18	--	0.04
Error (%)	181.9	0	24.3	4.3	42.8	--	15.4
		$K1_{(natural\ rainfall)} = C \times K2_{(nomograph)}$					
$C_{(Adjusted\ coefficient)}$	0.35	1.00	0.81	1.05	1.75	--	1.18

Table 5: Comparison of calculated and measured soil erodibility factor K by two different methods

4 Conclusion

The highest K values were found in the soils derived from Quaternary red clay when estimated by the Wishchmeier's nomograph among the seven soil types in subtropical China. In contrast the measured K values by the field plot method were the highest for the calcaric Regosol and cultivated Cambisol. In Subtropical China, Wischmeier's nomograph can be well applied to two of the seven soil types, with a relative error less than 5%, a relative error of 15.4% was found for one soil type; three soil types had significant differences with a relative error more than 24%, and one soil type could not be estimated by Wischmeier's nomograph.

Acknowledgments

This research was funded by the European Community (EC) and the National Natural Science Foundation of China.

References

Chen, F. Y. and Wang, Z.M. (1992): Application of the Universal Soil Loss Equation in Xiaoliang soil and water conservation station, Chinese bulletin of soil and water conservation **12(1)**: 23-41.

Lal, R. (1990): Soil Erosion in the Tropics. R. R. Connelley & Sons Company. 580 pp.

Ma, Z. Z. (1989): Discussion on application of satellite images to reckoning factors of Universal Soil Loss Equation (in Chinese), China Soil and Water Conservation **3**: 24-27.

Shi, D. M. (1983): Red Soils of China (in Chinese), Science Press, 137-153.

Shi, X. Z. and Yu, D. S. (1995): Study on soil erodibility in subtropical China by means of artificial simulated rainfall (in Chinese), Chinese J. of Soil and Water Conservation **9(3)**: 38-42.

Wischmeier, W. H., Johnson, C.B. and Cross, B.C. (1971): A soil erodibility nomograph farm land and construction sites. J. of Soil and Water Conservation **26(5)**:189-193.

Wischmeier, W. H. and Smith, D.D. (1978): Predicting rainfall erosion losses: A guide to construction planning. Agriculture Handbook 537. USDA, Washington, D. C. 58 pp.

Addresses of authors:
Shi Xuezheng
Yu Dongsheng
Xing Tingyan
Institute of Soil Science
Chinese Academy of Sciences
P.O. Box 821
210008 Nanjing, China
J. Breburda
Institute of Soil Science and Conservation
Justus-Liebig-University
Wiesenstrasse 3-5
D-35390 Giessen, Germany

Mechanics, Measurement and Modeling of Wind Erosion

D.W. Fryrear

Summary

Basic research on relationships between wind erosion mechanics and soil movement was conducted over fifty years ago. These relationships coupled with field erosion measuring equipment permit the verification of erosion models. Transport rates of 1231 kg/m-width of airborne soil were measured from a single storm from a 2.5 ha circular field. Average soil losses of 8.03 kg/m^2 were measured from the 2.5 ha field. Total measured soil losses for an entire erosion season varied from 0.1 to 30.9 kg/m^2 compared to erosion losses estimated with the Revised Wind Erosion Equation (RWEQ) of 0.1 to 34.8 kg/m^2 for fields 2.5 to 145 ha in size. The RWEQ is an empirical model requiring simple input data for soils, tillage, and crops. Physically based models including the Wind Erosion Prediction System (WEPS) and the Wind Erosion Assessment Model (WEAM) are under development, but the input data requirements will be much more extensive. When current models are fully operational, uses may select the most appropriate model to fit their objectives.

Keywords: Soil loss, transport rate, erosion losses, critical length, erosion control

1 Introduction

Wind erosion can be a major problem anywhere in the world if the soil surface is bare, wind velocities are high, and the soil surface is dry. The tremendous dust clouds generated when productive soils are eroded by wind may obscure the sun, damage crops, abrade paint, and impede air and automotive traffic. The valuable nonrenewable soil resource is degraded when fine soil particles are eroded by wind. Wind erosion is one of the basic geomorphological processes that have shaped eolian features on every continent, but the impact of wind erosion on soil productivity is very subtle. In the wind erosion process, fine soil particles are removed from the field and eventually, soils in the impacted areas become severely desertified and crop production declines. Wind erosion control becomes more difficult because the sand content of the surface soil increases. The combined effect of high winds, droughts, and erodible soils is amplified in soil loss by water. While not restricted to these arid regions, the severity of wind erosion is more pronounced in areas where annual rainfall is 250 mm or less (Boyadgiev, 1984).

Drier regions constitute 31.5% (46.1 million ha) of the total land of the world (Table 1) (Dregne, 1976). Africa has 17.7 million ha of arid soils (59% of total land area). The highest percentage of arid lands (82%) is in Australia. Asia has 14.4 million ha (33%) and North America 4.4 million ha (18%) of arid lands.

The erosion of soil by wind is only possible when the velocity of the wind at the surface of the soil exceeds the threshold velocity required to move the least stable soil particle. The detached particle may move a few millimeters before finding a more protected site on the landscape. The wind velocity required to move this least stable particle is called the static threshold (also called fluid threshold by Bagnold, 1941). If the wind velocity increases, soil movement will begin. If the velocity is sufficient the movement is sustained. This velocity is called the dynamic threshold (also called impact threshold by Bagnold, 1941).

	Arid-region soils	
Continent	Area in sq. km[a]	Percentage of continent
Africa	17,660	59.2
Asia	14,405	33.0
Australia	6,250	82.1
Europe	644	6.6
North America	4,355	18.0
South America	2,835	16.2
Total	46,149	

Source: From Dregne, 1976, Table 3.1.
[a] Figures in thousands.

Table 1: Distribution of arid-region soils by continent, excluding polar regions.

When soil movement is sustained, the quantity of soil that can be transported by the wind will vary as the cube of the velocity. Many transport equations have been reported that include a term for wind velocity and a single coefficient (Greeley and Iverson, 1985). Soil roughness, soil erodiblity, soil wetness, and quantity and orientation of crop residues are a few of the parameters that influence the single coefficient. As the wind velocity changes or any of the above factors are modified, soil erosion may increase or be controlled.

2 Mechanics

Laboratory wind tunnel tests have supported the theory that the transport capacity of the wind varies as the cube of the velocity (Fig. 1) (Bagnold, 1941; Chepil and Woodruff, 1963). With the development of field sampling equipment (Fryrear, 1986), it is possible to collect samples of airborne soil mass being transported from natural wind events. Field erosion data support that mass being transported does vary as the cube of the wind velocity (Fryrear and Saleh 1993).

Wind is the basic driving force when wind erodes soils. Since the capacity of the wind to transport soil varies as the cube of the velocity, reducing wind velocity at the soil surface is one method of controlling erosion. Detached particles may roll over the soil surface in a transport mode called surface creep. These particles never become airborne. Some detached particles move vertically from the soil surface, but are too large to be suspended in the wind stream, so these particles impact the soil surface with considerable energy dislodging many additional particles. Particles that return to the soil surface move in a transport mode called saltation. Smaller particles ejected from the soil surface may be suspended in the wind stream and can be transported great distances. This transport mode is called suspension. The size of particles in each transport mode will depend on the physical characteristics of the particle and the velocity and turbulence of the wind. The particle size and quantity distribution within each transport mode will vary throughout the erosion event.

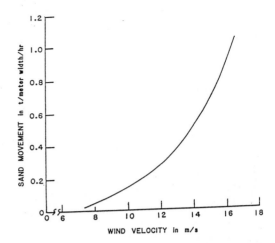

Fig. 1: Relationship between wind velocity and sand movement (Bagnold 1941).

As particle size increases, particle mass will prevent its movement by wind. As particles become very small (approach micron size), the cohesive forces between particles will cause them to coalesce into larger less erodible particles. The most erodible size depends on the velocity and turbulence of the wind, but usually the erodible particles are silt to very fine sand (Chepil and Woodruff, 1963).

If the erodible soil surface is covered with vegetation or residues from a previous crop, the surface is protected from the force of the wind and erosion is controlled. Unfortunately, in arid and semiarid regions of the world, rainfall may be insufficient to grow enough vegetation to protect the soil. Soils in these regions may erode until the surface becomes stable with a desert pavement surface. With deep sands, the surface may never stabilize and active dunes will dominate the landscape.

3 Control practices

Most control practices either reduce wind velocity at the soil surface or increase soil aggregate size. Standing vegetation will cover a portion of the erodible soil surface and also lower the wind velocity at the surface. The straw barriers in a checkerboard pattern used in China (Xu et al., 1982), and shelterbelts used in India (Gupta et al., 1981) are examples of practices used to reduce the wind velocity at the surface. From Fig. 1, it is apparent that a reduction in wind velocity can have a major impact on erosion.

Roughening the soil surface is a common practice in much of the Great Plains of the United States (Fryrear, 1984). A ridged soil surface may reduce soil erosion 90% compared to a smooth, flat surface (Fig. 2). The combination of a ridged surface and a cloddy soil surface is even more effective, but this benefit will only last until the surface is smoothed by rainfall. As soon as the bare surface loses its roughness, the field must be tilled again. Maintenance of soil roughness requires careful and continual management to be effective and is not suited for the sandy soils.

Growing vegetation and the residue from a previous crop are excellent methods of protecting the soil. Research has shown that covering even 30% of the soil surface with nonerodible plant materials will reduce soil losses 70% (Fig. 3) (Bilbro and Fryrear, 1994). Standing residues are about 6 or more times as effective in reducing soil erosion as the same quantity of residue flat on the soil surface (Fig. 4) (Bilbro and Fryrear, 1994).

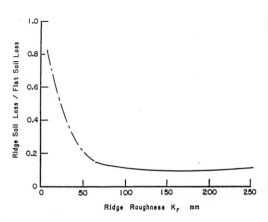

Fig. 2: Relationship between soil ridge roughness and soil loss ration (Fryrear 1985).

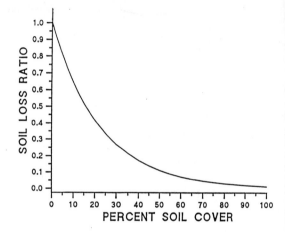

Fig. 3: Relationship between percent of the soil surface covered with non erodible materials (such as flat residues) and soil loss ratio. (SLR_f) = soil loss from partial cover divided by soil loss from bare soil.

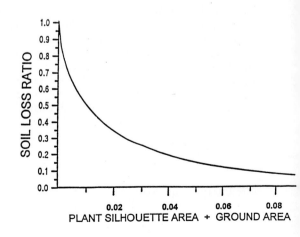

Fig. 4: Relationship between plant silhouette/unit ground area and soil loss ratio (Bilbro & Fryrear 1994).

4 Measurement

Methods of measuring soil erosion by wind have evolved from visual estimates of the quantity deposited at the edge of the field, to vertical slot samplers in laboratory wind tunnels and fields (Chepil, 1957), to the actual sampling of the material being transported by the wind at various heights (Fryrear, 1986). Improvements in field wind erosion measuring equipment continue. It is possible to detect the movement of sand grains with piezo-electric-quartz crystals (Gillette and Stockton, 1986). Data on initial soil movement is essential to determine the threshold velocities for different soils and differing surface conditions.

As the collection of wind erosion data became possible, it was necessary to develop methods of analyzing field erosion data. The ability to measure and analyze field erosion data has opened new areas of research in understanding and quantifying the wind erosion process (Fryrear et al., 1991; Vories and Fryrear 1991; Fryrear and Saleh, 1993). As scientists from other countries collect erosion data and develop other analytical procedures, the science and our understanding of wind erosion mechanics will expand.

5 Wind erosion models

Because it is now possible to measure wind erosion in the field, it is possible to check estimates of wind erosion from erosion models. The Wind Erosion Equation (WEQ) published by Woodruff and Siddoway (1965) has been widely used. This model assumes that the wind erosion process is similar to an avalanche of snow moving down the side of a mountain. However, unlike the snow, the quantity of material being transported by the wind will only increase until the wind becomes saturated. WEQ was developed to estimate annual soil erosion using single input values for weather, soil erodibility, and crop residues. As the need for more detailed erosion data grew, the time capabilities of WEQ were exceeded and a replacement model was in order.

Process and physically based models are being developed but have not been fully tested (Hagen 1991; Shao et al. 1996). These models will have tremendous capabilities for providing information on soil flux, surface changes during the erosion process, and dust emissions. The uncertainties for these models include the prediction of soil movement and dust emission for varying surface properties and the changes in surface properties due to erosion, weather, or disturbance.

The replacement model for WEQ is called the Revised Wind Erosion Equation (RWEQ) because it uses some of the same data as WEQ. However, with WEQ the basic relationship between field length and transport mass in RWEQ reflects that as the wind becomes saturated, any additional material picked up by the wind will result in the deposition of a portion of the original load. This is the basic process described by Bagnold (1941) and Chepil (1957), but is not a part of the physics in WEQ.

As the wind approaches its transport capacity, the larger particles may be deposited in favor of several smaller particles. The Revised Wind Erosion Equation incorporates erosion technology developed since 1965. RWEQ is structured to incorporate present technology, but is flexible to incorporate new technology in the future. Modern farming operations are very complex and RWEQ must permit the evaluation of systems that may include irrigation one year and not the next, multiple crops the same year, or different crops each year. More importantly, RWEQ is validated against actual field measurements of erosion.

The equation derived by Stout (1990) for transport at specific heights utilizes the self-balancing principle published by Owens (1964).

$$Q(x1)/Qc = 1 - e^{-(x/a)} \qquad [1]$$

where

Q(x1) = mass being transported at field length x in kg/m,
Qc = maximum transport capacity of wind over soil surface in kg/m,
x = field length in meters,
a = field length where Qx is 63.2% of Qc in meters.

This equation was used to describe the total transport mass from the soil surface to a height of 2 meters. Equation [1] assumes if the wind velocity is above threshold, the erosion process will detach and pick up soil particles until the wind attains its maximum transport capacity. As field length increases of the mass being transported in suspension will increase. The rate of increase in suspension is a function of many factors including the wind velocity, turbulence, surface conditions, soil texture, and field length. The field length required for the wind to become saturated will vary with the soil surface conditions and the velocity of the wind. Rarely does the wind become totally saturated before surface conditions or wind velocity changes.

From field observation, erosion does not begin immediately downwind from a noneroding boundary. At first, soil movement increases slowly but as the saltating particles abrade the surface, soil movement becomes more rapid. Based on the field erosion data and these observations, the basic relationship in equation [1] was modified to produce the „s" shaped curve in Fig. 5. The modified transport equation is:

$$Q(x)/Qc = 1 - e^{-(x/s)^2} \qquad [2]$$

where

Q(x) = transport at field length x, in kg/m,
Qc = maximum transport capacity, kg/m,
x = field length in meters,
s = field length where Q(x) is 63.2% of Qc.

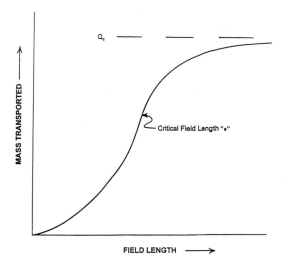

Fig. 5: Relationship between mass transported and field length (Equation 2), maximum transport capacity (Qc), and critical field lenth „a" where mass transport equals 63.2% of Qc.

From the upwind nonerodible boundary to the „s" field length, the capacity of the soil to emit soil particles will control the rate of increase in mass being transported. Beyond field length „s", the capacity of the wind to detach and transport soil will become the limiting factor until the wind becomes saturated (Qc).

To estimate soil movement for uninstrumented fields, a method of computing the empirical coefficients Qc and „s" had to be developed. To develop the equations for estimating Qc and „s", several erosion events were carefully selected based on the criteria there was no rainfall since the previous erosion event, and quality data were available on all parameters (Table 2). These data were used to determine the relationship between parameters Qc, „s", and soil erosion for a single storm.

From these analyses, the relationship between the input parameters for weather, soil erodibility, soil crust, soil roughness, and vegetation and Qc is:

$$Qc = 107.8 \, (WF \times EF \times SCF \times K' \times COG) \qquad [3]$$

where
Qc = maximum transport capacity, kg/m,
WF = weather factor from table 2, in kg/m,
EF = soil erodible fraction,
SCF = soil crust factor,
K' = soil roughness,
COG = vegetation factor including flat residues, standing residues, and canopy.

The „s" of the empirical relationship in [2] is also computed from the above input factor using the equation:

$$s = 146.86 \, (WF \times EF \times SCF \times K' \times COG) \qquad [4]$$

where
s = critical field length where Qx is 63.2% of Qc, in meters.

With the coefficients from equations [3] and [4] as input for equation [2], soil erosion can becomputed for any field size or shape. The WF has a time base, so erosion can be computed for any time interval when the remaining parameters are constant. The input data required to operate RWEQ includes factors for WEATHER, SOIL, CROP, FIELD, TILLAGE, and where applicable, IRRIGATION. Details on the components of each of these factors are available from the author. Within RWEQ, the WF is adjusted for air density, crop residues are decomposed (Steiner and Shomberg, 1994), and soil roughness is decayed by rainfall amount and intensity (Saleh, 1994).

The coefficients in equations [3] and [4] were developed from 10 single erosion events in 5 states. These same coefficients are used in equation [2] to estimate soil erosion for the entire erosion season from fields in other states and for erosion periods not included in the development of the basic coefficients in equations [3] and [4]. While RWEQ was developed to estimate soil erosion for entire seasons, not single events, the results in Table 2 illustrate that when good input data are available, the estimated erosion for single storms is very close to the measured values. For time periods longer than one day the WF is adjusted for rainfall/irrigation and snow cover and the wind factor is modified for winds parallel and perpendicular to the prevailing direction.

With RWEQ erosion is computed for a field by dividing the field into 200 equal width strips parallel to the dominant wind direction. The average field length is computed and used in equation [2]. From equation [3] the maximum transport capacity can be computed. Critical field length can be computed from equation [4]. The transport mass is important in evaluating potential plant injury from blowing sand. The critical field length is used to plan field barrier systems or field widths that will minimize soil erosion.

| Site[a] | Date | Factors[b] | | | | | Soil Loss[c] | | Q_c | Measured[d] |
		WF kg/m	EF	SCF	K'	COG	MSL ---kg/m²	Est ---	kg/m	s m
BS	17-04-95	14.2	.72	.77	1.00	.80	4.45	5.32	644	3
BS	27-01-90[e]	2.3	.64	.77	.95	.90	0.55	0.59	112	123
BS	29-01-90	2.8	.64	.77	.95	.90	0.80	0.71	133	88
BS	08-02-90	.6	.64	.77	.95	.90	0.15	0.15	96	289
BS	06-03-90	2.8	.64	.77	.95	.90	0.93	0.71	226	149
BS	29-03-93[f]	3.6	.77	.77	1.00	.96	2.46	1.24	402	84
MW	02-04-91	8.4	.79	.91	.82	.43	1.14	1.29	168	43
EK	09-03-92	41.9	.70	.65	.91	.65	8.03	6.82	1403	98
KM	13-03-93	15.3	.85	.90	.85	1.00	4.05	6.02	751	109
EC	12-03-91	179.9	.26	.21	.80	.48	2.14	2.28	648	179

a. Sites are coded Big Spring, Texas, (BS); Mabton, Washington, (MW); Elkhart, Kansas, (EK); Kennett, Missouri, (KM); and Eads, Colorado, (EC).
b. WF = Weather Factor
   EF = Soil erodible Fraction coefficient
   SCF = Soil crust coefficient
   K' = Soil roughness coefficient
   COG = Crop residue coefficient including the product of flat (SLR_F) and standing residues (SLR_s).
c. MSL = Measured average soil loss from 2.5 ha field.
   Est = Estimated average soil loss from 2.5 ha field.
d. Q_c = Maximum transport capacity over that surface.
   s = Critical field length.
e. Includes January 27th and 28th, 1990 wind data.
f. Includes March 28th and 29th, 1993 wind data.

Table 2. Input factors and resulting, soil loss measured and estimated values for 10 individual erosion events.

RWEQ estimates were tested against erosion measurements from fields that were much larger and different shapes than the original test sites (Table 3). Maximum time steps are fifteen days. Agreement between measured and estimated soil losses in Table 3 illustrates that RWEQ can provide accurate estimates of soil erosion from a variety of field conditions. The data in Table 3 do not represent the entire range of soils, crops, weather, and management systems used by farmers, but the data do represent typical conditions in the major wind erosion regions in the United States. When used with good weather, soil, crop, and management input data estimates of erosion with RWEQ agree closely with measured values. For one erosion at Elkhart, Kansas on March 9, 1992 a soil loss of 8.03 kg/m^2 was measured from the 2.5 ha field. Soil losses of 30.9 kg/m^2 have been measured for the entire erosion season at Crown Point, Indiana. As field measure-ment analyses are completed from fields in Fresno, California; Hawaii; Swan Lake, Minnesota; Fargo, North Dakota; Scobey, Lindsey, and Haver, Montana; and Prosser and Mabton, Washing-ton, they will be tested against estimates from RWEQ. In addition, erosion data are being collec-ted from arid rangelands and from agricultural lands in Canada and Australia.

Site	Soil Loss		Field	
	RWEQ	Measured	Shape	Size
	----------kg/m^2----------			ha
Crown Point, IN	34.8	30.9	circle	2.5
Big Spring, TX	17.2	17.1	circle	2.5
Sidney, NE	2.9	1.7	circle	2.5
Eads, CO	3.5	2.4	circle	2.5
Elkhart, KS	12.8	12.9	circle	2.5
Canada #1	17.0	15.2	circle	3.1
Canada #2	5.7	6.9	circle	3.1
TX #2	0.3	0.3	rectangle	55
TX #3	0.4	0.8	rectangle	41
TX #4	0.0	0.3	rectangle	145
Plains #1, TX	2.8	1.6	rectangle	145
Plains #2, TX	2.3	2.2	rectangle	73
Portales, NM	0.1	0.1	circle	59

Table 3. Comparison of estimated soil losses with RWEQ and measured soil losses.

6 Conclusions

Man has lived with the hazards of wind erosion for centuries, but the first basic research on wind erosion mechanics was conducted 50 to 60 years ago. With the development of good field erosion measuring equipment significant progress has been made to expand our understanding of wind erosion mechanics. Wind erosion has been measured in numerous fields to test a new wind erosion model with a variety of weather, crop, and soil conditions, and farming systems. With the measurement of field erosion, it is possible to establish base line conditions for future compari-sons. The field measurements verified that soil mass transport does vary as the cube of the wind velocity. Mass transport increases until the wind stream becomes saturated. The maximum transport capacity increases with wind velocity, and transport rates of 1403 kg/m have been measured from field sites. This level of transport is sufficient to destroy most crop seedlings within 15 minutes.

The effectiveness of crop residues in controlling soil erosion can be expressed as a percent of the soil surface that is covered. Standing residues present a silhouette to the wind that will reduce the wind velocity at the soil surface. Residues standing are more than six times as effective in reducing wind erosion as the same quantity laying flat on the soil surface. The combined effect of crop residues and soil roughness can be expressed as the product of the reduction coefficients.

Actual wind erosion measurements have verified that the Revised Wind Erosion Equation is suitable for a variety of soil, weather, crop and management conditions. Soil losses from a single storm were 8.03 kg/m² from a 2.5 ha circular field. Soil losses from an entire erosion season varied from 0.1 to 30.9 kg/m². Field sizes ranged from 2.5 to 145 hectares.

References

Bagnold, R.A. (1941): The physics of blown sand and desert dunes, London: Methuen, 265.
Bilbro, J.D. and Fryrear, D.W. (1994): Wind erosion losses as related to plant silhouette and soil cover. Agron. Jour. **86(3),** 550-553.
Boyadgiev, T.G. (1984): Report on the Modeling for Comilation of the Maps of Desertification Hazards in Africa and Soil Elements used in Assessing Desertification and Degradation in the World. FAO-UNEP.
Chepil, W.S. (1957): Width of Field Strips to Control Wind Erosion. Kan.Ag.Exp.Station Tech. Bull **92,** 16.
Chepil, W.S. and Woodruff, N.P. (1963): The physics of wind erosion and its control. In: A.G. Norman (ed.), Advances in Agronomy, Vol. **15,** 211-301. New York: Academic Press.
Dregne, H.E. (1976): Soils of Arid Regions. New York: Elsevier, p.237
Fryrear, D.W. (1984): Soil ridges-clods and wind erosion. Trans. ASAE. **27(2),** 445-448.
Fryrear, D.W. (1986): A field dust sampler. J. Soil and Water Conserv. **41(2),** 117-120.
Fryrear, D.W. and Saleh, A. (1993): Field wind erosion: Vertical Distribution. Soil Sci. **155(4),** 294-300.
Fryrear, D.W., Stout, J.E., Hagen, L.J. and Vories, E.D. (1991): Wind erosion: Field measurement and analysis. Trans. ASAE. **34(1),** 155-160.
Hagen, L.J. (1991): Wind erosion prediction system to meet user needs. J. Soil and Water Conserv. **46,** 105-111.
Gillette, D.A. and Stockton, P.H. (1986): Mass, Momentum, and kinetic energy fluxes of salting particles. In: Aeolian Geomorphology, Binghamton Symp. In Geomophol.: Int. Ser., no 17 edited by W.G. Nickling, Allen and Unwin, Boston., 35-56.
Greeley, R. and Iverson, J.D. (1985): Wind as a geological process on Earth, Mars, Venus, and Titan. New York: Cambridge Univ. Press.
Gupta, J.P., Aggarwal, R.K.and Raikhy, N.P. (1981): Soil erosion by wind from bare sandy plains in western Rajasthan, India. J. Arid Envir. **4,** 15-20.
Owens, P.R. (1964): Saltation of uniform sand grains in air. J. Fluid Mechanics **20(2),** 225-242.
Saleh, A. (1994): Measuring and predicting ridge-orientation effect on soil surface roughness. Soil Sci. Soc. Am. **58(4),** 1228-1230.
Shao, Yaping, Raupach, M.R. and Leys, J.F. (1996): A model for predicting aeolian sand drift and dust entrainment on scales from paddock to region. Aust. J. Soil Res. **34,** 309-342.
Steiner, J.L., Shomber, H.M., Douglas, C.L.and Black, A.L. (1994): Standing stem persistence in no-tillage small-grain fields. Agron. J. **86,** 76-81.
Stout, J.E. (1990): Wind erosion in a simple field. Trans. ASAE **33(5),** 1597-1600.
Vories, E.D. and Fryrear, D.W. (1991): Vertical Distribution of Wind-Eroded Soil over a Smooth Bare Field. Trans. ASAE 34(4), 1763-1768.
Woodruff, N.P. and Siddoway, F.W. (1965): A wind erosion equation. Soil Sc. Soc. Am. Proc. **29,** 602-608.
Xu, Junling, Pei Zhangquin, and Wang Renhua (1982): A research on the width of the protection belt of haif-hidden straw checkerboard barriers. J. Desert Res. **2(3),**16-22.

Address of author:
Donald W. Fryrear
USDA - ARS
P.O. Box 909
Big Spring, Texas 79721-0909
USA

Potential Hazard of Wind Erosion in Regions Affected by Chernobyl's Fallout

G.P. Glazunov

Summary
Factors affecting wind erosion in the western part of the Bryansk region of the Russian Federation have been investigated by both field and laboratory experiments. Soil blowing is possible due to a relatively high erodibility of arable light eutric podzoluvisols, eutric fluvisols and dystric histosols (light derno-podzolic, sod alluvial and peat-gley arable soils) by wind and the regional climate. This proved to be a factor in soil degradation. The biggest mean annual predicted amount of soil loss under present-day crop rotations (up to 9 $t \bullet ha^{-1} \bullet y^{-1}$) is characteristic for light eutric podzoluvisols (derno-podzolic light soils). The rate of ^{137}Cs depletion in these soils due to wind erosion is about 1.2% per year. Wind erosion should be taken into account in the Bryansk region and erosion control techniques should be implemented to protect soils and to prevent redeposition of radioisotopes.

Keywords: Soil, wind erosion, modeling, ^{137}Cs, Chernobyl

1 Introduction

A major part of the Bryansk region of the Russian Federation is affected by Chernobyl radioactive fallout. Arable lands of this region are subjected to water and wind erosion that causes detachment, transportation and accumulation of contaminated soil. The maximum amount of ^{137}Cs removal due to water erosion was estimated to be 20 times greater than for crops (Kuznetsov et al., 1995). Wind action is known to contribute significantly in ^{137}Cs transportation and accumulation (Aleksahin and Kornejev, 1992). However, the hazardous potential of wind erosion in this region is not as obvious as that of water erosion and has never been assessed before. Only limited information concerning wind erosion in the western part of the Bryansk region is available. Erosion is supposed to occur rarely but has not jet been studied systematically for wind erosion (Shandybin, 1987). Programs of this type are very expensive and currently are under way only in the USA (Hagen, 1991).

To estimate the importance of wind in the redistribution of contaminants, ^{137}Cs in particular, we investigated wind erosion in the Bryansk region. It included field measurements of the amount of soil transported by wind during separate erosion events; measurement of erodibility of local soils by wind using a laboratory wind tunnel; and assessment of potential soil loss by means of the wind erosion equation (Woodruff and Siddoway, 1965). The main aims were: 1) to estimate potential hazard of wind erosion in the western part of the Bryansk region; 2) to determine the existence of ^{137}Cs redistribution due to wind action; 3) to estimate the rate of ^{137}Cs reduction in soils as a result of wind erosion, and to check it against other factors.

2 Methods

We based our studies on the assumption that wind erosion can occur only if the wind is stronger than a threshold specific for the soil. There is no a standard methodology to determine the threshold velocity, so we used that of (Glazunov, 1983). The amount of soil loss depends on soil cloddiness and duration of wind. We used a flat sieve with 0.84 mm openings to determine soil cloddiness directly in the field. To trap blown soil, we used a box (353 486 300 mm) that was placed 100 m downwind from the edge of the field. The upper edge of the box was leveled with the soil surface. This simple method was shown to be reliable in the case of moderate winds (Uteshev and Semenov, 1967). The content of ^{137}Cs in soil probes was determined in the laboratory with the use of standard equipment (Aleksahin and Kornejev, 1992).

The amount of wind erosion, E, expressed in $t \cdot ha^{-1} \cdot y^{-1}$ was determined with the use of the wind erosion equation, $E = f(I',K',C',L',V)$, after some modification. Soil erodibility index I' determination was based on cloddiness of soil probes. Soil ridge roughness factor, K', as well, as slope factor, were equated to 1 (FAO, 1979). Climatic factor C' was determined with the use of wind erosion climatic erosivity (Skidmore, 1986). A simple empirical equation was developed to replace the chart with a movable scale, used for determining the factor of field length, because this chart does not allow one to take into account small amounts of soil loss, E3. Influence of vegetative cover was taken into account with the use of a monthly rating, similar to those, used in FAO methodology (FAO, 1979).

All independent variables of the wind erosion equation, with the exception of climatic erosivity, are deterministic. Wind erosion climatic erosivity is an expectation of soil blowing which is a random variable. A relative error of our method is equal to the ratio of the variance of this random variable and its expectation. Numerical integration for a limiting case of dry loose soil, absence of plant cover, and the biggest field size, gave 0,23 for the relative error.

Area of different soil types was taken from the map of soils of the Bryansk region (1:200,000 scale) and territories with different levels of ^{137}Cs contamination was taken from the "Map of radiation situation on the European part of the USSR for the December of 1990".

3 Results and discussion

Threshold velocities for major soil types of the Bryansk region, determined for the most hazardous spring period, are relatively low (Table 1). They are apparently changing over time. Having no data on these changes, we had to keep them constant as well, as for cloddiness. The latter is assumed by the wind erosion equation (Woodruff and Siddoway, 1965).

Threshold velocities were experimentally determined for the samples of the most widely spread soil series of the region. It was not possible to cover all the series represented on the soil map of the region (1:200,000 scale). Thus the data were used to arrange all soils into groups with equal threshold velocities (Table 2). In order not to overestimate the threat of wind erosion, the threshold velocities were taken as being equal to the upper limit of the range. Soils of the fifth group were excluded from the analyses because they are not subjected to wind erosion.

The majority of soils of the region in question falls into the first three groups. The state of these soils determines the possibility of wind erosion, which is very high, because 20-30% of the winds in spring exceed 5 $m \cdot s^{-1}$, and 5-10% exceed 9 $m \cdot s^{-1}$. Soils of the third group are moderately subjected to wind erosion because the probability of winds greater than 12 $m \cdot s^{-1}$ is only 0.5%.

Location	Soil	Friction velocity	Wind velocity at 10 m height
Lyshchichi	Eutric podzoluvisols (derno-podzolic)	0.2	5
Lyshchichi	Gleyic podzoluvisols (derno-podzolic gleyed)	0.5	12
Star. Bobovichi	Eutric podzoluvisols (derno-podzolic) before tillage	0.3	7
Star. Bobovichi	Eutric podzoluvisols (derno-podzolic) after tillage	0.2	5
Griva	Eutric podzoluvisols (derno-podzolic)	0.2	5
Griva	Blown material of eutric podzoluvisols (derno-podzolic soil)	0.1	2
Griva	Drifts of eutric podzoluvisols (derno-podzolic soil)	0.2	5
Popovka	Eutric podzoluvisols (derno-podzolic)	0.2	5
Popovka	Epigleyic podzoluvisols (surface gleyic derno-podzolic)	0.2	7
Star. Bobovichi	Eutric fluvisol (sod alluvial)	0.2	5
Kivai	Light greyzems (light gray forest)	0.4	9

Table 1: Threshold velocities, $m \cdot s^{-1}$

No	Soil	U^a
1	Eutric podzoluvisols (derno-podzolic), ameliorated dystric histosols (peat-gley, peat low moor), eutric fluvisols (sod alluvial, meadow alluvial), ameliorated silty dystric histosols (silty-peat-gley alluvial).	<=5
2	Epigleyic podzoluvisols (surface gleyic derno-podzolic)	6--7
3	Light greyzems (light gray forest)	8--9
4	Gleyic podzoluvisols (derno-podzolic gleyed)	10--12
5	Gully soils, dystric fluvisols (meadow swamp alluvial, silty-muck-gley alluvial)	>12

[a] Characteristic threshold velocity at 10 m height, $m \cdot s^{-1}$

Table 2: Wind erodibility groups of soils, represented on the 1: 200,000 scale soil map of the western part of the Bryansk region

Precise determination of probability of wind erosion events needs steady monitoring. Meteorological stations take into account only severe erosion events that cause visibility reduction. That is why the occurrence of wind erosion events is greater, than that recorded by meteorological stations. Our data support this statement (Table 3, 4). Those events that we observed in the field, were not registered at the station. All cases were caused by "fresh" to "strong" breezes. Wind velocities were higher than maximum daily velocities registered at the station. That is why wind velocity distributions are more useful for climatic erosivity calculations than averages alone.

The potential annual erosion, E, is always lower due to the protective role of vegetation. It also depends on the size of eroding fields. For the area under study, it was calculated on a regional level because data on field size distributions (factor L') and vegetation cover (factor V) were available only at the level of districts (Table 6).

Soil erodibility $E_3 = f(I',K',C')$ is the estimate of maximum possible soil loss under local conditions. It reflects differences between soils in cloddiness and resistance to wind action (Table 5). The higher is the resistance of soil to wind blowing, the lower are erosive work of the wind and the loss of soil. Time changes in E_3 for each soil are attributed to monthly changes in erosive work of the wind.

Data	1992					
	May 30	May 31	June 01	June 02	June 03	June 04
Wind speed, m•s-1	7	5	5	6	9	7
Air temperature, °C	17.3	18.3	18.8	17.4	16.1	15.2
Precipitation, mm	0	0	0	0	0	0
Dust storms	No	No	No	No	No	No

Table 3: Krasnogorsk meteorological station data records

Wind speed. m•s^{-1}	Aggregates size. mm								
	0--0.09	0.09--0.16	0.16-- 0.25--		0.315--	0.5--	0.63--	0.84--1	>1
	30.05.1992. 14—17. Lvshchichi.[a]								
6—9	60.0	26.1	11.4	0.5	1.7	Trace	Trace	Trace	
	30.05.1992. 18--20. Lvshchichi.[a]								
6—9	56.0	37.3	6.6	Trace	Trace	Trace	Trace	Trace	
	03.06.1992. 12--13. Griva.[b]								
6—9	88.7	9.4	1.8	0.2	Trace	Trace	Trace	Trace	
	04.06.1992. 14--16. Popovka.[a]								
6—15	7.3	30.0	50.3	8.3	2.0	2.0	2.5	0.2	

[a] Eutric podzoluvisols (derno-podzolic) and epiglevic podzoluvisols (surface glevic derno-podzolic)
[b] Eutric podzoluvisols (derno-podzolic)

Table 4: Particle size distribution of wind blown soil

Aggregate size distribution of the surface soil layer always differs from that of the soil particles trapped in the collector. The soil sampler used collects both saltating-size and surface creeping-size particles. Chepil (1945) found, that 62 to 97% of soil movement is in saltation and surface creep. That is why the total movement was greater than registered with the collector due to suspended particles.

Soil	Month					
	4	5	6	7	8	9
Eeutric fluvisols (meadow alluvial)	0.324	0.762	0.621	0.296	0.183	0.141
Eutric podzoluvisols (derno-podzolic)	1.24	2.91	2.37	1.13	0.701	0.539
Gleyic podzoluvisols (derno-podzolic gleyed)	0	0.036	0.036	0.059	1.13	0
Epigleyic podzoluvisols (surface gleyic derno-podzolic)	0	0.036	0.036	0.059	0.036	0
Light greyzems (light gray forest)	0	0.037	0.097	0.097	0.037	0.022
Ameliorated dystric histosols (peat-gley)	0	0.604	1.42	1.16	0.552	0

Table 5: Soil erodibility, $E_3 = F(I',K',C')$, $t \cdot ha^{-1}$

Even a slight increase in wind speed results in a noticeable change of the shape of size-distribution curve. A sharp peak appears, corresponding to fraction of particles, sized from 0.09 to 0.016 mm, highlighting that a considerable amount of saltating particles sized from 0 to 0.09 mm became suspended and, consequently, were subjected to much longer transportation by wind. In our case these particles were enriched with ^{137}Cs (Table 7).

The sorting action of wind has some consequences for the territory under study. First, mobility of ^{137}Cs enriched particles is higher than that of other particles. Thus, the range of their transportation by wind is greater, than that for a soil as a whole.

Second, as the amount of annual erosion does not exceed 1 t•ha^{-1}•y^{-1} (Table 6), it can be judged as relatively small. But in the case of local light-textured soils, this small amount causes more damage to the soil than in the case of heavy textured soils having uniform distribution of nutrients among aggregates of all sizes. This should be taken under control.

District	Soil loss	
	t•ha^{-1}•y^{-1}	t•y^{-1}
Krasnogorskij	0.98	28266
Gordeevskij	0.92	28146
Klintsovskij	0.69	28714
Novozybkovskij	0.94	32712
Zlynkovskij	0.74	18263
Klimovskij	0.69	50571

Table 6: Potential annual erosion, E, in the western districts of Bryansk region

Location. Soil	In aggregates of different size, mm					In arable layer
	<0.063	0.063--0.16	0.16--0.25	0.25--0.315	0.315--0.5	
Star. Bobovichi. Eutric podzoluvisols, sandy (derno-podzolic)	28.4	10.1	Not analyzed	Not analyzed	4.0	7.9
Star. Bobovichi. Dystric histosols (peat-gley) ameliorated	5.8	6.0	3.7	3.5	4.6	4.1
Griva. Eutric podzoluvisols (derno-podzolic)	5.2	3.4	3.2	3.0	5.2	3.3

Table 7: Contents of ^{137}Cs, Bq •kg^{-1} •10^3

Third, the amount of ^{137}Cs loss (Table 8) calculated on the basis of its average concentrations in the arable layer (taken from the map) appears to be underestimated. The difference depends on the intensity of soil blowing and the parameters of initial distribution of ^{137}Cs among fractions of aggregates. In fact, small particles contain 1.5—3.5 times as much ^{137}Cs, as the arable layer on average (Table 7). In case of sandy eutric podzoluvisols (derno-podzolic soils), the amount of ^{137}Cs removal should be multiplied by a factor of 3.5. Hence, the rate of ^{137}Cs depletion in these soils due to wind erosion, given an erodibility E$_3$ equal to 9 t•ha^{-1}•y^{-1}, will comprise 1.2% per year. Being 1.8 times lower than the rate of natural decay, this value is 2.4 times as big as the maximum rate of removal of ^{137}Cs with crops, which is 0.5% (Aleksahin and Kornejev,1992).

District	^{137}Cs loss	
	Bq•ha^{-1}•y^{-1}	Bq•y^{-1}
Krasnogorskij	3,1•10^8	9,0•10^{10}
Gordeevskij	2,5•10^8	7,6•10^{10}
Klintsovskij	1,7•10^8	7,2•10^{10}
Novozybkovskij	3,4•10^8	12•10^{10}
Zlynkovskij	1,3•10^8	3,3•10^{10}
Klimovskij	0,67•10^8	5,0•10^{10}

Table 8: Potential ^{137}Cs removal in the western districts of Bryansk region

4 Conclusions

Soils of the western part of the Bryansk region, contaminated to different extents with ^{137}Cs, are subjected annually to wind blowing due to their relatively low resistance to wind erosion and sufficient erosivity of the regional climate. An important consequence of this phenomenon is removal of small particles enriched with ^{137}Cs. Mean annual rate of soil loss due to wind erosion is equal to 1 t•ha^{-1}•y^{-1}, but the maximum rate, equal to erodibility, amounts to as much as 9 t•ha^{-1}•y^{-1}. The rate of ^{137}Cs depletion due to wind erosion can be 2.4 times as great as the maximum rate of removal with crops. Our results reinforce the statement that wind erosion is an important factor in the redistribution of ^{137}Cs in the western part of the Bryansk region and the vast expanses of neighbouring contaminated territories, and thus should be placed under control.

A very important question arises, concerning the deposition of blown soil enriched with ^{137}Cs. Wind action can be conceived as an action of a gigantic vacuum cleaner, collecting dust from extended territories and depositing it near to different obstacles, such as river valleys, gullies, forest fringes, and different constructions, thus generating new areas of secondary contamination. Accordingly, it seems worthwhile to trace the blown soil and to evaluate these deposits.

Acknowledgments

The author thanks Dr. A.D.Fless and Ye.N.Yesafova for their cooperation and Ye.L.Blochin for ^{137}Cs determination. The author is most grateful to the German Ministry of Environment, Bonn, for sponsoring his participation in the ISCO'96 Conference.

References

Aleksahin, R.M & Kornejev, N.A. (eds.) (1992): Agricultural radioecology, Ecology, Moscow.

A map of radiation situation on the European part of the USSR for the December of 1990 (Density of contamination with ^{137}Cs), (1991), GUGK, Minsk.

Chepil, W.S. (1945): Dynamics of wind erosion: 1. Nature of movement of soil by wind, Soil Science **60**, 305-320.

FAO (1979): A provisional methodology for erosion assessment, Rome.

Glazunov, G.P. (1983): Threshold velocity as an index of resistance of soils to wind erosion, Pochvovedenije **3**, 112-118.

Hagen, L.J. (1991): A wind erosion prediction system to meet user needs. Journal of Soil and Water Conservation **46(2)**, 106-111.

Kuznetsov, M.S., Pushkarjeva, M.M., Fless, A.D., Litvin, L.F., Blohin, Ye.L. Demidov, V.V. (1995): Prediction of water erosion intensity and radioisotopes migration in contaminated districts of the Bryansk region, Pochvovedenije **5**, 617-625.

Shandybin, A.I. (1987): Recommendations on lands of the river Desna basin protection from adverse effects by the example of Bryansk region, Bryansk.

Skidmore, E.L. (1986): Wind erosion climatic erosivity, Climatic change **9**, 195-208

Soil map of Bryansk region. Scale 1:200 000, (1988), GUGK, Moscow.

Woodruff, N.P., Siddoway, F.H. (1965): A wind erosion equation, Soil Science Society of America Proceedings **29(5)**, 602-608.

Uteshev, A.S., Semenov, S.Ye. (1967): Climate and wind erosion of soils, Kajnar, Alma-Ata.

Address of author:
G.P. Glazunov
Faculty of Soil Science
Moscow State University
119899 Moscow, B-234, Russia

A Wind Erosion Model and its Application to Broad Scale Wind Erosion Pattern Assessment

Y. Shao, R.K. Munro, L.M. Leslie & W.F. Lyons

Summary

Wind erosion is a complex process influenced by weather, soil and vegetation. For wind erosion assessment and prediction, it is necessary to understand the physical processes and to have reliable information about environmental susceptibility to erosion. In this paper we present an integrated system for wind erosion assessment and prediction, using a physically based wind erosion scheme with an atmospheric model and a GIS database. The system has been applied to the February 1996 dust storms in Australia. The importance of this system to land care practice lies in its capacity to identify areas and periods of wind erosion threat, and also to identify the environmental factors responsible.

Keywords: Wind erosion, weather prediction, Australia

1 Introduction

Wind erosion is a serious problem for a large proportion of Australian rural land. During erosion, fine soil particles rich in nutrients and organic matter can be carried away by wind over large distances, leading to irreversible soil degradation. According to Raupach et al. (1994), the 1983 Melbourne dust storm resulted in a loss of 2 million tonnes of top soil, including 3400 tonnes of nitrogen, 110 tonnes of phosphorus and a cost of 4.5 million Australian dollars. It was estimated that the dust storms of May 1994 resulted in a soil loss of between 10 and 20 million tonnes. In parts of Australia the rate of soil formation from bedrock is around 0.4 tonnes per hectare per year. In comparison, the rate of soil loss through wind erosion can be as high as several tonnes per year in some areas.

In the context of land care, the major tasks of wind erosion research are to quantify wind erosion risks on different scales of time and space, and to identify causative factors, so that strategic planning can be carried out and guidelines for wind erosion prevention measures established. It is also important to provide wind erosion forecasts for short term management of agricultural activities.

Wind erosion is a complex set of interacting physical processes governed by four main factors: climate; soil state; surface roughness; and land management practice. Wind erosion events are variable in space and intermittent in time, and they are often associated with meso-scale frontal systems under dry weather conditions. There have been some studies on wind erosion in Australia, including reports on dramatic wind erosion events; land degradation surveys; paddock-scale experiments (Leys and Heinjus, 1991); surveys based on meteorological records (McTainsh and Pitblado, 1987) and on calculated erosion indices (Burgess et al., 1989). These studies have shown, in a qualitative sense, that much of Australia is subject to severe wind erosion. There are also wind erosion assessment and prediction systems for other parts of the world (e.g. Gillette and Hanson, 1989). However, a quantitative wind erosion assessment and prediction system has not yet been accomplished.

In this paper, we describe an integrated system for wind erosion assessment and prediction (Shao et al., 1997). This system comprises a (mainly) physically based wind erosion scheme (Shao et al., 1996) with a high resolution atmospheric model (Leslie and Purser, 1991) and a detailed Geographic Information System (GIS) database. At present such a system is arguably the most powerful way of predicting a complex phenomenon such as wind erosion. The system has been applied to an investigation of the wind erosion events of February 1996 in Australia. This preliminary application shows that this kind of integrated system is effective. In this paper, we will present the results after a brief description of the system.

2 Wind erosion assessment and prediction system

2.1 Structure of the system

The centre-piece of the integrated system is a wind erosion scheme for the prediction of stream-wise sand flux Q and dust entrainment flux F from the knowledge of the wind, soil texture, soil structure, soil moisture and surface roughness elements. The wind erosion scheme is driven by the atmospheric forcing data obtained from the atmospheric model described below, and the land surface parameters derived from a detailed GIS database.

2.2 The atmospheric model

The atmospheric forcing data are obtained from a limited area atmospheric model developed at the University of New South Wales (Leslie and Purser, 1991). It has computational economy in terms of both storage requirements and algorithm efficiency. It is a two time-level scheme comprising a semi-Lagrangian advection step followed by a number of adjustments steps, using the forward-backward scheme. The temporal differencing is formally second-order, and the interpolations in the semi-Lagrangian step use bi-cubic splines.

The model has been tested extensively (e.g Leslie and Skinner, 1994). Standard statistical evaluation, averaged over 30 stations in the Murray-Darling Basin of Australia, has shown that for near-surface air temperature predictions, the rms error is 2.1 degrees K with a mean absolute error of 1.7 degrees K; for near-surface wind speed the rms error is approximately 3 ms^{-1}.

In this application, the limited area model is run continuously over the Australian region at 20km horizontal resolution and 31 levels vertically. In order to resolve the boundary layer, there are 10 levels from 850 hPa to the surface, with the lowest level at 1 m. The time covered by the simulations is the 60 day period January 1, 1996 to February 29, 1996. The model's initial and boundary conditions for this period are derived from the Australian Bureau of Meteorology's general circulation model.

2.3 Wind erosion scheme

The details of the scheme are described in Shao et al. (1996). In the scheme, wind erosion is modeled by considering the capacity of the wind in entraining and transporting aeolian particles, and the ability of the surface to resist wind erosion. The erosion model first describes the three physical components of the erosion process: saltation (hopping motion of soil particles), saltation bombardment for dust entrainment, and the sheltering effect of surface roughness elements. Saltation is the key process responsible for sand drift in the direction of wind, and the major mechanism responsible for dust entrainment (Shao et al. 1993a). The calculation of sand drift is based on Owen's (1964) results

$$Q \propto u_*^3 (1 - u_{*t}^2 / u_*^2) \qquad (1)$$

where u_* is friction velocity and u_{*t} is threshold friction velocity which is the minimum friction velocity required to initiate the erosion process. The dust emission rate is based on the experimental studies of Shao et al. (1993b) and can be described by

$$F \propto \beta Q \qquad (2)$$

where β is bombardment efficiency. Owen's theory has been extended to soils with multi-particle sizes, by assuming that the dependence of the saltation flux Q on friction velocity u_* and threshold friction velocity u_{*t} is not significantly altered by the presence of other particle sizes. Despite some limitations of this assumption, it remains a simple and sensible approach to saltation of multi-particle size soils. The dust entrainment results of Shao et al. (1993a) are also extended to multi-particle size soils. The sheltering effect of surface roughness elements is based on the theoretical work of Raupach (1992).

The resistance of the surface against wind erosion is characterised by u_{*t}. In the erosion scheme, u_{*t} is described by

$$u_{*t}(d_s; \lambda, w, c) = u_{*t}(d_s; 0,0,0) / R(\lambda) H(w) S(c) \qquad (3)$$

where d_s is particle size, λ is frontal area index, w top soil moisture and c a measure of surface crusting. R, H and S are empirical or semi-empirical functions. Embedded in u_{*t} is a complicated process and the resistance of the surface to wind erosion is manifested through this parameter in the model.

2.4 Verification of the wind erosion scheme for a single point

The performance of the scheme has been compared with observations from the Mendook area over a period of many weeks (Leys and McTainsh, 1996). In this experiment, five masts were set up around a paddock, each equipped with 6 Fryrear traps (Shao et al., 1993b), mounted at heights of 0.02, 0.25, 0.5, 1, 1.5 and 2 m. The aeolian particles accumulating in the samplers were collected at weekly intervals. Meteorological data were gathered by a weather station near one of the masts, which continuously recorded 12-minute averages of wind speed (at heights of 0.5, 1, 2 and 4 m), wind direction, temperature, relative humidity, atmospheric pressure, solar radiation and soil moisture content. For some periods, frontal area index was estimated from photographic images. Data obtained between October 22, 1990 to April 8, 1991 were used for comparison.

Figure 1 shows the predicted and observed streamwise sand fluxes. For the 20 weeks, reasonable agreement is found between the model and the observations for the first 6 weeks and the last 8 weeks. It is important to point out that the observation period is shortly after cultivation. The large erosion rate for the first 4 weeks can be attributed to several strong wind events and the relatively loose surface after cultivation. For the last 8 weeks, both observations and measurements showed weak erosion activity, consistent with the observed relatively frequent rainfall and low wind speeds. For the other 6 weeks (week 7 to week 12), however, the predicted wind erosion is higher than the observations, particularly for week 11. As discussed in Shao *et al.* (1996) this disagreement indicates the role of factors other than those already considered in the model. The omission of the evolution of the surface properties, particularly the response of the particle size distribution to the erosion process itself, possibly caused this disagreement.

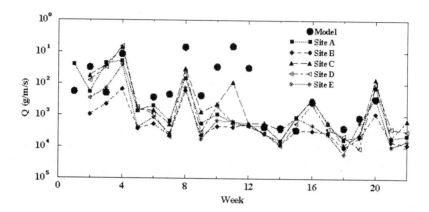

Figure 1: Comparison of predicted and observed weekly average streamwise sand flux Q at Mendook. Observations come from five different masts.

3 Atmospheric forcing and land surface parameters

Stationary parameters are derived from the GIS database and include vegetation height and frontal area index, soil types, surface areodynamic parameters, as well as the parameters for soil moisture modelling (Shao et al., 1996). Dynamic parameters, including wind speed and soil moisture, are predicted with prognostic models.

3.1 Vegetation

Information of vegetation type and height is obtained from the *Atlas of Australian Resources* (1988). Vegetation frontal area index is derived on the basis of Normalized Difference Vegetation Index (NDVI) from satellite imagery, using empirical relationships (e.g. McVicar et al., 1996). Vegetation type and land use data are taken into consideration in the empirical relationships between NDVI and frontal area indices.

3.2 Aerodynamic parameters

Aerodynamic parameters such as surface roughness length and zero-displacement height are estimated from leaf area index and vegetation height, according to Raupach (1994). For areas with no vegetation cover, constant roughness length is assumed.

3.3 Soil

According to the *Atlas of Australian Resources* (1980), Australian soils are classified into 28 soil classes. A soil particle size distribution and a description of surface crusting are assigned to each soil class. For major soil types, the particle size analysis is avaliable from the study of Leys and McTainsh (1996). For the purpose of this study, surface crusting is subjectively estimated from a general description of soil types, while 'random changes' (such as through ploughing) on surface crusting are ignored.

3.4 Surface friction velocity

Surface friction velocity u_* is obtained from predicted near surface wind speed U at a specified reference height z_U, using the logarithmic wind profile law

$$u_* = \kappa U / ln\{[(z_U - D)/z_0] - \Psi\} \tag{4}$$

where $\kappa = 0.4$ is the von Karman constant, D is the zero-displacement height, z_0 the aerodynamic roughness length of the surface and Ψ is a function which takes into account the effect of thermal stability on wind profile.

3.5 Soil moisture

Soil moisture is a major factor influencing u_{*t}. In contrast to vegetation cover, soil moisture in the (very) top soil layer fluctuates on a time scale of a few hours. Soil moisture simulation is in itself a difficult problem. In this study, the soil moisture model developed by Irannejad and Shao (1996) is used to simulate detailed soil moisture patterns (Shao et al., 1996).

4 Results

In early 1990s, Australia experienced one of the worst droughts this century. As a consequence, a considerable proportion of the land surface had little protective annual vegetation and the land surface was susceptible to wind erosion. The wind erosion assessment and prediction system has been applied to predict wind erosion events in Australia around 8 February 1996 and aspects of that period are shown in the following figures.

During the 1996 (southern hemisphere) summer, wind erosion was active in many parts of Australia. In New South Wales (NSW) for example, there were nine reported dust events in January 1996 and 10 dust events in February 1996. Surface weather maps show that on 7 February 1996, a cold front located over the Great Australian Bight was approaching the Australian continent. Within the next 12 hours, the main body of cold air invaded the continent with the cold front moving across South Australia and Victoria. As can be seen in Figure 2, the cold front was associated with strong winds (expressed here as u_*). On 9 February 1996, the cold front has moved further north and east. Strong winds covered a major part of central Australia.

Predicted sand drift intensity is as shown in Figure 2 for four different times (dust emission maps are also avaliable, but not shown). For 12.00 Universal Time Coordinate (UTC) 7 February 1996, wind erosion activities were limited. At 00.00 UTC 9 February 1996, wind erosion activities increased significantly in central Australia and some parts of western Australia, and further increased in intensity and extent during 9 February 1996. As the cold front moved further to the

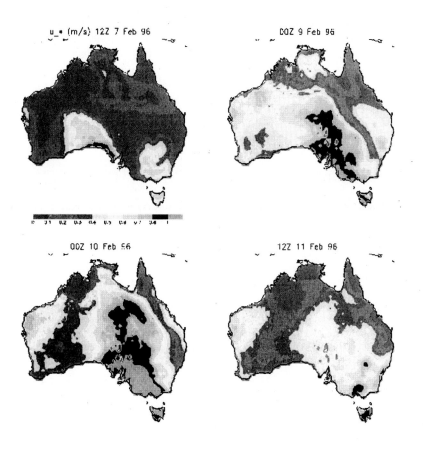

Figure 2: Surface friction velocity associated with a frontal system for the wind erosion period around 8 February 1996.

north east, and the wind speed reduced over the continent, wind erosion activity decreased significantly during the next day. These predictions are in agreement with meteorological observations. The total sand drift integrated over the period from 6 to 12 February 1996 indicated that central Australia, western Australia and the east coast experienced wind erosion. The worst affected areas are as shown in Figure 3 with the maximum total sand drift reaching 2000 kg/km.

A wind erosion model

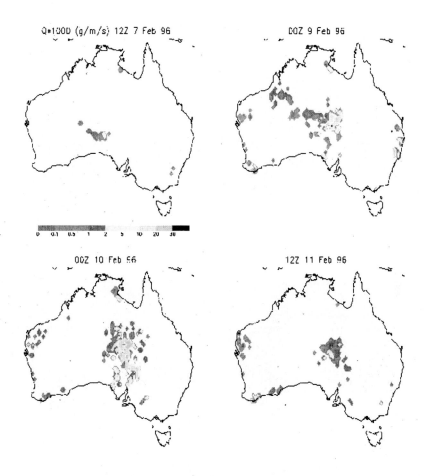

Figure 3: Predicted sand drift intensity for 12.00 UTC 7 Feb, 00.00 UTC 9 Feb, 00.00 UTC 10 Feb and 12.00 UTC 11 Feb 1996.

The model predicted that wind erosion would occur mainly in the central parts of Australia. The pattern of available soil moisture in the top 5cm soil layer over the Australian continent for the same four times described above exhibited a dramatic change. Around 12.00 UTC 7 February 1996, apart from the southern part of western Australia and the east coast of Australia, the majority of the continent was dry with soil moisture values close to air dry values ($w - w_{dry}$ close to zero). Rainfall associated with the cold front resulted in a shift in soil moisture distribution. By 12.00 UTC 11 February 1996, large parts of the eastern region were quite wet. The top soil layer can dry out very quickly in summer when evaporation is large. This variation in soil moisture in the very top soil has a strong influence on the threshold friction velocity for wind erosion, u_{*t}.

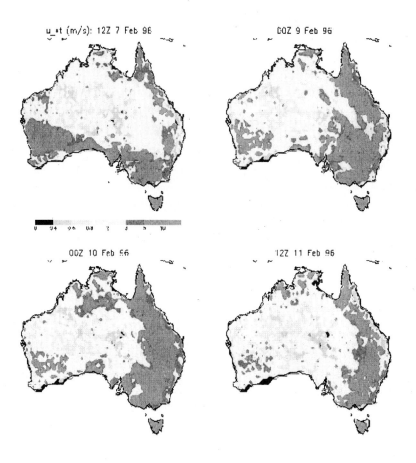

Figure 4: Predicted threshold friction velocity u_{*t} for 12.00 UTC 7 Feb, 00.00 UTC 9 Feb, 00.00 UTC 10 Feb and 12.00 UTC 11 Feb 1996.

The other important parameter influencing u_{*t} is the surface vegetation cover. The effects of soil moisture and surface vegetation cover influence the distribution pattern of u_{*t}, as shown in Figure 4. The figure shows that in large areas in western Australia and central Australia wind erosion risk is high. On 9 and 10 February, the strong winds resulted in friction velocity exceeding the threshold friction velocity for certain areas as can be seen from Figs. 2 and 4, resulting in substantial wind erosion.

5 Concluding remarks

We have presented an integrated system for wind erosion assessment and prediction, which couples a wind erosion scheme with an atmospheric model and a GIS database. We predict wind erosion events by considering both wind erosion erosiveness and wind erosion erodibility. Given the difficulties involved, the achievement of this study is very encouraging: the system predicted the 8 February 1996 wind erosion events well when compared with observations; the simulation also provided an interpretation of the environmental factors responsible for the erosion events.

The integrated system has several limitations:
(1) The present model does not satisfactorily describe the limitation of erodable substrate quantity on wind erosion. A possible consequence of this problem is that predicted wind erosion may be stronger than in reality;
(2) The concept of frontal area index is useful but may be too simplistic and additional parameters may be necessary to fully characterize the effect of surface roughness elements;
(3) Land surface parameters have not yet been accurately estimated for all areas.

References

Atlas of Australian Resources (1980): Volume 1, NATMAP, Canberra.
Atlas of Australian Resources (1988): Volume 6, NATMAP, Canberra.
Burgess, R.C., McTainsh G.H. and Pitblado J.R. (1989): An index of wind erosion in Australia. Aust. Geog. Stud. **27**, 98-110.
Gillette, D.A. & Hanson K.J. (1989): Spatial and temporal variability of dust production caused by wind erosion in the United States. J. of Geophy. Res. **94D**, 2197-2206.
Irannejad, P. and Shao, Y. (1996): The Atmosphere-Landsurface Interaction Scheme (ALSIS): Description and Validation. CANCES Technical Report, No. 2, University of New South Wales.
Leslie, L.M. and Skinner, T.C.L. (1994): Numerical experiments with the West Australian summertime heat trough. Wea. and Forec., **9**, 371-383.
Leslie, L.M. and Purser, R.J. (1991): High-order numerics in a three-dimensional time-split semi-lagrangian forecast model, Mon. Wea. Rev. **119**, 1612-1632.
Leys, J.F. and Heinjus, D.R. (1991): Simulated wind erosion in the South Australian Murray mallee. Research Report, Department of Conservation and Land Management of NSW, Australia, pp 38.
Leys, J. and McTainsh, G. (1996): Sediment fluxes and particle grain size characteristics of wind eroded sediments in south eastern Australia. Earth Surface Processes and Landforms, in press.
McVicar, T.R, Walker, J., Jupp, D.L.B., Pierce, L.L., Byrne, G.T., Allwitz, R. (1996): Relating AVHRR Vegetation Indices to In Situ Measurements of Leaf Area Index. CSIRO, Division of Water Resources, Technical Memorandum 96.5.
McTainsh, G.H. and Pitblado, J.R. (1987): Dust storms and related phenomena measured from meteorological records in Australia. Earth Surface Processes and Landforms **12**, 415-424.
Owen, R. P. (1964): Saltation of Uniform Grains in Air. J. Fluid Mech. **20**, 225-242.
Raupach, M.R. (1992): Drag and drag partitioning on rough surfaces. Boundary-Layer Meteorol. **20**, 225-242.
Raupach, M.R, McTainsh G.H. and Leys J.F. (1994): Estimates of dust mass in some recent major Australian dust storms. Aust. J. Soil and Water Cons. **7**, 20-24
Raupach, M.R. (1994). Simplified expressions for vegetation roughness length and zero-plane displacement as functions of canopy height and area index. Boundary-Layer Meteorol. **71**, 211-6.
Shao, Y., Raupach M.R. and Flindlater P.A. (1993a): Effect of saltation on the entrainment of dust by wind. J. Geophys. Res. **98**, 12719-12726.
Shao, Y., McTainsh G.H., Leys J.F. and Raupach M.R. (1993b): Efficiency of sediment samplers for wind erosion measurement. Aust. J. Soil Res. **31**, 519-532.
Shao, Y., Raupach M.R. and Leys J.F. (1996): Practical numerical model for estimation of sand drift and dust emission caused by wind erosion. Aust. J. Soil Res. **34**, 309-342.

Shao, Y., Leslie L.M., Munro R.K., Irannejad P., Lyons, W.F., Morison, R., Short, D. and Wood M.S. (1997): Soil moisture prediction over the Australian continent. Met. and Atm. Phys **63**, 195-215

Addresses of authors:
Y. Shao
Centre for Advanced Numerical Computation in engineering and Science
UNSW
Sydney, 2052, Australia
R.K. Munro
W.F. Lyons
National Resource Information Centre
P.O. Box E 11
Kingston ACT 2604, Australia
L.M. Leslie
School of Mathematics
UNSW
Sydney, 2052, Australia

Landslides in East African Highlands: Slope Instability and its Interrelations with Landscape Characteristics and Land Use

L.-O. Westerberg & C. Christiansson

Summary

The Nyandarua Range in central Kenya is a densely populated agricultural area. During the last 15 to 20 years mass movement activity has been reported to increase in the area. Due to the dense cultivation pattern, the mass movements constitute a serious problem to the farming population. This paper reports on studies of landslides in the Kangema Division, Murang'a District. We investigated some physical parameters of importance to slope instability and the extent to which human activities (land use, construction works, etc.) have contributed to the occurrence of mass movements. To this end, rainfall characteristics for the area were analysed. An inventory was conducted which identified 19 previously unsurveyed mass movement scars. Notes were made on type of mass movement, agro-ecological zone, soil type, slope form and angle and land use. Farmers were interviewed in order to establish the impact of the landslides on farming activities, and to find out how the landslides were triggered and how they have evolved subsequent to the initial movement. Landslide scars found in forested areas imply that natural conditions, such as soil physical parameters, climatic conditions and slope characteristics in themselves can cause slope instability. However, a few landslides were found to be caused by construction works and, possibly, soil conservation structures. Other slides were triggered following river-bank erosion. Upwelling groundwater was found to have preceded sliding in two cases, suggesting high porewater pressure conditions. The soil layering characteristics in andosols tend to favour shallow translational landslides, whereas nitisols tend to favour deep-seated rotational landslides. The translational landslides occurring in andosols appear to be more common than the rotational slides occurring in nitisols. On the other hand, the rehabilitation of translational slides is considerably faster.

1 Introduction

Mass movements occur on slopes throughout the world. In densely populated urban and agricultural areas, rapid mass movements can cause catastrophes and attract widespread public interest. The fact that mass movements also occur in large numbers in uninhabited and forested areas is seldom reflected upon. This, together with the fact that they are often related to extreme rainfall events, has led to the belief that mass movements are comparatively rare, accidental events with long return periods. However, mass movements may, especially in the humid and subhumid tropics, be one of the most important geomorphological agents (Temple and Rapp, 1972; Selby, 1993; Thomas, 1994).

The present paper is based on a study of landslides in the Kangema Division, Murang'a District, Kenya, that was conducted in October and November 1991. It describes the specific physical parameters of importance for the triggering of landslides in different agro-ecological zones. Furthermore, the paper discusses to what extent mass movements identified in the study are caused by natural processes vis-à-vis human activities. Finally the paper discusses whether certain soil conservation techniques are detrimental to slope stability, and the importance of mass movements as one, among many, slope-forming processes.

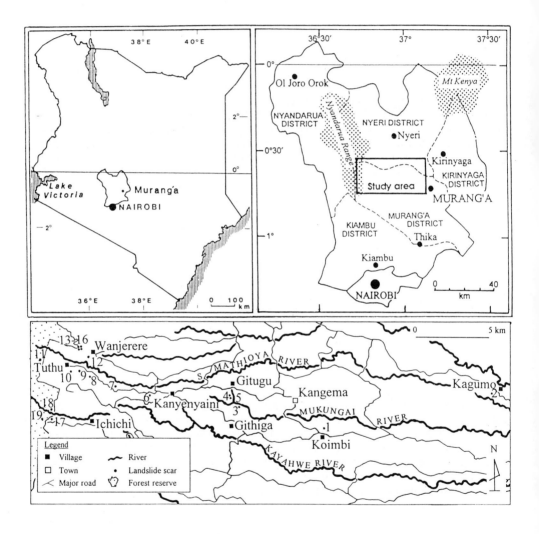

Figure 1: Kenya, Central Province and the study area.
Slide scars discussed in the text are indicated.

2 The study area

Kangema Division is located on the eastern foothills of the Nyandarua Range (Fig. 1). The population density in the Division is 495 people per km^2 (1989 census; Ministry of Planning and Economic Development, 1994). Hillsides are cultivated and grazed, up to 2300 m a.s.l. The mean annual precipitation ranges from about 1600 mm in the easternmost part of the division, to more than 2600 mm in the west (Jaetzold and Schmidt, 1983).

The geological strata are comprised of basalts, basaltic agglomerates and tuffs, associated with the formation of Gregory's Rift Valley (Thompson, 1964). The eastern slopes of the Nyandarua Range are characterised by deeply incised, fluvial valleys. Slope angles range from 10-30° in the lower part of the division, to 25-45° in the higher part. There are two dominant soil types in the division: nitisols in the east and andosols in the west (Sombroek et al., 1982). Deep-weathering is characteristic, and in general the soils have developed to a considerable depth.

3 Methods

An inventory of mass movement scars was made by car-surveying of the area. Extension officers and farmers were interviewed about mass movement incidences. For each scar found in the inventory, notes were made on location, agro-ecological zone, type of mass movement, probable development, approximate slope angle, slope form, the year of movement, and the land use in the immediate surroundings.

As all mass movements in the study area have occurred during the rainy seasons, rainfall data from Kanyenyaini Village (Fig. 1) were analysed.

4 Results

At least 50 mass movement scars of different ages are known in Kangema Division. The scars visited in this study are of a rather heterogeneous appearance, i.e. the location in relation to crests and watercourses varies, as does the steepness of the slopes on which they are situated. Also the morphometric type of movement varies between different areas in the division.

4.1 The mass movement inventory

During the inventory 19 previously unsurveyed mass movement scars were identified and investigated. Table 1 is a compilation of data about each of the scars surveyed.

4.2 Rainfall characteristics

The rainfall records from Kanyenyaini Village are here interpreted as representing the whole of Kangema Division. Annual records cover the period from 1940 to 1984, and daily records are from 1964 to 1984. Mean annual precipitation in Kanyenyaini is 2050 mm (Fig. 2) and the 80 %-probability value is 1560 mm. Precipitation occurs throughout the year, but April-May and October-November show marked peaks (Fig. 3). Dry spells are rare in the area, and only twice between 1964 and 1984 did a drought (< 50 mm per month) exceed 3 months. The greatest monthly rainfall during the period occurred in April 1978 (966.6 mm). The beginning of May 1978 was also very wet with 219.1 mm in 11 days. Several landslides are reported from that period (Kamau, 1980; Westerberg, 1989).

Number/ location	Map reference	Altitude (masl)	Agro-ecol. zone[a]	Movement type[b]	Land use[c]	Soil type[d]	Slope form[e]	Slope angle
1 Koimbi	S 0°41'15" E 36°59'12"	1680	MCZ	CSM (RLS)	G/C/F	N	∪	< 10°
2 Kagumo	S 0°41'00" E 37°05'40"	1440	MCZ	RLS	C	N	∪	15-20°
3 Githiga	S 0°41'30" E 36°55'55"	1740	CTZ	RLS (ESL)	C	N	∪/∩	10-20°
4 Gitugu	S 0°41'08" E 36°55'49"	1780	CTZ	RLS	G/F	N	∪	2-3°
5 Gitugu	S 0°41'08" E 36°55'49"	1780	CTZ	RLS (EFL)	C/G	N	∪	2-3°
6 Kanyenyaini	S 0°41'00" E 36°52'40"	2060	TDZ	RLS	C/B	N	∪	20-35°
7 Kiruri	S 0°40'45" E 36°51'30"	2080	TDZ	TLS	F/G/B	N/A	∪	30°
8 Tuthu	S 0°40'40" E 36°50'30"	2200	TDZ	TLS	F/C	A	—	30-40°
9 Tuthu	S 0°40'10" E 36°50'10"	2100	TDZ	TLS	G	A	∪	20-30°
10 Tuthu	S 0°40'10" E 36°50'00"	2100	TDZ	TLS	G	A	—	20-30°
11 Tuthu	c. S 0°40' E 36°48'40"	2300	AF	TLS	F	A	—	30-45°
12 Wanjerere	S 0°39'30" E 36°50'30"	2300	TDZ	TLS (EFL)	F/G/B	A	∪	35-40°
13 Wanjerere	ca. S 0°39-40' ca. E 36°50'	2360	AF	TLS	F	A	∪	50-60°
14 Wanjerere	ca. S 0°39-40' ca. E 36°50'	2360	AF	TLS	F	A	∪	40-50°
15 Wanjerere	ca. S 0°39-40' ca. E 36°50'	2340	AF	TLS	F	A	—	30-40°
16 Wanjerere	ca. S 0°39-40' ca. E 36°50'	2340	AF	TLS	F	A	—	30-40°
17 Ichichi	S 0°41'55" E 36°48'45"	2320	TDZ	TLS	F	A	∪	40°
18 Ichichi	S 0°41'55" E 36°48'40"	2300	AF	TLS	F	A	—	30-40°
19 Ichichi	S 0°41'40" E 36°49'00"	2340	AF	TLS	F	A	∪	35-40°

[a] MCZ=Main Coffee Zone; CTZ=Coffee-Tea Transition Zone; TDZ=Tea-Dairy Zone; AF=Aberdare Forest.
[b] RLS=rotational landslide; TLS=translational landslide; CSM=complex slope movement; ESL=earth slump; EFL=earth flow.
[c] C=cultivation; G=grazing; F=forest; B=building or construction activity.
[d] N=nitisol; A=andosol.
[e] ∪=concave slope profile; ∩=convex slope profile; —=rectilinear slope profile.

Table 1: Mass movements in Kangema Division (no. 1 is located in Kiharu Div.). Terminology after Varnes (1978). Agro-ecological Zones after Jaetzold and Schmidt (1983).

Landslides in East African highlands 321

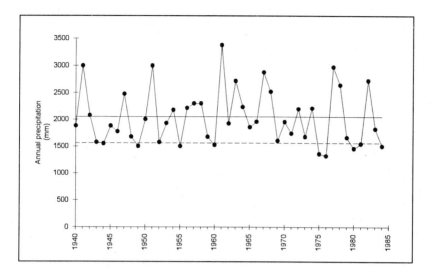

Figure 2: Annual rainfall at Kanyenyaini, 1940-1984. Mean annual rainfall (2050 mm) is indicated by a continuous line and the 80%-probability value (1560 mm) by a broken line.

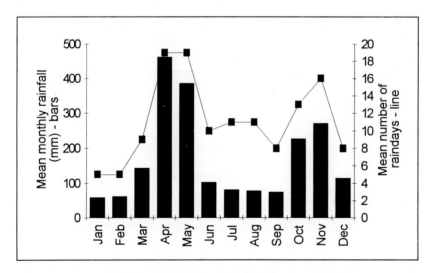

Figure 3: Mean monthly rainfall (histogram) and mean number of raindays (> 1 mm; lines) at Kanyenyaini. Based on records from 1964 to 1984.

Rainfall characteristics are often described by the mean intensity (I) of rainfall, i.e. rainfall per rainday. The intensity is calculated using the formula $I = P/N$, where P is the total amount of rainfall during a time period, and N is the number of raindays during the same period. In the very wet period in April and early May 1978, 1185.7 mm of rain fell during a 41-day period. The mean rainfall intensity per day during the period was about 29 mm, which is an amazingly high value. The highest daily rainfall recorded during the period was 123 mm.

Larsen and Simon (1993) estimated that low-intensity, long-duration storms giving some 2-3 mm/hour for about 100 hours, were sufficient to trigger landslides in Puerto Rico. The highest 100-hour (four-day) intensity in Kanyenyaini during April-May 1978 occurred over April 8-11(314 mm). No less than eight 100-hour periods gave more than 200 mm, and seventeen 100-hour periods gave more than 100 mm. A calculation of the mean intensity per hour during the 100-hour periods reveals that eight periods experienced an intensity of more than 2 mm/hour, and two periods more than 3 mm/hour, the highest being 3.3 mm/hour.

The highest daily rainfall intensities occur in April and May. The highest value during the 20 years of recorded data was in April 1966 (132.1 mm). The value corresponds to almost 20% of the total precipitation that month, and to almost 7% of the total rainfall during 1966. Rainstorms with an intensity of more than 100 mm occurred seven times during the period 1964-1984.

A recurrence interval diagram has been constructed (Fig. 4) which shows that storms with an intensity of 100 mm/day have a return period of between 2.5 and 3 years. Temple and Rapp (1972), in their report on landslides in the Uluguru Mountains in Tanzania, report 75 mm/day as a critical value for severe slide damage in that area. In Kanyenyaini, storms of that magnitude are likely to occur every year.

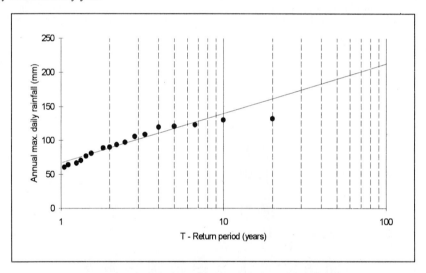

Figure 4: Maximum daily rainfall recurrence interval (cf. Temple & Rapp, 1972) at Kanyenaini, 1964-1984. The straight line equation has been calculated by regression analysis.

5 Discussion

The geologically recent evolution of the landscape in Nyandarua Range is characterised by fluvial processes. Valleys are deeply incised and slopes are shaped by denudational processes where mass movements have been important. A widespread conception among agricultural extension workers, that mass movements occur due only to human activities, is not entirely valid. It is a natural geomorphological process on slopes all over the study area. However, when occurring in densely populated agricultural areas, mass movements constitute a serious problem.

The present study discusses nineteen landslide scars in Kangema Division and surroundings. Eight of the landslides occurred in forested areas. Of these, two are situated close to a forest road, the construction of which may have disturbed slope stability. The other six scars in forested areas show no signs of human activities.

Of the nineteen scars discussed, six are situated in the nitisol zone, one in the transition zone between nitisols and andosols, and twelve in the andosol zone. The six landslides in the nitisol zone are all of the deep-seated rotational type, whereas the landslides in andosols are of the shallow, translational type. The nitisols, in general, are deeper and have more homogeneous soil profiles than andosols. Consequently, deep landslides tend to be more frequent in nitisols. Deep-seated rotational landslides seem to be related to high porewater pressure, which is likely to occur in concave, gently sloping areas, while shallow translational landslides are related to steepness of slope, i.e. rapid groundwater movements. Gently sloping areas are much more common in the nitisol zone, where infiltration of rainwater is further enhanced due to the practice of terracing coffee-plots.

An inventory made in 1985 (Larsson, 1986) indicates that landslides are common in both nitisols and andosols. The present study has identified more landslides in the andosol area than in the nitisol area. Observations of landslides not included in the present inventory, further supports the theory that landslides are more common in the andosol area. It is also likely that differences in slope morphology between the two areas have an impact on landslide occurrence (Westerberg, 1996a). However, though less frequent, landslides in nitisols constitute a more serious problem, due to the steepness of the scarps and the depth of the scars. The present study shows that landslides in andosols are more rapidly covered with vegetation. Eighteen months after the slide, top soil, although thin, had developed in the scar of landslide no. 12, and, in general, trees planted in scars in the andosol area establish quickly. It is likely that within another few years the andosol scars will not even be visible due to revegetation. In the nitisol landslides, on the other hand, remedial measures have frequently failed because of successive slumping from the scarps. Also rill and gully erosion have removed a substantial amount of soil from these deep-seated slides.

5.1 Causal factors

The central Kenya Highlands show many physical features which make the area highly susceptible to mass movement:
- The rainfall characteristics, with short return periods for intense rainstorms.
- The geological history, with bedrock and soil layers often parallel to the slopes.
- The closeness of the tectonically active Eastern Rift Valley.
- The soil characteristics, with deep-weathered homogeneous nitisols, and landslide-prone andosols.
- The slope characteristics, with very steep slopes in the andosol area and frequent concave slopes in the nitisol area where groundwater and surface water converge.

In addition, certain human activities are the direct causes of slope failure. Two of the slide scars were triggered in direct connection with excavation of building sites. The dumping of soil masses associated with building activities is another hazardous activity, and one of the landslides was triggered due to the weight of soil wastes added to the weight of water infiltrated in the soil. One landslide was triggered due to loss of stability at the base of the slope in connection with the excavation of a pond. Seven of the landslides are situated close to roads, and it is likely that construction actvities and drainage from the roads have had an impact on stability.

Soil conservation measures meant to increase the infiltration of rainwater may also induce slope instability. Four of the studied scars are located immediately below terraced coffee plots. At one of these scars the terraces are reverse, inward-sloping bench terraces, aiming to infiltrate as much rainwater as possible. Measurements indicate very high infiltration rates on the terraces. At another scar the terraces are horizontal, and also here the measurements indicate high infiltration. In an area where changes in pore water pressure may influence slope stability this is of course detrimental, and excessive infiltration on the terraces during high rainfall has contributed to instability in these cases.

The soil profile investigations, described more thoroughly in Westerberg (1996b), show that soil strata in the andosol area in the western part of the study area are roughly parallel to the slope. The concentration of groundwater flow to the boundary between the AB and C horizons is likely to have had a negative impact on slope stability. Several landslides in the andosol area, which have moved along the boundary between the AB and C horizons, are examples of this. Andosols have a reputation for being more resistant to water erosion than most other soils. However, due to the characteristic stratification and the thixotropic quality of some andosols, they are very susceptible to landslides (Van Wambeke, 1992).

6. Conclusions

The following conclusive points can be drawn from the study:

First, the eastern slopes of the Nyandarua Range are highly susceptible to mass movement processes. Annual rainfall is high and the return periods of high-intensity storms are comparatively short. The Range is, furthermore, part of a tectonically active zone, and geological strata are in places inclined roughly parallel to the slopes. The andosols in the western part of the area are characteristically landslide-prone, with tendencies of thixotropic behaviour when wet. Mass movements are thus natural processes in the area, and occur also in forested areas.

Second, in the Nyandarua Range intensive land use has led to clearing and cultivation of land less suitable for agriculture. Soil conservation projects implemented in the area have reduced the rate of surface erosion, but land degradation through mass movement continues and is reported to be increasing.

Third, landslides occurring in the nitisol area tend to be of the deep-seated rotational type, whereas andosols tend to favour shallow translational landslides. The rotational landslides are a more serious problem, due to the depth of the scars and the steepness of the slide scarps. Rills and gullies have developed in the nitisol landslides, which has further destabilised the slope so that mass movement processes have continued. Mass movement tends to be more common in concave areas due to converging water flow and high infiltration rates in such areas.

Lastly, it is likely that human activities have added to the growing mass movement problem. Activities that disturb slope stability include construction works, such as excavation of building sites, ponds and roads. Loading slopes with soil wastes and construction material may increase the shear stress acting on the soil mass. The increased infiltration caused by terraces and retention ditches may cause sufficiently high porewater pressure to trigger landslides. In particular, level and reversed coffee terraces show high infiltration capacity and seem to be an important causal factor in the destabilisation process.

Acknowledgements

The study has been conducted under the auspices of the Soil and Water Conservation Branch, Ministry of Agriculture, Kenya. Financial support was provided by the Swedish International Development Authority (SIDA). Thanks are due to the staff of the Regional Soil Conservation Unit (RSCU), Nairobi and to staff of the Department of Agricultural Engineering, University of

Nairobi. The staff of the Divisional Headquarters at Kangema Division are also gratefully acknowledged. Thanks also go to Mrs. Sara Cousins for text improvement.

References

Jaetzold, R. and H. Schmidt (eds.) (1983): Farm Management Handbook of Kenya, Vol. 2.' Ministry of Agriculture, Nairobi, Kenya, and German Agency for Technical Cooperation, 566-619.

Kamau, N.R. (1981): A study of mass movements in Kangema area, Muranga District, Kenya, (Unpublished project report, University of Nairobi).

Larsen, M.C. and Simon, A. (1993): A rainfall intensity-duration threshold for landslides in a humid-tropical environment, Puerto Rico, Geogr. Ann. 75 A (1-2), 13-23.

Larsson, M. (1986): Landslides in the mountain areas of Kenya: Comparative studies of different slopes within the Nyandarua Range, (Unpublished research proposal submitted to the Swedish Agency for Research Cooperation with Developing Countries, Dept. of Physical Geography, Stockholm University).

Ministry of Planning and Economic Development, Central Bureau of Statistics (1994): Kenya Population Census, 1989, Vol. 1, Nairobi.

Selby, M.J. (1993): Hillslope materials and processes, Oxford University Press, Oxford.

Sombroek, W.G., Braun, H.M.H. and van der Pouw, B.J.A. (1982): Exploratory Soil Map and Agro-Climatic Zone Map of Kenya, 1980, Exploratory Soil Survey Report No. E1, Kenya Soil Survey, Ministry of Agriculture, Nairobi.

Temple, P.H. and Rapp, A. (1972): Landslides in the Mgeta area, western Uluguru Mountains, Tanzania, Geogr. Ann. 54 A (3-4), 157-193.

Thomas, M.F. (1994): Geomorphology in the Tropics. A Study of Weathering and Denudation in Low Latitudes, John Wiley & Sons, Chichester.

Thompson, A.O. (1964): Geology of the Kijabe Area, Geological Survey of Kenya, Rep. No. 67, Nairobi.

Van Wambeke, A. (1992): Soils of the Tropics - Properties and appraisal, McGraw-Hill, New York.

Varnes, D.J. (1978): Slope Movement Types and Processes, in: R.L. Schuster and R.J. Krizer (eds.), Landslide Analysis and Control, Transportation Research Board, Special Report 176, 11--31.

Westerberg, L.-O. (1989): Rainfall characteristics, soil properties, land-use and landslide erosion in the Kanyenyaini area, Nyandarua Range, Kenya, (Unpublished MSc degree project report, Dept. of Physical Geography, Stockholm University).

Westerberg, L.-O. (1996a): Landform evolution through mass movement in the Kenyan Central Highlands - Towards a theory, EDSU Working Paper 32, Dept. of Physical Geography, Stockholm University.

Westerberg, L.-O. (1996b): Landslides in the Nyandarua Range, central Kenya. Natural prerequisites and consequences of human activities, EDSU Working Paper 34, Dept. of Physical Geography, Stockholm University.

Addresses of authors:
Lars-Ove Westerberg
Carl Christiansson
Evnironment and Development Studies Unit
Department of Physical Geography
Stockholm University
S-106 91 Stockholm
Schweden

Soil Degradation by Tillage Movement

D.K. McCool, J.A. Montgomery, A.J. Busacca & B.E. Frazier

Summary

Soil movement by tillage implements is an often overlooked but significant cause of soil degradation in the steep cropland of the Pacific Northwest, USA. Evidence is seen in ridgetops devoid of topsoil and in near-vertical embankments at the lower boundary of steep fields. Traditionally, the moldboard plow was the principal implement used for the first tillage operation after harvest, with the furrow turned downslope on steep slopes because of power requirements. Experiments were conducted to quantify soil displacement with the moldboard plow, and a relationship was developed between land slope and soil movement. Deposition was measured at the lower boundary of fields where water erosion was not involved. Calculated soil movement assuming typical tillage practices was approximated by the measured deposition. The results emphasize the magnitude and importance of tillage movement in land degradation in the region. Unless corrective measures are taken, exposed subsoil materials on ridgetops and other areas will be translocated to cover productive downslope topsoil.

Keywords: Soil movement, topsoil loss, tillage erosion, tillage translocation, tillage

1 Introduction

Tillage operations on slopes of steep cropland can move substantial quantities of soil, decreasing topsoil depth on ridgetops or at the upper boundary of tilled areas. This process is called tillage translocation, and is a form of soil erosion just the same as water or wind erosion. Evidence of tillage translocation is abundant in the steep non-irrigated wheatlands of the Pacific Northwest, USA, though the region was first intensively cultivated no earlier than about 1890. Here can be found hilltops denuded of topsoil, soil accumulated as eyebrows above grassed areas too steep to farm, and steep embankments formed along fence rows and other permanent features in the field. Moving a large mass of soil downslope exposes subsoil on top of the hills. In the Palouse region of the Pacific Northwest (central-southeast Washington), this subsoil is of a higher clay content than the topsoil, relatively low in organic matter, and has low permeability and undesirable structural characteristics. Kaiser (1961) reported the case of a hilltop cistern installed in 1911. Thirty-one years later more than 0.61 m (2 ft) of soil had been plowed away from it; in the next 17 years an additional 0.61 m (2 ft) was plowed away. The Soil Conservation Service (SCS) (USDA-ESCS, FS, SCS, 1979) reported examples of vertical banks of 0.41 to 0.91 m (16 to 36 in) depth formed within 10 to 30 years at the lower boundary of grass strips upslope of plowed fields, and the formation of banks or berms 1.2 to 3.1 m (4 to 10 ft) high at the lower boundary of plowed fields.

Tillage movement has been recognized for many years. Horner et al. (1944) reported "on steep slopes (above 20%) the tillage implements will not turn the furrow uphill, and all tillage gradually moves the topsoil down the slopes. During the past 50 years, it is likely that at least 6 in (15 cm) of topsoil has been moved a distance of not less than 25 ft (7.6 m)." Tillage translocation was largely ignored for many years. Only recently has there been renewed interest in this phenomenon.

In 1984, Alan Mace, an undergraduate student in the Agricultural Engineering Department at Washington State University, working with Dr. D. K. McCool's assistance, conducted an experiment to determine the effect of slope steepness on tillage movement with a moldboard plow. The experiment provided background for an entry in a student paper contest (Mace, 1984); the data is included as part of the present paper.

Lindstrom and colleagues (1990) conducted an experiment to quantify soil movement by tillage on different slope steepnesses. A moldboard plow with 46-cm share width and a disk were used on plots from 1 to 8% steepness with buried steel nuts as tracers. Soil loss was found directly related to slope. For contour tillage, soil movement up or downhill caused by the plow was related to slope as:

$$Y = 44.1 - 1.8\,S \tag{1}$$

where Y = net soil movement of the tilled depth, cm
  S = land slope, %. S is positive if the furrow slice is turned upslope, negative if turned downslope.

Lindstrom et al. (1992) reported a continuation of the previous study. The same 46-cm plow was used for all tests. The tracers were increased in number. The following equation was fit to the data from the plow:

$$Y = 44.3 - 1.12\,S \tag{2}$$

where Y = net soil movement, cm.
  S = land slope, %. S is positive if the furrow slice is turned upslope, negative if turned downslope.

A model was developed to calculate the effect on a specific site of long-term contour moldboard plowing. Lindstrom et al. (1992) observed that the effect of tillage translocation is to level the hillslope; soil is moved from convex and deposited in concave areas of the slope.

Govers and associates (1994) measured soil movement resulting from tillage by a moldboard and a chisel plow pulled up and downhill on slopes as steep as 25%. They found the following equation described up or downhill soil movement from the moldboard plow when pulled up or down the slope:

$$Y = 28 - 0.62\,S \tag{3}$$

where Y = net soil movement, cm.
  S = land slope, %. S is positive if the plow is pulled uphill, negative if pulled downhill.

The objectives of the current study were to quantify the distance soil is displaced up or down the slope when contour plowing with a conventional moldboard plow, and to compare computed soil movement from tillage with measured deposition at field boundaries.

2 Soil movement with moldboard plow

2.1 Procedure

Two different experiments are reported in this paper. The first, conducted in 1984, used buried, numbered, color-coded magnets placed in rows perpendicular to direction of tractor movement to track soil movement. The second, conducted in 1990, used buried, numbered, color-coded washers. In both experiments, metal detectors were used to locate the magnets or washers after contour plowing with a moldboard plow.

A six-bottom Case[1] pull-type moldboard plow with 40 cm (16 in) share spacing was used in both experiments. This furrow width is typical of those used in the region. In 1984 it was pulled by a John Deere Model 4240 front-wheel-assist 147 kw (110 hp) wheel tractor. In 1990, it was pulled by a Caterpillar track-laying tractor. The same field just north of Pullman, Washington, USA was used in both studies. The soil is a Palouse silt loam (fine-silty, mixed, Mesic, Pachic Ultic Haploxeroll). In 1984, the magnets were installed on slope steepnesses of 7, 23 and 31% and in 1990 the washers were installed on slope steepnesses of 4.5, 23 and 31%. Both experiments included contour plowing while turning the furrow slice upslope as well as downslope.

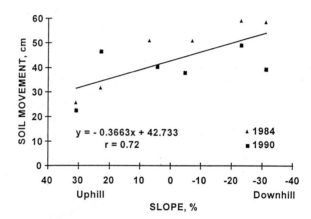

Fig. 1: Soil movement with 40-cm moldboard plow, plowing on contour with furrow turned uphill or downhill.

2.2 Results

Summarized data from the two studies is presented in Table 1 and Fig. 1. A linear fit of the combined data points yielded:

$$Y = 42.73 - 0.366\, S \qquad (4)$$

where Y = average movement of 40-cm-wide furrow slice from one plowing, cm.
 S = land slope, %. S is positive if the furrow slice is turned upslope, negative if turned downslope.
The correlation coefficient is 0.72.

Equation (4) indicates a furrow movement on zero slope approximately equal to the width of the furrow slice; at 30% slope, movement when turning the furrow downslope is about 125% of the plow share width, and movement when turning the furrow upslope is about 75% of the plow share width.

---

[1] Names of companies or commercial products are given solely for the purpose of providing specific information; their mention does not imply recommendation or endorsement by the U.S. Department of Agriculture over others not mentioned.

		Movement	
	Slope %	Furrow turned upslope cm	Furrow turned downslope cm
1984 Data	7	50.8	50.8
	23	31.5	59.0
	31	25.5	58.5
1990 Data	4.5	40.5	38.1
	23	46.6	49.3
	31	22.6	39.6

Table 1: *Soil Movement Upslope or Downslope With 40-cm Moldboard Plow*

The relationship indicates less influence of slope than that of Lindstrom et al. (1990, 1992). Their data was collected from land slopes no steeper than 8%. The assumed linear relationship between slope steepness and soil movement may be only an approximation, and speed of operation, moldboard shape, soil characteristics and soil moisture may influence the result. The relationship of Govers et al. (1994) is not comparable because the implement was pulled up and down the slope.

3 Tillage deposition at field border

3.1 Procedure

Field observations of soil deposition at lower boundaries of fields were made at three locations in the early summer of 1996. The sites were carefully selected just uphill from steep valley slopes that were in pasture and, due to the slope configuration, would never have been cultivated. Concave or concentrated flow areas were avoided; if possible, slightly convex areas were selected because they would have been less affected by water erosion, and any soil buildup at the field boundary could reasonably be assumed to be due to tillage movement.

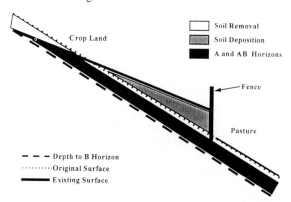

Fig. 2: *Tillage deposition measurements.*

The original soil surface was identified below the field boundary in the pastures (Fig. 2). The depth to the top of the B horizon was determined by augering below the field boundary in the pastures. Then the depths to the original soil surfaces and top of the B horizon were identified upslope of the field boundary by hand augering beneath the deposited material. Additional sample holes were drilled up the slope as necessary in order to establish where the original soil surface would have intersected the existing soil surface (point of zero deposited material). With these dimensions established, the cross-sectional area of the deposited soil was calculated.

Tillage removal downslope from a hill-top tree planting was estimated by projecting into the plowed field from the original, undisturbed soil surface and soil horizons in the trees. An additional removal site was located adjacent to a deep road cut through a ridge.

3.2 Results

Data from the measurements are presented in Table 2. The vertical cross-sectional area of deposits ranged from 2.2 to 6.0 m^2 (24 to 65 ft^2) on slopes from 20 to 26% (Sites 1, 4 and 5). The estimated vertical cross-sectional area removed downslope from the tree planting, Site 3, was 2.4 m^2 (25.5 ft^2) on a 24% slope. The depth of soil removed just downslope of the tree planting was approximately 60 cm. In the case of Site 2 where we were unable to determine a lower bounds, the depth of soil removed near the upper boundary was more than 90 cm.

	Slope %	Cross-Sectional Area m^2	Description
1	20	2.2	Deposition
2	-	-	Removal downslope from upper boundary. No way to determine lower bounds.
3	24	2.4	Removal downslope from hill-top tree planting
4	23	3.5	Deposition
5	26	6.0	Deposition

Table 2. Deposition or Removal at Field Boundary

4 Comparison of movement and deposition

The number of times any field has been plowed since removing the native vegetation could range from once per year to once every other year. In addition, there is no data on the number or type of non-inversion implements used, such as harrows, field cultivators or seeding drills. In most of the study area the primary rotation has been winter wheat alternating with summer fallow. Thus, most fields would have been plowed every other year for about 106 years.

Assuming a 15-cm deep downslope-directed moldboard plowing with shares spaced at 40 cm every other year since 1890 on sites 1, 4 and 5, and using the translocation estimated from equation (4), the estimated tillage translocation from the moldboard plow would be about 4 to 4.2 m^2 (0.15 m x 0.50 m x 53 yr.). These values fall within the range of measured deposition (Table 2) of 2.2 to 6.0 m^2. The removal downslope of the tree planting (site 3) is a special case. The period of tree planting was in the late 1930's to early 1940's. Thus, this movement has occurred only within approximately the past 55 years. However, this site is in an annual cropping zone, so it is possible the number of plowings was approximately equivalent to those for sites 1, 4 and 5. The estimated tillage translocation of 4.2 m^2 as calculated with equation (4) is 1.8 times the actual estimated removal of 2.4 m^2 on Site 3 (Table 2).

5 Discussion

With the measurements and assumptions in this set of experiments, measured tillage deposition and estimated tillage translocation were in reasonable agreement. However, unless there is an exact historical record of all tillage operations on a given field, it is doubtful that a more detailed comparison of calculated tillage movement and measured tillage deposition is possible. Also, the effect of land slope on tillage movement is affected by a number of factors. If relationships found by Lindstrom et al. (1990, 1992) were applied to this comparison, much greater tillage movement would be calculated and a poorer comparison with measured deposition would be seen. This indicates our general lack of understanding and data of tillage movement. Without better knowledge and understanding of the process, data and relationships are not readily transferable.

However, it is clear that tillage translocation at the rates measured and estimated in this study is sufficient to seriously degrade ridgetops by exposing subsoil materials that are of low productivity and have low infiltration and permeability. With time and additional downslope tillage, these ridgetop subsoil materials will be translocated to cover the topsoil on downslope areas.

The translocation data collected with the moldboard plow indicate it is possible to gradually move soil uphill if the furrow slice is always turned upslope, even on rather steep slopes. This requires adequate power to pull the plow at speeds sufficient to turn the furrow slice upslope on steep slopes. This will prevent additional soil degradation and may permit some level of recovery. The alternative is to use no-till or non-inversion tillage practices or to remove the degraded areas from crop production by planting permanent vegetation. If corrective measures are not taken the result will be gradual enlargement of areas of exposed nonproductive subsoil.

Acknowledgements

The authors wish to acknowledge the contributions to this project of Alan Mace who, as an undergraduate student working with Dr. McCool, collected the 1984 data, and Dr. John Simpson who worked with Dr. Montgomery in collecting the 1990 data.

References

Govers, G., VanDaele, K., Desmet, P., Poesen, J. and Bunte, K. (1994): The role of tillage in soil redistribution on hillslopes. European Journal of Soil Science **45**, 469-478.

Horner, G.M., McCall, A.G. and Bell, F.G. (1944): Investigations in erosion control and the reclamation of eroded land at the Palouse Conservation Experiment Station, Pullman, Washington, 1931-42. Tech. Bul. 860. Soil Cons. Serv., US Dept. Agr., Washington, DC.

Kaiser, V.G. (1961): Historical land use and erosion in the Palouse - a reappraisal. Northwest Science **35(4)**, 139-153.

Lindstrom, M.J., Nelson, W.W., Schumacher, T.E. and Lemme, G.D. (1990): Soil movement by tillage as affected by slope. Soil & Tillage Research **17**, 255-264.

Lindstrom, M.J., Nelson, W.W. and Schumacher, T.E. (1992): Quantifying tillage erosion rates due to moldboard plowing. Soil & Tillage Research **24**, 243-255.

Mace, A.G. (1984): Measurement and control of tillage erosion in the Palouse. ASAE Student Paper. Pacific Northwest Region. ASAE. 14 p.

USDA-ESCS, FS, SCS (1979): Palouse cooperative river basin study. USGPO. Washington, DC. 182 p.

Addresses of authors:
Donald K. McCool
USDA-ARS, Biological Systems Engr. Dept., Washington State University,
Pullmann WA 99164-6120, USA
J.A. Montgomery
Evnironmental Sciences Program, DePaul University,
1036 West Belden Avenue
Chicago IL 60614-3238, USA
A.J. Busacca
B.E. Frazier
Dept. of Crop and Soil Science, Washington State University,
Pullmann, WA 99164-6420, USA

Shifting Land Use and its Implication on Sediment Yield in the Rif Mountains (Morocco)

A. Merzouk & H. Dhman

Summary

A study was conducted to evaluate the effects of land use changes on soil erosion rate and sediment production in the Mediterranean mountains of Morocco. Spatial and temporal changes in land cover and use were analyzed for the 18 000 ha Tleta watershed using aerial photography (1976), SPOT (1990) and Landsat-TM (1996) imagery. The land use maps produced were integrated in a watershed resources geographic information system (GIS). The C-factor of the well known Revised Universal Soil Loss Equation (RUSLE) was derived and used to estimate the gross soil erosion. The change analysis showed that land use shifting through cultivation of forest and brushland has increased the area of cultivated land by 30% in twenty years. This expanded slope cultivation caused the C factor of the Tleta watershed to increase from 0.49 to 0.62. Gross soil loss and sediment yields were estimated using RUSLE and delivery ratio for the 1976 and 1996 land use patterns. The predicted sediment yields agreed with the measured values in the Ibn Battouta reservoir downstream of the watershed. The results indicate that the integration of RUSLE, GIS and remote sensing technology provides a powerful tool for watershed assessment and planning.

Keywords: Watershed, land use, erosion, sediment, Morocco

1 Introduction

The accelerating extension of rainfed agriculture onto increasingly steep and marginal lands, the historical neglect of mountain agriculture, and the paucity of input and techniques for sustainable mountain agriculture production and deforestation are leading to watershed degradation and soil loss far above sustainable levels in northern Morocco. The Rif mountains, representing less than 6% of the country, produce more than 60% of the total national sediment load (Merzouk, 1988). The only national study of soil erosion in Morocco, conducted in 1975, showed that 77% of the 22.7 million ha studied was susceptible to high or very high erosion risks. The extent and nature of soil erosion and its causing factors have been poorly documented, thus slowing down the watershed planning and assessment effort. Out of 78 major existing dams, less than 20 have a study and management plan, while their silting-up is progressing at the alarming rate of 60 Mm^3 per year (Boutaieb, 1988). This makes the effective management of the nation's watershed resources a priority of vital economic importance.

A systematic, cost-effective, multi-disciplinary approach to watershed management is urgently required to reduce upstream land degradation and the threat to Morocco's hydraulic infrastructures. This approach requires more advanced watershed planning and assessment technologies adapted to the Mediterranean mountains. In recent years, geographic information systems (GIS) and remotely sensed data have became useful spatial data source and handling tools in watershed assessment (Fedra 1993, Dhman, 1995, Kienzle 1993, Savabi et al., 1995). Research needs have been expressed by watershed planners for the development of methodologies applicable to the Mediterranean watersheds (FAO, 1993).

The primary objective of this project has been to evaluate the effect of land use changes on soil erosion and sediment production rates within a representative watershed in the Rif mountains (Morocco). A secondary objective was the setting up of a methodology integrating remote sensing and GIS techniques, and a spatially distributed erosion model for monitoring the dynamics of watershed replenishments and assessing the sediment sources in catchments.

2 Methods and materials

2.1 Watershed description

The Tleta watershed located upstream of the Ibn-Battouta dam in the northwestern part of the Rif mountains (Morocco). This 180 square kilometers basin is seasonally drained by Oued Sania, an effluent of Oued Mharher which drains towards the Atlantic Ocean. It is a well embanked basin with a culminating point at 600 m whereas the low altitudes near the dam are around 40 m asl, presenting a very rugged relief with steep slopes. The general litho-stratigraphic natures of the Rif mountains are well-represented in this watershed, namely by the dominance of very erodible geological formations made of flysh, limestone and marly material. Lithosols and regosols developed on sandstone and marly slopes are very dominant in the watershed and offer little resistance to water erosion. A Mediterranean climate with warm summers and mild winters prevails in the watershed. The watershed receives an average of 750 mm rainfall annually, concentrated in the period of October-January. The rainfall erosivity caused by the frequent storms of high intensity and short duration is high with an R factor of the Universal Soil Loss Equation valued around 200 5MJ mm/ha-h-yr (Dhman, 1995). Two-thirds of the watershed is presently cultivated with cereal crops with no soil conservation measures. This was the result of continuous process of converting the original cork oak forest to brushland then to cultivated slopes.

All the soil erosion factors are aggravated in this highly erodible watershed, making this land degradation process very threatening to the already limited land productivity and the storage capacity of the Ibn Battouta dam. The latest bathymetric survey of the sediment deposit showed an annual rate of 47.2 T/ha/yr (Merzouki, 1992).

2.2 Land use/cover change analysis.

Spatial and temporal changes in land cover and use were analyzed and mapped using aerial photography (1976), SPOT (1990) and Landsat-TM (1996) spring images. A computer image analysis system was used to overlay pixel to pixel the land use maps of 1976, 1990 and 1990 and to evaluate the temporal and spatial changes. From the digitized land use/cover information obtained with supervised classification and field observation, values of C-factor were assigned for each land cell (30 m x 30 m). The Revised Universal Soil Loss Equation, RUSLE (Renard et al., 1996), which is a widely used empirical and distributed erosion model, was the basis for the methodology used herein to evaluate the impact of temporal changes in land use on sediment yield for the different land use settings. The rainfall erosivity factor R was obtained from an iso-erodent map of the watershed published by Dhman (1995). The soil erodibility factor (K) was determined for each land cell from digitized soil survey map information using the RUSLE model. The topographical factor was derived from the digital elevation model (DEM) and the P factor was arbitrarily set to 1.0 since no conservation practices are established in the watershed. Quantity of sediment delivered to the river system was derived from the RUSLE computed soil loss using a delivery ratio (DR) as described by Hession and Shanholtz (1988). DR values were compiled for each land unit (modeling cells) using the DEM. A detailed description of the application of the model in the Moroccan conditions has been presented by Dhman (1995).

3 Results and discussion

The watershed land use map derived from the interpretation of the aerial photography (1/20 000) of 1977 was checked with the one established in 1977 upon the completion of the dam construction. It showed that half of the watershed was cultivated with cereal crops while the other half was still covered by forests and the so-called matoral which is represented by a brushland composed essentially of *Pistacia lentiscus, Olea europe, Chamerops humilis* and other Mediterranean shrubs. This brushland was used for grazing. Four major land use/cover were distinguished and used for the legend of the map (crop land, brushland/forest, bare soil and shrubland). The last unit represents the advanced stage of degradation of the natural plant cover and is represented mainly by *Chamaerops humilis* (Doum) and *Asphodelus microcarpa* and some annual plant species. Fig. 1 shows the 1976 land use map. The 1990 (July) SPOT-XS satellite image was interpreted and transformed into a land use map with a supervised classification by using the algorithm of maximum probability and ground information. The derived map was compared in 1991 with field data and presented an 82 % accuracy (Dhman, 1995). The map given in Fig. 1 documented the significant increase in the area cultivated (63%) that occurred during the last 20 years.

			Land use units as derived from the 1976 aerial photos			
			Crop land	Shrub/Doum	Brush/Forest	Bare soil
			51.31**	5.8	40.1	2.8
From A 1990 SPOT Image	Crop land	63.9**	51.3	3.9	8.7	-
	Shrub/Doum	16.6	-	1.9	14.7	-
	Brush/Forest	16.8	-	-	16.8	-
	Bare soil	2.8	-	-	-	2.8
From A 1996 LANDSAT TM image	Crop land	66.5**	51.3	1.2	14.0	-
	Shrub/Doum	11.8	-	4.6	7.2	-
	Brush/Forest	19.0	-	-	19.0	-
	Bare soil	2.8	-	-	-	2.8

*The table shows for example that the cultivated land representing 66.5% of the watershed area in 1996 cames from the total crop land of 1976 plus the area converted from Doum (matorral) and forest lands.
** in percent of the total water shed area (18 000 ha).

> Assigned C values (Dhman, 1995): C=0.8 for cereal cropping land; C=1.0 for bare soil; C=0.36 for shrubland (douml) and C=0.1 for forest and dense brushland.

Tab. 1: Crosstabulation table showing the magnitude of the land use shifting* as compared to the 1976 land use distribution.

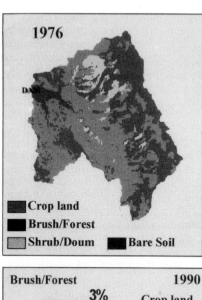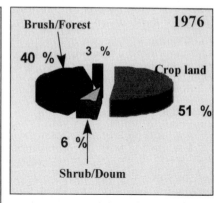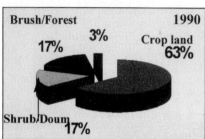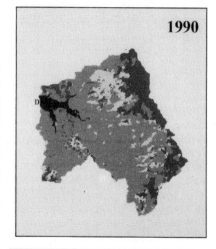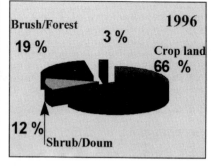

Fig. 1: Shifting land use pattern and the vegetal cover depletion in the telanta watershed between 1976 and 1996.

The Landsat-TM image acquired on 28 March 1996 shows that two-thirds of the watershed area is now cultivated (Fig. 1). The computer map overlay (pixel to pixel) allowed a precise change analysis between image pairs (1976-1990 and 1976-1996). Cross tabulation tables were obtained showing the frequency with which land use classes have changed. Fig. 1 and Table 1 present the results of the temporal and spatial changes. Crop land area has increased 30 % from 1976 to 1996 at the expense of the shrubland which represents the last stage of the forest and brushland degradation (Dhman, 1995). A computer image differencing technique produced the map of the area put under cultivation since 1977. The overlay of this map with the slope map showed that expansion of the cultivation took place mainly on steeper slopes. This extension of traditional, rainfed cereal cropping onto increasingly steep and marginal lands is the most serious problem facing watershed development in the Rif mountains. Mallyani (1988) has reported the same magnitude of land use transformation (1% of total area/yr) from natural forest/brushland to cropland, observed between 1963 and 1987 in a nearby watershed.

Fig. 2 presents the obtained digital maps representing the soil erosion factors of the RUSLE model. This GIS-based model estimated the gross soil loss in T/ha/yr for each land cell. Fig. 3 shows the soil loss rate distribution in the watershed for the 1996 land cover. A distributed delivery ratio used by Dhman (1995) allowed the GIS to estimate the sediment production rate from the gross soil loss map. The weighted average value for the DR was 13% which is very close to the 12% predicted delivery ratio for the entire watershed using the DR-watershed area curve proposed by the American Society of Civil Engineering. The 1996 land use average C value for the watershed was 0.62 and the corresponding average sediment yield was estimated at 34.4 ton/ha/yr. The model predicted a sediment yield with the 1977 land use C values (C=0.49) of 24.4 T/ha/yr. The model showed a very precise response of soil loss and sediment yield to land use changes. These results were compared with the published data of the sediment deposition survey of the reservoir which are 28.2 T/ha/yr in 1977 and 47.2 T/ha/yr 1996. Considering the contribution of the channel and river bed erosion that the model can account, for the results are very satisfactory.

4 Conclusions

The present study demonstrated the utility of the integrated remote sensing and GIS technology for the precise monitoring of land use dynamics in the Mediterranean mountains. The digital image differencing techniques were used to locate the forest and brushland that was cultivated during the last 20 years in the Tleta watershed (Morocco). Thus dense forest, low density forest, dense matorral (brushland), and low density matorral (shrubland) have regressed by 31% between 1977 and 1996. Logging, shift cultivation and over-grassing are the main causes of this change. The RUSLE model was easily linked to a GIS and used to estimate the soil loss and the sediment yield for each land use map (1977, 1990 and 1996). The RUSLE-GIS integration offered some sophisticated raster capabilities and modeling functions which facilitated the model applications in impact assessment.

Despite the satisfactory agreement between the predicted and the observed sediment yields, more work is needed on the proper selection of the input parameters corresponding to the semi-arid and humid Mediterranean mountains. This calls for more field research on the C (land use factor) and the K (soil erodibility) values. The present Tleta watershed resources GIS now has a sediment production spatial model that locates the critical areas to be corrected. It provides a useful tool to investigate the temporal variability of land use affecting the degradation of upstream land as well as the siltation of the dam. The watershed planner can use it for analyzing the various management scenarios and their impacts in order to formulate recommendations for optimal actions. This watershed GIS will be more powerful for decision making with the addition of a

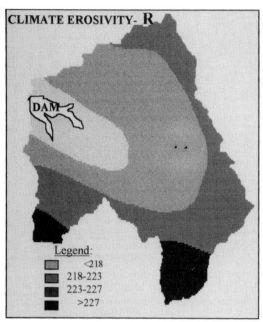
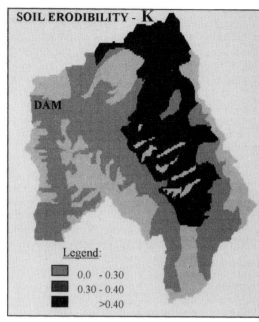
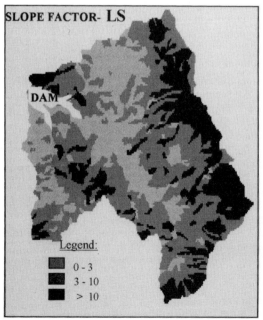
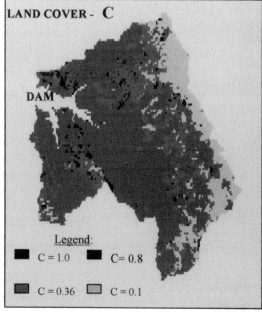

Fig. 2: Soil erosion factors used in the RUSLE model for the Telata watershed, Morcco.

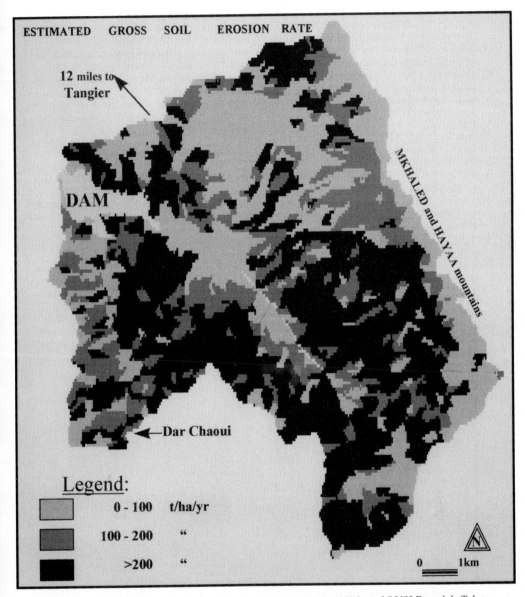

Fig. 3: Spatial distribution of soil loss rate as estimated with the GIS based RUSLE model; Telata watershed, Morocco

multicriteria analysis package. With the visual capabilities (maps, graphs, on screen animation) and the precise information stored, the GIS based model offers a useful tool to promote the communication between all the actors in watershed management (scientists, technicians, farmers, politicians and land use planners).

References

Boutaieb, M. (1988): Impacts économiques de l'envasement des barrages au Maroc. Actes du séminaire international sur l'aménagement des bassins versant. Direction des Eaux et Forêts, Rabat.

Dhman, H. (1995): Etablissement d'un SIG pour l'étude et la cartographie de l'érosion des sols: Application au bassin versant de Tleta, Tangérois. Mémoire de fin d'études. ENFI, Salé, Maroc.

FAO/MAMVA. (1993): Projet pilote d'aménagement des bassins versants au Maroc. Projet MOR/93/010. Rapport du projet, Rome.

Fedra, K. (1993): Models, GIS and expert systems: Integrated water resources models. HydroGIS 93: Application of Geographic Information System in Hydrology and Water Resources Management. In: Proceedings of the Vienna Conferences, April 1993. IAHS 211, 1993, 297-308.

Hession, W.C. and Shanholtz, V.O. (1988): A geographic information system for targeting nonpoint-source agricultural pollution. J. of Soil and Water Cons. Vol. 43, N° 3, May-June, pp.264-266.

Kienzle, S.W. (1993): Application of a GIS simulating hydrological responses in developing regions. HydroGIS 93: Application of Geographic Information System in Hydrology and Water Resources Management. In: Proceedings of the Vienna Conferences, April 1993. IAHS 211, 1992, 309-318.

Mallyani, M. (1988): Cartographie et evaluation de la degradation des types de peuplements entre 1963 et 1987 dans la foret de Mediar, Tanger. Mémoire de fin d'études en foresterie. IAV Hassan II, Rabat, Maroc.

Marzouki, T. (1992): Diagnostic de l'évasement des grands barrages marocains. La Revue Marocaine du Génie Civil, 38, Avril 1992.

Merzouk, A. (1988): L'érosion hydrique des sols dépricie leur productivité. ATTABEA: Revue des naturalistes enseignants, 55/87-88, Rabat, pp: 56-65.

Renard, K.G., Foster, G.R., Weesies, G.A., McCool, D.K. and Yoder, D.C. (1996): Predicting soil erosion by water: A guide to conservation planning with the Revised Universal Soil Loss Equation (RUSLE). USDA/ARS. AG. Handbook # 703. Washington., DC.

Savabi M.R., Flanagan, D.C., Hebel, B. and Engel, B.A. (1995): Application of WEPP and GIS -GRASS to a small watershed in Indiana. Journal of Soil and Water Conservation **50(5)**: 477-483.

Address of authors:
A. Merzouk
H. Dhman
Institut Agronomique et Vétérinaire Hassan II
BP: 6202-Institut
Rabat
Morocco

Optimisation of Soil and Water Conservation Techniques in a Watershed of the Tunisian Semi-Arid Region

L. Laajili Ghezal, T. Aloui, M.A. Beji & S. Zekri

Summary

Three approaches were used to relate erosion to soil productivity and soil and water conservation management and practices: a Productivity Index (PI), a conceptual model (GAMES) and a Decision Tree Approach (DTA). The PI indicates that the whole investigated watershed belongs to 3 productivity classes; average, poor, and extremely poor. The application of the model for a ten-year period shows that the calibration parameter α varied from 0.2 to 6.2 and that the predicted average gross erosion within the watershed is about 17 $t.ha^{-1}.yr^{-1}$. The scattering of the output data of the GAMES model using Principal Components Analysis (PCA) resulted in six groups of soil erosion being identified. Results obtained from the DTA approach show that 47% of the entire watershed is likely to suit reforestation and relatively intensive agriculture, 41% is suitable for grazing land, grazing forestry land and field cropping land when agricultural conditions are more appropriate. The remaining area (12%) is suitable for microcatchments.

Keywords: Modeling, erosion, catchment, land quality, information system, water harvesting.

1 Introduction

Soil degradation is one of the most important problems facing public authorities in Tunisia with water erosion being one of the major causes (Laajili Ghezal, 1988). Currently more than 325,000 ha are highly affected by erosion and 986,000 ha are fairly eroded. These figures represent 24% of the agricultural land in Tunisia. The Central Western part of the country, also known as the 'Dorsale', is the most influenced by erosion with some 563,000 ha found to be disturbed (DGPDIA, 1995). The Dorsale, a southwest-northeast oriented chain of montains, separates the country into two parts: one-third fairly rainy to the north and two-thirds an arid and semi-arid region to the south.

Facing the dual problem of increasing human needs and scarcity of natural resources, the central part of Tunisia, particularly the wadi Zioud site, has been subject to different investigations aiming at more rational land use. These investigations cover vegetation (Joffre, 1978), geology (Riaucourt, 1979), erosion and sedimentation (Delhoume, 1981; 1985, Ben Abdallah, 1983; Camus et al., 1988; Mansouri, 1991; Camus et al., 1990; 1992 and Laajili Ghezal et al., 1994), hydrology (Camus et al., 1982 and 1987), soil survey and mapping (Barbery and Delhoume, 1982). Based on these studies, the present paper tries to tackle land management so that soil and water resources could be used within the framework of sustainable agriculture.

Accordingly, a productivity index (PI) was developed. This is justified by the fact that relationships between soil properties and soil productivity are more and more needed, especially when these soils are affected by erosion. Different approaches which relate the properties of a soil to its productivity include: EPIC (Williams et al., 1982), ANSWERS (Beasly and Huggings, 1991), Soil Productivity Index (Neill, 1979; Kiniry, 1983; both in Gantzer and McCarty, 1987).

As defined in the Soil Survey Manual and reported by Pierce et al. (1983), soil productivity is the capacity of a soil for producing a specified plant or sequence of plants under a physically defined set of management practices. In this context, Neill's procedure which includes five soil parameters was used as such or modified for the purpose of quantifying the relationship between plant growth and soil attributes affected by soil erosion (Pierce et al., 1983; Frank and Zolman, 1985; Gantzer and McCarty, 1987).

The University of Guelph model for evaluating effects of Agricultural Management systems on Erosion and Sedimentation (GAMES) (Dickinson and Rudra, 1990), was applied to each parcel of land within the wadi Zioud watershed. The large variability observed in the output of the model was dealt with by mean of Principal Component Analysis (PCA). The limited number of clusters of cells having comparable characteristics thus obtained was analysed following a "Decision trees" procedure for the purpose of identifying technically more suitable management.

2 Materials and methods

2.1 Study area

The wadi Zioud watershed has a surface area of 774 ha. Its geological substratum is composed of marls and clays alternating with limestone outcrops of the lower and middle Cretaceous eras.

Rainfall is rather torrential and irregular: it fluctuates between a minimum of 127 mm.yr^{-1} and a maximum of 547 mm.yr^{-1}. Rainfall is relatively important in summer (18% of total rainfall) and stormy in this western part of the Dorsale as compared to the rest of the country (Baldy, 1965 in Delhoume, 1981). The mean minimum temperature of the coldest month (January) ranges between 1 and 3°C and the mean maximum temperature of the warmest month (July) is 34°C. The yearly evapotranspiration is about 2200 mm. Wind is rather erratic in terms of turbulence, gustiness and direction. The prevailing direction during the winter season is from the NW. The hot 'Sirocco' wind of the summer season is also fairly frequent. The accentuated topographic characteristics of the watershed, together with rainfall irregularity, explain the high variability of the Emberger quotient $Q = 2000P / (M^2-m^2)$, where P is the annual rainfall in mm, M the mean maxima of the warmest month and m is the mean minima of the coldest month. Thus climatic subdivisions based on Q values (LE Houerou, 1969) are high arid with temperate winters ($23 < Q \leq 35$), low semi-arid with temperate winters ($35 < Q \leq 45$) and high semi-arid with fresh winters ($45 < Q \leq 70$).

Two vegetative cover associations, one based on Pinus halepensis and Quercus Ilex (altitude greater than 1100 m) and the other on Pinus halepensis and Juniperus Oxycedrus at altitudes between 1000 and 1100 m correspond to the 'high semi-arid' subdivision. For the 'low semi-arid' subdivision, the Pinus halepensis and Juniperus oxycedrus association is observable down to 950 m altitude and that based on Pinus halepensis and Juniperus phoenicea down to 750 m. The Pinus halepensis and Genista microcephala (variety Capitellata) association prevails in the 'high arid' subdivision.

Most soils in the watershed are immature ones. Some of these cover eroded areas but most of them result from colluvium deposition on cracked flagstones or on marl layers. Calcimagnesimorphous soils on flagstones or altered limestone (Inceptisols) also appear in the watershed. Soils associations where denuded flagstone predominates are also a part of the soilscape of the watershed.

The present status of the watershed as a public nature reserve does not protect it sufficiently from human misuse essentially in terms of deforestation and overgrazing. Such practices have led to different advanced states of land degradation.

2.2 Productivity Index

Based on the fact that sustainable land use depends on soil characteristics, soil environment and land management practices, the productivity index used here is obtained by multiplying appropriate values of an organic matter content factor (O), a soil water-holding capacity factor (W), a soil effective depth factor (D), a soil textural class factor (T) and a slope factor (SP). The numerical values of the productivity index, like those of the five factors used in its computations, fall between 0 and 100:

$$(PI) = (SP) \times (D) \times (T) \times (W) \times (O) \tag{1}$$

where:
PI = Soil Productivity Index ($0 \leq PI \leq 100\%$);
SP = Slope parameter ($5 \leq SP \leq 100\%$);
D = Effective depth parameter ($10 \leq D \leq 100\%$);
T = Textural class parameter ($10 \leq T \leq 100\%$);
W = Soil water-holding capacity parameter ($10 \leq W \leq 100\%$);
O = Soil organic matter content parameter ($70 \leq O \leq 100\%$).

2.3 The GAMES Model

The Guelph model for evaluating the effects of Agricultural Management systems on Erosion and Sedimentation (GAMES) originally developed by Clark (1981, in Dickinson and Rudra, 1990), implemented and modified (Dickinson and Rudra, 1990), is used to describe and predict soil loss due to water erosion and subsequent delivery of transported solids from fields to streams in the wadi Zioud watershed. The discretization of a watershed into field-sized elements with homogenous characteristics of land use, soil type and class of slope was needed. The potential gross erosion for each cell was calculated by using the Universal Soil Loss Equation (Wischmeier and Smith, 1965):

$$E = R \times K \times LS \times C \times P \tag{2}$$

where:
E = computed soil loss per unit area, expressed in the units selected for K and the period selected for R;
R = the rainfall and runoff factor, expressed as the number of rainfall erosion index units;
K = the soil erodibility factor, expressed as the soil loss rate per erosion index unit for a specified soil;
LS = the topographic factor, expressed in terms of L, a slope-length factor, and S, a slope-steepness factor;
C = the cover and management factor which in turn can be related to land use;
P = the erosion control support practice factor.

The sediment delivered from each cell to the watershed's stream channels is determined by a calculated delivery ratio for each cell based on physical characteristics of the cell, i.e., roughness, slope and flow length. The general form of a "Micro Sediment Delivery Ratio" (Clark, 1981 in Dickinson and Rudra, 1990) is:

$$DR = f(V / L) \tag{3}$$

where:
DR = the Micro Delivery from a point on the watershed to the stream;
V = the average velocity of overland flow between the point and the stream;
L = the actual length of the surface drainage path from the point to the stream.

A simple formulation of this expression is:

$$DR = \alpha \times \theta \times [(1/n) \times (S^{1/2}) \times (Hc/L)]^{\beta} \qquad (4)$$

or

$$DR = \alpha \times \theta \times (t')^{\beta} \qquad (5)$$

where:
t' = inverse time component;
α = calibration parameter;
θ and β parameters dependent on individual cell characteristics;
Hc = hydrologic coefficient;
n = surface roughness coefficient;
L = length of cell overland flow path;
S = slope of land surface.

The total sediment load at the outlet of the watershed is calculated in GAMES by the expression:

$$SL = \sum_{i=1}^{N} Ai \times DRi$$

By minimizing the difference between the predicted total sediment load and the observed or measured total sediment load, the parameter alpha (α) can be calibrated for the watershed. The latter is obtained after having minimized the error of the following objective function Er:

$$Er = S_o - \sum_{i=1}^{N} A_i x D R_i \qquad (6)$$

where, Er is the error between observed and predicted downstream sediment loads for an assumed value of each of alpha (α), theta (θ) and beta (β). S_o is the observed sediment load attributed to field erosion immediately downstream of the watershed.

The gross erosion calculated, Ai - from the ith cell - is considered to be transported across each adjacent, connected downstream cell. Sediment deposition along the path from the (i+1)th cell down to the stream can take place. The amount of soil reaching the stream is dependent on the delivery ratio DRi calculated for the ith cell. In the stream, the DR is equal to 1. The sediment load SLi from the ith cell down to the stream is equal to:

$$SLi = Ai \times Dri \qquad (7)$$

and the total watershed sediment loading from all cells within the watershed is:

$$SL = \sum_{i=1}^{N} Si \qquad (8)$$

The watershed was divided into 20 and 33 land and stream cells respectively (Fig.1).

Soil and water conservation techniques, Tunisia   345

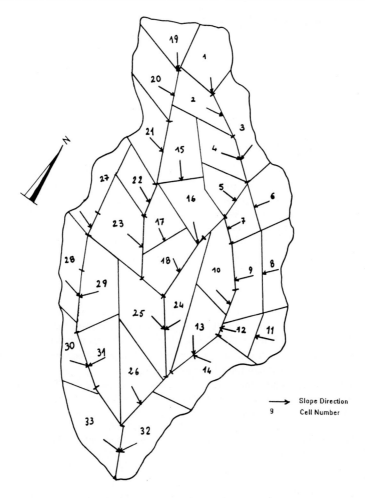

Fig. 1: Drainage network and composite overlay of the Wadi Zioud Watershed, scale 1:20 000. Adapted from Mansouri, R. (1991).

In feeding the model, the parameters K_i (Wischmeier and Smith, 1978 in Lal, 1988), C_i (Arnoldus, 1977; Masson, 1971) and n (Chow, 1959; Foster et al., 1980; both in Dickinson and Rudra, 1990) were used. Rainfall erosivity index was obtained from rainfall data of the Khanguet Zazia station, the closest to the watershed. Hydrological data for a 10 year period were used for the application of the model (Camus et al., 1982, 1987 and 1988; Ben Abdallah, 1983).

Information is fed to the model on yearly basis when applicable. Individual cell input files include: cell number, area, identification symbol (land or stream), length of flow path, slope of flow path, code of soil erodibility, code of cropping factor, number of subsequent cell, code of cropping practice and code of soil type.

The calibration parameter α is obtained, for each year, through the minimization of the difference between the predicted total sediment loads and the measured total sediment loads.

2.4 Principal component analysis (PCA)

The output of the GAMES model, i.e. erosion (E), sediment (SL) and topographic factor (LS), were analysed by PCA. This method permits, through transformations of the original variables called principal components, the representation of the data points into a lower dimensional space. This representation reveals the relationships which exist between the original variables as well as the clustering of the data points.

2.5 Decision tree approach

For a set of land units, specified by a number of parameters, a set of soil and water conservation managements or practices can be proposed. For suiting the appropriate management of an erosion-affected parcel of land, a decision tree procedure was followed. The latter, which takes into account soil-plant relationships, is based on slope, followed by outcrop of hard and/or marly substrata, soil depth and finally available water holding capacity.

The first decision criterion used is the slope with a threshold value of 10%. This allows the distinction between mechanical (S>10%) and biological (S≤10%) soil and water conservation management.

The percentage of outcrop (30%) enables the selection between earth and stone terraces, while the presence of marl substrata prevents the use of earth terraces.

The third criterion employed to decide about soil suitability for vegetation management (pasture, range management, field crops and fruit or forestry trees) is soil thickness. Based on the work of Barbery and Delhoume (1982) four classes of soil depth D were bilt: D≤12 cm (Pasture Management), 12<D≤23 cm (Pasture or Grazing or Range Management), 23<D≤40 cm (Field crops) and D>40 cm (Mountain Fruit Trees or Forestry Trees).

The last criterion used is soil available water holding capacity (AWHC). Calculation based on the product of soil water holding capacity and effective depth shows that AWHC ranges from less than 30 mm to more than 60 mm. Soil plots where AWHC is less than 30 mm are likely unable to sustain vegetation. Their management by the popular and efficient Tunisian soil and water conservation practice known as 'Meskat' or 'Jessour catchment' is recommended. The Meskat practice consists of maximizing runoff on area of low AWHC and directing the water thus harvested towards lower land areas (slope ≤10%) bounnded by earth terraces with a spillway. The other Tunisian soil and water conservation technique, known as 'Jessour' (Fig.2), harvests water by directed it towards micro-dams or sills built on appropriately chosen streams where the slope is greater than 15%.

Plots where AWHC is higher than 30 mm, shallow depth rooting plants (pasture), medium depth rooting plants (field crops) and even deep rooting plants where AWHC is around 60 mm could be considered essentially for soil and water conservation but also for biomass harvest and grazing where soil and topographic conditions are suitable. Finally, Jessours and low-lands of relatively good quality and suitable for the reception of upland harvested water could be considered for more intensive agriculture.

3 Results and discussion

3.1 Productivity index

Productivity indexes (PI) of thirty three plots, comprising the wadi Zioud catchment, were set against a scale of five productivity classes: class 1 (100-65), class 2 (64-35), class 3 (34-20), class

Fig. 2: Jessour systems in the south-east part of Tunisia.

4 (19-08) and class 5 (07-00) ranging from excellent to extremely poor (Anonymous). The results obtained showed that the whole catchment belongs to class 3 (average), 4 (poor) and 5 (extremely poor). This confirms field observations and soil analytical data (Barbery and Delhoume, 1982) showing that land of this area is average (6%), poor (27%) and essentially extremely poor or even useless (67%) as far as agriculture is concerned.

The PI as computed here involves just one point on the temporal axis. Therefore, erosion-caused changes in productivity with time are not taken into account. Owing to the low values of the PI's obtained, an estimation was needed of erosion within the watershed to plan for soil and water conservation management.

3.2 GAMES Model

The model was run in an analytical mode for a 10 year period. The calibration parameter, α, varied from 0.2 to 6.2 for the considered period (Table 1). The observed mean sediment load at the outlet of the watershed is 1913 ± 1696 (t.yr^{-1}). The predicted average gross erosion for the entire watershed is 17 ± 7 t.ha^{-1}.yr^{-1}, with a highest value of 29 t.ha^{-1}.yr^{-1} in 1989-90 and a lowest value of 7 t.ha^{-1}.yr^{-1} in 1983-84. The important variation in results could be attributed to the annual and seasonal variability and torrentiality of rainfall characterizing the Tunisian climate.

Year	α	So (t)	Ep (t.ha^{-1}.yr^{-1})	R
1977	3.6	2536	12	35.7
1978	0.3	274	11	41.0
1979	0.2	282	25	52.3
1981	3.6	3812	17	46.1
1982	1.9	2592	22	62.2
1983	3.5	1456	07	21.9
1984	1.4	812	09	27.5
1985	6.2	5478	14	43.7
1988	1.1	1144	17	41.0
1989	0.4	743	29	46.3
Mean	2.2	1913	17	41.8
Standard deviation	1.9	1696	07	11.6

α : Calibration parameter of the GAMES model;
So: Measured sediment load at the outlet of the watershed;
Ep: Predicted mean gross erosion;
R : Rainfall index of the USLE equation.

Table 1: Output data from the GAMES model for the 10 years period

The gross erosion obtained as output of the model showed that the eastern part of the watershed yielded a maximum amount of sediment downstream especially plots 8 (E=61 t.ha^{-1}.yr^{-1}, SL=8 t.yr^{-1}) and 14 (E=61 t.ha^{-1}.yr^{-1}, SL=11 t.yr^{-1}). Eroded plots are essentially used as pasture land except plots 7 and 9 which are cultivated with cereal.

3.3 Principal Component Analysis

The output of the GAMES model i.e., gross erosion (E), sediment load (SL) and topographic factor (LS) were analysed by the PCA procedure. The first two principal components explain 98.1% of the total variation between the plots. Figure 3 is a biplot of the data. The projections of the plot points onto the line connecting the origin and the E variable are good approximations of the values of the plots for that variable.

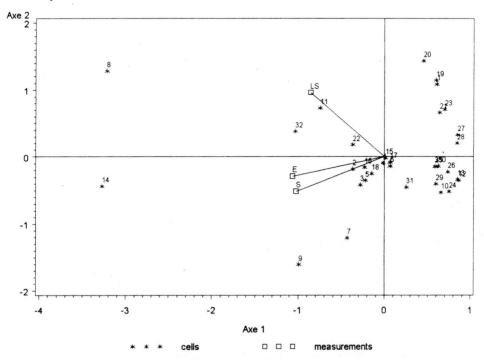

Fig. 3: Biplot of the first two principal components

There is high positive correlation between the three variables and especially E and SL. The 33 plots could be gathered into 6 clusters (groups), the characteristics of wich are summarized in Table 2 from best to worst (GA, GB, GC, GD, GE and GF) as far as soil erosion is concerned.

Group	E* (t.ha^{-1}.yr^{-1})	Percent of the watershed	S* (t.ha^{-1}.yr^{-1})	LS*	P*	L* %	LU (m)	K
GF	61	7	9.6	20.6	25.5	325	P	0.36-0.26
GE	31	6	3.5	13.1	19.5	325	PF	0.26
GD	31	5	6.3	4.7	10.0	365	C	0.41-0.31
GC	23	28	2.8	7.7	13.5	378	P,PF	0.26-0.36
GB	5	23	0.3	9.1	15.0	362	P	0.26-0.41
GA	12	31	1.3	4.3	9.4	361	PF	0.26-0.41

*: mean values

Table 2: Characteristics of the 6 clusters of the wadi Zioud watershed

For parcels number 8 and 14 (group F), both show comparable but highest values for the E, SL, LS, slope (S), land use (LU), cropping factor and Soil Conservation Service factor for soil type classification. The values of soil erodibility, K, although different, could be considered as low for one and average for the other. Slope length is important but this is also true for the others groups. However, its combination with high slope steepness values also explain differences between groups. With the exeption of the relatively high K values for parcels number 7 and 9 (group E), values for the rest of the parameters are lower than those of group F. Comparison of the latter group with group D shows that only slope steepness is higher and K values are lower.

The main cause of soil erosion is due to either grazing land (GF) or agricultural land (GE) within an arid region. In this respect GD compared in terms of gross erosion to GE - such as (GC, GB) - is still protected to some extent by remaining forest. This fact together with decreasing values of slope steepness, LS and E, explains the decreasing trand of soil erosion. Despite its relatively high LS and slope steepness, erosion of plots making group A yielded the lowest amount of soil loss. The reason is that vegetation cover (forest) in this area is still efficient.

The three first clusters are more likely to feed the watershed downstream with sediments and lose the greatest extent of land. Therefore, soil and water management and treatments of these plots should be a priority. Likewise, conservation and improvement of the remaining forest is necessary for soils of group B to maintain low soil losses.

3.4 Decision tree procedure

This procedure enables the association of specific soil and land water conservation techniques and land use to soil and soil environment characteristics. In the wadi Zioud watershed, three types of land uses capable of minimizing soil erosion clearly appear to suit the physical and biological conditions of the watershed and the users needs. Namely, these land uses are:

(a) relatively intensive agriculture including field crops and fruit trees in low land areas of AWHC higher than 60 mm and receiving 'Meskat' or 'Jessour' harvested water. Reforestation is also advisable in places of these same plots where erosion is more likely to take place and cause more soil degradation. Sixteen plots covering 47% of the total surface area of the watershed are likely to suit this type of land use.

(b) plots whose AWHC ranges from 30 to 60 mm have soil and soil environmental characteristics varying between poor and fair. Here, different land uses can be considered from grazing land and grazing forestry land, to field cropping land where agricultural conditions are best. Seventeen plots covering 41% of the total surface area of the watershed make this second class of land use.

(c) four microcatchments plots having 'Jessour' requirements and covering 12% of the total surface area of the watershed would be more appropriate to this kind of traditional soil and water conservation technique.

For soil and water conservation managements, the DTA shows that 36% of the total watershed could be treated by contour cultivation and vegetation management and practices corresponding to groups A and D. The total area of the watershed treated on earth terraces corresponds to 31% of the entire watershed. The remaining plots should be treated by stone terraces for groups B (14%), C (30%) and E (50%) and as microcatchments for groups B (15%) and C (20%).

Due to the fact that the parameters analyzed are physical nature without human impact taken into account, two major components should be included in future investigations for the purpose of improving the efficiency of the proposed management; user preferences and economic constraints. In fact, the rural population is the most concerned by management and the most involved in either implementing or inhibiting them. Thus, the opinion of the users is essential in the context being analyzed where management would be introduced on public property which, in practice and due to human pressure, is an almost free access domain. User opinion could either be obtained directly through surveys or observation via use of resources such as wood for fire, grazing, etc,.

Quantitative evaluation of population needs should be compared to resource availability after intervention. Economic constraints should be met for both public expenditures and farmer ability to participate in investment costs. If such considerations are accounted for, the process would not lead to an automatic choice. Rather, some compromises would be reached.

4 Conclusion

The use of a Productivity Index, GAMES model and a Decision Trees Approach to characterize soil degradation and to suggest technically more suitable management shows that the PI could be used as a land quality factor. Therefore, the quantification of gross erosion, sediment load and the characterization of slope steepness and length is necessary to optimize soil and water conservation measures and the GAMES model can be used for this purpose. The latter delineates critical sediment source areas within a watershed and can provide a means of evaluating soil conservation practices. However, management of parcels of land with relatively low topographic factors and sediment loads but with fairly high erosion (group A), needs more investigation.

The Decision Tree procedure based on land classification should be understood as an approach and needs to be better documented.

The application of such approaches necessitates specific investigations and assessments. However, we must note that the previous work was based on artificially-delineated land rather than naturally-bounded physical entities (e.g. soil series, geomorphologic units) or field farms, which necessitated manipulation of data.

Finally, due to the fact that the parameters analyzed are of a physical nature, two major components should be included in order to improve the efficiency of these measures; farmers preferences and economic constraints.

References

Arnoldus, H.M.J. (1977): Prediction des Pertes de Terre par Erosion en Nappe et en Griffe. In: Organisation des Nations Unies pour l'Alimentation et l'Agriculture (Editors), Aménagements des Bassins Versants. Cahier FAO: Conservation des Sols 1: 121-149.

Barbery, J. and Delhoume, J.P. (1982): Etudes en Milieu Méditerranéen Semi-aride- Carte Pedologique au 1:10 000 du Bassin Versant des Oueds Ed Dhiar et EZ Zioud (Djebel Semmama)- Tunisie Centrale. Division des Sols, Direction des Ressources en Eau et en Sol, Division des Sols, Ministère de l'Agriculture, République Tunisienne.

Beasley, D.B. and Huggings, L.F. (1991): ANSWERS User's Manual. Agricultural Engineering Dept, University of Georgia Coastal Plain Experiment Station Tifton, Ga. 31973-0748, USA.

Ben Abdallah, M. (1983): Recherches en Milieu Méditerranéen Semi-aride Djebel Semmama Tunisie Centrale (période 1980-1982). Direction des Ressources en Eau et en Sol, Ministère de l'Agriculture, République Tunisienne.

Camus, H., Dumas, R. and Ben Younes, M. (1982): Recherches en Milieu Semi-aride (période 1974-1977). Direction des Ressources en Eau et en Sol, Ministère de l'Agriculture, République Tunisienne.

Camus, H., R. Dumas, R. and Ben Younes, M. (1987): Analyse de l'Ecoulement sur le Bassin Versant de l'Oued El Hissiane (période 1977-1980). Direction des Ressources en Eau et en Sol, Ministère de l'Agriculture, République Tunisienne.

Camus, H., Dumas, R. and Ben Younes, M. (1988): Ecoulement et Erosion en Tunisie Centrale (période 1982-1986). Direction des Ressources en Eau et en Sol, Ministère de l'Agriculture, République Tunisienne.

Camus, H., Dumas, R. and Ben Younes, M. (1990): Ecoulement et Erosion sur le Bassin Versant de l'Oued El Hissiane (années 1987-1989). Direction Générale des Ressources en Eau, Ministère de l'Agriculture, République Tunisienne.

Camus, H., Ben Younes, M. and Chnina, M. (1992): Ecoulement et Erosion sur le Bassin Versant de l'Oued El Hissiane (Campagnes 1989-90 et 1990-91). Direction Générale des Ressources en Eau, Ministère de l'Agriculture, République Tunisienne.

Delhoume, J.P. (1981): Etudes en Milieu Méditerranéen Semi-aride- Ruissellement et Erosion en Zone Montagneuse de Tunisie Centrale (Djebel Semmama)- Résultats 1975 à 1979. Division des Sols, Direction des Ressources en Eau et en Sol, Division des Sols, Ministère de l'Agriculture, République Tunisienne.

Delhoume, J.P. (1985): Etude en Milieu Méditerranéen Semi-aride- Ruissellement et Erosion en Zone de Piedmont de Tunisie Centrale (Djebel Semmama)- Résultats 1976 à 1981. Direction des Sols, Ministère de l'Agriculture, République Tunisienne.

DGPDIA (1995): Etude sur la strategie des ressources naturelles. Volume 1: Bilan Stratégique des Ressources Naturelles. Ministère de l'Agriculture, République Tunisienne, SCET-Tunisie.

Dickinson, W.T. and Rudra, R.P. (1990): The Guelph Model for Evaluating Effects of Agricultural Management Systems on Erosion and Sedimentation, user's manual version 3.01. School of Engineering, Technical Report 126-86, University of Guelph.

Frank, R. R. and Zolman, M.G. (1985): Effect of Erosion on Soil Productivity-An International Comparison. J. Soil and Water Cons., 349-354.

Gantzer, C.J. and McCarty, T.R. (1987): Predicting Corn Yields on a Claypan Soil Using a Soil Productivity Index. Amer. Soc. of Agr. Engineers **30(5)** September-October.

Joffre, R. (1978): Notice des Cartes de la Vegetation du Bassin Versant de l'Oued El Hissiane (Centre Tunisien). Direction des Ressources en Eau et en Sol. Ministère de l'Agriculture. République Tunisienne.

Laajili Ghezal, L. (1988): Ruissellement et Erosion sur un Micro-bassin Versant: Modélisation et Simulation-Impact des Travaux de CES. Mémoire de fin d'Etudes du Cycle de Spécialisation Hydraulique et Aménagement Rural, Département de Génie Rural, Institut National Agronomique de Tunisie, Tunis, République Tunisienne, 174 pp.

Laajili Ghezal, L., Giraldez, J.V., Pontanier, R. and Camus, H. (1994): Estimation de l'Erosion en Zone Semi-Aride Tunisienne-Application à l'Aménagement Anti-érosif. In W.R.C., CIHEAM/IAM-B and IWRA (Editors), International Conference on Land and Water Resources Management in the Mediterranean Region : 519-533.

Lal, R. (1988): Erodibility and erosivity. In: R. Lal (Ed.), Soil Erosion Research Methods., USA, 141-160.

Le Houerou, H.N. (1969): La Végétation de la Tunisie Steppique. Annales Institut National de Recherches Agronomiques de Tunisie **42(5)**, 646 pp.

Mansouri, R. (1991): Validation d'un Modèle de Prediction du Ruissellement et des Sédiments dans le Bassin Versant de l'Oued Zioud en Tunisie Centrale. Mémoire de 3ème Cycle en Agronomie. Institut Agronomique et Vetérinaire Hassan II, Rabat, Royaume du Maroc, 152 pp.

Masson, J.M. (1971): L'Erosion des Sols par l'Eau en Climat Méditerranéen-Méthodes Expérimentales pour l'Etude des Quantités Erodées à l'Echelle du Champ. Thèse de Docteur-Ingénieur. Ecole Nationale Supérieure Agronomique de Nancy, Université des Sciences et Techniques du Languedoc, C.N.R.S.: AO 5445, République Française, 213 pp.

Pierce, F.J., Larson, W.E., Dowdy, R.H. and Graham, W.A.P. (1983): Productivity of Soils: Assessing Long-term Changes due to Erosion. J. Soil and Water Cons. **38**: 39-44.

Riaucourt, H. (1979): Recherches en Milieu Méditerranéen Semi-aride (Djebel Semamma- Tunisie Centrale)-Aperçu Géologique et Lithologique du Bassin Versant de l'Oued El Hissiane. Service Hydrologique, Division des Ressources en Eau, Direction des Ressources en Eau et en Sol, Ministère de l'Agriculture, République Tunisienne.

Williams, J., Dyck, P. and Jones, A. (1982): EPIC-a Model for Assessing the Effects of Erosion on Soil Productivity, User's manual.

Wischmeier, W.H. and Smith, D.D. (1965): Predicting Rainfall-Erosion Losses From Cropland East of the Rocky Mountains- Guide for the Selection of Practices For Soil and Water Conservation. Agricultural Handbook n°. 282.

Symbols and Abbreviations

Roman symbol	Description	Defining Eq. no. (if applicable)
A_i	Gross erosion from an individual cell	(6)
AWHC	Soil available water holding capacity	
C	Cover and management factor, in the USLE equation, related to land use	(2)
D	Effective depth parameter (10-D-100%)	(1)
DR	Micro Delivery from a point on the watershed to the stream	(3)
DR_i	Individual cell delivery ratio to the stream	(6)
DTA	Decision Trees Approach	
E	Computed soil loss per unit area, expressed in the units selected for K and the period selected for R	(2)
Er	Error between observed (measured) and predicted downstream sediment loads for an assumed value of α, θ and β	(6)
GA	Cluster of plots making group A using the PCA procedure	
GB	Cluster of plots making group B using the PCA procedure	
GC	Cluster of plots making group C using the PCA procedure	
GD	Cluster of plots making group D using the PCA procedure	
GE	Cluster of plots making group E using the PCA procedure	
GF	Cluster of plots making group F using the PCA procedure	
Hc	Hydrologic coefficient, an index of the probability and depth of overland flow	(4)
K	Soil erodibility factor, expressed as the soil loss rate per erosion index unit for a specified soil	(2)
L	Actual length of the surface drainage path from the point to the stream	(3)
LS	Topographic factor, expressed in terms of L, a slope-length factor, and S, a slope-steepness factor	(2)
n	Surface roughness coefficient	(4)
N	Total number of cells in the watershed	(6)
O	Soil organic matter content parameter (70-O-100%)	(1)
P	Erosion control support practice factor	(2)
PI	Soil Productivity Index (00- PI-100%)	(1)
R	Rainfall and runoff factor, expressed as the soil loss rate per erosion index units	(2)
S	Slope steepness of cell in percent	
SL	Total sediment load at the outlet of a watershed	(8)
SL_i	Sediment load from the ith cell	(7)
So	Observed sediment load, attributed to field erosion immediately downstream of the watershed	(6)
SP	Slope parameter (05-SP-100%)	(1)
T	Textural class parameter (10-T-100%)	(1)
t'	Inverse time component	(5)
V	Average velocity of overland flow between the point and the stream	(3)
W	Soil water-holding capacity parameter (10-W-100%)	(1)

Greek symbol	Description	Defining Eq. no (if applicable)
α	Calibration parameter	(4)
θ and β	Parameters dependent on individual cell characteristics	(4)

Addresses of authors:
L. Laajili Ghezal
T. Aloui
S. Zekri
Ecole Supérieure d'Agriculture de Mograne
1121 Mograne, Tunisie
M.A. Beji
Institut National Agronomique de Tunisie
43 Av. Charles Nicolle
1082 Tunis, Tunisie

Soil Erosion and Crop Productivity Research in South America

A. Tengberg, M. Stocking & S.C.F. Dechen

Summary

Progress in the quantification of the impact of erosion on crop productivity in South America within a major multinational initiative is reported. Eight different South American research groups are taking part in the experiments, and the sites encompass a broad range of soils from high exchangeable sodium soils to iron rich Oxisols and highly productive Mollisols.

Some of the highlight findings are erosion rates for different soil covers on different soils, the impact of erosion on *in situ* soil properties, erosion-yield relationships and nutrient losses following erosion. Combining erosion rates with erosion-yield relationships enabled the calculation of yield reductions with erosion over time and the 'half-life' of different soils. An important implication for land management is that the impact of erosion on yields is large for initial amounts of soil loss and progressively declines as the site becomes degraded.

Keywords: Soil erosion, crop productivity, soil nutrients, South America

1 Introduction

Erosion-induced loss in soil productivity is now recognised as one of the principal threats to agricultural sustainability (Pretty, 1995). Only if we can translate the dangers of erosion - and the benefits to conservation - into meaningful terms, do we have any possibility of assisting local people towards the goal of sustainability. This paper reports on progress in that direction in South America within a major multinational initiative promoted by the Food and Agriculture Organization (FAO) of the United Nations.

1.1 Experimental Design and Erosion-Productivity Network

Some fundamental gaps remain in our knowledge of erosion-induced loss in soil productivity. These were summarised in the report of FAO's Second Erosion-Productivity Network Workshop in Brazil in March 1996 (FAO, 1996):

a) *how much erosion will cause what level of change in soil properties and consequent decline in yields* - basic relationships are needed to quantify erosion impact;

b) *what are the measures of the specific impact of erosion which can be used by land managers and planners* - we need meaningful results on the financial impact of erosion and the benefit to society of soil conservation;

c) what is the nature of the impact process, or how can we predict the limiting factors of soil, crop, and climate that control the decline in productivity - crucial if any extrapolation of experimental results is to occur.

These are the same knowledge gaps identified more than a decade earlier (FAO, 1984), for which the FAO instituted a long-term cooperative research programme based around a standard design to collect consistent, comparative data. The original design (FAO, 1985) has been translated into Spanish (Stocking, 1985) and modified by two Brazilian researchers for South American conditions (da Veiga and Wildner, 1993). It has four principal objectives: (1) development of erosion-yield-time relationships; (2) identification of the variables which cause yield changes; (3) monitoring of possible explanatory variables; and (4) data acquisition on erosion impact. The recommended experimental procedure involves 27 soil loss and runoff plots, each about 50 m^2 in size. Plot treatments are as follows:

'N' PLOTS:- 12 plots are allowed to erode naturally. Plots are cleared of vegetation, and four different levels of erosion are achieved by covering the plots with different grades of artificial mesh. Each of the four treatments has three replicates.

'D' PLOTS:- 12 plots are artificially desurfaced to four different levels. The aim is to obtain the same range of responses to yield as will be obtained with the four 'N' treatments. This will usually mean that a greater depth of soil needs to be removed than erodes naturally. Each treatment has three replicates.

'C' PLOTS:- 3 remaining control plots are under continuous cropping or grass.

The experiment is designed to take four years. During the first two to three years, the N plots are allowed to erode until a range of levels of natural erosion has been achieved. Once this has occurred, the D plots are desurfaced, and all plots are planted with a standard crop. Differences in productivity between plots can then be assessed in relation to the amount of prior erosion. Throughout the experiment, seven groups of variables are recommended for monitoring: runoff and soil loss; physical and chemical characteristics of remaining soil; chemistry of eroded sediments and enrichment ratios; chemistry of runoff water; biological activity of the soil; climatic variables, principally rainfall and temperature; plant or crop measurements and indicators of growth stress.

Network research groups have autonomy to choose which variables will be monitored and which cropping systems to utilise in the comparative standard plantings. Groups which have published their results based on variations to the standard design include: Ethiopia (Tegene, 1992); Thailand (Krishnamra et al., 1990); and Kenya (Gachene, 1995). This paper presents an overview of the progress to date in South America where up to nine years of data are available at the longest-running sites and interest in erosion-productivity relationships is exceptionally high.

2 South American research sites

South American farming systems are characterised by diversity. From the highly productive commercial systems of Chile with *cero labranza* (no-till) through the agro-industrial production of the State of São Paulo and the small-farmer intensive systems using leguminous cover crops in Santa Catarina, Brazil, to the resource-poor subsistence farmers of Tarija in Bolivia, there is everywhere a growing realisation that soil degradation affects the economics of farming and, for some, it threatens the very livelihoods of large sections of the rural population. The imperative of sustainable agricultural production means that in this diversity of agroecologies we must develop the capability to quantify erosion-induced loss in productivity and predict the causative soil processes responsible. Table 1 lists those research groups which have accepted the challenge and joined with FAO in implementing, largely through their own resources, the erosion-productivity experimental design.

Country/State	Institution	Place & Mean Annual Rainfall	Soil Type, Slope & Crop	Status of Research
ARGENTINA	Instituto Nacional de Tecnologia Agropecuaria (INTA)	Estación Experimental Agropecuaria, Marcos Juáres, Córdoba, 885 mm	Mollisol (Luvic Phaeozem), 1%, soyabeans and wheat	est. 1993, soil loss and runoff data, yield data from desurfaced plots
BOLIVIA	Corporación Regional de Desarrollo de Tarija (CODETAR)	Coimata "Vivero Fruticola", 733 mm	Calcic Luvisol, 5%, maize	est. 1992, soil loss and runoff data
BRAZIL/ Paraná	Instituto Agronômico do Paraná (IAPAR)	IAPAR Station at Ponta Grossa, 1450 mm	Dystric Cambisol (Cambissolo álico), 12%, maize and soyabeans	est. 1994, soil loss and runoff data
BRAZIL/ Santa Catarina	Centro de Pesquisas para Pequenas Propriedades (CPPP), Empresa de Pesquisa Agropecuária e de Extensão Rural de Santa Catarina (EPAGRI)	(1) EPAGRI Station at Chapecó, 1850 mm (2) Colegio Agricola São José near Itapiranga, 1750 mm	(1) Rhodic Ferralsol, 16%, maize and soyabeans (2) Eutric Cambisol, 24%, maize	(1) est. 1989, data on soil loss, runoff, yield, nutrients in *in situ* soil, enrichment ratios (ERs) and crop nutrients (2) est. 1990, same as above but no ERs
BRAZIL/ São Paulo	Instituto Agronômico de Campinas (IAC)	(1) IAC, Campinas - natural erosion, 1410 mm (2) IAC, Mococa - desurfacing, 1300 mm	(1) Rhodic Ferralsol, 10%, maize (2) similar to (1)	est. 1987, data on soil loss, runoff, yield, nutrients in *in situ* soil, nutrient losses, ERs and crop nutrients
CHILE	Servicio Agricola y Ganadero (SAG), Ministry of Agriculture, Conceptión	Fundo Chequen, near Concepción 1112 mm	Orthic Luvisol, 17 %, maize	est. 1993, data on soil loss, runoff, and nutrient losses
COLOMBIA	Corporación Colombiana de Investigación Agropecuaria (CORPOICA)	CORPOICA Station at La Libertad, 2800-3000 mm	upper slope - Chromic Luvisol lower slope - Ferralic Cambisol, 4%, maize	est. 1995, no data available yet
PARAGUAY	Facultad de Ciencias Agrarias (FCA), Hohenau & Servicio de Extensión Agricola del MAG	FCA, Hohenau, 1700 mm	Chromic Luvisol, (Terra Roxa) 6%, maize	est. 1994, very limited amount of data on soil loss and runoff

Table 1: Research groups in South America participating in the network on erosion-induced loss in soil productivity.

The research sites encompass a broad range of soils from high exchangeable sodium soils on marginal, aridic conditions in Bolivia to the highly productive Mollisols of the Argentine *pampas* and the colluvial Inceptisols of SW Brazil. Reflecting their importance in South America, iron rich Oxisols (or Ferralsols) are well represented under the various soil classification systems. So, for example, the Chilean researchers have a baseline control of no-till while in Paraguay they compare results with maize-*Mucuna* intercropping widely planted by small farmers. All groups use greenhouse shade netting, locally called 'sombrite' to induce different levels of natural erosion. Eighteen per cent of sombrite gives about 40% mean effective cover, and 30% sombrite over 90% cover. Otherwise these on-station trials (except on Fundo Chequen, Chile, where the site is managed by a local farmer) follow closely the suggested design. Some have also instituted desurfacing comparison plots. Two sites, both in Brazil, have after nine years completed the experiment and the full results are in press and will be reported in due course.

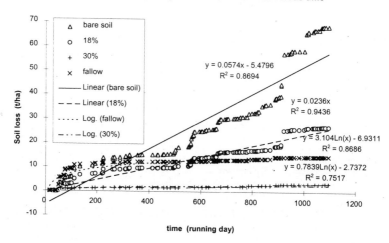

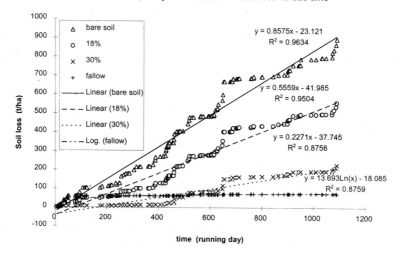

Figure 1: Soil loss-time relationships for a Cambisol and a Ferralsol, Santa Catarina State, Brazil.

3 Highlights of findings to date

3.1 Erosion-time relationships

Erosion rates for different soil covers are the first tangible results to emerge from the experiments and are now available for almost all the South American sites. Figure 1 depicts the relationships for a Ferralsol and a Cambisol in Santa Catarina State, Brazil, at two sites with an almost identical climate.

Soil loss (y) is linearly related to time (x) for non-vegetated plots (bare soil and different degrees of artificial cover) in the form:

$$y = Ax - B$$

where A represents the relative rate of erosion and B the time lag in the onset of erosion from the initiation of the experiment. On the fallow plots the best-fit relationship takes a logarithmic form.

At Chapecó cyclic oscillations around the mean erosion rate occur, which can be attributed to seasonal variations in rainfall amount and intensity. At Itapiranga, in contrast, there is a sudden inrease in erosion rate after approximately 2.5 years. Here, erosion has led to a progressive deterioration in soil surface structure which had originally been protected on this stony Cambisol and now generates substantially more runoff and sediment transport with a likely effect on soil productivity.

From the coefficients (A) in the equations in Fig. 1, erosion rates for the Ferralsol are much higher than for the Cambisol, despite the fact that the site at Itapiranga is steeper (24%) than at Chapecó (16%).

3.2 In situ soil changes

In Brazil, the Campinas group in São Paulo State has monitored erosion induced *in situ* soil changes on a Ferralsol for seven years. An increase in soil acidity and a decrease in organic C were the only changes that could be clearly attributed to erosion. In contrast, the Ferralsol at Chapecó, Santa Catarina State, showed significant declines in P, K and Ca+Mg, in addition to pH and organic C, and an increase in free Al after only three years of erosion. However, the Cambisol in Santa Catarina did not show any significant *in situ* soil changes after erosion. On the Mollisol in Argentina, desurfacing resulted in reductions in organic C, N and P. Initial indications are that there is a wide variety of important soil changes with erosion, depending on soil type, environment and cropping system. The most common, especially for iron-rich soils, appears to involve a complex chain reaction of reduced organic matter, increased leaching, soil acidity and consequent P-fixation and free aluminium - a dangerous combination but easily manageable (see last section).

3.3 Erosion-yield relationships

3.3.1 Natural erosion

To date, yield data from the naturally eroded plots is only available from Brazil (Chapecó, Itapiranga, Campinas). The best correlation between yield and soil loss is obtained when the data from the three sites are plotted together (Fig. 2). Two broad conclusions are drawn. First, the relationship between yield and soil loss is not site or soil specific at least under the conditions pertaining in Brazil. Instead, the farming system, which in this case is rain-fed cropping of maize with improved seeds, may be more important. Although, the soils that are included in the analysis - two Ferralsols and a Cambisol - do have some characteristics in common, such as high clay and Fe and low organic C, they react very differently in the erosion process. The Ferralsol at Chapecó has one of the highest measured erosion rates - nearly 300 tonnes per hectare per year. The Cambisol has less than a tenth of this rate. Secondly, of significant interest is the logarithmic form of the relationship, indicating that for these tropical soils there is an initial large decline in yields with the first five centimetres loss of topsoil. By contrast, further erosion has only a modest impact. This finding reinforces a conclusion from an earlier review of Alfisols (Stocking and Peake, 1986) which highlighted the utmost importance of protecting our good soils. Protection, or even attempts at rehabilitation, of severely degraded soils is costly and has very little potential yield benefit in the short term.

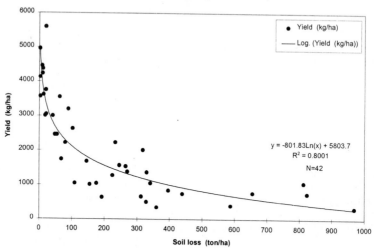

Figure 2: Erosion-yield relationships for rain-fed maize cropping, southern Brazil.

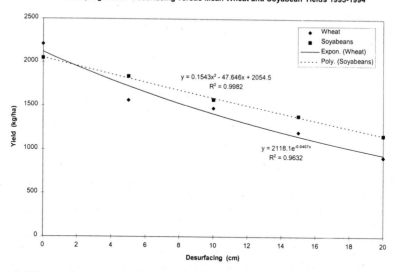

Figure 3: Wheat and soyabean yields (means for two years of cropping) versus simulated erosion on a Mollisol, Argentina.

When attempting to model erosion-yield-time relationships, site-specific factors such as soil, slope, cover and rainfall can be accounted for in the soil loss-time relationships (Fig. 1). Combining the soil loss-time relationships with the soil loss-yield relationship enables calculation of yield reductions with erosion over time. Thus, using the equations in Fig. 1 with the soil-loss yield relationship in Fig. 2, we can predict that without conservation, represented in the experiments by the bare soil treatment, the 'half-life' - that is, the time taken for yields to decrease to 50% of their original level on uneroded soil - of the Ferralsol is only one year and of the Cambisol three years. With good cover, represented by the 30% Sombrite, the half-life of the Ferralsol is three years and the Cambisol 50 years. If instead, we take a scenario for the time taken to reach a mere subsistence level of production of 1000 kg/ha, then the respective results are 1 and 19 years for no conservation and 3 and 332 years with good cover.

3.3.2 Simulated erosion.

Desurfacing experiments can also act as a means of transferring the erosion-productivity results to different farming and cropping systems. Figure 3 provides an example of the relationships found on a Mollisol by the Argentine group. It is evident that wheat is more sensitive to the impact of erosion than soyabeans. Similar findings have been reported from Chapecó, Brazil, where maize proved to be more sensitive to erosion than soyabeans. The experiment at Chapecó also showed the differential impact of erosion on different farming systems, with low input systems being far more sensitive to productivity losses.

3.3.3 Losses of nutrients in eroded sediment and runoff

The group in Chile has noted that considerably more nutrients are lost in the removed soil than in the runoff water. However, the losses seem to decrease with increasing cover. Similar observations have been made at Campinas, Brazil, where they note that this is particularly true for Ca and Mg, whereas there is little difference for the losses of K between sediment and runoff. Moreover, the groups at Chapecó and Campinas, Brazil, have reported decreases over time for nutrient enrichment ratios.

4 Implications for land management

It is too early to be definitive across all sites as to how land managers should respond to the challenges posed by erosion-induced loss in soil productivity. Nevertheless, certain consistent themes arise from our preliminary results which match findings elsewhere. In brief:

a) the impact of erosion on yields is large for initial amounts of soil loss and progressively becomes less as the site becomes degraded. The general exponential or logarithmic nature of the erosion-yield relationship is corroborated by our South American results to date. Protecting good quality soils must be a very high priority for food security;

b) erosion-yield-time relationships are variable between sites. Some soils have extremely high erosion rates but relatively low impact per tonne of soil loss (e.g. the Ferralsol at Chapecó); others present the opposite (the Cambisol at Itapiranga). Land managers will need to know not only the resilience of the site to soil erosion, but also the sensitivity of the soil to productivity loss;

c) soil changes with erosion indicate that no single soil variable explains fully the impact on soil productivity. Degradation, often led by decline in organic matter, results in a complex of soil physical and chemical changes including increasing acidity and declining plant-available water. In South America, the widespread use of no-till and cover crops is not just effective in reducing soil losses but also maintains a buffer of organic matter which, in turn, limits the chain reaction of declining productivity. Land managers have simple remedies against erosion-induced loss in soil productivity but the specific matching of technology to soil changes following erosion needs a knowledge of the causative variables.

and decision-making, the impact of erosion on farm economies and on national budgets needs to be assessed. If soil and water conservation techniques manage to prevent erosion this would minimise potential yield loss. Furthermore, the economic rationality of a farmer using a particular technique may be calculated by balancing the costs of land, labour and capital to install the measure against the benefits of the value of saved yield and any other incidental benefits such as reduced labour for weeding with a green cover crop. Already, the group in Argentina has measured an 80% yield reduction in wheat in 1993 with 20 cm artificial desurfacing. This amounts to a financial loss of US$190/ha - a total annual loss in farm income of more than US$600/ha. Such financial losses are clearly unsustainable. The corollary, however, is that investment in soil and water conservation should be well worth the effort and expense.

Acknowledgements

This paper is a contribution from FAO's Soil Conservation Programme, Land and Water Development Division, Rome, and the research reported here is partly funded by the Government of Japan. The authors thank FAO for its consistent support and are grateful to the Wenner-Gren Foundation for a post-doctoral Fellowship to the first author. Acknowledgement is also made to the research leaders at each site in Table 1.

References

da Veiga, M. and Wildner, L. do P. (1993): Manual para la Instalacion y Conduccion de Experimentos de Perdida de Suelos, Documento de Campo 1, Proyecto GCP/RLA/107/JPN, Organizacion de las Naciones Unidas Para La Agricultura y la Alimentacion, Santiago.

FAO (1984): Erosion and Soil Productivity: A Review. Consultants' Working Paper No. 1, Soil Conservation Programme, United Nations Food and Agriculture Organization, Rome.

FAO (1985): Erosion-Induced Loss in Soil Productivity, Consultants' Working Paper No. 2, Soil Conservation Programme, United Nations Food and Agriculture Organization, Rome.

FAO (1996): Erosion-Induced Loss in Soil Productivity. Second Workshop: Preparatory Papers and Country Report Analyses, Chapecó, Santa Catarina, Brazil, 3-7 March 1996, Food and Agriculture Organization of the United Nations.

Gachene, C.K.K. (1995): Effect of Soil Erosion on Soil Properties and Crop Response in Central Kenya, Reports and Dissertations No.22, Department of Soil Sciences, Swedish University of Agricultural Sciences, Uppsala.

Krishnamra, J., Somsopon, U. and Srikhajon, M. (1990): Erosion Induced Loss in Soil Productivity, Department of Land Development, Bangkok, Thailand.

Pretty, J.N. (1995): Regenerating Agriculture: Policies and Practice for Sustainability and Self Reliance, Earthscan, London.

Stocking, M.A. (1985): Perdida en la productividad del suelo a causa de la erosion: un disino de investigacion. Programa de Conservacion de Suelos, Servicio de Recursos, Manejo y Conservacion de Suelos, Direccion de Fomento de Tierras y Aguas, FAO, Roma.

Stocking, M. and Peake, L. (1986): Crop yield losses from the erosion of Alfisols, Tropical Agriculture (Trinidad) 63, 41-45.

Tegene, B. (1992): Effects of erosion on properties and productivity of eutric nitisols in Gununo area, southern Ethiopia. In: H. Hurni and K. Tato (eds.), Erosion, Conservation and Small-scale Farming. Geographica Bernensia, Berne, Switzerland, 229-242.

Addresses of authors:
Anna Tengberg*, Michael Stocking
Overseas Development Group, School of Development Studies, University of East Anglia
Norwich NR4 7TJ, U.K.
*current address: Göteborg University, Department of Earth Science, Section of Physical Geography
Box 460, SE-405-30 Göteborg, Sweden
Sonia C.F. Dechen
Instituto Agronômico de Campinas, Seção de Conservação do Solo, Av. Barão de Itapura 1481,
Caixa Postal 228, 13001-970 Campinas SP, Brazil

The Effect of Soil Erosion on Soil Productivity as Influenced by Tillage: With Special Reference to Clay and Organic Matter Losses

A. Moyo

Summary

Sheet erosion is still a major threat to agricultural production in Zimbabwe. To combat this type of erosion, conservation tillage systems (mulch ripping and tied ridging) have been developed and are being evaluated in their effectiveness to sustain agricultural production through maintenance of soil organic matter and clay content.

Run-off and soil loss as well as organic matter and clay losses differed significantly between treatments ($P<0.01$) with the two conservation tillage treatments having the least losses. Yields differed significantly only when the seasonal rainfall distribution and amount were poor (1994/95), but when rainfall distribution and amount were normal (1993/94) there was no significant difference between treatments.

The productivity of the soil is best maintained by mulch ripping and tied ridging through retention of clay and organic matter components, while these are lost in great quantities under conventional tillage and bare fallow.

Keywords: Conservation tillage, sheet erosion, soil productivity

1 Introduction

In Zimbabwe rill and gully erosion have been largely controlled through mechanical conservation structures (Elwell, 1984). Sheet erosion, however, is still a major threat to agricultural production (Braithwaite, 1976; Elwell, 1984). To combat this type of erosion, conservation tillage systems were developed and are still being evaluated. Their aim is to reduce run-off and sheet erosion in small-scale agriculture. Research work has, therefore, concentrated on evaluating tillage systems on their soil and water conservation merits without reference to their effect on the productivity of the soil. Thus productivity decline due to the associated losses of plant nutrients, organic matter and fine soil particles has not been a primary focus for research to date.

Clay and organic matter maintain and improve soil productivity (Stocking, 1983). In sandy soils these soil constituents promote a good aggregate structure, retard erosion, retain moisture and provide a suitable habitat for microfauna (Follet, et.al., 1987; Stocking & Peake, 1987). Their removal by erosion is coupled with the removal of plant nutrients and degradation of the soil. Reduced soil organic matter and clay content cause a drop in fertility, an increase in bulk density and result in a reduced storage capacity of plant available water, a fact particularly evident with the sandy soils in Zimbabwe's communal areas.

2 Materials and methods

2.1 Site

The research work was carried out in the semi-arid region of southern Zimbabwe which is characterised by erratic and unreliable rainfall both between and within seasons (Thompson, 1967; Anon, 1969). The average annual rainfall ranges between 450 and 650mm (Thompson and Purves, 1981). The soils are inherently infertile granitic sands of low pH (4-5), low organic matter (0,8%) and low clay (4-6%) contents (Thompson, 1967; Vogel, 1993). The rooting depth is shallow owing to the stone line occurring in the 50 to 80 cm depth (Vogel, 1993).

2.2 Tillage treatments

Four different tillage systems were evaluated:
- Conventional tillage (CT): the land is ox-ploughed to 25 cm using a single-furrow mould-board plough and thereafter harrowed with a spike harrow
- Mulch ripping (MR): crop residues from the previous season are left to cover the ground and only rip lines are opened between the mulch rows, 25 cm deep, using a ripper tine
- Tied ridging (TR): ridges are constructed at a 1% slope and are 0.9 m apart. Crossties (about two thirds of the ridge height) are built in the furrows at an interval of 1 m. The ridges are maintained for several years through re-ridging so as to maintain their correct size and shape
- Bare fallow (BF): ploughing is done using a tractor disc plough and disc harrow and left bare and weed-free throughout the season (Elwell & Norton 1988; Working document, 1990; Vogel, 1993).

All treatments were laid out at 4.5% slope. Tillage operations were done across the slope. Maize (*Zea mays* L.) was grown at a population of 36,000 plants/ha on all treatments except for BF.

2.3 Erosion measurements and soil sampling

Soil and run-off assessments were from 30 x 10 m erosion plots with 5 m buffer strips on either side and from 150 x 4,5 m with two guard rows above and below for the TR treatment. Surface run-off and soil loss were collected in 1500 litre conical tanks installed downslope of the plots.

Soil sampling on the trial plots was done at the end of each season, at a depth of 25 cm. The sediments were thoroughly mixed and a sample taken by driving a hollow plastic tube through the "sediment profile". The suspended load was pumped into plastic containers, left to stand for 3 days and the settled material at the bottom of the container was sampled. All soil samples were air dried and texture (using the hydrometer method) and organic matter content (using the Walkley and Black method) were determined.

3 Results and discussion

3.1 Soil loss and run-off

The effect of soil and water conservation through conservation tillage treatments was apparent in both years (Table 1). MR is most effective as it reduces both the raindrop impact on the soil as well as the run-off volume and velocity. It must be noted, however, that the effect of mulch is cumulative and becomes most noticeable after a number of years. TR was slightly less effective because it only reduces run-off and not raindrop impact. The overall difference of soil loss between treatments was highly significant at $P < 0.01$ for both years. When CT was contrasted with the mean of TR and MR, the difference was highly significant ($P < 0.01$). No significant difference was found between the two conservation tillage treatments, when tested both within the group and independently.

Tillage system	Run-off mm	Soil loss kg/ha	Yield t/ha
1993/94 seasonal rainfall = 483mm			
Bare fallow	123	81817	-
Conventional till	95	34303	2.4
Mulch rip	5	169	2.6
Tied ridge	16	1537	3.0
ANOVA	***	***	ns
1994/95 seasonal rainfall = 384mm			
Bare fallow	65	43525	-
Conventional till	49	6824	0.9
Mulch rip	4	83	2.2
Tied ridge	4	137	1.1
ANOVA	***	***	**

Significance level: *** < 0.01; ** < 0.025; * <0.05; ns = not significant

Table 1: Run-off, soil loss and yield for different tillage systems for two seasons

As expected run-off follows the same trend as soil losses, since the latter is a direct function of raindrop impact and run-off. For both seasons the run-off from MR treatment was maintained at less than 1% of total rainfall, ranged between 1 and 3% for TR, 13 and 20% for CT and 17 and 26% for the bare plot. The difference between the treatments overall was highly significant ($P < 0.01$) in both years. The difference between MR and TR was not significant.

The soil and run-off losses from the two conservation treatments are minimal. From the conservation point of view, both treatments appear to be sustainable. Elwell (1975) quotes soil losses less than 5t/ha, on sandy soils, to be tolerable in relation to the soil life span estimates. The very high soil losses with CT will eventually lead to productivity loss as the soil depth is limited due to the depth (50-80 cm) of the stone line. Furthermore, the soils become shallow, have less organic matter and clay contents thus reducing plant available water and nutrients. Plant nutrient losses can be replenished by the addition of fertilisers or manure, however, under rain-fed condition, losses in plant available water cannot be redressed. Therefore, physical properties (e.g. water holding capacity) altered by soil erosion, are the most longterm yield-limiting factors (Lowery & Larson, 1995).

3.2 Yield

There were no significant treatment differences in terms of yield for the 1993/94 season. In 1994/95, however, due to the very poor rainfall distribution that occurred, yields differed significantly (Fig. 1 and Table 1). Yields obtained under CT and the conservation treatments were significantly different at $P < 0.05$, while the two conservation treatments differed significantly at $P < 0.025$ (Table 1). Yield differences between TR and CT are relatively small compared to between TR and MR. Whilst the MR treatment could sustain the severe dry spells that occurred throughout the season, the crops under CT and TR were severely affected. The positive performance of MR is mainly due to the reduced evaporative losses as the ground is covered with mulch. Moyo and Hagmann (1994) found lower evaporative losses with MR as compared to TR and CT, which resulted in higher soil moisture contents throughout the season under MR.

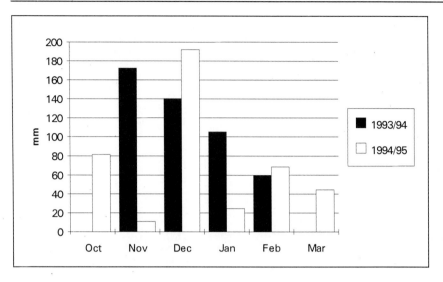

Figure 1: Rainfall pattern for two seasons, 1993/94 and 1994/95

3.3 Particle size distribution of the sediments

Erosion results in losses of topsoil and exposure of the subsoil. Because the Makoholi 5G soils have a greater clay content in the subsoil, particle size distribution of the soil after erosion (25 cm depth) is not a reliable indicator for sheet erosion. After the selective removal of clay particles in the uppermost topsoil a replenishment from the subsoil occurs. The treatments thus show a minimal change in the mechanical composition of the soil, compared to virgin land, with no significant differences between treatments.

The average mechanical composition of the sediments collected in the erosion studies is shown in Table 2 for each tillage treatment. The sediments consist of a higher clay (2.23 - 3.78 times) and silt (3.26 - 5.20 times) content as compared to the original soil. This finding confirms the hypothesis that erosion, especially on sandy soils, is a selective process that depletes the soil of its fine soil particles and leaves less productive coarse sand and gravel behind (Hudson, 1961; Follet et.al. 1987).

Treatment	Clay %	Silt %	Fine Sand %	Medium Sand %	Coarse Sand
Bare fallow	12	18	36	14	20
Conv. Till	10	18	33	17	22
Mulch rip	18	29	29	10	16
Tied ridge	15	25	24	15	21
Average	14	23	31	14	20

Table 2: Mechanical composition of sediments from 4 tillage systems

The conservation tillage systems have a higher percentage of clay and silt in their sediments, although the total amount of the clay/silt lost is only a fraction of the amount lost from conventional tillage systems (Tables 1 and 2). The difference in the amounts of clay lost from each treatment is highly significant ($P < 0.01$) because with conservation tillage treatments there is virtually no soil loss from fields except for the little that is held in suspension. With CT, however, the high run-offs carry greater loads of coarse materials from the field. Furthermore, the amounts of run-off from the conserved lands is minimal, thus also resulting in minimal losses of soil.

3.4 Organic matter loss with sediments

The original organic matter content of the virgin soils at the start of cultivation in 1988 averaged approximately 0.8%. After continuous cultivation for five years the organic matter content was found to have declined by 50% for the BF; 25% for CT; 5% for MR and 17.5% for TR. It is clear that with continuous cultivation the organic matter status of a soil decreases and to a more significant degree if no plant residues are left in the field, e.g. BF. For the CT, the little addition of organic matter through root mass leads to a somewhat better but still poor maintenance of organic matter. TR combines this effect with that of soil conservation to give even less losses. However, the best effect is achieved under MR, where roots together with plant residues and soil conservation effects contribute to a much improved maintenance of organic matter with only a 5% loss. The mineralisation of organic matter after cultivation normally occurs, however, with MR the mineralisation rate is slowed down due to lower soil temperatures and reduced cultivation.

Organic matter loss is a direct function of soil loss (Follet et.al. 1987), i.e. the higher the soil losses, the higher the organic matter losses. The same trend which was established for soil loss and run-off is true for the organic matter loss, also resulting in highly significant differences between treatments (at $P < 0.01$). CT did not differ significantly with BF but the mean of the two treatments differed significantly with the mean of TR and MR ($P < 0.01$). This shows that the two conservation tillage treatments are very effective in conserving and/or maintaining soil organic matter. As it is important to show which of the two conservation treatments performs better than the other, an independent t-test was carried out, i.e. disregarding the other two treatments altogether and comparing only the two conservation treatments. The results showed a significant difference at $P < 0.05$, in favour of the MR treatment. This shows that MR is by far the most effective treatment for sustaining productivity and its greatest advantage lies in the high infiltration rate. This leads, therefore, to less run-off and improved soil structure (Moyo and Hagmann, 1994).

The concentration of organic matter in sediments is higher for conservation tillage systems (3.53 - 5.02%) than for CT and BF (0.44 - 0.89). This is because most of the soil lost under the conservation tillage systems is in suspended form (Table 3) and the soil organic matter is generally associated with the finer and more reactive clay and silt fractions of the soil (Follet et al., 1987).

Treatment	OM content (%)		Soil loss (kg/ha)		OM loss (kg/ha)		Enrichm. ratios	
	Sludge	Susp	Sludge	Susp	Sludge	Susp	Sludge	Susp
BF	0.23	2.59	74535	7282	168	189	0.43	4.80
CT	0.35	3.27	27974	6329	99	207	0.51	4.81
MR	0.00	5.02	0	169	0	8	0.00	6.28
TR	0.42	4.56	380	1157	2	53	0.67	7.24

Table 3: Differences in the organic matter contents of sludge and suspended soil.

The proximity and concentration of soil organic matter near to the soil surface (< 25 cm), and its close association with plant nutrients in the soil, makes erosion of soil organic matter a strong indicator of overall plant nutrient losses resulting from erosion (Follet et.al., 1987). Thus the effectiveness of the two conservation tillage treatments is appreciated based on the insignificant levels of organic matter loss with eroded sediments, compared to the conventional treatments.

4 Conclusion

From this study MR and TR proved to be the most effective tillage techniques in reducing soil and water losses and therefore, organic matter and clay particle losses to negligible levels. While MR reduces evaporative losses, these are increased as a result of ridging (Moyo and Hagmann, 1994) thus resulting in lower yields by the later. The main factor affecting productivity (after seven years of cultivation) was found to be the plant available water more than fertility, as fertility loss is redressed through fertiliser application. This is why there were no significant yield differences between treatments when rainfall amount was high.

MR therefore, is the most viable technique for soil and water conservation as well as for sustainable production, while TR should be combined with mulch for better yields. The high soil and water losses under CT warrant a gradual change to conservation tillage techniques so as to maintain and improve soil productivity.

Acknowledgements

The author would like to sincerely thank Prof. Horst Mutscher, Dr. Horst Vogel, Mr. J. Hagmann, Mr. Mike Connolly for the editorial comments and Mr. Kennerd Masunda for helping with the laboratory analysis.

References

Anon. (1969): Guide to Makoholi Experiment Station. DR&SS, Salisbury.
Braithwaite, P.G. (1976): Conservation tillage - Planting systems. Rhod. Farmer **10**.Salisbury.
Elwell, H.A. (1975): Conservation implications of recently determined soil formation rates in Rhodesia. Sci. Forum, Salisbury.
Elwell, H.A. (1984): Sheet erosion from arable lands in Zimbabwe. Prediction and control. Harare Symposium pp 429-438.
Elwell, H.A. and Norton, A.J. (1988): No-till tied ridging. A recommended sustained crop production system. IAE, Harare.
Follet, R.H., Gupta, S.C. & Hunt, P.G. (1987): Conservation Practices: Relation to the management of plant nutrients for crop production. In SSSA special publication No. 19. Soil fertility and organic matter as critical components of production systems.
Hudson, N.W. (1961): An introduction to the mechanics of soil erosion under conditions of subtropical rainfall. Proceedings and Transactions of the Rhodesia Scientific Association **XLIX**, Part 1 15-25, Salisbury.
Lowery, B. & Larson, W.E. (1995): Erosion impact on soil productivity. SSSA **59** (3) 647-648.
Moyo, A. and Hagmann, J. (1994): Growth-effective rainfall in maize production under different tillage systems in semi-arid conditions and shallow granitic sands of southern Zimbabwe. In H.E. Jensen, P. Schonning, S.A. Mikkelsen and K.B. Madsen (eds.), ISTRO, Proceedings of the 13th International Conference **1** 475-480, Aalborg.
Stocking, M. (1983): Field and laboratory handbook for soils in Development: Manuals and Reports Series No. **17**. Norwich.
Stocking, M. and Peake, L. (1987): Erosion-induced loss in soil productivity: trends in research and international cooperation. In: I. Pla Sentis (ed.), Soil conservation and productivity: Proceedings IV International Conference on Soil Conservation, 399-438. Maracay.
Thompson, J.G. (1967): Report on the soils of the Makoholi Experiment Station. DR&SS, Salisbury.
Thompson, J.G. and Purves, W.D. (1981): A guide to the soils of Zimbabwe. Technical Handbook No. **3**. DR&SS, Harare.
Vogel, H. (1993): An evaluation of five systems for small holder agriculture in Zimbabwe. Der Tropen Landwirt **94**: 21-36.
Working document (1990): Working document for the on-station component of the project, Conservation Tillage for Sustainable crop production Systems (CONTILL), IAE, Harare.

Address of author:
A. Moyo, AGRITEX / GTZ, Contill Project, Box 790, Masvingo, Zimbabwe

Erosion Intensity and *Crotalaria juncea* Yield on a Southeast Brazilian Ultisol

A.A.C. Salviano, S.R. Vieira & G. Sparovek

Summary

Soil erosion reduces crop yields due to modification of its physical, chemical and biological properties as well as reduction of rooting depth. The purpose of this study was the assessment of the relationship between soil thickness and productivity of *Crotalaria juncea* on a naturally eroded area. The field experiment was carried out in 1994/5 in Piracicaba (Brazil). Soil erosion was evaluated through soil thickness, defined as the depth from the surface down to the C horizon over a grid of 50 x 70 m. Severe erosion, with the exposure to the surface of the C horizon, reduced yield by 65 % of the maximum. Soil thickness over 60 cm did not directly affect productivity, and thickness under 20 cm was strongly correlated to low or very low productivity.

Keywords: Soil erosion, crop productivity.

1 Introduction

Soil erosion may reduce crop yields due to modification of its physical, chemical and biological properties as well as reduction on rooting depth. Reductions in crop yields were found to be most closely associated with reduced soil depth, although this variable is closely related to many others which also change (Sessay & Stocking, 1995). In a five year experiment, Mokma & Sietz (1992) found that corn yields in a severely eroded soil phase were 21% less than on slightly eroded phase. Monreal et al. (1995) found, through linear function, that each Mg of soil loss reduces wheat yield to 0.89 kg ha^{-1}.

The relationship between erosion and productivity is not yet well understood. Each method, used in this kind of experiment, has inherent strength, weakness, and bias which can result in the measured soil productivity response attributed to erosion being potentially confounded with other variables such as landscape position, soil formation and management (Olsen et al., 1994). Despite spatial heterogeneity, due to those variables, application of fertilizer-nutrient inputs is usually done in an uniform manner (Cassman & Plant, 1992). The use of mean soil test to formulate fertilizer recommendation may result in over and/or under fertilization of large areas, reducing its efficiency, and increasing potential for contamination of surface and ground water (Mulla et al., 1992).

The purpose of this research was to study the relationship between soil erosion and productivity of *Crotalaria juncea* (Crotalaria), using regression analyses, Tukey's test and geostatistical tools for mapping on a naturally eroded area.

2 Materials and methods

The study was carried out in 1994 on a field located in Piracicaba (Brazil) that was used for sugar cane production for about 50 years. The last sugar cane ratoon was eliminated by disking and incorporated in the surface soil. The crop was *Crotalaria juncea* (Crotalaria), a green manure usually used before sugar cane replanting. Soil erosion damage (rill and interrill processes) caused by unsuited cover management is widespread in the experimental site.

The dominant soil is a highly erodible Ultisol on which an area (50 × 70 m) was sown at the rate of 30 kg.ha^{-1} and resown 15 days later. This area was divided in 140 sampling points according to a grid of 5 × 5 m. Soil erosion was evaluated through soil thickness (ST), defined as the depth from the surface down to the C horizon. The ST was determined by auguering in the center of each sampling site. After soil sampling, 4 Mg.ha^{-1} of lime were uniformly applied. At early flowering, plots of 5 m^2 of the aboveground part of the plants were collected in each sampling point for determining the dry matter (DM) production.

Means, median, standard deviation, coefficient of variation, maximum and minimum values, coefficients of skewness, kurtosis were calculated. The sampling points were grouped in relation to the soil thickness ranges of 0.00-0.20 m, 0.21-0.40 m, 0.41-0.60 m, 0.61-0.80 m, 0.81-1.00 m and 1.01-1.20m. The mid-value of yield from each group of soil thickness was plotted against the mean soil thickness and a quadratic equation was fitted to the data by regression analysis. The yield values from the thickness groups were submitted to analyses of variance and Tukey's test.

Semivariograms were used to examine the spatial dependence and kriging was applied to interpolate points of ST and DM for a grid of 1 × 1 m. For DM and ST, mapping and overlapping of the attribute data were grouped in 5 classes. For DM classes statistical limit criteria were used ($\overline{X} - \sigma$; $\overline{X} - \frac{1}{2}\sigma$; $\overline{X} + \frac{1}{2}\sigma$ and $\overline{X} + \sigma$ from the kriged data), and layer criterion were applied for ST classes. Classes of DM and ST were ranked from A to E as shown in Table 1, represented by maps and overlay matrices.

Classes	Classification	DM Mg.ha^{-1}	ST m
A	very high	> 11,33	> 0.80
B	high	9,25 - 11,33	0.60 - 0.80
C	middle	7,11 - 9,25	0.40 - 0.60
D	low	6,03 - 7,11	0.20 - 0.40
E	very low	≤ 6,03	≤ 0.20

Table 1: Area classes and respective value.

Variables	Number samples	Mean	Standard deviation	Maximum	Minimum	Kurtosis	Skewness	C.V.
ST (m)	140	0.77	0.31	1.20	0.0	2.67	-0.70	40.6
DM (Mg ha^{-1})	140	8.25	2.80	22.54	1.08	7.43	0.33	34.0

Table 2: Descriptive data statistics.

3 Results and discussion

DM was non-normally distributed while ST presented coefficients of skewness and kurtosis very close to normal function (Table 2); DM and ST exhibited a moderate coefficient of variation, 34% and 41%, respectively.

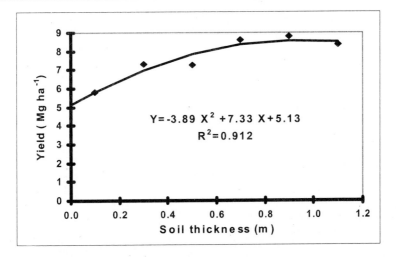

Figure 1- Effect of soil thickness on crotalaria yield.

Crotalaria yields were affected by reduction in soil thickness. Quadratic regression (Fig. 1) suggests that soil thickness only has a direct relation to DM production up to 0.60 m. Beneath this ST limit, water-holding capacity and rooting depth could be restrictive. Monreal et al. (1995) and Sessay & Stocking (1995) have also found a close relation between soil thickness and crop yield. For values of ST above 0.60 m, crotalaria yields were not directly affected by soil thickness which is probably most affected by soil erosion. Only the group with an ST of 0.00-0.20 m had DM yield significantly different from the others (Tukey 5%). In this case, the usual practice of fertilization may counterbalance the productivity losses due to limited soil thickness. Mokma & Sìetz (1992), working with corn, also found significant difference in crop productivity, but only under severe soil erosion. In our study in the lower soil thickness values (corresponding to high erosion values), crotalaria yields were 65 % of the highest yields, indicating a relatively low decrease in productivity. This may be related to lime application prior to seeding, and to the crops' capacity of symbiotic nitrogen fixation compensating for soil N deficiencies. Wani et al. (1994) have found that biological fixation of N by legumes can be used as the exclusive source of N for barley production on low fertility soil without sacrificing yield or soil quality.

Variables	Classes					
	0.00-.20m	0.21-0.40m	0.41-0.60m	0.61-0.80m	0.81-1.00m	1.01-1.20m
ST						
DM	5.787$_a$	7.135$_b$	7.135$_b$	8.594$_b$	8.797$_b$	8.353$_b$

Table 3: Crotalaria yield (Mg ha^{-1}) for soil thickness class.

Variables	Model.	Nugget	Sill	Range (m)
DM	spherical	1.53	7.50	27
ST	spherical	160	960	20

Table 4: Summary semivariogram results.

Semivariograms showed spatial dependence (Table 4). The range is the measure of the maximum distance over which properties remain spatially correlated (Mulla et al., 1993). The ranges of influence for ST and DM measured were 20 and 27 m, respectively. These semivariograms were best represented by the spherical model.

Classes	A	B	C	D	E
DM	12.0	25.4	33.2	11.3	18.1
ST	44.7	26.4	16.6	8.2	4.1

Table 5: Area percentage of class versus variables

Soil thickness

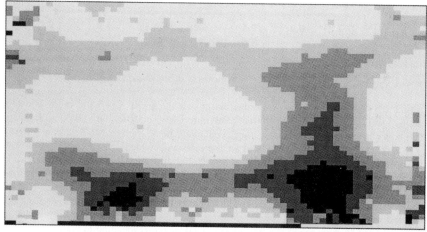

Crotalaria yield

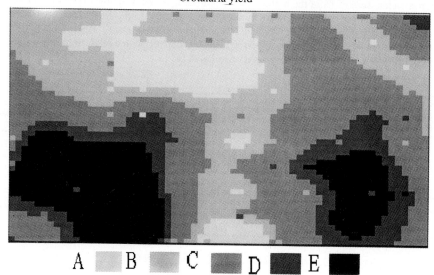

Figure 2: Maps of classes (area 50x70m) crotalaria yield and soil thickness.

On the maps the percentage of classes (Table 5 and Figure 2) had a variation of distribution from 4.1% for E class of the ST (<0.20m), representing linear erosion gullies of the area, to 44.7 % for A class of the ST (>0.80 m). DM exhibited a more homogeneous spatial variation. Mapping of these attributes is very important for land use planning allowing a more accurate recommendation of soil liming and fertilization. The use of the mean value from a soil test to formulate fertilizer recommendation may result in low efficiency of the applied fertilizers, increasing costs and increased risks of environmental contamination (Mulla et al., 1992).

			ST				
		class	A	B	C	D	E
DM	class	A	6.5*	5.2*	0.1	0.0	0.1
		B	14.5*	8.2*	2.4	0.3	0.1
		C	14.7*	7.3*	7.2*	3.1	1.0
		D	5.4*	2.0*	1.8	1.1	1.1
		E	3.6*	3.7*	5.2*	3.7*	1.9*

(*) Classes used to overlay maps.

Table 6: Overlay crotalaria yield with soil thickness.

The overlay of DM and ST is shown in Table 6 and Figure 3. The A class of DM (12 %) was mostly concentrated on classes A (6.5%) and B (5.2%) of ST indicating that 97% of the most productive area was associated to ST > 0.60 m. Conversely, E (1.9%) and D classes (1.0%) of DM was correlated to E class of ST (4.1%). 70% of the E class of ST was associated with areas of low and very low crotalaria yields.

4 Conclusions

- Severe erosion, with exposure to the surface of the C horizon, reduced crotalaria yield to 65 % of the maximum yields.
- Soil thickness over 60 cm did not directly affect crotalaria productivity.
- Soil thickness under 20 cm was strongly correlated to low or very low crotalaria productivity.
- Mapping the area may help in understanding soil erosion effects.

References

Cassman, K.G. and Plant, R.E. (1992): A model to predict crop response to applied fertilizer nutrients in heterogeneous fields. Fertilizer Research **31**: 151-163.
Mokma, D.L. and Sietz, M.A0. (1992): Effects of soil erosion on corn yields on Marlette soils in south-central Michigan. J. Soil and Water Cons. **47**: 325-327.
Monreal, C.M., Zentner, R.P. and Robertson, J.A. (1995): The influence of management on soil loss and yield of wheat in Chernozemic and Luvisolic soils. J. Can. Soil Sci. **75**: 567-574.
Mulla, D.J. (1993): Mapping and managing spatial patterns in soil fertility and crop yield. Procee. Soil Specific Crop Management. A workshop on research and development issues. Minneapolis, MIN. 15-26.
Mulla, D. J., Bhatti, A.U., Hammond, W.M. and Benson, J.A. (1992): A comparison of winter wheat yield and quality under uniform versus spatially variable fertilizer management. Agriculture, Ecosystems and Environment. **38**: 301-311.

Olson, K.R., Lal, R. and Norton, L.D. (1994): Evolution of methods to study soil erosion-productivity relationships. J. Soil and Water Cons. **49(6):** 586-590.

Sessay, M. F. and Stoking, M.A. (1995): Soil productivity and fertility maintenance of a degraded Oxisol in Sierra Leone. In: Sustainable Land Management in African Semi-arid and Subhumid regions, Dakar, Proceedings. Dakar, Senegal, 1995. P. 189-201.

Wani, S. P., McGill, W.B., Haugen-Kozyra, K.L., Robertson, J.A. and Thurston, J.J. (1994): Improved soil quality and barley yields with fababeans, manure, forages and crop rotation on a Gray Luvisol. J. Can. Soil Sci. **74:** 75-84.

Addresses of authors:
Adeodato Ari Cavalcante Salviano
Federal University of Piauí
64.049-550 Teresina, Brazil
Sidney Rosa Vieira
Agronomic Institute of Campinaas
13.001-970 Campinas, Brazil
Gerd Sparovek
University of São Paulo
13.418-900 Piracicaba, Brazil

Impact of Soil Erosion on Soil Productivity and Crop Yield in Tanzania

F.B.S. Kaihura, I.K. Kullaya, M. Kilasara, J.B. Aune, B.R. Singh & R. Lal

Summary

Major agricultural soils in three distinct eco-regions representing sub-humid/semi-arid, sub-humid and humid eco-regions were selected for studies on soil erosion and crop productivity relationships in Tanzania. Nutrient contents significantly decreased with increased surface erosion in most soils. The levels for pH, Organic carbon (Oc), total nitrogen (Tot-N) were higher in slight and least eroded classes for all seven soils. Phosphorus (P) was extremely low in all soils in the sub-humid and semi-arid/sub-humid eco-regions. Bulk density significantly increased with severity of erosion, and available water capacity significantly decreased with increased severity of erosion. Soil pH, texture and organic matter variously and significantly influenced crop yield. Maize and cowpea yield was significantly influenced by the extent of erosion. On average 64 kg/ha maize grain and 34 kg/ha cowpea grain was lost per cm topsoil loss for the studied soils. Cowpea was relatively less sensitive to changes in topsoil depth as compared to maize.

Keywords: Erosion, productivity, eco-regions, management, yield.

1 Introduction

Soil productivity is the capacity of a soil to produce a particular crop or sequence of crops under a specified management system (NSESPCRP 1981). It depends on factors such as soil management practices and inherent soil characteristics which include surface soil depth, plant nutrient availability, soil organic carbon content, available water capacity and rooting depth (McCormack et.al. 1982). Soil erosion depletes soil productivity, but the relationship between erosion and productivity is soil, crop and eco-region specific. Until the relationship is adequately developed, selecting management strategies to sustainable crop production is difficult (Lal 1987). Among the most important chemical and nutritional constraints aggravated by soil erosion are low cation exchange capacity (CEC) and low nutrient content (Lal 1988). Important among physical properties are bulk density, rooting depth and surface soil depth (Vogel 1994).

Research data on the impact of erosion on soil properties and its effect on crop yield is grossly missing in tropical Africa including Tanzania. Here a study was conducted on six major agricultural soils in three different agro-ecological zones in Tanzania to evaluate past erosion impact on soil physical and chemical properties and its effect on crop yield.

2 Materials and methods

Three locations, Sokoine University of Agriculture Farm (Morogoro), Agricultural Research Institute Mlingano (Tanga), and Agricultural Research Institute Lyamungu (Kilimanjaro), were selected to represent three district agro-ecological regions, namely sub-humid/semi-arid, sub-humid and humid eco-regions, respectively.

Agro-ecological zone characteristics (De Pauw, 1984) were used to characterize eco-regions in terms of geology, hydrology, climate, vegetation, present land use, predominant cropping system and length of growing season. Within each eco-region, two major agricultural soils with known history of land management were selected. Within each soil type evidence of past erosion as evidenced by changes in surface soil colour, surface stoniness and thickness of the Ap horizon were used to demarcate areas for detailed studies. For each soil type, mini-pits (30 cm x 60 cm) were excavated to 50 cm depth at 5 m grid points and described in terms of A horizon thickness, colour, consistence, and texture of the A horizon and the immediately underlying horizon. Impermeable layers at deeper depths were checked by augering. For each major soil, profiles were described following FAO (1988a) guidelines for soil profile description and classified using both the USDA system of soil classification (Soil Survey Staff 1992) and the FAO revised legend of the soil map of the world (FAO 1988b). Based on surface soil depth, erosion classes were established and plots identified for erosion productivity relationships studies in each erosion class.

Soil samples were collected, air dried and sieved to pass through a 2 mm sieve. Texture, pF at saturation, field capacity and permanent wilting point, infiltration rate and bulk density were determined following methods outlined in Klute (1986). Available water capacity was calculated as the difference between moisture content at field capacity and permanent wilting point. Soil pH, organic carbon, total nitrogen, available phosphorus, cation exchange capacity, exchangeable bases and base saturation were also determined according to methods outlined in Page et al. (1986). Maize (*Zea mays* var TMV-1) was used as a test crop during the long rains growing season while Cowpea (*Vigna unguiculata* var *Tegemeo*) was used as a test crop during the short rains. Relationships between soil properties and crop performance were evaluated by regression analysis techniques.

3 Results and discussion

Major soils studied included Nitisols in the humid eco-zone, Ferralsols and Lixisols in the sub-humid zone, Cambisols, Alfisols and Luvisols in the sub-humid/semi-arid zone. Nitisols in the Kilimanjaro eco-region had the deepest surface soil depth up to 40 cm, while Cambisols in the Morogoro eco-region were 9 cm and shallowest of all soils.

Deep horizons were rated as least eroded while very thin horizons were rated as severely eroded. Kaihura and Kaitaba (1995), reported an average surface soil thickness of 15 cm from 44 study sites under agricultural production in Tanzania.

Table 1 shows texture and chemical properties for composite surface (0-20 cm) and sub-surface (30-40 cm) soil samples from profiles of each major agricultural soil. The soils at all sites contained less clay in the upper 20 cm suggesting existence of preferential removal of the fine particles from the surface through erosion under the common farmer management practices. Soil pH was in the favourable range for high availability of most nutrients. Organic carbon and total nitrogen were high in the surface horizon and decreased suddenly with depth beyond 20 cm. With the exception of the Nitisols soils had extremely low levels of available phosphorus in both horizons. However, these low levels of available phosphorus at Mlingano and Morogoro did not influence maize growth. Phosphorus nutrition and dynamics in these soils need further investigations. Cation exchange capacity is low to medium for most soils while Ca is in the adequate range. Potassium levels are in the high range for most soils. The results in Table 1 indicate the need to restore surface soil productivity lost through removal of fine earth by erosion. Within each erosion class (1=severe, 2 = moderate, 3 = slight, 4= least) a trend of change in soil chemical properties with various degrees of erosion were examined. (Table 2).

Soil/Site*	sand	silt	clay	pH(H$_2$O)	O.C.	N	Av-P	CEC	Ca	Mg	K	Na	BS
	......%......			....%....	-mg/kg			Cmol/kg......%					
Humic Nilisol													
K.Boro	30	26	44	7.0	2.4	0.21	38	ND	14.0	2.9	3.9	0.5	ND
	22	22	56	6.6	0.8	0.11	29	ND	10.3	1.7	1.7	0.5	ND
Humic Nilisol													
XHelena	44	22	34	6.0	2.0	0.14	7.6	ND	6.7	1.6	4.3	0.2	ND
	36	22	42	6.3	1.6	0.13	14	ND	6.3	2.4	1.3	0.5	ND
Rhodic Ferralsol													
ML1	26	8	66	6.6	2.7	0.22	4.0	15.7	9.8	3.4	1.3	0.6	96
	24	4	72	6.5	0.9	0.07	1.0	12.4	3.5	1.7	0.9	0.4	52
Haplic Lixisol													
ML2	48	8	44	6.5	2.3	0.19	4.0	14.0	9.6	2.7	0.8	0.6	98
	34	5	61	6.4	0.5	0.07	4.0	8.5	4.4	2.5	0.8	0.4	96
Eutric Cambisol													
MS1	75	6	19	6.5	1.2	0.11	<1	19.5	8.0	3.1	1.1	0.3	64
	68	7	25	7.0	0.5	0.07	<1	25.7	20.0	3.3	0.4	0.3	93
Chromic Luvisol													
MS3	76	3	21	6.3	1.0	0.12	1.0	12.0	4.0	1.9	5.3	0.2	95
	67	3	30	6.6	0.6	0.07	<1	13.2	4.0	3.4	0.5	0.3	83
Haplic alisol													
Mindu	70	5	25	5.8	1.2	0.11	<1	9.6	4.0	2.7	1.2	0.1	83
	60	5	35	5.8	0.5	0.08	<1	11.5	2.0	2.9	0.6	0.3	50

*K.Boro = Kirima Boro, X.Helena = Xeno Helena, ML1= Mlingano 1, ML2 = Mlingano 2, MS1 = Misufini 1, MS3 = Misufini 3, ND = Not determined.

Tab. 1: Soil properties for composite surface and sub-surface soils from representative sites profiles.

Differences with changes in soil depth were significant at Mlingano and Morogoro (Misufini 1 and 3, Mindu) soils and not significant for Kilimanjaro (Kirima-Boro, Xeno-Helena) soils. Since soils in Kilimanjaro are volcanic, they are deeper and higher in inherent fertility and less sensitive to erosion effects in the short run as compared to the soils in the other two eco-regions

In most cases significant decreases in nutrient contents with increased surface erosion were observed. The levels for pH, Oc, Tot-N and available phosphorus were higher in slight and least eroded classes for all seven soils. The changes in soil chemical properties with severity of erosion are in agreement with those of Langdale et al. (1997) and Thomas et al. (1989). The results suggest that in these organic carbon, total nitrogen and to some extent CEC are the most negatively affected parameters by erosion. Phosphorus did not show any definite trend for Misufini 3 and Mindu sites probably due to the extremely low levels in these soils.

Bulk density was observed to increase significantly with severity of erosion at most sites (Tab. 3). Available water capacity significantly decreased with increased severity of erosion at all sites. The increase in bulk density with increased erosion at Mlingano sites follow the trend as reported by Frye et al. (1982). preferential removal of fine particles with erosion reduces soil fertility as well as an increase in bulk density due to increased compaction, less root penetration and lower microbiota populations. Erosion also reduces the available water capacity in the soil. Erosion therefore had a negative impact on soil productivity at all tested soils.

Site	Erosion Class	pH(H$_2$O)	OC %	Total N %	Av-P mg/kg	CEC Cmol/kg
Kirima B	1	5.2	2.0	0.18	26.8	25.3
	2	5.3	2.0	0.19	27.7	25.0
	3	5.3	2.2	0.20	38.4	32.9
	4	5.5	2.2	0.20	41.2	28.1
	LSD(0.05)	NS	NS	NS	6.35*	NS
Xeno.H	1	4.9	1.8	0.21	35	26.4
	2	4.9	1.8	0.21	34	26.7
	3	4.8	1.9	0.23	36	28.4
	LSD(0.05)	NS	NS	NS	NS	NS
Mlingano 1	1	6.0	1.9	0.18	1.26	14.75
	2	6.1	2.0	0.19	1.04	12.12
	3	6.4	2.3	0.20	1.69	15.02
	4	6.1	2.3	0.21	1.56	14.21
	LSD(0.05)	0.17*	0.03*	0.02*	0.35***	0.86*
Mlingano2	1	6.2	2.1	0.18	0.98	13.52
	2	6.2	2.2	0.18	2.34	14.02
	3	6.5	2.6	0.21	3.75	14.64
	LSD(0.05)	0.19**	0.17**	0.02**	1.02***	0.85*
MS 1	1	7.4	1.0	0.09	7.00	21.8
	2	7.3	1.5	0.12	11.00	21.3
	LSD(0.05)	NS	0.25**	0.03*	NS	NS
MS 3	2	6.0	1.3	0.11	3.00	17.3
	3	6.2	1.4	0.11	6.00	21.8
	4	5.9	1.5	0.11	3.00	22.4
	LSD(0.05)	0.58**	0.15*	NS	2.00*	3.43*
Mindu	1	6.8	1.2	0.18	2.00	20.4
	2	6.8	1.2	0.12	3.00	20.8
	3	6.6	1.4	0.14	3.00	23.3
	LSD(0.05)	NS	0.12*	NS	0.58*	NS

NS = Non significant, * = Significant at 5%, ** = Significant at 1%, *** = Significant at 0.1%.

Table 2: Impact of past erosion (measured by erosion classes) on selected chemical properties of the major agricultural soils in the three eco-regions.

4 Surface soil depth and crop yield.

Maize yield trend with changes in surface soil depth varied amongst eco-regions. Maize yield in the Morogoro eco-region was significantly influenced by the extent of erosion. Topsoil depth explained 57 to 81 % of the variation in maize yield at the Morogoro sites and was associated with 58 to 130 kg maize grain loss for every centimetre topsoil loss. Maize yield decreases with decreasing topsoil depth was also significant for Mlingano 2 and Kirima-Boro and was associated with 101 and 23 kg maize grain loss per centimetre eroded soil respectively. Maize loss at Mlingano 1 and Xeno-Helena were 9 and 54 kg per centimetre topsoil loss respectively but non significant. The low significance for the latter sites is probably due to other factors that influence maize yield besides topsoil depth. On average 64 kg of maize grain is lost per cm topsoil for the studied soils. Decrease in maize yield with decrease in topsoil depth was also reported by Vogel (1994). The effect of erosion on cowpea yield was also tested on each site. Changes in topsoil depth significantly reduced cowpea yield. Average cowpea grain loss was 34 kg/ha per centimetre topsoil loss. Table 4 summarises average grain yield losses for maize and cowpeas as a result of erosion.

Site	Erosion class	Sand %	Silt %	Clay %	BD Mg/cm*	AWC mm/m
KB	1	33	18	49	1.2	107
	2	31	19	50	1.2	124
	3	35	20	45	1.0	124
	4	38	18	44	1.1	129
	LSD(0.05)	2.7*	NS	2.7*	NS	NS
Xeno-Helena	1	51	14	35	0.97	118
	2	50	13	37	0.87	125
	3	54	14	32	0.97	128
	LSD(0.05)	NS	NS	NS	NS	NS
Mlingano1	1	54	8	38	1.39	ND
	2	57	9	34	1.29	ND
	3	53	11	36	1.30	ND
	4	57	11	32	1.21	ND
	LSD(0.05)	1.47***	1.37***	1.52***	0.06**	ND
Mlingano2	1	44	09	47	1.46	ND
	2	43	09	48	1.42	ND
	3	45	07	48	1.41	ND
	LSD(0.05)	NS	NS	1.8*	0.03*	ND
MS 1	1	52.5	16.1	31.4	1.47	103
	2	51.0	14.8	32.5	1.48	98
	LSD(0.05)	NS	1.18*	0.92**	NS	NS
MS 2	2	50.2	12.5	37.2	1.4	090
	3	53.8	12.4	33.7	1.42	122
	4	50.1	13.2	36.7	1.35	126
	LSD(0.05)	2.85**	NS	2.42**	0.06*	23.5*
Mindu	1	41.3	16.3	41.6	1.39	93
	2	41.5	11.8	36.1	1.41	93
	3	36.6	13.3	49.8	1.35	127
	LSD(0.05)	2.85**	2.32	4.2**	NS	20.5*

ND = Not determined, NS = Non significant, * = Significant at 5%, ** = Significant at 1%, *** = Significant at 0.1%, BD = Bulk density, AWC = Available water capacity.

Tab. 3: Impact of past erosion on selected physical properties for the seven soils in different eco-zones.

Site		Maize yield reduction per cm eroded soil	Cowpea yield reduction per cm roded soil	R^2	Level of significance
Kirima Boro		23	-	0.14	P<0.01
		-	7	0.09	P<0.05
Xeno Helena		54	-	0.06	NS
		-	29	0.17	P<0.05
Mlingano	1	9	-	0.01	NS
Mlingano	2	101	-	0.23	P<0.001
		-	66	0.13	P<0.05
Misufini	1	75	-	0.57	P<0.001
Misufini	3	130	-	0.68	P<0.001
Mindu		58	-	0.81	P<0.001
Av. Yield loss		64	34		

Table 4: Maize and cowpea grain loss (kg/ha) per centimetre topsoil loss for the tested major agricultural soils in Tanzania

The results, however, indicate that cowpea is less sensitive to changes in topsoil depth than maize. Cowpea is known to be a more tolerant crop to drought than maize. Since decrease in surface soil depth is associated with decrease in available water capacity, low sensitivity for cowpea to changes in surface soil depth could be accounted for by its drought tolerance.

5 Soil properties and crop yield

Influence of soil properties on maize yield was assessed by stepwise multiple regression analysis. Erosion effects on soil properties variously affected maize yield at different sites. In the sub-humid/semi-arid eco-region, surface soil depth, soil pH and soil texture accounted for 69% of the variation in maize yield across sites. Soil pH and available water capacity explained 82% of the variation in yield at Misufini 3. Topsoil depth alone explained 80% of the variation in maize yield at the Mindu site. Decrease in pH negatively affected phosphorus availability at Misufini 1. In the sub-humid eco-region, texture and bulk density were negatively correlated with maize yield at Mlingano 1 and explained 44% of maize yield while topsoil depth organic carbon and nitrogen explained 46% of the variation in yield at Mlingano 2. Effects of soil properties on maize yield were various but non significant for soils of Kilimanjaro eco-region. The results indicate that the soils in the sub-humid/semi-arid eco-region are most sensitive to erosion, Nitisols in the humid eco-erogion being least sensitive.

6 Conclusions

Studies on the impact of past soil erosion on soil productivity were carried out. The study was conducted with the assumption that variations in surface soil depth are mainly due to different magnitudes of past in-situ soil erosion. The methodology used ensured that the soils are of the same parent material, same soil series on the same landscape units. Observed differences in erosion impacts could be attributed to differences in soils, management practices and diferences in micro-relief.

Decrease in surface soil depth significantly decreased soil pH, organic carbon and total nitrogen for the most eroded soil classes at all sites. Available phosphorus was very low except for Nitisols but did not affect crop yield.

Decrease in surface soil depth significantly increased soil bulk density and reduced available water holding capacity respectively. Erosion effects on soil properties had a direct and negative impact on crop yield. Cowpea was more tolerant to erosion effects than maize.

The results suggest that:
* Soil erosion has economic implications from farm to national level. The impact is more serious under minimum/low input peasant agriculture.
* Economic implications may be used as a guide to establish policies on sustained use of the land for present and future generations.
* Management interventions for soil productivity enhancement are soil and eco-region specific.
* Phosphorus availability in the studied soils needs further investigations.

References:

De Pauw (1984): Soils, physiography an agro-ecological zones of Tanzania. Ministry of Agriculture, Dar es salaam. Food and Agriculture Organisation of the United Nations. Rome.

FAO (1988a): Guidelines for soil profile description. Third edition. Food and Agriculture Organization of the United Nations. Rome.

FAO (1988b): FAO/UNESCO Soil Map of the World, Revised Legend. World Resources Report 60, FAO, Rome. Reprinted as Technical Paper 20, ISRIC, Wageningen, 1989.

Frye, W.W. Ebelhar, S.A., Murdock, L.M. and Blevins, R.L. (1982): Soil erosion effects on properties and productivity of two Kentucky soils. Soil Sci. Soc. Am. J. 1051-1055.

Kaihura, F.B.S. and Kaitaba, E. (1995): Low input land management technologies and their impact on surface soil fertility. Southern Africa Development Cooperation-Land and Water Management Research Programme Annual Scientific Conference, Lusaka, Zambia. 16-20 October 1995.

Klute, A. (ed). (1986): Methods of soil analysis Part 1. Physical and mineralogical methods. Second Edition. ASA, SSSA, Madison, W1.

Lal, R. (1987): Effects of soil erosion on crop productivity. CRC Critical reviews in Plant Sciences. **5(4):** 303-367. CRC Press, Inc. Boca Raton, Florida.

Lal, R. (1988): Monitoring soil erosion's impact on crop productivity. In: R. Lal. (ed.), Soil erosion research methods. Soil and Water Conservation Society, Iowa, USA, p. 187-200.

Langdale, G.W., Box, J.E. jr., Leonard, R.A., Barnett, A.P. and Fleming, W.G. (1979): Corn yield reduction on eroded southern piedmont soils. J. Soil and Water Conservation **34**: 226-228.

McCormack, D.E., Young, K.K. and Kimberlin, L.W. (1982), Current criteria for determining soil loss tolerance. ASA special publ. No. 45. Soil SC. Soc. Am. Madison, Wisconsin.

National Soil Erosion-Soil Productivity Research Planning (NSESPRP) (1981): Soil erosion effects on soil productivity: A perspective. J. Soil and Water Conservation **36** (2): 82-90.

Page, A.L., Miller, R.H. and Keeney, D.R. (eds) (1986): Methods of soil analysis Part 2. Chemical and microbiological properties. Second Edition, ASA, SSSA. Madison, Wisconsin, W1.

Soil Survey Staff, (1992): Keys to soil Taxonomy, 5th edition, SMSS technical monograph No. **19**. Blacksburg, Virginia, Pocahontas Press, Inc, 556 pp.

Thomas, P.J., Simpson. T.W. and Baker, J.C. (1989): Erosion effects on productivity of Cecil soils in the Virginia piedmont. Soil Sci. Soc. Am. J. **53**: 928-933.

Vogel, H. (1994): Conservation tillage in Zimbabwe, Evaluation of several techniques for the development of sustainable crop production systems in smallholder farming. African studies series A **11**, Berne, 150, pp.

Addresses of authors:
F.B.S. Kaihura
Agricultural Research Institute Ukiriguru
P.O. Box 1433
Mwanza, Tanzania
I.K. Kullaya
Agricultural Research Institute Lyamungu
P.O. Box 3004
Moshi, Tanzania
J.B. Aune
Centre for International Environment and Development Studies - Noragric
P.O. Box 5001
1432 Aas, Norway
B.R. Singh
Department of Soil Science
NLH
1432 Aas, Norway
R. Lal
School of Natural Resources
The Ohio State University
210 Kottman Hall 2021 Coffey Road
Columbus, Ohio 43210, USA

Soil Loss and Soil Conservation Measures on Steep Sloping Orchards

Chia-Chun Wu & A-Bih Wang

Summary

Heavy and concentrated rainfall, brittle rocks, unstable stratum, steep terrain, intricate land usage and management, as well as diversified cropping system makes soil erosion problems in Taiwan unique. In order to deal with soil loss and land degradation problems, Taiwan's first pilot soil conservation demonstration project was established in 1952. However, the increasing pressure on land resources makes the soil loss assessment on steep slopes more important than ever.

On-site measurements on soil losses from 60%-steepness orchards were monitored for four consecutive years. The bare plot had the highest soil loss with an average of 365.2 tons/ha. Vegetation, grass strip, surface mulch, and hillside ditches were found to be effective on steep slopes. These conservation measures can greatly reduce soil losses and conserve surface runoff.

By feeding the Universal Soil Loss Equation (USLE) with a local rainfall erosivity index and an estimated soil erodibility index, the estimated annual soil loss from a 60%-steepness bare slope was found to be 73 times higher than the actual average field measurements. Slope length and slope steep-ness factors are the variables most likely to over-estimate soil loss.

Keywords: Soil loss, soil conservation, sloping orchards, Taiwan

1 Introduction

The use of slopelands for agriculture has a very long history in Taiwan, starting from rice paddies and fruit trees to teas, vegetables, and betel nuts. The migration of agriculture from lowlands to highlands is mainly due to an increase in population and expansion of urban and industrial areas. Agricultural production may partially fulfill the demands of the markets. However, the loss of fertile top-soils and off-site environmental impacts caused by soil erosion are major problems that Taiwan is trying to resolve.

Taiwan is a mountainous island with 74% classified as slopelands and mountains (CAPD, 1980). Sedimentary, volcanic, and metamorphic rocks form the mountain ranges. Rainfall in Taiwan is abundant yet concentrated. The average annual rainfall approaches 2,430 mm, while about 78% of the annual precipitation falls in the summer from May to October especially during the typhoon season. Hence, steep terrain, brittle rocks, unstable stratum, and concentrated rainfall make the island of Taiwan less sustainable due to erosion.

In order to better manage the land resources, the Taiwan government established its first soil conservation demonstration project in 1952. A number of soil conservation practices have been developed to suit the local needs. They include hillside ditches, bench terraces, vegetative barriers, grassed waterways, cover crop, and mulch (Liao, 1980; Lee, 1984). However, experience in soil conservation for agricultural lands were mainly in the scope of mild and flat slopes. Therefore, the objectives of the study were (1) to measure the soil losses from 60% steepness orchards, and (2) to assess the applicability of the selected conservation practices on steep sloping orchards.

2 Experiments and measurements

2.1 Site description and experimental setup

The outdoor experimental site is located in the Campus of the National Pingtung University of Science and Technology, Taiwan. The entire site occupies 0.20 ha with an average slope of 60% (31°). The soil is clas-sified as sandy clay loam based upon United State Department of Agriculture (USDA) standards. This experimental site had been used for growing mango before it was converted into a research site. All the mango trees and ground vegetation were removed at the beginning of site preparation. Minor grading was applied to the slope to achieve uniformity.

Seven small plots were installed on the experimental site. Each plot had a slope length of 22.1 m, a width of 4 m across the plot, and a slope steepness of 60%. Brick walls were used to partition the plots, and a set of stilling basins was constructed at the downslope end of each plot to collect surface runoff and eroded soil.

Photo 1: Layout of the experiment site.

Several conservation measures commonly used in Taiwan were followed in this study. They include Bahia grass (*Paspalum notatum Flugge*) overall vegetation in litchi orchard (treatment A), bare check plot (treatment B), Bahia grass overall vegetation (treatment C), Bahia grass overall vegetation plus the hillside ditch in litchi orchard (treatment D), clean-cut litchi orchard (treatment E), Bahia grass strips plus grass residue mulch between strips in litchi orchard (treatment F), and clean-cut betel nut plantation. The entire experimental site is illustrated in Photo 1, in which all treatments were arranged sequentially from left to right with treatment A on the far-left end of the Photo.

At the end of a rainy season, exposed soil was re-plowed to a depth of 30 cm to break up the possible surface crusts so that possible surface effects left over from the previous rainy season would not be carried over in the next rainy season. Air-dried soil stored in the warehouse was added to the re-plowed plots. For those plots receiving fresh soils, manual grading was applied to achieve a uniform slope.

2.2 Measurements:

Measurements conducted in this study include daily precipitation, surface runoff, and soil loss. Daily precipitation was measured in 15-minute intervals using a 0.5-mm tipping bucket rain gauge. A data logger was attached to the rain gauge to record the measurements.

Surface runoff and soil loss were measured on a single storm basis. During a storm event, surface runoff and eroded soil were collected in the stilling basin as the lump sample. At the end of the storm event or the next day, total volume of surface runoff was first measured. For the measurement of soil loss, three small samples of suspended and deposited sediment with known volumes were taken, respectively, from the stilling basin. Volumetric method was used to calculate the total dried soil loss for each treatment.

To prevent the stilling basin from overflowing, a set of triangle weirs was used to split the excess surface runoff into two uneven portions. Only one-sixth of the excess runoff was kept in the spare tank, whereas the rest of the excess runoff was discarded.

Obser-vation Year	Effective Precip. (mm)	Treatment						
		A	B	C	D	E	F	G
		Total soil loss (kg/m^2)						
1992	1360	0.00	28.80	0.00	0.00	22.11	0.08	20.46
1993	765	0.00	25.89	0.00	0.00	24.89	0.00	28.82
1994	2413	0.00	44.17	0.01	0.01	25.88	0.00	22.92
1995	1260	0.01	36.65	0.01	0.01	19.74	0.00	15.95
Average		0.00	36.52	0.01	0.01	23.53	0.02	21.61
		Total surface runoff (m^3/m^2)						
1992	1360	0.000	0.072	0.000	0.004	0.063	0.004	0.074
1993	765	0.000	0.125	0.000	0.002	0.059	0.000	0.189
1994	2413	0.002	0.448	0.004	0.009	0.227	0.002	0.513
1995	1260	0.003	0.166	0.007	0.007	0.090	0.002	0.126
Average		0.002	0.256	0.003	0.007	0.137	0.002	0.283

Remarks: Treatment A - Litchi + Bahia grass vegetation
Treatment B - Bare check plot
Treatment C - Bahia grass vegetation
Treatment D - Litchi + Hillside ditch + Bahia grass vegetation
Treatment E - Clean-cut Litchi orchard
Treatment F - Litchi + Bahia grass strips + Grass residue mulch
Treatment G - Clean-cut betel nut plantation

Table 1 - Summary of annual soil loss and surface runoff

3 Results and discussions

Only annual soil loss and surface runoff from four consecutive years are summarized in Table 1. The effective precipitation in Table 1 refers to the total amount of precipitation that actually caused soil erosion. The average soil loss was calculated by weighting the effective precipitation to the soil loss in the same year, adding all the weighted soil loss together, and dividing by the total effective precipitation. The average surface runoff was calculated in the same manner.

The field observations indicate that treatments A, C, and F produce the least amount of surface runoff. The reasons are mainly due to, (1) the presence of grass roots that serve as additional paths for surface runoff to infiltrate, (2) the absorption capability that grass residue possesses, and (3) the additional depression that grass strips provide. On the basis of the laboratory experiments, air-dried grass residue has the capability to absorb as much as 4.8 times its own weight of water. This trapped water will be gradually released over a period of seven days (Wu et. al., 1995). Hillside ditch (treatment D) is equally effective in controlling soil loss and conserving surface runoff. However, the amount of surface runoff produced from treatment D is a little bit greater than those from treatments A, C, and F. This was mainly due to the experimental design. A PVC pipe was installed in the hillside ditch to convey the surface runoff generated from the upper slope to the stilling tank. Because of the lack of storage capability, PVC pipe cannot retain any surface runoff.

From the view point of cost effectiveness, a combination of grass strip and residue mulch is probably the best choice for conserving soil and water resources on steep sloping orchards. The grass strip not only protects soil from erosion, but also serves as a barrier to trap the eroded sediment. Residue mulch applied between grass strips has multiple functions. They include, (1) protecting soil from erosion, (2) conserving surface runoff for crop intake, (3) becoming part of the organic resources as they decompose, and (4) exerting least competition in water and nutrient consumption between main crops and themselves.

The hillside ditch has its own advantages and disadvantages. It has been proven effective in soil and water conservation. It also serves as a linkage road for farming operation. However, excavation and backfill processes are needed in the installation of hillside ditches. These processes not only raise the costs, but also disturb the top soil. Hence, a hillside ditch may not achieve its full potential in erosion control before the disturbed soil becomes stable.

The soil loss from treatment G (clean-cut betel nut) is slightly less than that from treatment E (clean-cut litchi). However, the surface runoff is in the reverse order. The main reason is due to crop height and the characteristics of the crop canopy. The average crop height for a 7-year-old betel nut tree can reach 4 m, whereas the average crop height for a 7-year-old litchi tree with two trims annually can only reach 2 m. Hence, the average fall height of the intercepted raindrops is less for litchi trees. In addition, betel nut tree canopies are concentrated at one height, whereas litchi tree canopies are scattered in several heights. Therefore, raindrops intercepted by litchi canopies can be further shattered into smaller sizes. Smaller drop sizes, shorter fall height, and larger canopy cover allow the bare soil in clean-cut litchi orchards to experience less impact energy. Because of the greater raindrop impact, apparent surface crusting was observed in the clean-cut betel nut plantation. The surface crusting has made the bare soil in clean-cut betel nut plantation become less erodible and less permeable than the clean-cut litchi orchard.

The soil loss from treatment B (bare plot) is the highest among all treatments, as expected. As indicated in Table 1, the average annual soil loss from a 60%-steepness bare plot may reach 36.5 kg/m^2. By ignoring the factors affecting sediment delivery, the average annual soil loss will reach 365 tons/ha, which is equivalent to the loss of 2.4 cm of top soil. The surface runoff from treatment B is slightly less than that from treatment G. The reason causing this unexpected result can only be explained by human error.

The average annual soil loss from treatment B (bare plot) was used to compare with the estimated soil loss obtained by the USLE. In order to obtain the representative rainfall erosivity index (R-index), actual drop size measurements were conducted for four consecutive years to capture the drop size distribution reflecting local rainfall characteristics. A set of rainfall kinetic energy equations were also obtained as follows (Wu and Wang, 1996):

$$E = 0.119 + 0.0873 \log_{10} I; I < 4mm/h$$

$$E = 0.276 - \frac{0.521}{I} + 1.146 \exp(-I); I \geq 4mm/h$$

The rainfall erosivity index for year 1993, 1994, and 1995 is 8,800, 36,406, and 10,982 MJ·mm/ha·hr·yr respectively. By taking the average, the annual rainfall erosivity index for the experiment site is taken to be 18,729 MJ·mm/ha·hr·yr. The soil erodibility index (K-index) is estimated to be 0.078 tons/ha·yr · ha·hr·yr/MJ·mm using Wischmeier and Smith's (1978) nomograph. Using LS equation presented by Wischmeier and Smith (1978) and taking the C and P factors to unity, the estimated annual soil loss from a 60%-steepness bare slope is found to be 26,683.8 tons/ha. It is 73 times higher than the actual field measurement!

4 Conclusions

Vegetation, grass strip, surface mulch, and hillside ditches are found to be effective not only on mild slopes but also on steep slopes. These conservation measures can greatly reduce soil losses and conserve surface runoff. The combination of grass strip and residue mulch between strips is the most cost-effective practice for steep sloping orchards. Without any conservation practices, the average annual soil loss from a 60%-steepness bare slope may reach 36.5 kg/m^2.

By feeding the USLE with a local rainfall erosivity index, the slope length and slope steepness factors are probably the most likely variables causing the serious over-estimation in soil loss. Hence, further research on slope steepness factors on steep slopes is needed.

References

Council for Agricultural Planning and Development (CAPD) (1980): Preliminary report on regional agricultural development planning in the Taiwan area, 8p.
Lee, S.W (1984): Soil conservation and slopeland development in Taiwan. In: S.-C. Hu (ed.), Proc. Sino-Korea Bilateral Symposium on Soil and Water Conservation of Sloped Farm Land, National Science Council, R.O.C., 29-42.
Liao, M.C. (1980): Soil conservation research and development in Taiwan, J. Agric. Assoc. of China **112**, 1-22 (in Chinese)
Wischmeier, W.H. and Smith, D.D. (1978): Predicting rainfall erosion losses - A guide to conservation planning, US Department of Agriculture, Agricultural Handbook Number 537, 58p.
Wu, Chia-Chun, Chen, C.-H. and Tsou, C.-C. (1995): Effects of different mulching materials on soil moisture variation and erosion control for steep sloping lands, Journal of Chinese Soil and Water Conservation **26(2)**, 121-133. (In Chinese).
Wu, Chia-Chun and A.-Bih Wang (1996): Drop size characteristics and erosive kinetic energy of natural rainstorms in Pingtung Laopi area, Journal of Chinese Soil and Water Conservation **27(2)**, 151-165. (In Chinese).

Address of authors:
Chia-Chun Wu
A-Bih Wang
Department of Soil and Water Conservation Technology,
National Pingtung University of Science and Technology, Pingtung, Taiwan

The Extent of Soil Degradation on Hilly Relief of Western Lithuania and Antierosion Agrophytocenoses

B. Jankauskas & G. Jankauskiene

Summary

Soil degradation on the hilly-rolling relief of Western Lithuania is a result of both natural and Antropogenic erosion. The thickness of the lost soil layer varies from a few millimetres to 107 centimetres. Soil loss was found to be 28.8-82.4 under potatoes, 11.7-26.4 under spring barley and 3.5-8.3 m^3 ha^{-1} under winter rye. This was a result of accelerated water erosion on slopes of 2-5^0 and 5-10^0. The quick-growing grasses completely stopped soil erosion even on a slope of 10-14^0. In turn the natural fertility of slightly, moderately and severely eroded soils decreased by 22, 39 and 62 %, respectively.

The ingenious application of antierosion agrophytocenoses led to a decrease in soil losses on 2-10^0 slopes of arable land by 77-81% compared to the losses on field crop rotation, and by 22-24% in comparison to grain-grass crop rotation.

Keywords: Hilly relief, soil erosion, crop rotation, antierosion agrophytocenoses.

1 Introduction

There are about 45% of Aquic Glossoboralfs (USDA 1994) or of Soddy Podzolic (classification of V.V. Dokuchaev school) soils (SP) in Lithuania (Vaitiekunas 1981). The degree of lessivation and podzolization in Western Lithuania is higher than in the other parts, and therefore there is a significant amount of moderately (SP_2) and strongly (SP_3) lessivated-podzolized soils. About 53% of terrain in Western Lithuania are on hilly-to-rolling relief (Aleknavichius et al., 1989). The parent rock is rather special on the Zhemaichiai upland (Western Lithuania). The last glacier (12,000 BP) left a thin layer of glacial clay loam moraine on the old bases of the upland. This moraine is erodible (Kudaba 1983). The abundance of precipitation, and its intensity, induces water erosion as well as the processes of lessivation and podzolization. The average annual precipitation is 858 mm in the central part of the Zhemaichiai upland and 750-800 mm in the lower parts (average annual precipitation in Lithuania: 626 mm).

2 Materials and methods

We investigated the Aquic Glossoboralfs (Soddy Podzolic) loamy sand and clay loam soils on hilly-rolling relief of the Zhemaichiai upland. Twenty-three longitudinal relief profiles, 87 soil profiles, and 69 boreholes to a depth of 160 cm were described and characterised by data of agrochemical analysis of 647 soil samples. For determining the degree of erosion, we used an approach (Table 1) that we modified for Soddy Podzolic moderately and strongly podzolized soils (Jankauskas 1995).

Degree of erosion	Initials	Ploughed out horizon[a] on the soils:			Thickness of lost soil layer cm[b]	Correspondence to slope gradient
		SP_3	SP_2	SP_1		
Slight	$N_1(S)$	Top E[c]	E	EB	5-20	$2-3^0$
Moderate	$N_2(M)$	E	E, EB	Bt1	20-40	$3-7^0$
Severe	$N_3(V)$	EB	Bt1	Bt2	40-60	$7-10^0$
Very severe	$N_4(E)$	Bt1	Bt2	C(BC)	>60	$>10^0$

[a] Soil horizons under arable layer. Their identities were established by colour of subsoil, which is ploughed out by ploughing to 25 cm depth: E - light gray, EB - light gray with brown patches, Bt1 - brown with light gray patches, Bt2 - brown, C(BC) - brown (calcareous).
[b] It was established by difference in depths of a calcareous horizon between non-eroded and eroded soils.
[c] Top part of E horizon.

Table 1: Dependence of degree of erosion of Soddy Podzolic soils on the level of their lessivation-podzolization and thickness of lost soil layer

Barley (*Hordeum vulgare L.*), grain, and straw gross yield on growing stage of ripeness was harvested on the plain tops of hills (conditionally non-eroded soil), on slopes of $2-5^0$ (slightly eroded), $5-10^0$ (moderately eroded), $10-14^0$ (severely eroded) and on the deposit footslopes.
 The field experiments comparing erosion-preventione crop rotations on slopes of various gradients have been carried out since 1982. Four crop rotations of the following structure were compared:
I. The field crop rotation: tilled crop 17%, grain crop 50%, mixture of clover-timothy (CT) (*Trifolium pratense L.-Phelum pratense L.*) 33% (1. Winter rye (*Secale cereale L.*), 2. Potato, (*Solanum tuberosum L.*), 3-4. Spring barley, 5-6. CT);
II. The grain-grass crop rotation: grain crop 67%, CT 33% (1. Winter rye, 2-4. Spring barley, 5-6. CT);
III. The grass-grain I crop rotation: grain crop 33%, CT 67% (1. Winter rye, 2. Spring barley, 3-6. CT);
IV. The grass-grain II crop rotation: grain crop 33%, mixture of orchard grass-fescue red (*Dactylis glomerata L.-Festuca rubra L.*) (OF) 67% (1. Winter rye, 2. Spring barley, 3-6. OF).
 The field experiments were carried out on slopes of $2-5^0$, $5-10^0$ and $10-14^0$. Instead of the field crop rotation, perennial grasses of multiple composition for long-term use were grown on a slope of $10-14^0$. This grass mixture included timothy common, fescue red, clover white (*Trifolium repens L.*), bluegrass Kentucky (*Poa pratensis L.*) and trefoil birdsfoot (*Lotus corniculatus L.*) (20% of each). The soil was an eroded sandy loam with pH_{KCl} of 5.3-5.8, base saturation percentage of 67.9-85.5%, mobile P_2O_5 of 42-114, mobile K_2O of 158-176 mg kg^{-1} of soil, and humus 1.96-2.69%. Optimum ground and fertilizer treatments were used according to soil features. The yields of different crops were evaluated by the amount of metabolizable energy according to the data of chemical analyses of each crop production. The losses of soil by water erosion were established using the method of rills' volume measurement (Sobolev 1961). The annual precipitation during the period of study was from 635 to 1075 mm.

3 Results and discussion

The degree of soil erosion on the investigated slopes was determined (Table 1) by comparison of the thickness of the lost soil layer, slope gradient and different thickness of genetic horizons (according to the degree of lessivation-podzolization). Among 155 investigated soil profiles or boreholes, 116 eroded plots were discovered. Very severely and severely eroded soils were discovered on 49 plots; 29 plots were moderately eroded; 18 plots were slightly eroded and 20 plots contained eroded-deposited soil. On the tops of small hills and on the slopes of 5-10^0, the thickness of the lost layer of soil was from 69 to 107 cm. On the gently-sloping plateau the thickness of the lost layer was 25 cm. The thickness of soil layers up to the calcareous horizon was from 103 cm to 185 cm. The thickness of the deposit layer of soil (sediments) in footslopes or on hollows was 31-162 cm.

The deterioration of the agrophysical properties of loamy sand and clay loam soils (usually having A, E, B and C horizons) were ascribed due to soil erosion (Table 2). Dry bulk density and percentage of clay-silt and clay fractions increased, while the percentage of total porosity and water field capacity decreased.

Strong acidity (3.8-5.6 pH$_{KCl}$) and low stock of nutrients (10-96 mg kg^{-1} P$_2$O$_5$, 57-162 mg kg^{-1} K$_2$O) on E, EB and B horizons of eroded Soddy Podzolic soils is a characteristic feature of agrochemical properties.

Degree of soil Erosion	Percentage of fraction		Dry bulk density (g cm^{-3})	Percentage of	
	clay-silt	Clay		total porosity	water field capacity
Loamy sand soil					
Non-eroded	17	8	1.42	44	31
Slight	28	8	1.44	44	30
Moderate	31	14	1.53	42	27
Severe	39	28	1.58	39	25
Clay loam soil					
Non-eroded	32	15	1.54	41	27
Slight	38	19	1.57	40	25
Moderate	44	24	1.65	37	22
Severe	49	35	1.69	36	22

Table 2: Agrophysical characteristics of the surface horizon of eroded SoddyPodzolic soils

Parts of relief	Degree of soil erosion	Average of 3 year green weight of top from:					
		48 investigated plots			75 investigated plots		
		yield	extra yield	rel[a]	yield	extra yield	rel[a]
		(t ha^{-1})			(t ha^{-1})		
Plains	Non-eroded	19	-	100	21	-	100
Slopes of 2-5^0	Slight	15	4	78	16	4	78
Slopes of 5-10^0	Moderate	11	8	60	13	8	61
Slopes of 10-14^0	Severe	7	12	38	-	-	-
Footslopes	Deposit soil	19	-	102	21	0	100
LSD$_{05}$		1.1			1.1		

[a]In relative numbers
LSD$_{05}$ Least significant difference

Table 3: Dependence of barley yield upon slopes gradient and degree of soil erosion

The extent of soil degradation due to soil erosion is evident in the natural fertility of the soil (Table 3). The soil fertility had decreased by 22%, 39% and 62% on slightly, moderately and severely eroded slopes, respectively.

The extent of water erosion is revealed by data from the field experiments. Grasses (mixtures of CT, OF and multiple composition) completely stopped water erosion even on the slopes of 10-14^0, while rates of erosion under winter rye, spring barley and potato on slopes of 2-5^0 and 5-10^0 were, respectively, 3.5-8.3, 11.7-26.4 and 28.8-82.4 m^3 ha^{-1}.

Antierosion agrophytocenoses (the erosion-preventive crop rotations) are the most important part of the conservation cropping system on arable land. This system was prepared according to the research data of the Kaltinenai Research Station (Jankauskas 1996). The erosion-preventive grass-grain crop rotations were most productive under an optimum ground and fertilizer treatment, (Fig. 1). The amount of metabolizable energy accumulated by the erosion-preventive grass-grain crop rotations was 14.1-32.7% higher than in the field crop rotation, and 11.8-27.7% higher than in the grain-grass crop rotation. The perennial grasses of long-term use on the slope of 10-14^0 were more productive than the grass-grain crop rotation.

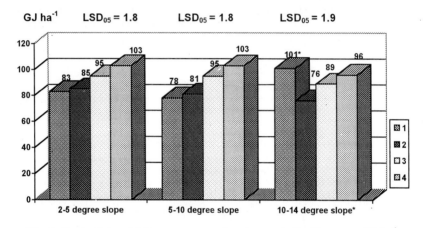

Fig. 1: Productivity of different crop rotations in metabolizable energy GJ ha^{-1}.
Crop rotations: 1. Field, 2. Grain-grass, 3. Grass-grain I, 4. Grass-grain II.
* Perennial grasses of long-term use were grown on a slope of 10-14° instead of the field crop rotation.

The part of perennial grasses in composition of crop rotations determined the total erosion-resisting capability of crop rotations. The annual losses of soil by water erosion under the erosion-preventive grass-grain crop rotations on slopes of 2-5^0 and 5-10^0 were 1.9-3.8 m^3 ha^{-1}. It decreased by 77-81% in comparison to the field crop rotation, where it reached 8.2-19.5 m^3 ha^{-1}. The loss of soil under the grain-grass crop rotation was 6.2-15.3 m^3 ha^{-1}, or decreased by 22-24%. Therefore even the grass-grain crop rotations could not completely halt soil erosion. The annual rates of soil loss due to water erosion under the grass-grain crop rotations were 6.1-6.2 m^3 ha^{-1} on the slope of 10-14^0. We find however, that grassing on slopes of 10^0 and over is especially important. Grass stands of high fertility for early, medium and late hay-making or grassing were recommended (Jankauskas 1996; Norgailiene and Zableckiene 1994), and erosion-preventive tillage with antierosion liming-fertilizing must be practised on slopes of 2-10^0.

4 Conclusions

1. The different rates of soil loss were affected due to present land use by water erosion. The quick-growing perennial grasses completely stopped soil erosion.
2. The natural fertility of slightly, moderately and severely or very severely eroded soils decreased by 22, 39 and 62%, respectively.
3. The antierosion agrophytocenoses (the grass-grain crop rotations) decreased the erosion risk by reducing the soil loss. This alternative is the most important management system on hilly-rolling relief of Western Lithuania.
4. The amount of metabolizable energy accumulated by the erosion-preventive grass-grain crop rotations under an optimum ground and fertilizer treatment was 14.1-32.7% higher than in the field crop rotation and 11.8-27.7% higher than in the grain-grass crop rotation.

Acknowledgements

The author thanks the Organising Committee of ISCO for covering the Conference participation and travelling costs.

References

Aleknavichius, P., Gogelis, A., Jasinskas, J. Juodis, J., Pakeltis, V., Skuodziunas, V., Pakutinskas, J., Kaushyla, K. and Poshkus, B. (1989): The land cadastre, Mokslas, Vilnius, 727 pp (In Lithuanian).
Jankauskas, B. (1995): Comparison of methods for determination of degree of erosion of Soddy Podzolic soils with different degree of podzolization. In: Leuchovius T., Hallerstrom B. and Lazauskas S. (eds.), Proceedings of the fourth regional conference on mechanisation of field experiments, Uppsala, IAMFE, 51-56.
Jankauskas, B. (1996): Soil erosion, Margi rashtai, Vilnius, 168 pp (In Lithuanian with comprehensive summary in English).
Kudaba, Ch. (1983): Uplands of Lithuania, Mokslas, Vilnius (In Lithuanian).
Norgailiene, Z. and Zableckiene, D. (1994), Selection of different manurity of moving and grassing grass-stands for erodible clay loam soils, Agricultural sciences, 3, 72-76 (In Lithuanian).
Sobolev, S.S. (1961): Soil conservation and increasing of fertility, The publishing house of Academy sciences of SSSR, Moscow, 230 pp (In Russian).
USDA (1994): Keys to soil taxonomy, Sixth edition, Washington, 306 pp.
Vaitiekunas, J.-K. (1981): Soil and soil resources, in Fedoseva M. and Fetisova N., ed., Atlas of Lithuanian SSR, VGKV, Moscow, 89-96 (In Lithuanian and in Russian).

Address of authors:
B. Jankauskas
G. Jankauskiene
Kaltinenai Research Station of the Lithuanian Institute of Agriculture
LT-5926 Kaltinai, Silale District
Lithuania

Degradation of a Sandy Alfisol and Restoration of its Productivity Under Cotton/Maize Intensive Cropping Rotation in the Wet Savannah of Northern Cameroon

Z. Boli Baboule & E. Roose

Summary

On the sandy alfisols of the cotton belt of Northern Cameroon, conventional intensive cropping based on soil tillage and mineral fertilizers application does not prevent soil degradation or free farmers from shifting cultivation. The aim of this study was to characterize degradation of these soils under intensive cropping, and to develop ways for restoring the productivity of these degraded soils that have been rendered unsuitable for intensive cultivation. The study was based on evaluation of soil constituents, structure and functions as affected by time. Run-off plots were used in two locations at different stages of soil degradation.

Degraded soils were characterized by a low organic carbon content (around 0.2%) in its upper 10 cm, a high erodibility, and a high compaction of the upper layers when not recently tilled. Restoration of productivity on these soils should start by aerating and stabilizing the compacted layers. Only then can fertilizer application and other improvements be profitable. Soil mulching after tillage and incorporation of farm manure, adoption of conservation tillage practices and short duration fallows, allow restoration of productivity on these degraded soils.

Keywords: Northern Cameroon; sandy alfisols; conventional vs no-tillage; soil compaction erosion, restoration of productivity.

1 Introduction

In the wet Sudanese savannah of Northern Cameroon, cotton (*Gossypium hirsutum*) and maize (*Zea maidis*) are grown in an intensive cropping system with conventional soil tillage, mineral fertilization, application of insecticides and improved varieties. However, this intensive cropping system does not prevent soil productivity from declining. Lands where inputs are no longer profitable are abandoned and new fields are opened from savannah bush fallow. In a previous survey carried out in the main area of intensification of the cotton belt of Cameroon, Boli et al. (1991) noticed that soil degradation and erosion were major causes of the continuous yield decline.

To better understand how disfunctioning of the "soil - plant - atmosphere system" occurs, an experimental design with 57 run-off plots of 100 to 1000 m² was set up and observed both on a new and an old field at Mbissiri, from 1991 to 1994. Experiments presented in this paper had two objectives : (1) characterization of sandy alfisol degradation under different tillage systems and, (2) restoration of productivity of soil degraded by conventional tillage.

2 Material and methods

2.1 Climate

The region falls within the wet Sudanese zone. Its annual rainfall varies from 1000 to 1500 mm. The length of the rainy season is six to seven months, from April to October (Suchel, 1988).

2.2 Soils

Sandy alfisols are the most representative soils for rainfed farming in this region (Brabant and Gavaud, 1985). The Mbissiri soils are developed on a sandstone and have less than ten per cent clay content in the upper 15 cm, a poor organic matter content (1.2%) and a low cation exchange capacity (1.5 to 3 meq/100g). Depth of arable soils above parental rock or ferruginous hard pan varies with the position on the slope (from 0 to 100 cm). Average slope is two per cent.

2.3 Field methodology approach

Soil constituents, structure and functions are compared for two situations: (1) recent fields and (2) degraded old fields. In each field, conventional and conservation tillage are compared. The basic experimental unit is a run-off plot of 100 m² (50 × 20). Three of the four cropped blocks have 16 plots each. Not all kinds of treatment are replicated in the third block located on degraded soil. The evolution of the constituents is evaluated by comparing the mineral and organic fractions and the pH of the 0 - 10 cm soil layer between February 1991 (t_1) and February 1995 (t_5). For the evolution of the soil structure, the status of the soil surface is considered after one or several rainfall events occurring after tillage. The soil water content at the pF values of 1.8 and 4.2 also gives an idea of the change in soil structure. The determination of the soil bulk density after the first two cropping cycles, the visual estimation of macroporosity and the counting of roots of the cultivated plants in the 0 - 40 cm layer were also considered. Run-off, soil loss and crop yield are considered in this study as descriptors of soil functions (Boli, 1996).

2.4 Treatments

Two groups of treatment are tested: (1) the conventional tillage group and (2) the no-tillage group (Table 1). Mineral fertilizers were applied to cotton and maize.

Conventional tillage	No-tillage
• Control, which is the intensive cropping system: soil ploughing, sowing, weeding and moulding up six weeks after sowing • Bare fallow control • Ploughing after application of 3 t/ha of goat farm manure, then sowing ... • Ploughing after two years rest (without ploughing) on the degraded soil	• No-tillage on soil that has not been loosened (new and old fields) • No-tillage on degraded field after loosening • No-tillage on degraded field after two years of fallow (one plot of bush fallow and another one of improved fallow with Calopogonium mucunoïdes)

Table 1: Tested groups of treatment

3 Results

3.1 Rainfall

The rainfall characteristics of the four years are indicated in Table 2. The low number of days with run-off in at least one plot in 1991 is due to late planting (July 10) as compared to the other three years (June 20). The year with the highest total rainfall (1992) was also the year with the highest soil losses. The driest year (1993) recorded the most aggressive single rainfall event.

Year	Total annual rainfall (mm)	Total rainfall cycle (mm)	Number of days with run-off	Maximum $I_{max}30'$ observed (mm/h)	RuSA index	Rainfall series occurrence
1991	1209	673	13	-	419*	yes
1992	1570	1184	24	97	785	yes
1993	1072	772	19	117	496	no
1994	1352	1073	23	91	433	yes
*estimated						

Table 2: Main rainfall characteristics of the 4 cropping cycles

3.2 Evolution of soil organic carbon content

Figure 1 indicates the evolution of the soil organic carbon content in the upper 10 cm layer of the savannah soil (SAV), bare soil (NUE), the control (TRM), and no-tillage (ZT) for new and old fields. Four years after land clearing, the soil organic carbon content has dropped down to the level of 0.3 % found on the degraded soil after many years of continuous cultivation.

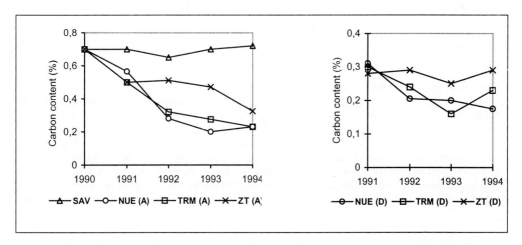

Figure 1: Evolution of organic carbon content in the 0 to 10 cm soil layer as affected by soil treatment between February 1990 and February 1994

3.3 Changes in soil structure

The evolution of the structure of the topsoil is determined through the evaluation of soil crusting (Table 3) and the soil water content at the specific values of pF (Table 4). When conventionally tilled, 90 % of sandy soils surface can be crusted after 50 to 100 mm rainfall, depending on the rainfall intensity. The general tendency is a decrease in water content at pF 1.8 and an increase of the one at pF 4.2.

Period	Treatment	Crusted surface (%)		Total rainfall in the period (mm)	
		New field	Old field	New field	Old field
First rain After Ploughing	Conventional tillage	85.6	56.1	35.0	12.5
	No-tillage	2.1	1.2		
Before Moulding Up	Conventional tillage	92.8	84.2	214.6	219.4
	No-tillage	18.3	20.9		

Tab. 3: Frequencies of crusted soil surfaces as affected by cultural practices and rainfall distribution

pF	Savannah		Conventional tillage				No-tillage			
			New field		Old field		New field		Old field	
	T1	T5	T1	T5	T1	T5	T1	T5	T1	T5
1.8	13.7	11.5	15.0	11.9	15.0	10.9	18.5	12.2	15.9	13.2
4.2	2.1	2.2	2.6	2.7	3.0	2.3	4.1	5.2	2.3	3.2

Tab.4: Soil water content at pF 1.8 and 4.2 as affected by treatment and time in new and old fields

3.4 Soil functions

Among other variables, run-off, soil loss and crop yields demonstrate how soil response to climate and management is affected by tillage (Tables 5, 6 and 7). Fields degraded by many years of continuous cropping and conventional tillage lose two to three times more soil than new fields. The effect of farm manure on yields of degraded fields is reduced. Under no-tillage, soil erosion is low (six to 16 times less than with conventional tillage), irrespective of the degradation stage. Degraded soils are more compacted under no-tillage systems and soil compaction/crusting appears as a limiting factor for the productivity of these systems, especially when it rains frequently during the first two months after sowing. Run-off and soil losses increase with the frequency of soil tillage. Compacted soils can be aerated by deep ploughing or by a fallow, even of short duration (2 years).

	Conventional tillage			No-tillage			
	Erosion (t/ha/yr)	Yield control (t/ha)	Yield Manure* (t/ha)	Erosion (t/ha/yr)	Bulk density (kg/l)	Number roots 40cm	Yield (t/ha)
New fields	6-16	C = 2.1 M = 4.5	C = 2.9 M = 7.0	1.3	1.4 - 1.6	88	C = 2.1 M = 2.2
Old fields	18 - 38	C = 1.8 M = 5.0	C = 2.1 M = 5.7	2.2	1.6 - 2.0	32	C = 1.4 M = 2.2

* control + 3t/ha of goat manure before ploughing   C: cotton   M: maize (dry grain)

Table 5: Comparative soil behaviour in new and old fields

Treatment	Run-off (%)		Erosion (t/ha/yr)	
	1993	1994	1993	1994
Conventional tillage since 1991	34	35	15 + 7*	25 + 8*
Conventional tillage since 1993 after two years of no-tillage	12	23	3 + 1*	7 + 4*
No-tillage since 1991	5	7	0.9 + 0.1*	1,7 + 0*

* coarse sediments + suspensions

Table 6: Effect of the frequency of soil tillage on run-off and soil loss in the degraded lands

Treatment	Cotton yield 1993 (t/ha)	Maize grain yield 1994 (t/ha)
No-tillage (control)	1.4	2.8
No-tillage after loosening	1.9	3.6
No-tillage after two years fallow	2.1	3.6

Table 7: Improvement of crop productivity under no-tillage systems on degraded soils

4 Discussion

Soil carbon content appears to be the best indicator of soil degradation. It drops from 0.7 % in the savannah to 0.3 % in the cultivated field, within three years after clearing a thirty years old savannah fallow. Later on, the carbon content hardly changes. Similarly, the surface status of sandy soils changes rapidly before reaching a new equilibrium. But it varies widely from one location to another as the rainfall characteristics change.

The texture of the mineral fraction, however, does not show a clear change either in newly opened savannah or in older fields. Cropping systems that leave the soil uncovered after tillage and sowing are exposed to rapid soil surface crusting. As the surface becomes crusted, water infiltration and aeration of underlying layers are limited. The macroporosity of these layers decreases. It was noticed that the macroporosity generated by soil tillage was fugacious and soil compaction fast, except where goat manure was incorporated to the soil.

The high cohesion of degraded soil under no-tillage systems appeared more harmful to crop yield than to water infiltration. Run-off, soil loss and crop yield are directly related to soil instability.

Various definitions have been given to soil degradation. Many of them refer to crop production (Riquier, 1977 ; Oldeman, 1988 ; Piéri, 1989 ; IFAD, 1992 and Roose, 1994). It follows from our observations that soil degradation has many causes and appears under different forms. However, one can observe a convergence of its effects on the macro-porosity and water distribution in the soil, resulting in reduced water infiltration. Soil degradation appears as the action by which physical, chemical and biological processes destabilise soil surface layers, resulting in a reduction of water and air flows between the atmosphere and the soil.

A comparison of soil functions of new and degraded fields (Table 5) reveals significant differences in soil loss between conventional and no-tillage systems. Degraded soils are characterized by a high instability to water under conventional tillage and a high compaction under no-tillage. Besides these two characteristics associated with a low organic carbon content, degraded fields are not necessarily unproductive. Good yields comparable to those obtained in new fields were recorded when some conditions were met. These included soil stabilization and aeration prior to any other chemical or physical improvement and a favourable rainfall pattern for no-tillage systems.

Run-off and soil loss from degraded fields could be reduced considerably by changing from conventional tillage to mulching, fallowing or no-tillage. Crop productivity of degraded soils under no-tillage could be improved by deep ploughing or by fallowing (two years bush or improved fallows). While some authors have indicated that short duration fallows have no potential in improving chemical soil fertility (Piéri, 1989; Hien and Sédogo, 1991 ; Roose, 1994), this study clearly shows the important role of a two years fallow in aerating and stabilizing the soil surface layers (Boli, 1996). It confirms results obtained by Morel and Quantin (1972) in Central African Republic and by Valentin (1989) in the northern part of Ivory Coast. We observed that a factor like grass roots plays an important role in soil stability. However, such a factor is seldom used by other authors.

When aeration and stabilization conditions of a degraded soil are met, the next steps for soil productivity restoration could be effective. These mainly consist in applying major and minor nutrients through mineral and organic fertilizers in order to meet crop needs and control soil acidity. In addition, crop rotations need to be respected as degraded soils were heavily attacked by *Striga hermontica* (120 stands/m² against 0.4 in the new fields; Boli et al., 1991).

5 Conclusions

It follows from this study that soil degradation is mainly related to tillage systems. Conventional tillage results in soil instability and increased water erosion. Soil erosion in turn affects soil productivity by depleting soil fertility and reducing plant growth and biomass production. On the other hand, applying no-tillage to degraded soil, compaction of the surface layers is the main constraint to crop productivity.

Finally, when a degraded soil still has an acceptable depth (> 50 cm) above the hard rock or ferruginous pan, it may regain its productivity without rebuilding up its initial texture or organic matter content. But it will be necessary to aerate and stabilize the surface layers before any other chemical or physical improvements.

References

Brabant, P. and Gavaud, M. (1985): Les sols et les ressources en terre du Nord-Cameroun (Provinces du Nord et de l'Extrême-Nord). Cartes à 1/500000è; éd. ORSTOM Paris, 285p.
Boli, Z., Bep à Ziem and Roose, E. (1991): Enquête sur l'érosion pluviale dans la région de Tchollire (Sud-Est Bénoué), Nord-Cameroun. Bull. Rés. Erosion n° **11**: 127-138.

Boli Baboulé, Z. (1996): Fonctionnement des sols sableux et optimisation des pratiques culturales en zone soudanienne humide du Nord-Cameroun (expérimentation au champ en parcelles d'érosion à Mbissiri). Thèse de doctorat en sciences de la terre à l'Université de Bourgogne, 344p.

Hien, V. and Sédogo, M.P. (1991): Etude des effets de la jachère de courte durée sur la production et l'évolution des sols dans differents systèmes de culture du Burkina Faso. In La Jachère en Afrique de l'Ouest; pp.221-232.

IFAD (1992): Soil and water conservation in the Sub-Saharan Africa. Towards sustainable production by rural poor, 110 p.

Morel, R. and Quantin, P. (1972): Observations sur l'évolution à long terme de la fertilité des sols cultivés à Grimari (Rép. Centrafricaine). Agron.Trop., **27, 6**: 667-739.

Oldeman, L.R. (1988): Directives pour une évaluation générale de l'état de la dégradation des sols par l'homme. Document de travail et première épreuve GLASSOD, n°88/4, 12p.

Piéri, C. (1989): Fertilité des terres de savanes. Bilan de trente ans de recherche et de développement agricole au sud du Sahara. Min. Coopération et du développement, 444 p.

Riquier, J. (1977): Philosophy of the world assessment of soil degradation and items for discussion. Assessing soil degradation, FAO Soils Bulletin n°34, sp.

Roose, E.J. (1994): Introduction à la gestion conservatoire de l'eau, de la biomasse et de la fertilité des sols (GCES). Bulletin Pédologique FAO n° **70**, 420 p.

Suchel, J.B. (1988): Les climats du Cameroun. Thèse de doctorat d'Etat. Tome I pp.1-443; Tome III pp.791-1188.

Valentin, C. (1989): Etat de dégradation de deux terroirs Sénoufo - Nord Côte D'Ivoire. Abidjan IIRSDA, 9p. multigr.

Wischmeier, W.H. & Smitt, D.D. (1978): Predicting rainfall erosion losses. A guide to conservation planning. US Department of Agriculture; Agriculture Handbook N°537: 58p.

Addresses of authors:
Z. Boli Baboulé
IRA
BP 2123
Yaoundé (Messa), Cameroon
E. Roose
ORSTOM – Réseau Erosion
BP 5045
34042 Montpellier, France

Catchment Discharge and Suspended Sediment Transport as Indicators of the Performance of Physical Soil and Water Conservation in the Ethiopian Highlands

U. Bosshart

Summary
The Anjeni Research Unit is typical of the high-potential, intensely cultivated, cereal belt of the north-western Ethiopian Highlands, which are seriously endangered by soil degradation. In 1986, the entire catchment (113 ha) was treated mainly with graded *Fanya juus*. However, as of 1990, when supervision and pressure from governmental institutions disappeared due to dramatic political changes, the farmers partially halted maintenance work and began to remove some of the graded structures on their fields.

Analysis and interpretation of the hydro-sedimentological characterisation of the Minchet catchment, based on a data base compiled over a period of 10 years, clearly reflect the above-mentioned changes in soil conservation (Bosshart, forthcoming). The hypothesised impact of physical soil and water conservation measures on catchment discharge, and particularly on suspended sediment transport, can be identified. During the period of well-maintained, highly efficient physical soil and water conservation measures (1986-1990), suspended sediment transport continuously decreased in a significant way, finally even falling below average tolerable soil loss rates. Simultaneously, a slightly insignificant reduction in catchment discharge occurred, assuming steady precipitation. Since 1990, suspended sediment transport has increased again, with a trend towards the previous values before conservation, when the farmers reduced maintenance and even started to remove conservation structures. The numerous complex reasons for this problematic development were not only limited to technical problems related to the conservation. In particular, they were related to socio-economic and institutional constraints. This highlights the importance of multi-disciplinary approaches to establish the necessary preconditons for successful application of essential physical soil and water conservation.

Keywords: River discharge, Ethiopia, physical soil and water conservation, socio-economic constraints, suspended sediment transport, watershed

1 Introduction

The Anjeni Research Unit, which comprises the Minchet catchment, is one of eight microcatchments selected as Research Units of the Soil Conservation Research Programme (SCRP)[1]. For the Minchet catchment the standard, multi-disciplinary SCRP research programme has been operational since 1984. The standard research set-up accommodates different levels of survey within the catchments. Plot experiments, for example, provide highly accurate data base on soil erosion processes and the inter-relationships of subsystems of factors over a small area with well-defined conditions. On the other hand, the construction of gauging stations at the outlets of the catchments allows monitoring and assessment of discharge and suspended sediment transport from the catchment area. The latter two factors can both be considered as an integral reflection of soil degradation and the impact of soil and water conservation at the catchment level (Bosshart 1996; Bosshart, forthcoming; Hurni 1986, 9; SCRP 1991a, 7; SCRP 1991b, 1).

The Anjeni Research Unit is typical of the high-potential, intensely cultivated, ox-plough cereal belt of the north-western Ethiopian Highlands. According to Thornthwaite's climatic classification, the Minchet catchment lies in the humid areas; in the Breitenbach diagram, it is part of the highland forest zone; and according to Hurni's classification, it represents the wet Weyna Dega to wet Dega agroclimatic zone. Minchet catchment is situated at 37°31" east and 10°40" north at an altitude between 2,405 and 2,500 m a.s.l. The catchment has a hydrological surface of 113.4 ha. Three geomorphological units (plateau, slope and foothill zone/valley bottom) form a moderated undulated topography (Photo 1). The most important soils are Nitosols, Regosols and Phaeozems. The soils on the ridges are strongly to extremely degraded, while the soils on the slopes and the valley bottom are slightly to moderately degraded. The local drainage network consists of Minchet River, which starts as a perennial spring, located in a fingered gully head in the upper part of the catchment, and some adjoining (artificial) waterways. The mean land cover pattern (1984-1991) exhibits the very intensive agricultural use. The mean proportions of the individual categories on the total catchment area are 67% cultivated land, 20% grazing land, 5% fallow land and 8% forest, bush, settlements, roads and tracks. A remarkable increase in the amount of cultivated land, which peaked in 1988 and 1989, can be seen in the data (Breitenbach 1961, 1963; Bosshart 1996; Hurni 1986, 9; Krauer, forthcoming; Kefeni Kejela 1995, xi, 34-37, soil map; Ludi 1994, 77, 19; SCRP 1991a, 55; SCRP 1991b, 4, 9, 49; SCRP 1988, 35; SCRP 1986, 10; Thornthwaite 1948; Werner 1986, 11f).

Traditional shallow drainage ditches running diagonally across the slope are used on untreated fields or in combination with physical soil and water conservation (Million Alemayehu 1992, 1, 9, 24; Werner 1986, 14). In early spring 1986, the entire Minchet catchment was systematically treated under the supervision of the Ministry of Agriculture, mainly with graded *Fanya juus*[2]

---

[1] The Soil Conservation Research Programme was initiated in 1981 to provide the Ethiopian soil conservation programme with basic data on soil erosion and conservation, and to assess the effectiveness of soil and water conservation measures under the specific conditions of the Highlands of Ethiopia and Eritrea. By 1993, eight micro-catchments had been selected as Research Units (Bosshart 1996; Bosshart, forthcoming; Kebede Tato, in Hurni 1986:VII; SCRP 1982:9; SCRP 1991a:7; SCRP 1991b:1).

[2] A graded *Fanya juu* («Throw uphill» in Swahili) is an embankment along the contour, made of soil and/or stones, with a basin at its lower side. The latter acts as a slightly graded drainage ditch going sideways towards a waterway or river with a gradient of up to 1%. Graded *Fanya juus* are about 50-75 cm high and have a base width of 100-150 cm. The ditch is about 50 cm deep and 50 cm wide. The space between the ditch and the berm is at least 25 cm. The spacing between two graded *Fanya juus* is determined by the slope gradient and the depth of reworkable soil. Tied ridges placed in the drainage ditch about every 10 m serve to prevent flow from becoming too rapid and creating linear erosion, and to provide small basins for water storage. About every 50 m, a gap can be left open to allow ploughing oxen to cross and reach their land. Graded *Fanya juus* retain small amounts of runoff above their wall and they drain excess runoff from heavy storms through the ditch below, which would cause overflow and downslope destruction on level structures. Some of the soil eroded between two *Fanya juus* is deposited above the wall, some is deposited in the ditch, while the rest is drained sideways during heavy storms and lost from the land. Whenever the ditch is full of sediment, the graded *Fanya juus* must be raised, keeping the gradient of the drainage ditch

Photo 1. Minchet Catchment (Photograph by Urs Bosshart, 12 September 1992). The view is to the north. The gauging station is located to the right of the SCRP Research Station, below the forested spot in the middle ground.

(Fig. 1), supplemented with grass strips, cut-off drains, waterways and area closure associated with afforestations. These measures took account of the "traditional" problem of waterlogging on flatter slopes, which the farmers had already tackled with drainage ditches (Ludi 1994: 6; SCRP 1991b: 9). In 1990, when supervision and pressure from governmental institutions disappeared due to the dramatic political changes, the farmers partially halted maintenance work and began to remove some of the graded structures on their fields. The numerous complex reasons for this problematic development were not only limited to technical problems related to the conservation structures (e.g. loss of arable land due to the construction of graded Fanya juus, spacing that was too narrow for ox-ploughing and threshing). In particular, they were related to socio-economic and institutional constraints, e.g. lack of cultivable land due to population growth, unsatisfactory cost-benefit-ratio, differences in perception, social prestige, lack of motivation and conviction on the part of the farmers, insufficient funds, limited working capacity, time constraints, and uncertainty about land tenure (Herweg 1993: 394, 397, 402, 405; Tsehai Berhane-Selassie 1994: 16f).

2 Hydro-sedimentological characterisation

The hydro-sedimentological characterisation of the Minchet catchment is based on 10 years of measurements (1984 to 1993). This period is obviously short for any statistically sound interpretation, as long-term observations of 15 to 20 years are needed for the spatio-temporal characterisation of hydro-meteorological patterns (Krauer 1988: 38f; WMO 1987: 2, 38). However, with regard to comprehensive and multi-disciplinary research on soil erosion in micro-catchments, 10 years of data are a good basis for a sound hydro-sedimentological characterisation of the Minchet catchment, particularly given existing infrastructure and security.

---

going sideways, which is difficult to manage. This will produce well-developed a bench terraces in the course of years (Hurni 1986:44, 48f).

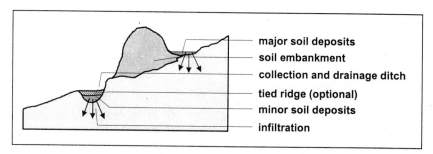

Figure 1: Cross-sectional sketch of a graded Fanya juu (according to Hurni 1986, 44).

The depth of water is continuously recorded with float-activated recorders at the Minchet gauging station. In addition, sediment sampling is carried out if suspended sediment transport is visible, i.e. if the river water changes colour from "clear" to "brown". Single point 1-litre samples are manually taken near the water surface at 10-minute intervals during the rising and the recession limbs of the hydrograph. Rainfall is continuously recorded by an autographic rain gauge (Bosshart 1996; Bosshart, forthcoming; Krauer 1988).

The mean annual precipitation for the Minchet catchment, 1,616 mm (Fig. 2, Table 1), is very high, with a low coefficient of variation (Cv) of 14%. The remarkable mean annual rainfall erosivity amounts to 620 N/h, with a slightly higher coefficient of variation (23%). The mean annual catchment runoff is 731 mm. The coefficient of variation (10%) is also very low. Despite the comparatively high drainage ratios, on average 45%, waterlogging problems often occur on arable land due to the intense torrential rains. The mean annual suspended sediment yield is 25.2 t/ha (Cv: 64%). This value is very high compared with a maximum tolerable annual soil loss rate[3] of 10 to 16 t/ha. The enormous susceptibility of soils to erosion is also highlighted by measurements on cultivated test plots, where soil losses up to 200 t/ha·year were observed (Herweg, forthcoming). Regarding the coefficients of variation, the very high figure for the suspended sediment yield is striking, particularly compared with the very low values for precipitation and catchment runoff. The first is obviously related to the varying impact of the physical soil and water conservation measures implemented, with three periods to be distinguished. Thus, the calculated mean can only give a hint. On the other hand, the low coefficients of variation for precipitation and catchment runoff show that these means are fairly useful for forecasting, as the annual data are only slightly scattered around the means.

The graph of mean monthly values of precipitation, catchment runoff and suspended sediment yield (Fig. 3, Table 2) shows clear extended unimodal regimes, in terms of seasonality. Two periods can be distinguished: *Firstly*, the interval from November to April with little or no precipitation, low discharge and very rare and low suspended sediment transport; and *secondly*, the extended "rainy season" from May to October, which contributes an average of 77% of the mean annual precipitation, 88% of the mean annual discharge, and 99% of the mean annual suspended sediment transport. Despite these clear regimes, reference should be made to the occasionally high coefficients of variation of the mean monthly data, mainly for precipitation and suspended sediment transport. Regarding precipitation, a clear intra-annual distribution can be seen. The lowest coefficients of variation exist, on one hand, during the major rainy season, i.e. these

---

[3] Hurni 1983, quoted in SCRP 1988:53. This figure is a site-specific value. Therefore, it is definitely not comparable with the area loss (i.e. the suspended sediment yield). But a comparison gives a hint about processes within the catchment. The value given is based on a rough estimation of the annual soil formation rate. Soil formation is related to subsoil that is less fertile than the eroded surface soil, which contains more organic matter.

Indicators of soil and water conservation performance, Ethiopia

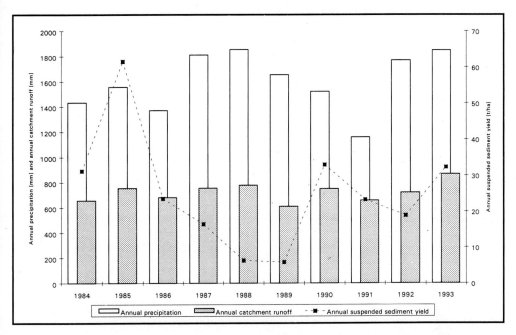

Figure 2: Minchet Catchment - Annual precipitation, catchment runoff and suspended sediment yields between 1984 and 1993.

	Precipitation	Discharge				Drainage Ratio	Suspended Sediment		
			Volume	Specific	Runoff		Rate	Yield	Concent.
	[mm]	[l/s]	[m³]	[l/s·km²]	[mm]	[%]	[t]	[t/ha]	[g/l]
1984	1432[a]	40.3[a]	744529[a]	35.5[a]	657[a]	45.9[a]	3540	31.22	4.75[a]
1985	1556	27.2	856278	23.9	755	48.5	6980	61.55	8.15
1986	1371	24.5	773368	21.6	682	49.8	2669	23.54	3.45
1987	1811	27.2	857791	24.0	756	41.8	1868	16.48	2.18
1988	1854	27.9	883431	24.6	779	42.0	717	6.32	0.81
1989	1654	22.0	692539	19.4	611	36.9	663	5.85	0.96
1990	1522	27.0	851005	23.8	750	49.3	3736	32.95	4.39
1991	1159	23.6	745400	20.8	657	56.7	2634	23.22	3.53
1992	1768	25.8	817317	22.8	721	40.8	2140	18.87	2.62
1993	1848	21.2	983139	27.5	867	46.9	3660	32.28	3.72
MEAN	1616	26.3	828919	23.2	731	45.2	2861	25.23	3.45
STD	238	2.7	84868	2.4	75	6.0	1821	16.06	2.20
Cv [%]	14.8	10.2	10.2	10.2	10.2	13.2	63.7	63.7	66.1
MAX	1854	31.2	983139	27.5	867	56.7	6980	61.55	8.15
MIN	1159	22.0	692539	19.4	611	36.9	663	5.85	0.81
Days[b]	183	365	365	365	365	--,--[c]	89	89	--,--[c]

a) All data are given for 1984, despite some missed rain and discharge due to delays in starting the monitoring in April and June, respectively. The data marked with a) were not used to calculate statistical measures and deduced factors. However, as suspended sediment transport is empirically limited to the rainy season, the data collected are considered to be representative of the annual total for 1984 (according to SCRP 1986:15, 60, 62).
b) Mean number of days with rainfall, discharge and suspended sediment transport.
c) Data not available or not calculated.

Table 1: Minchet Catchment - Annual precipitation, discharge and suspended sediment transport between 1984 and 1993.

Blume, Eger, Fleischhauer, Hebel, Rej & Steiner (Editors): Towards Sustainable Land Use

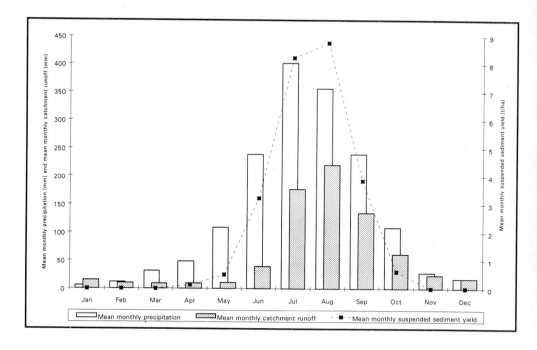

Figure 3: Minchet Catchment - Mean monthly precipitation, catchment runoff and suspended sediment yield data for Minchet Catchment.

	Precipitation[a]	Discharge[b]				Drainage	Suspended Sediment[b]		
			Volume	Yield	Runoff	Ratio	Rate	Yield	Concent.
	[mm]	[l/s]	[m³]	[l/s·km²]	[mm]	[%]	[t]	[t/ha]	[g/l]
Jan	6	6.4	17305	5.7	15	260.4	0.0	0.00	0.00
Feb	12	4.7	11412	4.2	10	86.8	0.0	0.00	0.00
Mar	31	3.8	10256	3.4	9	28.8	0.0	0.00	0.00
Apr	48	4.1	10739	3.7	9.5	19.6	14.7	0.13	1.37
May	109	4.6	12233	4.0	11	9.9	56.3	0.50	4.61
Jun	239	17.3	44762	15.2	40	16.5	365.4	3.22	8.16
Jul	402	75.0	200917	66.2	177	44.1	932.6	8.22	4.64
Aug	356	93.3	250009	82.3	221	61.9	993.3	8.76	3.97
Sep	240	59.0	152961	52.0	135	56.3	435.8	3.84	2.85
Oct	109	26.0	69516	22.9	61	56.3	69.1	0.61	0.99
Nov	28	10.4	26887	9.2	24	84.5	0.2	0.002	0.01
Dec	17	7.4	19678	4.5	17	98.5	0.6	0.005	0.03

a) The monthly means of precipitation for January to March are based on the data collected from 1985 to 1993. However, data collection started in April 1984, and therefore monthly means for April to December are based on the data collected from 1984 to 1993.

b) The monthly means of discharge and suspended sediment transport for January to May are based on the data collected from 1985 to 1993. However, data collection started in June 1984, and therefore monthly means for June to December are based on the data collected from 1984 to 1993.

Table 2: Minchet Catchment - Mean monthly precipitation, discharge and suspended sediment transport data for Minchet Catchment.

monthly means show less uncertainty. On the other hand, the months with minor rains are characterised by monthly means with higher coefficients of variation, i.e. the monthly data are largely scattered around these means. Regarding suspended sediment transport, the same intra-annual distribution of the coefficients of variation can be seen. This points to the only occasional suspended sediment transport in April, May, November and December. This is related to "favourable" conditions, i.e. the coincidence of high rainfall erosivity and low vegetation cover. Finally, the mean monthly discharge values have the smallest coefficients of variation, which are even rather constant.

The hydrographs clearly show a perennial flow behaviour for Minchet River, while the sedigraphs exhibit a limited period of sediment transport during the major rainy season, with sporadic occurrence whenever the conditions are favourable. High flow is of little importance for annual discharge, as the five highest quick flows contribute an average of only 7% to the annual discharge. High-yielding events, however, are rather important for suspended sediment transport, as the five highest-yielding events, an average of 6.5% of the events per year, contribute an average of 39.5% of the annual suspended sediment rate.

3 Impact Study

Miscellaneous indicators of the impacts of soil and water conservation measures on catchment discharge and suspended sediment transport are found in the comprehensive hydrosedimen-tological characterisation. Further indicators were found in a statistical and process-based analysis from which three examples are taken: double mass analysis, t-tests, and a comparative analysis of monthly mean suspended sediment concentration (Bosshart, forthcoming).

Fig. 4 shows the double mass plots for monthly catchment runoff, and suspended sediment yield data for the period January 1984 to December 1993. Double mass plots help to detect inhomogeneities in data sets that can be interpreted, in our case, as a modification of the suspended sediment transport pattern, possibly due to an impact of physical soil and water conservation measures. With regard to the hypothesised impacts of physical soil conservation measures on suspended sediment transport, a general negative curvilinear trend is interpreted as resulting from an on-going reduction of the suspended sediment transport, while catchment runoff remains stable.

Straight lines are visually fit to the data to distinguish periods for varying conditions of physical soil and water conservation measures. The pairs of boxes, containing dates and connected by lines, mark the temporal grouping of the data set.

Fig. 4 exhibits the mean suspended sediment concentration, i.e. the ratio of suspended sediment yield and catchment runoff, for the accumulated data at any time of observation. In this scatter plot three sections can clearly be distinguished, which are obviously related to the physical soil and water conservation measures. Each period can be roughly approximated by a straight line with a different gradient[4]. The first section represents the period before graded *Fanya juus* were systematically implemented on arable land in the entire Minchet catchment, with high mean suspended sediment concentrations. The second section characterises the period of well-maintained, highly-effective graded structures until 1990, when there was an obvious reduction of the mean suspended sediment concentration. The last section represents the time since the farmers reduced maintenance and even started to remove graded structures. This resulted in a renewed increase in mean suspended sediment concentrations.

---

[4] Statistical tests showed that the linear regression analysis carried out to prove the identified inhomogeneity was statistically unsuitable to describe the data sets with linear models. Therefore, causally based interpretations are given for the identified inhomogeneities, as there was no evidence of change in the monitoring set-up.

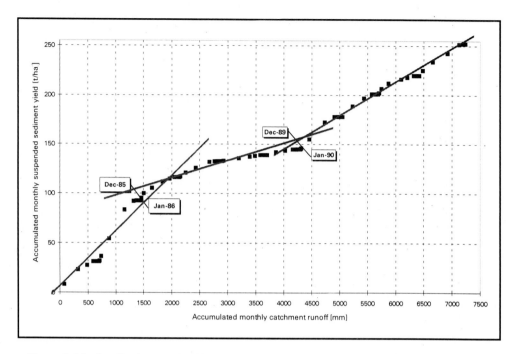

Figure 4: Minchet Catchment - Double mass plots of the monthly catchment runoff and suspended sediment yield data for the period 1984 to 1993.

Thus, the evident change of the gradient can be related to changes in the decisive conditions of physical soil and water conservation measures. The reduction of suspended sediment concentration can be related to the implementation of conservation structures[5]. The subsequent drastic increase in suspended sediment concentration can be related to an observed and reported reduction in maintenance, and even the removal of conservation structures (Herweg, forthcoming; Ludi 1994, 199).

Considering the periods of differences in suspended sediment yields, preferable t-tests were carried out to determine whether the differences are significant and attributable to identified changes in physical soil conservation measures implemented, or merely to random fluctuations. Table 3 shows the calculated mean and standard deviation of suspended sediment yield data for the different periods distinguished. The test results show that the mean monthly suspended sediment yields and the mean monthly suspended sediment concentration for the period of well-maintained, efficiently working physical soil and water conservation measures are statistically smaller at a 95% significance level than for the preceding period of observation.

---

[5] The modification of the gradient could also be linked to a modification of the land use pattern that obviously took place. However, the proportions of cultivated land, the land most susceptible to soil erosion, generally increased, with a peak in 1988 and 1989. This trend was obviously the opposite of that displayed by the suspended sediment transport, particularly for the period 1986 to 1989, when conservation structures were well-maintained and working effectively. In addition, the amount of pasture, i.e. the most effective cover unit against soil erosion, simultaneously decreased in a dramatic way.

	N	q_s [t/ha]		C_s [g/l]	
		mean	std	mean	std
1st Period	19	4.88	7.96	3.99	5.15
2nd Period	48	1.09	1.81	1.43	3.29
3rd Period	48	2.24	3.87	1.90	2.74
Notation	1st Period:	June 1984 to December 1985			
	2nd Period:	January 1986 to December 1989			
	3rd Period:	January 1990 to December 1993			
	N:	number of observations (months)			
	Q_s:	monthly suspended sediment yield			
	C_s:	monthly suspended sediment concentration			

Table 3: Mean and standard deviation of monthly suspended sediment yield and suspended sediment concentration data for different periods.

Figure 5 shows a comprehensive trend in the monthly mean suspended sediment concentration from April to October for the period 1984 to 1993[6]. The year before conservation measures were implemented, 1985, was chosen as a "year of reference", i.e. the data for this year are considered as baseline data. Thus the monthly mean suspended sediment concentration data for the other years show the percentage of change with reference to the data for 1985. The data presented as columns are linked with lines indicating the kind of change. Diverging lines indicate a comparative increase in the monthly mean suspended sediment concentration, converging lines a decrease.

These changes, outlined below and marked with capital letters, which refer to Figure 5, are attributable to the conservation measures (maintenance or deconstruction), but also to exceptional rainfall erosivity rates after periods of human-induced physical soil detachment caused by ploughing of fields. These observations highlight the importance of physical soil and water conservation measures, as high erosivity rates at the beginning and end of the major rainy season occur regularly, and a protective vegetation cover is lacking, particularly at the beginning of the rainy season.

(A) There is a general reduction in the monthly mean suspended sediment concentrations (MSSC), except for May and October. In May, a comparatively high monthly rainfall erosivity was recorded for a period with unconsolidated soil, resulting from the construction of conservation measures. In October, moderate rainfall erosivity was also recorded for a period of optional ploughing for the second barley crop.

(B) A further general decrease in the MSSC can be observed, except for April. Unconsolidated soil material, resulting from a comprehensive maintenance campaign to deal with the completely silted drainage ditches, was assumed to be most susceptible to the high monthly rainfall erosivity.

(C) A further general reduction in the MSSC can be seen, except for September, where they remained approximately stable, and in October. The latter is characterised by the highest monthly rainfall erosivity ever recorded in October, for a period of optional ploughing.

(D) The MSSC are generally steady or show a slight decrease. The lack of response in April to a comparatively high rainfall erosivity is remarkable, especially as it was observed in most of the other years.

(E) A first general increase in the MSSC occurred, particularly in May. This was apparently related to the highest monthly rainfall erosivity observed in May, the start of neglect of maintenance, and some removed conservation structures.

(F) Generally the MSSC remained approximately constant, except for May.

---

[6] Suspended sediment transport was never observed in January, February and March. Suspended sediment transport, which rarely occurred in November (1986, 1991) and December (1989), was excluded as it was being very low and might have overloaded the Figure.

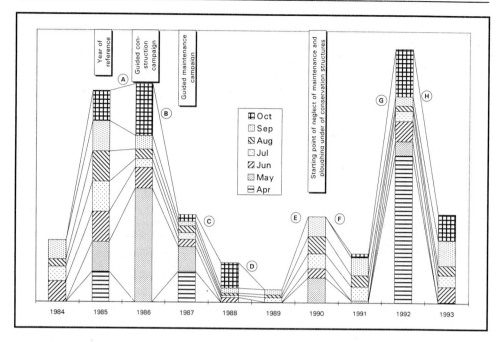

Figure 5: Minchet Catchment - Comparative trend of monthly mean suspended sediment concentrations for the period 1984 to 1993.

(G) A slight general increase in the MSSC occurred, with two remarkable exceptions. Firstly, the highest monthly rainfall erosivity was recorded in April, which coincided with a second period of intensive removal of conservation measures. Secondly, the highest monthly rainfall erosivity was observed in October, during a period of optional ploughing.

(H) A general decrease in the MSSC took place, except in September, which had the second highest monthly rainfall erosivity observed.

4 Conclusions

The hypothesised impact of conservation measures on catchment discharge and particularly on suspended sediment transport can be clearly identified for Anjeni catchment. During the period of well-maintained, highly efficient conservation measures, suspended sediment transport continuously decreased in a significant way, finally even falling below average tolerable soil loss rates. Since 1990, suspended sediment transport has increased again, with a trend towards the previous values before conservation, when the farmers reduced maintenance and even started to remove conservation structures.

References

Bosshart, U. (1996): Measurement of River Discharge for the SCRP Research Catchments: Gauging Station Profiles, Soil Conservation Research Project, Research, No. 30, Berne.

Bosshart, U. (forthcoming): Catchment Discharge and Suspended Sediment Transport as Indicators of Soil Conservation in the Minchet Catchment, Gojam Research Unit, A case study in the north-western Ethiopian Highlands, Soil Conservation Research Project, Research Report, Berne.
Breitenbach, F. von (1961): Forest and Woodlands of Ethiopia, Ethiopian Forestry Review, 1/1961, 5-16.
Breitenbach, F. von (1963): The Indigenous Trees of Ethiopia, 2nd edition, Ethiopian Forestry Association, Addis Abeba.
Herweg, K. (1993): Problems of Acceptance and Adaptation of Soil Conservation in Ethiopia, in Acceptance of Soil and Water Conservation Strategies and Technologies, E. Baum, P. Wolff and M. A. Zöbisch (eds.), Topics in Applied Resource Management 3, Witzenhausen, 391-411.
Herweg, K. (forthcoming): Soil loss and runoff (working title), in Soil Conservation Research Project (SCRP), forthcoming, SCRP - Data Bank Report (working title), Berne
Hurni, H. (1983): Soil Formation Rates in Ethiopia (with 8 maps, scale 1:1 mio), Working Paper 2, Ethiopian Highlands Reclamation Studies, Addis Abeba.
Hurni, H. (1986): Guidelines for Development Agents on Soil Conservation in Ethiopia, Community Forests and Soil Conservation Development Department (CFSCDD), Addis Abeba.
Kefeni Kejela (1995): The soils of the Anjeni Area - Gojam Research Unit, Ethiopia (with 4 maps), Soil Conservation Research Project, Research Report 27, Berne.
Krauer, J. (1988): Rainfall, Erosivity & Isoerodent Map of Ethiopia, Soil Conservation Research Project, Research Report 15, Berne.
Krauer, J. (forthcoming): Spatio-Temporal Perspectives on the Soil Resources of the Ethiopian Highland with Reference to Rural Environmental Development (working title), Geographica Bernensia, African Studies, Berne.
Ludi, E. (1994): Handlungsspielraum kleinbäuerlicher Familien und Handlungsstrategien zur Verbesserung ihrer Lebenssituation, Ein Fallbeispiel aus Anjeni, Äthiopien, Unveröffentlichte Diplomarbeit am Geographischen Institut der Universität Bern, Bern.
Million Alemayehu (1992): The Effect of Traditional Ditches on Soil Erosion and Production, on-farm trials in Western Gojam, Dega Damot Awraja, Soil Conservation Research Project, Research Report 22, Berne.
SCRP, Soil Conservation Research Project (1982): First Progress Report (Year 1981), H. Hurni, Soil Conservation Research Project, Volume 2, Berne.
SCRP (1986), Fourth Progress Report (Year 1984): H. Hurni, Soil Conservation Research Project, Volume 5, Berne.
SCRP (1988): Fifth Progress Report (Year 1985), H. Hurni and M. Grunder, Soil Conservation Research Project, Volume 6, Berne.
SCRP (1991a): Seventh Progress Report (Year 1987), M. Grunder and K. Herweg, Soil Conservation Research Project, Volume 8, Berne.
SCRP (1991b): Eighth Progress Report (Year 1988), K. Herweg and M. Grunder, Soil Conservation Research Project, Volume 9, Berne.
Thornthwaite, C. W. (1948): An approach toward a rational classification of climate, Geographical Review, Vol. XXXVIII, 55-94.
Tsehai Berhane-Selassie (1994): Social Survey of the Soil Conservation Areas Dizi, Anjeni and Gununo (Ethiopia), Soil Conservation Research Project, Research Report 24, Berne.
Werner, Ch. (1986): Soil Conservation Experiments in the Anjeni Area, Gojam Research Unit, Ethiopia, Soil Conservation Research Project, Research Report 13, Berne.
WMO, World Meteorological Organization (1987): Tropical Hydrology, Operational Hydrology, Report No. 25 (WMO Publication No. 655), Geneva.

Address of author:
Urs Bosshart
Federal Department of Public Economy,
Federal Office of Agriculture
Mattenhofstrasse 5,
CH-3003 Bern, Schweiz

Organic Matter Stability and Nutrient Availability under Temperate and Tropical Conditions

H. Tiessen, E. Cuevas & I.H. Salcedo

Summary

Soil organic matter (SOM) provides plant nutrients to low-input agriculture. Nitrogen and phosphorus release depend on the mineralisation of SOM, while cation exchange and soil physical properties depend on the maintenance of SOM levels. Research and modelling of SOM transformations in temperate soils has advanced to a point where prediction of SOM levels under different management and climate is possible, but SOM transformations in tropical soils deviate from these patterns. Examples are presented from one temperate and two tropical sites to illustrate differences in SOM transformations. Organic matter has a higher decomposition rate in the tropics, as might be expected at the higher temperatures. In addition, tropical soils have a lower ability to stabilise the residues of SOM turnover in the long term. This lower stability is seen in the much lower radiocarbon ages of tropical SOM. As a result of faster SOM turnover, organic nutrient supply is short-lived, and biological short-term transformations of organic residues are more important for optimal management of fertility in the tropics than in temperate soils.

Keywords: Radiocarbon age, soil organic matter models, phosphorus, nitrogen, slash-burn agriculture.

1 Introduction

The ecological impact of high input agriculture in the developed world and the continuing reliance of developing countries on minimal-input production has rekindled an interest in the role of soil organic matter (SOM) in soil quality and fertility. Soil organic matter buffers nutrient supply and stabilises soil structure. It also represents a large pool of sequestered C, which maintains lowered atmospheric CO_2 levels. The N and P supplying abilities of SOM are related to its transformations, whereas C sequestration, stabilisation of physical soil properties, and contributions to cation exchange capacity rely on SOM stability. In order to optimize SOM function in soils, a balance between turnover and stability must be managed. Disturbance of soil carries the risk of SOM degradation and nutrient loss. On the other hand, any agricultural, grazing or forestry production depends on nutrient mobilisation for crop or pasture uptake and subsequent export. The sustainability of a production system depends on the efficiency of nutrient mobilisation for crop use, so that economic returns may be obtained, and on the avoidance of nutrient loss, so as to minimise required inputs and environmental degradation. Maintaining a balance between mobilisation and conservation is the essential task in "sustainable land use".

Many low-input agricultural systems rely entirely on the decomposition of SOM and biomass to release nutrients for crop uptake. In the North American Great Plains, for instance, crop nutrition during the initial years of cultivation relied heavily on mineralising SOM (Tiessen et al., 1982). This permitted adequate crop production without supplemental fertilisation in the semi-arid Canadian prairies for some 50 years. A similar reliance on mineralized nutrients under tropical semi-arid conditions of northeastern Brazil limits continuous land use to only 5 years before land is abandoned in a cycle of shifting cultivation (Tiessen et al., 1992). Even shorter cycles of shifting cultivation of 2-4 years are typical for extremely humid rainforest sites (Uhl, 1987). The viability or durability of agriculture without supplemental fertilisation appears to be related to the stability of soil organic matter at those sites (Tiessen et al., 1994a). In order to be able to predict soil fertility and "life span", it is therefore important to understand the controls on SOM turnover and stabilisation.

2 Organic matter turnover

Much of the understanding of SOM transformations, plant production and soil fertility has been developed on temperate soils (Parton et al., 1987). Soil organic matter is not uniform, and different SOM components turn over or mineralize at different rates, from rapidly decomposing recent litter to more inert mineral-associated humus. The relative amounts of SOM in pools with different stabilities are estimated in SOM models. Soil texture, climate and management determine the SOM distribution among different pools. The resulting computer models such as the CENTURY model are able to predict SOM levels under different conditions, and also provide a spectrum of nutrient availabilities associated with SOM transformations (Parton et al., 1989). Organic matter with slow turnover is implicated in the long-term supply of nutrients. Nutrients tied up in rapidly mineralizing SOM are quickly exhausted upon disturbance. Application of models that are successful in temperate zones to nutrient-limited tropical soils is meeting with only limited success (Gijsman et al., 1996). The following is an attempt to reconcile some of the information available on SOM production and turnover in tropical and temperate soils in order to advance our understanding of SOM transformations in tropical soils.

SOM builds up as a result of the incorporation of plant residues into the soil. Plant litter has an approximate C: N: P ratio near 500: 10: 0.6. When decomposers mineralize plant litter, CO_2 is respired and nutrients are concentrated in the decomposer biomass with an approximate C: N: P ratio of 50: 10: 1. The residue of this process, i.e. the soil organic matter, typically has an intermediate C: N: P ratio of 100: 10: 1. Carbon turnover therefore results in N and P release as the nutrient contents of SOM change during transformations.

In a mature primary ecosystem, the biomass production and sequestration of nutrients is closely balanced with decomposition and nutrient release. Knowledge of primary production therefore provides an opportunity to estimate the flow-trough of C and associated nutrients in the system. The input of organic matter into the soil is determined by the litter production. Based on the linear simplification of a first order decay for SOM, the theoretical mean residence time (MRT) of SOM can be calculated from an equation used to describe decay:

$$k[SOM] = \frac{(\Delta[SOM])}{(\Delta t)}$$

and

$$MRT = -\frac{(1)}{(k)}$$

where k is the reaction constant for the decomposition of [SOM] and t is the time, here measured in years. Since decomposition is unknown but litter production is equal to decomposition under steady state, ecological productivity data can be related to the turnover of SOM.

In a cold temperate grassland most of the primary production is transformed to litter within the annual cycle of seasons. Above-ground productivity for prairie near Saskatoon, Canada, has been estimated between 1.3 Mg ha^{-1} y^{-1} (Redmann et al., 1993) and 2.5 Mg ha^{-1} y^{-1} (Redmann, 1991). Below-ground allocation of photosynthate is about 50% of the photosynthesis, giving a total for plant production in the prairie ecosystem a little below 5 Mg ha^{-1} y^{-1} or about 2 Mg C ha^{-1} y^{-1}. Under this productivity regime, a sandy Mollisol had a total C content of 88 Mg ha^{-1} with a ^{14}C-derived MRT of 330 y (Tiessen et al., 1994a). Adding to the SOM the standing stock biomass of approximately 10 Mg C ha^{-1}, gives a total C reservoir in the ecosystem near 100 Mg ha^{-1}. The annual turnover is therefore 2 % of the stock, giving an MRT of around 50 y, or only one-sixth of the MRT determined by ^{14}C dating. When only below-ground production and stocks are considered (to avoid above-ground material that might decompose without entering the SOM), MRT is raised to 90 y, still much below the ^{14}C value. Since some of the above ground litter enters the soil compartment, the most realistic estimate for MRT will be between 50 y and 90 y, say 70 y.

The large difference between the MRT calculated from annual production and consequent decomposition as compared to the ^{14}C-based MRT is related to the soil's ability to stabilise SOM differentially. To account for such differential stabilisation, the CENTURY model assigns a portion of the soil's C to a stable pool with a MRT of 1500 y, while most of the C turns over much more rapidly. This model has been successful in predicting SOM levels in many temperate ecosystems, but when CENTURY was used to model SOM in a tropical soil, it apparently produced the best results with the stable pool "turned off" (Veldkamp, 1994).

A turnover estimate from production data for the tropical semiarid site examined by Tiessen et al. (1992, 1994a) is necessarily based on more tenuous information since production and biomass data for Caatinga vegetation are scarce. Schacht et al. (1989) estimated a total litter production equivalent to 1.5 Mg C ha^{-1} y^{-1}. Tiessen et al. (1997), reviewing the available data arrived at a total (above and below ground) biomass estimate of 10 to 20 Mg C ha^{-1}. Since above and below ground biomasses are approximately equal, one might assume similar production rates, giving a total C addition of 3 Mg C ha^{-1} y^{-1}. Total soil C at the site under native conditions was 34 Mg ha^{-1} (Tiessen et al., 1994a). Excluding long-lived woody above-ground biomass from the stock, total C in the system is about 45 Mg ha^{-1} which, at a production rate of 3 Mg, gives a MRT of 15 years. The ^{14}C-derived estimate for MRT of the SOM was between 20 and 50 y. The much closer fit between ^{14}C- and input- derived MRTs in this tropical soil than the above temperate example indicates that little C is in the stable pool. This substantiates Veldkamp's (1994) experience that the turnover of SOM under tropical conditions is best described without a stable SOM pool. What is responsible for this difference between temperate and tropical conditions?

The ratio of input-derived MRTs between the temperate and tropical semi-arid sites for below ground materials is near 5 (70/15). Jenkinson and Anayaba (1977) reported a 5 times faster decomposition rate of organic residue for Nigeria than for the U.K. This is an indication that both litter and the active C pools in the tropics cycle some 5 times faster than in temperate soils (under comparable moisture regimes, humid in Jenkinson and Anayaba's study, semi-arid in our sites). This accelerated turnover may well be a simple temperature (Q_{10}) effect on decomposition rate.

The comparisons of production- and ^{14}C- derived MRTs indicate that there is also a "stabilisation" factor in addition to the Q_{10} effect. The difference between the two MRT estimates is a factor of about 5 in Canada, and half that in Brazil. A search for resistant C in the tropical soil, showed only one somewhat stable fraction, associated with silt-sized aggregates, with a ^{14}C MRT of 100y (Shang and Tiessen, submitted to Soil Science). Coarse clay-associated SOM, in Saskatchewan on the other hand has been dated older than 1000 y. This indicates that the tropical soil provides less long-term stabilisation of C.

Applying the same reasoning to the more humid Amazon forest site (Tiessen et al., 1994a), OM stabilisation is even less. The site at San Carlos had a total biomass of 150 Mg C ha^{-1} with and annual litter production of 10 Mg C ha^{-1} y^{-1} and a SOM (including litter mat) pool of 60 Mg C ha^{-1}. In this case, much of the biomass C is in long-lived boles, so that a more relevant figure might be the below-ground plus litter-layer biomass of 30 Mg C ha^{-1} (Medina and Cuevas, 1989). The C in SOM plus below-ground biomass has a calculated MRT of only 9 years. The ^{14}C MRT of mineral associated SOM in this soil was 30 to 50 y but this fraction accounted for only 15% of SOM (Tiessen et al., 1994a). The remainder was too labile to be dated using ^{14}C. In this Amazonian soil the "soil stabilisation" of introduced C is therefore imperceptible, and SOM turnover is largely determined by biotic processes of production and decomposition. The experience of local slash-burn agriculture confirms this: land is usually abandoned when the litter mat has disappeared, 3 years after the burn.

What are potential mechanisms for the soil "stabilisation factor", and why are there such differences between temperate and tropical soils? Clay content certainly plays a role. Heavier textured Amazonian soils than that at San Carlos retain much more organic C under cultivation (Sanchez et al., 1983). In a subtropical soil, C associated with clay particles had 10 times greater turnover times (although this was still only 59 y) than SOM in coarser fractions (Bonde et al., 1992). Expanding layer clays, which are more common in temperate soils are more effective in sorbing organic materials than 1:1 clays, and a cation exchange with polyvalent cations aids sorption (Harter, 1977). In Ca-rich soils, clay-Ca-SOM complexes are formed that are in part responsible for soil aggregation (Oades, 1988) and that provide long term stability to SOM in temperate soils (Anderson and Paul, 1984). But recent ^{14}C dates from a Kenyan Vertisols were also much younger than those of temperate soils (Warren and Meredith, 1997), indicating that tropical soils containing expanding clays perhaps do not provide substantially more long-term stability for SOM. Transition metals form strong complexes with organic materials (Harter, 1977). High Al concentrations have been implicated in very strong SOM stabilisation in a temperate soil (Carballas et al., 1980), and once formed, the SOM-Al complexes were not disrupted by liming which precipitates free Al as hydroxides (Condron et al., 1993). Iron stabilised SOM in the Amazonian soil inside lateritic concretions to such an extent that ^{14}C age was 4600 y in a soil that otherwise contained essentially only modern material (Tiessen et al., 1994a), but this may be an effect of physical obstruction inside Fe-rich nodules. In summary, the available evidence suggests that cation suite, mineral type, particle size distribution and biological activity all affect SOM stabilisation. Most of these factors differ substantially between tropical and temperate soils, and the mechanistic details that could explain the SOM lability observed in the tropics have not yet been synthesised. The high cost of ^{14}C dating has precluded a more systematic study of these factors responsible for long-term SOM stabilisation.

3 Organic nutrient cycling

After some 50 years of cultivation, the cold temperate soil from Canada requires N and P fertilisation for adequate crop production. It's SOM loss has levelled off, crop production with fertilizer applications is economically viable, and environmental impact is minimal as long as adequate ground cover is maintained to prevent erosion losses of SOM and nutrients. The average ^{14}C age of the organic matter in cultivated sites is nearly double that of the native site (Martel and Paul, 1974). Currently, the remaining SOM is therefore more stable than under native conditions, providing a buffer resistant to further degradation.

At the tropical sites, SOM degradation led the process of fertility decline terminating in land abandonment. The Oxisol in semi-arid NE Brazil was examined trough a complete cycle of slash burn of the native thorn forest, followed by cultivation, abandonment and regrowth (Tiessen et al., 1992). After clearing the native semi-arid thorn forest, six years of cultivation with minimal fertilization resulted in the mineralisation of 30% of the soil's C, N and organic P. In this nutrient-poor Oxisol, crop production depended on organic nutrient supply, and the site was abandoned to bush fallow after 6 years.

Differences in soil fertility between different stages of the shifting cultivation cycle were examined in detail (Salcedo et al., 1997). Soils from the native thorn forest, from a recently slashed and burnt site and from a site abandoned after 6 years of manual cultivation to cassava were compared. The effect of burning was an increase in soil nutrient content, followed by decreases of N and organic P by 20%, available P by 70%, and exchangeable Ca by 60%, Mg by 70% and K by 45% during subsequent cultivation. These changes affected plant production, which was examined in a subtractive fertilisation greenhouse trial and an incubation experiment for C and N mineralisation.

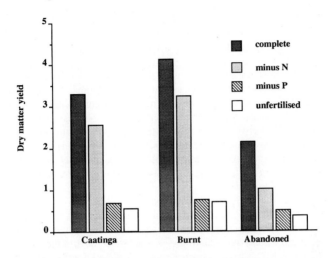

Figure 1: *Plant dry matter yields in a greenhouse subtractive fertilisation trial on soils native, burnt and abandoned plots of a semi-arid Oxisol (Tiessen et al., 1997).*

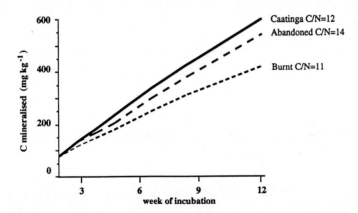

Figure 2: *Mineralisation of C and C:N ratios of mineralised soil organic matter during incubation of soils from native, burnt and abandoned plots of a semi-arid Oxisol (Tiessen et al., 1997).*

Without remedial fertilization highest dry matter yields were obtained in the burnt, and only half that yield in the abandoned soil. The relative yield ratios were not changed by nutrient addition, although total yields were increased by a factor of 5. In all soils, the most severe nutrient limitation was for P followed by N (Fig. 1). Carbon mineralisation over a 12 week incubation was greatest for forest, intermediate for the abandoned and least for the burnt soil whereas N removed in 13 leachings was greatest from the forest less from the burnt and least from abandoned soil (Fig. 2). The SOM in the abandoned soil was more labile than that of the two other soils but also had a higher C:N ratio, so that nutrient release was less. Nutrient limitations were the main reason for curtailing production on this Oxisol, and organically-held nutrients played a crucial role in supplying crop demands. Nevertheless, remedial fertilization was unable to fully compensate for the yield decline in the abandoned soil. The observed yield reductions, which were responsible for land abandonment were therefore not explained entirely by the factors examined. It is apparent that SOM plays a vital role in soil quality at this site, but not all of the productivity decline can be attributed to nutrient availabilities. Perhaps (micro-) biological interactions with SOM abundance play a role.

The highly leached rainforest soil is an extreme example for the fragility of an ecosystem. Several studies have examined the finely-tuned organic matter and nutrient cycles that maintain the rainforests on a soil with extremely low nutrient availability (Cuevas and Medina, 1986; Medina and Cuevas, 1989). Organic matter in the forest litter mat plays the main role in conserving and recycling nutrients. Nutrient recycling in the intact forest depends largely on biological processes, such as the retranslocation prior to leaf abscission (Cuevas and Medina, 1990), and the reutilisation of nutrients from litter fall by a dense above-ground root mat. Comparisons between sites less than 10 elevation metres apart showed that forest adaptation was extremely fine-tuned to edaphic conditions (Tiessen et al., 1994b). Litter fall on Oxisols under mixed forests was 50% greater than on lowland sites with quartz sands. Associated N fluxes were 400% greater on Oxisols, while P fluxes were similar to those on sands, indicating that forest on Oxisol is limited by P availability and forest on sand by N availability and that the vegetation has adapted its nutrient uptake and recycling to the dominant nutrient limitations (Table 1). These intricate mechanisms of forest adaptation and biotic recycling of nutrients cannot be achieved under agricultural use. In the Oxisol, 60% of the soil's C, 65% of the N and 50 % of the organic P have a mean residence time of less than 4 y. The destruction of the litter layer is therefore sufficient to reduce nutrient availability to levels too low for agricultural use within 4 to 5 y of land clearing. The end of the cultivation cycle coincides with the disappearance of the root and litter mat, indicating that the mineral soil contributes little to sustained crop nutrition.

Soil	litter production per nitrogen ($g\ g^{-1}$)	Litter production per phosphorus ($g\ g^{-1}$)
Oxisol	63	3600
Quartz sand	140	2000

Table 1: Nutrient use efficiency in the production of leaf litter (expressed as the mass of litter produced per mass of N or P contained in the litter) on an Oxisol and a quartz sand at San Carlos de Rio Negro (Cuevas and Medina, 1986).

In conclusion, low-fertility tropical soils appear to rely to a greater extent on organic nutrient supply than their temperate counterparts. Organic matter turns over rapidly in the tropics, and is also stabilised to a lesser extent. The SOM-associated nutrient supplies are rapidly exhausted. Management of tropical soils therefore needs to maintain adequate soil organic matter levels which hold nutrients in a form protected from leaching, excessive sorption or volatilisation, but mineralisable at rates sufficient for crop, pasture or forest supply. In the absence of substantial mineral stabilisation, it is likely that management of short-term organic matter supply, for instance through mulches and agroforestry, will provide the most appropriate options for sustainable management.

References

Anderson, D.W. and , E.A. (1984): Organomineral complexes and their study by radiocarbon dating. Soil Sci. Soc. Am. J. **48**, 298-301.

Bonde, T.A., Christensen, B.T. and Cerri, C.C. (1992): Dynamics of soil organic matter as reflected by natural 13C abundance in particle size fractions of forested and cultivated Oxisols. Soil Biol. Biochem. **24**, 275-277.

Carballas, M., Cabaneiro, A., Guitan-Ribera, F. and Carballas, T. (1980): Organo-metallic complexes in Atlantic humiferous soils. Ann. Edafol. Agrobiol. **39**, 1033-1043.

Condron, L.M., Tiessen, H., Mª Trasar-Cepeda, C., Moir, J.O. and Stewart, J.W.B. (1993): Effects of liming on organic matter decomposition and phosphorus extractability in an acid humic Ranker soil from northwestern Spain. Biology and Fertility of Soils, **15**, 279-284.

Cuevas, E. and Medina, E. (1986): Nutrient dynamics within amazonian forests I. Nutrient flux in fine litter fall and efficiency of nutrient utilization. Oecologia **68**, 466-472.

Cuevas, E. and Medina, E. (1988): Nutrient dynamics within amazonian forests II. Fine root growth, nutrient availability and leaf litter decomposition. Oecologia **76**, 222-235.

Cuevas, E. and Medina, E. (1990): Phosphorus/nitrogen interactions in adjacent Amazon forests with contrasting soils and water availability. In: H. Tiessen, D. López-Hernández and I.H. Salcedo (eds.), 1991 Phosphorus Cycles in Terrestrial and Aquatic ecosystems Regional Workshop 3: South and Central America. Saskatoon: Saskatchewan Institute of Pedology, University of Saskatchewan, pp. 84-94.

Gijsman, A.J., Oberson, A., Tiessen, H. and Friesen, D.K. (1996): Limited applicability of the CENTURY model for highly weathered tropical soils. Agronomy Journal, **88**, 894-903.

Harter, R.D. (1977): Reactions of minerals with organic compounds in the soil. In: J.B. Dixon and S.B. Weed (eds.), Minerals in soil environments. Soil Sci. Soc Am. Madison, USA, pp. 709-739.

Jenkinson, D.S. and Anayaba, A. (1977): Decomposition of carbon-14 labelled plant material under tropical conditions. Soil Sci. Soc. Am. J. **41**, 912-915.

Martel, Y.A. and Paul, E.A. (1974): Effects of cultivation on the organic matter of grassland soil as determined by fractionation and radiocarbon dating. Can. J. Soil Sci. **54**, 419-426.

Medina, E. and Cuevas, E. (1989): Patterns of nutrient accumulation and release in Amazonian forests of the upper Rio Negro basin. In: J Proctor (ed.) Mineral nutrients in tropical forest and savanna ecosystems. Blackwell Scientific Publications, Oxford, U.K. p. 217-240.

Oades, J.M. (1988): The retention of organic matter in soils. Biogeochemistry 5, 35-70.

Parton, W.J., Schimel, D.S., Cole, C.V. and Ojima, D.S. (1987): Analysis of factors controlling soil organic matter levels in Great Plains grasslands. Soil Sci. Soc. Am. J. 51, 1173-1179.

Parton, W.J., Cole, C.V., Stewart, J.W.B., Ojima, D.S. and Schimel, D.S. (1989): Simulating regional patterns of soil C, N and P dynamics in the US central grasslands region. p. 99-108. In: M. Clarholm and L. Bergström (eds.), Ecology of arable land. Kluwer Academic Publishers, Dordrecht, Netherlands.

Redmann, R.E. (1991): Primary productivity. Chapter 5 In: R.T. Coupland (ed.) Ecosystems of the World 8A, Natural Grasslands. Elsevier, Amsterdam, pp. 75-93.

Redmann, R.E., Romo, T.J. and Pylypec, B. (1993): Impacts of burning on primary productivity of Festuca and Stipa-Agropyron grasslands in Central Saskatchewan. Am. Midl. Nat. **130**, 262-273

Salcedo, I.H., Tiessen, H. and Sampaio, E.V.S.B. (1997): Nutrient availability in soil samples from shifting cultivation sites in semiarid NE Brazil. Agriculture Ecosystems and Environment (in press).

Sanchez, P.A., Villachica, J.H. and Bandy, D.E. (1983): Soil fertility dynamics after clearing tropical rainforest in Peru. Soil Sci. Soc. Am. J. **47**, 1171-1178.

Schacht, W.H., Mesquita, R.C.M., Malechek, J.C. and Kirmse, R.D. (1989): Response of caatinga vegetation to decreasing levels of canopy cover. Pesp. Agropec. Bras. **24**, 1421-1426.

Tiessen, H., Stewart, J.W.B. and Bettany, J.R. (1982): Cultivation effects on the amounts and concentrations of C, N and P in grassland soils. Agronomy Journal, **74**; 831-835.

Tiessen, H., Cuevas, E. and Chacon, P. (1994a): The role of soil organic matter in sustaining soil fertility. Nature, **371**, 783-785.

Tiessen, H., Chacon, P. and Cuevas, E. (1994b): Phosphorus and nitrogen status in soils and vegetation along a toposequence of dystrophic rainforests on the Upper Rio Negro. Oecologia, **99**, 145-150.

Tiessen, H., Salcedo, I.H. and Sampaio, E.V.S.B. (1992): Nutrient and soil organic matter dynamics under shifting cultivation in semi-arid Northeastern Brazil. Agriculture Ecosystems and Environment, 38, 139-151.

Tiessen, H., Feller, C., Sampaio, E.V.S.B. and Garin, P. (1997): Carbon sequestration and turnover in semi arid savannas and dry forest. Climate Change. (in press).

Uhl, C. (1987): Factors controlling succession following slash-and-burn agriculture in Amazonia. J. Ecol. **75**, 377-407

Veldkamp, E. (1994):Organic carbon turnover in three tropical soils under pasture after deforestation. Soil Sci. Soc. Am. J. **58**, 175-180.

Warren, G. and Meredith, J. (1997): Isotopic composition of organic C. In: L. Bergström and H. Kirchman (eds.), Carbon and nutrient dynamics in natural and agricultural tropical ecosystems. CAB International, Wallingford, U.K. in press

Addresses of authors:
H. Tiessen
Dept. of Soil Science
University of Saskatchewan
51 Campus Drive
Saskatoon S7N 5A8, Canada
E. Cuevas
Centro de Ecologia
Inst. Venezolano de Investigaciones Cientificas
Apt. Postal 21827
1020A Carcas, Venezuela
I.H. Salcedo
Depto. De Energia Nuclear
Universidade Federal de Pernambuco
Ave. Prof. Luiz Freire 1000
50730 Recife, PE Brazil

Soil Organic Matter and Sustainable Land Use

M. Körschens

Summary
Sustainable land use involves optimal soil carbon content which is sufficient for high biomass production and yet not too high to induce nitrogen and carbon losses. Investigations should include the mineralisable part of carbon and nitrogen. The evaluation of data for 20 long-term experiments in central Germany show that the mineralisable C and N in the soil have a relatively narrow ecological optimum in arable soils which ranges between 0.2 to 0.6 % C or 0.02 to 0.06 % of N in the topsoil. Below these values, soil fertility, yield and CO_2 binding are insufficient; above these values, environmentally dangerous losses occur.

Keywords: Sustainable land use, soil organic matter, nitrogen balance, long-term experiments, fertilisation

1 Introduction

Sustainable land use requires the maintenance and improvement of soil fertility as well as avoiding environmental pollution. The mineralisable part of soil organic matter (SOM) can be influenced by management systems, especially by organic and mineral fertilisation.

The carbon content (also soil organic matter content) is an important factor which influences the concentration of trace gases such as CO_2, N_2O CH_4 in the atmosphere. Despite intensive humus research in the last few decades the question on the optimal humus content is still not settled. In most investigations the total carbon content is taken into account, which may lead to incorrect conclusions, because the mineralisable part of carbon is important as well. This portion can, for example, amount in a sandy soil roughly 50 %, and in a loamy soil only 25 % of the total carbon content. To solve this question, long-term experiments are necessary which, whenever possible, have reached a steady state.

2 Methods

The impact of management on the mineralisable carbon content in the soil was quantified by using the results of numerous long-term experiments, listed in Table 1 and of other long-term experiments in Bad Lauchstädt (Germany). Optimal soil carbon and nitrogen contents were estimated using the diffence between control plot and the carbon content in the plot with highest organic and mineral fertilisation.

According to the results of previous studies, the carbon content of the control plots which have not received any fertiliser for decades is regarded as a criterion for the inert-or stable-carbon (C_i), i.e. carbon which is not involved in transformation processes. The carbon input by roots on loamy soils is less than 0.1 % C and is not taken into account. On very light sandy soils there is no plant growth on control plots and thus no carbon input.

No.	Experimental sites	Initial year	clay content %	mean temp °C	precipitation mm	Author
1	Groß Kreutz (P 60)	1959	2	8,9	537	Asmus, 1990a
2	Groß Kreutz (M 4)	1967	2	8,9	537	Asmus, 1990b
3	Bad Salzungen	1966	2	7,7	600	Ansorge et al., 1992
4	Thyrow (soil fert.)	1938	3	8,6	520	Schnieder, 1990
5	Thyrow	1937	3	8,6	520	Schnieder, 1990
6	Ascov (Denmark)	1894	4	7,7	790	Christensen, 1989
7	Müncheberg	1963	5	8,2	521	Rogasik, 1995 unpubl.
8	Skierniewice (Poland)	1923	5	7,9	520	Mercik, 1993
9	Dülmen	1958	5	9,7	878	Wollring, 1993 unpubl.
10	Berlin-Dahlem	1923	5	9,2	549	Krzysch et al., 1992
11	Rostock	1953	6	8,4	599	Reuter, 1990
12	Spröda	1966	6	8,3	540	Ansorge et al., 1992
13	Seehausen (crop rot.)	1958	8	9,0	556	Leithold, 1992
14	Seehausen (comb.)	1967	8	9,0	556	Hülsbergen, 1992
15	Halle	1878	8	9,2	501	Stumpe et al., 1990
16	Ascov (Denmark)	1894	12	7,7	790	Christensen, 1989
17	Bernburg	1910-1962	16	8,8	474	Wabersich, 1967
18	Bad Lauchstädt	1902	21	8,7	484	Körschens et al., 1994
19	Grignon (France)	1875	22	11	640	Houot et al., 1995
20	Järna (Sweden)	1958	30	6	550	Pettersson et al., 1992

Experiments without specification of country are located in Germany

Table 1. Overview of the evaluated long-term fertilisation experiments.

3 Results and discussion

The differences in the carbon and nitrogen content between the control plots and the differently fertilised plots of long-term experiments are taken as experimental proof for the mineralisable SOM.

Figure 1 shows the results of 20 long-term experiments. The Corg content of the control plots (C_i) lies between 0.34 and 2.45 % with a mean value of 0.85 %. The difference between the control plot and the treatments with the highest fertilisation (C_m) ranges from 0.17 to 0.66 % and is 0.37 % on average.

In Fig. 2 the C_{org} and N_t contents of the treatments without any fertiliser, with exclusive mineral fertilisation and with combined fertilisation, are given for selected long-term experiments. The C_i content ranges from 0.56 to 2.45 %, the exclusively mineral fertilisation causes an increase of 0.08 to 0.16 %, and the influence of the additional manure fertilisation shows an increase of 0.37 % on average (0.16 %-0.50 %).

At the same time, the close connection between the clay content and the C_i content is evident, the latter rising significantly with the clay content.

From these and previous investigations (Körschens et al., 1995) it can be concluded, that the mineralisable carbon and nitrogen fall within a relatively narrow optimal range, which under the conditions of central Germany and comparable locations amounts to 0.2 to 0.6 % C and 0.02 to 0.06 % N, respectively. Lower contents of mineralisable carbon and nitrogen are insufficient for soil fertility and yield.

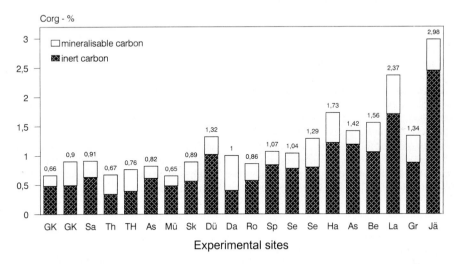

Figure 1: Content of inert carbon (C_i) and mineralisable carbon (C_m) in selected long-term experiments (see Table 1)

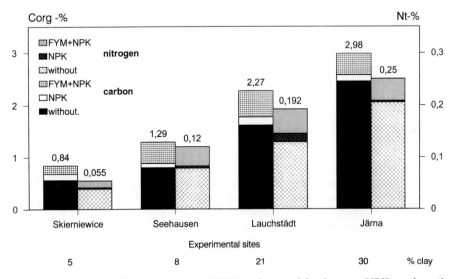

Figure 2: Influence of farmyard manure (FYM) and mineral fertilisation (NPK) on the soil carbon and nitrogen content in selected long-term experiments

When stipulating optimal soil carbon contents a second factor is of importance. The relation between soil carbon content and physical soil properties is almost functional. According to Körchens and Waldschmidt (1995), an increase of the carbon content by 0.1 % results in:
- an increase in hygroscopicity of   0.06   to 0.08 mass percent
- an increase in water capacity of   0.4   to 0.6 mass percent
- a decrease of dry matter density of 0.004 to 0.005 g/cm^3
- a decrease of bulk density of   0.006 to 0.008 g/cm^3.

The humus content plays an important role on light sandy soils, such as maintaining a sufficient water and sorption capacity. On an ideally textured Haplic Chernozem, this component is of much less importance.

Nevertheless, when transferring these results into practice one should consider, that the C and N content of the soil changes very slowly. It takes many decades until a new steady state has been established after a change of the management system. The necessary time period depends on the initial level. Figure 3 shows the results of a model experiment on a Chernozem illustrating the influence of black fallow on the soil carbon content under largely differing initial carbon contents. At a high initial level the C content decreases by 0.57 %, at a low initial level the decrease is only some 0.18 %. After 40 years the new steady state, which is to be expected at a content of 1.5 % of C (according to the C_i in this location), has not yet been achieved.

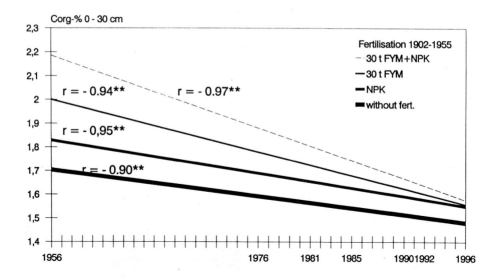

Figure 3: Dynamics of soil carbon content depending on initial level, investigated in a model experiment with Chernozem under bare fallow

In 1978 a part of the Static Experiment at Bad Lauchstädt was reorganised in a way that the different SOM levels of the fertilisation variants after seventy years were used as new initial levels. On this basis it was possible to test the effect of different organic and mineral fertilisation treatments on the Corg content depending on the initial level (Fig. 4).

The omission of any fertilisation at a high initial level leads to a decline of the Corg of 0.14 % (from 2.25 % to 2.11 %). In case of unchanged fertilisation the initial level remains unchanged. Combined organic/mineral fertilisation at a low initial level leads to an increase of 0.33 % C_{org}

(from 1.74 % to 2.02 %). A continued omission of fertilisation did not cause any further decline of the low C_{org} level. However it is obvious that after 16 years the new steady state has not yet been reached.

Sustainable land use requires high and increasing yields per land unit to provide enough food for the steadily increasing population in the world.

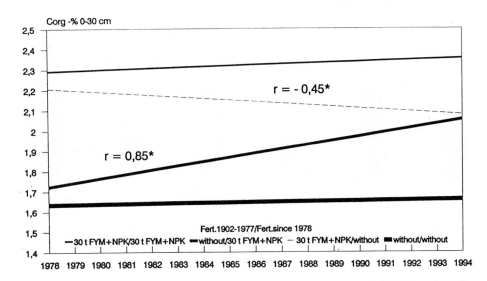

Figure 4: C dynamics depending on initial level and fertilisation in the Static Experiment Bad Lauchstädt after extension of experimental question (crop rotation: potatoes, winter wheat, sugar beet, spring barley)

The following discussion includes the comparison of the yields of the main variants of the Static Experiment Bad Lauchstädt for the years 1906 - 1915 and those for the years 1986 - 1995.

Winter wheat (Fig. 5 as one example) shows a mean yield increase of 91 % for all fertilised variants. During the first period, there were no significant differences between the yields of all fertilised treatments. The yields of the control variant increased by nearly 2 t/hectare, the main reason of this being the atmospheric deposition of N, the improved productivity of the sorts and the application of agrochemicals. Moreover, wheat is capable of taking up N from depths of down to 2 meters. A yield advantage resulting from organic fertilisation or from the higher soil C content in the organically fertilised variants can, as in the first period, not be proved.

Spring barley, with a yield increase of 103 % in the two investigated periods, shows a stronger reaction to organic manure application than winter wheat. Here, the yield increase is 9 %.

Summarizing it can be detected from the figures that there has been a considerable yield increase within a period of 80 years, which is highest for cereals amounting to 90-100 %. This is the result of progress in plant breeding, which for cereals is more advanced than for other crops. It is also a result of improved management of fertilisation and the application of pesticides, especially during the last decades. The same reasons can explain the significant rise in the yields of the non-fertilised variants. Here, in addition, the N input from the atmosphere plays an important role. It has been steadily rising over the past 50 years and is now about 50 kg/hectare per year (Körschens et al., 1995, Russow et al., 1995). The soil-improving effect of the organic substance, i.e. the non- nutritional effect is deduced from a comparison of the optimum exclusive mineral

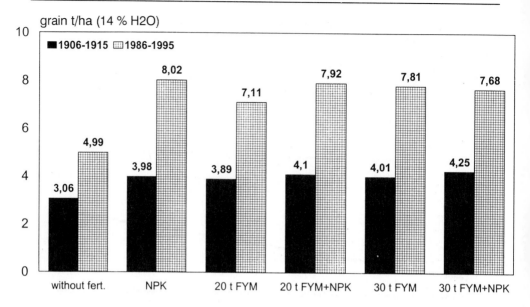

Figure 5: Yields of winter wheat of selected treatments and periods in the Static Experiment Bad Lauchstädt ((farmyard manure (FYM) application every 2nd year))

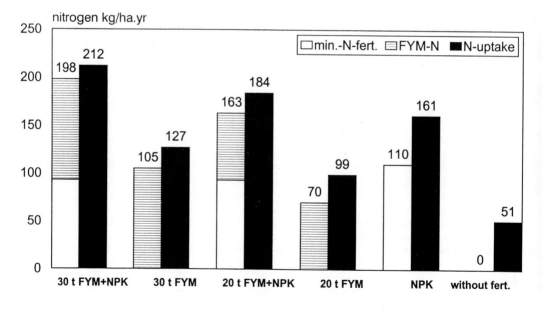

Figure 6: Nitrogen balance of the Static Experiment Bad Lauchstädt (Mean values of all crops in the period 1985 - 1994)

fertilisation with the optimum combination of organic/mineral fertilisation. Under these local conditions this effect makes up a yield increase of 5 % for the crop studied. According to a previous analysis of 1088 test years of long-term experiments, the soil improving effect of the SOM on the yield is up to 6 % in clay soils and up to 10 % in sandy soils.

According to the results presented and the preconditions required for sustainable land use, all fertilised treatments meet the requirements regarding absolute yields and the development of yields.

Nevertheless, the demand for increasing yields is accompanied by the demand for decreasing environmental impacts. In Figure 6, the N balance of the examined plots is presented. In this case it is evident that the experiment has reached the steady state. This means that carbon and nitrogen contents remain constant and that the difference between input and output in case of a positive balance must consequently result from additional sources. In case of a negative balance, gaseous losses and/or leaching occurs. The balance of the treatment with exclusive mineral fertilisation shows the largest positive difference between output and input. About 51 kg/ha nitrogen from additional sources could be used for plant production. In comparison with all other treatments, except for the control plot, mineral fertilisation shows the highest N gain and thus the lowest N losses.

Mineral fertilisation has proved to have the lowest N losses. Given optimum management of N fertilization, a large amount of the atmogenic N input can be used for plant production under given local conditions.Similar conclusions have been drawn from the results of the long-term experiment in Prague-Ruzyne and on the Albic Luvisol in Thyrow (Körschens & Ritzkowski 995).

4 Conclusions

1. In the soil C and N have a relatively narrow ecological optimum range, which is between 0.2 and 0.6 % of mineralisable C or 0.02 and 0.06 % of N. Below these values, soil fertility, yield and CO_2 sequestration are insufficient; above these values, environmentally dangerous losses occur.
2. The application of mineral fertiliser N within the limits of today's knowledge is positively influencing the C and N balances and is, by means of higher yields and CO_2 sequestration, leading to a decrease in environmental pollution.
3. The aim must be to organize agriculture in a way that it becomes efficient, ecologically justifiable and economic.
4. The influence of field manure on yield is only 6 - 10 % . This is not very impressive.

References

Ansorge, H., Pößneck, J. (1992): Untersuchungen über den Einfluß einer langjährig differenzierten organischen Düngung auf die Wirkung der mineralischen N-Düngung und den Boden auf drei Standorten. In: UFZ-Umweltforschungszentrum Leipzig-Halle GmbH (ed.) Symposium Dauerfeldversuche und Nährstoffdynamik 9. bis 12. Juni 1992 in Bad Lauchstädt, DS Druck-Strom GmbH, Leipzig, 53-55.

Asmus, F. (1990a): Versuch P 60 Groß Kreutz - Prüfung verschiedener Möglichkeiten der organischen Düngung. In: Akademie der Landwirtschaftswissenschaften (ed.), Dauerfeldversuche, Terra-Druck GmbH Olbernhau, 231-243.

Asmus, F. (1990b): Versuch M 4 Groß Kreutz - Wirkung organischer und mineralischer Düngung und ihrer Kombination auf Pflanzenertrag und Bodeneigenschaften. In: Akademie der Landwirtschaftswissenschaften (ed.), Dauerfeldversuche, Terra-Druck GmbH Olbernhau, 245-250.

Christensen, P.T. (1989): Askov 1894 - 1989. Research on Animal Manure and Mineral Fertilizers, Proceedings of the Sanborn Field Centennial, Papers presented June 27, 1989 at Jesse Wrench Auditorium University of Missouri-Columbia SR-415, 28-48

Houot, S., Chaussod, R. (1995): Impact of agricultural practices on the size and activity of the microbial biomass in a long - term field experiment, Biol Fertil Soils, **19**, 309 –316

Hülsbergen, K.-J. (1992): Erträge, Stickstoff- und Kohlenstoffbilanz im Kombinationsversuch Seehausen. In: UFZ-Umweltforschungszentrum Leipzig-Halle GmbH (ed.), Symposium Dauerfeld-versuche und Nährstoffdynamik 9. bis 12. Juni 1992 in Bad Lauchstädt, DS Druck-Strom GmbH, Leipzig, 108-113.

Körschens, M., Stegemann, K., Pfefferkorn, A., Weise, V., Müller, A. (1994): Der Statische Düngungs-versuch Bad Lauchstädt nach 90 Jahren, B. G. Teubner Verlagsgesellschaft Stuttgart-Leipzig

Körschens, M., Ritzkowski, E.-M. (1995): Problems of Sustainable Soil use. Plant Microbe Interaction in Sustainable Agriculture. R. K. Behl, A. L. Khurana u. R. C. Dogra (eds.), CCS HAU, Hisar & MMB, New Delhi, 1 - 10

Körschens, M. & Waldschmidt, U. (1995): Ein Beitrag zur Quantifizierung der Beziehungen zwischen Humusgehalt und bodenphysikalischen Eigenschaften, Arch. Acker- Pfl. Boden., **39**, 165 - 173

Körschens, M., Müller, A. & Ritzkowski, E.-M. (1995): Der Kohlenstoff-haushalt des Bodens in Abhängig-keit von Standort und Nutzungsintensität, Mitt. Deutsche Bodenkundl.Gesellsch., **76**, 847-850

Krzysch, G., Caesar, K. (1992): Einfluß von langjährig differenzierten Bewirtschaftungsmaßnahmen und Umweltbelastungen auf Bodenfruchtbarkeit und Ertragsleistung eines lehmigen Sandbodens, TU Berlin, Schriftenreihe des Fachbereiches Internationale Agrarentwicklung, Berlin **141**, 1-327

Leithold, G. (1992) Zur Dynamik des Humus- und N-Haushaltes im Fruchtfolge-Düngungsversuch See-hausen. In: UFZ-Umweltforschungszentrum Leipzig-Halle GmbH (ed.), Symposium Dauerfeld-versuche und Nährstoffdynamik 9. bis 12. Juni 1992 in Bad Lauchstädt, DS Druck-Strom GmbH, Leipzig, 78-82.

Mercik, St. (1993): Seventy years of fertilizing experiments in Skierniewice, Warsaw, 31-54

Pettersson, B. D., Reents, H.J., v. Wistinghausen, E. (1992): Düngung und Bodeneigenschaften, Ergebnisse eines 32jährigen Feldversuches in Järna, Schweden, Schriftenreihe: 2, Institut für biologisch-dynami-sche Forschung, Darmstadt, 1-59

Reuter. G. (1990): Dauerversuch Hu1 bzw. Hu1To9 in Rostock. In: Akademie der Landwirtschafts-wissenschaften (ed.) Dauerfeldversuche, Terra-Druck GmbH Olbernhau, 319-322.

Russow, R., Faust, H., Dittrich, P., Schmidt, G., Mehlert, S., Sich, I. (1995): Untersuchungen zur N-Transformation und zum N-Transfer in ausgewählten Agrarökosystemen mittels der Stabilisotopen-technik, in: Körschens, M., Mahn, G. (ed.) Strategien zur Regeneration belasteter Agrarökosysteme des mitteldeutschen Schwarzerdegebietes, B. G. Teubner Verlagsgesellschaft Stuttgart - Leipzig, 131 - 166

Schnieder, E. (1990): Die Dauerversuche in Thyrow, 205-229, in:Akademie der Landwirtschaftswissen-schaften (ed.) Dauerfeldversuche, Terra-Druck GmbH Olbernhau

Stumpe, H., Garz, J., Hagedorn, E. (1990): Die Dauerdüngungsversuche auf dem Versuchsfeld in Halle. In:Akademie der Landwirtschaftswissenschaften (ed.) Dauerfeldversuche, Berlin, 25-71

Wabersich, R. (1967): Der Bernburger Dauerdüngungsversuch. 2. Mitt.:Die Kohlenstoffverhältnisse, Thaer-Arch. Berlin 11, **9**, 859-869

Address of author:
Martin Körschens
Centre for Environmental Research Leipzig-Halle
Department of Soil Science
Hallesche Straße 44
D-06246 Bad Lauchstädt, Germany

Influence of Organic Matter and Soil Fauna on Crop Productivity and Soil Restoration after Simulated Erosion

G. Sparovek

Summary

Soil and crop yield restoration due to the application of organic matter and chemical fertilizers, and soil fauna (earthworm, *Pontoscolex corethrurus*) activity was evaluated on an acid, low fertile Oxisol (Rhodic Kandiudox) after surface horizon removal (simulated erosion). Following soil removal under natural conditions (control treatment), the soil became unproductive. The organic matter restored crop productivity to normal levels immediately (first crop after removal) indicating high erosion tolerance under high input agriculture systems. The addition of chemical fertilizer led to intermediate recovery. Soil structure restoration was strongly correlated to earthworm activity.

Keywords: Soil restoration, crop productivity, soil structure, *Pontoscolex corethrurus*

1 Introduction

Soil erosion may cause the exposure of subsurface soil layers with less available nutrients and organic matter and adverse physical properties reducing crop productivity. The influence of soil erosion on crop productivity and the possibility of soil restoration was the subject of several studies (McCool, 1985; Blum & Wenzel, 1989; Larson et al., 1990; Lal & Stewart, 1992). However data for tropical conditions are limited and incomplete.

Soil restoration after intense erosion processes or other forms of soil degradation require organic matter and nutrient addition and the action of soil microbes and soil fauna. In tropical environments the geophagous earthworms, specially *Pontoscolex corethrurus* (Oligochaeta) are significant in soil structure building. Barois and Lavelle (1993) demonstrated that *P. corethrurus*, after consuming soil material, disperse it into a suspension that is reaggregated into stable and nutrient rich aggregates. This process is important for the reformation of soil structure. The rehabilitation of soil structure due to soil fauna action was also suggested by Shipitalo and Protz (1988) and by Marinissen and Dexter (1990).

The purpose of this study was to evaluate soil and crop yield restoration due to the action of organic matter, chemical fertilizers and soil fauna on an acid, low fertile Rhodic Kandiudox following surface horizon removal.

2 Materials and methods

The experiment was carried out on a cultivated acid, low fertile Oxisol (Rhodic Kandiudox) located in Piracicaba (Brazil). Table 1 shows some chemical attributes and clay content determined prior to soil removal.

Depth cm	pH	OM g kg^{-1}	P mg kg^{-1}	K	Ca	Mg	Al	CEC	BS	Al sat.	Clay
				----------mmol$_c$ kg^{-1}----------					----------%----------		
0-20	5.3	25	3	1.8	28	15	4	101	44	10	56
20-40	4.8	21	2	0.7	15	10	13	112	23	35	61
40-60	4.6	19	2	0.6	15	5	20	128	16	50	61
60-80	4.6	17	2	0.4	12	4	19	126	13	53	60
80-100	4.6	17	2	0.4	9	3	21	124	10	62	61

Table 1: Chemical properties and clay content from the Rhodic Kandiudox used for the simulated erosion experiment, prior to soil removal (average of 12 samples)

The pH was measured in a 0.01 M CaCl$_2$ solution; Organic Matter (OM) was oxidized by Na$_2$Cr$_2$O$_7$ in a H$_2$SO$_4$ solution and quantified by colorimetry; P, K Ca and Mg was extracted with ion exchange resin, and P was quantified by colorimetry and K, Ca and Mg by atomic absorption spectroscopy; Al was extracted with a KCl 1M solution and quantified by acid basic titration; Cation Exchange Capacity (CEC) was measured at pH = 7.0, BS = base saturation; Al sat = aluminum saturation in effective CEC and Clay quantified by pipette method and dispersed with NaOH 0.1N.

A surface soil layer of 50 cm (all A horizon + part of B horizon) was removed mechanically with a caterpillar in order to assure the complete elimination of earthworms in the remaining site. The plots were isolated with a plastic film down to 1.5 m in order to avoid earthworm migration. After soil removal the following treatments were applied: **Control** (50 cm surface soil removal) **Lime+P** (Control + 5 Mg dolomitic lime ha^{-1}, 1 Mg thermophosphate ha^{-1} applied 20 days before planting) **Lime+P+Worm** (Lime + P + inoculation with 100 *P. corethrurus* adults m^{-2} 10 days before panting) **Lime+P+Manure** (Lime + P + 100 Mg stabilized cattle manure ha^{-1} incorporated 20 days before planting), **Lime+P+Manure+Worm** and **Manure+Worm**. Experimental plots of 2x3 m with 3 replications were randomly distributed. Winter wheat was sown on July 10, 1995 and harvested on September 23, 1995 and the summer green manure *Crotalaria juncea* was sown on December 15, 1995 and harvested on March 23, 1996. The yield values are expressed in dry matter (DM) and correspond to the wheat grains and the total above ground mass of the *Crotalaria juncea*. The earthworm population was estimated in one point per plot 125 days after inoculation. Simultaneously, 1.0 kg of soil material was collected at the depths of 0-10; 10-20; 20-30; 30-40 and 40-50cm and dry sieved in fractions of >9.15; 9.15-4.00 and 4.00-2.00 mm. The aggregates built by the action of earthworms (casts) had a different shape and color than the pedological soil aggregates, and were visually separated and weighted.

3 Results

Table 2 shows the yield values from wheat (grain production) and *Crotalaria juncea* (above ground mass).

	Wheat		Crotalaria juncea	
	Grains		Above ground mass	
Treatment	Mg (DM) ha^{-1}		Mg (DM) ha^{-1}	
Control	0.00	C	0.3	C
Lime+P	1.33	B	5.1	B
Lime+P+Worm	0.99	B	4.8	B
Lime+P+Manure	2.95	A	15.7	A
Lime+P+Manure+Worm	2.99	A	16.4	A
Manure+Worm	2.99	A	14.4	A

Different letters in the columns indicate different values with 1% of probability (Tukey).

Table 2. Wheat and Crotalaria juncea yields.

All treatments with manure application had approximately the same yield indicating that organic matter addition was the most significant factor affecting productivity restoration. The absolute values of productivity are comparable to the normal average yield values on this soil without erosion or soil removal, denoting the complete restoration of the productivity in the first crop even after the complete removal of the surface layer (complete removal of A horizon + part of B horizon). After soil removal the control treatment became unproductive due to the exposure of the B horizon with low organic matter, low nutrient contents and high acidity. The treatments with chemical fertilizer application showed an intermediate trend. It is probable that the incorporation of crop residues will have the same effect on crop productivity as manure application, but will take longer.

In high input agricultural systems, which are common in this region, the addition of fertilizers and the production of organic matter as crop residues are intense. The results shown indicate that under these conditions this soil is very resistant against productivity loss due to soil erosion, consequently having high erosion tolerance values. This erosion tolerance is expected to be higher than the usually observed erosion rates. The control treatment will probably remain unproductive once the crop residue production is very low, insufficient to increase soil organic matter.

The earthworm activity represented by its increase rate and cast formation as aggregates greater than 2.0 mm are represented in Table 3.

	Population after 125 days[a]		Increase rate	Casts > 2mm
Treatment	Individuals m^{-2}		Ind. m^{-2} day^{-1}	Cast (g)/100g soil
Control	0	B	0.00	0.0
Lime+P	0	B	0.00	0.0
Lime+P+Worm	120	B	0.16	0.6
Lime+P+Manure	0	B	0.00	0.0
Lime+P+Manure+Worm	240	B	1.12	0.1
Manure+Worm	2,800	A	21.60	3.9

[a] Initial inoculation was of 100 *P. corethrurus* adults m^{-2}
Different letters in the columns indicate different values with 1% of probability (Tukey).

Table 3. Earthworm (Pontoscolex corethrurus) activity and cast formation.

The earthworm increase rate was significantly higher in the treatment **Manure+Worm**. The earthworm population was not eliminated by lime and chemical fertilizer application but showed low increase rates under these conditions. Soil structure building due to soil fauna activity was also observed. In the period considered, 3.9% of the aggregates were built due to the action of the earthworms in the **Manure+Worm** treatment. Taking into account the short period of time considered in this experiment (125 days from earthworm inoculation) the structure formation by *P. corethrurus* was very significant.

4 Conclusions

1. Organic matter restored crop productivity immediately, even after the complete removal of the surface horizon.
2. The soil became unproductive without organic matter addition.
3. Chemical fertilizer addition partially restored crop productivity. Complete soil restoration may be dependent on the organic matter addition via crop residue incorporation.
4. Soil fauna activity (earthworm, *Pontoscolex corethrurus*) had a major impact in the rehabilitation of soil structure.

Acknowledgments

The author wishes to thank the Staden Institute and the Martius Foundation for supporting my participation in the 9th Conference of the International Soil Conservation Organization in Bonn (August 1996).

References

Barois, I. and Lavelle, P. (1993): Changes in respiration rate and some physicochemical properties of a tropical soil during transit through Pontoscolex corethrurus. Soil Biol. Biochem., **18**, 539-541

Blum, W.E.H. and Wenzel, W.W. (1989): Bodenschutzkonzeption. AV-Druck, Wien.

Lal, R. and Stewart, B.A. (1992): Soil Restoration. Advances in Soil Science, vol. **17**.

Larson, W.E., Foster, G.R., Almaras, R.R. and Smith, C.M. (eds) (1990): Proceedings of soil erosion and productivity workshop. Minnesota. Published by University of Minnesota, Saint Paul. 142 pages.

Marinissen, J. C. Y. and Dexter, A. R. (1990): Mechanisms of stabilization of earthworm cast and artificial casts. Biol. Fert. Soils, **15**, 163-167.

McCool, D.K. (1985): Erosion and Productivity. ASAE Publication 8-85, Michigan.

Shipitalo, J. M. and Protz, R. (1988): Factors influencing the dispersibility of clay in worm casts. Soil Sci. Soc. Am. J., **52**:152-157.

Address of author:
Gerd Sparovek
University of São Paulo, ESALQ
13.418-900 Piracicaba (SP), Brazil

Quantifying Soil Fertility Changes and Degradation Induced by Cultivation Techniques in the Nigerian Savanna

J.O. Agbenin

Summary

Investigations of the impact of cultivation techniques on gains and losses of soil chemical fertility in the Nigerian savanna are limited. The study quantifies the area-based changes in soil mass, amounts of organic carbon (OC), nitrogen (N), phsophorus (P) and potassium (K), and the exchangeable cations induced by cultivation techniques in a savanna ferruginuous soil (Typic Haplustalf). Soil samples were taken from 0-5 cm, 5-20 cm and 20-30 cm from three cultivation systems: no-tillage (NT); minimum tillage (MT) in which ridges are made with traditional hoes; and conventional tillage CT involving ploughing and harrowing. Averaged across sampling depths, amounts of OC, P, K and exchangeable cations were not significantly different between cultivation treatments. However, in relation to pre-tillage status, more losses of soil material, OC, total P and basic cations occurred in MT and CT than NT. Net gains of total K, exchange K and Ca were made in NT. The variations in soil fertility between cultivation treatments were related to tillage-induced soil losses and re-distribution, organic matter loss and possible differential leaching regimes. The inherent chemical fertility of savanna soils will be best maintained or preserved under no-tillage system.

Keywords: Exchange cations, gains, losses, organic matter, soil fertility, tillage

1 Introduction

The savanna agroecosystems are unique in terms of their climatic and edaphic environments. The soils are mostly sandy, coarse-textured, poorly aggregated with low amounts of organic matter (OM) and occur under a climatic environment characterized by high intensity rainfall with exceptionally high kinetic energy load (Kowal and Kassam, 1978). Under continuous cultivation now prevalent in the savanna, there is a range of soil risks. These risks are further exacerbated by unsuitable tillage techniques such as mechanical tillage.

One profound effect of tillage in soils is reduction of OM, total P and cation exchange capacities (Lal, 1976; Aina, 1979). In western Nigeria, untilled Alfisols have four times the OM content of tilled Alfisols (Aina, 1979). In view of the importance of OM in soil fertility, nutrient retention and cycling, tillage practices that enhance OM build-up are likely to improve soil productivity. Zero-tillage leads to more OM accumulation near the surface than conventional tillage (Edwards et al., 1992).

In evaluating the impact of tillage on soil fertility in most Nigerian soils, nutrient concentrations under different tillage practices were compared (Aina, 1979, Ike, 1986). Direct comparison of concentrations of nutrients in a given sampling depth does not accomodate possible differences in soil mass between tillage systems, and consequently possible differences in masses of nutrient elements. For example, soil-A having a bulk density of 0.96 Mg m^{-3} and 250 mg kg^{-1} soil from 0-5 cm depth does not contain more P than soil-B with a bulk density of 1.36 Mg m^{-3} and 200 mg kg^{-1} soil. The amount of P in soil-A will be 120 kg ha^{-1} compared to soil-B having 170 kg ha^{-1} even though the concentration of P in soil-A was more than soil-B. In this study, amounts of OC, total N, P and K, and the exhangeable cations per unit area (hectare basis) under three tillage systems were determined. The objective was to quantify the area-based changes in soil fertility induced by three tillage techniques in a savanna Alfisol.

2 Materials and methods

The tillage experiment for this report was carried out at the Institute for Agricultuaral Research Farm at Samaru (11° 11'N 7° 38'E) in the northern Guinea savanna zone of Nigeria. The annual precipitation ranges from 880 to 1200 mm. Soils at the experimental site are Isohyperthermic Typic Haplustalf. The tillage experiment was initiated in 1977, and consisted of nine tillage treatments with four replicates. However, for the present study, three tillage experiments, representing the common tillage systems in the region, with three replicates were sampled for analysis. The tillage systems were:

i. No-tillage (NT): No soil disturbance other than that involved in seed-drilling. Residues of previous year's crops were left as surface mulch, and weeds were controlled with the herbicide glyphosate applied at 2.2 L ha^{-1} before seeding.

ii. Minimum tillage (MT): The traditional tillage system in the savanna in which old ridges from previous year cropping were split by forming new ridges in a furrow between two ridges. The previous year's crops were removed after harvest. Weeds were controlled by hoe-weeding. Depth of soil disturbance was about 15 cm.

iii. Conventioanl tillage (CT): The soil was disc ploughed, harrowed and ridged. Crop residues from previous year were removed. Depth of ploughing was up to 20 cm.

The experimental fields were cropped to cotton and maize in rotation and fertilized with 33 kg ha^{-1} as calcium ammonium nitrate [(CaNH$_4$(NO$_3$)$_3$], and 9 kg P ha^{-1} as single superphosphate. Prior to the initiation of the tillage experiment at the study site, soil samples were collected at 0-5 cm, 5-20 cm and 20-30 cm depth to characterize the initial nutrient status of the soil. In 1984, seven years after the initiation of the tillage experiment, twenty core samples were taken from 0-5 cm, 5-20 cm and 20-30 cm and bulked for each of the sampling depths. Bulk density for each of the sampling depths was measured by pushing a metal cylinder into the soil. Both ends of the excavated cylinders were trimmed and dried at 105 °C to a constant weight. Bulk density was determined as the weight of soil divided by volume of the metal cylinder.

2.1 Soil and data analysis

Soil pH was determined in 1:2 soil:water suspension. Organic C was determined by dichromate oxidation (Nelson and Sommers, 1982). Exchangeable cations were displaced with 1.0 M NH$_4$OAc (Thomas, 1982). Exchangeable Na and K were determined by flame photometry, and Ca and Mg by atomic absorption spectrophotometry. Total P and K were determined by digestion with sulphuric and perchloric acids. Total P in digests was determined colorimetrically by the method of Murphy and Riley (1962) while total K was determined by flame photometry. Total N was determined by the regular Kjeldahl method (Bremner and Mulvaney, 1982).

In estimating the area-based changes in the soil chemical fertility data (net gains and losses of soil and nutrients), the entire sampling depth was considered since tillage could have re-packed and re-distributed soil particles within the profile. The mean bulk density averaged across the sampling depths (0-30 cm) was not significantly different between tillage treatments. This is expected because bulk density of tilled soil usually returns to its untilled status in the savanna after few weeks because of the high kinetic energy load of the rains (Kowal and Kassam, 1978; Adeoye, 1986). The amounts of OC, N, P, K and the exchangeable actions were determined by multiplying their concentrations by the entire 30 cm depth and the dry bulk density, and extrapolated to a hectare basis. Table 1 gives the bulk density, OC, total P, K and exchangeable cations of the soil prior to the imposition of the tillage treatments.

Soil Depth (cm)	Bulk density ($Mg\ m^{-3}$)	Soil mass	Organic C	Total P	Total K	Exch. Na	Exch. K	Exch. Ca	Exch. Mg
		------ $Mg\ ha^{-1}$ --------		-------------------------- $kg\ ha^{-1}$ ----------------------------------					
0-5	1.44	720	7.06	130	292	11	84	884	207
5-20	1.43	2140	10.5	183	743	64	150	1352	482
20-30	1.43	1430	4.23	95	819	20	123	1195	412
Mean[a]	1.43	1667	7.84	145	693	41	130	1222	413

[a] Means were weighted by multiplying each value with respective sampling depth, summing up and dividing by total sampling depth

Table 1: Bulk density, soil mass and amounts of OC, P, K and exchangeable cations at three sampling depths prior to tillage

To estimate the area-based gains and losses of nutrients, the following calculation was done: if $X\ kg\ ha^{-1}$ was the amount of total P in soil prior to tillage, and $Y\ kg\ ha^{-1}$ was the amount of total P after tillage, net gain or loss of P induced by tillage was given as $\Delta P\ (kg\ ha^{-1}) = Y-X$ after correcting for fertilizer-P input during the cultivation treatment. Critical to this calculation is the assumption that pre-tillage soil fertility was minimally affected by temporal changes under native vegetation. It is unlikely that a period of seven years could significantly alter pedogenesis under native vegetation, moreso in these highly weathered soils where the factors of soil formation are at a steady state. However, it cannot be entirely discounted that the gains and losses of different nutrient elements, as calculated above, might be little affected by temporal changes in the soil properties.

To determine the tillage-related factors/processes responsible for variations in the soil chemical fertility, the data were subjected to factor analysis. Factor analysis allows subtle relationships in a multivariate dataset to be identified. Such relationships reflect the correlation of each variable with mutually orthogonal factors (Johnson, 1978). Varimax rotation was employed to reduce the number of variable loading highly on any one factor in order to simplify the interpretation. The factor analysis was performed on Statview 11 (Abacus Concepts, 1987).

3 Results and discussion

The mass of soil in each sampling depth did not differ significantly between tillage treatments (Table 2) apparently because of similarity in soil bulk density under the different tillage systems. The effect of cultivation on bulk density of savanna soils is usually short-lived as bulk density returns to its pre-tillage status after few weeks (Adeoye, 1986). Tillage effects on OC, N, P, K and the exchangeable cations were apparent at 0-5 cm depth (Table 2). The higher amounts of OC, N and P in NT than MT and CT at 0-5 cm depth might be explained by surface mulch of crop residues on NT, and rapid OM decomposition in MT and CT. Aina (1979) and Lal (1976) reported higher OC of zero-tilled soils than conventionally tilled Alfisols in western Nigeria. Since tillage stimulates oxidation of OM, build-up of P and N will be limited as these nutrients are cycled through soil organic matter (SOM).

Soil depth (cm)	Tillage treatm.	Bulk density Mg m^{-3}	Soil mass ---- Mg ha^{-1} ----	Org. C	Total N	Total P	Total K	Exch. Na	Exch. K	Exch. Ca	Exch. Mg
							-- kg ha^{-1} --				
0-5	NT	1.34	670	6.56	938	169	551	9	89	934	139
	MT	1.26	630	3.47	ND	88	290	8	42	678	94
	CT	1.23	615	3.63	480	62	287	8	58	599	128
LSD =	0.05	0.10	NS	1.51	ND	39	116	NS	28	NS	NS
5-20	NT	1.44	2160	10.1	ND	301	888	23	180	1785	439
	MT	1.39	2085	9.92	ND	341	1452	24	144	2083	462
	CT	1.36	1950	11.8	ND	197	1019	24	129	1898	432
LSD =	0.05	NS	NS	NS	-	90	NS	NS	NS	NS	NS
20-30	NT	1.44	1440	5.47	1094	146	1259	17	146	1306	458
	MT	1.39	1390	4.89	ND	158	1015	16	81	1210	565
	CT	1.48	1430	6.58	1201	217	873	31	98	1411	388
LSD =	0.05	NS	NS	NS	-	NS	NS	10	NS	NS	155

Table 2: The pH, amounts of soil, OC total N, P and K, and the exchangeable cations of the soils on area basis after seven years of tillage oeration

The higher amounts of exchangeable cations in 0-5 cm depth of NT than MT and CT cannot only be explained by reference to higher OM accumulation furnishing the soil with basic cations, but also related to differential trapping of harmattan dusts in the savanna. Harmattan dusts, which are deposited on the soil surface in the savanna annually between November and February, are usually laden with primary weatherable minerals rich in basic cations. Analysis of harmattan dusts showed, on the average, that concentration of Ca = 39 mg g^{-1} dust, Mg and Na = 15 mg g^{-1} dust and K = 30 mg g^{-1} (Tiessen et al., 1991). Estimates of harmattan dust deposition are as high as 990 kg ha^{-1} yr^{-1} in northern Nigeria (McTainsh, 1980). Thus harmattan dusts are a major source of the high base status of surface soils of west African savanna soils (Tiessen et al., 1991; Moberg et al., 1991). The probability of trapping and retention of harmattan dusts by surface much provided by previous years's crop residues in NT is likely to be greater than MT and CT without surface mulch. This pre-disposes the soil to water and wind erosion. Below 0-5 cm depth, CT had the highest OC content (Table 2) due to deep incorporation of organic residue during ploughing. The amounts of OC, N, P, K and the exchangeable cations were, however, not significantly different between tillage systems when considering the whole or entire sampling depth (Table 3).

3.1 Area-based losses and gains

In assessing the area-based changes in soil fertility, the entire sampling depth (0-30 cm) was considered because of the tendency of tillage to re-distribute soil particles and OM through residue incorporation. Furthermore, this depth represents the zone of maximum root growth and distribution. About 77 % of maize root and 61 % of sorghum root, two major cereals cultivated in the savanna, are concentrated within 30 cm of soil depth (Nofziger, 1974). In relation to pre-tillage status, losses of soil particles occurred in all tillage treatments (Table 4). The degree of soil losses followed the order CT > MT > NT consistent with the degree of soil exposure, disturbance and pulverization. The net losses of soil in MT and CT were probably caused by water and wind erosion because of exposure. This contrasts with NT with surface mulch of previous year's crop residues curtailing soil detachment, water flow and soil transport (Angels et al., 1984; Mostaghimi et al., 1991).

Tillage treatm.	Bulk density ($Mg\ m^{-3}$)	Org. C ($Mg\ ha^{-1}$)	Total N	Total P	Total K	Exch. Na	Exch K	Exch. Ca	Exch. Mg
					$kg\ ha^{-1}$				
NT	1.42	7.48	1113	208	931	17	147	1407	374
MT	1.37	6.08	ND	196	919	16	89	1324	376
CT	1.39	7.32	898	159	727	21	95	1301	299
$LSD^a_{0.05}$	NS	NS	-	NS	NS	NS	NS	NS	NS

[a] NS = not significant; ND = not determined.

Table 3: *Effect of tillage bulk density, amounts of OC, P, K and exchangeable cations for the whole sampling depth (0-30 cm) on area basis.*

Tillage treatm.	Soil particles ($Mg\ ha^{-1}$)[a]	Org. C	Total N	Total P	Total K	Exch. Na	Exch K	Exch. Ca	Exch. Mg
					$kg\ ha^{-1}$				
NT	-3.30	-390	-178	-0.60	+236	-24	+16.9	+9.0	-37.2
MT	-56.7	-1800	ND	-15.1	+229	-25	-41.6	-76.2	-37.2
CT	-113	-550	-379	-55.5	+35	-20	-35.1	-88.0	-116

[a] Minus and plus signs indicate net losses and gains respectively

Table 4: *Area-based gains and losses of soil particles, OC, total P, K and exchangeable cations of the whole sampling depth (0-30 cm) after seven years of tillage.*

The higher loss of OC in MT and CT than NT is related to rapid oxidation in tilled soils. All tillage systems incurred net loss of N and P in the order CT > MT > NT consistent with the order of soil loss (Table 4). Ryden et al. (1973) showed that P loss was related to the amount of silt and clay-sized particles transported by overland flow to which P might be intimately bound. The net gain of total K in all tillage treatments is partly related K-input from harmattan dusts which are usually rich in mica and K-feldspars (Moberg et al., 1991). If the estimate of annual deposition of harmattan dusts in northern Nigeria given by McTainsh (1980) is correct, it will translate into K-input of 29.7 $kg\ ha^{-1}\ yr^{-1}$ from harmattan dusts.

The soil under MT and CT incurred net losses of exchangeable Ca, Mg, K and Na due to soil loss and possibly leaching. It seems, from the results of this study, that the variations in OC, P, K and the exchangeable cations under different tillage operations can be explained by soil and OM losses and redistribution, and possible differential leaching regimes. This seems supported by the result of factor analysis in which only two factors accounted for variations in the soil fertility criteria examined in the study (Table 5). Soil properties having high loading on the same factor are said to be covariant, hence one can infer from the association of soil properties the physico-chemical interactions that can be mechanistically interpreted. The fact that OC, P and the exchangeable cations loaded highly on Factor 1 (Table 5) suggests that this factor consists of prcocesses associated with soil particle and OM losses, re-distribution and oxidation in the different tillage systems. For instance, loss of OM and soil particles will not only affect N and P, but also the cation exchange properties and base status of soils. Jones (1973) showed that OM contributed more than 80% of CEC of west African savanna soils. Exchange Mg and pH loaded more on Factor 2 than Factor 1 suggesting that this factor is related to differences in leaching between tillge systems consistent with the negative loading of soil pH.

Variable	Factor laoding		Final Communality estimate
	Factor 1	Factor 2	
Organic C	0.9	---	0.8
Total P	0.0	---	0.7
Total K	---	0.8	0.8
pH	---	-0.9	0.9
Exch. Na	0.7	---	0.6
Exch. K	0.8	---	0.6
Exch. Ca	0.9	---	0.9
Exch. Mg	0.5	0.8	0.8

Table 5: Results of factor analysis of soil properties for all tillage systems with varimax rotation

4 Conclusions

The study indicated that all tillage systems, compared to original soil conditions, resulted in loss of soil and OM with MT and CT losing 57 and 113 Mg ha^{-1} of soil particles, respectively. This contrasts with 3.3 Mg ha^{-1} loss of soil under NT. Losses of OC and other nutrient elements were higher in CT and MT than NT. Losses of OC and other nutrient elements were higher in CT and MT than NT. The variations in area-based losses and gains of nutrient elements and OC between tillage systems were related are influenced by factors affecting OM addition, oxidation and re-distribution processes, in addition to leaching regimes. The results of this study would suggest that the inherent fertility of savanna soils will be best maintained or preserved under a no-tillage system. The acceptability and adoption of no-tillage technology in the savanna must, however, await the development of simple weed control technologies at the farmer level.

References

Abacus Concepts (1987): Statview 11. Abacus Concepts, Berkely, CA
Adeoye, K.B. (1986): Physical changes induced by rainfall in the surface layer of an Alfisol, Northern Nigeria. Geoderma: 59-66
Aina, P.O. (1979): Soil changes resulting from long-term management practices in western Nigeria. Soil Sci. Soc. Am. J. **43**: 173-179.
Angel, J.S., McCLung, G., McIntosh, M.S., Thoma, P.M. and Wolfe, D.C. (1984): Nutrient losses in runoff from conventional and no-till corn watershed. J. Environ. Qual. **13**: 431-435.

Bremner, J.M. and Mulvaney, C.S. (1982): Nitrogen - total. In: Methods of soil analysis. Part 2. Chemical and Microbiolgical Properties. pp 595-624 (A.L. Page, R.H. Miller and D.R. Keeney eds.). SSSA and ASA, Madison, WI.

Edwards, J.H., Wood, C.W., Thurlow, D.L. and Ruf, M.E. (1992): Tillage and crop rotation effects on fertility status of a Hapludult. Soil Sci. Soc. Am. J. **56**:1577-1582.

Ike, I.F. (1986): Soil and Crop responses to different tillage practices in a ferruginuous soil of the Nigerian savanna. Soil Till. Res. **6**: 261-272.

Johnston, R.J. (1978): Multivariate statistical analysis in geography. Longmans, New York

Jones, M.J. (1973): The organic matter content of savanna soils of west Africa. J. Soil Sci. **24**: 42-53.

Kowal, J.M. and Kassam, A.H. (1978): Agricultural ecology of savanna. Clarendom Press, Oxford.

Lar, R. (1976): No-tillage effects on soil properties under different crops in western Nigeria. Soil Sci. Soc. Am. J. **40**: 762-768

McTainsh, G. (1980): Harmattan dust deposition in northern Nigeria. Nature **286**: 587-588.

Moberg, J.P., Esu, I.E. and Malgwi, W.B. (1991): Characteristics and constituent composition of harmattan dust in northern Nigeria. Geoderma **48**: 73-81.

Mostaghimi, S., Younos, T.M. and Tim, U.S. (1991): The impact of fertilizer application techniques on nitrogen yield from two tillage systems. Agric. Ecosyst. Environ. **36**: 13-22.

Murphy, J. and Riley, J.P. (1962): A modified single solution for the determination of phosphorus in natural waters. Anal. Chim. Acta. **27**: 31-36

Nelson, D.W. and Sommers, L.E. (1982): Total carbon, organic carbon and organic matter. In: Methods of soil analysis. Part 2. Chemical and Microbiolgical Properties. pp 539-579 (A.L. Page, R.H. Miller and D.R. Keeney eds.). SSSA and ASA, Madison, WI.

Nofziger, D.L. (1974): Root growth of maize and sorghum at Samaru. Mimeo. Institute for Agricultural. Research, Ahmadu Bello University, Zaria, Nigeria.

Ryden, J.C., Syers, J.K. and Harris, A.F. (1973): Phosphorus in run-off and streams. Adv. Agron. **25**:1-45.

Thomas, G.W. (1982): Exchangeable cations. In: Methods of soil analysis. Part 2. Chemical and Microbiological Properties. pp 159-165 (A.L. Page, R.H. Miller and D.R. Keeney eds.). SSSA and ASA, Madison, WI.

Tiessen, H., Hauffe, H and Mermut, A.R. (1991): Deposition of harmattan dust and its influence on base saturation of soils in northern Ghana. Geoderma **49**: 285-299.

Address of author:
John O. Agbenin
Department of Soil Science
Ahmadu Bello University
Zaria, Nigeria

Soil Microbial Biomass and Organic Carbon in Reforested Sites Degraded by Bauxite Mining in the Amazon

E.S. Costa, R.C. Luizão & F.J. Luizão

Summary

In eastern Amazon, parameters related to the soil organic matter such as litter mass, microbial biomass and organic carbon were measured at sites which were reforested after mining. Treatments of the same age were compared with the control (primary forest): two 1-year-old reforested sites, with and without topsoil replacement before the plantation; two 11-year-old sites, one with natural regeneration and the other one reforested.

Litter accumulation was lower in both 1-year-old sites than in the control. The 11-year-old sites had similar litter mass in the wet and dry season but both were greater than the control, confirming that their litter production may be higher than that of the climax forest. As expected, sites with topsoil replacement showed higher soil organic carbon than sites without topsoil. Soil organic carbon in the 11-year-old sites were similar though lower than in the control, suggesting that the recovery of the organic carbon stock needs more than a decade. There were no differences in microbial biomass between the 1-year-old sites with and without topsoil. Differences in soil microbial biomass were found neither between the two 11-year-old sites nor between each one and the control, indicating that microbial biomass were fully recovered within that time.

Keywords: Microbial biomass, soil rehabilitation, reforestation, bauxite mining

1 Introduction

During the last 30 years, there has been a considerable increase in prospecting of minerals and on the establishment of industrial mining projects, especially bauxite, in the Brazilian Amazon (Santos, 1987; Fernandes & Portela, 1990). According to Fernandes & Portela (1990) in Pará state only there are 2.5 billions tons of bauxite yet to be exploited. Though industrial mining is still little developed, mining activities can greatly affect the soil ecosystem. Removal of the vegetation and soil excavation usually causes soil impoverishment, erosion and soil toxicity, thus affecting the flora, fauna, water quality, and people of the region in many ways (Griffith, 1980).

In 1987, with the support of the Economic European Community (EEC) and the Deutsche Gesellschaft für Technische Zusammenarbeit (GTZ), a program was started to encourage mining companies to adopt a policy of rehabilitation of degraded areas following bauxite exploitation in Eastern Amazon. Sites have been filled back with all the layers of material which were taken out to reach the bauxite and the topsoil was replaced before either letting the site for natural regeneration or introducing tree plantations.

The impact of the absence of the organic layer on soil processes is still not well understood. The aim of this study was to investigate which parameters are related to soil organic matter, such as litter layer, soil microbial biomass and soil organic carbon that could be used as an index of soil rehabilitation in reforested bauxite mining sites in Eastern Amazon.

2 Materials and methods

2.1 Study site

The study was carried out in Saracá Mine located 65 km north-west of the town of Oriximiná in Pará State, Brazil (1° 45' S and 56° 30' W). The local climate is classified as Aw in the Köppen system (RADAMBRASIL, 1976) with a mean annual temperature (1984-1994) of 26.5 °C with distinct dry (June-November) and wet (December-May) seasons. The mean monthly temperature in 1995 ranged from 25 °C (January) to 28 °C (October). The total precipitation in 1995 during the study period was 1969 mm. The predominant soils at Saracá Mine are Oxisols (RADAMBRASIL, 1976) with low fertility, very acid, but rich in iron and aluminum. The floristic composition of the primary tropical rain forest has been described by Higuchi et al. (1982) as being relatively homogenous. The main families are the Leguminosae (56 species), Lauraceae (14 spp), Sapotaceae (13 spp), Annonaceae (11 spp), and Moraceae (10 spp).

2.2 Experimental design, soil and litter layer sampling

Among the various treatments of the rehabilitation program, five sites were selected for this study: 1) 1-year-old reforested sites with topsoil replacement; 2) 1-year-old reforested sites without topsoil replacement; 3) 11-year-old natural regeneration; 4) 11-year-old reforested sites; and 5) primary forest (control). For each treatment, three sites of 50 m^2 were selected. The soil and the litter layer were collected three times in 1995 in order to cover seasonal variability: in April/95 (wet season); August/95 (dry season); and November/95 (the transitional period between the dry season and the wet season). Each soil sample consisted of five sub-samples of the top soil (0-10 cm), pooled as a composite sample. The litter layer (one sample composed of 15 sub-samples was randomly collected in each plot using a 20×20 cm wooden frame.

2.3 Soil particle fractions and chemical properties

Organic carbon was measured by an adjusted Walkley Black method (Jackson 1958); microbial biomass contents by fumigation-extraction (Vance et al., 1987); and soil pH in water and KCl suspension, using a digital pH-meter. The soil nutrients analyzed were: total N and total P, both using auto-analyzer (SFA2); extractable K, Ca and Mg were analyzed by atomic absorption spectrophotometry (Anderson & Ingram, 1989).

2.4 Statistical analysis

The mean values for treatments and seasons were compared using an one-way analysis of variance (ANOVA), followed by a Tukey test (considered to be significantly different where $p < 0.05$), and correlation tests.

3 Results and discussion

3.1 Soil particle fraction and chemical properties

With the exception of coarse silt which was significantly higher ($p < 0.05$) in the 1-year-old reforested sites without topsoil replacement, none of the treatments showed a granulometric composition different from the control forest (Tab. 1). Obviously, soil excavation affected the soil texture.

Soil moisture increased from 17% in dry season in the 1-year-old reforested sites without topsoil replacement to 52% in the forest control and, as expected, the moisture content reflected the vegetation cover in each site. Thus, all sites showed moisture contents lower than the forest control. In both wet and dry seasons, the sites without topsoil, with a lower vegetation cover, showed lower moisture content than the plot with topsoil replacement (data not reported).

Soil pH in KCl raised from 3.7 in wet season in the primary forest to 5.4 in dry season in the 1-year-old reforested sites without topsoil replacement. This can be associated with the fertilization which was done before the reforestation. Another factor possibly contributing to the higher pH may have been the lower deposition of litter which, depending on the phenolic concentrations, can cause soil acidification (Kuiters 1990).

Plots	Particle fractions (%)				
	Clay	Fine silt	Coarse Silt	Coarse Sand	Fine sand
RWTS1	85,0 ±1,15 a	2,4 ±0,84 a	1,8 ±0,31 a	4,8 ±0,36 a	3,2 ± 0,75 ab
RWTH1	65,7 ±2,22 b	14,1 ±1,81 a	4,2 ±0,36 b	9,1 ±0,62 a	6,6 ±0,57 a
NR11	69,5 ±1,06 ab	17,5 ±0,95 a	1,9 ±0,47 a	3,5 ±0,46 a	1,9 ±0,30 b
REF11	72,2 ±7,63 ab	14,5 ±7,05 a	1,2 ±0,47 a	4,4 ±0,88 a	2,7 ±0,74 ab
PF	70,9 ±1,30 ab	13,1 ±5,06 a	1,1 ±0,40 a	6,1 ±2,54 a	3,6 ±1,77 ab

Means followed by similar letters are not significantly different ($p < 0.05$) according to the Tukey test

Tab. 1: *Soil particle fractions from upper layer (0-10 cm) in 1-year-old reforested plots with (RWTS1) and without (RWTH1) topsoil replacement; in 11-year-old natural regeneration (NR11); in 11-year-old reforested plots (REF11); and in primary forest (PF) in Saracá Mine, Porto Trombetas-PA. Values are represented by means ± SE (n = 3).*

Plots	Soil nutrients									
	Wet					Dry				
	N	P	K^+	Ca^{++}	Mg^{++}	N	P	K^+	Ca^{++}	Mg^{++}
RWTS1	0,08 b	138 a	0,09abc	0,46 c	0,16cd	0,12ab	139 a	0,06 cd	0,39 bc	0,12 cd
	(±0,02)	(±7,94)	(±0,02)	(±0,07)	(±0,03)	(±0,02)	(±5,50)	(±0,01)	(±0,12)	(±0,02)
RWTH1	0,02 b	88 b	0,01 c	0,03 c	0,01 d	0,01 b	84 b	0,01 d	0,07 c	0,03 d
	(±0,01)	(±0,10)	(±0,01)	(±0,01)	(±0,01)	(±0,01)	(±4,50)	(±0,01)	(±0,02)	(±0,17)
NR11	0,10 b	150 a	0,11 ab	1,53 ab	0,38ab	0,17ab	160 a	0,10abc	1,79 ab	0,39 ab
	(±0,03)	(±3,26)	(±0,02)	(±0,35)	(±0,08)	(±0,01)	(±5,00)	(±0,01)	(±0,58)	(±0,06)
REF11	0,23 a	168 a	0,16 a	2,34 a	0,49 a	0,12ab	166 a	0,15 a	2,47 a	0,59 a
	(±0,02)	(±6,74)	(±0,01)	(±0,24)	(±0,04)	(±0,04)	(±6,74)	(±0,03)	(±0,45)	(±0,10)
PF	0,21 a	146 a	0,06 bc	0,07 c	0,11cd	0,27 a	153 a	0,13 ab	0,14 c	0,18bcd
	(0,01)	(±8,67)	(±0,03)	(±0,23)	(±0.00)	(±0,02)	(±6,75)	(±0,01)	(±0,02)	(±0,05)

Means followed by similar letters are not significantly different ($p < 0.05$) according to the Tukey test

Tab. 2. *Soil nutrient concentrations of the top layer (0-10 cm) in 1-year-old reforested plots with (RWTS1) and without (RWTH1) topsoil replacement; in 11-year-old natural regeneration (NR11); in 11-year-old reforested plots (REF11);and in primary forest (PF) in Saracá Mine, Porto Trombetas-PA, in the wet and dry seasons. Values are represented by means ± SE (n = 3).*

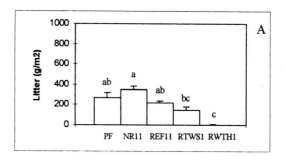

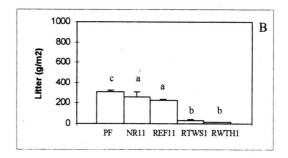

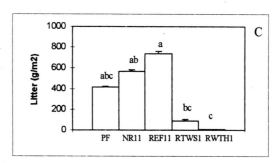

Figure 1: Soil litter layer (g/m²) in primary forest (PF), 11-year-old natural regeneration (NR11); 11-year-old reforested plots (REF11), in 1-year-old reforested plots with topsoil (RWTS1), and 1-year-old reforested plots without topsoil replacement (RWTH1) in the A) wet season, B) dry season and C) transitional period. Values are represented by means ± SE (n = 3).

The 1-year-old reforested sites without topsoil replacement showed significantly lower ($p < 0.05$) concentrations of N and P than the control forest in the wet and dry season, while concentrations of K were lower only in the dry season (Tab. 2). On the other hand, the N of the control forest was significantly higher ($p < 0.05$) than in the 1-year-old site with topsoil reposition in the wet season, but K was significantly higher ($p < 0.05$) in the forest control in the dry season. Concentrations of P in the 1-year-old reforested sites with topsoil replacement were significantly higher ($p < 0.05$) than the 1-year-old reforested sites without topsoil replacement in wet and dry seasons. Both 1-year-old reforested sites showed significantly lower concentrations of P than the forest control. In both 11-year-old sites, concentrations of Ca and Mg were significantly higher (p < 0.05) than in the primary forest in the wet season, but in the dry season, both were higher for Ca (Tab. 2). The higher concentrations of Ca and Mg in the 11-year-old sites may still be an effect of liming (site REF11) and of burning (NR11 sites).

3.2 Litter layer

It was evident that the vegetation cover and the soil litter layer were greater in the 1-year-old reforested site with topsoil replacement, but the difference for the sites without topsoil was not significant. The absence of significance can be related to the high heterogeneity of the study sites. The lower litter layer was visible when compared to the forest control, especially in the sites without topsoil (Fig. 1).

The 11-year-old natural regeneration and the 11-year-old reforested sites did not show significant different litter mass in both wet and dry seasons, but in the transitional period the litter layer in the reforested sites was significantly higher ($p < 0.05$) than in the natural regeneration in the same period. When compared with the control forest in both wet and dry seasons, the 11-year-old natural regeneration and 11-year-old reforested sites showed significantly higher ($p < 0.05$) litter mass.

In contrast to other studies which reported the highest litterfall in the dry season (Dantas & Phillipson, 1989; Luizão, 1989), for this study the highest litter layer mass was reported in the transitional period (between the wet and the dry season). The most likely reason was a long dry spell prior to the sampling of the transitional period. Also the tree diversity with their distinct phenology for different tree species in the study sites may have influenced the observed pattern. The greatest litter accumulation occurred in the 11-year-old natural regeneration, which can be attributed to the highest concentration of *Cecropia* trees which have coarse leaves difficult to decompose, and to a possible coincidence of higher litterfall of other pioneer tree species.

Plots	Organic Carbon (%)			Organic Matter (%)		
	Wet	Dry	Transitional	Wet	Dry	Transitional
RWTS1	2,29 ±0,23 c	2,95 ± 0,40 b	2,35 ± 0,16 b	3,94 ±0,40 c	5,08 ±0,69 b	4,04 ±0,28 d
RWTH1	0,05 ±0,05 d	0,18 ± 0,06 c	0,16 ± 0,00 c	0,09 ±0,09 d	0,31 ±0,10 c	0,28 ±0,01 c
NR11	2,98 ±0,10abc	3,74 ± 0,17 b	3,48 ± 0,13 b	5,14 ±0,17 abc	6,46 ±0,29 b	6,34 ±0,22 b
REF11	3,51 ±0,14 ab	3,95 ± 0,18 b	3,23 ± 0,21 b	6,06 ±0,24 ab	6,82 ±0,31 b	5,57 ±0,37 b
PF	3,92 ±0,39 a	6,51 ± 0,16 a	5,07 ± 0,53 a	6,77 ±0,68 a	11,23 ±0,48 a	8,75 ±0,92 a

Means followed by similar letters are not significantly different ($p < 0.05$) according to the Tukey test

Table 3: Organic carbon and organic matter from upper layer (0-10 cm)) in 1-year-old reforested plots with (RWTS1) and without (RWTH1) topsoil replacement; in 11-year-old natural regeneration (NR11); in 11-year-old reforested plots (REF11); and in primary forest (PF) in Saracá Mine, Porto Trombetas-PA, in the wet season, dry season and transitional period. Values are represented by means ± SE (n = 3).

3.3 Soil organic carbon and organic matter

The carbon and organic matter showed the highest concentration ($p < 0.05$) in the control forest and the lowest concentration in the 1-year-old reforested sites without topsoil (Tab. 3). Clearly, the removal of the topsoil reduced the input of organic material resulting in smaller quantities of organic substrates. The organic carbon at the 11-year-old natural regeneration and at the 11-year-old reforested sites did not differ in all samplings, but both showed lower values than for the control forest in the dry season and in the transitional period, suggesting that the time needed for re-building the soil organic carbon stock is longer than a decade.

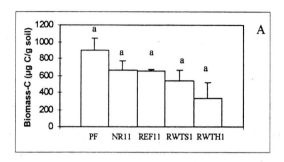

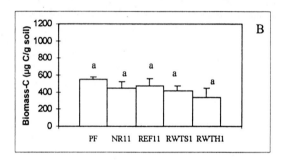

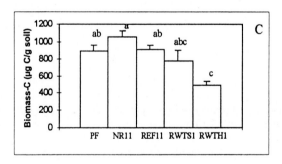

Figure 2: Soil microbial biomass (µgC/g soil) from upper layer (0-10 cm) in primary forest (PF), 11-year-old natural regeneration (NR11); in 11-year-old reforested plots (REF11); in 1-year-old reforested plots with topsoil (RWTS1) and in 1-year-old reforested plots without topsoil replacement (RWTH1) in the A) wet season, B) dry season and C) transitional period. Values are represented by means ± SE (n = 3).

3.4 Soil microbial biomass

The most important contribution of soil organic matter to the soil fertility and cropping production is its capacity to supply nutrients for plant growth (Bonde & Robertson 1991; Henrot & Robertson 1994). Based on this assumption, the initial hypothesis of the present study was that the sites receiving topsoil replacement would have higher concentrations of organic matter and consequently higher soil microbial biomass. Surprisingly, at any time, there were no differences in the microbial biomass between the two 1-year-old reforested sites, with and without topsoil replacement (Fig. 2). However, when compared with the control forest, the 1-year-old plot without topsoil showed significantly lower ($p < 0.05$) microbial biomass in the transitional period.

Despite the few significant differences between the treatments, the results in this study were sensitive enough to show significant differences ($p < 0.05$) between seasons in the soil microbial biomass, which confirmed findings by Luizão et al. (1992), who reported strong seasonal fluctuation (along the year) for soil microbial biomass in mature forest and young pasture in Central Amazon. Microbial biomass followed the same pattern observed for organic carbon, of which microbial biomass is a small part. However, the lack of significance between the two 1-year-old treatments emphasize the high plasticity of microbial population, suggesting that during one year the survived or reintroduced microorganisms became adapted to the new environment. In addition, microbial biomass differed between the treatments of 1-year-old without topsoil and the forest control which points to less microbial activity process at sites with reduced input of organic matter. In contrast, no significant differences were observed between the 11-year-old sites of natural regeneration and reforestation, or between these two treatments and the forest control in all samplings. This indicates that one decade after reforestation the soil microbial biomass is fully recovered.

4 Conclusions

In this study we found that among the 1-year-old treatments the organic carbon content of the soil was the best indicator of soil organic matter recovery. The replacement of the topsoil layer and of its nutrients greatly affected the vegetation growth and hence the organic matter production.

Among the 11-year-old treatments it was observed that after one decade organic carbon was still lower than for native forest, whereas the microbial biomass approached a similar level independent of the treatment.

Acknowledgements

Deutsche Gesellschaft für Technische Zusammenarbeit (GTZ), Instituto Nacional de Pesquisas da Amazônia (INPA), and Conselho Nacional de Desenvolvimento Científico e Tecnológico (CNPq).

References

Anderson, J.M. and Ingram, J.S.I. (1989): Tropical Soil Biology and Fertility. A Handbook of Methods. CAB-International, 171 p.

Bonde, T.A. and Robertson, K. (1991): Soil organic carbon and nitrogen dynamics changes after clearfelling and cultivation of a tropical rainforest soil. In: Bonde, T. A. (Ed.). Size and Dynamics of Active Soil Organic Matter Fraction as Influenced by Soil Management. Linköping Studies in Arts and Science **63**: 1-35.

Dantas, M. and Phillipson, J. (1989): Litterfall and litter nutrient content in primary and secondary Amazonian 'terra firme' rain forest. Journal of Tropical Ecology **5**: 27-36.

Fernandes, F.R.C. and Portela I.C.M.H.M. (1990): Recursos minerais da Amazônia (alguns dados sobre situação e perspectivas). Projeto Especial-DTA (Desenvolvimento da Tecnologia Ambiental). CNPq - CETEM 23p.

Griffith, J.J. (1980): Recuperação conservacionista de superfícies mineradas: uma revisão de literatura. Sociedade de Investigações Florestais, Viçosa. Boletim Técnico 2, 51p.

Henrot, J. and Robertson, P.G. (1994): Vegetation removal in two soils of the humid tropics: effect on microbial biomass. Soil Biology & Biochemistry 26: 111-116.

Higuchi, N., Jardim, F.C.S. and Barbosa, A.P. (1982): Inventário Florestal no Rio Trombetas (INPA/ SHELL-ALCOA). ALCOA MINERAÇÃO S. A. Apêndice. INPA, Manaus/AM, 13, 49p.

Jackson, M.L. (1958): Soil Chemical Analysis Contable & Co. Ltd., London.

Kuiters, A.T. (1990): Role of phenolic substances from decomposing forest litter in plant-soil interactions. Acta Botanica Neerlandica 39, 329-348.

Luizão, F.J. (1989): Litter production and mineral element input to the forest floor in a Central Amazonian forest. GeoJournal 19: 407-417.

Luizão, R.C.C., Bonde, T.A. and Rosswall, T. (1992): Seasonal variation of soil microbial biomass - the effect of clear-felling a tropical rainforest and establishment of pasture in the Central Amazon. Soil Biology & Biochemistry 24: 805-813.

RADAMBRASIL (1976): Levantamento de Recursos Naturais Vol. 10; Folha S.A. 21 - Santarém. Ministério das Minas e Energia; Departamento Nacional de Produção Mineral,. Rio de Janeiro/RJ, 510 p., 7 mapas.

Santos, B.A. (1987): Carajás: patrimônio nacional. In: MCT-CNPq (ed.). A questão mineral na Amazônia: seis ensaios críticos. Recursos Minerais. Estudos e Documentos 5. Brasília, 93-131.

Vance, E.D., Brookes, P.C. and Jenkinson, D.S. (1987): An extraction method for measuring soil microbial biomass C. Soil Biology & Biochemistry 19:703-707.

Addresses of authors:
E.S. Costa
Divisão do Curso de PG-Ciências de Florestas Tropicais
Do Instituto Nacional de Pesquisas da Amazônia (INPA)
Cx. Postal 478
69011-970 Manaus / Amazonas
Brazil
R.C. Luizão
F.J. Luizão
Departamento do Ecologia (INPA)
Cx. Postal 478
69011-970 Manaus / Amazonas
Brazil

Impact of Intensive Agriculture on Changes in Humus Status and Chemical Properties of Arable Luvisols

L. Reintam

Summary

An increase in the depth of the A-horizon from 15–20 cm up to 25–35 cm was the result of deep ploughing over 20–30 years of intensive management. The total deepening of the humus horizon sometimes led to a decrease in the amount of organic carbon.

Against the background of relative stability of the C:N ratio (8.5–10.5) and total soluble fractions, an increase in humus fulvicity as well as in iron activity was characteristic of Luvisols. This was related to the formation of R_2O_3-fulvates and also with the fixation of fulvic compounds in the structure of clay minerals and humic acids with inactive sesquioxides. Compared with the other soils, an increase in the degree of humification as well as the formation of active humic acids were found only in Luvisol sections on calcareous till. A slight tendency to a change in base saturation was accompanied by an evident increase in nonsiliceous iron compounds and iron activity as well as in the enlargement of the ratio between humus acids bound with mobile sesquioxides, and earth alkalines.

Keywords: Luvisols, humus status, humus quality, iron relationships

1 Introduction

Mechanical and chemical degradation of soils is of global importance and has been the subject of worldwide study (Oldeman et al., 1990). It tends to be due to quite intensive management in agriculture, the impact of which on the resilient tolerance of soils as well as on the development of sustainability in economy and ecology needs to be ascertained (Blum, 1995).

In the 1980's a rather good level of crop (2.5–3 tons per ha of cereals, 15–20 tons per ha of potatoes, nearly 5 tons per ha of perennial grasses) and animal production (more than 4 tons of milk per cow) was achieved by mechanization and chemization of agriculture in Estonia. These activities in their turn under the condition of quite intensive management gave rise to certain degradation of soils (Nuudi and Reintam, 1989) as well as to the worsened changes in humus relationship (Reintam, 1993).

Due to the complicated polygenetic and polyphasic nature, a buffer capacity is characteristic of any soil under any natural and human impacts. To ascertain the possible trends of changes in humus status and some chemical properties of arable Luvisols special investigations were carried out.

2 Material and methods

In the course of large-scale soil mapping and general pedogenetic investigations, Calcaric Luvisols (LVc) on yellowish-grey (y-g) and reddish-brown (r-b) calcareous tills, Chromi-Calcaric Luvisols (LVxc) on reddish-brown calcareous till and Stagnic Luvisols (LVj) on bisequal reddish-brown noncalcareous till, were sampled by their entire sections during 1956–1967. Recurrent investigations were carried out in 1986 in the localities where primary studies had been made prior to the intensive use of heavy machinery and chemicals under conditions of large-scale farming. This procedure was once again repeated in 1996 after the partial privatization and changes in land use and management in the early nineties.

The recurrent samplings of Luvisols in two replications for humus and chemical investigations were carried out to a depth of 30 (or 40) cm where obvious changes in their morphology had occurred. Luvisol section (A-EL-Bmt-C and A-Baf-ELg-Bt-C in calcaric and stagnic formations, respectively) was preserved in all cases, in spite of the development of a humus horizon at depth.

The composition of humus was determined by acid-alkaline treatment using the volumetric method after Tyrin-Ponomareva (Ponomareva, 1957). The results were expressed in percentages of organic carbon. Humic acids bound with mobile sesquioxides were classified as active, while those bound with clay minerals were termed as inactive. Their percentage of total humic acids was calculated whereas the rest of humic acids belong to those bound with alkaline earths. The ratio between the percentage of humic and fulvic acids (Ha:Fa) represents the integral parameter for the humus maturity and mobility. Ratio between the 1st and 2nd fractions (1st fr:2nd fr) characterize the relations of brown humic-fulvic complexes bound with mobile sesquioxides to grey (black) ones bound with calcium and magnesium. The supply of organic carbon as one of the features of soil humus status was calculated at the level of smoothed bulk density (Kitse, 1979), that of organic carbon energy was determined according to Volobuyev's (1974) criteria.

Exchangeable aluminium after Daikuhara-Sokolov, base saturation after Kappen, nitrogen after Kjeldahl, nonsiliceous dithionite-soluble iron after Coffin, oxalate-soluble amorphous sesquioxides after Tamm and iron activity after Schwertmann, were determined (Arinushkina, 1970; Zonn, 1982).

3 Results and discussion

An increase in the depth of A-horizon from 15–20 cm (1950's–1960's) up to 25–35 cm (in the 1980's) was the result of deep ploughing during 20–30 years of extensive collective management. In spite of the changes in the land use system after the restoration of independence, this situation has continued, because deep ploughing is characteristic of farming practices. In most cases perennial grasses sown in the 1980's have not been ploughed and continue their natural impact on the progress of the humus profile. The total deepening of the humus horizon resulted in the dilution of topsoil with subsoil poor in humus, but rich in amorphous sesquioxides and/or ferrous-ferric neoformations formed due to compaction and changes in perched water status in Stagnic Luvisol sections. This sometimes led to a decrease in the percentage of organic carbon (Table 1), but to a lesser degree than in Rendzic Leptosols, Cambisols and/or Planosols (Reintam, 1993). An evident tendency to an increase in humus accumulation during the last decade is likely due to the permanent grass sward, but also to manuring under conditions of private management.

According to the C:N ratio there is a trend for a slight worsening in the maturity of humus. It may occur due to the deficiency of organic nitrogen and weakening of nitrogenous bridges in phenolic molecules of humus acids rich in carboxydes (Flaig, 1971) which can take place even under conditions of good nutrition and tillage (Clapp et al., 1990). Simultaneously, with the enlargement of the C:N ratio, an increase in the importance of both free fulvic acids and fulvates bound in the structure of clay minerals (hydrolysate of 0.5 M sulphuric acid) can be mentioned. The total soluble fraction was relatively stable or increased within the last decade. Except for Calcaric Luvisols on yellowish-grey till, it is accompanied by a decrease in the degree of humification (total percentage of humic acids) and an increase in the activity of humic acids. This can be interpreted as a frequent phenomenon developed in the human-impacted unstable situation where the intersystem mobilization of soil mineral resources is possible only against the background of changes in the quality of organic agents of pedogenesis.

Soil	Year	Depth, cm	C	N	C/N	Total soluble	Humic acids	Free fulvic acids	Hydrol. 0.5 M H_2SO_4	% of total humic acids	
			%			% of organic C				active	inactive
LVc y-g	1	0-25	1.42	0.14	10.1	63	18	3	9	73	19
		25-40	0.40	0.04	10.0	70	15	5	13	50	33
	2	0-10	1.53	0.15	10.2	57	21	5	11	81	19
		10-25	1.38	0.15	9.2	60	22	5	5	86	14
		25-40	1.13	0.16	7.1	55	21	6	8	77	16
	3	0-10	1.94	0.16	12.1	72	30	8	13	41	29
		10-25	1.58	0.16	9.9	77	19	8	14	82	11
		25-33	1.61	0.14	11.5	76	19	9	9	67	15
		33-40	0.52	0.06	8.7	100	6	9	1	67	0
LVc r-b	1	5-15	1.15	0.11	10.5	58	23	4	8	59	15
		23-33	0.25	0.03	8.3	56	10	11	11	32	16
	2	0-15	1.02	0.12	8.5	52	17	5	6	56	14
		15-32	1.67	0.16	10.4	44	17	4	7	50	21
		32-40	0.52	0.05	10.4	39	8	12	6	0	25
	3	0-15	1.50	0.12	12.5	71	13	6	10	94	5
		15-30	0.89	0.10	8.9	98	14	13	7	90	2
		30-35	0.47	0.06	7.8	90	9	27	4	62	13
LVxc	1	5-15	1.14	0.13	8.8	63	19	3	5	46	18
		23-30	0.25	0.03	8.3	64	16	12	8	50	25
	2	0-15	1.09	0.11	9.9	69	30	5	9	54	12
		15-30	1.13	0.12	9.4	72	31	7	12	48	6
		30-40	0.28	0.03	9.3	100	31	19	6	40	20
	3	0-15	0.98	0.08	12.3	78	16	12	6	93	3
		15-30	1.00	0.10	10.0	84	19	15	7	92	4
		30-40	0.48	0.08	6.0	97	19	22	10	55	22
LVj	1	0-20	1.16	0.12	9.7	67	26	9	6	81	10
		20-50	0.28	0.05	5.6	73	13	21	6	67	19
	2	0-15	0.94	0.09	10.4	78	21	11	9	47	37
		15-30	0.93	0.09	10.3	50	14	6	6	64	14
	3	0-15	1.30	0.10	13.0	60	14	2	6	93	1
		15-30	1.65	0.09	18.3	59	16	5	5	86	1

Table 1: Humus characteristics in 1956–1967 (1), 1986 (2) and 1996 (3)

Against the background of these changes in the percentage and composition of humus (Table 1), in connection with the deepening of humus horizon, the organic carbon supply in most cases and its energy sources increased (Table 2). A decrease in carbon and its energy sources was found in Calcaric Luvisols on reddish-brown till within the last decade under conditions of cereal monoculture (LVxc) or fallowed grass sward rich in dandelions and renoved of legumes (LVc, r-b). Such a situation tends to be rather common in a great number of arable soils used by new private farmers and/or cooperatives slow to organize normal organic management.

Soil	Year	Depth, cm	Ha/Fa	1st fr./2nd fr.	Org. C in the layer of 30 cm	
					supply kg·m^{-2}	energy 10^3 kcal
LVc y-g	1	0-25	0.5	1.9	5.12	20.5
		25-40	0.4	0.6		
	2	0-10	0.8	2.5	5.94	23.8
		10-25	0.6	1.0		
		25-40	0.8	1.9		
	3	0-10	0.9	2.8	7.06	28.2
		10-25	0.4	1.9		
		25-33	0.4	1.7		
		33-40	0.1	0.4		
LVc r-b	1	5-15	0.8	3.0	4.21	16.8
		23-33	0.3	1.3		
	2	0-15	0.6	2.9	6.05	24.2
		15-32	0.8	2.4		
		32-40	0.3	0.4		
	3	0-15	0.3	1.8	5.40	21.6
		15-30	0.2	0.8		
		30-35	0.1	0.4		
LVxc	1	5-15	0.5	1.9	4.55	18.2
		23-30	0.4	1.7		
	2	0-15	1.0	1.9	4.83	19.3
		15-30	1.1	0.9		
		30-40	0.5	0.4		
	3	0-15	0.4	1.9	4.45	17.8
		15-30	0.4	2.9		
		30-40	0.2	1.2		
LVj	1	0-20	0.8	4.7	3.90	15.6
		20-50	0.2	1.6		
	2	0-15	0.4	1.1	3.89	15.6
		15-30	0.5	3.8		
	3	0-15	0.4	4.2	6.12	24.5
		15-30	0.4	2.2		

Table 2: Humus quality indices in 1956–1967 (1), 1986 (2) and 1996 (3)

The widespread public opinion just like a total decrease in soil humousness is typical of the present arable soil probably proceeds from the frequency of the phenomenon in the topsoil, and being inadvertent to the status in the subsoil deepened and enriched with humus. A predominant increase in the supplies and energy of organic carbon was nevertheless characteristic of most Luvisols and is worthy of special attention in the interpretation of their contemporary status and productive efficiency.

Intensive collective as well as private farming management resulted in an increase in humus fulvicity of Luvisols. The ratio between R_2O_3- (1st fraction) and Ca- (2nd fraction) humic-fulvic complexes has been more or less stable (Table 2). This was related to the formation of the fulvates of mobile sesquioxides and also with the fixation of fulvic compounds in the interlayer structure of clay minerals (hydrolysate of 0.5 M sulphuric acid) and humic acids with sesquioxides. An increase in the ratio of the $1^{st}:2^{nd}$ fraction in the deepened humus horizon of Stagnic and Chromi-Calcaric Luvisols seems to be an indirect result of machinery degradation. Under the influence of seasonal perched water on compacted solum the alternation of reduction-oxidation processes took place leading to the accumulation of ferric-ferrous humic-fulvic complexes (Brinkman, 1979) and an evident increase in the iron oxides (Table 3).

Soil	Year	Depth, cm	pH		Al^{3+} mg·kg^{-1}	V	Oxalate-soluble oxides (%)		Nonsil. oxide Fe	Fe activity
			H_2O	KCl		%	Fe	Al	%	%
LVc y-g	1	0-25	6.3	5.4	0	85	0.48	0.39	0.88	54
		25-40	6.4	5.5	0	86	0.51	0.39	0.81	63
	2	0-10	6.6	5.7	3	75	0.35	0.28	0.70	50
		10-25	6.6	5.8	4	78	0.61	0.31	0.99	62
		25-40	6.7	5.8	5	87	0.45	0.35	0.76	59
	3	0-10	6.7	6.3	2	80	0.55	0.31	0.83	66
		10-25	6.7	6.3	1	80	0.49	0.33	0.54	91
		25-33	6.9	6.4	1	83	0.52	0.33	2.29	23
		33-40	6.8	6.4	1	85	0.34	0.29	0.48	71
LVc r-b	1	5-15	6.4	5.5	4	80	0.29	0.26	0.76	38
		23-33	6.4	5.5	4	83	0.38	0.28	0.79	48
	2	0-15	6.8	5.9	3	75	0.46	0.31	0.60	77
		15-32	7.1	6.4	1	92	0.45	0.38	0.82	55
		32-40	6.8	6.2	3	82	0.64	0.34	1.57	41
	3	0-15	5.9	5.6	4	75	0.39	0.27	0.67	58
		15-30	6.4	5.7	2	90	0.32	0.10	0.54	59
		30-35	5.9	5.5	3	85	0.35	0.12	1.38	25
LVxc	1	5-15	6.1	5.2	2	66	0.34	0.28	0.68	50
		23-30	5.8	4.9	5	74	0.30	0.28	0.60	50
	2	0-15	6.2	5.4	5	69	0.50	0.24	0.79	64
		15-30	5.9	5.3	7	70	0.38	0.18	0.69	55
		30-40	5.8	5.0	6	75	0.38	0.18	0.48	79
	3	0-15	6.2	5.8	3	70	0.36	0.21	0.55	65
		15-30	6.2	5.7	3	70	0.41	0.19	0.53	77
		30-40	6.0	5.6	2	75	0.40	0.20	0.44	91
LVj	1	0-20	5.4	4.5	82	48	0.41	0.40	0.67	61
		20-50	5.5	4.6	60	47	0.43	0.40	0.63	68
	2	0-15	6.9	6.4	15	90	0.38	0.39	0.80	47
		15-30	6.7	6.3	13	88	0.51	0.34	1.04	49
	3	0-15	6.4	6.0	3	93	0.54	0.31	0.70	77
		15-30	6.6	6.1	1	85	0.55	0.33	0.59	93
		30-40	6.3	5.9	2	70	0.49	0.37	0.84	58

Table 3: Some chemical characteristics in 1956–1967 (1), 1986 (2) and 1996 (3)

Oxalate-soluble (amorphous) aluminium was rather stable in time and space. Intermediate intensification of amorphous iron oxide formation has stabilized within the last decade simultaneously with a decrease in acidity and with an increase in base saturation. A slight tendency to a decrease in base saturation accompanied by an obvious increase in iron compounds and iron activity was characteristic of collective farming management of Luvisols on calcareous tills. A decrease in the application of fertilizers resulted in the contemporary increase in base saturation and immobilization of exchangeable aluminium. It is possible that the joint impact of air pollution (acid precipitation) and acid fertilization on the exchange complex of soil has weakened within the last decade.

Although changes in the described humus and iron status are far from being dangerous to the normal functioning of the arable plant-soil system, only a tendency to an increase in fulvicity of humus and to a decrease in humification can demonstrate certain irregularities under conditions of intensive agriculture. The transformation of fulvic acids into humic ones is evidently disturbed, though a cereal monoculture as well as a frequent fallow situations ensured the progress of pedogenesis characterized by mobile fulvic humus and high activity of iron. That is why stability and/or an increase in the sources of organic carbon and nonsiliceous iron cannot make amends for a deterioration in humus qualitative properties.

4 Conclusions

Luvisols were relatively stable with respect to human impact during intensive management both in large-scale collectives and new private farming. However, stability and/or an increase in quantitative characteristics (sources of organic carbon and energy, increase in humus horizon, etc.) cannot make amends for deterioration in some qualitative indices such as high mobility of humus, great fulvicity, relative low degree of humification, variable in time active acidity, and mobility of sesquioxides. To some extent, these phenomena have been caused by machinery compaction, deficiency of fresh organic sources and surplus of mineral nitrogen in cooperative farming management, but also by a monoculture and unregulated turnover of substances and energy after privatization of family farms.

References

Arinushkina, E.V. (1970): Handbook for Soil Chemical Analyses. (in Russian) University Press, Moscow, 488 pp.
Blum, W.E.H. (1995): A Concept of Sustainability, Based on Soil and Soil Functions. State-of-the-Art-Lecture, Internat. Congress on Soils of Tropical Forest Ecosystem, Balikpapan/ Indonesia, 24 pp.
Brinkman, R. (1979): Ferrolysis, a Soil-Forming Process in Hydromorphic Conditions. Pudoc, Wageningen, 105 pp.
Clapp, E., Hayes, M., Malcolm, R., Chen, Y. and Dowdy, R. (1990): Characterization of humic substances from soils managed by different nitrogen, tillage, and maize residue treatments. In:Trans. of 14th Internat. Congr. of Soil Sci. Kyoto, Japan, **II**, Comm. II, pp. 405–406.
Flaig, W. (1971): Organic compounds in soil. Soil Science **111**, 1:19–33.
Kitse, E. (1979): Soil Moisture. (Mulla vesi in Estonian) Valgus, Tallinn, 142 pp.
Nuudi, T. and Reintam, L. (1989): Soils are the basis of life, but they are in danger. Estonian Nature **4**: 210–217, 270.
Oldeman, L.R., Hakkeling, R.T.A. and Sombroek, W.G. (1990): World Map of the Status of Human-Induced Soil Degradation. GLASOD, ISRIC, UNEP. 27 pp. and maps.
Ponomareva, V.V. (1957): To the method for the study of humus after I.V.Tyurin's scheme. (in Russian) Pochvovedenie (Soil Science), **8**: 66–71.
Reintam, L. (1993): Humus relationship as a representation of soil resilience in extensive agriculture. In: Ü. Mander, J. Fenger, T. Oja and M. Mikk (Eds): Energy, Environment and Natural Resources Management in the Baltic Sea Region. 4th Internat. Conf. on System Analysis. Copenhagen: Nordiske Seminar- og Arbejdsrapporter, **653**: 409–415.
Volobuyev, V.R. (1974): Introduction to the Energetics of Pedogenesis.(in Russian) Nauka, Moscow, 128 p

Address of author:
Loit Reintam, Institute of Soil Science and Agrochemistry, Estonian Agricultural University
Viljandi Road, Eerika, EE-2400 Tartu, Estonia

Degradation of Dryland Ecosystems: Assessments and Suggested Actions to Combat It

A. Ayoub

Summary

This paper gives an account of several attempts by UN organizations, particularly UNEP, in the area of desertification assessment and monitoring. Approximately 70 % of the susceptible drylands are undergoing various forms of land degradation. Overgrazing of the rangelands by livestock is the most widespread cause of degradation in the drylands. The UN Convention to Combat Desertification recommended that efforts should move from global assessment to national and local level assessments, and focus should shift from physical parameters and directed more to people issues.

Keywords: Drylands, desertification, assessment, physical parameters, overgrazing.

1 Introduction

The World Atlas of Desertification (UNEP, 1992), defines drylands as those areas other than polar and sub-polar regions in which the ratio of annual precipitation to potential evapotranspiration falls within the range from 0.05 to 0.65. This definition has also been adopted by the UN Convention to Combat Desertification. It reads: "desertification means land degradation in arid, semi-arid and dry sub-humid areas resulting from various factors, including climatic variations and human activities".

Accordingly, drylands occupy 6150 million ha (41% of the world's land area). Land degradation is one of the major problems of dryland ecosystems, accompanied by rainfall variability both in time and space. This coupled with the inherent ecological fragility of drylands weakens the resilience of the ecosystem.

The United Nations held a Conference on Desertification (UNCOD) in Nairobi in 1977 to address the 1968-1973 drought in the African Sahel, which was followed by another meeting in 1984 as a result of the 1979 - 1984 drought in the same region. In those meetings it was acknowledged that any attempt to attain sustainable development for such dry regions should recognize the links between poverty, inequality and environmental degradation. The principal causes of desertification were identified to be: (a) natural vulnerability of the ecosystem to degradation; (b) over-exploitation of the biotic resources; (c) unsustainable land use; and (d) political unrest. UN Plan of Action to Combat Desertification (UNPACD) was developed for arresting desertification and restoring productivity of these lands, but external funding was not forthcoming. Having been designated as the intergovernmental body to prepare for the UNCOD, UNEP has focused substantial efforts on the issue of desertification and its control. UNEP has carried out various studies on desertification at national, regional and global levels and has built up a comprehensive data base on its various facets.

UNEP, FAO and UNESCO have collaborated on desertification issues since the mid 1970's when the first world map of desertification was produced for UNCOD in 1977. In the early 1980's, FAO and UNEP developed the first methodology for the assessment of desertification, which since has been modified to be more appropriately applicable to global needs. UNEP and the International Soil Reference and Information Centre (ISRIC) cooperation has been particularly rewarding in the development of the Global Assessment of Soil Degradation (GLASOD) and the Soils and Terrain (SOTER) data base, which provided and continues to provide considerable information for the database of the World Atlas of Desertification.

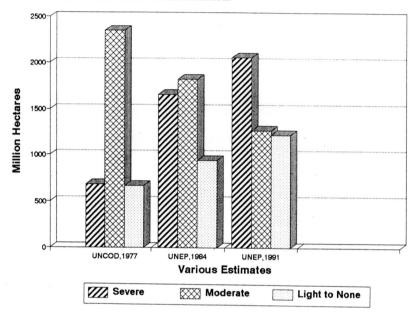

Figure 1: Estimates of global drylands degradation

2 Assessments of desertification/dryland degradation

There are considerable differences in opinion on the extent, severity and trends of land degradation in the drylands. Several global or regional attempts of land degradation/desertification assessments have been attempted (Fig. 1). The first attempt was done by FAO/UNESCO/WMO (1977) for the UNCOD. This assessment revealed the following data:
- The area threatened at least moderately by desertification within the drylands amounted to about 3.0 billion hectares.
- More than 100 countries were affected by desertification the inhabitants of which constituted more than 15 % of the world's population. The population of areas undergoing severe desertification totaled 78.5 million, and that the annual rate of land degradation in arid and semi-arid areas alone amounted to a total of 5.825 million hectares, broken down as follows: irrigated lands 0.125, rainfed croplands 2.5, and rangelands 3.2.
- The annual loss of productive capacity (income foregone) due to desertification amounted to US$ 26 billion. The annual cost of land reclamation measures was estimated to be US$ 388 million and the annual benefit of land reclamation measures amounted to more than twice that figure.

- A twenty-year worldwide programme to arrest further desertification required approximately US$ 4.5 billion a year or US$ 90 billion in total, of which developing countries in need of financial assistance would require US$ 2.4 billion a year or US$ 48 billion for twenty years.

The general assessment of the status and trend of desertification was undertaken by UNEP and FAO in 1984 (FAO/UNEP, 1984). The main findings arising from the 1984 assessment showed that:
- The scale and urgency of the problem of desertification as presented to UNCOD in 1977 were confirmed;
- Desertification continued to spread since 1977, efforts were too modest to be effective;
- Land degraded to desert-like conditions continued to increase at the rate of six million hectares annually;
- Areas affected by at least moderate desertification totalled about 3.48 billion hectares; 3,100 million hectares of rangelands, 335 million hectares of rainfed croplands, and 40 million hectares of irrigated lands.
- Rural populations in the areas severely affected by desertification numbered 135 million people;
- Projections to the year 2000 indicated that desertification in rangelands would continue to increase at existing rates. In rainfed croplands it would accelerate to critical proportions; in irrigated lands, it would probably remain stable, with gains balancing losses and with possible local improvements;
- The cost of losses due to desertification was estimated as five times the cost of halting desertification.

The third assessment of the world status of desertification undertaken by UNEP in 1990-91 (UNEP, 1991) was based on the GLASOD (1990) and Dregne (1991). The first data set (Dregne, 1991) was produced by the International Centre for Arid and Semi-Arid Land Studies (ICASALS) of Texas Tech University, USA. The second data set related to soil degradation within drylands of the world, based on the World Map of the Status of Human-Induced Soil Degradation (GLASOD, 1990) prepared by the International Soil Reference and Information Centre (ISRIC) and UNEP in 1990. Although the two data sets were different, they were interrelated. The major difference between the two global figures were in the status of degradation in the rangelands of the dry regions.

The 1991 assessment shows the following:
- About 43 million hectares of irrigated lands or 30% of their total area in the world's drylands (145 million hectares) were affected by various processes of degradation, mainly waterlogging, salinization and alkalinization. It would be safer to assume that the situation did not change appreciably since the 1984 assessment, but remained unsatisfactory with a tendency towards worsening. Soil scientists have established that the world is losing about 1.5 million hectares of irrigated lands annually due to various processes of soil degradation, mostly salinization and mainly in the drylands. It would thus be safe to assume that about 1.0 to 1.3 million hectares of irrigated land are currently lost every year throughout the world's drylands; this is presumably compensated by irrigating the best rainfed croplands and rangelands, whose area, how-ever, decreases accordingly.
- Nearly 216 million hectares of rainfed croplands or about 47 per cent of their total area in the world's drylands (457 million hectares) are affected by various processes of degradation, mainly water and wind soil erosion, depletion of nutrients and physical deterioration. It shows some decrease in comparison with the 1984 assessment.
- About 3,333 million hectares of rangeland or nearly 73 per cent of its total area in the world's drylands (4,556 million hectares) are affected by degradation, mainly by degradation of vegetation, which, on some 757 million hectares, is accompanied by soil degradation, mainly erosion. It shows an increase of some 233 million hectares in comparison with the 1984

assessment. As in the case of irrigated lands, it would be safer to assume that the situation did not change appreciably during this period and remained very unsatisfactory with a tendency towards worsening.

Desertification assessment reported in the World Atlas of Desertification produced by UNEP (UNEP, 1992) was based on the GLASOD (1990) database, which shows that just over one billion hectares are affected by human-induced soil degradation. Areas affected by soil degradation are 43 million ha of irrigated lands, 215 million ha of rainfed croplands and 757 million ha of rangelands. Soil degradation is a serious phenomenon in the drylands because of the slowness of recovery following disturbance. For example salts, once accumulated, tends to remain *in situ*.

Despite the inaccuracy of the available data, both the 1991 and 1992 assessments very definitely show a dramatic decline in the land resources of the world's drylands. Approximately 70% are undergoing various forms of land degradation. It is difficult at this stage to predict definite trends, but the process, if unabated, may lead to very serious socio-political and economic consequences world-wide, largely in developing countries.

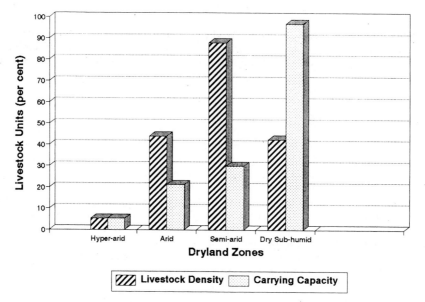

Figure 2: Indicative potential carrying capacity and livestock density in Africa drylands.

3 Causes of rangeland degradation

Overgrazing of the rangelands by livestock is the most widespread human-induced cause of degradation in the drylands. Figure 2 shows the potential carrying capacity and livestock density in different vegetation zones of the drylands in the Sudano-Sahelian zone of Africa. The potential carrying capacity for each vegetation zone was estimated using the mean annual rainfall and land quality (soils and terrain characteristics). Temporal vegetation variability was also considered. Calculations of livestock units were done following Le Houerou and Grenot (1986) in Thomas and Middleton (1994).

In the hyper-arid zone, livestock, which are mostly camels and goats, survive on zerophytic shrubs and ephemeral grasses. As the carrying capacity of this zone is very limited, only less than

4% of livestock in the African drylands survive here. At the fringes of this zone better winter grazing of camels is possible: the animals subsist on juicy plants without water. In the arid and semi-arid zones the animal density is far above the carrying capacity and these are the areas where most of the land degradation takes place. These zones are overgrazed and highly denuded of their tree cover. Uncontrolled communal grazing and deforestation for fuelwood and domestic use accentuated by frequent drought exposed the soils to severe water and wind erosion. The semi-arid zone is mostly inhabited by cattle-owning nomads who graze their herds on stabilized sand-dunes. Grazing is good during the rains, but as the dry season proceeds, the nomads move with their cattle to the more reliably-watered rangeland. This zone usually carries the highest animal density which exceeds the carrying capacity by about 3-fold. Animal densities in the dry sub-humid zone are usually below the carrying capacity. Parts of this zone could be without livestock, particularly cattle, due to the presence of the tsetse fly (*Glossina moritans*).

Overgrazing around settlements is often related to the sedentarization of nomadic herders. The settlement of these former nomads has meant that their herds have been concentrated onto grazing around their new homes. Drought conditions have also forced herders to concentrate their animals in areas where drinking water was available causing the disappearance of herbaceous cover in many places. This is especially the case around boreholes which provide drinking water for humans and animals year round with consequent sheet erosion and windblown loss of topsoil and reactivation of ancient sand dune deposits. Rangeland destruction is also due to fire incidents which have an effect equal to overgrazing. Infestation of the area with grasshoppers, rats, and locusts produces a similar situation to overgrazing.

The Sudano-Sahelian region of Africa and some other drylands of the world are experiencing, since the late 1960s, a comparable or even worse environmental disaster that occurred in the "Dust Bowl" in the 1930's in the semi-arid region of the United States. The Dust Bowl experience was a pivotal event in the formation of dryland management practices. The United States government mounted an unprecedented reclamation programme to restore productivity in the dust bowl. This effort involved the creation of new government agencies such as the Soil Conservation Service, new legislative initiatives, new funding mechanisms, the formation of farmer cooperatives, intensified agriculture research and extension, and building national highways to serve the affected areas. These actions have been credited with reversing the course of desertification in the western United States. It is expected that this type of programme could have similar beneficial effects on the problems of land degradation in the Sudano-Sahelian region and elsewhere. Drylands contain large areas of high quality soils which when effectively managed can sustain more than double its current population.

4 The Post-Agenda 21 and CCD Efforts

UNEP and UNDP/UNSO are jointly initiating a programme to develop desertification indicators in response to Agenda 21 (UN, 1993) and the Convention to Combat Desertification (UN, 1994). The new approach has to move from global assessment to national and local level assessments. Integration of key social, economic and demographic factors with the full range of physical factors and indicators, as appropriate to the local situation, is essential (Berry, 1996). Past assessments have focussed on the global picture and the physical parameters; future assessments will be much more integrated and directed more to people issues and be linked from local to national levels. The prime purpose of this activity is, therefore, to put the priority assessment at the local level to be undertaken with the help of the local people.

Rangelands desertification indicators are not well developed. There exists a great deal of confusion between vegetation degradation and soil degradation and the relative importance of each in determining the extent and severity of desertification at global and national levels. Appropriate indicators of rangelands degradation should show the status of the problem at a given time, the trend of the severity with time, and could also lead to the prediction of the impacts of desertification. This will help policy makers appreciate the significance of rangelands desertification to the livelihood of those who live in the desertification-prone areas, and the consequent impacts on national economy and social and political stability of the country. The indicators could also provide a means of comparing the severity of the problem from one country to another. However, in defining such indicators distinction should be made between short-term cyclic episodes of degradation and the longer-term trends (Berry, 1996).

The World Commission on Environment and Development (Brundtland, 1989) recommended that environmental policies in the drylands have to focus in the short-term on repairing damages previously caused and in the long-term to prevent damage and to reduce the adverse effects of human activities, and to actively promote a socio-economic policy that expands the base of sustainable development.

Agenda 21, on the other hand, recommended implementation of urgent direct preventive measures in drylands that are vulnerable but not yet affected, or only slightly desertified drylands by introducing (i) improved land-use policies and practices for more sustainable land productivity; (ii) appropriate, environmentally sound and economically feasible agricultural and pastoral technologies; and (iii) improved management of soil and water resources. It further recommended that accelerated afforestation and reforestation programmes be carried out, using drought-resistant, fast-growing species, in particular native ones, including legumes and other species, combined with community-based agroforestry schemes. In this regard, creation of large-scale reforestation and afforestation schemes, particularly through the establishment of green belts, should be considered, bearing in mind the multiple benefits of such measures. It also recommended urgent implementation of direct corrective measures in moderately to severely desertified drylands with a view to restoring and sustaining their productivity.

5 Conclusions

It is clear from the various global assessments of desertification that a dramatic decline in the land resources of the world drylands is taking place. This process, if unabated, may lead to very serious socio-political and economic consequences world-wide, largely in developing countries. Overgrazing of the rangelands by livestock and cropping without appropriate nutrients input are the most widespread human-induced causes of degradation in the drylands. Past assessments of desertification have focussed on the global picture and the physical parameters. Future assessments should be much more integrated and directed more to people issues.

References

Berry, L. (1996): Dryland Assessment and Monitoring for Sustainable development. A draft joint UNDP/UNEP proposal. UNSO/UNDP, New York.

Brundtland, G.H. (1987): Our Common Future, Report of the World Commission on Environment and Development, University of Oxford.

Dregne, H.E (1991): Desertification Costs; Land Damage and Rehabilitation. International Centre for Arid and Semiarid Land Studies, Texas Tech University, Lubbock, USA.

FAO/UNESCO/WMO (1977): World Map of Desertification, United Nations Conference on Desertification, 29 August - 9 September 1977, Nairobi, Kenya.

FAO/UNEP (1984): Map of Desertification Hazards, United Nations Environment Programme, Nairobi.

Le Houerou, H.N., and Grenot, C.J. (1986): The Grazing Lands Ecosystems of the African Sahel: State of Knowledge. In R.T. Coupland (ed) Ecosystems of the World. Vol. 8: Natural Grasslands. Elsevier, Amsterdam.

GLASOD (1990): World Map of the Status of Human-Induced Soil Degradation. An Explanatory Note, UNEP/ISRIC, Wageningen.
Thomas, D.S.G. and Middleton, N.J. (1994): Desertification: Exploding the Myth. Wiley & Sons, England.
UNEP (1984): General Assessment of Progress in the Implementation of the Plan of Action to Combat Desertification 1978-84, Nairobi.
UNEP (1991): Status of Desertification and Implementation of the United Nations Plan of Action to Combat Desertification. Report of the Executive Director to UNEP/GCSS. III/3, UNEP, Nairobi, Kenya.
UNEP (1992): World Atlas of Desertification, Edward Arnold, London.
UN (1993): Report of the United Nations Conference on Environment and Development, Volume 1: Resolutions Adopted by the Conference, New York.
UN (1994): United Nations Convention to Combat Desertification in those Countries Experiencing Serious Drought and/or Desertification, Particularly in Africa, UNEP/IPA, Nairobi.

Address of author:
A. Ayoub
United Nations Environment Programme
P.O. Box 47074
Nairobi
Kenya

The Impact of Desertification on Sahelian Ecosystems
A Case Study from the Republic of Sudan

M. Akhtar

Summary

The natural grasslands of the Sahel in the Republic of Sudan form the economic backbone for a majority of the existing animal husbandry systems. With respect to the availability and the distribution of local animal feed resources in the eastern Sahel in humid and in dry years of the eighties and the early nineties estimations indicated alarming changes related to the quality and quantity of the herb-layer. These changes are principally not correlated with the short-term rainfall induced fluctuations of the biomass production and of the floristic composition.

Based on long-term field surveys in the Sahel, geomorphic, morphodynamic, climate, vegetation, land use and desertification maps were delineated. Field results were supplemented by the interpretation of satellite images (MSS, TM, NOAA) and aerial photographs. Soil analyses (e.g. soil seed reserves) and vegetation mappings show that the severity of site degradation or desertification differs locally, and is controlled by the seasonal availability of the water resources and the soil conditions.

Keywords: Sahel, land use rights, vegetation, desertification.

1 Causes of land degradation and desertification in NE Sudan

Until the early 1970s, the uncontrolled and unlimited accessibility to vegetal resources was prevented by a hierarchy of coercive measures enforced on a tribal basis. These measures varied according to the annually alternating availability of the water resources and biomass. The renunciation of the exclusive rights systems of resource utilization in 1971, i.e. the introduction of an open access system to the Sahel of the Sudan, encouraged maximum use of such resources (Kirk, 1994, p. 133). Especially in years with sub-optimal rainfall events, interventions in traditional land use regulations favoured the concentration of different land use practices within the remaining favourable localities, i.e. those with sufficient water and pasture resources.

Recording the current pattern of pasture utilization of the different livestock keeping systems in wet and lean years permitted the precise identification of agro-pastoral core areas in the eastern Sahel of Sudan (Fig. 1). In dry years, when the annual rainfall only amounts to 30% of the long-term average, it is the dry season pastures south of the 15° latitude which could be discerned as areas with a fairly reasonable grazing potential. According to Pflaumbaum (1994, p. 78) these areas had a 'theoretical' carrying capacity of 16.1 ha / Tropical Livestock Units in the vegetation period of 1991/92 which followed the sub-optimal rainfall of 1991. However, the unilateral promotion of land cultivation (mechanization of the rainfed agriculture and the implementation of irrigation schemes) to the eastern Sahel in this century has intruded upon the former dry season grazing areas.

ISBN 3-923381-42-5
© 1998 by CATENA VERLAG, 35447 Reiskirchen

Accordingly, in the dry year of 1991, fundamentally important dry season or dry year pastures of the Butana region (north-eastern Sudan) were reduced in size. This left a maximum 70 km wide strip of natural grasslands for mobile animal husbandry (Fig. 1). The mechanized rainfed agriculture bordering on the south of these pastures, aggravates or even inhibits the southward movements of nomadic pastoralists with their animal herds as long as the crops have not been harvested. The unrestricted accessibility to the natural pastures has supported the decision-making processes of farmers from agricultural areas to also invest in livestock keeping. This has added to the ecological stress due to competition arising with the more traditional livestock keeping systems for the meagre resources in dry years. The regular overstocking of dry season pastures, and the uncontrolled exploitation of the vegetal resources is thus the major cause of desertification.

The environmental effects caused by increased plant cover exploitation and the ecologically required regeneration phase in the Sahel have been accelerated by the recent dry periods of the 1980s and the early 1990s.

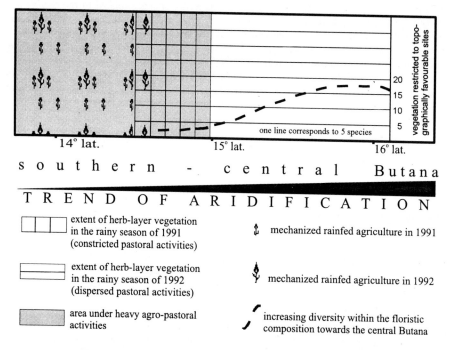

Fig 1: Agro-pastoral core area in the Butana region / NE Sudan

2 Major indicators and effects of desertification

Prior to the development and the implementation of rehabilitation measures at different scales, it was necessary to study the ecological impact of contemporary resource utilization practices. The surveillance of the plant cover over several years enabled the identification of a vegetal degradation sequence which could be typical for the Sahel.

The decrease in floristic diversity is a major feature of the frequently over-stocked pastures. Further investigations into the specific characteristics of the plants, i.e. their grazing values, indicated that during the past decades there has been a gradual decline in the availability of perennial and dry season grazing plants in north-eastern Sudan. Harrison (1955) defined the important dry season grazing forb, *Blepharis edulis*, as the climax vegetation of the Butana region. Instead, annual grasses and forbs prevail today in the frequently utilized pastures of this region. Due to various noxious attributes, these increasers in the herb-layer are not, or only to a limited extent, available for grazing throughout their life cycle (e.g. *Aristida spp.*, *Urochloa trichopus*, *Ocimum basilicum*) (Akhtar, 1995).

Presently, there is a remarkable correlation between the seasonal availability of water resources, and the distribution pattern of important dry season perennial grazing plants. An abundant occurrence of the highly cherished and perennial *Blepharis edulis* is restricted to localities with a water deficiency throughout the dry season. Thus the drop in the pasture quality of the agro-pastoral core zones in northern Sudan is principally not the outcome of the severe 1984/85 and 1990/91 drought events. Rather, it is the result of resources mismanagement during the last 25 years.

The decline in the perennial plant cover has also affected the morphodynamics of the Sahelian zone. The compact cotton soils (Table 1) which dominate the vast and featureless plains of the eastern Sahel are not prone to marked aeolian erosion, even in areas with a considerable vegetal degradation. However, in case of an overall decrease in the plant cover, surface horizons of sandy soils are subject to escalated erosion and redeposition.

The multi-temporal comparison of MSS, TM and NOAA satellite images and field studies in the Qoz areas (old dune belt) of central Sudan have revealed that the natural plant cover has been cleared extensively in favour of rainfed cultivation. During the recent droughts, the absence of rainfed cultivation, and the degradation of the natural vegetation has accelerated the pace of new sand formations developing, and encroaching upon the south, at a rate of more than 30 km in six years near the passage of Qoz el Harr (Akhtar et al., 1992). This is a hazard to the arable lands in the south.

Investigations into seed reserves (vegetal regeneration potential) of the different soils documented a discrepancy between the seed amounts of sandy and clayey soils (Table 1). Under similar climate and land uses, the sandy soils revealed little seed reserves when compared with the clay dominated cotton soils.

Morphological Unit	Granulometric conditions (% weight)	Sampling: (6 samples)	General seed, fruit contents	Other characteristics:
Plains in the Nubian Sandstone	clay: 20 % silt: 25 % sand: 53 % gravel: 2 %	600 ml from surface up to 5 m depth	approx. 20	about 50 % *Aristida* seeds
Basement peneplain	clay: 30 % silt: 50 % sand: 20 %	600 ml from surface up to 5 m depth	approx. 110	seed bank indicates diversity

Source: Akthar, 1995, pp. 127-137.

Table 1: Characteristics of the seed budget of different soils sampled in different morphological units.

Due to their visibly higher erosion and redeposition rates, it is possible that sand soils generally have a lower seed budget than the cotton soils. But the marked homogeneity of the meagre seed reserves in sandy soils under frequently utilized grasslands, and the dominance of the fairly undesired *Aristida spp.* seeds indicate the effects of improper grazing management on the regeneration potential.

3 Consequences for management

By identifying the environmental effects of resource mismanagement in the eastern Sahel, one can promote measures towards the ecological readjustment of present land uses. It is evident that years with below-average rainfall events especially require control measures over the meagre resources. Traditional control measures, which have been abolished, revealed that in dry years the seasonal restrictions on accessibility to regional water resources, by means of introducing exclusive rights of utilization, do curb uncontrolled exploitation, hence, preventing land degradation.

Long-term and interdisciplinary field investigations have emphasized that major corrective measures (Table 2) to the present land uses are possible, and would promote sustainable development in the eastern Sahel.

Proposed measures	Effects
Pushing back the northern boundary of the mechanized rainfed cultivation (this was also proposed in the Bill of 1991 in Sudan)	Especially in dry seasons or in lean years more grazing areas will be available, reducing the risk of overstocking and tribal conflicts
In dry years, the introduction of controlled access regulations to the water resources in the former tribal areas. The development of such regulations should consider traditional tribal land use rights, i.e. reintroducing some exclusive rights of utilization	In times of limited resources, and a reduced regeneration potential of the plant cover, this measure would prevent allochthonous tribes and farmers from unlimited access to the free of charge pasture resources in the former tribal areas
In lean years, a comparatively free availability of crop residues from the cultivated lands for the mobile animal husbandry systems	A free of charge access to these crop residues existed up to the early eighties. Its partial reintroduction would relieve the strain on the limited and vulnerable pastures
Implementation of desertification monitoring measures in frequently utilized areas	Regular ground checks and the implementation of satellite images and aerial photographs enable the immediate recording of degradation processes in the plant cover and the human induced changes to the morphodynamics

Table 2: Remedial action plans to combat desertification, and to promote sustainable land uses in eastern Sahel

References

Akhtar, M. (1995): Degradationsprozesse und Desertifikation im semiariden randtropischen Gebiet der Butana / Rep. Sudan. Dissertation an der Georg-August-Universität zu Göttingen. In: H.S.H. Seifert, P.L.G. Vlek and H.-J. Weidelt (eds.), Göttinger Beiträge zur Land- und Forstwirtschaft in den Tropen und Subtropen, **105**, 166 pp.

Akhtar, M., Mensching, H.G., Pflaumbaum, H. & Pörtge K.-H. (1992): The natural potential of a transition zone - examples from different ecosystem units of the Sahel, Republic of Sudan. Z. Geomorph. N.F., suppl.-Bd. **91**, 175-184.

Harrison, M.N. (1955): Report on Grazing Survey of the Sudan, (Dept. of Animal Prod.), Khartoum.Unpubl.

Kirk, M. (1994): Property rights in Resources, Pluri-activities and Socio-economic Differentiation: Livestock Production in the Butana Today. Interdisciplinary Research on Animal Production in the Sahel. Symposium 28.-29. Oktober 1993 Göttingen. In: H.S.H. Seifert, P.L.G. Vlek and H.-J. Weidelt (Editors), Göttinger Beiträge zur Land- und Forstwirtschaft in den Tropen und Subtropen, **98**, 125-149.

Pflaumbaum, H. (1994): Futterressourcen und Tragfähigkeit. In: H.G. Mensching and H.S.H. Seifert (Editors), Projekt Endbericht: Tierhaltung im Sahel - Rezente Entwicklungen und Perspektiven in der Rep. Sudan. Göttinger Beiträge zur Land- und Forstwirtschaft in den Tropen und Subtropen, **99**, 64-83.

Address of author:
Mariam Akhtar, Schillerpromenade 26, D-12049 Berlin, Germany

Concepts, Assessment and Control of Soils Affected by Salinization

I. Szabolcs[†]

Summary

The problem of salt-affected soils has gained ever-increasing importance in science, technology, ecology and economics alike during the last decades. This is expected given that more and more territories were found to be salt-affected in various regions and by the pressing demands for the production of foodstuffs and raw materials in many countries on the one hand and the conservation and protection of the natural environment on the other. Soil salinization is closely associated to these often conflicting requirements and has become a global problem, even one of the most important issues in many countries, as manifested by recent land utilization programmes taking into account protection of the natural environment.

Keywords: Alkalinity, degradation, land use, salinity, salt-affected soils, sodicity

1 Short review of the definition and extension of salt-affected soils

All soil formations in which water-soluble salts play a dominant role, deciding their physical, chemical, and biological properties, belong to the family of salt-affected soils in spite of the fact that, depending on the type of the salts, they can be very diverse in their appearance, morphology, properties, as well as the possibilities of their utilization.

In different soil classification systems, salt-affected soils are represented differently and appear on different taxonomical levels (Kovda, 1947; Richards, 1954; Szabolcs, 1992; Keys to Soil Taxonomy, 1992 [see Table 1], Abrol et al., 1988). However, for several technical and practical reasons, a grouping system, shown in Table 2, is recommended as its application has numerous advantages.

In Table 2 the different groups of salt-affected soils are listed according not only to the chemistry of salts causing their formation but also to the environment where they appear, and their properties adversely affecting the biota as well as to the possibilities of their sustainable management.

Salt-affected soils cover roughly 10% of the surface of the continents (Szabolcs, 1989). More than a hundred countries have salt-affected soils occupying different proportions of their territory. In many of them a great part of the utilizable land is covered by such soils causing great, often principal problems, for national production and economy. Salt-affected soils are particularly extended in dry areas, in many developing countries of Asia, Africa, and Latin America, hindering agri-, sylvi-, and horticulture and reducing the potential of food production. Salt-affected soils may be found not only in the vast territories of deserts and semi-deserts but also occur frequently in fertile river valleys, lowlands, foothills, sea shores and other areas where all natural conditions, apart from salinity, would be favourable for production.

Order	Suborder	Great Goup
Alfisols	Aqualfs	Natraqualfs
	Boralfs	Natriborals
	Udalfs	Natrudalfs
	Ustalfs	Natrustalfs
	Xeralfs	Natrixeralfs
Aridisols	Argids	Natrargids
	Orthids	Salorthids
		Gypsiorthids
Entisols	Aquents	Sulfaquents
Inceptisols	Aquepts	Sulfaquepts
Mollisols	Aquolls	Natraquolls
	Borolls	Natriborolls
	Ustolls	Natrustolls
	Xerolls	Calcixerolls
		Natrixerolls

Tab. 1: Salt-affected soils in the hierarchy of the US Soil Taxonomy

Type of salt-affected soils	Electrolyte(s) causing salinity and/or alkalinity	Environment	Properties adversely affecting the biota	Methods for reclamation
Saline	Sodium chloride and sulphate (in extreme cases nitrate)	Arid and semi-arid	High osmotic pressure of soil solution, toxic effect of chlorides	Removal of excess salts
Alkali	Sodium ions capable of alkaline hydrolysis	Semi-arid. semi-humid, and humid	High (alkali) pH, poor water physical conditions, Ca deficiency	Lowering or neutralizing the high pH by chemical amendments
Magnesium	Magnesium ions	Semi-arid and semi-humid	Toxic effect, high osmotic pressure, Ca deficiency	Chemical amendments, Leaching
Gypsiferous	Calcium ions (mainly $CaSO_4$)	Semi-arid and arid	Low (acidic) pH nutrient deficiency	Alkaline amendments
Acid sulphate	Ferric and aluminium ions (mainly sulphates)	Seashores and lagoons with heavy, sulphate containing sediments, diluvial inland slopes and depressions	High acidity and the toxic effect of aluminium Nutrient deficiency	Liming

Tab. 2: Grouping of salt-affected soils and their environmental relations

Soils affected by salinization

In Fig. 1 the distribution of salt-affected soils is shown. Salt-affected soils occur practically in all climatic belts, from humid tropics to beyond the polar circle. They can be found in different altitudes, from territories below sea level, e.g. the district of the Dead Sea, to mountains rising over 5,000 metres, such as the Tibetan Plateau or the Rocky Mountains. The rather widespread opinion, sometimes voiced not only in the mass media but even in scientific circles, which limits the occurrence of salt-affected soils to desert conditions, is simply incorrect. On the contrary, more and more salt-affected soils are being discovered e.g. in the tropical belt of Africa, Latin America and even in arctic regions, particularly the territory of the Antarctic. Accordingly it is necessary to broaden our sight whenever and wherever salt-affected soils are concerned in association with environment and production.

Fig. 1: The global distribution of salt-affected soils

Whenever and wherever salt-affected soils are studied and methods are elaborated for their reclamation and utilization, the environmental and economic conditions should first be taken into consideration in order to find proper methods which can guarantee sustainable management. The properties of salt-affected soils are, as a rule, in close correlation with the geochemical and geomorphological conditions of their environment.

2 Natural and secondary (man-made) salt-affected soils; the effects of irrigation

Most salt-affected soils developed through natural geological, hydrological and pedological processes and a great number of them have existed for millennia. However, Man, interfering with natural processes, created salt-affected soils in many parts of the world. It is well-known that in ancient times large irrigated territories turned into wastelands (in Mesopotamia, the valleys of the

Yangtze and the Hwang Ho in China, the Nile Valley in Egypt, etc.) due to the improper methods of primitive irrigation. It is less-known that the deforestation and overgrazing entailed by these ancient civilizations also contributed to the salinization of vast expanses of land in many regions.

It is a tremendous waste that neglected or abandoned irrigation systems are rather frequent and account for a very high percentage of all those existing. According to the estimates of the FAO (Food and Agricultural Organization of the UN) and the UNESCO (United Nations Educational, Scientific and Cultural Organization), as much as half of all the current irrigation systems of the world are more or less under the influence of secondary salinization, alkalization and waterlogging. This phenomenon is very common not only among old irrigation systems but also in areas where irrigation has been recently introduced.

According to the estimates of the above-mentioned agencies, as well as of the Subcommission on Salt-Affected Soils of the International Society of Soil Science, 10 million hectares of irrigated land are abandoned yearly as a consequence of the adverse effects of irrigation, mainly secondary salinization and alkalization (Szabolcs, 1989).

At present no continent is free from the very serious occurrences of this phenomenon. In Argentina 50% of the 40,000 ha of land irrigated in the 19th century are now salinized. In Australia, secondary salinization and alkalization take place in the River Murray valley, and in Northern Victoria 80,000 ha have been affected. The situation is no better in Alberta, Canada. Similar processes have been recorded in the northern states of the USA, where irrigation was introduced much later than in the dry West. It should be noted that the above-mentioned and many other irrigated regions are far from being arid and the majority of salts accumulating are sodium salts capable of alkaline hydrolysis, not the neutral sodium salts that are familiar to those who study desert and semi-desert areas.

As a rule, secondary saline and/or alkali soils were abandoned, remained salt-affected, and so it went on for many thousands of years. Secondary salt-affected soils have thus accumulated and now account for a great part of today's salt-affected soils on all the continents. Such soils are seldom ameliorated and where they are, the pace of amelioration cannot keep up with accelerating secondary salinization. In fact the ratio of the former to the latter is negligible.

Although salt-affected soils have been reclaimed on a large-scale in many countries, no data is available on the results of these projects which suggests that they can hardly be remarkable. We cannot expect this situation to change in the foreseeable future.

In Fig. 2 the global development of irrigation and secondary salinization of soils is charted also indicating the envisaged increase of irrigated territories and the estimated increase of those, secondary soil salinization by the beginning of the 21st century.

Fig. 2 clearly shows that the trends of increased irrigation and progressive salinization are nearly parallel; the increase of secondary salinization and alkalization even surpasses that of irrigation. It should be noted that the territory of secondary salinized soils is larger than the territory of irrigated land. This is because the former includes all those which were affected by irrigation in the past even if they have not been irrigated for centuries. This fact and because secondary salinization, induced by irrigation, influences to an ever-growing extent the larger areas surrounding the irrigation systems, results in a sharp increase of the secondary salinization of soils which accelerates with the further extension of irrigated agriculture.

3 Anthropogenic processes, other than irrigation, causing secondary salinization

Improper irrigation and drainage are presumed to be primarily responsible for secondary salinized and alkalized soils. However there are other anthropogenic effects which also lead to the intensification or initiation of secondary salinization and/or alkalization.

The most important are as follows:

Soils affected by salinization

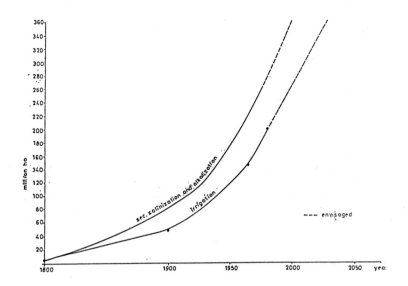

Fig. 2: Global development of irrigation and secondary salinization of soils

(1) Deforestation.
   Under arid and semi-arid conditions, sometimes even in non-arid conditions, intensive deforestation often results in changes in the water economy of the landscape and consequently in the appearance of bare land in place of forests. This is especially frustrating when the target has been to establish plantations.
(2) Overgrazing.
   The overgrazing of pastures is another factor which often leads to the intensification of salinization. As a consequence of overgrazing, the original balance and metabolism of compounds between the natural plant cover and the soil will alter.
(3) Changes in cultivation pattern.
   Similar problems may arise in the wake of changing the type of cultivation e.g. by turning natural meadows into arable land or utilizing them for plantation. The latter methods of cultivation often drastically change the water and nutrient balance in the soils, depleting them and initiating salinization.
(4) Depletion of fresh water layers near the surface of the soil and their replacement with saline ground water from deeper layers.
   This may happen due to irrigation (tube wells), or by the over-utilization of fresh water reserves for drinking by people and animals, for household purposes, etc. All of these courses of action further the salinization of both soil and water.
(5) Utilization of biomass in arid areas for fuel.
   The practice of burning scarce wood and shrub reserves is widespread in many dry countries of Africa, Asia and Latin America. Due to the mineralization of organic matter, considerable amounts of water-soluble salts accumulate in the soil layers and cause salinization in many places.
(6) Chemical contamination.

In intensive farming, and particularly in greenhouses, large amounts of mineral fertilizers have been applied, particularly in the last decades. Part of the fertilizer residues accumulates in the soil and causes salinization such as observed in Japan.

4 Principles and possibilities of the assessment and sustainable management of salt-affected soils

The problem of managing salt-affected soils in a sustainable way is very peculiar. By sustainable management of a soil we usually mean the application of high or low technology production systems with considerable output and the maintenance of good environmental conditions.

The management of salt-affected soils cannot be approached in this way as most of them cannot, and must not, be reclaimed or utilized for agricultural production in the foreseeable future.

A great part of the salt-affected soils of the world should maintain their present state without any attempt to utilize them where impractical costly and risky. The majority of saline deserts belong to such regions. In some places the natural halophyte vegetation of saline soils can be utilized (Ayoub 1992, Squires and Ayoub 1994).

The utilization and management of salt-affected soils, where and when possible, first needs a holistic approach, and a consideration of all the major imminent aspects and properties. It should be decided which part of the salt-affected territories can be reclaimed and/or managed, under which circumstances, with which environmental consequences, and a costs-benefit analysis should be prepared.

It is also of primary importance to decide the side-effects of amelioration and the management of salt-affected soils on the surrounding territories, water, air and biosphere.

Such considerations should be underlain by the general principles of pedology, biology, hydrology, etc. However, they should always be concrete and site-specific. This is a principal question because it often seems to be the easier way to adopt and transfer available methods without thorough analysis of local circumstances. It is not only necessary but also rewarding to carry out feasibility studies, preliminary surveys and proper planning before projecting and executing soil utilization programmes in salt-affected areas.

There are also many examples of the utilization of saline soils in arid areas, where the introduction of proper irrigation and drainage techniques ensures a good return on the investment with reliable and steady yields. Such methods are well-known and are widely used in Israel, Egypt, India and the western states of the U.S.

In parallel with applying the mentioned methods of amelioration, due attention should be paid to the maintenance of irrigation and drainage systems and the control of short- and long-term salinization processes.

The ways to solving this problem are either:
(a) to control or mitigate the processes of salinization, i.e. to keep the salt balance at its present level; or
(b) to reduce the processes of salinization, i.e. to remove part of the salts from the soil layers.

The proper strategy of this kind of combat against salinization depends on local circumstances, techniques of irrigation and drainage, and on the requirements of production.

In irrigated areas, or in areas to be irrigated, the study and control of alkalinity and/or salinity must be started well before putting the irrigation and drainage systems into operation; even before making plans for their construction. Such an approach is necessary because, with a proper survey, it can be decided whether the territory is reasonably suitable for irrigation. By studying the salinity problems of the areas where irrigated farming is envisaged, considerable worry can be avoided. During the preliminary survey, the soils, environmental conditions (climate, hydrology, etc.), as well as the possible sources of irrigation, and the quality of both ground and irrigation water, should be studied.

In irrigated areas it is essential to construct and employ up-to-date monitoring systems for the observation and control of salinity conditions and their changes enabling the anticipation of possible consequences.

In Table 3 a scheme is given of the methods recommended for the control of salinity and alkalinity in irrigated areas. Table 3 shows that a comprehensive survey and monitoring system is necessary both before and during irrigation.

(A) Before construction of irrigation system		
Preliminary survey		
<u>Soils</u>	<u>Landscape</u>	<u>Planned irrigation</u>
genetics spatial distribution typology and properties salinity/alkalinity	climate geomorphology hydrology hydrogeology	available irrigation water (quality and quantity) ground water depth & quality technology of irrigation salt tolerance of crops
(B) During irrigation		
<u>Monitoring</u>		
salinity and alkalinity of soil and ground water table chemical composition of ground water chemical composition of irrigation water water filtration physical soil properties toxic elements, if any, in soil and water		

Tab. 3: Scheme of methods recommended for the control of salinity and alkalinity in irrigated areas

Even when operating a proper monitoring system, unexpected changes may happen. It is highly necessary to recognize the early warning signals of adverse processes as soon as they appear in the soils and waters so that we can undertake in due time effective measures of precaution and correction. Unfortunately we still do not have sufficient knowledge of such early warning signals and of the monitoring systems to identify them.

References

Abrol, I.P., Yadav, J.S.P. and Massoud, F. (1988): Salt-affected Soils and their Management. FAO Soil Bulletin **39**, Rome.

Ayoub, A.T. (1992): Some features of salt tolerance in Sinna in Sudan. Intl. Workshop on Halophytes. Nairobi, 297-301.

Keys to Soil Taxonomy (1992): USDA, SMSS Technical Monograph No. 19. 5th ed. Pocahontas Press, Inc., Blacksburg, Virginia.

Kovda, V.A. (1947): Origin and Regime of Salt-affected Soils. (Russian), Vols. 1 and 2, Izd. Ak. Nauk. SSSR, Moscow.

Richards, L.A. (1954): Diagnosis and Improvement of Saline and Alkaline Soils, USDA Handbook No. 60., U.S. Department of Agriculture, Washington D.C.

Squires, V.A. and Ayoub, A.T. (1994): Halophytes as a Resource for Livestock and for Rehabilitation of Degraded Lands. Kluwer.
Szabolcs I. (1989): Salt-affected Soils. CRC Press, Boca Raton, Fl.
Szabolcs I. (1992): Salinization of Soil and Water and its Relation to Desertification, Desertification Control Bulletin, UNEP, Nairobi, No. 21, 32-37.

Address of author:
Istvan Szabolcs[†]
Research Institute for Soil Science and
Agricultural Chemistry
Hungarian Academy of Sciences
P.O.B. 35
H-1525 Budapest, Hungary

Quantifying Soil Salinity in a Dynamic Simulation of Crop Growth and Production

J. Barros & P. Driessen

Summary

A methodology is tested for quantifying (changes in) soil salinity in land-use systems with sunflower. Dynamic simulation of crop growth and crop production potential requires that soil attributes be parameterised for analysis of the soil's water budget during the cropping period. The total soil moisture potential is composed of matric and osmotic components. Both are calculated for a sequence of short time intervals in the growing season of the crop. The matric soil potential is quantified by combining water budget calculations with soil-specific water retention characteristics. A salt budget is calculated alongside the water budget by accounting for the salt load of each water flux in or out of the rooted soil compartment. The osmotic potential is quantified using generally valid transfer functions that relate the electric conductivity of the soil moisture with osmotic potential and salt concentration.

Keywords: Soil salinity, crop growth, model, simulation.

1 Introduction

In many agricultural lands, especially in arid zones, soil salinity builds up because of input of salts with irrigation water, lateral seepage or capillary rise. The present work discusses a methodology for quantifying physiological drought induced by salinity and its effects on crop growth. The model used is a comprehensive, deterministic crop production model, developed for dynamic simulation of the sufficiency of essential 'land qualities' in rigidly defined production situations (Driessen and Konijn, 1992).

Field experimentation was carried out at the experimental farm of the Institute for Natural Resources and Agrobiology of Seville (IRNAS) in Coria del Rio, Spain. Selected land-use systems with sunflower were monitored in 1993 and 1994.

2 Materials and methods

A water budget study keeps track of all water fluxes in or out of the rooted soil compartment. Lateral seepage may be ignored in the (flat and level, alluvial) study area but precipitation, irrigation, evaporation, transpiration, runoff and capillary rise or deep, percolation/drainage (can) play a role. Changes of soil salinity are made visible by accounting for the salt load of each water flux.

The electric conductivity of the groundwater (EC_w, in $dS.m^{-1}$) and of the irrigation water (EC_i) are inputs in the model just as the initial electric conductivity of the soil's saturation extract (EC_e). The electric conductivity of the *actual* soil solution (EC(day)) is the 'dependent', i.e. calculated, state variable value that expresses the momentary salinity level of the rooted soil compartment.

Uptake of water by plant roots is only possible after the total force with which water is retained by the soil (i.e. the total soil moisture potential, PSItot) is compensated by the crop. The total soil moisture potential is composed of a matric component (PSI) and an osmotic component (OP). The relation between OP (in atm) and the shift in freezing point caused by dissolved salts (DELFRPNT, in °C) is given by Thorne and Peterson (1954) as:

$$OP = 12.06 * DELFRPNT - 0.21 * DELFRPNT^2 \tag{1}$$

Weast (1975) lists electric conductivity values and corresponding osmotic potentials for aqueous solutions of sodium chloride. Working out the relations between electric conductivity (EC, in $dS.m^{-1}$) and DELFRPNT, the following equation was constructed:

$$EC = 28.26 * DELFRPNT - 2.33 * DELFRPNT^2 \tag{2}$$

Combining equations (1) and (2) yields the approximate relation between OP (in hPa) and EC:

$$OP = 463 * EC \tag{3}$$

A relation between EC and sodium chloride concentration (Co in $g.l^{-1}$) reads:

$$EC = 1.464 * Co \tag{4}$$

Relations (3) and (4) hold for solutions of pure sodium chloride. Saline soils may contain a variety of salts. The impression exists however that the coefficient 1.464 in equation (4) does not change much between the $NaCl - Na_2SO_4$ dominated saline soils that were examined in the present study.

The total quantity of salt in the root zone (SOILSalt) is calculated by adding to the quantity of salt in the original rooted soil compartment (SaltSURFACE) all salt that is added through root growth (SaltDEEP), irrigation (SaltIRRIG) and capillary rise (SaltRISEN) and by deducting any quantity of salt leached out of the rooted soil compartment with deep percolation (SaltPERCED):

$$SOILSalt = SaltSURFACE + SaltDEEP + SaltIRRIG - SaltPERCED + SaltRISEN \tag{5}$$

The salt load (in $kg.ha^{-1}$) of each flux is calculated for every time interval (Dt, in days) in the growing cycle of the crop.

Leaching water (D, in $cm.Dt^{-1}$) is assumed to have the salt content of the soil moisture. The quantity of salt leached, per hectare and per interval is approximated as:

$$SaltPERCED = D * Dt * (EC(day) / 1.464) * 100 \tag{6}$$

where 100 is a factor to satisfy units.

Likewise, salt influx with capillary rise (CR, in $cm.Dt^{-1}$), expressed in kg per hectare and per interval, amounts to:

$$SaltRISEN = CR * Dt * (EC_w / 1.464) * 100 \tag{7}$$

Irrigation water is a further source of salts. Not all irrigation water enters the soil; some of it may be discharged as surface runoff. Salt added with effective irrigation water inputs (IE, in $cm.Dt^{-1}$) amounts to:

$$\text{SaltIRRIG} = \text{IE} * \text{Dt} * (\text{EC}_i / 1.464) * 100 \tag{8}$$

Salt that was already present in the root zone of the soil (SaltSURFACE, in kg.RD^{-1} ha^{-1}) is assumed to be fully dissolved in the soil moisture stored over the (calculated) depth of the rooted surface compartment (oldRD, in cm):

$$\text{SaltSURFACE} = (\text{SMPSI} / \text{BD}) * \text{oldRD} * (\text{EC(Day)} / 1.464) * 100 \tag{9}$$

where SMPSI is soil moisture fraction (cm^3.cm^{-3})
       BD   is soil bulk density (g.cm^{-3})

Salt added with root growth (SaltDEEP, in kg.ha^{-1}) amounts to:

$$\text{SaltDEEP} = (\text{SMPSI} / \text{BD}) * (\text{RD} - \text{oldRD}) * (\text{EC(Day - 1)} / 1.464) * 100 \tag{10}$$

The total quantity of salt in the rooted surface compartment can now be calculated (SOILSalt; equation (5)) and the state variable EC can be adjusted for use in the calculations for the next time interval:

$$\text{EC(day)} = (\text{SOILSalt} / ((\text{SMPSI} / \text{BD}) * \text{RD} * 100)) * 1.464 \tag{11}$$

Bulk density (BD) is used to convert volumetric water fraction (SMPSI) to gravimetric water content. It may be input from file or calculated from total soil porosity (SM0 in cm^3.cm^{-3}, specified in the soil data file) and the specific density of the solid soil phase (SDSP, approximately equal to 2.6 g.cm^{-3} in mineral soils):

$$\text{SM0} = 1 - \text{BD} / \text{SDSP} \tag{12}$$

Groundwater is the most important source of irrigation water in the study area. Its electric conductivity (EC$_w$) varied between 2.11 and 1.74 dS.m^{-1}. The electric conductivity of the soil saturation extract (EC$_e$) varied between 0.17 and 0.24 dS.m^{-1}.

3 Results and discussion

Parameterisation of a sunflower crop and characterisation of the soil water balance permits the calculation of potential crop yield and production at PS1 and PS2 levels, provided that adequate daily weather data are available. Results were generated for scenarios with emergence on day 85 (26th of March) and a sowing density of 5 kg.ha^{-1}.

Fig. 1 shows yield potentials calculated for the period between 1972 and 1994. The curve 'Yield-PS1' represents the yield component of the 'constraint-free' biophysical production potential (PS1). Its variation reflects the effect of variable environmental conditions on crop production. Calculated yield potentials varied between 4.8 t.ha^{-1} and 3.9 t.ha^{-1}.

The curve 'Yield-PS2' represents the yield component of the water-limited yield potential (PS2). The effect of water scarcity reduces yields to less than 1.0 t.ha^{-1}. Water is the main limiting crop production factor in Andalusia.

Fig. 2 illustrates this limitation by showing precipitation during the crop growing period (curve 'PREC') alongside the water requirement for unlimited crop production (curve 'TWR').

Results generated for the two production situations are listed in Table 1.

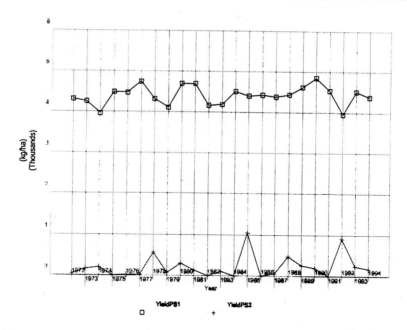

Fig. 1: Calculated potential sunflower yields for PS1 and PS2 production situations.

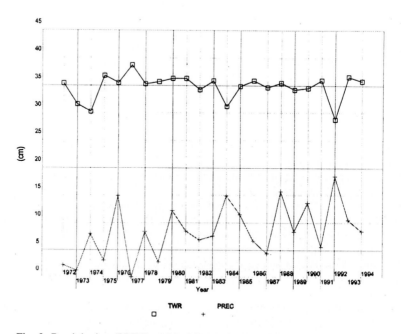

Fig. 2: Precipitation (PREC) and (total) water needs (TWR) of sunflower.

Soil salinity in a dynamic simulation

The difference between required and available quantities of water has to be bridged by supply of water, e.g. through irrigation or capillary rise, both of which add salts to the (rooted) surface soil.

Precipitation during the crop growing period varied between 18.5 cm and no precipitation at all. Water requirements were calculated to meet the maximum consumptive water use with the soil moisture tension kept at a constant 1000 hPa (pF 3). In some years, crop failure by water shortage was observed as early as 11 days after emergence. On average, the first water shortage occurred around day 48 after emergence. The calculated cumulative water requirement varied between 38.7 cm and 28.8 cm, with an average of 34.8 cm.

Year	LPG	PS1 LAD	LAImax	LAIf	PS2 LAD	LAImax	LAIf	CFW
1972	121	299	5.6	0.0	83	2.5	0.0	48
1973	118	253	5.0	0.3	57	1.7	0.0	46
1974	115	229	4.7	0.3	91	3.0	0.0	57
1975	119	258	4.9	0.2	*			(11)
1976	112	219	4.8	0.5	18	0.6	0.0	20
1977	119	247	4.6	0.4	46	1.3	0.0	45
1978	117	213	4.0	0.5	108	2.7	0.0	57
1979	110	217	4.5	0.6	77	2.5	0.0	52
1980	108	183	3.8	0.9	102	3.2	0.0	61
1981	110	197	4.3	0.7	84	2.8	0.0	59
1982	106	167	3.8	0.9	*			(13)
1983	114	201	4.2	0.4	89	2.9	0.0	59
1984	115	186	3.6	0.9	*			(37)
1985	112	203	4.1	0.8	129	3.5	0.2	64
1986	111	223	4.9	0.5	*			(38)
1987	105	190	4.1	1.0	54	1.8	0.0	46
1988	108	207	4.3	0.9	69	2.1	0.1	43
1989	104	196	4.4	0.9	73	2.1	0.0	48
1990	106	199	4.3	1.0	63	2.2	0.0	23
1991	106	216	4.7	0.7	*			(34)
1992	104	180	3.8	0.6	54	1.3	0.2	11
1993	112	230	4.7	0.5	131	3.9	0.0	63
1994	106	206	4.3	0.6	111	3.4	0.0	57
Year	LPG	LAD	LAImax	LAIf	LAD	LAImax	LAIf	CFW
maximum:	121	299	5.6	1.0	131	3.9	0.2	64
minimum:	104	167	3.6	0.0	18	0.6	0.0	11

where
LPG   is length of plant growth (d).
LAImax   is maximum LAI in the crop cycle ($m^2.m^{-2}$).
CFW   is first day after beginning of water shortage.
LAD   is leaf area duration ($m^2.d$).
LAIf   is LAI at maturity ($m^2.m^{-2}$).
*   crop fails.

Table 1: Some crop parameters values for PS1 and PS2 production situations.

A number of alternative irrigation schedules, with different applications of water, and different levels of water stress, were evaluated. The electric conductivity of the irrigation water (EC_i) was set to the same value as that of the groundwater ($EC_i = EC_w = 2.0$ $dS.m^{-1}$) and the initial electric conductivity of the saturated soil (extract), EC_e, was set to 0.20 $dS.m^{-1}$. All scenarios were run with weather data recorded at Coria del Rio in 1993. The initial matric suction (on day 96) was 1000 hPa, and 5 kg of (dry) seeds were sown per hectare. Table 2 shows the results obtained.

Row 'TDM/cm appl' in Table 2 can be seen as an expression of water use efficiency. It is clear that the relation between water use efficiency and water application is not one of inverse proportionality even though the general trend is that water use efficiency decreases with increasing rates of water application.

Rows 'CFW' in Table 2 indicate the first day in the crop cycle at which the crop experiences water shortage under a particular scenario. Rows 'SMPSI' give the soil moisture content at crop maturity. Runs with single applications of 13 cm of water resulted in (predicted) crop failure because of extended periods of soil saturation.

TDM	7061	6083	7582	8852	8336	9083	9633
SO	1901	1839	2826	2582	2672	2606	3228
CFW	63	57	54	67	61	69	67
EC	3.99	3.63	4.46	8.29	5.85	5.36	4.28
Ec_e	1.14	0.96	1.19	2.32	1.58	2.20	1.14
SMPSI	0.13	0.12	0.13	0.13	0.13	0.19	0.13
Irrigation schedule (Day and cm applied)	50 5 55 5 60 5 65 5 70 5 75 5 80 5 85 5	50 3 55 3 60 3 65 3 70 3 75 3 80 3 85 3	50 2 55 2 60 3 65 4 70 5 75 6 80 5 85 4	40 5 50 7 60 8 70 10 80 8 90 4	40 5 50 7 60 10 70 10 80 10 90 8	40 5 50 7 60 10 70 10 80 10 90 8	50 5 57 7 64 10 71 10 78 10 85 8
number, gift	8 40	8 24	8 22	6 40	4 50	6 50	6 50
SO/TDM	0.27	0.30	0.37	0.29	0.32	0.29	0.34
TDM/gift	1.8	2.5	3.4	2.2	1.7	1.8	1.9
TDM	10100	10120	10065	9404	10029	8845	6264
SO	4390	4464	4538	3228	3464	2553	996
CFW	54	54	54	54	54	54	54
EC	6.10	8.29	8.67	6.44	4.60	3.53	3.77
Ec_e	3.41	3.78	2.43	1.76	1.23	0.93	2.11
SMPSI	0.26	0.21	0.13	0.13	0.13	0.12	0.26
Irrigation schedule (Day and cm applied)	55 10 65 10 75 10 85 10 95 10	55 8 65 12 75 12 85 12 95 6	60 10 70 12 80 12 90 6	55 10 65 12 75 12 85 6	55 10 70 10 85 10	55 10 75 10	55 10 85 10
number, gift	5 50	5 50	4 40	4 40	3 30	2 20	2 20
SO/TDM	0.43	0.44	0.45	0.34	0.35	0.29	0.16
TD/cm applied	2.0	2.0	2.5	2.4	3.3	4.4	3.1

where:
TDM   is total dry matter (kg.ha^{-1}).
SO   is storage organ dry matter (kg.ha^{-1}).
CFW   is first day in the crop cycle with water shortage.
EC, EC_e and SMPSI are values at crop maturity.
Irrigation schedule is defined by the dates and amounts of single applications (Day and cm water applied); and by the number of applications and the total amount given (number, amount).

Table 2: Salinity risks calculated for various irrigation regimes.

Irrigation strategies can be tested for an unlimited number of scenarios; the results presented in Table 2 refer to a specific soil type. Irrigation and leaching requirements can be examined by defining a permissible level of soil salinity and analysing a number of prospective irrigation schedules, land uses (different crops and/or fallow), or depths of the groundwater.

Ayers and Westcot (1985) state that sunflower is moderately sensitive to soil salinity. This relative tolerance is reflected by its threshold salinity level, that is said to correspond with EC_e- values between 1.3 and 3.0 dS.m^{-1}. EC_e is the electric conductivity measured in the extract of a saturated soil paste at 25 °C. The real soil water content is not that of a saturated paste but (normally) less and varies strongly between sites and between years. Table 2 suggests that the actual soil salinity level (EC) may be 2 to 4 times greater than EC_e, and even more when the soil dries out. It is conceptually better to relate yield depression by excess electrolytes in the soil solution not only with EC_e but also with soil moisture regime. In this study we have expressed EC as the electric conductivity at actual soil moisture content. Note that this is only permissible if salt contents are low or moderate and solubility products of the dissolved salts are high. At high contents of dissolved salts ionic imbalances would poison the crop and reactions to increased soil moisture potentials would become unpredictable.

$$EC = EC_e * SM0 / SMPSI \qquad (13)$$

Table 3 shows results generated for scenarios with 40 cm of irrigation water (2 dS.m^{-1}) and 6 applications (5 cm on day 40, 5 cm on day 50, 8 cm on day 60, 10 cm on day 70, 8 cm on day 80 and 4 cm on day 90 in the crop cycle). Emergence was on day 96 in the year, the seed application rate was 5 kg.ha^{-1}). The results were generated with weather data recorded at Coria del Rio, in 1992 and 1993.

Table 3 suggests that soil salinisation varies strongly between years, even with the same rates of irrigation water application. Note also that soil salinity build-up is not (always) correlated with the precipitation sum. It is clearly more telling to analyse processes than interpret lumped water budgets.

Year	1972	1975	1978	1981	1984	1987	1990	1993
tdm	5732	7549	7694	8752	8439	10177	8603	8852
so	307	1569	1850	3445	2878	4721	3402	2582
CFW	64	72	75	64	67	60	57	67
EC	4.14	5.99	6.11	8.82	5.66	10.91	9.18	8.29
Ec_e	1.72	1.78	1.66	2.98	1.53	4.49	3.84	2.32
SMPSI	0.19	0.14	0.13	0.16	0.13	0.19	0.20	0.13
PREC	2.2	0.0	8.0	3.7	13.7	5.7	10.0	10.5
so/tdm	0.05	0.21	0.24	0.39	0.34	0.47	0.40	0.29
SO/TDM	0.47	0.50	0.52	0.53	0.54	0.52	0.53	0.51
so/SO	0.05	0.24	0.31	0.54	0.47	0.81	0.52	0.40
WUE	1.4	1.9	1.9	2.2	2.1	2.5	2.2	2.2

where
PREC   is amount of precipitation during the growing season (cm).
so/tdm   is ratio of 'so' and 'tdm' (-).
SO/TDM   is 'so/tdm' but for PS1 (-).
so/SO   is ratio of actual and potential storage organ dry masses (-).
WUE   is water use efficiency, with ratio of 'tdm' and water application set to 4000 m^3.ha^{-1} (kg.m^{-3}).

Table 3: Salinity risks calculated for different years.

4 Conclusions

The methodology presented allows one to evaluate alternative irrigation scenarios (with different irrigation schedules, with free or forced drainage, etc.). Leaching requirements can be studied for different combinations of water quality and land use (different crops, fallow).

Soil salinity level, an important indicator of land use sustainability, can be 'previewed' for irrigation strategies with different (timing and amounts of) water applications, and different levels and timing of water stress.

The electric conductivity of actual soil moisture is a much more meaningful indicator of soil salinity than the electric conductivity of a 'normalised' saturation extract because these values have practical significance for the identification of (approximate) threshold salinity levels of field crops.

References:

Ayers, R.S. and Westcot, D.W. (1985): Water Quality for Agriculture, F.A.O. Irrigation and Drainage Paper **29**, Rev.1, Rome.

Driessen, P.M. and Konijn, N.T. (1992): Land Use Systems Analysis, Dept. of Soil Science & Geology, W.A.U., Wageningen.

Thorne, D.W. and Peterson, H.B. (1954): Irrigated Soils: Their Fertility and Management, The Blakiston Company, Philadelphia.

Weast, R.C. (Ed.) (1975): Handbook of Chemistry and Physics, Chemical Rubber Publishing Company, Cleveland.

Addresses of authors:
Jorge Barros
Rua Gil Vicente 14, 2 FR
2780 Oeiras
Portugal
Paul Driessen
Department of Soil Science
Wageningen Agricultural University
P.O. Box 37
6700 AA Wageningen
The Netherlands

Relations Between Soil Salinity and Water Quality as well as Water Balance in Yinbei Plain, PR China

G. Xie, J. Breburda, A. Kollender-Szych & A. Battenfeld

Summary

The Yinbei alluvial plain, situated in the northern part of the Ningxia Hui Autonomous Region, suffers from strong salinization and alkalization. Occurence and development of salinization are related to water quality and salt balance. Due to frequent irrigation and insufficient drainage the water and salt balance of the region shows a disequilibrium. 14 % of the salts carried into the area mainly by irrigation water are accumulated in the soils. This implies that drainage or leaching ought to be ensured, drainage equipment controlled, or soil salinity will reach toxic levels soon.

Keywords: Yinbei plain, soil salinity, water balance

1 Introduction

The Yinbei alluvial plain is situated in the northern part of the Ningxia Hui Autonomous Region, PR China. It has an area of 3,700 km^2, with a comprehensive gravity irrigation system and drainage network. Owing to arid climatic conditions, the rainfall cannot meet crop water requirements. Agriculture totally depends on irrigation supplied by water from the Qintongxia dam on the Yellow River. The volume of water used from the Yellow River in the region is about 2.5 billion m^3 per year. The plain suffers from strong salinization and alkalization of the soil because evapotranspiration exceeds precipitation. It has been estimated that due to the elevated salt content on about 40% of the irrigated farmland, crop yields are depressed by approximately 50%. Another 800 km^2 of uncultivated land is saline. The extension of salinization and alkalization in the region hinders or prevents agricultural production. It is known that the quality and movement of water play a very significant role in the occurrence and development of soil salinization (Breburda, Kollender-Szych 1992; Kollender-Szych et al.. 1995; Xie 1993, 1995). The aim of this study was to determine the properties of salt affected soils, ground water and irrigating water, the relations between soil salinization, and quality and balance of water in the plain.

2 Materials and methods

For the test, 62 water samples from the Yellow River, irrigating canals, draining canals and groundwater, 6 profiles of salt affected soils, saline soils and sodic soils, as well as 40 soil samples of upper horizons of the salt affected soils were used. The pH value was determined electrometrically in 0.02n CaCl$_2$-extracts of soil samples (1:5) with a pH-Meter according to Schlichting et al. (1995). The measurement of electrical conductivity (EC) of the soil samples was carried out in 1:5-aqueous extracts of soil samples using an EC-Meter type CD M2. The concentrations of the ions: Na^+, Ca^{2+}, Mg^{2+}, Cl^-, SO_4^{2-}, HCO_3^-, CO_3^{2-} in the soil samples were carried out in 1:5-aqueous extracts. The concentrations of the cations: Na, Ca, Mg were determined by an Atomic Absorption Spectrometer (AAS-Perkin Elmer). The measurement of Cl^- concentration was carried out using a Cl-electrode. The concentrations of HCO_3^-, CO_3^{2-} and SO_4^{2-} were determined by titration.

Numerous schemes have been proposed for classifying water with respect to their suitability for irrigation and the interaction between the physico-chemical properties of the soil and the water, as well as their hazards of salinity, sodicity and toxicity. These have ranged from general schemes designed for average use conditions to specific water quality ratings, based on a given crop in a specific region. For evaluation of the irrigating water and ground water, the following criteria for evaluating water quality, based on electrical conductivity (EC) and the sodium adsorption ratio (SAR) according to Richards (1954), were used: Salinity classes: C1 - low salinity water, C2 - medium salinity water, C3 - high salinity water, C4 - very high salinity water. Sodium classes: S1 - low sodium water, S2 - medium sodium water, S3 - high sodium water, S4 - very high sodium water. In addition, the water was assessed based on residual sodium carbonate (RSC) value according to Wilcox (1968): B1 - low RSC water (RSC<1.25 $mmol_c \cdot l^{-1}$), water probably safe for irrigation, B2 - medium RSC water (RSC 1.25-2.5 $mmol_c \cdot l^{-1}$), water marginally suitable for irrigation, B3 - high RSC water (RSC>2.5 $mmol_c \cdot l^{-1}$), water unsuitable for irrigation and based on soluble sodium percentage (SSP) values according to Christiansen-Weniger (1967): A1 - low SSP water (SSP<60%), water suitable for irrigation, A2 - high SSP water (SSP>60%), water that can lead to alkalization of the soil (Landon, 1984).

3 Results and discussion

3.1 Salt content and ionic composition of soils

Estimates of salinity and sodicity of soils are important because of their effects on crop yields Shainberg et al., 1984). These effects arise when salt concentrations in the soil are high because the resultant salt concentrations in the soil solution reduce or even reverse the flow of water into the plants by osmosis. In addition, some ions (notably Na^+, Cl^-, SO_4^{2-}) are specifically toxic for certain crops. The effects of high sodium concentration are also noticeable in its deleterious effects on soil structure. The pH-value, salt content and ion composition of soils in the plain are shown in Table 1. It was found that the soil salt content in the plain is very high (0.06-7.94%). The soils contain sufficient soluble salts to interfere with growth of most crops. The soluble salts are mainly composed of ions: SO_4^{2-} (0.11-33.55 $cmol_c \cdot kg^{-1}$ soil), Cl^- (0.2-21.86 $cmol_c \cdot kg^{-1}$ soil) and Na^+ (0.13-44.47 $cmol_c \cdot kg^{-1}$ soil). These ions are specifically toxic for some crops.

	pH	salt content %	Na^+	Ca^{2+}	Mg^{2+} $cmol_c \cdot kg^{-1}$	HCO_3^-	SO_4^{2-}	Cl^-
n	43	43	43	43	43	43	43	43
$\bar{x}$	7.81	0.57	4.59	1.05	1.01	0.97	3.25	2.28
s	0.62	1.36	10.36	1.04	1.31	0.42	7.25	5.09
min.	7.13	0.06	0.13	0.28	0.16	0.34	0.11	0.20
Max.	10.1	7.94	44.47	5.14	6.85	2.29	33.55	21.86

$\bar{x}$: mean
s: standard deviation

Table 1: pH, salt content and ionic composition in the upper horizons (0-20cm)

It is necessary to point out that the ionic composition of soil soluble salts in the region varies among different soil samples. These differences are closely related to the salt content of the soils. The regression relation between the ionic composition and the salt content of the soil is shown in Table 2. It can be found that if the salt content of the soil is less than 0.2%, the salt is mainly composed of Na^+, Ca^{2+}, Mg^{2+} and HCO_3^-. If the salt content ranges from 0.2% to 0.4%, it is mainly composed of Na^+, SO_4^{2-}, HCO_3^-. If the soil salt content is above 0.4%, the salt is mainly composed of Na^+, SO_4^{2-} and Cl^-.

Ions	correlation	regression
Na^+	r = 0.5509**	Y1 = -10.3470+6.4619ln(X)
Ca^{2+}	r = -0.5178**	Y2 = 36.1463-3.9224ln(X)
Mg^{2+}	r = -0.4438*	Y3 = 27.9026-2.9296ln(X)
HCO_3^-	r = -0.6913*	Y4 = 50.8232-6.0908ln(X)
SO_4^{2-}	r = 0.4584*	Y5 = 0.6359+3.2804ln(X)
Cl^-	r = 0.5319**	Y6 = -5.1611+3.2006ln(X)

**: p = 0.01
*: p = 0.05
X: salt content in mg/100g soil
Y1-Y6: $mmol_c\%$ of ions Na^+, Ca^{2+}, Mg^{2+}, HCO_3^-, SO_4^{2-}, Cl^-

Table 2: Correlation and regression of the ionic composition and the salt content in the upper horizons (0-20cm)

3.2 Salinity and sodicity of the irrigation water and its quality evaluation

The salinization and alkalization problem that develops in irrigation agriculture, arises from the chemical composition of the irrigation water and groundwater (Breburda 1980, 1981, 1985). It is recognized that the interaction between the physico-chemical properties of the soil and the irrigation water, as well as ground water, is a very important parameter in evaluating the suitability of water for irrigation (Fang 1989). The analyzed region is the floodplain of the Yellow River. Much of the flood plain is irrigated by diverting Yellow River water. Some of the alluvial plain has been irrigated for as long as 2000 years. The salt content in the water is 382.6-480.0mg/l and it is mainly composed of three positive ions: Na^+ (1.33 $mmol_c \cdot l^{-1}$), Ca^{2+} (3.02 $mmol_c \cdot l^{-1}$), Mg^{2+} (1.95 $mmol_c \cdot l^{-1}$) and three negative ions: Cl^- (1.07 $mmol_c \cdot l^{-1}$), SO_4^{2-} (0.99 $mmol_c \cdot l^{-1}$), HCO_3^- (3.98 $mmol_c \cdot l^{-1}$) (see Table 3).

	PH	Salt content $mg \cdot l^{-1}$	Na^+	Ca^{2+}	Mg^{2+}	HCO_3^-	SO_4^{2-}	Cl^-
					$mmol_c \cdot l^{-1}$			
n	8	8	8	8	8	8	8	8
$\bar{x}$	7.82	445.14	1.33	3.02	1.95	3.98	0.99	1.07
s	0.32	32.12	0.56	0.33	0.22	0.38	0.29	0.16
min.	7.40	382.60	0.43	2.49	1.68	3.37	0.75	0.92
max.	8.30	480.00	2.10	3.55	2.35	4.46	1.63	1.34

$\bar{x}$: mean; s: standard deviation

Table 3: pH, salt content and ionic composition of the irrigating water

Irrigation water with a mean EC value and low SAR, RSC and SSP values can be used for irrigation on almost all soils with little danger of the development of harmful levels of exchangeable sodium and with little likelihood that a salinity problem will develop. However, some leaching is required and sodium sensitive crops may accumulate injurious concentrations of sodium (see Table 4).

sample	EC $\mu S \cdot cm^{-1}$	SAR	RSC $mmol_c \cdot l^{-1}$	SSP %	EC & SAR	RSC	SSP
Q1	610.2	0.84	-0.67	19.8	C2 - S1	B1	A1
Q2	569.8	0.67	-0.91	16.7	C2 - S1	B1	A1
Q3	629.9	1.01	0.27	23.3	C2 - S1	B1	A1
Q4	661.2	0.99	-0.74	22.1	C2 - S1	B1	A1
S1	649.8	1.45	-0.91	30.4	C2 - S1	B1	A1
S2	641.3	1.41	-1.56	28.2	C2 - S1	B1	A1
S3	519.9	0.71	-0.70	18.5	C2 - S1	B1	A1
S4	599.6	0.28	-1.58	7.0	C2 - S1	B1	A1

A1: low SSP water (SSP<60%), water suitable for irrigation
B1: low RSC water (RSC<1.25$mmol_c \cdot l^{-1}$), probably safe for irrigation
C2: water of medium salinity
S1: water of low sodium concentration

Table 4: Evaluation of irrigating water

3.3 Relations between soil salinity and groundwater

Besides the irrigation water, physico-chemical properties and the depth of groundwater exert a great influence. The salt content in the groundwater varies from 429 to 38,643mg/l. It is mainly composed of the cation Na^+ (0.7-532.2$mmol_c \cdot l^{-1}$), and the anions: Cl^- (0.8-242.6$mmol_c \cdot l^{-1}$), SO_4^{2-} (1.2-332.2$mmol_c \cdot l^{-1}$) (Table 5). However, the ionic composition varies with the salt concentration of the groundwater: if the salt concentration is less than 1g/l, the salt is mainly composed of Mg^{2+} and Ca^{2+}, HCO_3^-. If the salt concentration ranges from 1-2g/l, the salt is mainly composed of Na^+, SO_4^{2-} and HCO_3^-. If the salt concentration is above 2g/l, the salt is mainly composed of Na^+, SO_4^{2-} and Cl^- (Table 6). Most of the groundwater having a high or very high EC value and medium SAR value, may produce harmful salinization and will present an appreciable sodium hazard in fine textured soils (Table 7).

	pH	salt content $mg \cdot l^{-1}$	Na^+	Ca^{2+} $mmol_c \cdot l^{-1}$	Mg^{2+}	HCO_3^-	SO_4^{2-}	Cl^-
n	40	40	40	40	40	40	40	40
$\bar{x}$	7.5	3,148.2	33.9	5.9	10.3	8.6	19.0	20.2
s	0.25	6,166.0	87.7	5.9	11.9	3.4	72.4	42.5
max.	8.1	38,643.0	532.2	25.9	77.0	18.5	332.2	242.6
min.	7.1	429.9	0.7	0.4	0.9	3.5	1.2	0.8

$\bar{x}$: mean, s: standard deviation

Table 5: pH, salt content and ionic composition of the groundwater

Ions	Correlation coefficient	regression equation
Na^+	r = 0.7482**	Y1 = -44.8283+9.2028ln(X)
Ca^{2+}	r = -0.5792**	Y2 = 48.2242-5.0087ln(X)
Mg^{2+}	r = -0.4596*	Y3 = 45.7069-3.9667ln(X)
HCO_3^-	r = -0.8104**	Y4 = 79.9741-8.2894ln(X)
SO_4^{2-}	r = 0.4352**	Y5 = -3.1334+2.4787ln(X)
Cl^-	r = 0.7323**	Y6 = -25.9436+5.5833ln(X)

**: p = 0.01   X: salt concentration in $mg \cdot l^{-1}$
*: p = 0.05   Y1-Y6: $mmol_c$% of ions Na^+, Ca^{2+}, Mg^{2+}, HCO_3^-, SO_4^{2-}, Cl^-

Table 6. Correlation and regression of the ionic composition and the salt concentration of the groundwater

The salt content in the upper layers of the soils correlates positively with the salt concentration of the groundwater and negatively with the depth of the groundwater. The content of Na^+ in upper soil layers shows a positive correlation with the Na^+ concentration of the groundwater and a negative correlation with the depth of the groundwater. The SO_4^{2-} content of the top layer of the soils correlates positively with the SO_4^{2-} concentration of the groundwater and negative with the depth of groundwater. The content of Ca^{2+} in the top layers correlates positively only with the Ca^{2+} concentration in the groundwater. The contents of Mg^{2+} and Cl^- correlate negatively only with the depth of the groundwater. The HCO_3^- content in the top layers shows no significant correlation with either the HCO_3^- concentration and depth of the groundwater (Table 8). It is evident that the groundwater is an important factor leading to land salinization. Owing to an excess of intake over drainage, the groundwater level is near to the ground surface, and salt-containing groundwater evaporates quickly: a significant amount of water and salt moves to the soil surface by evaporation and causes soil salinity.

3.4 Relations between soil salinity and water balance

The regional salinity problems are associated with many factors, including climatic and hydrogeological factors, irrigation practices, etc. (Wang 1990). However, all these factors might be contributed due to water and salt balance or water movement. Soluble salts move with the water in which they are dissolved. The movement of water causes movement of salt. Therefore, salts are transported from places where water is infiltrating towards places where it evaporates. They accumulate in places where water evaporates. The regional water balance equation can be expressed as:

$$Qc = (Q+P+G1+G2+Gt)-(q+E+G3+G4)$$

where Qc is the change of storage during a period including surface storage, groundwater storage and storage in the unsaturated zone. Q is the input amount of water taken from the Yellow River through the canals. P is the input amount of precipitation. G1 is the input amount of surface water flow. G2 is the input amount of interflow. Gt is the input amount of pumpage water. The q is the output, the discharge of the drains. E is the output of evapotranspiration. G3 is the output, groundwater discharge directly into the river. G4 is the output, the amount of water used by industry and the residents.

Sample	EC µS·cm⁻¹	SAR	RSC mmol$_c$·l⁻¹	SSP %	EC & SAR	RSC	SSP
ZW1	3,849.8	3.72	-25.86	30.6	C4 - S2	B1	A1
ZW2	1,912.3	6.59	-3.40	61.5	C3 - S2	B1	A2
QT1	3,894.8	6.60	-16.60	49.0	C4 - S3	B1	A1
QT2	6,299.2	12.79	1.45	80.1	C4 - S4	B2	A2
QT3	1,193.7	1.61	-1.63	27.4	C3 - S1	B1	A1
QT4	1,212.0	1.58	-1.84	26.7	C3 - S1	B1	A1
LW1	4,919.8	10.57	-27.20	56.3	C4 - S4	B1	A1
LW2	1,965.7	5.99	-2.06	58.4	C3 - S2	B1	A1
YN1	562.4	0.93	-0.90	23.2	C2 - S1	B1	A1
YN2	1,144.5	1.51	-1.09	26.4	C3 - S1	B1	A1
YN3	546.9	0.43	-1.75	11.9	C2 - S1	B1	A1
YN4	1,175.7	1.51	-1.63	26.1	C3 - S1	B1	A1
YJ1	1,250.0	3.56	-0.89	48.8	C3 - S1	B1	A1
YJ2	1,057.4	1.26	-1.41	23.4	C3 - S1	B1	A1
YJ3	3,311.3	4.16	-6.43	38.1	C4 - S2	B1	A1
HL1	1,688.7	1.28	-6.38	18.8	C3 - S1	B1	A1
PL1	2,305.9	3.28	-9.12	36.2	C4 - S1	B1	A1
PL2	2,341.1	4.12	-7.52	42.8	C4 - S2	B1	A1
PL3	1,178.3	3.10	-2.55	44.8	C3 - S1	B1	A1
PL4	54,278.7	74.67	-95.50	83.9	C4 - S4	B1	A2
PL5	2,242.7	5.81	-3.30	55.3	C3 - S2	B1	A1
PL6	3,284.6	7.18	-6.71	55.6	C4 - S2	B1	A1
PL7	2,341.1	4.04	-7.93	42.1	C4 - S2	B1	A1
PL8	1,349.8	3.13	-3.08	43.2	C3 - S1	B1	A1
PL9	7,359.4	14.59	-18.50	66.2	C4 - S4	B1	A2
PL10	9,597.8	31.23	-6.99	83.4	C4 - S4	B1	A1
PL11	1,150.2	2.46	-3.89	38.1	C3 - S1	B1	A1
PL12	2,270.8	3.03	-6.01	34.5	C4 - S1	B1	A2
PL13	18,523.6	111.10	5.25	96.9	C4 - S4	B3	A2
PL14	6,085.5	8.79	-77.9	51.6	C4 - S3	B1	A1
PL15	1,950.2	8.77	1.80	71.6	C3 - S3	B2	A1
PL16	3,155.2	11.42	1.31	72.6	C4 - S4	B2	A2
PL17	2,688.4	3.85	-5.05	39.3	C4 - S2	B1	A1
SZ1	4,597.9	12.50	1.27	69.1	C4 - S4	B2	A2
SZ2	4,056.6	11.77	-2.29	69.1	C4 - S4	B1	A2
SZ3	3,837.2	6.95	-9.98	51.7	C4 - S4	B1	A1
SZ4	1,406.1	1.50	-4.17	23.4	C3 - S1	B1	A1
SZ5	2,889.5	7.57	1.78	60.5	C4 - S3	B2	A2
SZ6	1,861.6	2.17	-6.72	28.5	C3 - S1	B1	A1
SZ7	1,983.9	3.61	-2.71	40.7	C3 - S1	B1	A1

Table 7: Evaluation of the groundwater

Concentration of salt & ions in the soil	multiple-R with X_1 and X_2	Partial-R with X_1	with X_2
Na^+	0.7221**	0.4935*	-0.6816**
Ca^{2+}	0.6725**	0.5251**	-0.3331
Mg^{2+}	0.5570**	0.2608	-0.5432**
HCO_3^-	0.0731	0.0603	-0.0606
SO_4^{2-}	0.8111**	0.6200**	-0.5932**
Cl^-	0.7269**	0.3276	-0.7205**
Salt	0.8735**	0.5698**	-0.8378**

**: p = 0.01   n = 36 X_1: salt concentration in mg·l^{-1} of the groundwater
*: p = 0.05 X_2: depth of the groundwater in cm

Table 8: Correlation of the salt content (mg/100g soil) in the upper horizons of the soil (0 - 20cm) and the salt concentration X_1 and the depth X_2 of the groundwater

The regional salt balance depends on the water balance. It is important for all the processes which contribute to inflow, outflow and changes in the salt composition of the soil. This can be expressed mathematically as a mass conservation equation for the region:

$$SQc = (SQ+SP+SG1+SG2+SGt)-(Sq+SG3+SG4)$$

where SQc is the salt balance. SQ is the quantity of salts transported into the region by irrigation water. SP is the quantity of salts derived from precipitation. SG1 is the quantity of salts transported by surface water flow. SG2 is the quantity of salts transported by horizontal inflow of groundwater. SGt is the quantity of salts transported by pumpage water. Sq is the quantity of salts transported by drains. SG3 is the quantity of salts transported by horizontal outflow of groundwater into the river. SG4 is the quantity of salts transported by water used by industry and the residents.

The results of the water and salt balance are shown in Table 9. Due to frequent irrigation and insufficient drainage, the water and salt balance shows a disequilibrium. The total input of salts by irrigation per year amounts about 1,5 million tons. Therfrom, 14% (210.8 million kg) were accumulated in the soil. Under the assumed conditions, this equation states that in order to maintain the salt balance, the amount of salt added during irrigation must be equal to the amount drained. This implies that drainage or leaching ought to be ensured in irrigating agriculture, or soil salinity will ultimately reach toxic levels (Xie 1995).

← Legend of Table 7:
A1: low SSP water (SSP<60%), water suitable for irrigation
A2: high SSP water (SSP>60%), water can lead to alkalization
B1: low RSC water (RSC<1.25 mmol$_c$·l^{-1}), probably safe for irrigation
B2: medium RSC water (RSC 1.25-2.5 mmol$_c$·l^{-1}), marginally safe for irrigation
B3: high RSC water (RSC>2.5 mmol$_c$·l^{-1}), unsuitable for irrigation
C2: water of medium salinity
C3: water of high salinity
C4: water of very high salinity
S1: water of low sodium concentration
S2: water of medium sodium concentration
S3: water of high sodium concentration
S4: water of very high sodium concentration

On the basis of the above relations between soil salinity and irrigation water, groundwater as well as the water balance, recommended measures of water control for the prevention and reclamation of salt affected lands, as well as water and salt balance in the Yinbei alluvial plain, should include:
(1) Drainage: Many factors have proved that drainage by drain-ditches and draining-shafts can effectively lower the groundwater table. The increase of artificial discharge of groundwater by pumpage or draining management causes groundwater discharge instead of evaporation.
(2) Reducing the seepage of irrigation canals: The seepage of irrigation canals is very serious and common. It results in an obvious rise of the groundwater table. A realized and effective method to solve this problem is to cover the bottom and the sides of the irrigation canals by cement board.

		water-balance		salt-balance		
		$10^6 m^3$/year	%		10^6 kg/year	%
Input	Q	2,536.0	75.9	SQ	1,128.5	75.2
	P	665.0	19.9	SP	21.3	1.4
	G1	19.6	0.6	SG1	15.7	1.1
	G2	21.3	0.6	SG2	21.3	1.4
	Gt	103.0	3.0	SGt	324.2	21.6
	Total	3,344.9	100.0	total	1,500.0	100.0
Output	Q	1,167.0	34.9	Sq	1,258.0	97.6
	E	2,108.0	63.1	SE	0.0	0.0
	G3	0.8	0.2	SG3	2.5	0.2
	G4	64.4	1.9	SG4	28.7	2.2
	Total	3,340.2	100.0	total	1,289.2	100.0
balance	Qc	+4.7			+210.8	

Q:	water from the channels		SQ:	salts in irrigating water
P:	precipitation water		SP:	salts from precipitation
G1:	surface water flow		SG1:	salts in surface water flow
G2:	interflow		SG2:	salts in horizontal inflow of groundwater
Gt:	pumpage water		SGt:	salts in pumpage water
q:	discharge of the drains		Sq:	salts in the draining water
E:	evapotranspiration water		SE:	salts in evaporated water
G3:	groundwater discharge		SG3:	salts transported by horizontal outflow of groundwater into the river
G4:	water used by industry			
Qc:	change of storage during a period		SG4:	salts transported by industrial water

Table 9: Water- and salt-balance

References

Breburda, J. & Kollender-Szych, A. (1992): The soil of the Yinchuan plain and some results of investigations on salt-affected soils. Proceedings of the international symposium on strategies for utilizing salt-affected lands. Bangkok, 106-120

Breburda, J. (1980): Salz- und Alkaliböden in Indien und ihre Verbesserung. Z. f. Kulturtechnik und Flurbereinigung, **21**, 366-373

Breburda, J. (1981): Landwirtschaft in der Volksrepublik China heute. DLG-Mitteilungen, **21**, 1148-1158

Breburda, J. (1985): Reclamation of saline-alkali soils in Soviet Central Asia and in Armenia. Applied Geography and Development, **25**, 103-105

Christiansen-Weniger, F.(1967): Bewässerung und Bodenversalzung im Nahen Orient. Landwirtschaft im Ausland, **1**, 3-5

Fang, S.X. (1989): Analysis of the water and salt balance in groundwater in Yinbei irrigation region. Diss. Water Resources Research and Documentation Centre, Italian University for Foreigners, Perugia, Italy

Kollender-Szych, A., Breburda, J. & Xie, G. (1995): Degradation of soils owing to salinization and desertification in Ningxia, PR China. Zesz. Probl.Post. N. Roln. **418, II**, 805-814

Kollender-Szych, A.(1993): The soils of the Yinchuan flood plain with special emphasis on the salt affected soils of Tong Dong. Giess.Abhdlgn. zur Agrar- und Wirtschaftsforschung des europäischen Ostens. **185**, 125-142

Landon, J.R. (ed.) (1984): Soil and water salinity and sodicity. In: Booker Tropical Soil Manual - A handbook for soil survey and agricultural land evaluation in the tropical and subtropics, Pitman Press

Richards, L.A. (1954): Diagnosis and improvement of saline and alkali soils. United States Salinity Laboratory Staff. Department of Agric.

Schlichting, E., Blume, H.-P. & Stahr K. (1995): Bodenkundliches Praktikum, Blackwell Wissenschaftsverlag, Wien

Shainberg, I. & Shalvelet, J. (1984): Soil salinity under irrigation, Springer-Verlag Berlin, Heidelberg, New York, Tokyo

Wang, J.Z. (1990): The soils in Ningxia, Verlag Ningxia Reming, Yinchuan

Wilcox, I.G (1968): The quality of water for irrigation use. US Dept. Agric. Techn. Bull. **962**

Xie, G. (1993): Soil salinization and its control on the Ningxia flood plain. Giess. Abhandlgn. zur Agrar- und Wirtschaftsforschung des europäischen Ostens. **185**, 81-85

Xie, G.(1995): Versalzung und Alkalisierung der Böden der Yinbei-Ebene/China, Diss. Giessen, Berichte aus der Umweltwissenschaft. Shaker-Verlag, Aachen

Address of authors:
G. Xie
J. Breburda
A. Kollender-Szych
A. Battenfeld
Department for Continental Agriculture and Economic Research
Otto-Behaghel-Str. 10D
D-35394 Giessen, Germany

Soil Salinization in the Central Chaco of Paraguay: A Consequence of Logging

M. Nitsch, R. Hoffmann, J. Utermann & L. Portillo

Summary

In the Central Chaco and its eastern margins, highly saline groundwater with EC's of up to 80,000 µS/cm is threatening agricultural land use through salinization of the soils. In field studies the effect of deforestation on the groundwater table and electrical conductivity of groundwater and soil, respectively, were investigated. The studies confirm the pattern of damage as effect of salinization resulting from rigorous deforestation. From the studies a "critical groundwater table" of about 2 m is derived as a threshold value in order to avoid further damage in the topsoils created by clearcutting.

Keywords: Soil salinization, logging, Paraguay, critical groundwater table, electrical conductivity

1 Introduction

The Chaco Boreal is a physiographic region located in the western part of Paraguay, extending west and north to Bolivia and continuing south towards northern Argentina, and is composed of unconsolidated sediments of Quaternary age. It is characterized by its abundant vegetation still virgin on a large scale, varying from extensive pasture grounds to homogeneous forests with abundant and dense shrub stratum and low vegetation (Mitlöhner, 1990). The climatic differences from the semi-arid western Chaco to the sub-humid eastern Chaco and the flooded plains from the Pilcomayo River have created great physiographic divisions.

The soils have developed according to the climate, the parent material, and the distance of the saline groundwater to the soil surface. In the investigated area, clayey soils are typical; in depressions and along temporary rivers where the saline groundwater is close to the surface, Solonchaks and Gypsisols are found. The average yearly precipitation varies from 500 mm on the Paraguayan - Bolivian border to 1,400 mm around the Paraguayan River eastwards. The average annual temperature is 25°C, with a maximum of 45°C and a minimum of just below 0°C. It has prevailing strong north and south winds and a relative humidity of between 20% and 65%. The potential evapotranspiration is 1,300 - 1,500 mm/year, due to high temperatures and minimal seasonal precipitation.

In the Central Chaco region pasture is expanding exclusively at the expense of forests and woodlands. The spreading of agricultural land use through deforestation of the natural vegetation is carried out without taking into account the agricultural potential of soils. In particular in the area east of Loma Plata, the combination of heavy soils with highly saline groundwater (Godoy & Paredes, 1994, 1995) is a great threat to the soils by salinization after deforestation (Nitsch, 1994).

Clearcutting initiates various forms of salinization. The first and immediately visible one is initiated by levelling and spreading the salty ash material of the burnt vegetation. The second is the removal and distribution of the material of termite mounds which is rich in salt. The third form of salinization is initiated through the rising saline groundwater table or capillary fringe into the effective rooting[1] zone of the plants or even to the soil surface caused by rainwater which infiltrates through cracks and macro pores.

In the study area large areas are already lost or have to be considered as potentially threatened by salinization due to surface near groundwater tables and high salinity of the groundwater. Three locations with groundwater tables of 1, 2 and 3 m, respectively, were chosen, to study the depth of the groundwater table and the electrical conductivity of the groundwater and of the soil as a function of precipitation events and land use.

2 Locations and methods of investigations

The three locations of investigation are: **Lindendorf** with a groundwater table at 3 m, **Santa Sofía** at 2 m and **Laguna Porá** at 1 m. The depth between 1 and 3 m was chosen, because in these heavy textured soils the capillary rise of the groundwater was expected to be around 2 m. These three locations have in common a natural thornbush vegetation and one nearby situated pasture, all on silty clay soils. In this paper results are discussed based on two measuring periods of about 4 months in 1994 and 1995 at the location **Santa Sofía.** Santa Sofía lies about 25 km northeast of the Trans-Chaco Road and about 30 km southeast of the township of Loma Plata. According to the FAO soil classification system (FAO-Unesco, 1990) the soil under thornbush vegetation was classified as a gypsic Solonetz (SNy) and the soil under pasture vegetation as a gleyic Luvisol (LVg).

During the measuring period 1994 the precipitation was measured by rain gauges, whilst in 1995, the precipitation data were taken from the nearest station. The electrical conductivity of the groundwater was measured in two depths: the first directly at the groundwater table surface and the second 50 cm deeper. The electrical conductivity of the groundwater is measured in µS/cm at 25°C. The electrical conductivity of the soil was measured by means of soil salinity sensors, which were installed at various depths, beginning at 10 cm, 20 cm, and then increments of 20 cm down to a depth of 220 cm, just above the groundwater table. Measurements of the soil electrical conductivity are related to a satured soil extract and are expressed in mS/cm at 25°C.

3 Results

3.1 Groundwater - thornbush site

On February 16, 1995, the groundwater table at the thornbush site was at 2.62 m below the soil surface (Fig 1) and had an electrical conductivity at the surface of 55,800 µS/cm and 50 cm below of 56,900 µS/cm. On the Feb. 19 and also on the Feb. 22 it rained 33 mm. The groundwater table rose to a level of 2.04 m. The electrical conductivity changed only little, at the groundwater surface to 50,400 µS/cm and 50 cm below to 54,500 µS/cm. In the following drier period until April 10, the groundwater table sunk again to 2.29 m, without greater influence on electrical conductivity. Heavy rainfall in the period from April 17 to May 17, a total of 297 mm, caused the groundwater table to rise to 1.25 m on May 22. The electrical conductivity changed to 36,000 µS/cm at the groundwater surface and to 51,500 µS/cm at 50 cm depth (Fig. 1). The soil surface was, in contrast to the pasture site, not flooded.

---

[1] The effective rooting zone as a soil parameter is the potential depth of plant available water, that can be taken up by annual agricultural crops.

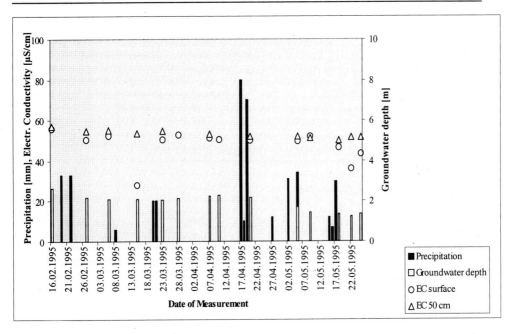

Fig. 1: Precipitation, depth of the groundwater table and the electrical conductivity of the groundwater for the period 16.02.1995 to 25.05.1995 (Santa Sofía, thornbush)

Date	10.02.95	14.02.95	16.02.95	27.02.95
Time	11.00	14.45	13.40	14.30
Groundwater Table [m]	2.61	2.62	2.62	2.18
Soil depth [cm]	Electrical Conductivity [mS/cm 25°C]			
10	2.2	1.5	1.2	1.5
20	2.1	1.5	1.5	1.5
40	4.0	7.6	7.8	11.5
60	6.3	22.0	20.5	24.5
80	9.0	24.0	24.0	40.0
100	12.7	34.0	35.0	38.5
120	21.0	45.5	45.5	28.0
140	10.5	32.0	40.0	38.0
160	13.5	38.0	40.0	45.5
180	18.0	40.0	48.0	45.5
200	28.0	45.5	50.0	40.0
220	29.0	28.0	28.0	34.0

Precipitation: 08.02.95 - 55 mm,   19.02.95 - 33 mm,   22.02.95 - 33 mm

Tab. 1: Precipitation, depth of the groundwater table and electrical conductivity of the soil for the period 08.02.1995 to 27.02.1995 (Santa Sofía, thornbush)

3.2 Soil - thornbush site

The electrical conductivity of the soil (Table 1) clearly reveals the relationship between rainfall and dryer periods in the form of changing salt concentrations in the soil. After a period of heavy rainfalls, measurements on the Feb 16,1995 show a zone "free" of salts down to 20 cm with conductivity values of about 1.5 mS/cm. Below this zone the conductivity values increase from 7.6 mS/cm to 46 mS/cm at a depth of 200 cm. Approximately one month later, on the March 15, 1995, after a longer drier period, the soil conductivity values in the upper zone from 0 - 20 cm reached a value of 15.8 mS/cm, due to the increased capillary rise of saline groundwater. Periodical rainfalls did not influence the electrical conductivity significantly (Table 1).

3.3 Groundwater – pasture site

On Feb 16, 1995, the groundwater table depth at the pasture site was 2.56 m with an electrical conductivity of 33,100 µS/cm. Rainfall of 33 mm on Feb 19 and 22, respectively, resulted in a groundwater rise to 1.79 m, with an electrical conductivity of 2,800 µS/cm. In the following drier period the groundwater table fell to 2.29 m with a slightly increasing electrical conductivity up to 8,300 µS/cm. Inspite of a complete flooding of the pasture site caused by 297 mm rain in the period from April 17 to 19, 1995, the groundwater table showed only little change. This phenomenon is caused by swelling of the soil closing cracks, fissures and macro pores, thus impeding infiltration of the rain into the soil (Fig. 2). Between May 12, 1994 to May 22, 1995, an oscillation of the groundwater table of 1.29 m was observed.

Date	08.02.95	09.02.95	14.02.95	16.02.95	19.02.95	22.02.95	27.02.95	28.02.95
Time		8.45	14.10	15.45			13.10	8.45
Groundwater table [m]		2.57	2.58	2.56			2.31	
Soil depth [cm]			Electrical Conductivity [mS/cm 25°C]					
10			1.3	1.3			1.2	1.3
20			4.3	12.5			1.2	1.3
40			1.3	7.0			14.6	13.6
60				9.2			30.0	28.0
80				7.0			38.0	37.5
100				17.2				44.0
120			18.7	38.0				19.2
140			17.8	27.0				35.0
160			11.7	20.0				34.0
180			12.0	19.0				36.0
200			12.9	17.5				40.0
220			12.2	18.8				32.0

Precipitation: 08.02.95 - 55 mm,   19.02.95 - 33 mm,   22.02.95 - 33 mm

Tab. 2: Precipitation, depth of the groundwater table and electrical conductivity of the soil for the period 08.02.1995 to 28.02.1995 (Santa Sofia, pasture)

3.4 Soil - pasture site

The soil sensor measurements of the pasture site reveal comparable results to those of the thornbush site. Measurements on Feb 16, 1995, show a zone nearly free of salts in the upper 20 cm with electrical conductivities below 2 mS/cm. From 20 cm downwards to a depth of 120 cm, a continuous increase of the conductivity to ~ 40 mS/cm could be observed. After two heavy rainfall events with 33 mm each on Feb 19 and 22, 1995, the electrical conductivity above the groundwater table increased strongly (Table 2). This is a result of quickly infiltrating rain water through cracks and macro pores as long as the soil remains dry, thus lifting the groundwater table and also the capillary fringe. After wetting of the top soil and swelling of the clay mineral particles, cracks and coarser pores are closed, so that the infiltration rate is reduced to nearly zero, causing the already mentioned flooding of the area.

4 Discussion

4.1 Precipitation, groundwater table and electrical conductivity of the groundwater

Data for the selected measuring period show the close relationship between precipitation, groundwater table and the electrical conductivity of the groundwater. After heavy rains the saline groundwater table and the capillary fringe, respectively, rise into the rooting zone or even to the soil surface. Depending on the duration of conditions favouring high infiltration rates and the salt resistance of the grasses, the effect can be slight damage or a total loss of the pasture. Under natural conditions the movement of water (and salt) is regulated by the vegetation adapted to these conditions. The higher transpiration of forests compared to pastures keeps the ground-water table at a lower level.

Referring to this vegetation effect on groundwater level it is repeatedly suggested to leave plots or stripes of the natural forest as a buffer zone, so that water could be pumped from the grassland to keep the groundwater table at a safe distance to the rooting zone. With respect to the hydraulic conductivity characteristics of the heavy textured soils in the Chaco, such an appraoch does not seem very promising to reduce the risk of soil salinization under pasture because the stripes would have to be kept at intervals too small for agricultural land use. From our own observations and Australian publications (Gillard et al., 1989; Williamson, 1990) it can be concluded, that even larger natural forest areas cannot pump enough water from the neighboured pasture areas in heavy clayey soils.

4.2 Precipitation, groundwater table and electrical conductivity of the soil

<u>Thornbush site:</u> The canopy of the vegetation prevents the direct impact of rain drops on the soil surface and the dense rooting system prevents (quick) percolation. With time a nearly salt free zone developed in the uppermost soil horizons. With the beginning drought the capillary rising water is absorbed and consumed by the fine root system. Little or no moisture will reach the upper soil horizons or even the surface which results in only little change of the salt content of the soil.

<u>Pasture site:</u> After periods of drought, rainwater infiltrates very quickly through cracks, fissures and macro pores to the groundwater table (during the 1994 drought deep clefts of about 1.30 - 1.40 m depth and 3 to 5 cm width were observed). The groundwater table rises and lifts the saline zone in the soil; the soil gets wetter and the electrical conductivity in the upper soil horizons increases (Fig. 2).

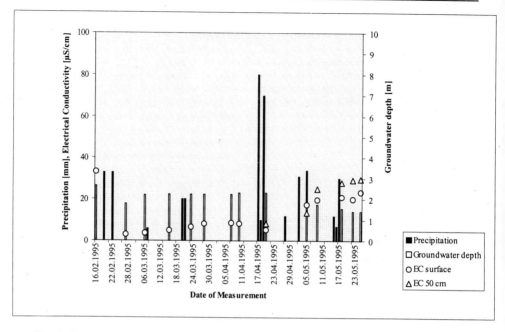

Fig. 2: Precipitation, depth of the groundwater table and the electrical conductivity of the groundwater for the period 16.02.1995 to 25.05.1995 (Santa Sofia, pasture)

4.3 Calculation of a "critical groundwater table" [2]

From soil characteristics, field observations and measured soil parameters it is concluded that a "critical groundwater table" must be in the range of about 2 m. This assumption is confirmed by calculating a "critical groundwater table" (CGT) for silty clay soils according to Equation (1):

$$\text{CGT} = \text{eRD} + \text{CR at } 0.3 \text{ mm/d} = 5 + 13\ (16)\ \text{dm} = 18\ (21)\ \text{dm} \qquad (1)$$

with: **CGT**   = the "critical groundwater table" in dm,
        eRD   = the effective rooting depth in dm, which was found to be about 5 dm on the pasture site with Buffalo- and Estrella-grasses
        CR   = the capillary rise in dm.

The soil texture for this location was determined as a silty clay with a capillary rise between 13 and 16 dm at a rate of 0.3 mm/day (AG Boden, 1994). The rate of <0.3 mm/day seems to do no harm to the vegetation at this location.

---

[2] The "critical groundwater table" with regard to the risk of salinization of the top soil shall be that distance of the groundwater table to the soil surface, at which the capillary rise from the groundwater reaches the rooting zone of cultivated plants. If the distance between the groundwater table and the soil surface is smaller than the calculated value, roots are harmed by the high salt contents in the soil.

5 Conclusions

- Clearcutting of the natural thornbush vegetation causes the initial damage by salinization when salty ash of the burnt vegetation and/or salty termite mound material is spread during the cutting and levelling process.
- Then damage after clearcutting and cultivation is caused by the capillary rise of saline groundwater where the critical groundwater table is less than 2 m below surface. The evaporation from this unsheltered land is much higher than under forest cover, and therefore a permanent transport of salts towards the effective rooting zone or even to the soil surface takes place.
- The same effect is to be observed, when on bare land (mainly pastures) rain water percolates to the groundwater table via preferential flow through cracks, fissures and quickly draining pores. Subsequently the capillary zone or the salty groundwater itself is raised into the rooting zone (affecting the plant roots) or even to the soil surface.
- Rainshowers wetting the soil cause a swelling of clay minerals in soils under pasture, thus resulting in a sealing of the soil surface. In addition grazing cattle compact the soil surface and in doing so also reduce pore space and infiltration. This leads to a sometimes longer-lasting flooding of the land with increased surface run-off and erosion. Because of the sealing of the soil surface, only little salts are washed out.

6 Recommendations

1. A change of the present land use (clearcutting for the extension of agricultural land) can only be recommended if the groundwater table is kept deeper than 2 m (not taking into account the annual oscillation of the groundwater table, which was observed to exceed 1 m during a season).
2. Potentially salt-endangered areas should be defined to keep them out of agricultural use.
3. Where the land is already damaged, salt adapted plants should be introduced to lower the groundwater table to or below the critical level.

References

AG Boden (Ad-hoc-Arbeitsgruppe Böden der Geologischen Landesämter und der Bundesanstalt für Geowissenschaften und Rohstoffe der Bundesrepublik Deutschland) (1994): Bodenkundliche Kartieranleitung, 4.Aufl., 392 S., 33 Abb., 91 Tab, Hannover.
FAO-Unesco (1990): Soil map of the world - revised legend. Rome.
Gillard, P., Williams, J. & R. Moneypenny (1989): Tree clearing in the semi-arid tropics. - Agricultural Science 2, 34 - 39.
Godoy, E. & Paredes, J.L. (1994). Las Aguas Subterráneas del Chaco Boreal - Central Sudamericano. II Congreso Latinoamericano de Hidrología Subterránea. 7 a 11 Nov 94. p. 641 - 656. Santiago - Chile.
Godoy, E. & Paredes, J.L. (1995): Aquíferos Potenciales del Paraguay. Anais 1º Mercosur de Aguas Subterráneas. 03 a 06 Sep 95. p. 24 - 37. Curitiba - Brasil.
Mitlöhner, Ralph (1990): Die Konkurrenz der Holzgewächse im regengrünen Trockenwald des Chaco Boreal, Paraguay. Göttinger Beiträge zur Land- und Forstwirtschaft in den Tropen und Subtropen. Göttingen.
Nitsch, M. (1994): Versalzungsgefährdung von Böden im östlichen Zentralchaco als Folge nicht angepaßter Rodungsmaßnahmen. Unpublished Report Federal Institute for Geosciences and Natural Resources, Hannover.
Williamson, D.R. (1990): Salinity - an old environmental problem. - In: Year Book Australia 1990, p. 202 - 211. (AGPS: Canberra).

Addresses of authors:
M. Nitsch
J. Utermann
Federal Institute for Geosciences and Natural Resources, Stilleweg 2
D-30655 Hannover, Germany
R. Hoffmann
Proyecto Sistema Ambiental del la Región Oriental del Paraguay, Calle Ciencias Veterinarias
San Lorenzo, Cassilla de Correo 1859
Asunción, Paraguay
L. Portillo
Subsecretariá de Estado de Recursos Naturales y Medio Ambiente, Dirección de Ordenamiento Ambiental
Ruta Mcal. Estigarribia km 10.5, Cassilla de Correo 2461,
San Lorenzo, Paraguay

Changes of Soil Qualities under Pivot Irrigation In the Bahira Region of Morocco: Salinization

M. Badraoui, B. Soudi, A. Merzouk, A. Farhat & A. M'Hamdi

Summary

In the arid and semi-arid areas of Morocco, irrigation is necessary to obtain good and sustainable agricultural production. In the last 10 years, pivot irrigation, along with other irrigation techniques, has been introduced in arid regions on range lands where an uncertain rainfed agriculture is practiced. The objectives of this study were i) to evaluate changes of selected soil characteristics caused by pivot irrigation in the Bahira region of central Morocco and ii) relate the rates of soil change to water quality.

Forty-nine pivot plots were selected ranging in size from 25 to 50 ha, and irrigated from 1 to 10 years. Both irrigated and control (non-irrigated) plots were sampled for determining the major soil characteristics. Irrigation water quality was also determined for each plot.

Irrigation caused a significant positive effect on the annual variation rate of EC, ESP, exchangeable sodium and magnesium, and permeability. The effect of irrigation was not significant on other soil characteristics (pH, organic matter, calcium carbonate equivalent, cation exchange capacity, and exchangeable calcium). Even with good water quality and good soil permeability, irrigation caused rapid salinization and alkalinization of the soils under high evapotranspiration. For pivot center units having the same soil type, salinization increased as a function of increased irrigation water salinity and the number of years under irrigation. Sodification of the soil increased with the increase of irrigation water SAR in only clayey soils. No relationship was found between ESP and permeability.

The number of years necessary to reach critical EC values for cereal crops is variable from one pivot unit to another. Desalinization measures should be undertaken to prevent building up of salts in the soils.

Keywords: Irrigation, soil quality, water quality, salinization, Morocco

1 Introduction

Good soil and water properties are essential for sustainable land use under intensive cropping. A soil management system is sustainable only when it maintains, or can improve, both soil and water quality (Larson & Pierce, 1991). In Morocco, many studies have shown that irrigation leads to soil and water degradation (Mathieu & Ruellan, 1980; Baaki, 1987; Badraoui & Merzouk, 1994). Irrigation water quality associated with soil texture and drainage largely influences the rate of soil degradation, especially salinization (Umali, 1993).

In the arid and semi-arid areas of Morocco, irrigation is necessary to obtain good and sustainable yields. In the last 10 years, pivot irrigation, along with other irrigation techniques, has been introduced in arid regions on range lands where an uncertain rainfed cereal production is practiced. Pivot irrigation has only been possible in areas where large undergroundwater reservoirs are available. Undergroundwater of Bahira reservoir has a medium to high salinization risk (EC> 0.75 dS/m) and low to medium alkalinization risk (SAR ranges from 1.6 to 4.8 $(meq/l)^{1/2}$. Up to 1995, 20,380 ha were irrigated using this technique. The government has a program for 25,000 ha to be irrigated by the year 2000.

Monitoring of soil quality under pivot irrigation is a basic means of evaluating the sustainability of this land management system. The objectives of this study were i) to assess changes in selected soil characteristics of an irrigated area in the Bahira region of northwestern Morocco, following 10 years of irrigation and ii) to relate the rates of soil quality changes to irrigation water quality.

2 Materials and methods

The Bahira plain is a large sedimentary closed depression of 1667 km² located between the Haouz of Marrakech, the Tadla plain, and the Rehamna plateau in the northwestern part of Morocco. It has an arid mediterranean climate. The mean rainfall is 225 mm/year, the a mean maximum temperature is 34.4°C, and the mean annual evapotranspiration is above 2400 mm/year (Watts & El Mourid, 1988). Apart from the central part of the depression which contains saline and hydromorphic soils, the most representative soils are typic xerochrepts and haploxerolls.

Pivot irrigation was introduced to the Bahira region in 1985. At present, 3096 ha are under irrigation using this technique, which increased cereal yields from less than 1 t/ha to 6 t/ha. A program of 5,000 ha is being carried out for this region.

Forty-nine pivot units, ranging in size from 25 to 50 ha and irrigated for a period of time ranging from 1 to 10 years, were selected for this study. Both irrigated and control (non-irrigated) plots were sampled to a depth of 40 cm. Three compound samples, made of 20 sub-samples each, were considered representative of each plot. The major soil characteristics (particle size, pH, EC, organic matter, CEC, and exchangeable Ca^{++}, Mg^{++}, and Na^+) were analysed using standard analytical techniques. The 49 pivot units sampled are irrigated by 29 wells having a water flow ranging from 12 to 60 l/s. Irrigation water quality was evaluated by determining EC, pH, Na^+, Ca^{++}, Mg^{++}, Cl^-, and NO_3^-. The annual quantity of water used for the irrigation of each plot was evaluated. Because the period of time under irrigation is different from one pivot unit to another, a normalized indicator of soil quality changes under irrigation was used for comparison. Thus, an annual variation rate (AVR) was calculated for each soil property X, as: AVR(X)=(Irrigated value - Control value)/Number of irrigation years.

Statistical comparison between means of irrigated and control plots for each soil property was parformed using the least significant difference test (LSD) at 5 % level of significance (Gomez and Gomez, 1984).

3 Results and discussions

According to the standards contained in Richards (1954), the undergroundwater of the Bahira reservoir has a medium to high salinization risk and a low to medium sodification risk depending on the well. The ECw (water) ranged from 1.16 to 5.47 dS/m and SAR (Sodium Adsorption Ratio) varied from 1.59 to 4.02 $(meq/l)^{\frac{1}{2}}$. With these values, no reduction of the infiltration rate is likely to occur (Rhoades, 1982). Other analytical results are reported in Table 1. Chlorine is the major anion present in irrigation water. The overuse of nitrogen fertilizers causes nitrate pollution of underground water (Farhat, 1995). Eleven of the 29 wells have more than 50 mg/l of nitrate.

Parameter	Minimum	Maximum	Mean	SD	Units
pH	6.65	8.02	7.23	0.38	-
ECw	1.16	5.47	1.87	0.93	dS/m
SAR	1.59	4.02	2.77	0.79	$(meq/l)^{1/2}$
Na	3.58	17.54	6.91	3.67	meq/l
Cl	3.58	17.54	11.75	5.49	meq/l
NO_3	14.24	79.36	46.3	18.40	mg/l

Table 1: Ranges in undergroundwater quality in the Bahira region for 29 wells.

Irrigation caused significant changes in several soil qualities (Table 2). Irrigation-induced soil salinity was the most significant effect. For non-irrigated control plots, soil salinity (ECs) varied between 2 and 48 dS/m. It increased after irrigation, but this increase was only significant in 24 plots out of the 49 studied. A decrease in ECs, but not signficant, was found in 5 plots presenting sandy soils. The AVR-ECs ranged from 0.64 to 1.25 dS/m/year (Fig. 1). The large variations of AVR-ECs for plots irrigated using waters of similar quality (1.2<ECw<1.8 dS/m) may be due to differences in soil texture, the amount of water used over the period of irrigation, the salinity of the soil prior to irrigation, and to the frequency and intensity of rainfall which can cause some salt leaching.

Soil quality	Number (1)	Significant effect (2)	Impacts (3)
pH	49	4	4 -
ECs	49	24	24 +
OM	49	3	1 - 2 +
CEC	49	11	3 - 8 +
Na	49	15	1 - 14 +
Ca	49	3	2 - 1 +
Mg	49	4	1 - 3 +
ESP	49	16	1 - 15 +
Infiltration rate	29	5	5 +

(1) : Number of plots analysed
(2) : Number of plots in which irrigation caused significant effect
(3) : (-) decrease; (+) increase

Table 2 : Impacts of pivot irrigation on selected soil qualities in the Bahira region, Morocco

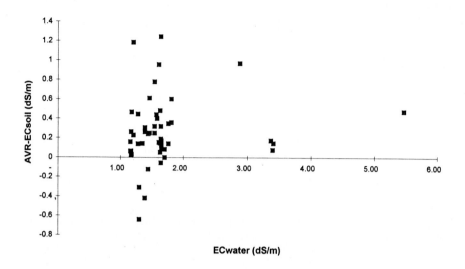

Fig. 1: Relationship between the ECwater and the annual variation rate (AVR) of soil salinity (ECsoil) in the Bahira region, Morocco

Water EC influences the accumulation of salts in the soil profile, but more importantly is the quantity of salt added (Q) over time by irrigation water. This quantity varied from 2 to 10.3 tons/ha/year depending on the salt content in water and the amount of water used for irrigation. AVR-ECs is best related to Q than to ECw (Fig. 2). Despite the relatively good water quality (ECw < 2 dS/m) and the good infiltration rate (> 2 cm/h) of the soils in the Bahira region (Farhat, 1995), irrigation induced salt accumulation in the soils.

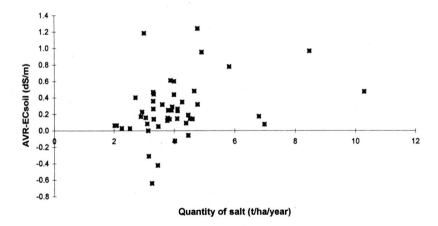

Fig. 2: Relationship between the quantity of salt added to the soil by irrigation water and the AVR-ECs in the Bahira region, Morocco

Sodium accumulation in the soil was significant in 14 plots out of the 49 studied (Table 2). As expected, this increase was more important in clayey soils than in medium textured or sandy soils. The AVR-Na ranged from - 0.2 to 0.76 meq/100g of soil and was related to the SAR of irrigation water. Negative values were not significant (except in one case). In parallel, the AVR-ESP increased linearly with the increase of AVR-Na (Fig. 3). Dispersion of the clays by sodium accumulation in Bahira soils did not significantly reduce the infiltration rate. On the contrary, the infiltration rate generally increased. Three facts could explain this observation; i) intensive cropping under pivot irrigation is associated with better tillage practices than to rainfed agriculture, ii) the clay mineralogy of the soils is mixed with low amounts of smectites and high quantities of illites and kaolinites (Badraoui, unpublished data), and iii) as shown by ECw and SAR values, no reduction of infiltration was predicted. The effect of salt concentration was relatively more important in maintaining the stability of soil agregates than the dispersive effect of sodium.

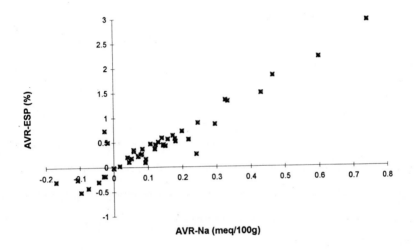

Fig. 3: Relationship between AVR-Na (soil) and AVR-ESP in the Bahira region, Morocco

Irrigation had almost no effect on the other soil characteristics such as pH, organic matter, Ca, and Mg.

It is clearly shown that irrigation in the Bahira region induced salinization of the soils. The annual rate of salinization is site-specific and depends on the water quality. Values of AVR-ECs could be used to assess the sustainability of the system. The regional agricultural development service in the Bahira region is now asking the following question : for how long would cereal production under pivot irrigation system be sustainable? Taking into account the critical salt tolerance threshold for wheat (7.4 dS/m for 10% reduction and 13.0 dS/m for 50% reduction of yield), the present salinity of the soil and the annual increase in soil salinity under each pivot unit, we have calculated that the number of years necessary to have a yield reduction is under 5 years for 10 pivot units.

4 Conclusion

Monitoring of soil quality under irrigation in the Bahira region demonstrated that salinization is a serious problem for the sustainability of cereal production. Desalinization measures should be undertaken to prevent building up of salts in the soil. Optimization of water-use efficiency is an important parameter to consider.

References

Baaki, M. (1987): Effet de differentes eaux d'irrigation sur la salinité et la sodicité d'un sol du Tadla. MS thesis, IAV Hassan II, Soil Science Dept., Rabat, Morocco.

Badraoui, M. & Merzouk, A. (1994): Changes of soil qualities under irrigation: the effect of salt accumulation on water retention by vertisols. In CIEHAM-IAM-B ed, Advanced course on farm water management techniques. Morocco 7-22 May 1994, p. 145-155.

Farhat, A.(1995): Effets de l'irrigation par pivot sur la qualité des sols dans la Bahira: situation actuelle et perspectives de développement. MS thesis, I.A.V. Hassan II, Soil Science Dept. Rabat, Morocco.

Gomez, K.A. and Gomez, A.A. (1984): Statistical procedures for agricultural research. » An International Rice Research Institute book », Second Edition, John Wiley & Sons, pp. 187-193.

Larson, W.E. and Pierce, F.J. (1991): Conservation and enhancement of soil quality. In evaluation for sustainable land management in the developing world. Vol 2: Technical papers. Bangkok, Thailand : International Board for Soil Research and Management, 1991, IBSRAM Proceedings No 12 (2).

Mathieu, C. and Ruellan, A. (1980): Evolution morphologique des sols irrigués en région méditerranéenne semi aride. Cahier ORSTOM, Série pédo. **13**, 3-25.

Rhoades, J.D. (1982): Reclamation and management of salt-affected soils after drainage. Proc. First Annual Western Provincial Conf. Rationalization of Water and Soil Res, and Management. Lethbridge, Alberta, Canada, Nov. 29-Dec. 2, 123-178.

Richards, L.A. (1954): Diagnosis and improvement of saline and alkali soils. U.S.S.L., USDA, Handbook 60.

Umali, D.L. (1993): Irrigation-Induced Salinity. A Growing problem for development and the environment. Wold Bank Technical Paper 215, 78.

Watts, D.G. and El Mourid, M. (1988): Régimes et probabilités des précipitations dans des régions semi-arides et céréalières du Maroc occidental. Centre aridoculture-INRA Maroc (ed.) Projet INRA/MIAC/USAID No: 608-D136.

Addresses of authors:
M. Badraoui
B. Soudi
A. Merzouk
A. Farhat
A. M'Hamdi
Institute of Agronomy and Veterinary Medicine
BP. 6202
Rabat-Instituts
Rabat, Morocco

Drainage for Soil Conservation

M. M. Moukhtar, M. H. el Hakim & A.I.N. Abdel Aal

Summary

Some areas on the northern periphery of the Nile Delta are clay, salt- affected soil with, poor productivity. The soil is threatened by shallow highly saline groundwater. We present here some soil hydropedological data and discuss the use of local experience for simple non-costly management. The field experiment included 20 and 40 m spacing treatments. Moling was introduced later at 0.45 m depth and 2.0 m apart. Results indicated that the static water table was variable and affected by environmental activity in the surrounded areas. Soil salinity was strongly related to water table depth and its salinity. Severe salinization occurs in top soil layers under fallow periods. Draw down rate was improved after moling and consequently salinity of upper layers decreased to less than 4 dS/m as long as drainage works efficiently.

Keywords: Soil conservation, drainage, salt-affected, heavy textured soil, clay, low permexable.

1 Introduction

Soil degradation as a result of water logging and salinity is a worldwide phenomena in irrigated agricultural lands in arid and semi-arid regions. Causes of salinization are different and so are either preventive or remedial measures. The problem becomes more complicated in clay soils because salts might be found in micropores, and more serious in the presence of a shallow saline groundwater. Water management in such lands reqirees considerable attention, especially in countries with limited land resources and an increasing population. Local experience can be very important in dealing with such problems.

Many agricultural lands of fluvio-marine sediments in the northern periphery of the Nile Delta are considered problem soils (Moukhtar et al. 1990a, b; 1995). They are salt- affected and have poor productivity. Moreover, they are assumed to lie in a zone of hydrostatic piezometric pressure (Amer and Ridder 1989). The control of these detrimental conditions could improve their potential productivity (Ritzma 1994; Moukhtar et al. 1995). The purpose of the present work is to discuss results of hydropedological studies of the deteriorated soils in the northeastern Nile Delta, and to present local experiencefor their rational use of such soil.

2 Materials and methods

The site of the study is a representative area of deteriorated soils of the northeastern Delta (Fig. 1), located on the farm of El Serw Research Station near lake Manzala. The area is at 0 msl. The clay content reaches 63.5% up to 90 cm depth. Soils are low permeable and the average hydraulic conductivity is 0.0669 cm/day. A permanent saline groundwater table (average 25 dS/m, mainly

NaCl) is the main source of soil salinization. Conditions affecting soil salinity are shown in representative profiles in a large area of the farm with non-adequate field drainage. This area is bounded by a main drain in the north, a main irrigation canal in the south, an open collector drain in the east and an irrigation sub-canal in the west.

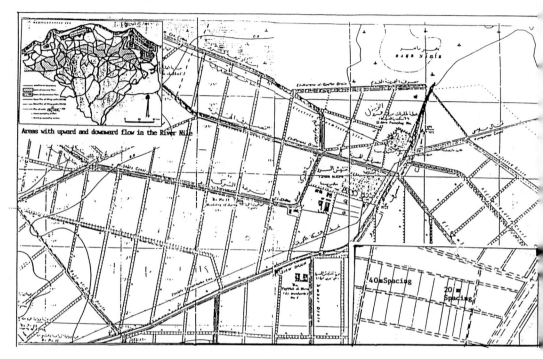

Fig. 1: The site and experimental field in the Northeastern Delta of Egypt.

The importance of drainage for soil conservation was studied in an experimental drainage field of 20 feddans (8 ha). The design involved 2 drain spacing treatments 20 and 40 m at 90 cm depth (usually adopted by the farmers) and separated by buffer zones. The effect of spacing treatments on hydrological and soil conditions was tested along 4 cropping seasons. Also moling 45 cm depth and 2 m apart, introduced in the field in conjunction with field drain ditches was tested in the following 3 cropping seasons. Crops tested in winter were clover and wheat and, in summer, sorghum. Water table depth was measured in observation wells placed at the centre of the plots of each spacing treatment during several irrigation intervals in each cropping season. Water samples and soil salinity in saturation extract were determined according to Richards (1954).

3 Results and discussion

3.1 Hydropedological features for deteriorated soils and classification

The soil classification of the study area is Aquic chromuderts, fine, montmorillontic, thermic (Abdel-Aal, 1995) according to Soil Survey Staff (1990). Data for some soil profiles in the area are presented in table (1) and shows the variable conditions affecting soil salinity.

Data shows that soil salinity increases with soil depth since salts in the top soil are more or less effectively leached upon irrigation. But severe soil salinization occurs even in the top soil under fallow condition. Data indicates that the top-soil salinity reached a value of 40 dS/m in a field left fallow for several seasons, with a groundwater salinity of 59 dS/m at 102 cm depth. If returned to cultivation, its desalinization will be difficult (Moustafa et al. 1990; Michaelsen et al. 1993). It is worthy to mention that soil adjacent to the main drain is protected from degradation as shown from the relative low salinity especially in the upper 60 cm.

Generally, salt content in the soil is strongly related to the water table depth and its salinity (Moukhtar et al. 1990$_b$; Ismail 1996). The dominant salt either in groundwater or in soil is sodium chloride, magnesium ions exceed calcium. On the other hand, the deepest water table is always found adjacent to the main drain and the shallowest adjacent to the main irrigation canal.

Profile	Adjacent to the Main Canal	Adjacent to the Main Drain	In Between	Fallow
Soil Layer (cm)				
0-30	5.5	3.7	5.6	40.0
30-60	5.9	4.7	6.1	16.4
60-90	15.0	6.1	7.6	18.3
90-120	22.6	6.7	21.1	41.7
Water Table:				
Salinity	22.3	15.4	18.1	59.3
Depth (cm)	53	100	78	102

Table 1: Soil salinity (EC,dS/m) and water table depth and salinity in some representative soils.

3.2 Drainage and soil conservation

The main purpose of drainage is not only to remove of excess water but to prevent soil degradation. Local experience showed that the salinity of the falling water table varied according to its position during the irrigation intervals. For the soils under study, the saline groundwater table should first be considered as to its effect on the upper soil profile until 50-60 cm depth, which includes the rootzone and which must be preserved from water logging and salinity. Through enhancing the downward water movement in the first days after irrigation, excess salts could be removed from the rootzone and the fresh irrigation water will constitute a temporary front separating the saline groundwater from the rootzone. On the other hand, the saline groundwater increases the drainage requirements which means that the rate of water table drawdown should increase.

The water table recession after irrigation was rather slow both for 20 or 40 m spacing in the experimental field. As shown in Fig. 2, the rate of water table drawdown in the rootzone did not exceed 6 cm/day in the first season. However much improvement in water table recession occurred when moling was introduced. Water table dropped to about 50 and 40 cm in 5 days in treatments with 20 and 40 m spacing respectively. The rate of water table drawdown values were around that required (8 - 10 cm/day), values being higher in 20 m spacing treatment. Previous studies by El-Hakim et al (1990) showed that a rate of water table drawdown of 8-10 cm /day under a saline groundwater of 20-30 dS/m must be obtained in the upper 60 cm.

The beneficial effect of moling is to avoid the harmful stagnation of irrigation water and dissolved salts around the rootzone. The downward water movement is enhanced through cracks and fissures developed by the mole plough blade and water is evacuated partly through the mole drains.

Desalinization which occured as a result of open drains alone and in combination with moling is shown in Fig. 3 which represents the electric conductivity (EC) of the upper soil layers (up to 60 cm) and of the lower layers (60 -120 cm) in both drainage treatments.

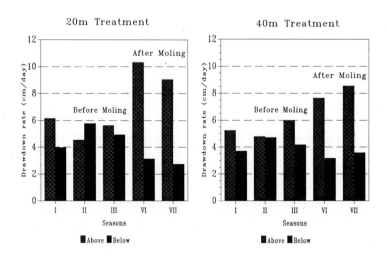

Fig. 2: Rate of drawdown after irrigation in the consecutive seasons in 20 and 40 metre spacing treatment before and after moling (above and below mole depth).

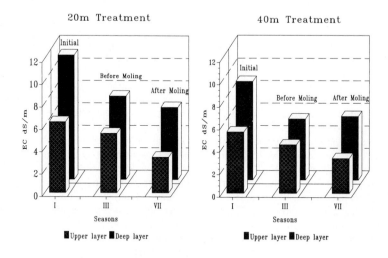

Fig. 3: Mean EC in upper layers (0-60 cm) and deep layers (60-120 cm) in seasons before and after moling under both drainage treatments.

Only after moling the EC of the upper soil layers in both treatments decreased to a value of 4 dS/m. Obviously moling does not interfere in the desalinization of the deeper soil layers mole depth. Therefore shallow rooted crops are preferred under these soil conditions. Sorghum and wheat crops were quite successful in the field under study. Irrespective of moling, rice crop is always included in the crop rotation since the ponded water conditions are effective in controlling the salinity of the soil profile and a good yield can be obtained.

4 Conclusion

Moling combined with open field drains can be highly recommended as an auxiliary drainage treatment in low level clay salty soils with a saline water table to raise soil productivity. It is a low-cost measure needing no advanced machinery and it can be adopted by small farmers instead of using narrow drain spacing which wastes agricultural land. In addition this type of soil should not be left fallow even for a short period, otherwise salinization quickly occurs. Delay in irrigation should be avoided. Also, shallow root crops are preferred. Water submerged crops i.e. rice, amshout should be included in agricultural rotation. Care for tillage operations should be taken.

References

Abdel-Aal, A. I. N. (1995): Macro and Micromorphological Studies on the Water Table Affected Layer in some Soils of Egypt, Ph.D. Thesis, Fac. Agric., Cairo University.

Amer, M. H. and Ridder, N.A. (1989): Land drainage in Egypt, DRI, Cairo, Egypt.

El-Hakim, M. H., Moukhtar, M.M., Moustafa, M.K.H. & El-Sheikh, M.B. (1990): Groundwater salinity variations during irrigation intervals in northeastern Delta, Egypt. J. Appl. Sci. **5**,140-149.

Ismail, M. I. (1996): Effect of Water Table on Soil Properties, M.Sc. Thesis, Fac. Agric., El Mansoura University.

Michaelsen, J., Moustafa, M.K.H., Gunther, D., Moukhtar, M.M., Widmoser, P. and Woldemichael, A. (1993): Hydraulic conductivity and leaching behaviour in response to pore system dynamics in swelling clay soil of the Northern Nile Delta, Egypt. J. Appl. Sci. **5** 343-354.

Moukhtar, M.M., El-Hakim, M.H., Abdel Mawgoud, A.S. and Ismail, M.I. (1995): Field experiment for restoring the productivity of a salty clay soil with saline groundwater, International Symp. on Salt-Affected Lagoon Ecosystems [ISSALE-95], Valencia (Spain): 188-191.

Moukhtar, M. M., Moustafa; M.K.H., Abdel Mawgoud, S.A. and Ismail, M.I. (1990_a): Field testing for water table control in coastal north east Delta, Egypt. J. Appl. Sc. **5**, 132-139.

Moukhtar, M. M., El Hakim, M.H. and Moustafa, M.K.H. (1990_b): Soil conditions and adequate water table depth in northeastern Delta, Symp. On Land Drainage for Salinity Control in Arid and Semi-Arid Regions. **2**,102-112, Cairo, Egypt.

Moustafa, M. K. H., Moukhtar, M.M. and El Gayer, A. (1990): Preferential water flow and salt leached in salty clay soil. Egypt. J. Appl. Sci. **5**, 228:235.

Richards, L.W. (1954): Diagnosis and improvement of saline and alkali soils, U.S.D.A. Agric. Handbook No. 60.

Ritzma, H.P. (1994): Drainage principles and Application, ILRI Publication 16, Wageningen, The Netherlands.

Soil Survey Staff (1990): Keys to Soil Taxonomy. United States Department of Agriculture, Soil Conservation Service, Sixth Edit ion.

Addresses of authors:
Mohamed M. Moukhtar
Soils, Water and Environment Research Institute, Agric. Research Centre
3, Str. no. 41, Madient El-Tahrir, Imbaba, Giza, Egypt
Mahida M. El Hakim
Aly I.N. Abdel Aal
Soil, Water and Environment Research Institute, El Gamma St., Giza, Egypt

Part II: Growing Impacts of Industrialized Agriculture and Urbanization on Soils

Introduction

H.-P. Blume

During the last century new forms of soil degradation such as compaction and several forms of contamination with harmful substances have become increasingly important (Blume, 1992). Reasons for these developments include growing populations and increasing industrialization will be described in the following chapters as well as possibilities for protecting soils against such problems.

Soil compaction is induced by the use of heavy field machinery in crop production as well as in fruit plantations and forests. It decreases the porosity of soils, especially the coarse pores. This takes place not only in the top soil but also further down. Worse plant rooting, water stagnation and oxygen defiency are normal consequences. Possible strategies against soil compaction are lighter machineries with broad wheels; wet soils should not be passed at all. Rehabilitation strategies against light compaction can be to promote biological activity, especially loosening by earthworms, e.g. by organic fertilizers. Stronger forms of compaction can be overcome by deep loosening or deep ploughing. But these forms of rehabilitation are only successful under special soil and climate conditions.

In many cases soil compaction is accompanied by surface sealing. But numerous cultivations or lack of organic fertilizers can also lower the organic matter contents of the topsoil and can promote surface crusting and sealing, especially of silty soils, too. In some soils subsoil hardening can become a problem for root penetration, e.g. in volcanic ash soils or in Ferralsols, which must be solved by loosening.

High quantities of organic or anorganic fertilizers, especially nitrogen and pesticides, can pollute groundwater and can influence the latter and even damage soil life. Commercial mineral fertilizers are required to supplement the nutrient supply of arable land to optimize crop yields and quality, and to avoid a loss of soil productivity. Negative effects on the soils have only been brought to light where either the chemical forms or the application rates deviated from advisory recommendations. It is quite clear for example that physiologically acid fertilizers reduce the base saturation of soils which need liming. Groundwater pollution by nitrates and air pollution by laughing gas and other nitrous oxides are more severe problems but can be induced by organic fertilizers as well. A decrease of fertilizer intensity, an orientation after the nutrient uptake by plants, and dating after plant growth are possible strategies to lower these problems.

Groundwater pollution by pesticides is a more severe problem of industrialized agriculture. One possibility is to seek non-chemical means of attacking wild plants and parasites, but calamities cannot be avoided under all circumstances. In these cases there are possibilities to look for special ingredients with which the risks can be minimized under special site conditions.

Especially in highly industrialized countries acid rainfall, e.g. caused by using fossil energy, like coal or fuel, pollutes soils, increases soil acidity, weathering and loss of nutrients, especially in forests. Some papers show strategies of reclamation of acidified soils.

Sewage sludge and composted municipial wastes can be good organic fertilizers for arable land in principle. But they lead to the risk of soil pollution by heavy metals and harmful organic chemicals. Meanwhile many countries have laws to avoid soil pollution by harmful substances. This requires measurements of total and of mobile metals and organics in the sludge and the wastes as well as in the soils. Methods should be internationally standardized.

Many soils of urban industrial agglomerations are polluted by harmful organics and heavy metals: this is the case because many of them are polluted by traffic or industry but others are formed from so-called technological substrata such as wastes, sludges, slags, mortar and bricks, and muds which can be enriched with heavy metals. In many cases soils of such substrates differ in their quality for plant growth and as a buffer against groundwater pollution from those of natural substrata.

Many reclamation problems of soils affected by mining industries and disposal sites have to be solved, especially in urban industrial agglomerations. Soil physical problems like compaction and sealing as well as erosion have to be solved in addition to soil chemical problems, e.g. strong acidity caused by sulfide-enriched coal-mined lands. Soil quality standards should be formulated to focus land reclamation agencies on potential problems, supported by sustained post-project inspection, and enforcement should be the best way of securing long-term productivity of reclaimed lands.

In the following chapters the problems are described and ideas for their solution are suggested.

References

Blume, H.-P. (ed.) (1992): Handbuch des Bodenschutzes. 2. Aufl.; ecomed, Landsberg

Address of author:
Hans-Peter Blume
Institute of Plant Nutrition and Soil Science
Christian-Albrechts-University
D-24098 Kiel, Germany

Soil Compaction:
A Global Threat To Sustainable Land Use

B.D. Soane & C. van Ouwerkerk[†]

Summary

Soil compaction arising from the use of field machinery in crop production is implicated in a loss of soil quality, reduced efficiency of agricultural crop production and deterioration in the quality of the environment. The opportunities available for both reducing the incidence of soil compaction and the rehabilitation of compacted soils are considered and compared in terms of technical, economic and sustainability criteria.

Keywords: Soil compaction, machinery, axle load, contact pressure, rehabilitation, environment

1 Introduction

Today we face two dominant global sustainability challenges: (1) environmental degradation problems including soil degradation; (2) declining per capita food production problems. Soil compaction has a dominant impact in both challenges. Compaction is a process of densification in which porosity and permeability are reduced, strength is increased and many changes are induced in the soil fabric and in various behaviour characteristics (Soane and Van Ouwerkerk, 1994a; Campbell, 1994). The anthropogenic factors considered here induce compaction as a result of loads applied to the soil surface (Koolen, 1994; Horn and Lebert, 1994; Gupta and Raper, 1994). Compaction problems arise in agricultural crop production (Raghavan et al., 1990; Van Ouwerkerk, 1991; Soane and Van Ouwerkerk, 1994 a,b; Håkansson and Voorhees, 1997), and also in forest crops (Wronski and Murphy, 1994). Proposals for further research on soil compaction were set out by Van Ouwerkerk and Soane (1994).

2 Machinery use and soil compaction

The continuing reduction in the number of agricultural workers and the amalgamation of land holdings in many countries have been accompanied by increases in the power, working width, mass, and often the ground contact pressure of field machinery (Fig. 1). The mass of agricultural machines has increased by a factor of 3 to 4 during the past 3 decades (Horn, 1994). The average mass of tractors in the U.S.A. has increased from 2.7 Mg in 1948 to 6.8 Mg in 1994 but specialist tractors and six-row sugar beet harvesters may reach to 30 Mg and 38 Mg, respectively.

The spatial and temporal distribution of vehicle traffic in the field can be quantified in terms of: overall coverage, rut length, traffic intensity, traffic effect, load index, mobility index, mechanization degree and compaction risk factor (Kuipers and Van de Zande, 1994). Traffic treatments used in experiments should be quantified in a uniform manner, e.g. Mgkm ha^{-1} (Arvidsson and Håkansson, 1996). Traffic distribution varies widely, depending on the type of machinery and the field operation undertaken. Operations of particular importance are the transport of bulky amendments such as slurry (Douglas et al., 1994), in-furrow tractor wheel traffic during mouldboard ploughing (Tijink et al., 1995), seedbed preparation operations, and harvest traffic.

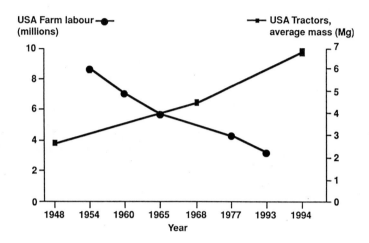

Fig. 1: The relationship between the decline of farm labour and the average mass of tractors in the U.S.A.

The primary characteristics of tyres which affect the incidence of compaction are: (1) size (diameter, width); (2) load (static, dynamic); (3) ground contact pressure (average, maximum, hard ground, soft ground); (4) inflation pressure; (5) forward speed; (6) tread type; (7) slip; (8) number of passes (Tijink, 1994). The characteristics of tracks which relate most closely to their ability to compact soil are: (1) size (contact length, width); (2) ground contact pressure (average, max.); (3) load; (4) grouser type; (5) vibration (Tijink, 1994). The intensity of compaction in a field operation is controlled by the interactions of: (1) the compaction capability of the running gear; (2) the soil compactibility at the time; (3) the distribution of wheel traffic.

3 Compaction as a threat to the environment

Soil compaction may influence the environment at a distance from the immediate location at which the compaction occurs, sometimes many km from the site, such as pollution of surface and ground waters, and, sometimes affecting factors of global importance such as the composition of the atmosphere and soil resources (Soane and Van Ouwerkerk, 1995; Lal et al., 1995; Horn et al., 1995; Van Ouwerkerk, 1995).

The energy required for fracturing a compacted soil by primary or secondary cultivation is greater than that for a non-compacted soil (Dickson and Ritchie, 1996a), and the additional fuel consumed will result in additional CO_2 emission to the atmosphere. Compacted soils at high water contents are likely to become anaerobic which will accentuate the risks of increased emission of N_2O, a powerful greenhouse gas (Hansen et al., 1993; Douglas et al., 1994; Soane and Van Ouwerkerk, 1995; Lal et al., 1995). Normal traffic over grassland soils resulted in peak N_2O

emissions 2 to 3 times larger than from soil receiving low ground pressure traffic or zero traffic (Douglas et al., 1994). High soil water content and hence low air content, following compaction can also lead to local production of methane, an important greenhouse gas, whereas in untrafficked parts of the field, methane oxidation may remain dominant (Hansen et al., 1993; Lal et al., 1995). Although ammonia is not a greenhouse gas, its increased volatilization to the atmosphere after the application of slurry to compacted soils of reduced infiltration rate, represents a loss of available nitrogen to subsequent crops, as well as a contribution to pollution from acid rain.

The role of soil quality and soil resources in relation to local, regional and global environmental problems was emphasized by Yaalon (1996). Soil degradation, including soil compaction, is recognised as one of the most serious threats to sustainable land use. Traffic-induced compaction influences the stability of soil resources because of the increased risks of runoff and erosion due to reduced infiltration rate (Horton et al., 1994). The tendency for considerably larger amounts of fertilizers to be applied to compacted soils than to non-compacted soils leads to increased risks of loss of environmental quality due to increased emissions of N_2O and increased loss of fertilizer nutrients into surface waters. On grassland soils, additional applications may amount to about 50-100 kg N ha^{-1} y^{-1} (Douglas, 1994), with corresponding N losses to the atmosphere and to ground waters. The biological activity of soils, an essential component in soil quality and stability, is seriously affected by compaction (Brussaard and Van Faasen, 1994; Whalley et al., 1995).

4 Compaction as a threat to crop production

Soil compaction decreases the yield and quality of crop plants in all parts of the world (Soane and Van Ouwerkerk, 1994b). Crop performance, usually quantified by yield, bears a quadratic relationship to soil compactness, provided a wide enough range of soil conditions is tested. The optimum value of compactness and the shape of the yield/compactness relationship depend on many factors such as soil type, crop type (perhaps variety), soil wetness and crop nutrition. The occurrence of an optimum has been widely established for both temperate (Boone and Veen, 1994; Lipiec and Simota, 1994; Lindstrom and Voorhees, 1994) and tropical crops (Kayombo and Lal, 1994). Crop species, as well as some varieties (Braunack, 1994), show distinctly different sensitivities to compactness (Alakukku and Elonen, 1995; Dickson and Ritchie, 1996b). The optimum compactness relationship is dependent on the interaction of a number of mechanisms influencing crop responses adversely at both high and low compactness. These mechanisms interact markedly with soil type and weather conditions (Soane and Van Ouwerkerk, 1994a). Where soil water contents tend to be high, the dominant mechanism is associated with deficient aeration (Stępniewski et al., 1994), whereas under drier conditions the dominant mechanisms are increases in soil strength (Guérif, 1994) and limitations to the supply of nutrients (Boone and Veen, 1994).

5 Machinery options to reduce soil compaction

5.1 Reduction of axle load

Axle load is considered to be the dominant factor affecting the transmission of vertical stress into the lower parts of the profile, resulting in the creation of persistent, unfavourable conditions in subsoils (Håkansson and Petelkau, 1994). Van den Akker (1994) used a model to predict whether a subsoil would be overloaded by certain surface loads by taking into account the measured strength characteristics of the subsoil. To investigate the need for such restrictions in axle load, an experimental programme involving 26 experiments in Europe and North America was established (Håkansson, 1994; Håkansson and Reeder, 1994) in which a common 10 Mg single axle load treatment was applied on a single occasion. The control treatment was a 5 Mg axle load or less. Results varied greatly at different sites. At clay contents of 10% or less, yield depressions were very slight, even in the first year. In a few cases, high axle loads tended to outyield normal axle loads (Melvin et al., 1994), perhaps related to water supply during the growing season. Where high axle loads are applied for several years consecutively, crop yield reductions are more marked (Arvidsson and Håkansson, 1996; Lal, 1996). Axle loads can be reduced by reducing the size and mass of vehicles and by fitting more axles but the economic consequences may not always be acceptable (Vermeulen and Perdok, 1994).

5.2 Reduction of ground contact pressure

A reduction of ground contact pressure below running gear, usually through increased contact area, has a dominant effect on reducing compaction within the topsoil (Vermeulen and Perdok, 1994; Tijink et al., 1995). Possible options include the use of: (1) extra tyres (duals, triples); (2) wide tyres with low inflation pressure (flotation); (3) reduction of vehicle weight; (4) increase of contact area by reduction of tyre inflation pressure; (5) replacement of tyres with tracks. Many manufacturers now produce wide section tyres and, in addition, have reduced the minimum allowed inflation pressure for standard tyres. Automatic control of inflation pressure to permit adjustment of contact area to suit ground conditions (Tijink, 1994) is technically very attractive but not yet readily available. The adoption of low ground pressure tyres, having a very much greater width than conventional tyres, results in much more soil area being exposed to traffic and hence in higher average compactness in the topsoil, whereas under conventional tyres, soil compactness may show greater variability and greater maximum values (Dickson and Ritchie, 1996b).

Yield responses of a wide range of crops to low ground pressure tyre systems have been reported by Vermeulen and Perdok (1994), Tijink et al. (1995), Dickson and Ritchie (1996b). Low ground pressure options for running gear are comparatively easy to obtain and to fit, with relatively modest extra costs, in contrast to zero-traffic systems (Chamen et al., 1994b). Crop responses to the use of tracks were reported by Erbach (1994) and Melvin et al. (1994).

5.3 Proposals for standards for vehicle running gear

Following extensive evidence, particularly in Eastern Europe, for severe compaction problems where heavy conventional vehicles are used, standards have been proposed for axle loads and ground pressures of running gear of agricultural vehicles, especially with a view to minimising compaction at depths below 50 cm depth. Such recommendations usually specify maximum axle loads of 4-6 Mg and maximum ground contact pressures of 100-200 kPa (Rusanov, 1994; Håkansson and Petelkau, 1994). The variation in such proposals is attributable to differing soil type, soil water content and season affecting soil compactibility. Certain countries, such as Russia, Ukraine and Sweden, have already accepted the desirability of such standards (Håkansson and Medvedev, 1995).

5.4 Zero traffic

In U.S.A. terminology (Taylor, 1994), the term "controlled traffic" refers to the complete and permanent separation of the crop zone from traffic lanes, while in European terminology (e.g. Chamen et al., 1994b), the term "zero traffic" is preferred and will be used here. Machinery options for zero traffic systems include: (1) conventional tractors with extended axles up to about 3 m track width; (2) wide wheel-track gantries up to about 12 m width (Chamen et al., 1994a; Taylor, 1994; Torbert and Reeves, 1994).

Marked decreases in soil compactness and changes in many other properties occur where zero traffic systems are employed (Unger, 1994). Crop responses to zero traffic may be large, small, negligible or negative, depending on the relative soil compactness in the treatments compared and on weather conditions. Where intensive cultivation is practised, zero traffic may result in sub-optimum soil compactness. Among advantages of zero traffic are the absence of recompaction after deep cultivation (Daniel et al., 1994), improved crop uniformity, reduced nitrogen fertilizer requirement (Dickson and Ritchie, 1996b), increased uptake of fertilizer nutrients (Torbert and Reeves, 1994), increased precision of planting and pesticide applications, increased timeliness of operations, and reduced cultivation requirement and energy consumption (McPhee et al., 1995 a,b; Dickson and Ritchie, 1996a; Chamen and Longstaff, 1995). Disadvantages of zero traffic systems include the high capital cost of equipment, the loss of cropped area and the unreliability of yield increases. These inhibit the commercial adoption of zero traffic systems for general cropping, although zero traffic may be commercially acceptable for specialist crops (Chamen et al., 1994a).

6 Soil management and rehabilitation

The compactibility of a field soil is dependent partially on its inherent texture and partially on transient properties, such as structure, surface trash, organic matter content, calcium status and water content. Options are available to farmers to manage these transient soil properties to reduce compactibility at the time of traffic (Larson et al., 1994).

Soils which have been damaged by traffic-induced compaction can, within certain limits, be rehabilitated in a number of ways to improve crop production (Horn, 1994; Larson et al., 1994). Ploughing and other cultivations are widely and regularly practised on compacted topsoils but may not eliminate residual effects, such as greater surface roughness, coarser clod structure or more cohesive or smeared furrow slices, necessitating additional secondary cultivation.

Subsoil management has particular importance after deep compaction (Jayawardane and Stewart, 1994). Large areas in the former USSR have persistent subsoil compaction deeper than 80 cm (Libert, 1995). Subsoil cultivation is not universally adopted due to: (1) recompaction (natural and traffic-induced; (2) high cost; (3) special equipment not universally available; (4) high draught requirement; (5) need for subsoils to be dry to encourage maximum fracture; (6) transitory benefit, no benefit or even reductions in crop yield (Larson et al., 1994; Melvin et al., 1994). Subsoils, after loosening, are prone to rapid recompaction (Horn, 1994), with a loss of vertical macropores which are essential for drainage and aeration (Kooistra and Tovey, 1994). Careful traffic control after subsoil cultivation is essential to reduce recompaction (Daniel et al., 1994; Melvin et al., 1994).

Segmental subsoil cultivation (slotting/slitting) is intended to loosen only a part of the subsoil, leaving strips where no loosening occurs (Jayawardane and Stewart, 1994; Horn, 1994). The unloosened segments retain sufficient strength to support loads applied by traffic on the surface, while the movement of roots, water and air into the loosened slits is facilitated. In-row subsoil cultivation, a form of slit tillage, is favoured for wide-row crops (Torbert and Reeves, 1994), with no wheel traffic being allowed over or along the loosened slot, thus minimising recompaction.

Biological activity such as plant roots, fungi and soil fauna, especially earthworms, can rehabilitate high strength subsoils (Brussaard and Van Faasen, 1994; Kayombo and Lal, 1994; Whalley et al., 1995), although mechanical loosening may be needed first (Horn, 1994). With a high population of earthworms and other burrowing animals, subsoils may exhibit suitably high strength to bear transmitted loads from the surface, well developed vertical porosity and free internal drainage.

7 Economic considerations

The economic consequences to farmers of soil compaction problems include: (1) reduced income due to lower yields and poorer quality of products; (2) increased costs due to additional cultivation and fertilizer requirements; (3) costs associated with additional erosion and runoff; (4) penalties due to reduced timeliness. There are corresponding, but largely unquantified, economic consequences to society arising from associated environmental degradation. Losses of crop production attributable to compaction problems in the U.S.A. have been estimated to amount to at least 1 billion US$ per year, while in Russia the estimated losses of grain and fodder production amount to 8 and 35%, respectively.

Reductions in energy and cultivation costs affect the economics of alternative machine systems, such as zero traffic (Chamen et al., 1994b) and low ground pressure running gear systems (Tijink et al., 1995). The overall effects of compaction on the economics of crop production at farm level were studied by Eradat Oskoui et al. (1994) by quantifying the loss of earnings and increased production costs imposed by compaction on a whole-farm situation.

8 Conclusions

8.1. Crop responses to changes in soil compactness are a threat to the efficient production of high quality food, fibre, timber and other products but show complex interactions with soil type, soil wetness and nutrient availability, which have not been fully investigated.

8.2. The environmental implications of soil compaction may be sufficiently important to justify the adoption of regulatory standards for vehicles and running gear.

8.3. The use of low ground pressure systems, especially in the context of reduced axle loads, offers encouraging prospects for reducing soil compaction problems. Zero-traffic systems have also potential advantages but crop yields are often not sufficiently great or reliable to offset the higher capital costs and the loss of cropped area.

References

Alakukku, L. and Elonen, P. (1995): Long-term effects of a single compaction by heavy field traffic on yield and nitrogen uptake of annual crops. Soil Tillage Res **36**, 141-152.

Arvidsson, J. and Håkansson, I. (1996): Do effects of soil compaction persist after ploughing - results of 21 long-term field experiments in Sweden, Soil Tillage Res **39**, 175-197.

Boone, F.R. and Veen, B.W. (1994): Mechanisms of crop responses to soil compaction. In: B.D. Soane and C. van Ouwerkerk (eds.), Soil Compaction in Crop Production. Developments in Agricultural Engineering **11**, Elsevier, Amsterdam, 237-264.

Braunack, M.V. (1994): Tillage and traffic for sustainable sugarcane production. Proc. 13th Int. Conf., Int. Soil Tillage Research Organization, Aalborg, Denmark, Vol. 2, 769-775.

Brussaard, L. and Van Faassen, H.G. (1994): Effects of compaction on soil biota and soil biological processes. In: B.D. Soane and C. van Ouwerkerk (eds.), Soil Compaction in Crop Production. Developments in Agricultural Engineering **11**, Elsevier, Amsterdam, 215-235.

Campbell, D.J. (1994): Determination and use of soil bulk density in relation to soil compaction. In: B.D. Soane and C. van Ouwerkerk (eds.), Soil Compaction in Crop Production. Developments in Agricultural Engineering 11, Elsevier, Amsterdam, 113-139.

Chamen, W.C.T. and Longstaff, D.J. (1995): Traffic and tillage effects on soil conditions and crop growth on a swelling clay soil. Soil Use Manage **11**, 168-176.

Chamen, W.C.T., Dowler, D., Leede, P.R. and Longstaff, D.J. (1994a): Design, operation and performance of a gantry system: experience in arable cropping. J Agric Eng Res **59**, 45-60.

Chamen, W.C.T., Audsley, E. and Holt, J.B. (1994b): Economics of gantry- and tractor-based zero-traffic systems. In: B.D. Soane and C. van Ouwerkerk (eds.), Soil Compaction in Crop Production. Developments in Agricultural Engineering 11, Elsevier, Amsterdam, 569-595.

Daniel, H., Jarvis, R.J. and Aylmore, L.A.G. (1994): Hardpan amelioration and redevelopment in loamy sand under a zero traffic system. Proc. 13th Int. Conf., Int. Soil Tillage Research Organization, Aalborg, Denmark, Vol. 1, 61-66.

Dickson, J.W. and Ritchie, R.M. (1996a): Zero and reduced ground pressure traffic systems in an arable rotation. 1. Cultivation power requirement. Soil Tillage Res 38, 71-88.

Dickson, J.W. and Ritchie, R.M. (1996b): Zero and reduced ground pressure traffic systems in an arable rotation. 2. Soil and crop responses. Soil Tillage Res 38, 89-113.

Douglas, J.T. (1994): Responses of perennial forage crops to soil compaction. In: B.D. Soane and C. van Ouwerkerk (eds.), Soil Compaction in Crop Production. Developments in Agricultural Engineering 11, Elsevier, Amsterdam, 343-364.

Douglas, J.T., Crawford, C.E. and Clayton, H. (1994): Soil compaction control, slurry disposal and nitrous oxide fluxes on grassland for silage. Proc. 13th Int. Conf., Int. Soil Tillage Research Organization, Aalborg, Denmark, Vol. 1, 25-30.

Eradat Oskoui, K., Campbell, D.J., Soane, B.D. and McGregor, M.J. (1994): Economics of modifying conventional vehicles and running gear to minimize soil compaction. In: B.D. Soane and C. van Ouwerkerk (eds.), Soil Compaction in Crop Production. Developments in Agricultural Engineering 11, Elsevier, Amsterdam, 539-567.

Erbach, D.C. (1994): Benefits of tracked vehicles in crop production. In: B.D. Soane and C. van Ouwerkerk (eds.), Soil Compaction in Crop Production. Developments in Agricultural Engineering 11, Elsevier, Amsterdam, 501-520.

Guérif, J. (1994): Effects of compaction on soil strength parameters. In: B.D. Soane and C. van Ouwerkerk (eds.), Soil Compaction in Crop Production. Developments in Agricultural Engineering 11, Elsevier, Amsterdam, 191-214.

Gupta, S.C. and Raper, R.L. (1994): Prediction of soil compaction under vehicles. In: B.D. Soane and C. van Ouwerkerk (eds.), Soil Compaction in Crop Production. Developments in Agricultural Engineering 11, Elsevier, Amsterdam, 71-90.

Håkansson, I. (ed.) (1994): Subsoil Compaction By High Axle Load Traffic. Special Issue, Soil Tillage Res 29, 105-304.

Håkansson, I. and Reeder, R.C. (1994): Subsoil compaction by vehicles with high axle load - extent, persistence and crop response. Soil Tillage Res 29, 277-304.

Håkansson, I. and Petelkau, H. (1994): Benefits of limited axle load. In: B.D. Soane and C. van Ouwerkerk (eds.), Soil Compaction in Crop Production. Developments in Agricultural Engineering 11, Elsevier, Amsterdam, 479-499.

Håkansson, I. and Medvedev, V.M. (1995): Protection of soils from mechanical overloading by establishing limits for stresses caused by heavy vehicles. Soil Tillage Res 35, 85-97.

Håkansson, I. and Voorhees, W.B. (1997): Soil compaction. In: R. Lal, W.E.H. Blum, C. Valentin and B.A Stewart (eds.), Methods of Assessment of Soil Degradation. Advances in Soil Science, CRC Publ., Boca Raton, 167-179.

Hansen, S., Maechlum, J.E. and Bakken, L.R. (1993): N_2O and CH_4 fluxes in soil influenced by fertilization and tractor traffic. Soil Biol Biochem 25, 621-630.

Horn, R. (1994): Stress transmission and recompaction in tilled and segmentally disturbed subsoils under trafficking. In: N.S. Jayawardane and B.A. Stewart (eds.), Subsoil Management Techniques. Lewis Publ., Boca Raton, 197-210.

Horn, R. and Lebert, M. (1994): Soil compactability and compressibility. In: B.D. Soane and C. van Ouwerkerk (eds.), Soil Compaction in Crop Production. Developments in Agricultural Engineering 11, Elsevier, Amsterdam, 45-69.

Horn, R., Domżal, H., Słowińska-Jurkiewicz, A. and Van Ouwerkerk, C. (1995): Soil compaction processes and their effects on the structure of arable soils and the environment. Soil Tillage Res 35, 23-36.

Horton, R., Ankeny, M.D. and Allmaras, R.R. (1994): Effects of compaction on soil hydraulic properties. In: B.D. Soane and C. van Ouwerkerk (eds.), Soil Compaction in Crop Production. Developments in Agricultural Engineering 11, Elsevier, Amsterdam, 141-165.

Jayawardane, N.S. and Stewart, B.A. (eds.) (1994): Subsoil Management Techniques. Lewis Publ., Boca Raton, 256 pp.

Kayombo, B. and Lal, R. (1994): Responses of tropical crops to soil compaction. In: B.D. Soane and C. van Ouwerkerk (eds.), Soil Compaction in Crop Production. Developments in Agricultural Engineering **11**, Elsevier, Amsterdam, 287-316.

Kooistra, M.J. and Tovey, N.K. (1994): Effects of compaction on soil microstructure. In: B.D. Soane and C. van Ouwerkerk (eds.), Soil Compaction in Crop Production. Developments in Agricultural Engineering **11**, Elsevier, Amsterdam, 91-111.

Koolen, A.J. (1994): Mechanics of soil compaction. In: B.D. Soane and C. van Ouwerkerk (eds.), Soil Compaction in Crop Production. Developments in Agricultural Engineering **11**, Elsevier, Amsterdam, 23-44.

Kuipers, H. and Van de Zande, J.C. (1994): Quantification of traffic systems in crop production. In: B.D. Soane and C. van Ouwerkerk (eds.), Soil Compaction in Crop Production. Developments in Agricultural Engineering **11**, Elsevier, Amsterdam, 417-445.

Lal, R. (1996): Axle load and tillage effects on crop yields on a Mollic Ochraqualf in northwest Ohio. Soil Tillage Res **37**, 143-160.

Lal, R., Fausey, N.R. and Eckert, D.J. (1995): Land use and soil management effects on emissions of radiatively-active gases from two soils in Ohio. In: R. Lal, J. Kimble, E. Levine and B.A. Stewart (eds.), Soil Management and Greenhouse Effect. Advances in Soil Science, CRC Publ., Boca Raton, 41-60.

Larson, W.E., Eynard, A., Hadas, A. and Lipiec, J. (1994): Control and avoidance of soil compaction in practice. In: B.D. Soane and C. van Ouwerkerk (eds.), Soil Compaction in Crop Production. Developments in Agricultural Engineering **11**, Elsevier, Amsterdam, 597-625.

Libert, B. (1995): The Environmental Heritage of Soviet Agriculture. CAB International, Wallingford, 228 pp.

Lindstrom, M.J. and Voorhees, W.B. (1994): Responses of temperate crops in North America to soil compaction. In: B.D. Soane and C. van Ouwerkerk, eds., Soil Compaction in Crop Production. Developments in Agricultural Engineering **11**, Elsevier, Amsterdam, 265-286.

Lipiec, J. and Simota, C. (1994): Role of soil and climate factors in influencing crop responses to soil compaction in Central and Eastern Europe. In: B.D. Soane and C. van Ouwerkerk (eds.), Soil Compaction in Crop Production. Developments in Agricultural Engineering **11**, Elsevier, Amsterdam, 365-390.

McPhee, J.E., Braunack, M.V., Garside, A.L., Reid, D.J. and Hilton, D.J. (1995a): Controlled traffic for irrigated double cropping in a semi-arid tropical environment: Part 2. Tillage operations and energy use. J Agric Eng Res **60**, 183-189.

McPhee, J.E., Braunack, M.V., Garside, A.L., Reid, D.J. and Hilton, D.J. (1995b): Controlled traffic for irrigated double cropping in a semi-arid tropical environment: Part 3. Timeliness and trafficability. J Agric Eng Res **60**, 191-199.

Melvin, S.W., Erbach, D.E. and Cruse, R.M. (1994): Effect of axle load and subsoiling on maize yields on three midwestern U.S. soils. Proc. 13th Int. Conf., Int. Soil Tillage Research Organization, Aalborg, Denmark, Vol.1, 179-187.

Raghavan, G.S.V., Alvo, P. and McKyes, E. (1990): Soil compaction in agriculture: A view towards managing the problem. Adv Soil Sci **11**, 1-36.

Rusanov, V. A. (1994): USSR standards for agricultural mobile machinery: permissible influences on soils and methods to estimate contact pressure and stress at a depth of 0.5 m. Soil Tillage Res **29**, 249-252.

Soane, B.D. and Van Ouwerkerk, C. (eds.) (1994a): Soil Compaction in Crop Production, Elsevier, Amsterdam, 662 pp.

Soane, B.D. and Van Ouwerkerk, C. (1994b): Soil compaction problems in world agriculture. In: B.D. Soane and C. van Ouwerkerk (eds.), Soil Compaction in Crop Production. Developments in Agricultural Engineering **11**, Elsevier, Amsterdam, 1-26.

Soane, B.D. and Van Ouwerkerk, C. (1995): Implications of soil compaction in crop production for the quality of the environment. Soil Tillage Res **35**, 5-22.

Stępniewski, W., Gliński, J. and Ball, B.C. (1994): Effects of compaction on soil aeration properties. In: B.D. Soane and C. van Ouwerkerk (eds.), Soil Compaction in Crop Production. Developments in Agricultural Engineering **11**, Elsevier, Amsterdam, 167-189.

Taylor, J.H. (1994): Development and benefits of vehicle gantries and controlled-traffic systems: In: B.D. Soane and C. van Ouwerkerk (eds.), Soil Compaction in Crop Production. Developments in Agricultural Engineering **11**, Elsevier, Amsterdam, 521-537.

Tijink, F.G.J. (1994): Quantification of vehicle running gear. In: B.D. Soane and C. van Ouwerkerk (eds.), Soil Compaction in Crop Production. Developments in Agricultural Engineering **11**, Elsevier, Amsterdam, 391-415.

Tijink, F.G.J., Döll, H. and Vermeulen, G.D. (1995): Technical and economic feasibility of low ground pressure running gear. Soil Tillage Res **35**, 99-110.

Torbert, H.A. and Reeves, D.W. (1994): Traffic and tillage system effects on N fertilizer uptake and yield for cotton and wheat. Proc. 13th Int. Conf., Int. Soil Tillage Research Organization, Aalborg, Denmark, Vol. 2, 839-844.

Unger, P.M. (1994): Controlled traffic effects on soil density, penetration resistance, and hydraulic conductivity. Proc. 13th Int. Conf., Int. Soil Tillage Research Organization, Aalborg, Denmark, Vol. 1, 1-6.

Van den Akker, J.J.H. (1994): Prevention of subsoil compaction by tuning the wheel load to the bearing capacity of the subsoil. Proc. 13th Int. Conf., Int. Soil Tillage Research Organization, Aalborg, Denmark, Vol. 1, 537-542.

Van Ouwerkerk, C. (ed.) (1991): Soil Compaction and Plant Productivity. Special Issue, Soil Tillage Res **19**, 95-362.

Van Ouwerkerk, C. (ed.) (1995): Soil Compaction and the Environment. Special Issue, Soil Tillage Res **35**, 1-113.

Van Ouwerkerk, C. and Soane, B.D. (1994): Conclusions and recommendations for further research on soil compaction in crop production. In: B.D. Soane and C. van Ouwerkerk (eds.), Soil Compaction in Crop Production. Developments in Agricultural Engineering **11**, Elsevier, Amsterdam, 627-642.

Vermeulen, G.D. and Perdok, U.D. (1994): Benefits of low ground pressure tyre equipment. In: B.D. Soane and C. van Ouwerkerk (eds.), Soil Compaction in Crop Production. Developments in Agricultural Engineering **11**, Elsevier, Amsterdam, 447-478.

Whalley, W.R., Dumitru, E. and Dexter, A.R. (1995): Biological effects of soil compaction. Soil Tillage Res **35**, 53-68.

Wronski, E.B. and Murphy, G. (1994): Responses of forest crops to soil compaction. In: B.D. Soane and C. van Ouwerkerk (eds.), Soil Compaction in Crop Production. Developments in Agricultural Engineering **11**, Elsevier, Amsterdam, 317-342.

Yaalon, D.H. (1996): Soil science in transition: Soil awareness and soil care research strategies. Soil Sci **161**, 3-8.

Addresses of authors:
B.D. Soane
Formerly: Scottish Centre of Agricultural Engineering
SAC, Bush Estate
Penicuik, EH26 0PH, UK
C. van Ouwerkerk[†]
Formerly: Research Institute for Agrobiology and Soil Fertility (AB-DLO)
P.O. Box 129
9750 AC Haren Gn, Netherlands

Assessment, Prevention and Rehabilitation of Soil Degradation Caused by Compaction and Surface Sealing

R. Horn

Summary

Soil strength depends on internal parameters and kind, intensity and number of loading events. The more aggregated soils are, the stronger they are, but exceeding the precompression stress value as a material function of each soil horizon results in soil deformation, either by divergency (volumetric strain) or by shear processes. In both cases soil strength can be either increased due to compaction or reduced due to kneading. However, each stress applied will always be transmitted 3-dimensionally and results in the creation of major or minor principle stresses and octahedral shear stress as well as mean normal stresses. If the applied soil stress cannot be attenuated by internal soil strength, further deformation by volumetric strain occurs which again differs depending on soil aggregation, water content, humus content and type of humic substances. In very weak soil, too intensive soil homogenization by ploughing and chiseling also leads to slaking and hard-setting processes. As a consequence, root penetration, gas exchange and water infiltration are hindered. Overcompacted soils can be rehabilitated by deep loosening or partial ploughing if the pore water pressure is small enough (pF > 3) in the deeper soil horizons. Nevertheless tillage operations have to be reduced or even neglected for the next 3 - 5 years and all consecutive mechanical stressing of the rehabilitated site must be adjusted to the new and much lower soil strength. Soil compaction results in a further reduction of the water infiltration rate and in more pronounced water erosion in sloping areas and the formation of deep gullies starting from the top soil. The more intensive soils are homogenized during, e.g., tillage processes, or in combination with increasing slope angle, particle immersion and a reduced shear strength inbetween the single particles or aggregate fragments results in a more intensive gully formation up to the depth of the plough pan layer. If the capillary rise is prevented or reduced due to a too small hydraulic conductivity, and/or if the shear strength inbetween single particles is too small compared to the dragging forces, soil will be detached by wind erosion even in slopeless areas.

Keywords: Soil strength, strain, stress, shear parameters, water and wind erosion, soil sealing, soil rehabilitation, sustainable agriculture

1 Introduction

It is well known in the literature that soils as three phase systems undergo intensive alteration in their physical, chemical, and biological properties both during natural soil development as well as during anthropogenic impact processes such as soil ploughing, sealing, soil erosion by wind and water, soil amelioration, soil material excavation and refilling of devastared land. In agriculture soil compaction, as well as soil erosion by wind and water, are classified as the most harmful processes which do not only end in a reduction of site productivity but are also responsible for groundwater pollution, gas emission and higher requirement of energy input in order to gain a comparable crop yield.

These interrelationships have very often been described (Soane and Ouverkerk 1994) and quantified (Hakansson et al., 1988). In Germany up to 40 % of yield decline has been reported by Werner et al. (1992) because of too intensive wheeling and there are several indications that in total about 33 Mha of arable land are already completely devastated by soil compaction only in Europe (Oldeman 1992). Furthermore several papers report more pronounced soil erosion by wind and water because the seedbed preparation results in a very weak soil „bed" with hardly no internal soil strength, and which can also be more easily transported both by wind or water. In addition, Boone and Veen (1994), Horn et al. (1984), Lipiec and Simota (1994), Stepniewski et al. (1994) argue that due to a complete homogenisation of the seedbed pore system, gas, water and heat transport into and out of the soil are also prevented.

In addition, the hardsetting and soil crust formation because of slaking must be considered in combination with stress-affected reduced soil strength, e.g. because of tillage effects and/or because of less effective chemical, biological and physical boundings.

Consequently, a closer look is required at all these processes and reactions in order to
- understand and predict the formation of a site and land use specific soil strength, stress distribution and stress attenuation,
- reduce soil erosion by smaller mechanical loading and applied dynamic forces,
- understand physical processes on macro- and microscales such as particle arrangement and formation of aggregates as well as the corresponding functioning of pore systems.

Furthermore, the possibilities to improve soil structure and to affect soil physical and chemical properties by tillage operations and natural processes must be considered in order to deal adequately with the environmental and ecological problems.

2 Mechanical aspects

2.1 Soil compaction processes and soil strength

Soil strength can be quantified by stress strain measurements defined by the precompression stress value (Horn 1988). Based on soil mechanics theory, this value defines the stress range with complete elastic deformation behaviour, while beyond this strength value mainly plastic soil deformation occurs in the virgin compression line (i.e. the stress range above the precompression stress).

The precompression stress value is no material constant, but defined in dependence of the actual pore water pressure and remaining pore continuity and the actual structure (defined as aggregation) for given internal parameters. In comparison with data for the completely homogenized samples (refilled and dried with the same pore water pressure value at the given bulk density), the effect of soil aggregation on strength can also be easily quantified.

The strength values differ for various soil types, and they depend on texture, structure, pore water pressure, organic matter and bulk density. At a given soil texture, soil aggregate formation always results in a strength increase compared to the coherent one, but at a given aggregate type, increasing clay content results at first in a strength increase and only after exceeding approx. 40% clay, the strength values get smaller because of water content or pore water pressure effects (Horn 1981). Furthermore, this method and/or these results can also be used to detect anthropogenic processes such as plough pan formation or the determination of geological processes during glacial times. This method is also applicable to quantify long-term tillage effects on soil strength as well as used to quantify the effect of pore water pressure on soil strength. In Fig 1. an example is given about soil type dependent strength values.

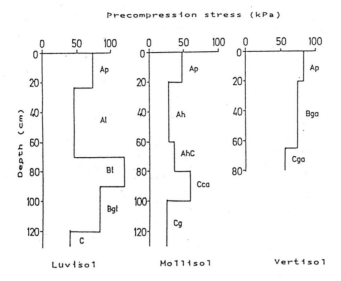

Fig.1: Precompression stress values in different horizons of 3 soil types at a pore water pressure value of -6kPa (from Horn 1988)

The effect of clay migration on strength decline in the Al horizon and strength increase due to aggregate formation in the Bt horizon of the Luvisol derived from loess can be detected as well as the strength increase due to Ca precipitation in the corresponding horizon of the Mollisol. In all 3 soil types the parent material is always weakest. Anthropogenic processes such as the yearly ploughing procedure creates a very strong plough pan layer with precompression stress values similar to the ground contact pressure of tractor tires which are preserved even for decades irrespective of the changes in soil management. It could be shown that even after more than 30 years of no tillage, the plow pan layer can still exist (Horn, 1986).

2.2 Effect of kind of loading on soil strength and shear processes

Each soil deformation can either be induced by divergency processes such as expansion or soil compaction and by shear processes. Both processes are time dependent and react according to the effective stress equation (Terzaghi and Jelinek, 1954) via mechanical and hydraulic properties. During short-term loading, clay soils for example with low hydraulic conductivity can be stronger than sandy or well-structured soils having the same bulk density, load and pore water pressure because of the incompressibility of water at given hydraulic properties. In sandy soils the settling process is dominated by timeless settlement while in clay soils the precompression stress can be even doubled, if a static compression is applied only for a short time (less than a second) compared to long-lasting compression. In homogenized and unsaturated clay or loamy soils, the positive pore water pressure is created during soil compaction and soil shear and may induce a short-term strength increase. Such higher strength, however, does not help at all if long-term stability of soils is being discussed and if dynamic loading effects are considered.

The proportion of soil deformation by compaction and/or shearing depends on the internal strength as well as on the applied stress and kind of stresses. Repeated wheeling or loading and unloading leads not only to a destruction of existing structure and the creation of platy structure at the transition from the overconsolidated to the less intensive deformed soil, but it also results in an altered proportion of elastic to plastic deformation, and in completely altered ecological properties.

In addition it has to be pointed out that during static loading the rearrangement of particles may occur primarily in the virgin compression load range, but the single aggregates in the soil volume may persist if their internal strength exceeds the overall applied external load. Consequently each external load always diminishes at first the interaggregate pore system while the aggregate shape and strength and their physical/chemical properties persist. During dynamic (shear dependent) processes, however, not only the inter-aggregate pore system, but also the intra-aggregate pores and the total aggregates can be destroyed or rearranged as soon as the corresponding partial strength values are smaller compared to the shearing forces. It is well known that the shear parameters of single aggregates exceed by far those values of the structured bulk soil and only after exceeding this internal strength, the data for the different components of the bulk soil become the same as for the homogenized material (Baumgartl, 1991).

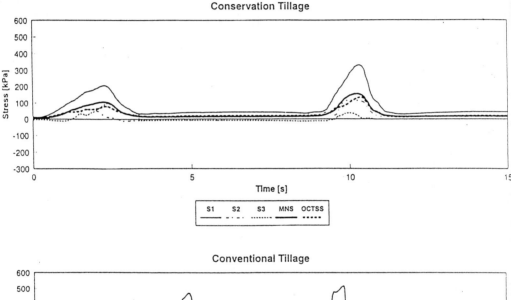

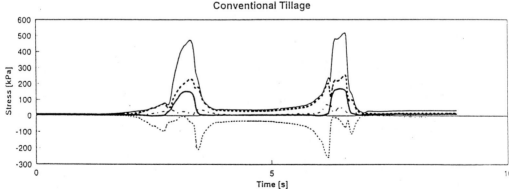

Fig. 2: Effect of wheeling on stress distribution in the A-horizon of argillic cambisol derived from loess under conservation and conventional tillage.

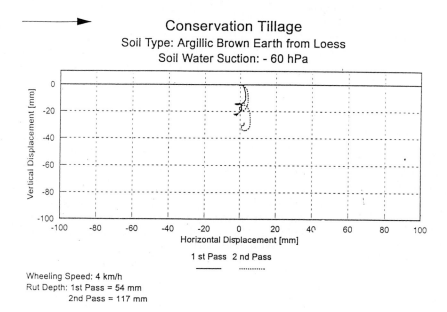

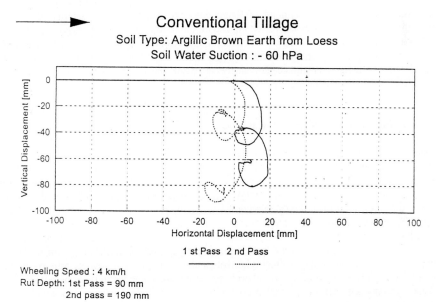

Fig. 3: 2-dimensional soil deformation during wheeling at a depth of 10 cm for conventional and conservation tillage treatments

2.3 Stress distribution in soils

Any load applied to the soil is transmitted in the soil 3-dimensionally via the solid, liquid and gaseous phase. Soil stress determination requires stress state transducer (SST) systems (Horn et al., 1992) if the effective stress equation as well as the stress components are to be quantified. At a given tecture, stresses will be transmitted to deeper depths if th soil is
- wet / moist
- less aggregated
- less dense

and if the contact area at the same contact area pressure ist larger. The bigger the load at given properties the deeper the stress propagation. The pattern of the stress propagation equipotential line differs depending on soil strength, kind of loading, contact area and contact area pressure and when an estimated concentration factor is used. The stronger the soil is when a given external stress is applied, the smaller the value of this factor. It varies from 2 - 3 in soils with high strength and up to 9 in weak, wet and loose soils. If the normal stress in soils has been determined as a function of the number of wheeling events, divergency and shear processes result in a strength increase at first in the top soil. During the following loading events, the stronger top soil horizon attenuates the stress applied and spread it more laterally too. The remaining vertical stress component deforms the deeper and still weak soil horizons even if the movement of the stronger top soil is reduced to the elastic component. Thus, additional soil loadings induce an increase in the precompression stress values of the deeper soil horizons. Owing to progressive stress attenuation this effect fades out at greater depths.

How far various tillage systems can cause altered stress distribution patterns (e.g. at a given depth) is shown in Fig 2. At a given depth of 30 cm, wheeling always results in internal soil stress, but the kind and intensity of stresses created depend on the farming system. For argillic cambisols derived from loess under long-lasting conservation tillage treatment (thirty years), only very small major principal stresses and small octahedral shear stress and mean normal stress values are created. However conventional tillage plots at the same site show very pronounced stress peaks under the front and rear tyre of a 10 Mg tractor.

2.4 Stress/strain processes in agricultural soils

With respect to changes in soil function, not only 3-D stresses, but also volumetric strain has to be determined by a special Displacement Transducer System (DTS) because both stress and strain tensors can only be quantified if 3-D sensor systems are placed in the soil volume. The sensor technique is described by Kühner et al. (1994). If only the 2-dimensional deformation during the passage of a tractor (axle load: 5 Mg) is recorded, it can be seen that under conservation tillage only a slight vertical deformation occurs of max. 2 cm during the passage of a tractor which, in addition, has only a slight horizontal displacement. The same tractor creates a very pronounced vertical and horizontal displacement of up to 8 cm (Fig. 3). In addition it must be pointed out that plastic as well as elastic deformation creates horizontal cracks and, in combination with excess soil water which cannot be drained off in very short times, also a kneading due to shearing occurs.

These consequences can be derived from fig. 4, where the effect of repeated wheeling on changes in the ratio of the stress components at 30 cm depth at a pore water pressure of - 30 kPa is shown in a Luvisol derived from loess. Repeated wheeling during a single day at constant water content induces a relative increase in the vertical principal stress S1, compared with the 2 horizontal stress components S2 and S3. Thus, the intensive stress concentration in the vertical direction resembles an increased concentration factor value which again points to weaker soil. Consequently mechanical soil loading can result either in a more intensive soil compaction, which coincides with an intensive increase in precompression stress value up to depths of > 80 cm by using normal agricultural machinery, but it can also cause a complete homogenization due to dynamic loading, especially under moist or wet conditions. The latter process results in very small precompression stress values and the formation of soil samples which show normal shrinkage behaviour. In addition, these two different soil deformation processes lead to bulk density values which cannot be compared at all and which cannot be used for any prediction of ecological parameters.

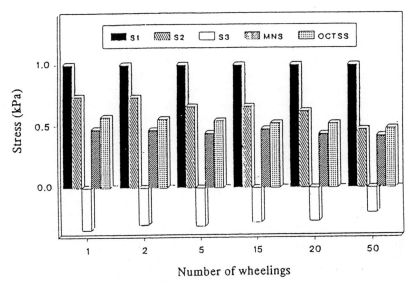

Fig. 4: *Effect of repeated wheeling on changes in the ratio of the stress components at 30 cm depth at a given pore water pressure of -30 kPa (after Semmel, 1993)*

3 Effect of compaction or wheeling on soil sealing

According to Mullins and Ley (1995) soil sealing can be created by hard setting due to a structure breakdown of weakened soil aggregates during wetting and then setting to a hard structured mass during drying. It can be further induced by slaking as a consequence of fragmentation that occurs when aggregates which are predamaged by tillage or shearing, are suddenly immersed in/or placed in contact with water. Such slaking process occurs especially if aggregates are not strong enough to withstand these stresses resulting from rapid water uptake. These stresses are produced by differential swelling (Emerson, 1977) and embedded air, the rapid release of heat, wetting, and the mechanical action of moving water. It is also affected by antecedent pore water pressure, rate of wetting, the concentration of organic matter and clay mineralogy. Both hard setting and slaking are induced by a too intensive seed bed preparation which in itself results in an intensive decreasing biological activity, rapid decline in organic material and chemical leaching because of an improved accessibility of particle surfaces for percolating water. Soil strength due to biological activity is also reduced because increasing tillage intensity reduces earth worm abundance, number of earth worm channels, and net consumption of organic matter, etc. In combination this results in a lower site productivity, higher susceptibility for compressibility during moist conditions, more pronounced water as well as wind erosion. If the chemical aspects: filtering and buffering are also considered, both are reduced resulting in a more pronounced pollution of surface water. (So et al., 1995)

4 Rehabilitation of soil degradation

The variety of possible methods to rehabilitate compacted soils or single soil layers is site and use dependent and can either result in the need to completely homogenize soils up to depths of more than 1 m or a partial reloosening by slit ploughing (Reich et al., 1985), slotting (Blackwell et al., 1989), various kinds of deep loosening (Schulte-Karring, 1988), or under extreme conditions by dynamite explosion or air pressure application. The various techniques, however, require a very intensive predetermination of internal soil strength and the judgement of subsequent land use. Irrespective of these variations in techniques, soil loosening always results in a steep decline in internal soil strength which leads to a higher susceptibility for further soil compaction. If after such loosening process soil treatment is continued as before, even worse ecological properties must be anticipated. Horn (1994) and Horn et al. (1997) have defined those changes induced by variations in deep loosening techniques in detail.

With respect to biological processes for soil reloosening no effects during the short run can be detected. The effectivity of biological processes has to be based on primarily mechanical reloosening techniques and a reduced application of machinery after soil reloosening processes in order to support reaggregation by physical and biological processes, gaining strength increase. Physical as well as mechanical properties also clearly underline the necessity to consider the various components of internal soil strength and to derive from those results also the time required for soil rehabilitation. For the Browncoal mining area in Germany (Rheinbraun), even after more than 30 years no soil strength recovery occurred, but in case of a non site-specific soil refilling and levelling system, a very pronounced and irreversible soil degradation could be detected (Lebert and Horn, 1995; Jayawardane and Stewart, 1994).

5 Effect of soil deformation on water and wind erosion

5.1 Water erosion

As soon as the infiltrability for water is less than the rainfall intensity, surface water runoff will occur in sloping areas. How far divergency and shear processes can affect water erosion is shown in Fig. 5. Both processes alter pore systems and the pore functions including their tortuosity.

If soil strength is exceeded by external loading, destruction of the inter-aggregate pore system occurs as well as a more pronounced homogenization of the pores with respect to quantity and diameter. During the first step of soil compaction very dense and less conductive intra-aggregate pores define the hydraulic site properties, while the inter-aggregate pore system is destroyed. The very dense platy structure results in a drastic decline of the hydraulic conductivity (up to several orders of magnitude) which also leads to more-pronounced surface water runoff. Erosion starts at the soil surface and creates even deeper gullies if shear strength inbetween the mineral particles is smaller than the velocity-dependent hydraulic stress of the running water.

If during stress application both the inter-aggregate pore system and the aggregates themselves are destroyed, the hydraulic conductivity is even more reduced, because homogenous soil samples are more susceptible to soil compaction. Consequently, pronounced surface water runoff occurs and forms gullies at the soil surface. For addition, subsurface horizontal water and soil flux must be considered as consequence of divergency processes.

Shear processes during seed bed preparation result in a strength decline of the complete A horizon and in isotropic but very tortuouse pore systems. Water infiltration is reduced, and the strength decline due to swelling increased, which finally leads to particle immersion and to a lateral soil volume movement on top of the plough pan layer in sloping areas. Soil gully formation therefore coincides with the depth of tillage operations during seed preparation.

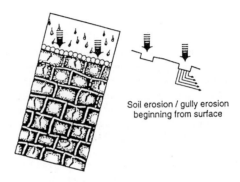

Fig. 5: Soil deformation processes (divergency/compaction; shearing) as causes for water erosion

5.2 Wind erosion

As soon as the top soil is very weak with a very small shear strength between single particles, and the water saturation is low so that the pore water pressure does not strengthen the soil via the waterfilled pores, each wind velocity creates a corresponding particle mobility, especially if the soil surface is flat and the threshhold gradient is rather small. Consequently, too-intensive tillage preparation, a bare soil surface, and an uneven distribution of water and gas-convecting pores result in a more-pronounced soil loss by wind erosion. The measures to prevent soil loss by wind erosion are well-defined and described in several textbooks (Morgan 1996). However, the effects of soil texture and structure on pore systems and pore functions with respect to wind erosion are not fully understood, especially if the effect of surface roughness on the threshhold gradient is also included. How far the shear resistance between single aggregates and in the total soil or the arrangement of particles in the top soil can affect this threshhold gradient is completely unknown, as well as the influence of various kinds of tillage treatments on kind of wind profiles and turbulence effects. Several equations exist on wind erosion processes, but the effect of water menisci forces on soil strength and the effect of wheeling and compaction on reduced or increased susceptability to wind erosion is not finally formulated, and no real mechanical relationships are defined on this. Instead of regression analyses, more physically-based research is required in order to predict the effect of soil deformation on erodibility by wind.

6 Conclusions

The determination of soil strength and the ecological consequences of such processes require mechanical and physical methods. The determination of stress/strain behaviour results in precompression stress values as a measure for internal soil strength. Soil deformation after exceeding the precompression stress value is induced by divergency and shear processes but depend on internal as well as external parameters. Soil compaction can be defined by reduction of inter-aggregate pore systems, while single aggregates remain stable or at highest soil stresses by a complete homogenization and deterioration of the total structure system. Stresses will always be transmitted to deeper depths, but the pattern of the equipotential lines depends on internal as well as external parameters. Short term strength increase can be explained by time-dependent settlement processes which again are induced by hydraulic time-dependent flux processes or movement of particles. Soil deformation results in an increased susceptability to water and wind erosion as well as soil slaking and hardsetting. All processes always result in worse ecological soil properties. Soil rehabilitation requires the application of mechanical energy, followed by biological processes, but the applicability of consecutive tillage techniques have to be adjusted to the reduced soil strength in order not to degrade soil more intensely. Soil rehabilitation requires several decades to centuries as it can also be derived from former soil genetic descriptions and developments. Soil deformation results in an increased susceptibility to water and wind erosion, but the erosion patterns differ depending on the hydraulic properties of the corresponding soil horizons. The prevented water infiltration in the top soil leads to lateral water movements and to corresponding shear processes starting from the top soil. If the total top soil is homogenized, an additional load reduction due to immersion has to be considered, resulting in a very pronounced deep gully erosion up to the deeper plough pan layer. Wind erosion depends on shear forces between single particles or in the bulk soil and it is affected by the kind of pore system. At sites where the capillary rise is reduced due to a low hydraulic conductivity, the top soil dries out and becomes more susceptible to wind erosion than at sites with a high capillary rise of water. In the latter case due to the formation of water menisci forces, the shear resistance between particles is increased and results in less-intensive soil loss by wind erosion. Nevertheless, further research is necessary to deal with the mechanical aspects of this problem.

References:

Baumgartl, T. (1991): Spannungsverteilung in unterschiedlich texturierten Böden und ihre Bedeutung für die Bodenstabilität. PHD Thesis Schriftenreihe Institut für Pflanzenern. u. Bodenkunde CAU Kiel., 128 pp.

Blackwell, J., Horn, R., Jayawardane, N., White, R. and Blackwell, P.S. (1989): Vertical stress distribution under tractor wheeling in a partially deep loosened typic Palesutalf. Soil Till. Res. **13**, 1-12

Boone, F.R. and Veen, B.W. (1994): Mechanisms of crop response to soil compaction. 237-264. In: Soane, B. und C. van Ouwerkerk, 1994: Soil compaction in crop production. Development in Agricultural Engineering **11**, Elsevier, Amsterdam, 167-189

Emerson, W.W. (1977): Physical properties and soil structure. In: J.S. Russell and E. L. Greacen (eds.): Soil factors in crop production in a semiarid environment. Univ. of Queensland Press, Brisbane, 78-104.

Hakansson, I., Voorhees, W.B. and Riley, H. (1988): Vehicle and wheel factors influencing soil compaction and crop response in different traffic regimes. Soil Tillage Res. **11**, 239-282

Horn, R., (1981): Die Bedeutung der Aggregierung von Böden für die mechanische Belastbarkeit in dem für Tritt relevanten Auflastbereich. Schriftenreihe des FB 14, TU Berlin, Heft 10, 200 S.

Horn, R. (1986): Auswirkungen unterschiedlicher Bodenbearbeitung auf die mechanische Belastbarkeit von Ackerböden. Z. Pflanzenern. Bodenk. **149**, 9-18

Horn, R. (1988): Compressibility of arable land. In: J. Drescher, R. Horn and M. de Boodt (eds.): Impact of water and external forces on soil structure. CATENA SUPPLEMENT **11**, 53-71

Horn, R.(1994): Stress transmission and recompaction in tilled and segmently disturbed soils under trafficking. Advances in Soil Science, Hersg.: N. Jayawardane und R. Stewart, Lewis Publ., 53-87.

Horn, R., Semmel, H., Schafer, R., Johnson, C. and Lebert, M. (1992): A stress state transducer for pressure transmission measurements in structured unsaturated soils. Pflanzenern. u. Bodenkde. **156**, 269-274

Horn, R., Taubner, H., Wuttke, M. und Baumgartl, T. (1994): Soil physical properties in processes related to soil structure. Soil Tillage Res. **30**, 187-216

Horn, R., Kretschmer, H., Baumgartl, T., Bohne, K. and Neupert, A. (1997): Soil mechanical properties of a partly relosened (slip ploughed system) and a conventionally tilled overconsolidated gleyic Luvisol derived from glacial till. International Agrophysics (in press).

Jayawardane, N. and Stewart, E. (1994): Advances in Soil Science. Lewis Publishers, 250 pp.

Kühner, S., Baumgartl, T., Gräsle, W., Way, T., Raper, R. und Horn, R. (1994): Three dimensional stress and strain distribution in a loamy sand due to wheeling with different slip. Proc.13th Int. ISTRO Conf. Aalborg, 1994, 591-597

Lebert, M. and Horn, R. (1995): Bodenmechanische Aspekte bei der Rekultivierung von Braunkohletagebaugebieten. Report No. 4, Eigenverlag Firma Rheinbraun, Köln, 132 pp.

Lipiec, J. and Simota, C. (1994): Role of soil and climate factors influencing crop responses to compaction in Central and Eastern Europe. In: Soane and v. Ouwerkerk (eds.) (1994): Soil compaction in Crop production. Developments in Agricultural Engineering **11**, Elsevier, Amsterdam, 365-390

Morgan, R. P. C. (1996): Soil Erosion and Conservation, Longman Publ., 198 pp

Mullins, C.E. and Lei, G.J. (1995): Mechanisms and characterization of hard setting soils. 157 - 176. In: H.B. So, G.D. Smith, S. R. Raine, B.M. Schöfer and R.J. Loch (eds.): Sealing, crusting and hard setting soils: productivity and conservation. ASSI Publisher.

Oldeman, L.R. (1992): Global extent of soil degradation. Proc. Symp. Soil Resilience and Sustainable Landuse, Budapest, Hungary

Reich, J., Unger, H., Streitenberger, H., Mäusezahl, C., Nussbaum, S. and Stewart, P. (1985): Verfahren und Vorrichtung zur Verbesserung verdichteter Unterböden. EB DDR Nr. 233915, also described in: Reich, J., Streitenberger, H. and Romanesko, N. (1991): Agrartechnik **14**, 57-62

Schulte-Karring, H. (1988): 150 Jahre Technik der Tieflockerung. Eigenverlag Landeslehr- und Versuchsanstalt f. Landwirtschaft, Bad Neuenahr Ahrweiler

Semmel, H. (1993): Auswirkungen kontrollierter Bodenbelastungen auf das Druckfortpflanzungsverhalten und physikalisch-mechanische Kenngrößen von Ackerböden, PHD CAU Kiel, in: Schriftenreihe des Instituts für Pflanzenernährung u. Bodenkunde, Heft 26, 183 pp.

So, B., Smith, G.D., Raine, S.R., Schöfer, B.M. and Loch, R.J. (eds.) (1995): Sealing, crusting and hard setting soils: productivity and conservation. ASSI Publisher. 527 pp.

Soane, B. und van Ouwerkerk, C. (1994): Soil compaction in crop production. Development in Agricultural Engineering **11**, Elsevier, Amsterdam, 662 pp.

Stepniewski, W., Glinski, J. and Ball, B.C. (1994): Effects of soil compaction on soil aeration properties. In: Soane, B. und C. van Ouwerkerk, 1994: Soil compaction in crop production. Development in Agricultural Engineering **11**, Elsevier, Amsterdam, 167-189

Terzaghi, K. and Jelinek, P. (1954): Theoretische Bodenmechanik, Springer Verlag, Berlin

Werner, D., Roth, D., Reich, J., Mäusezahl, C., Pittelkow, U., Steinert, P. (1992): Verfahren der Unterbodengefügemelioration mit dem Schachtpflug B 206 A. Publisher: Landwirtschaftliche Untersuchungs- und Forschungsanstalt Thüringen, 92 pages

Address of author:
Rainer Horn
Institute for Plant Nutrition and Soil Science
Olshausenstraße 40
D-24118 Kiel, Germany

The Impact of Soil Seals and Crusts on Soil Water Balance and Runoff and their Relationship with Land Management

C.J. Chartres & G.W. Greeves

Summary

The term 'soil seal' generally refers to a surface layer of soil with significantly reduced porosity and permeability resulting from rapid wetting of dry soil, raindrop impact, deposition of fine soil material, chemical dispersion, or to some combination of these processes. Subsequent drying of the soil can result in a layer with significantly increased strength that is commonly referred to as a soil crust. Seals or crusts can limit plant production by restricting infiltration of rainfall or irrigation water, by physically restricting seedling emergence and early root growth, or by restricting the gas exchange necessary for soil aeration. Soil seals that decrease infiltration rate will result in increased potential for surface runoff and for water erosion and other associated adverse off-site impacts. Prediction of the impacts of seals or crusts on the soil water balance is necessary for guiding improved land management.

Soil surface hydraulic properties, surface roughness and depressional storage are critical factors in determining potential infiltration and run-off. Each of these can be significantly affected by surface sealing processes, but incorporating adequate description of their effects in water balance models can be problematic. Particular attention is given to modeling of the impact of sealing and crusting processes on soil hydraulic properties.

Recent studies on measurement and prediction of the impact of cultivation and other land management practices in south-eastern Australia on the soil water balance are summarised. The results demonstrate that long term cultivation management practices have diminished organic carbon contents of topsoils and increased susceptibility to dispersion. However, a range of surface and subsurface properties including horizon depth, surface and subsurface hydraulic conductivity and macroporosity need to be taken into account as well as crusting if runoff is to be predicted. The practical significance of potential reductions in infiltration and increases in runoff due to declining surface hydraulic conductance during surface sealing under rainfall is illustrated for cultivated soil from Cowra in New South Wales, Australia using the SWIM model (Soil Water Infiltration and Movement) together with published soil and climatic data.

Keywords: Predictive modelling of soil water, crusting, sealing, erosion, soil hydraulic properties.

1 Introduction

The aim of this paper is to indicate how cultivation management practices influence seal and crust development and thus the soil water balance including the generation of runoff. Whilst the development of soil seals and crusts may be of major importance in the generation of runoff, the impact of other land management practices on soil properties and the nature and properties of soil horizons themselves may also be of considerable significance in runoff generation. The paper advocates a simple modeling approach, for use in data-sparse environments where empirical experimental data may be unavailable, to estimate potential runoff risk given different management/soil property combinations. Some examples of the impact of sealing and crusting on the water balance of soils and erosion risk are also given.

Soil seals and crusts are but one form of degraded soil structure. However, because they occur at the soil/atmosphere interface they impact significantly on water and gaseous exchange between the two media and on plant germination and development. Consequently, they are probably the most significant form of soil structural degradation in terms of both lost crop and pasture yields and environmental degradation impacts both on and off-site. The term 'soil seal' generally refers to a surface layer of soil with significantly reduced porosity and permeability resulting from rapid wetting of dry soil, raindrop impact, deposition of fine soil material, chemical dispersion or some combination of these processes. Subsequent drying of the soil can result in a layer with significantly increased strength that is commonly referred to as a soil crust.

The following sections describe the nature of crusts and seals and their impact on soil hydrology, relate different cultivation management practices to soil structural stability and then demonstrate the role of models to describe how soils respond to rainfall with and without sealing and crusting.

2 Crust types and genesis

Sumner (1995) defined four types of crusts that occur in various locations worldwide; these include:
a) chemical crusts - composed of precipitated salts and commonly found in arid environments;
b) structural crusts - caused by raindrop impact on a wide range of soil types;
c) depositional or sedimentary crusts - formed by the transport and deposition of suspended material; and
d) cryptogamic crusts - formed by the development of mosses, liverworts, lichens and algae on natural and degraded soil surfaces.

To these should be added erosional crusts as defined by Valentin and Bresson (1992).

Soil micromorphological studies have been instrumental in developing our understanding of both the properties of crusts and seals and in their classification as is shown in Table 1. This table indicates a further differentiation of crust types based on descriptive and inferred genetic properties. As noted by Sumner (1995) structural crusts are probably of most widespread occurrence and significance, although as shown in Table 1, they can form as a result of a number of physical processes which occur following rain impact. The general process of structural crust formation involves a number of related and independent processes. As a raindrop hits the soil surface, Moss (1991) demonstrated that it initiates a "seismic" wave that compacts the grains due to the collapse of pores. This process in silty soils can result in extremely dense packed surface layers in its own right. In other materials the combination of the energy input from raindrops and the extremely low electrical conductivity of the rainwater itself leads to slaking, collapse of soil aggregates and, in some cases, dispersion. Dispersed fine particles may then migrate either upwards or downwards leaving variably sorted layers of material, which give the crust a characteristic laminar appearance. Generally, under cultivated conditions, crusting processes lead to a reduction in relief of the microtopography and thus a concomitant reduction in the ability of the surface's water detention capacity. This process has been well described using microtopo-graphers and laser scanners by several authors including Huang and Bradford (1992). Thus as time goes on during both single events and through a season, the combination of seal and crust formation and reduced microtopography lead to a greater potential for runoff to occur. In most systems it is usual to find depositional crusts in topographic lows associated with structural and/or erosional crusts on the topographic highs. It is to be stressed that sealing and crusting is a dynamic process which is highly variable through time and space.

CRUST TYPE	SUBCLASS	DIAGNOSTIC MICROMORPHOLOGICAL FEATURES
Structural crusts	Slaking	Reduced aggregate size, increased microporosity, no textural separation of skeleton and plasma
	Swelling	Banded skeleton grains within superficial parts of clods
	Infilling	Textural separation, net-like infillings of silt grains
	Coalescing	Porous, coalescence of aggregates decreasing with depth
	Sieving	Surface skeleton grains overlying translocated clay
Rain impact crusts		Compacted silt layers
Erosional crusts		Poorly oriented fine particles, absence of relationship between layer thickness and surface microtopography
Depositional crusts	Run-off	Silty surface seals, loose aggregates
	Ponding	Clay and silt laminae
Cryptogamic crusts		Mosses, lichens, liverworts and algae, subsurface hyphae

Tab. 1: Crust types and diagnostic features after Chartres et al. (1994).

Rain impact seals or crusts have been found to significantly reduce water infiltration in many field studies (e.g. Freebairn and Gupta, 1990; Loch and Foley, 1994) with steady infiltration rates being reduced to 20 mm h^{-1} or below for many soils and to less than 10 mm h^{-1} for some. Laboratory rainfall simulations where hydraulic gradients across the surface are measured or controlled (e.g. McIntyre, 1958; Bosch and Onstad, 1988) can be used to assess surface hydraulic conductivity decline associated with sealing. McIntyre reported a reduction by a factor of 2000 in saturated hydraulic conductivity while Bosch and Onstad reported final conductivities ranging from 13.7 mm h^{-1} down to 0.7 mm h^{-1} due to raindrop impact crusting. Whilst these reported seal conductivities are sensitive to the assumed seal thickness, their magnitudes are sufficiently low to generate run-off even under relatively low intensity rainfall.

Crusts formed by other mechanisms can differ in surface permeability from the values quoted above for rain impact crusts. Crust morphology can be related to the dominant processes of crust formation and Casenave and Valentin (1992) have used this in their quantitative runoff capability classification system for semi-arid areas in West Africa.

Soil management practices clearly influence the infiltration rates of A horizons from a range of soils as indicated in Figure 1 which relates to soils with texture contrast characteristics (Alfisols) in south-eastern Australia (Geeves et al., 1995a). The data presented in this study encompassed both crusting and non-crusting sites. Part of the aim of the current investigation is to determine the impact that crusting has on the soils when different cultivation management practices are used. The soils in question all have potentially dispersive A horizons (Figure 2) as demonstrated using the method of Rengasamy et al. (1984), in which the sodium adsorption ratio (SAR) is plotted against total cation concentration. These data suggest that even when soils have low SAR or exchangeable sodium percentages, they may be potentially dispersive under low total soluble cation concentrations (low electrolyte concentration of dissolved salts). Such conditions prevail in many surface horizons in south-eastern Australia. This dispersibility is accentuated by low organic carbon contents which further limit the development of stable aggregates. As can be seen in Fig. 3, organic carbon contents are 2% or less in all but the woodland and low and medium intensity grazing sites. Thus the maintenance of stable aggregate structures is of critical concern to management from the point of view of maintaining aggregate stability and reducing seal and crust development.

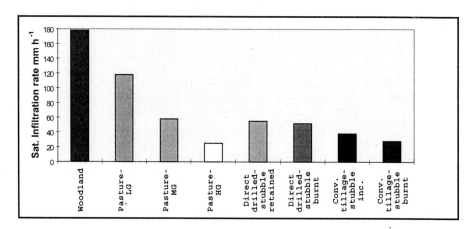

Figure 1. The impact of different management practices on saturated infiltration rates in soils from south-eastern Australia (after Geeves et al., 1995a). LG, MG and HG represent light, medium and heavy grazing respectively.

Whilst studies of the surface structural properties of soils in south-eastern Australia have demonstrated the propensity of most soils to disperse and to deteriorate structurally under cereal cultivation, also of concern is our ability to predict both potential run-off losses and deep drainage from these soils. These factors are of major concern with respect to the key land degradation processes of erosion, salinisation and acidification. Consequently, we have embarked on a number of model based predictive studies, which are discussed in the following sections.

3 Modeling runoff generation and the impact of crusts

There is considerable literature on modeling approaches to determining run-off and soil erosion. These approaches are usually either event-based, or based on continuous simulation. Both approaches have advantages and disadvantages with respect to data requirements and their flexibility for use under different environments from those for which they were developed.

Approaches to quantitatively describing infiltration have generally been empirical models, such as the equations of Kostiakov (1932) and Horton (1940) and the SCS curve number method (McCuen, 1982); models based on approximations of the Richards' equation (Marshall and Holmes, 1979) such as the model of Green and Ampt (1911) and the equation of Philip (1957) or models based on solution of the Richards' equation (e.g. Ross, 1990a). A common feature of many approaches for incorporating effects of surface sealing into infiltration models in each of the categories has been exponential functions describing decline in both infiltration (e.g. Morin and Benyamini, 1977) and surface hydraulic conductivity.

Straightforward approaches are generally available to incorporate sealing effects in empirical infiltration models providing measured infiltration values are available. However, extrapolating beyond the conditions under which the measured data were collected can be misleading.

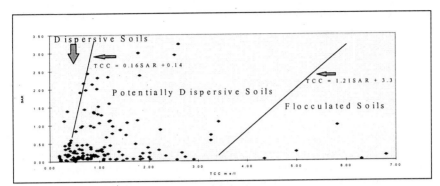

Fig. 2. Plot of sodium adsorption ratio vs. total cation concentration for samples from the 0-5 cm layer of soils in the south-eastern Australian wheatbelt (data from Geeves et al., 1995b; regression lines showing dispersive, potentially dispersive and flocculated soils from Regasamy et al., 1994)

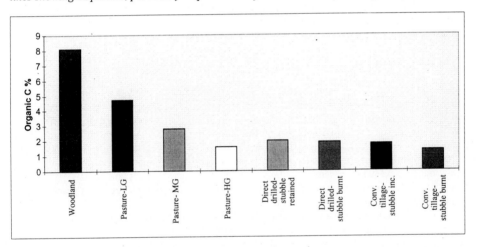

Figure 3. Differences in organic carbon contents in soils under different management practices in south-eastern Australia (data from Geeves et al., 1995b).

Models of the second type have been widely used and methods have been developed for incorporating effects of surface sealing. For instance, Moore (1981) developed a modification to the two stage Green and Ampt equation (Mein and Larson, 1973); Brakensiek and Rawls (1983) presented a modification to the multi-layer model and Rawls et al. (1990) developed an empirically evaluated crust factor for use with the single layer Green and Ampt model.

A number of attempts have been made to describe the flow of water through surface seals in terms of the physically based Richards' equation. These attempts involve characterising the change in soil water characteristic and unsaturated hydraulic conductivity function for the sealing layer (e.g. Mualem and Assouline, 1989; Romkens et al., 1990). In other cases, less physically based descriptions of surface seals as infinitely thin layers of declining hydraulic conductance have been added on top of more physically based models of water flow through the soil profile (e.g. Ross, 1990b).

Despite the availability of theoretical advances and improvements in electronic computing that have increased the availability of models of the third kind, difficulties in obtaining the data necessary for appropriate parameterisation have often meant that routine applications of infiltration models to agricultural management problems have not explicitly considered the impact of sealing/crusting events on run-off.

The magnitude of reductions in infiltration and increases in runoff caused by surface sealing at any particular time and location will depend on various rainfall factors (e.g. intensity, duration and raindrop kinetic energy), soil factors (e.g. antecedent water content, surface depressional storage, surface detention, surface cover or protection, underlying soil hydraulic properties) and site factors, which may influence site hydrology (e.g. slope angle and length) as well as the nature of the seal formed. Water balance simulation using both historical rainfall records (e.g. Cresswell et al., 1992) and designed rainfall events (e.g. Bristow et al., 1995) has been used to gain insight into the complex relationship combining these highly dynamic factors. As has been shown above, seal and crust formation in most soils depend upon the soil properties themselves and on the degree of crop cover. In this part of the paper two modeling applications are presented. The first relates to the use of a run-off index which is useful in the assessment of soil structural degradation and the second to the prediction of run-off during the early Australian winter when soils have been cultivated and sown, but crop cover is minimal. Both these applications used the Soil Water Infiltration and Movement Model (SWIM) (Ross, 1990a,b).

3.1 The run-off index

Cresswell et al. (1995) have demonstrated that run-off generation is dependent upon not only the properties of the soil A horizon and its surface condition, but also on the properties of the underlying horizon and antecedent soil moisture conditions. They developed a run-off index, which predicts the run-off from a 1-in-20 year rainfall event using SWIM. The run-off index does not consider the impact of sealing and crusting *per se*, but is included here to emphasise the significance of other important factors which generate run-off and potential erosion prior to the presentation of data which focus specifically on the impact of seals and crusts. It also is instrumental in demonstrating the complex linkage between cultivation management practices and hydraulic properties of the surface layers of the soil.

The run-off index is parameterised using soil layer depths, saturated hydraulic conductivity and the soil water characteristic for each layer. Campbell's method (1985) is used to predict unsaturated hydraulic conductivity. The index also assumed that initial water potential is -1.0m (i.e. the drained upper limit) throughout the profile, drainage occurs due to gravity at the base of the profile, a constant surface detention of 2.0 mm and that no further surface sealing occurs in a given event. The results of some runs using this approach are reproduced in Table 2. These show saturation excess is an important run-off mechanism in texture contrast soils in south eastern Australia. This excess is controlled not simply by soil structural condition, but by a number of interacting factors including depth of each soil layer, layer properties and rainfall characteristics. Thus, as the results demonstrate, tillage and other land management practices including stubble retention, can lead to significant improvements in soil structural condition which improve infiltration and reduce run-off. However, on some soils with shallow A horizons and underlying B horizons with poor conductivity (e.g. Wellington), well maintained surface structure was not sufficient to reduce run-off because of the low conductivity of the underlying horizon.

Whilst the results (Table 2) indicate that run-off generation is dependent upon a number of weather factors and soil properties, the adverse impact of sealing and crusting were not considered in detail. Furthermore, although rainfall intensities are less in the winter months in southern Australia, the late May to July period coincides with maximum bare and often cultivated soil. The data presented in the following section consider the potential impact of winter rainfall intensities on uncrusted and crusted soils using a similar modeling approach.

Site location	Land Use	Hydraulic conductivity ($\Psi_m = -10$ mm) Surface mm hr-1	Subsurface mm hr-1	A horizon macro-porosity %	Rainfall in 30 minutes	Runoff Index
St Arnaud	Traditional tillage [a]	9	3	11	27.1	6
Wellington	Direct drill [b]	21	1	7	34.4	16
Wellington	Woodland+ traffic	45	2	9	34.4	8
Quandialla	Pasture	32	7	9	30.7	3
Harden	Traditional tillage	18	17	9	29.7	0
Reefton	Traditional tillage	13	12	17	30.3	3
Reefton	Woodland	18	9	10	30.3	1
Cowra	Direct drill	81	8	14	31.8	0

[a] Tradition till crops sown after multiple passes with disc or tined implements - stubble incorporated or burnt.
[b] Direct drill crops sown after a single pass with a tined implement - stubble retained.

Table 2. Modeled run-off at a range of sites with different soil management practices and hydraulic properties (after Cresswell et al., 1995).

3.2 The effect of surface sealing on the water balance of a cropping soil

A similar approach to that of the run-off index is used here to give a simple illustration of the significance of potential effects of changes in surface hydraulic properties on the water balance when structural surface seals form on cultivated cropping soil from Cowra in New South Wales (NSW), Australia.

Geeves et al. (1995b) measured the rate and degree of decline in surface hydraulic conductivity under artificial rainfall for soil under long-term Traditional Tillage at Cowra. One dimensional water balance simulations using SWIM are used here to illustrate the practical significance of the measured surface sealing under relevant climatic conditions.

The water balance was simulated for a fully disturbed layer of soil overlying a settled soil profile for the month of June over the years from 1949 to 1958. Winter cereal crops are commonly planted at Cowra during late May or early June and planting commonly involves complete soil disturbance. Even under very conservative tillage treatments this one period of full soil disturbance is considered necessary to prevent disease caused by Rhizoctonia and other root pathogens. Potential for structural surface sealing is generally greatest when rain falls on loose tilled soil and so the time of sowing provides a convenient time to illustrate effects of sealing on the soil water balance. Underlying soil hydraulic properties were measured by Geeves et al. (1995b). Historic pluviograph and evaporation records for the site were provided by the NSW Department of Land Water Conservation. Daily evaporation was calculated from long-term mean monthly pan evaporation adjusted using a pan factor of 0.7. Antecedent soil water conditions were not monitored in the field but were derived for this illustration by simulating the water balance over the preceding two month period using the measured rainfall and evaporation. Surface depressional storage was arbitrarily set constant at 2 mm for both tillage treatments. In reality, surface depressional storage is likely to be greater than 2 mm for the Cowra soil following crop sowing and is likely to decline rapidly with rainfall. Run-off parameters were arbitrarily set to ensure that surface water in excess of surface depressional storage was rapidly removed as run-off. Gravitational drainage was assumed at the bottom boundary of both soil profiles throughout all simulations.

Year	Max.intensity (mm h^{-1})	June rain	With sealing Runoff	Evap	Drainage (mm)	Without sealing Run-off	Evap	Drainage
1949	32.5	64.1	6	21	37	0	21	43
1950	10.6	77.4	16	21	40	0	21	55
1951	9.2	74	14	19	22	0	19	36
1952	35	96	29	22	73	0	22	103
1953	4.7	18	0	14	0	0	14	0
1954	21.7	42	2	16	9	0	16	11
1955	66.7	59	2	22	29	0	22	31
1956	10	89	20	21	44	0	21	63
1957	22.5	18.9	0	7	0	0	7	0
1958	56.9	85	12	43	12	8	12	47

Table 3. The effect of surface sealing on simulated water balance for traditionally tilled Cowra soil for the month of June during 1949-1958.

Table 3 shows the resulting water balance for the month of June over the period 1949-1958. Comparison of simulated runoff with and without simulated surface sealing clearly indicates that surface sealing, to the extent measured by Geeves et al. (1995a), can significantly increase simulated runoff. Failure to account for this degree of surface sealing could potentially lead to errors in the runoff term of the water balance of up to 30% of total rainfall. In this Mediterranean winter-cereal cropping environment, increased surface run-off is of concern to farmers. It can lead to a reduced amount of soil water that may limit crop growth later in the growing season (French and Schultz, 1984) and also to significant soil loss through sheet and rill erosion (e.g. Hairsine et al. 1993).

The simulations presented here use field measured inputs but the output terms of the water balance listed in Table 3 have not been validated by field measurement. Furthermore they are subject to various simplifying assumptions including assumptions that:
a) the water balance model realistically represents the physical processes actually occurring;
b) the measured hydraulic properties are representative of the areas under the tillage treatments; and
c) that the effects of soil spatial heterogeneity and three dimensional soil water flow do not dominate.

Subject to these limitations, the simulations presented here indicate that structural surface sealing can have a significant effect on the water balance of cropping soils even under the relatively low intensity mid-winter rainfalls observed in this region. Any attempt to simulate the water balance for this case that does not account for surface sealing is prone to major underestimations of runoff.

The predictions made above relate to soils which are used for extensive cropping. Management of soil structural properties relies on the application of correct management practices with respect to minimal cultivation, stubble retention and rotations to overcome disease rather than the application of expensive chemical amendments to overcome sealing and crusting.

4 Conclusions

The current results demonstrate that long-term changes in soil surface properties caused by differences in management practices can result in very different soil hydrological responses to rainfall events when all relevant properties are taken into account. In soils in which management practices have degraded soil structure, soil sealing and crusting can have a very marked impact on rainfall infiltration even when rainfall intensities are relatively low. Data from the south-eastern Australian wheatbelt show that infiltration rates are significantly reduced by management practices which reduce organic carbon. In these conditions even soils with low sodium adsorption ratios show a propensity to seal and crust. This is important because the soils in question are devoid of cover in late Autumn and early winter and crusting can occur under relatively low rainfall intensities. Simple 1D modeling approaches demonstrate that management practices which reduce cultivation, promote organic matter retention and aggregate stability will significantly reduce potential run-off by minimising seal/crust development potential. However, as well as management practices and crusting, other factors including horizon depths and hydraulic properties also play an important role in influencing soil behaviour under heavier rainfall events or during prolonged wet spells. We would conclude, that although this paper has demonstrated the complex interaction of factors that may promote sealing and crusting, considerably more work is required to extrapolate the techniques described to a wider range of soils and antecedent moisture conditions.

References

Bosch, D.D. and Onstad, C.A. (1988): Surface seal hydraulic conductivity as affected by rainfall. Trans. A.S.A.E. **31**, 1120-1127.

Brakensiek, D. L. and Rawls, W.J. (1983): Agricultural management effects on soil water processes, II: Green and Ampt parameters for soil crusting. Trans. A.S.A.E. **26**, 1753-1757.

Bristow, K.L., Cass, A., Smettem, K.J.R. and Ross, P.J. (1995): Water entry into sealing, crusting and hardsetting soils: a review and illustrative simulation study. In Sealing, Crusting and Hardsetting Soils: Productivity and Conservation. Proc. 2nd International Symposium on Crusting and Hardsetting Soils, 183-204. ASSSI, Brisbane.

Campbell, G.S. (1985): Soil physics with BASIC, Transport models for soil-plant systems. Elsevier, New York. pp 150.

Casenave, A. and Valentin, C. (1992): A runoff capability classification system based on surface features criteria in the arid and semi-arid areas of West Africa. J. Hydrol. **130**:213-249.

Chartres, C.J., Bresson, L.-M., Valentin, C. and Norton L.D. (1994): Micromorphological indicators of anthropogenically induced soil structural degradation. Proc. XV Congress of Int. Soc. Soil Sci., Vol. 6a, 206-228. ISSS. Acapulco, Mexico.

Cresswell, H.P., Smiles, D.E. & Williams, J. (1992): Soil structure, soil hydraulic properties and the soil water balance. Aust. J. Soil Res. **30**, 265-83.

Cresswell, H.P., Geeves, G.W. and Murphy, B.W. (1995): Describing structural form with a mechanistically based runoff index. In Sealing, Crusting and Hardsetting Soils: Productivity and Conservation. Proc. 2nd International Symposium on Crusting and Hardsetting Soils, 253-258. ASSSI, Brisbane.

Freebairn, D.M. and Gupta, S.C. (1990): Microrelief, rainfall and cover effects on infiltration. Soil & Tillage Research. **16**, 307-327.

French, R.J. & Schultz, J.E. (1984): Water use efficiency in a Mediterranean-type environment. I. The relation between yield, water use and climate. Aust. J. Agric. Res. **35**, 743-64.

Geeves, G.W., Cresswell, H.P., Murphy, B.W., Gessler, P.E., Chartres, C.J., Little, I.P. & Bowman, G.M. (1995a): The physical, chemical and morphological properties of soils in the wheatbelt of southern NSW and northern Victoria. NSW Department of Conservation and Land Management /CSIRO Aust. Division of Soils. Occasional report.

Geeves, G.W., Hairsine, P.B. and Moore, I.D. (1995b): Rainfall-induced aggregate breakdown and surface sealing on a light textured soil. In Sealing, Crusting and Hardsetting Soils: Productivity and Conservation. Proc. 2nd International Symposium on Crusting and Hardsetting Soils, 145-150. ASSSI, Brisbane.

Green, W.H. and Ampt, G.A. (1911): Studies on soil physics, 1: The flow of air and water through soils. J. Agric. Sci. **4**(1), 1-24.

Hairsine, P., Murphy, B., Packer, I. and Rosewell, C. (1993): Profile of erosion from a major storm in the south-east cropping zone. Aust. J. Soil & Water Cons. **6**, 50-55.

Horton, R.E. (1940): An approach toward a physical interpretation of infiltration capacity. Soil. Sci. Soc. Am. Proc. **5**, 399-417.

Huang, C.H. and J. M. Bradford (1992: Applications of a laser scanner to quantify soil microtopography. Soil Sci. Soc. Am. J. **56**: 14-21.

Kostiakov, A.V. (1932): On the dynamics of the coefficient of water-percolation in soils and on the necessity for studying it from a dynamics point of view for the purposes of amelioration. Trans. Sixth Comm. Int. Soc. Soil Sci., Part A., 17-21.

Loch, R.J. and Foley, J.L. (1994): Measurement of aggregate breakdown under rain: Comparison with tests of water stability and relationships with field measurements of infiltration. Aust. J. Soil Res. **32**, 701-720.

Marshall, T.J. and Holmes, J.W. (1979): Soil physics. Cambridge University Press, Cambridge.

McIntyre, D. (1958): Permeability measurements of soil crust formed from raindrop impacts. Soil Sci. **85**: 185-189.

McCuen, R. (1982): A guide to hydrologic anlysis using SCS methods. Prentice-Hall, Englewood Cliffs, N.J.

Mein, R.G. and Larson, C.L. (1973): Modeling infiltration during a steady rain. Water Res. Res. **9**, 2, 384-394.

Moore, I.D. (1981): Effect of surface sealing on infiltration. Transactions of the A.S.C.E. **24** (6), 1546-1552.

Morin, J. and Benyamini, Y. (1977): Rainfall infiltration into bare soils. Water Res. Res. **13**, 813-817.

Moss, A.J. (1991): Rain-impact soil crusts I. Formation on a granite derived soil. Aust J. Soil Res. **29**: 271-289.

Mualem, Y. and Assouline, S. (1989): Modeling soil seal as a non-uniform layer. Water Resources Research **25** (10), 2101-2108.

Philip, J.R. (1957): The theory of infiltration. 1. The infiltration equation and its solution. Soil Sci. **83**, 345-357.

Rawls, W.J., Brakensiek, D.L., Simanton, J.R. and Kohl, K.D. (1990): Development of a crust factor for a Green-Ampt model. Trans. of the A.S.A.E. **33**, 1224-1228.

Rengasamy, P., Greene, R.S.B., Ford, G.W. and Mehanni, A.H. (1984): Identification of dispersive behaviour and the management of red brown earths. Aust. J. Soil Res. **22**, 413-431.

Romkens, M.J.M., Prasad, S.N., and Parlange, J.-Y. (1990): Surface seal development in relation to rainstorm intensity. CATENA SUPPLEMENT **17**, 1-11.

Ross, P.J. (1990a): Efficient numerical methods for infiltration using the Richards' equation. Water Resour. Res. **26**, 279-90.

Ross, P.J. (1990b): SWIM: A simulation model for soil water infiltration and movement. Reference manual, CSIRO Division of Soils, Townsville, Australia.

Sumner, M.E. (995): Soil Crusting: chemical and physical processes. The view forward from Georgia, 1991. In Sealing, Crusting and Hardsetting Soils: Productivity and Conservation. Proc. 2nd International Symposium on Crusting and Hardsetting Soils, 1-14. ASSSI, Brisbane.

Valentine, C. and Bresson, L-M. (1992): Morphology, genesis and classification of surface crusts in loamy and sandy soils. Geoderma **55**, 225-245.

Addresses of authors:
C.J. Chartres
CSIRO Division of Soils, GPO Box 639, Canberra, ACT 2601, Australia,
current address:
Australian Geological Survey Organisation, GPO Box 378, Canberra 2601, Australia
G.W. Geeves, NSW Dept. Land and Water Conservation, PO Box 445, Cowra, NSW 2794, Australia

A Finite Element Model of Soil Loosening by a Subsoiler with Respect to Soil Conservation

A. M. Mouazen & M. Neményi

Summary
A three dimensional, non-linear, finite element analysis was conducted for soil cutting processes by a medium-deep subsoiler. Two sandy soils were considered as elastic-perfectly plastic materials of non-linear behaviour. The Drucker-Prager model was adopted. Two different aspects were taken into consideration during model construction; soil strength and soil-tool interaction (external friction). The effect of a hard pan existing during soil tillage by deep tools was investigated assuming highly dense layer (hard pan). Higher draught and vertical forces were calculated from the model including hardpan than those calculated from the model free of hard pan. Soil displacement and soil stress distribution in the soil structure due to the subsoiler load were demonstrated.

Keywords: Subsoiling, finite element method, forces, reduced tillage.

1 Introduction

The running tillage systems tend to reduce the soil inverting pattern of mouldboard ploughs. Instead, soil loosening by lifting is now widely spread. Furthermore, the work done by different chisels and subsoilers can be successfully combined with other tillage tools for soil preparation. Soil compaction by heavy agricultural machinery during the cropping season negatively affects soil moisture content and soil productivity. However, layers of compacted soil can be built up by the action of tillage tools, particularly where the same tool (mouldboard plough or heavy disc harrow) is used at the same cultivating depth in successive seasons (Trouse, 1985). The major problem of using heavy disc harrows as a primary and secondary tillage tool simultaneously (conservative tillage type) is the formation of a hard pan (disc pan). Hard pans, however, greatly restrict vertical root growth. This reduces water and nutrients extraction from deeper strata under the hard pan. Consequently, hard pans cause yield reduction in situations of moisture deficit (Stafford and Hendrick, 1988). In areas of high precipitation, the creation of hard pans accelerates soil erosion by decreasing infiltration rates and increasing surface runoff and soil loss.

On the other hand, subsoiling is one common practice being used to disrupt hard pans and provide a pathway for roots to the less dense horizons. It is a conservative tillage type, since subsoilers neither mix nor turn the surface soils. In addition, a considerable surface residue is left undisturbed.

The finite element method (FEM) is being used to study soil cutting and tillage. The strategy of this paper is to apply the FEM to study soil cutting processes by a medium-deep subsoiler in terms of soil protection and energy saving. On the basis of the Drucker-Prager material model, a commercially available COSMOS/M FEM program was used to accomplish the analyses.

ISBN 3-923381-42-5
© 1998 by CATENA VERLAG, 35447 Reiskirchen

The goals of the paper are to: develop a FEM model of subsoiler breaking of hard pan by which the draught and vertical forces needed to eliminate a hard pan layer can be predicted, model subsoiler cutting of uniformly compacted and loose sandy soils using the FEM, and to introduce a theoretical concept of combining heavy disc harrows with subsoilers.

2 Theory

2.1 Mechanical principles of elastic-perfectly plastic material behaviour

Agricultural soils such as bulk materials suffer plastic deformations after a given external load. Hence, considering the soil as an elastoplastic material, the resulting strain rates can be divided into elastic and plastic deformations as:

$$d\varepsilon = d\varepsilon^e + d\varepsilon^p \tag{1}$$

where: $d\varepsilon$ = incremental total strain, $d\varepsilon^e$ = incremental elastic strain and $d\varepsilon^p$ = incremental plastic strain.

According to the generalised Hook's law of elasticity, the incremental elastic strain can be related to incremental stress as:

$$d\sigma = D^e d\varepsilon^e = D^e \left(d\varepsilon - d\varepsilon^p\right) \tag{2}$$

where: $d\sigma$ = incremental stress and D^e = elastic material matrix.

In another form, Eqn. (2) can be expressed as a function of the incremental total strain only, as follows:

$$d\sigma = (D^e - \frac{1}{d} D^e \frac{\partial g}{\partial \sigma} \frac{\partial f}{\partial \sigma} D^e) d\varepsilon \tag{3}$$

where:

$$d = (\frac{\partial f}{\partial \sigma} D^e \frac{\partial g}{\partial \sigma}) \tag{4}$$

g = plastic potential function and f = yield function.

The yield function of the Drucker-Prager elastoplastic material model can be written as:

$$f = 3\alpha\sigma_m + \overline{\sigma} - k = 0 \tag{5}$$

where: α, k = material parameters, σ_m = mean principal stress that can be expressed as:

$$\sigma_m = \frac{1}{3}I_1 = \frac{1}{3}\left(\sigma_x + \sigma_y + \sigma_z\right) \tag{6}$$

$\overline{\sigma}$ = effective stress that might be related with the second derivative stress invariant as:

$$\overline{\sigma} = J_2^{1/2} \tag{7}$$

But:

$$J_2 = \frac{1}{2}[(\sigma_x - \sigma_m)^2 + (\sigma_y - \sigma_m)^2 + (\sigma_z - \sigma_m)^2] + \tau_{xy}^2 + \tau_{yz}^2 + \tau_{xz}^2 \tag{8}$$

where: τ = shear stress, I_1, J_2 = first stress invariant and second deviatoric stress invariant, respectively, and σ = compressive stress.

The material parameters (α, k) included in Eqn. (5) have to be calculated as a function of the soil shear strength coefficients (soil cohesion c and soil internal friction angle ϕ) that can be obtained experimentally by the standard triaxial compression apparatus. It is worth noting that soil strength coefficients are plastic parameters, since soil is plastically deformed at peak shear stress.

2.2 Material and geometrical non-linearity

Two sources of non-linearity appear when a soil is under an external load; material and geometrical non-linearity. Material non-linearity can be fully described by a stress-strain relationship. More than a single stress-strain coefficient is required to fully represent the mechanical behaviour of any material under a general system of changing stresses (Duncan and Chang, 1970). For soil material, these coefficients are the Young's modulus of elasticity and Poisson's ratio that can be determined based upon the standard triaxial compression test as well.

In order to deal with material non-linearity, an incremental analysis technique was used. Inside each, the Newton-Raphson iteration method was applied. The geometrical non-linearity was solved considering the small strain assumption (Chen and Mizuno, 1990).

2.3 Finite Element Formation

The basic FEM equation can be formulated by transforming the non-linear equations of equilibrium for the elastic-plastic problems to the incremental form of linearized FEM equations (Chen and Mizuno, 1990). The simplified incremental form of the basic equation is:

$$[K]\{dU\} = \{dR\} \tag{9}$$

where: $[K]$ = total stiffness matrix, which relates a vector $\{dR\} = [dR_1, dR_2, ..., dR_n]^T$ of the total load increments to a vector $\{dU\} = [dU_1, dU_2, ..., dU_n]^T$ of the total displacement increments.

The incremental strain $\{d\varepsilon\}^{n+1}$ for the element can be evaluated from the kinematic condition:

$$\{dU\}^{n+1} = [K]^{-1}_{n+1/2}\{dR\} \quad \text{in a structure level} \tag{10}$$

$$\{d\varepsilon\}^{n+1} = [B]\{dU\}^{n+1} \quad \text{in an element level} \tag{11}$$

where: $[B]$ = the transformation matrix.

The stress increment $\{d\sigma\}^{n+1}$ can be obtained by using the current constitutive matrix as:

$$\{d\sigma\}^{n+1} = [D^{ep}]^{n+1/2}\{d\varepsilon\}^{n+1} \tag{12}$$

where: $[D^{ep}]$ = elastoplastic material matrix.

Eight-node, solid elements, were selected to represent the soil and subsoiler. The subsoiler was determined as a rigid body. The subsoiler is geometrically described in Bánházi et al. (1984). The chisel and the shank enclose 23° and 90° with the horizon, respectively. The thickness of the chisel cutting edge is 6 cm, whereas it is 3.6 cm of the shank.

To investigate the interaction and sliding characteristics of the soil-subsoiler system, the Coulomb's criterion of friction was adopted:

$$\tau_s = \sigma_n \tan\delta \tag{13}$$

where: $\tan\delta$ = coefficient of external friction, τ_s = shear strength of the interface, σ_n = normal stress on the interface and δ = angle of soil-subsoiler friction.

Interactions between the soil and subsoiler were studied so that an interface, two-node, gap elements were inserted between the soil and front edges of the subsoiler. Soil adhesion, which may exist in some cases but not in the case of sandy soil under dry conditions, was ignored (Mouazen and Neményi, 1996). The subsoiler was forced to move 10 cm in the travel direction as was followed by Araya and Gao (1995). This displacement was set at all interface nodes.

Three model variations, comprising three tillage problems, were hypothesised:
First, a highly compacted hard pan layer of 4.5 cm thick was incorporated into a loose body at 30 cm depth, directly beneath the tilling depth of mouldboard ploughs or heavy disc harrows, as

shown in Fig. 1. It also illustrates how the heavy disc harrow cultivates the upper layer down to 30 cm leaving a more dense hard pan layer to be disrupted later on by means of the subsoilers coming behind.

Second, a homogenous loose soil model was simulated in order to compare the results calculated from it with that calculated from the hard pan model.

Third, the entire soil body was assumed homogenous and highly compacted. The above-mentioned three variations were separately analysed. Considering two sandy soils, the parameters incorporated into the FEM analyses are summarised in Table 1. Parameters of the soil (2), either dense or loose readings, were used to analyse the first and second model variations. The parameters of soil (1), were used to analyse the third model variation.

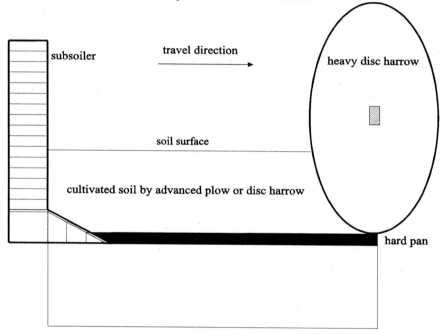

Fig. 1: Subsoiler cutting of hard pan created by advanced heavy disc harrow

Property	Symbol	Soil		
		(1)[a]	(2)[b]	(2)[c]
Compaction status	-	Compacted	Compacted	Loose
Cohesion (kPa)	C	9,130	1,000	1,000
Bulk density (kg/m^3)	ρ	1430,000	2210,000	1800,000
Young's modulus (kPa)	E	83360,000	82430,000	34900,000
Poisson's ratio	ν	0,248	0,280	0,280
Angle of soil internal friction (deg)	ϕ	23,800	36,500	30,400
Angle of soil metal friction (deg)	δ	15,000	17,000	15,000

[a] Araya and Gao (1995).
[b, c] Duncan and Chang (1970).

Table 1: Soil properties for non-linear finite element analysis

3 Results

3.1 Loose soil models (with and without hard pan)

Since soil loosening by chisels, cultivators or different subsoilers is of the lifting type, the upward soil displacement can be considered as an indicator of soil loosening quality. The maximum upward soil displacement predicted from the FEM model, was 13.2 cm and 12.9 cm of soil with and without hard pan, respectively. The alternation of the structural unit size of the hard pan layer caused 0.3 cm more upward soil movement. Alternatively, loosening and distortion of the hard pan might have caused the higher upward soil movement due to an increase in the overall soil volume over that of the model free of hard pan.

On the other hand, the highest soil shear and compression stresses were concentrated about the chisel, particularly at the tip. A high shear stress concentration about the chisel proves that soil breaking in shear takes place at the chisel depth. Meanwhile, the soil part in front of the subsoiler shank is lifted up by means of the inclined chisel, the shank in turn continues the job by cutting the upwards lifted soil. However, the maximum shear stress at the chisel tip of the hard pan model was 165 kPa, whereas it was 91.6 kPa for the model without hard pan. This means that higher shear stress need to be generated to overcome the high shear strength value of the hard pan layer. On the other hand, maximum compression stresses, at the chisel tip, were 1410 kPa and 992 kPa with and without hard pan, respectively. Accordingly, higher draught and vertical forces must proceed from the FEM model with hard pan.

The total calculated draught and vertical forces calculated from the FEM model are illustrated in Fig. 2. The total draught forces increased with subsoiler progress in the travel direction till they achieved peak values of 3259 N and 3115 N of the models with and without hard pan, respectively (Table 2). The total maximum vertical forces were 1486 N and 1130 N, respectively.

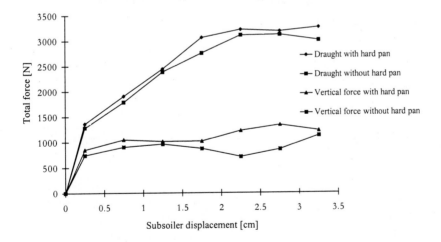

Fig. 2: Comparison of draught and vertical force with and without hard pan

An extra draught of 144 N as well as extra downward vertical force of 356 N are attributed to the higher stiffness of the hard pan layer than that of the identical loose one. Hence, about a 5% extra draught of the subsoiler total draught is required to break the hard pan. Since draught is directly related to energy requirements of tillage tools, 5% higher energy is consumed to eliminate a hard pan layer of 4.5 cm thick.

Draught force	Depth (cm)	Speed (m/s)	Value (N)
Measured in field [a]	40,000	0,550	9810,000
Measured of compacted soil [b]	34,000	0,115	8750,000
Predicted of compacted soil [c]	34,000	0,066	7835,000
Predicted of loose soil [d]	34,000	0,066	3115,000
Predicted of loose soil with hard pan [e]	34,000	0,066	3259,000

[a] Bánházi et al. (1984).
[b] Authors' measurement of draught.
[c, d, e] Authors' predictions of draught.

Table 2: A comparison of measured and predicted draught

3.2 Totally compacted soil model

The highest calculated shear and compression stresses were again generated for the whole chisel, particularly at the tip, like those of the loose soil models. The high stresses extended here to prevail over the entire soil bulk in front of the subsoiler from the chisel tip up to the soil surface. This can be attributed to the high stiffness of the whole soil body. However, maximum draught and vertical forces calculated from the FEM model were 7835 N and 5013 N, respectively. Bánházi et al. (1984) reported that the field measured draught force of the subsoiler studied herein was 9810 N, performing their measurements in compacted soil down to 40 cm depth. The estimated total draught was compared with the measured one by Bánházi et al. (1984). As shown in Table 2, the estimated draught from the FEM model was 2000 N less than the field measurement. A lower draught estimation of 2000 N might be attributed either to a bigger cultivating depth of 40 cm or higher speed of 0.55 m/s of the field subsoiling.

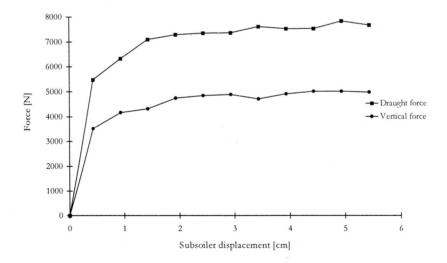

Fig. 3: Total draught and vertical force with subsoiler displacement

Figure 3 illustrates a comparison of total draught and vertical forces. Note that soil failure occurs at about 5 cm subsoiler displacement, because maximum draught refers to the force required for soil failure.

On the other hand, in order to accomplish further verification of the predicted results, a laboratory soil bin experiment was carried out at the Technical University of Munich, Institute of Agricultural Machinery. An identical subsoiler to the one incorporated into the FEM model was manufactured at the Pannon Agricultural University, Faculty of Mosonmagyaróvár and transferred to Munich University. The subsoiler was mounted on a dynamometer wagon, which pulled it with 0.115 m/s speed. The test was preceded by soil preparation, which consisted of soil loosening and compaction. The draught force was 8750 N measured at 34 cm cultivating depth. As compared to the soil bin measurement, 900 N less draught was calculated by the FEM model. In other words, the relative prediction error of the draught was about 10%. This assures the accuracy of the model predictions. Moreover, the lack of available experimental data reported for a similar tillage combinator to the one proposed here (subsoiler and heavy discs) - tilling in soils involving hardly condensed layers - opens the way for further conclusions based only upon the results predicted from the FEM analyses.

4 Conclusions

Three FEM analyses were conducted of soil tillage by a medium-deep subsoiler. The increase in the unit volume size of the hard pan layer developed an additional upward soil displacement of 0.3 cm. Moreover, higher shear and compression stresses were generated about the chisel, particularly at the tip of cultivated soil with a hard pan, than that of a cultivated one free of hard pan. For primarily cultivated soil including hard pan the total computed draught was 3259 N, whereas it was 144 N less for cultivated soil without hard pan. The maximum estimated draught of fully compacted soil was 7835 N. A comparison between the FEM computation and the laboratory soil bin measurement of draught showed good agreement.

It is common tillage practice to perform subsoiling, once every 2 or 3 seasons, in order to disrupt hard pans in the areas where mouldboard plough or heavy disc harrow are successively employed to the same depth. Usually, the subsoiler runs separately. If the heavy disc harrow and subsoiler are incorporated into one tillage machine (Fig. 1), two objectives can be successfully achieved; energy savings and restricting wheel traffic. The upper 30 cm compacted soil will be loosened by the heavy disc harrow. The shortcoming of the disc harrow in condensing the soil beneath its heavy mass will be avoided by means of subsoilers coming behind. This would eliminate one extra machine run besides decreasing the subsoiler specific energy consumption due to the reduction in its specific draught force, from 7835 N to 3259 N.

It is clear that the optimal tillage system by which hard pans can be avoided is by combining a heavy disc harrow with a subsoiler (Fig. 1). It is worthwhile incorporating the negative effect of heavy disc harrows in creating hard pans with the positive effect of subsoilers in avoiding them. One row of subsoilers can be installed behind two rows of heavy discs. However, further empirical investigation and field testing will be needed to study such a combinator.

References

Araya, K. and Gao, R. (1995): Non-Linear Three-Dimensional Finite Element Analysis of Subsoiler Cutting with Pressurised Air Injection, J. Agric. Engng. Res. **61**(2), 115-128.

Bánházi, J., Jóri, J.I. and Soós, P. (1984): Középmélylazító Szerszámok Összehasonlító Vizsgálata (Investigation and Comparing of Soil Cultivation with Medium-Deep Subsoilers), Hungarian Academy of Science, Budapest.

Chen, W.F. and Mizuno, E. (1990): Nonlinear Analysis in Soil Mechanics, Elsevier Science Publisher, Amsterdam.

Duncan, J. and Chang, C.Y. (1970): Nonlinear Analysis of Stress and Strain in Soils, J. Soil Mech. Foundations Div., ASCE, **96**(SM5), 1629-1653.

Mouazen, A.M. and Neményi, M. (1996): Two-Dimensional Finite Element Analysis of Soil Cutting by Medium-Deep Subsoiler, Hungarian agric. Engng. **9/96**, 32-36.

Stafford, J.V. and Hendrick, J.G. (1988): Dynamic sensing of soil pans. Trans. Am. Soc. Agric. Engs. **31**(1), 9-13.

Trouse, A.C. (1985): Development of the Controlled Traffic Concept, 'Soil Dynamic as Related to Tillage Machinery System', Proceeding of International Conference on Soil Dynamics, Auburn, Alabama.

Symbols and Abbreviations

FEM	finite element method,
$d\varepsilon$	incremental total strain,
$d\varepsilon^e$	incremental elastic strain,
$d\varepsilon^p$	incremental plastic strain,
D^e	elastic material matrix,
D^{ep}	elastoplastic material matrix,
$d\sigma$	incremental stress,
σ_m	mean principal stress,
σ_n	normal stress,
$\underline{\sigma}_{x,y,z}$	compressive stresses,
σ	effective stress,
τ	shear stress,
τ_s	shear strength of soil-tool interface,
f	yield function,
g	plastic potential function,
α, k	material parameters of the Drucker-Prager yield function,
I_1	first stress invariant,
J_2	second deviatoric stress invariant,
$[K]$	total stiffness matrix,
$\{dU\}$	total displacement increment,
$\{dR\}$	total load increment,
$[B]$	transformation matrix,
δ	angle of soil-metal friction,
ρ	soil bulk density,
E	Young's modulus,
ν	Poisson's ratio,
c	soil cohesion strength and
ϕ	soil internal friction angle.

Address of authors:
Abdul Mounem Mouazen
Miklós Neményi
Pannon Agricultural University
Department of Agricultural and Environmental Engineering
9200 Mosonmagyaróvár, Hungary

Influence of Establishment and Utilisation Techniques of Hillside Pastures on Soil Physical Characteristics

A.F. Ferrero & G. Nigrelli

Summary

Results are presented of a trial involving oversowing, traditional sowing, and the utilization of herbage production by haymaking plus rotational cattle grazing or by grazing alone in hillslope pastures. The production and botanical composition changes of the sward were checked at every regrowth. Some chemical, physical and hydrological properties of the soil were determined annually and over the grazing season.

After three years, no noticeable different effects emerged between the two utilisation techniques. Sown swards had higher total herbage production with better content of forage species than permanent oversown pastures, which yielded more in summer. At the end of the third year, a soil pan formed on all plots. Differences in the physical and hydrological properties of the soil became negligible. Oversowing of permanent sward allowed satisfactory herbage yields, reducing machinery traffic.

Keywords: Hillslopes, pasture, establishment, utilisation, soil compaction

1 Introduction

Hill country has always played an important role in Italian agriculture and has been intensively cultivated, but with major changes occurring to the local ecosystem.

The deruralisation process of the post-WWII years strongly reduced manpower availability which, due to environmental conditions and farm structure, has been only partially replaced by mechanisation. Where cash crops are cultivated, farms still are viable, though arable, meadow and coppice land are being abandoned. In addition, the disappearance of hill livestock breeding has lowered national meat production.

In order to correctly manage the land a sustainable agriculture system must be maintained on hillslope areas.

To increase the farm income, it has been proposed to use these marginal plots for semi extensive breeding of livestock within the integrated management of agroforestry resources (Talamucci, 1979).

2 Motivation and aims

A 15 - year experiment with free-range cattle rearing in a hilly area in Piedmont (N.W. Italy) showed a positive interaction between pasture and coppice woodland when used together (Ferrero and Lisa, 1983). Operating conditions are, however, different from those of traditional stock farms, since frequent rotations of livestock on fenced plots are necessary, even small, throughout the grazing season.

The research confirmed the problems involved in the establishment of long-term pastures on hillslopes by ploughing and reseeding, as well as the difficulties of ensuring over time, suitable quantitative and qualitative characteristics of the sward. Moreover it was not possible to match the concurrent needs of the livestock to those of the pasture sward by continuous adjustment of the stocking rate.

It is generally accepted that the intensity and frequency of defoliation and also repeated trampling are the main causes of the sward decay (Charles, 1979).

It is difficult to contain the adverse effects of trampling on the soil since animals have to be rotated on the same paddock several times, even when soil bearing capacity is low. It has been estimated that at the end of the grazing season, in conditions similar to ours, every part of the sward had been trodden eight to ten times (Edmond, 1958).

As part of the experiment outlined above, a paddock-trial was developed to assess the effects of the traffic of agricultural machinery and animals on sloping pastures with reference to the establishment technique. The major aims of the research were to quantify the effects of compaction on sloping pastures, relating them to the changes in the sward composition due to different establishment and utilization techniques.

We also wished to determine a bearing capacity threshold for each paddock - i.e., the minimum soil resistance level below which the hoof sinks - to help the cattle manager to decide when to move the animals into the pasture.

3 Experimental procedures

The farm, 29 ha, 450 m elevation, comprises 8 ha of forage crops split into 9 plots. The climate has cold winters and summer droughts (mid-July to mid-September). Rainfall averages 841 mm, but can vary greatly from year to year.

The herd consists of 22 Piedmontese cattle, a local beef breed.

Two paddocks were used, occupied by long-term poor quality meadows with species of the *Arrenatherum* and *Festuco-Brometus* association, one with a slight slope, the other sloping from 15% to 50%. The soils are mainly of loam silt texture.

Half of these paddocks was ploughed and, at the end of the summer, sown with a mixture of grasses and legumes (S). Following close grazing, the other half was oversown (0) with white *(Trifolium repens L.)* and red clover *(Trifolium pratense L.)*, ryegrass *(Lolium perenne L.)* and cocksfoot *(Dactylis glomerata L.)*, using a seeder fitted with discs.

Inside each plot control areas (C) with permanent enclosures were located. On half of each renewed pastures the herd was rotated four to five times from April to late October (G): on the other half, haymaking of the first regrowth was performed in May and three to four grazing rotations were carried out (HG). A 52 kW four-wheel tractor was used for mowing and baling, and a 33 kW tractor to turn and ted the hay.

Changes in ground cover of the various species and the bare areas were recorded each season. Herbage production was sampled at each regrowth, just before grazing, and the botanical composition of the dry matter yield (DM) was determined.

Soil physical and hydrological measurements were carried out with the methods and devices already experimented in woodland grazing tests (Ferrero, 1989).

At the start and three years after, the soil texture and chemical characteristics at three depth intervals - from 0 to 250 mm - were assessed. Soil cores for bulk density and the hydrological characteristics determinations were taken on the trial pasture, at the beginning, the end, and at selected times through the grazing season, and also the cone resistance was measured. On some undisturbed samples (200 cm^3 in volume) the water release curve - weight versus time - was determined by a method developed at our institute.

4 Results and discussion

Total annual herbage production in the first three years revealed no significant differences between systems of utilisation, inside each establishment technique (Table 1).

Herbage yield (DMtha^{-1})	Oversown pasture			Sown pasture		
Year	C*	G	HG	C*	G	HG
I Annual	9.8a	9.6a	9.4a	9.2a	9.1b	8.9b
I Summer	3.9a	3.8a	3.6a	3.2b	3.0b	2.7b
I Legumes	1.2a	1.3a	1.2a	0.8b	0.9ab	0.8b
II Annual	10.8ab	10.6b	10.3b	11.4a	11.2a	10.7ab
II Summer	4.3a	4.1a	4.2a	4.5a	4.2a	4.0b
II Legumes	1.3	1.2	1.3	1.3	1.2	1.2
III Annual	9.9a	9.5b	9.6ab	10.2a	10.1a	9.8a
III Summer	4.1a	3.7b	3.9a	4.3a	4.0a	3.9a
III Legumes	1.2	1.2	1.1	1.1	1.2	1.1

*C = Control plots untrampled
a,b = means in the same rows with different letter differ at 0.05 level

Table 1: Annual herbage production, Summer yield and its legumes content (DMtha^{-1}) from oversown or sown hillslope pastures under grazing (G) or hay plus grazing (HG) utilisation regime

Spring yields in the first year were considerably lower in the sown swards where, due to the limited carrying capacity of soil, the haymaking and the first spring grazing were deferred. In the second and third year, the annual herbage productions from these swards were higher. In the third and fourth regrowth oversown sloping pastures yielded more. The average total stocking on the trial pastures, expressed as animal unit grazing days (AUGd) - one AU = 500 kg of liveweight - ranged from 530 to 630 AUGdha^{-1}; the lower values were recorded in sown hay paddocks (SGH).

In the second year, utilisation techniques began to exercise a significant influence on botanical composition, especially on that of (S) sward: Spring mowing followed by grazing (SHG) fostered the formation of a more compact sward, with higher content of grasses and restrained the development of poor forage species (Fig.1).

There was a significant higher presence of legumes on oversown surfaces; a contribution was also given by natural reseeded lucerne *(Medicago sativa L)*. Legumes sustained the yield of the third and fourth regrowth, over the drought period, increasing by up to 25% the herbage yield with respect to sown pastures.

The soil chemical and physical characteristics examined did not differ greatly in relation to the utilisation system inside the same sward establishment technique (Table 2). In the third year, comparison between oversown and reseeded pastures and between the latter and the untrampled control (C) revealed significant differences: the C/N ratio and bulk density in the sown sward were higher than the control, but intermediate with respect to the oversown pasture. Saturation capacity (SC) did not differ, though it was lower, in the top layer of HG plots. Field capacity (FC), in the top 100 mm of the oversown sward was significantly higher.

The above spring measurements showed negligible differences between the compared treatments. In Autumn, greater differences were recorded and the influence of pasture management and establishment technique became evident.

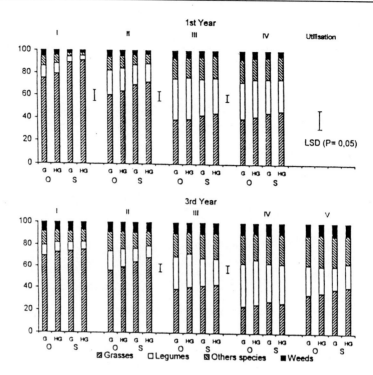

Fig. 1: Botanical composition (% DM) of the herbage production before each utilization, in the oversown (O) and in the sown hillslope pastures (S), under rotational grazing (G) or under haymaking plus grazing regime (HG) in the first and third year of trials.

Soil properties	Oversown pasture				Sown pasture			
	Grazing		Hay+Grazing		Grazing		Hay+Grazing	
	A	B	A	B	A	B	A	B
pH (water)	7.1	7.4	7.2	7.3	7.5	7.4	7.6	7.5
Organic matter (gkg^{-1})	48a	39	46a	37	41a	34	38b	32
C/N ratio	9.7	9.1a	9.5	9.0a	9.2	8.6ab	9.1	8.4b
Bulk density (gcm^{-3})	1.40	1.38	1.41	1.39	1.40	1.36	1.42	1.38
Sat. capacity (%, v/v)	47.2	48.4	44.8	49.5	45.7	47.4	44.6	50.5
Field capacity (%, v/v)	41.0a	39.1	38.2a	40.2	38.6a	39.2	36.5b	41.8

A = 0-70 mm depth   B = 80-150 mm depth
a,b = means in the same row, for the same depth, with different letters, differ at 0.05 level

Table 2: Soil chemical and physical properties determined at the end of the third year of trial, in oversown or sown hillslope pastures under different utilisation regime

From these measurements, however, it was not possible to identify from the outset the formation of the compact "floor" found in all plots at 80 to 100 mm depth.

Penetration resistance measurements up to 250 mm depth revealed a different trends (Fig. 2), if carried out at the end of the grazing season, with soil water content near FC. Despite high spatial variability, they revealed different soil responses to the sward implantation technique used; the highest values being recorded in sown and hay swards (SHG).

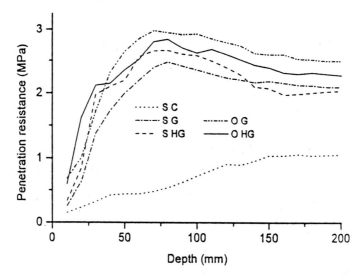

Fig. 2: Penetration resistance in the hillslope sown pastures (S), in oversown old pastures (O) and in the control untrampled (SC), measured after three years of utilization by rotational grazing (G) or by haymaking plus grazing (HG).

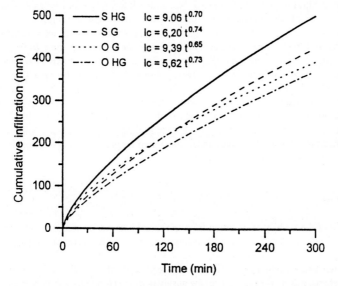

Fig. 3: Water cumulative infiltration in the soil of the hillslope sown pastures (S) and of the oversown old pasture (O) after three years of utilization by grazing (G) or by haymaking plus grazing (HG).

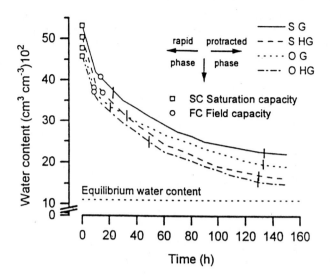

Fig. 4: Desiccation trend of soil samples from oversown old pastures (0) and from the newly sown pastures (S) after three years of rotational grazing (G) or haymaking plus grazing (HG).

The average penetration resistance trend over the first three years, showed that in spring differences were smaller. It would thus appear that soil compaction induced by traffic and trampling is largely improved in winter by rain and snow cover. This is in agreement with the findings reported by (Tanner and Mamaril, 1959).

The water infiltration measured "in situ" was much lower in the newly implanted pasture than in the untrampled control (C), but higher than that of the oversown pasture (Fig. 3).

The water release curve - weight versus time - of undisturbed soil samples, saturated from below, confirmed that soil in sown pastures has a higher water content at saturation. In this soil, a first, more rapid phase of desiccation up to about FC was recorded, as well as an intermediate one - to about 45% of the FC - slower than the long-term oversown pastures. Conversely the latter showed a more linear desiccation trend (Fig. 4).

This trend, if confirmed on other soil with different granulometric composition and on samples taken in different periods of the season, might enable us to identify water release behaviour models typical for pastures subjected to different levels of trampling.

The indications achieved from the model, integrated by penetration resistance data, could be used by the breeder for pasture utilisation management.

5 Conclusions

The utilisation of the herbage production from trial paddocks was carried out following ordinary farm management. Thus we feel that the conclusions drawn from the results presented are applicable to other farms.

It has been shown that in hill country the technique of sward establishment, as opposed to utilisation systems, has marked effects on the survival and performance of species, in both altering the botanical composition and influencing annual and seasonal yield.

The results highlight that, if carefully managed, trampling has a substantial impact on pasture, so the herbage utilisation techniques do not affect soil physical properties in notably different ways. Further controls however should be performed over longer periods of time.

The experiment has confirmed (Bryan, 1985) that hillside long-term pasture can be improved by surface oversowing and topdressing with productive results not markedly lower than those obtainable with reseeding by ploughing.

In addition, oversown pastures exhibited greater elasticity of utilisation from the very first year. Due to the higher presence of legumes, they showed more productive regrowths in Summer droughts. Finally, oversowing holds promise for reducing inputs while maintaining acceptable levels of production on hillslope long-term pastures.

Aknowledgements

We are grateful to our colleagues in the Hill Farming Section G. Benvegnù, S. Parena, L. Sudiro, who helped in farm work and data collection.

Symbols and Abbreviations:

AUGd	Animal unit grazing days
CA	control area untrampled
DM	dry matter
FC	field capacity
G	rotational grazing
HG	haymaking plus grazing
O	oversowing on sod
S	conventional sowing
SC	saturation capacity
SHG	sown pasture under hay and grazing regime.

References

Bryan, W.B. (1985): Effects of sod-seedings legumes on hill land pasture productivity and composition. Agron. J. **77**, 901-905.
Charles, A.H. (1979): Treading as a factor in sward deterioration. Charles A.H., Haggar R.J. Editors, New York.
Edmond, D.B. (1958): The influence of treading on pasture, a preliminary study. New Zeal. J. Agric. Res. **1**, 319-328.
Ferrero, A.F. (1989): Effect of compaction simulating cattle trampling on soil physical characteristics in woodland. Soil Tillage Res. **19**, 319-329.
Ferrero, A.F. & Lisa, L. (1983): Integration of beef cattle grazing with coppice on hillands Proc. Int. Hill Lands Symp. Foothill for Food and Forests, Corvallis (Oregon, USA), 375-378.
Talamucci, P. (1979): Rivalutazione produttiva delle aree marginali. Italia Agr. **116**, 93-106.
Tanner, C.B. & Tamaril, C.P. (1959): Soil compaction by animal traffic. Agron. J. **51**, 329-331.

Addresses of authors:
A.F. Ferrero
G. Nigrelli
Institute for Agricultural Mechanization
National Research Council
Strada delle Cacce, 73
10135 Torino, Italy

Subsoil Amelioration

H. Schulte-Karring, D. Schröder & H.-Chr. Von Wedemeyer

Summary

After removal of the top soil at a winter construction site, the subsoil had suffered extreme compaction over a pipeline route distance of 16 km caused by heavy transportation and construction vehicles.For amelioration, the soil was loosened to a depth of 70 to 80 cm using new technology, the subsoil was fertilised and green manure crop planted to stabilise the new soil structure.After periods of one and two years following amelioration the soil physics and biology were examined.

The results obtained confirmed the checks carried out in connection with other tests over the past 35 years showing that the preservation of loosened arable land is less a function of the type of soil than the degree of loosening and subsequent treatment, i.e. the mechanical pressure exerted by agricultural equipment or construction vehicles.

Keywords: Soil compaction, pipeline construction, restratoration, successfull amelioration

1 Introduction

Research work in the field of mechanical deep loosening began in 1954 with a dissertation at the Institute of Soil Science in Bonn. The investigations were continued at the State Teaching and Testing Establishment (SLVA) in Ahrweiler in 1959 and have been pursued there up to the present day (Schulte-Karring,1957).

The task in Ahrweiler at the time was to find an effective means of remedying an area of about 300 hectares of compacted soils. We considered the technique used world-wide until then - systematic pipe drainage - to be wrong because it did not tackle the cause of waterlogging, namely soil compaction. Moreover, in an area with an average annual precipitation of only 560 mm, the rainwater had to be preserved and not drained. Our aim was to eliminate the subsoil compaction and create a new, efficient soil structure which was
* aerated,
* supportive of growth,
* easily permeable for roots and
* capable of storing water. (Schulte-Karring, 1964, 1980)

2 Technical developments

Although it proved possible to convince the relevant authorities that our approach was right, there was a lack of equipment for carrying out deep loosening on a wide scale (Schulte-Karring, 1967). The development led from rigid lift loosening implements, which loosen the soil by lifting action, to lift loosening devices with direct drive, and then to implements that loosen the compacted soil by a cutting and breaking mode (Fig.1). Subsoil fertilisation technology was also developed further.

In this context, the problem of distributing fertiliser in the subsoil had to be solved. After experimenting with mechanical aids, compressed air and blowers, use of an impeller blower finally proved successful. But this device was only suitable for powdery fertilisers (Schulte-Karring, 1995).

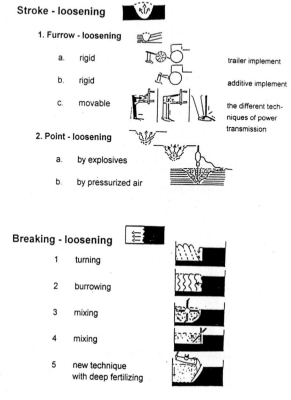

Fig. 1: *Possibilities of deep loosening*

3 Experiments and results

The technical developments were accompanied by experiments and evaluations. The objective was to discover the optimum method of deep loosening and ascertain the impact on the soil, root and plant growth, soil life and especially the water balance.

Another important subject of investigation was the sustainability of deep loosening, i.e. the durability of amelioration. The experiments initiated 36 years ago are still available for this purpose. The results were evaluated in close collaboration with the universities of Bonn (Schröder and Scharpenseel, 1975, Martinovic et al., 1983) and Trier (Schröder and Schulte-Karring, 1984).

The experiments related to physical, micromorphological and microbiological processes. They showed that proper mechanical deep loosening ensures that the compacted soil remains aerated, supportive of growth, easily permeable for roots and capable of storing water for a long time.

The greatest success is achieved wherever the greatest compaction, i.e. the largest damage, has occurred and that the duration of the amelioration effects depends less on the type of soil than on its subsequent use. For example, incorrect treatment of arable soils quickly results in renewed compaction in the top layer of the subsoil. By contrast, in the case of soils exposed to less or no mechanical pressure, the deep loosening is largely or entirely preserved once the soil has settled.

Subsoil amelioration

Experiments	part	pore total %		coarse pores > 50 µm %		water permeability (kf cm/s * 10⁻⁴)		infiltration (mm/sec.)	air permeability (ka µ²)		shear strength (kp/cm²)	
		30-40 cm	50-60 cm	30-40 cm	50-60 cm	36-40 cm	55-60 cm	35-40 cm	36-40 cm	55-60 cm	35-40 cm	55-60 cm
Ahrweiler	not loos.	43,0	39,4	7,1	4,6	4,8	8,6	1,6	4,8	0,9	305	950
I	loosened	43,2	43,6	11,4	7,4	15,0	50,0	18,5	11,0	25,0	290	190
24 years ago	loosened	+0,2	+4,2	+4,3	+2,8	+10,2	+41,4	+16,9	+6,2	+24,1	-15	-760
Kreuznach	not loos.	43,3	46,3	5,5	0,9	24,7	2,0	0,0	12,7	2,2	0	0
II	loosened	44,7	46,6	4,7	5,9	17,1	35,2	0,0	7,7	10,5	0	0
21 years ago	loosened	+1,4	+0,3	-0,8	+5,0	-7,6	+33,2	+0,0	-5,0	+8,3	0	0
Ahrweiler	not loos.	40,7	42,8	5,4	4,1	0,6	0,5	1,7	4,2	0,4	1050	1212
III	loosened	44,3	46,4	6,9	6,9	17,0	24,0	1,6	1,5	7,1	437	300
20 years ago	loosened	+3,6	+3,6	+1,5	+2,8	+16,4	+23,5	-0,1	-2,7	+6,7	-613	-912
Ahrweiler	not loos.	47,3	44,3	9,3	7,5	8,3	2,3	3,5	8,5	4,4	731	758
IV	loosened	47,9	48,8	12,7	11,6	54,0	47,0	11,5	14,0	68,0	630	438
20 years ago	loosened	+0,6	+4,5	+3,4	+4,1	+45,7	+44,7	+8,0	+5,5	+63,6	-101	-320
Ahrweiler	not loos.	46,3	45,7	10,4	9,7	35,0	3,7	0,09	16,0	20,0	410	316
V	loosened	47,4	50,5	9,3	10,8	3,1	130,0	0,02	4,4	57,0	446	543
20 years ago	loosened	+1,1	+4,8	-1,1	+1,1	-31,9	+126,3	-0,07	-11,6	+37,0	+36	+227
Ahrweiler	not loos.	44,9	42,7	9,9	8,5	6,0	6,8	0,03	5,1	3,2	640	700
VI	loosened	47,8	50,5	12,4	13,6	19,0	120,0	0,09	9,7	13,0	280	400
12 years ago	loosened	+2,9	+7,8	+2,5	+5,1	+13,0	+113,2	0,06	+4,6	+9,8	-360	-300

scientific interpretation: D. Schröder - University Bonn and Trier

Tab. 1: Soil physical properties, 12 – 24 years after amelioration

	depth cm	total pore volume %		coarse pores > 50 µm %		dry - weight (g/cm³)		water permeability (geom.avg.kf - cm/d)		air permeability (geom.avg.ka - µ²)		resistance to penetration (k/Pa)	
		loos.	not loos.	loos.	not loos.	loos.	not loos.	loos.	not loos.	loos.	not loos.	loos.	not loos.
Ahrweiler	40	41,6	40,9	6,4	5,7	1,55	1,58	24,5	52,1	2,5	3,9	2698	2682
1	60	42,0	38,1	4,1	1,2	1,54	1,65	14,1	9,6	1,1	0,5	2489	3353
with draining	80	37,6	35,0	2,2	0,0	1,69	1,76	1,9	22,8	0,2	0,3	2633	3982
Ahrweiler	40	42,5	39,6	4,4	2,8	1,50	1,60	12,5	21,7	1,9	0,8	2847	2197
2	60	44,0	42,7	4,7	3,2	1,50	1,60	5,9	2,4	0,4	0,2	1814	2858
without draining	80	39,8	39,4	5,6	1,7	1,60	1,60	2,6	2,6	0,6	0,0	3003	4095
Draining-ditch	40	45,0	41,5	14,2	5,4	1,45	1,53	83,3	5,5	9,2	1,4	2649	2955
	60	46,6	39,5	15,3	2,6	1,41	1,59	163,2	2,4	17,0	0,2	2042	2874
at right angle to the ploughing direction	80	46,6	39,5	15,5	3,8	1,41	1,61	404,0	6,5	23,0	0,4	1310	3645

Soils: Gray Plastosols - Pseudogleys; stagnation layer from 40 cm; kaolinitic clays about 30%, silt about 60%
Location: Ahrweiler - 100 m a.s.l., annual precipitation 560 mm, average temperature 8,5° C
Equipment: lift loosening device, attached rigid, size of share 120 * 300 mm
Experiments: Ahrweiler 1 = 60 cm maximum depth of loosening —— Ahrweiler 2 = 80 cm maximum depth of loosening
Scient.Interpret.: D. Schröder - University Trier

Tab. 2: Soil physical properties, 31 years after amelioration

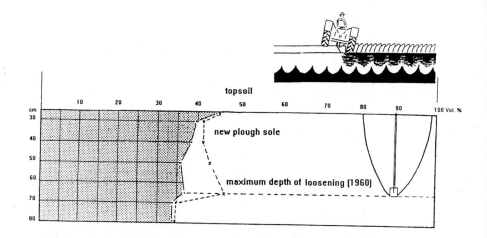

Fig. 2: Typical pore space-diagram of arable land intensively cultivated, three decades after amelioration. (Gray Plastosol-Pseudogley with stagnation at a depth of 40 – 50 cm)

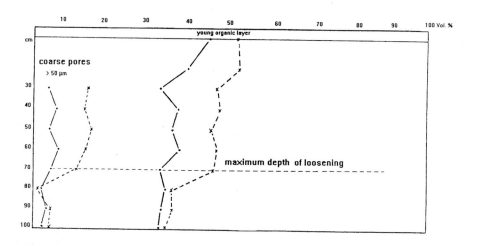

Fig. 3: Typical pore space-diagram of soils without tillage, e.g. forest subsequent agricultural land use, 10 years after amelioration. (Gray Plastosol-Pseudogley with stagnation at a depth of 50–60 cm)

This applies, for example, when using the soil for perennial plants, sports facilities or silviculture (Table 1+2, Figs. 2+3).

The enhancement of supercapillary pores over 50 µm by mechanical loosening is considered particularly beneficial for aeration and water absorption. Even after 31 years, the enlargement of this coarse pore volume is detectable in the deep loosening that was then carried out to a depth of only 60 cm.

4 Areas of Application

After 1965, deep loosening was employed in 283 land management projects in Rhineland-Palatinate. The impact of this soil amelioration was also checked (Schröder and Schulte-Karring, 1984). The deep loosening performed on a large scale in diverse soils has led to few complaints, thus indicating that success has been achieved in the form of long-lasting amelioration (Schneider, 1987).

Initially, mechanical deep loosening was carried out only in soils used for agricultural purposes. In the 1960s, this amelioration method was also employed in Ahrweiler vineyards. In viticulture, however, deep drainage ploughing is only possible where a conventional tractor can be driven. The new MM (Mehrzweck-Melioration) technology with the advantage of a strong pushing force made it possible to cope with inclines up to 45% (Schulte-Karring, 1976).

In other areas of application, such as fruit, asparagus and hop growing, deep loosening is also becoming more widespread either to prepare or to enhance plantations. Deep loosening with the aid of the new MM technology avoids damage to roots and leads to even growth, even where soils differ greatly (Schulte-Karring, 1979, 1985).

This amelioration technique is also used after construction measures or on earthwork that was wet during filling and levelling.

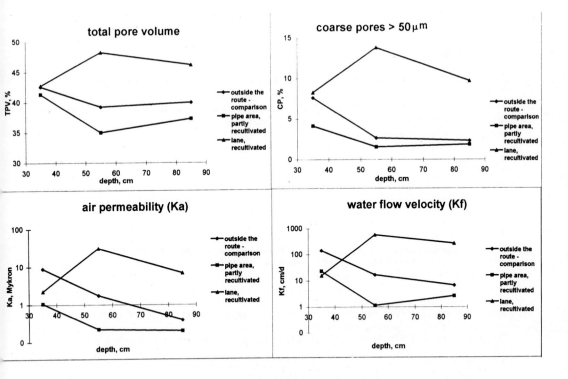

Fig. 4: Pore volume (total and >50 μm), air permeability (Ka) and water flow velocity (Kf) of a highly compacted route soil one year after recultivation. Soil: deep sandy loam. (Location I (Cölbe, July 1994)

While deep loosening is now a generally accepted means of improving growth conditions, aeration, soil activation and water absorption, reducing erosion damage or enhancing filtering functions, it is still largely unknown in silviculture. A missed opportunity were two cases of major storm damage in the 1980s. At the time, technology was already available for cutting the waste wood, including stumps, and for subsequent deep loosening that avoids clogging. Furthermore, prior to EU-wide forestation on formerly compacted arable soils, soil preparation would have been essential to improve deep rooting and the water balance (Schulte-Karring, 1992).

A special challenge was posed by the recultivation of a pipeline route that had been greatly impaired by unfavourable weather conditions. Over a distance of 16 km, the soil at the winter construction site consisted of viscous mud, with highly compacted soil underneath. Deep soil damage had occurred above all on the track next to the pipeline trench. After preparatory work, such as clearance, levelling and connection of the drainage system, the compacted soil was subjected to deep loosening and deep fertilisation. This was followed by stabilisation of the loosening with the aid of a deep-rooted mixed-seed green manure, mulching of the subsequently lush vegetation produced by favourable nutrient and humidity conditions, and then final loosening to a depth of 50 cm, after the top-soil had been carefully spread in order to preserve the structure of the loosened subsoil (v.Wedemeyer, 1978, 1993).

This recultivation was carried out in 1993. The effects of the amelioration were checked in 1994 and 1995. The findings of the physical and chemical investigations show that, through proper use of the technology created, even extreme soil damage can be remedied within a short space of time (Schulte-Karring and v. Wedemeyer, 1996, Fig.4)

5 Conclusions

Deep loosening is increasingly being used in all areas of application. In silviculture, however, where compacted soils in our region are frequently encountered, the method of amelioration is virtually unknown. No doubt forest experts and silviculturists will come to accept that forest vegetation needs specific growth conditions if it is to flourish and that technology is on hand for remedying deficiencies of compact soils.

References

Martinovic, Lj., Mückenhausen, E. and Schröder, D. (1983): Einfluß mechanischer und pneumatischer Tieflockerung auf drei Bodentypen. Z. f. Kulturtechnik und Flurbereinigung 24: 213-223.

Schneider, R. (1987): Auswirkungen von Meliorationsmaßnahmen in Flurbereinigungsverfahren der Nordosteifel auf verschiedene Standorteigenschaften. Diplomarbeit Universität Trier.

Schröder, D. and Scharpenseel, H.W. (1975): Infiltration von Tritium-markiertem Wasser in zwei tiefgelockerten Graulehm-Pseudogleyen. Z. f. Pflanzenern. u. Bodenkunde 4/5: 483-488.

Schröder, D. and Schulte-Karring, H. (1984): Nachweis 20-jähriger Wirksamkeit von Tieflockerungen in lößbeeinflußten Graulehm-Pseudogleyen. Z. f. Pflanzenern. u. Bodenkunde 147: 540-552.

Schulte-Karring (1957): Die Untergrundlockerung allgemeindichter Böden und ihre Auswirkung auf Boden und Pflanzenertrag. Diss. Bonn.

Schulte-Karring, H. (1964): Neue Wege der Melioration strukturkranker Böden. Zweijahresbericht der LLVA Ahrweiler (1961-1963), pp. 32-69

Schulte-Karring, H. (1967): Die technischen Probleme der Untergrundlockerung. Landtechnik 4: 96-104.

Schulte-Karring, H. (1970): Die meliorative Bodenbewirtschaftung. Warlich, Ahrweiler, 170 pp.

Schulte-Karring, H. (1976): Bodenschäden und Maßnahmen zu ihrer Behebung - aufgezeigt an Beispielen aus dem Weinbau Südafrikas. Der Deutsche Weinbau 25: 941-944.

Schulte-Karring, H. (1979): Der Einsatz des Ahrweiler Meliorationsverfahrens im Landschafts- und Gartenbau. Neue Landschaft 3: 180-184.

Schulte-Karring, H. (1985): Die Gefügemelioration durch Tieflockerung - Einsatz und Auswirkung des Ahrweiler Verfahrens in verdichteten Böden unter besonderer Berücksichtigung des Gemüse-, Obst- und Weinbaus. DVWK-Schriften, Teil IV, 144-270.

Schulte-Karring, H. (1992): Wirksame Waldbodenmelioration oder so weitermachen wie bisher? Allgemeine Forstzeitschrift **10**: 538-541.

Schulte-Karring, H. (1995): Die Unterbodenmelioration - Technik und Einsatz. H. Schulte-Karring, Bad Neuenahr, 136 pp.

Schulte-Karring, H. and v.Wedemeyer, H.-C. (1996): Subsoiling in the case of extreme damage. Sonderdruck anläßlich der 9. ISCO Tagung in Bonn, Ruhrgas AG, Essen.

v. Wedemeyer, H.-C. (1978): Darstellung der landwirtschaftlichen Folgeschadensentwicklung nach der unterirdischen Verlegung einer Erdgasleitung am Beispiel eines Projektes am Niederrhein. Z. 3 R international **5**: 331-335.

v. Wedemeyer, H.-C. (1993): Flur- und Folgeschadensregulierung beim Rohrleitungsbau. Schriftenreihe des Hauptverbandes der landwirtschaftlichen Buchstellen und Sachverständigen e.V.. Heft 92, 2. Auflage. Pflug und Feder, St. Augustin, 35 pp.

Addresses of authors:
H. Schulte-Karring
Johannisberg 13
D-53474 Bad Neuenahr, Germany
D. Schröder
Universität Trier
FB VI - Bodenkunde
D-54286 Trier, Germany
H.-Chr. von Wedemeyer
Ruhrgas AG
Huttropstraße 60
D-45138 Essen, Germany

Reclamation of Hardened Volcanish Ash Soils in the Central Mexican Highlands – Their Productivity and Erodibility

J. Baumann, G. Werner & W. Moll[†]

Summary

Nutrient supply of 25 treated Tepetates in the Central Mexican Highlands were investigated and maize yield potential (productivity) was calculated with the QUEFTS-model. From 1991 to 1994, runoff and soil loss were measured on two recently treated Tepetates, three hardened Tepetates and one Chromic Cambisol. The erodibility was calculated as average soil loss per EI_{30}-erosivity unit according to the USLE. The nutrient supply for Ca, Mg and K is high, whereas P, N and organic C amounts low to very low. After 7 to 10 years of cultivation the available P-amounts indicated a medium level of P-supply, whereas N-supply is still very low. The annual erosivity (EI_{30}) varied from 110 to 390 N/h. Soil loss occurred predominantly during high erosive storms and as a consequence of erosive rainfall accumulation over a period of several days. K-factors for Tepetates were between 0.32 and 0.38. They indicate a medium to high erodibility compared with the Cambisol having a K-factor around 0.09.

Keywords: Tepetates, reclamation, erosivity, erodibility, productivity

1 Introduction

In view of the need to extend arable land in the near future, the rehabilitation of degraded lands in the tropics is an important task. In the state of Tlaxcala, Central Mexican Highlands, more than 15 % of the surface is covered by indurated volcanic ash soils. Soils developed in Toba-sediments are often characterized by pedogenetic silica-acid enrichment in the subsoil. The formation of so called "Tepetates" results from the exposure of the silica-rich subhorizon at the surface after erosion of the topsoil; the exposed subsoil hardens irreversibly. For about 20 years the government of Tlaxcala has been treating Tepetates by deep-loosening under a land reclamation program. The program is restricted only to mechanical measures. The deep loosened sites are left immediatly for cultivation to the small farmers without further recommendations for adequate management of these labile soils. Under traditional management maize yields are very low within the first years of agricultural use. At some fields corn yields are below 1 t/ha.

If the technical measures are not performed in a correct way, concerning size and inclination of prepared terraces, and if farmers do not take into consideration effective soil erosion control measures after deep loosening, erosion can lead to great damage during the first or second rain period.

The aim of this study was to evaluate fertility status of treated Tepetates and the erodibility of recently deep-loosened Tepetates in comparison to a widespread Cambisol.

2 Site description

The state of Tlaxcala (3914 km²) is situated in the eastern section of the Central Mexican Highland. The southern area of the state forms part of the valley of Puebla-Tlaxcala. It is separated from the valley of Mexico by the Sierra Nevada. The northern boundary of the valley is marked by the blocks of Tlaxcala and Huamantla which cross the state in a west-east direction and where the studied sites are located. The blocks reach heigths of about 2550 - 2700 m a.s.l.. Widespread soil types are Vertic Cambisols and Chromic Cambisols, both with duric phases in the subhorizon (Werner, 1988). They are associated with Duri-/Fragipans (Tepetates), formed by erosion of the topsoil of the Cambisols (Aeppli, 1973). About fifteen percent of the area is covered by Tepetates (Baumann, 1996). The climate is characterized by a summer rainy season extending from May to October. The average annual rainfall is about 600 - 700 mm. With a mean annual temperature of 15 - 17°C and 5 - 6 humid months, the climate is semiarid (Lauer and Stiehl, 1973).

3 Materials and methods

For chemical analysis 25 treated Tepetates in the "block of Tlaxcala" were examined. The number of years of cultivation ranged from 0 (recently deep loosened) to 10 years. The following chemical soil analyses were undertaken:
- pH-values were measured potentiometrically using a glass electrode in deionised H_2O and 1 M KCl-suspension. The proportion soil : solution was 1 : 2.5.
- Exchangeable cations were extracted by percolation with 1 N NH_4-acetate (pH 7.0) and measured by AAS. Soil samples of 2.5 g were placed in percolation tubes. After a contact time of 2 hours with NH_4-acetate the samples were percolated with 100 ml of extracting solution. The percolation time was 4.5 hours.
- Total carbon was determined by burning with oxygen and measuring the CO_2 conductometrically with the Wösthof gas analyser. Organic carbon has been evaluated by subtracting the carbon encountered within the carbonates from total carbon.
- Carbonates were determined by measuring CO_2-volume after destroying with HCl in the Scheibler apparatus.
- Total nitrogen was determined according the Kjeldahl method (Reeuwijk, 1986).
- Plant available phosphorus was determined by the Olsen-Method ($NaHCO_3$-extractable P) (Olsen et al., 1954; Houba et al., 1989).
- Phosphorus adsorption was determined according to Blakemoore (Mizota and Reeuwijk, 1989; Reeuwijk, 1986).
- Nutrient reserves were estimated by a 1 M HCl-extract according to Knickmann (1955), modified by Moll (1965). 50 ml of 1 M HCl were added to 5 g of soil. The samples were heated to 90°C. The contact time was three hours. K, Ca, Mg and Na were measured using AAS. Phosphorus was measured colorimetrically as Molybdovanadophosphoric acid (Jackson, 1958).
- Calculation of pore size distribution and available water capacity was based on the determination of water content at different water retentions (Hartge and Horn, 1989), using pressure plate extractors (Soil Moisture Inc.).
- Aggregate stability was determined by the wet sieving method as "ΔMWD[1]" (Hartge and Horn, 1989) and by the percolation method (Becher and Kainz, 1983).

Erosion studies were undertaken on three sites in the "Block of Tlaxcala" from 1991 to 1994. On two sites, Tlalpan and El Carmen, recently treated Tepetates were examined. On the third site,

---

[1] "ΔMWD" = "Mean Weighted Diameter"

Matlalohcan, the erodibility was assessed of a Chromic Cambisol which is widespread in the area of Tlaxcala. Runoff and soil loss were measurd on erosion plots under natural rainfall. Rainfall was measured with 24-hour recording rain gauges. The scale of the register-stripes was 16 mm/h and 8 mm/1 mm rain amount. Erodibility was calculated as the average soil loss per EI_{30}-erosivity unit according to the USLE (Wischmeier and Smith, 1978). The following plots were installed: Tlalpan, 2 plots on treated Tepetates „TR"/bare fallow, "TR"/maize (22 m × 2 m, Wischmeier standard) and one plot on Tepetate "T" (3 m × 2 m). El Carmen, 2 plots on treated Tepetates "TR"/fallow, "TR"/maize (22 m × 2 m, Wischmeier standard), and one plot on Tepetate "T" (3 m × 2 m). Matlalohcan, 2 plots on a Chromic Cambisol: "CAM"/bare fallow, "CAM"/natural vegetation (22 × 2 m, Wischmeier standard) and 1 plot on Tepetate "T" (10 × 2 m) (Table 5).

4 Results and discussion

The **results of chemical analyses** are given in Tables 1 and 2. The pH-values showed a slightly acidic to slightly alkaline soil reaction. These are favourable conditions for the cultivation of most agriculture crops. Soil-pH decreased slightly with time since cultivation. The total sum of exchangeable cations showed a strong correlation with pH-values (r = 0.843***). The soils contain medium to high amounts of exchangeable cations. So they are well supplied with Ca, Mg and K. Above that they have high to very high reserve amounts (HCl-extractable) of these nutrients. Problematic is the very low nitrogen and phosphorous content. The mean N and P amounts of recently treated Tepetates are only 540 kg N/ha (0.018 %, soil depth: 0-30 cm, bulk density: 1.2 g/cm^3) and 2.1 kg/ha (0.7 mg/kg) available P (P-Olsen) respectively. Therefore, mineral N and P fertilizer and/or organic manure applications are necessary. The amounts of soil nitrogen and available phosphorouse increased with time since cultivation. After 7 - 10 years of cultivation the mean level of nitrogen reached 1600 kg/ha, which is still very low. However, there is a great variation of N-status (690 kg/ha - 3840 kg/ha) between the individual fields. It can be concluded that management greatly varies.

Years since cultivation		pH		Exchangeable Cations meq/100g soil					P-Olsen mg/kg	C_{org} %	N_t %
		KCl	H$_2$O	Ca	Mg	K	Na	total			
0 (n = 5)	Mean[1]	6,4	7,9	10,4	8,5	1,8	0,7	21,4	0,9	0,02	0,02
	STD[2]	0,5	0,3	1,5	1,1	0,5	0,1	2,4	0,3	0,01	0,004
1 (n = 1)		7,0	8,0	8,6	5,7	2,0	0,3	16,1	1,1	0,1	0,02
2 (n = 1)		6,6	8,0	12,7	9,5	2,9	0,6	25,7	2,9	0,21	0,02
4 (n = 5)	Mean	5,9	7,3	9,0	7,1	1,5	0,4	18,3	4,6	0,33	0,05
	STD	0,8	0,8	2,7	1,9	0,5	0,2	5,7	2,6	0,30	0,02
7 (n = 9)	Mean	5,5	6,8	7,4	5,7	1,2	0,3	14,5	5,3	0,37	0,06
	STD	0,4	0,4	1,6	1,7	0,7	0,1	3,4	2,6	0,23	0,03
10 (n = 4)	Mean	5,4	6,8	7,7	5,1	1,8	0,3	14,8	8,3	0,47	0,06
	STD	0,6	0,6	1,5	0,7	1,0	0,1	2,9	0,7	0,12	0,08

[1] arithmetic mean   [2] standard deviation

Table 1: pH-values, exchangeable cations, P-Olsen, organic carbon, total nitrogen of 25 treated Tepetates in the block of Tlaxcala (topsoil, 0 - 30 cm).

	P-adsorption %	HCl-extractable nutrient reserves mg/100g soil				CaCO$_3$ %
		P	K	Mg	Ca	
Mean	11,0	173	19738	31252	35743	1,4

Table 2: *P-adsorption, HCl-extractable nutrient reserves and CaCO$_3$ content of 25 treated Tepetates in the block of Tlaxcala (topsoil, 0 - 30 cm).*

In comparison to N-status the available phosphorous amounts after 7 - 10 years of cultivation indicate a medium level of P-supply. The mean values for P-Olsen were 15.9 kg/ha (5.3 mg/kg) and 24.9 kg/ha (8.3 mg/kg), respectively. The results are in the same range as those obtained by Etchevers et al. (1991). Because of the very low P-adsorption between 7 % - 20 % (mean value 11 %) of the Tepetates, improvement of P-status can easily be reached by application of soluble mineral fertilizers. Investigations of P-supply for different soils in the state of Tlaxcala from Cruz et al. (1991, 1992) showed that 73 % of the soils contained high to very high (P-Olsen > 12 mg/kg resp. > 24 mg/kg) available phosphorus amounts. Comparable investigations from Gonzales (1975) fifteen years ago showed a reversed situation; 50% of the examined soils contained less than 5 mg/kg of P-Olsen. The improvement of P-status can be explained by increasing amounts of P-fertilizer application within the last twenty years (Cruz et al., 1992).

The theoretical maize yield potential was estimated using the QUEFTS model (Quantitative Evaluation of the Fertility of Tropical Soils; Janssen et al. 1986, 1990; Smaling, 1993). Because of the partly empirical character of the model the system is only applicable to soils that have a pH (H$_2$O) in the range 4.5 – 7.0 and values for organic carbon, P-Olsen and exchangeable potassium below 70 g/kg, 30 mg/kg and 30 mmol/kg, respectively. The results are presented in Table 3. The official annual yield evaluations conducted by the Agricultural Department in the state of Tlaxcala resulted in average maize yields for rainfed conditions between 1300 kg/ha and 2100 kg/ha for the period from 1985 to 1991. The mean yield for the seven year period was 1800 kg/ha/yr. The results show that primarily insufficient nitrogen and organic matter contents limited the productivity of treated Tepetates (Table 3). Maintenance and improvement of soil productivity and yield increase in semiarid regions require additional inputs of organic substances (Turenne, 1989; Moll, 1980). A lasting elevation of humus level seems to be possible by continous application of manure or compost (Jones, 1971 cit. in Müller-Sämann and Kotschi, 1994; Pichot, 1983 cit. in Turenne, 1989). Good results are obtained by combined applications of mineral and organic fertilizers (Mokwuney, 1980). Marquez et al. (1992) showed that additional application of 40 t/ha of organic manure increased crop yield significantly in comparison to fields with only mineral fertilization.

Years since cultivation	maize yield kg/ha	pH (H$_2$O)	C$_{org}$ g/kg	P-Olsen mg/kg	K$_{sorb}$ mmol/kg
7	986	6,7	2,5	5,5	9,1
7	995	6,9	2,0	6,8	11,7
7	1304	6,5	3,3	3,3	10,2
10	1349	6,4	3,5	8,8	8,3
4	2586	6,2	6,3	7,0	7,6
7	2720	6,3	7,2	4,5	5,6
7	2770	6,5	6,5	7,7	7,2
10	2989	5,7	8,2	5,7	4,6

Table 3: *Maize yield potential of 8 treated tepetates according to QUEFTS in the block of Tlaxcala.*

Water supply: Soil productivity depends largely on soil water capacity. In semiarid regions climatic constraints such as extreme variability of rainfall (both within seasons and between years), high energy of erosive storms and high evapotranspiration rates, limit crop production. Periods of excesses or deficits of available soil water during the growing season often reduce crop yields. To minimize negative effects for plant growth sufficient storage capacity for plant available water is needed. The mechanical loosening primarily eliminates the physical resistance for root penetrating and provides new pore space for root growth. The effect on pore size distribution was primarily an increase of about 15 % - 20 % of the volume of large pores (> 30 m), whereas the volume of medium and fine pores slightly decreased. The treated Tepetates are characterized by a favourable available water capacity ranging from 14 % to 20 % (pore volume between pF 4.2 and pF 2.0). Because of the low average loosening depth of about 45 cm the total available water capacity is only 65 mm to 90 mm. Investigations from Sivakumar et al. (1987) on Indian Alfisols showed that the risk of yield depression caused by short drought periods within the rainy season were strongly correlated with soil depth and total water storage capacity. Virmani et al. (1980) reported from investigations in semiarid regions of India which showed that the length of effective plant growing periods could be extented from 105 days up to 133 days if water storage capacity increased from 50 mm up to 150 mm. Therefore, for further reclamation measures it is strongly recommended to increase the average loosening depth.

Erosion studies: Soil loss in t/ha on the different erosion plots, annual precipitation and annual erosivity measured in the four year period from 1991 to 1994 are presented in Table 4. The annual rain erosivity (EI_{30}) varied from 110 N/h to 370 N/h. In spite of these low to medium values, individual high erosive storms occured. These storms are characterized by high EI_{30}-values and/or high intensities (rain amount: 62 mm, EI_{30}: 155 N/h, peak intensity: 162 mm/h). For storms with rain amounts > 15 mm the peak intensities varied from 11 mm/h to 162 mm/h. The highest 30-minutes-intensity of an individual storm was 91 mm/h. Soil loss occurred mainly during high erosive storms and as a consequence of erosive rainfall accumulation over a period of several days. At Tlalpan three rain accumulations in 1991 caused 92 % (maize), 82 % (bare fallow) and 69 % (tepetates) of total annual soil loss. Over a period of one week five storms each of 20 mm and more were registered, although in the entire rainy season only seven storms occurred greater than 20 mm. The five storms caused 64 % (maize), 58 % (bare fallow) and 49 % (tepetate) of soil loss. Nonetheless there is a high "interstorm variability" in the sense of Romero Diaz et al. (1988) regarding the erosive efficacy of individual storms.

Site	plots/rainfall	1991	1992	1993	1994	units
Tlalpan	TR /bare fallow	128	106	109	71	t/ha
	TR/maize	26	2	0,5	0,9	t/ha
	T (Tepetate)	43	57	37	32	t/ha
	Rainfall	803	793	663	719	mm
	EI_{30}	301	196	151	390	N/h
El Carmen	TR/bare fallow	90	104	76	63	t/ha
	TR/maize	24	5	25	0,8	t/ha
	T (Tepetate)	7	8	11	9	t/ha
	Rainfall	779	758	642	627	mm
	EI_{30}	157	110	257	149	N/h
Matlalohcan	CAM/bare fallow	49	43			t/ha
	CAM/maize	0,3	0,1			t/ha
	T (Tepetate)	8	9			t/ha
	Rainfall	775	678			mm
	EI_{30}	301	196			N/h

Table 4: soil loss (t/ha), rainfall, annual rainfall-erosivity (EI_{30}) on reclamated tepetates (TR), tepetates (T) and a Chromic Cambisol (CAM) at three sites in the block of Tlaxcala.

Site	soil type	K-factors[1]	number of erosive events (n)	ΔMWD	percolation stability[2]
Tlalpan	TR	0,38	n = 117	3,38	307
El Carmen	TR	0,32	n = 129	3,00	176
Matlalohcan	CAM	0,09	n = 81	1,29	1392
Tlalpan	T	0,13	n = 140		
El Carmen	T	0,03	n = 124		
Matlalohcan	T	0,02	n = 92		

[1] US-units: [(ton · acre · hour)/(acre · tonf · inch)], conversion factor to SI units [(t · h)/(ha · N)]: 1,317 (FOSTER et al. 1981).
[2] percolation rate in ml after ten minutes of percolation.

Table 5: K-factors and aggregate stability of two reclamated tepetates (TR), a Chromic Cambisol (CAM) and three tepetates (T) in the block of Tlaxcala.

The K-factor determined after 4 years are presented in Table 5. According to the classification system of Goldsmith (1977) the treated Tepetates are characterized by a medium to high erodibility, whereas the K-value for the Cambisol indicates a low erodibility. In view of the time-consuming erosion measurements under natural rainfall conditions, many attempts were undertaken to estimate erodibility of soils using simple methods. Various authors reported that aggregate stability is often a good index to describe the susceptibility of soils to erosion (literature cited in Baumann 1996). The results indicated a good relation between erodibilitiy and aggregate stability. Here, the results of the wet sieving method as well as the percolation method showed a significantly higher aggregate stability of the Chromic Cambisol in comparison with the two Tepetates (Table 5). Further investigations on the relation between erodibility and aggregate stability should be done with the aim of estimating relative erodibility of the volcanic ash soils in the state of Tlaxcala.

5 Conclusions

Suitable use of treated Tepetates requires the development of adequate land management systems. A guideline therefore may be the World Bank Technical Paper 221 (Srivastava, 1993). These Tepetates offer several advantages for agricultural crop cultivation. They have medium to high amounts of exchangeable cations, favourable soil-pH and very low P-adsorption rates. At the same time they have various disadvantages such as very low N and P contents recently after loosening. Here, a lasting increase of humus content and nitrogen supply must be a major goal. Although they have a medium to high field capacity, the plant available water content is limited by a low medium loosening depth. The treated Tepetates show a medium to high erodibility. Therefore, effective erosion control measures are indispensible. The following measures are advisable:
- terraces must be well designed such as reverse slope bench terraces or ridge type terraces with a slightly inward slope. To stabilize the ridges they should be planted with perennial grass and Maguey plants (Agave).
- loosening depth should be extended to about 70 cm (100 mm - 140 mm field capacity).
- legumes should be integrated in the crop production system.
- organic manures such as animal waste or compost should be used to increase humus content and to improve soil physical conditions such as soil structure and infiltration.

Long term advice is very important for the small farmers about appropriate cropping systems, erosion control measures and fertilization. Otherwise, economical investments will never be amortised.

References

Aeppli, H. (1973): Barroböden und Tepetate - Untersuchungen zur Bodenbildung in vulkanischen Aschen unter wechselfeuchtem gemäßigtem Klima im zentralen Hochland von Mexiko, Dissertation, Justus-Liebig-Universität Gießen.

Baumann, J. (1996): Die Wirkung der Tieflockerung auf Ertragspotential und Erodierbarkeit verhärteter Vulkanascheböden im Staat Tlaxcala/Mexiko. Boden und Landschaft, Bd. 10, JLU Giessen, 208 S..

Becher, H.H. and Kainz, M. (1983): Auswirkungen einer langjährigen Stallmistdüngung auf das Bodengefüge im Lößgebiet bei Straubing, Z. Acker u. Pflanzenbau 152, 152-158.

Cruz, L., Etchevers, J., Rodriguez, S., Galvis S., A. and Lopez, R.M. (1992): Levantamiento nutrimental de los suelos de Tlaxcala, Agrociencia 3, 157-162.

Cruz, L., Etchevers, J.D. and Rodriguez, J. (1991): Situación del fósforo en los suelos de Tlaxcala, Terra 9, 204-209.

Etchevers, J.D., Zebrowski, C., Hidalgo, M. and Quantin, P. (1991): Fertilidad de los tepetates: situación del fósforo y del potassio en tepetates de los Estados de México y Tlaxcala (México). 1er Símposio Internacional - Suelos volcánicos endurecidos, Resumenes Ampliados, CP, Montecillo, pp. 74-77.

Foster, G.R., McCool, D.K., Renard, K.G. and Moldenhauer, C.W. (1981): Conversion of the universal soil loss equation to SI metric units, J. Soil Water Cons. 36, 355-359.

Goldsmith, P.F. (1977): A practical guide to the use of the universal soil loss equation. Unpublished BAI Techn. Monogr., pp. 34. In: Landon, J.R. (1991): Booker tropical soil manual. A handbook for soil survey and agricultural land evaluation in the tropics and subtropics. Longman, Hong Kong.

Gonzales, D.R. (1975): Predicción de la respuesta del maíz a la fertilización fosfatada en el estado de Tlaxcala, basada en la disponibilidad de fósforo del suelo y otras variables del sitio, Tesis de maestría en Ciencias, Colegio de Postgraduados. Chapingo, México.

Hartge, K.H. and Horn, R. (1989): Die physikalische Untersuchung von Böden. Stuttgart.

Houba, V.J.G., Lee, J.J., Novozamsky, I. and Walinga, I. (1989): Soil and Plant Analysis, a series of syllabi. Part 5: Soil Analysis Procedures, Department of Soil Science and Plant Nutrition, Wageningen Agricultural University. Wageningen, Niederlande.

Jackson, M.L. (1958): Soil chemical analysis, 498 pp., Prentice-Hall, Englewood Cliffs, N.J.

Janssen, B.H., Guiking, F.C.T., Eijk, D. van der, Smaling, E.M.A. and H. Reuler (1986): A new approach to evaluate the chemical fertility of tropical soils. Trans. XIII Congr. ISSS, Hamburg, Vol. III, pp. 791--792.

Janssen, B.H., Guiking, F.C.T., Eijk, D. van der, Smaling, E.M.A., Wolf, J. and Reuler, H. (1990): A system for quantitative evaluation of the fertility of tropical soils (QUEFTS). Geoderma 46, 299-318.

Knickmann, E. (1955): Die Untersuchung von Böden. Methodenbuch Bd. 2, Neumann-Verlag, Radebeul und Berlin.

Lauer, W. and Stiehl, E. (1973): Hygrothermische Klimatypen im Raum Puebla-Tlaxcala (Mexiko), Erdkunde 27, 230-234.

Marquez, A., Zebrowski, C. and Navarro, G.H. (1992): Alternativas agronómicas para la recuperación de tepetates. Terra, 10, 465-473.

Mizota, C. and Reeuwijk, L.P. (1989): Clay mineralogy and chemistry of soils formed in volcanic material in diverse climatic regions. ISRIC, Soil Monograph 2, Wageningen, Niederlande.

Mokwunje, U. (1980): Interactions between farmyard manure and fertilizers in savanna soils, FAO Soils Bulletin 43, 192-200.

Moll, W. (1980): Der Verwitterungszustand des Bodens in den Tropen als begrenzender Faktor der Nutzung, Giessener Beitr. Entwicklungsf., Reihe I, 6, 87-97.

Müller-Sämann, K.M. and Kotschi, J. (1994): Sustaining Growth: soil fertility management in tropical smallholdings. Weikersheim (Markgraf).

Olsen, S.R., Cole, C.V., Watanabe, F.S. and Dean, L.A. (1954): Estimation of available phosphorus in soils by extraction with sodium bicarbonate. USDA circ. 939.

Reeuwijk, L.P. van (Ed.) (1986): Procedures for soil analysis, ISRIC Technical Paper No. 9, Wageningen, Niederlande.

Romero-Diaz, M.A., Lopez-Bermudez, F., Thornes, J.B., Francis, C.F. and Fisher, G.C. (1988): Variability of overland flow erosion rates in a semiarid mediterranean environment under mattoral cover, Murcia, Spain, CATENA SUPPLEMENT 13, 1-11.

Sivakumar, M.V.K., Singh, P. and Williams; J.H. (1987): Agroclimatic aspects in planning for improved productivity in Alfisols, In: Alfisols in the Semi-Arid Tropics. Proc. Consultants Workshop State of the Art and Manag. Altern. Optim. Prod. SAT Alf., 1-3 Dec. 1983, ICRISAT, 15-30.

Smaling, E.M.A. (1993): The soil nutrient balance: An indicator of sustainable agriculture in Sub-Saharan Africa, The Fertilizer Society, Paper read at the international conference in Cambridge on 8-9 december 1993.

Srivastava, J.P., Tamboli, P.M., English, J.C., Lal, R. and Stewart, B.A. (1993): Conserving soil moisture and fertility in the warm seasonally dry tropics. World Bank Technical Paper No. 221, Washington, D.C..

Turenne, J.F. (1989): Soil organic matter and soil fertility in tropical and sub-tropical environments. In: Maltby, E. and Th. Wollerson (eds.), Soils and their management - A sino-european perspective. Elsevier, London and New York.

Virmani, S.M., Sivakumar, M.V.K. and Reddy, S.J. (1980): Climatological features of the SAT in relation to the farming system program. Proc. Intern. Workshop Agroclimatological Research Needs of the Semi-Arid Tropics, 22-24 Nov. 1978, ICRISAT, Hyderabad, A.P.,India. pp. 5-16.

Werner, G. (1988): Die Böden des Staates Tlaxcala im zentralen Hochland von Mexiko. Das Mexiko-Projekt der Deutschen Forschungsgemeinschaft, XX, 207 S., Wiesbaden (Steiner).

Wischmeier, W.H. and Smith, D.D. (1978): Predicting rainfall erosion losses. A guide to conservation planning. USDA Agriculture Handbook No. 537, 58 pp., Washington.

Address of authors:
J. Baumann
G. Werner
W. Moll[†]
Justus-Liebig-Universität Giessen
Tropical Research Center
Schottstraße 2
D-35390 Giessen, Germany

Use of Soil Amendments to Prevent Soil Surface Sealing and Control Erosion

D. Norton & K. Dontsova

Summary

Surface sealing leads to low water infiltration rates producing runoff and erosion even in low intensity rainfall events. Structural instability and the low electrolyte concentration of rainwater leads to breakdown of aggregates and dispersion of colloids producing low steady state infiltration rates (I_s). It has been hypothesized that for some soils exchangeable Mg (ExMg) may behave similar to soils with Na causing dispersion and increased surface sealing. In order to study this effect, a rainfall simulator study was conducted to measure infiltration and erosion of five soils from the midwest USA. Small interrill plots (0.14 m^2) were packed with the sieved soil and brought to saturation from below for 2 hours, and subjected to deionized rainfall at a rate of 64 mm/hr at 5% slope with -5 cm tension for one hour. Samples of infiltrating water and runoff were collected at 5-minute intervals. Soil loss was determined by measuring gravimetrically the sediment concentration in the runoff. Four replications for each soil were performed. In addition to the control, each soil was surface amended with anionic polyacrylamide (PAM) at 20 kg/ha, 5,000 kg/ha fluidized bed combustion bottom ash (FBCBA) and PAM plus FBCBA.

All soils rapidly developed surface seals when untreated at this rainfall intensity except a sandy soil. Three of the soils had I_s of <7mm/hr. The type of clay minerals did not seem to have a significant effect on this process. A fine-silty soil with a low ExMg content had a I_s of >7x that of a similar soil with a Mg:Ca ratio >1. The total soil loss (TSL) was also significantly less for the low Mg soil. Amending the soils with PAM and FBCBA increased the I_s and reduced the TSL for all soils except the sandy soil. The effect was greatest when both PAM and FBCBA were applied to the soil. Amending soils to increase I_s and reduce runoff and TSL seems to be a viable conservation practice.

Keywords: Clay dispersion, clay flocculation, infiltration, runoff polyacrylamide, gypsum

1 Introduction

Farmers in the USA and Canada have recently started using yield monitors equipped with Global Positioning System (GPS) and computers to map yields across a field (Lachapelle, et al., 1994). Their crop yields were highly variable although the crop may have appeared similar. Some of the more progressive farmers have performed grid sampling for soil fertility analyses and found that the soil test results have a negative correlation with crop yield. Many are convinced that the yield variation and negative correlations of fertility with yield are due to water entry problems related to poor soil structure or quality. Water entry problems due to soil structural instability and surface sealing causes runoff on elevated landscape positions and ponding in lower landscape positions of the late-Wisconsinan glacial till plain of the USA cornbelt. Lack of water in the higher positions and too much water in the lower positions leads to moisture stresses on plants reducing yields in both areas, thus significantly reducing the overall yield of the field. The runoff production from poor water entry also produces soil loss and causes adverse environmental problems off-site.

It has been long known that soils with even low amounts of sodium and sodium absorption ratios (SAR) can be dispersive and have low water infiltration rates (Singer, et al., 1982). Recently, the effect of exchangeable magnesium on dispersion and surface sealing has received attention. Keren (1991) found that soils containing montmorillonite were more susceptible to surface sealing and had lower water intake rates with Mg on the exchange complex than the same soil saturated with Ca even in the presence of carbonates. He attributed this effect to the greater hydration radii of Mg ions on the external surfaces of clay tactoids.

Many soils from the eastern USA that are low in Na experience water entry problems with or without montmorillonite (Norton et al., 1993, Norton, 1995). This poor water intake rate causes runoff even with low intensity rainfall and results in poor water use efficiency. This can occur under both rainfed and irrigated farming. The low electrolyte content of rainwater enhances dispersion and surface sealing which can be eliminated by a surface application of gypsum or gypsum like materials (Reichert and Norton, 1994) by releasing electrolytes to the rainwater. Although the effect of electrolytes on reducing chemical dispersion and surface sealing is well known, its relative effect versus that of the Ca:Mg ratio is not known. The objective of this study is to determine the relative importance of Ca:Mg ratios versus the effect of electrolyte release on surface sealing and water entry for some soils common to the USA cornbelt and to evaluate the efficiency of two soil amendments on reducing surface sealing and controlling erosion, and to test if the Ca:Mg on the soil exchange complex affects the surface sealing process.

2 Materials and methods

Five soils were chosen for study that represented a range of soils found within the corn belt of the USA. These soils (Table 1) varied in texture, mineralogical and chemical composition. The soils were characterized for chemical, physical and mineralogical composition using the procedures detailed in Reichert and Norton (1996). Soil aggregates were measured using the Yoder apparatus to compute median diameter and diameter at 50 percent passing and a Griffith fall velocity tube (Hairsine and McTainsh, 1986). The aggregates were measured in the Griffith tube both saturated and air-dry to compute a "slaking index" (SI).

Soil	Classification in US Taxonomy
Blount	fine, illitic, mesic Aeric Ochraqualf
Catlin	fine-silty, mixed, mesic Typic Hapludoll
Chelsea	mixed, mesic Alfic Udipsamments
Fayette	fine-silty, mixed, mesic Typic Hapludalf
Miami	fine-silty, mixed, mesic Typic Hapludalf

Table 1: Soils selected for study.

The infiltration, runoff and soil loss was measured on small pans 0.14 m^2 using the procedures described in Reichert and Norton (1996). The treatments included in this study were: control, 20kg/ha anionic polyacrylamide (PAM, American Cyanamide Magnifloc 836A) surface applied in solution, 5MT/ha fluidized combustion bottom ash (FBCBA) surface applied dry granules, and PAM+FBCBA surface applied at the same rates. Mean values for infiltration, runoff and soil loss measurements for treatments were compared using Duncan's multiple range test with P=0.05.

3 Results and discussion

Particle size distribution for the soils is given in Table 2. The Fayette had the highest clay content followed by Blount, Miami, Catlin, the later two with very high silt contents. The Chelsea soil is a sandy soil. The clay mineralogies (not shown) for Catlin, Miami and Chelsea were mixed and nearly identical, whereas, Blount was illitic and Fayette, smectitic.

Soil	Sand	Silt	Clay	Texture Class	MWD	D_{50}	Wet V_{50}	Dry V_{50}	SI
	------------%-------------				---mm---		--cm/s--		
Blount	30.6	44.0	25.4	L	1.1	0.2	8.0	9.7	0.82
Catlin	23.4	58.5	18.1	SiL	1.5	0.7	8.1	8.4	0.96
Chelsea	95.5	2.0	2.5	FS	0.5	0.2	1.8	ND*	ND
Fayette	5.7	65.2	29.1	SiC	0.5	0.1	8.8	8.1	1.08
Miami	10.4	71.1	18.5	SiL	0.5	0.1	8.3	3.4	2.44

*ND = Not determinable and not applicable.

Table 2: Results of the soil textural and aggregate size analyses.

The soils exhibited a considerable difference in aggregation (Table 2). The Catlin soil had the greatest mean weight diameter (MWD) and diameter at 50 percent passing (D_{50}) with Miami and Fayette the least. This indicates that the Catlin had the most stable aggregates of all the soils. The sealing index (SI) which is defined as the ratio of the wet to dry fall velocity at 50 percent mass (V_{50}) was least for the Blount and greatest for the Miami (Table 2).

The chemical properties of the soils also varied widely. The main differences (Table 3) were in the cation exchange capacity (CEC), organic carbon (OC) and the Mg/Ca ratio of the exchange complex. The Catlin had the lower Mg/Ca ratio (Table 3) and Miami by far the greatest with Fayette intermediate. Miami had the greatest total CEC followed by Fayette, Catlin, Blount and Chelsea. Catlin had the greatest OC content followed by Miami, Blount, Fayette and Chelsea.

Soil	H	K	Na	Ca	Mg	CEC	Mg/Ca	pH	OC
	---------------------mol_c/kg---------------------							-%-	
Blount	5.7	0.30	<0.01	15.4	3.9	25.3	0.25	6.5	1.45
Catlin	3.6	0.29	<0.01	21.2	3.4	28.5	0.16	7.0	5.48
Chelsea	3.7	0.16	<0.01	1.0	0.2	5.1	0.20	5.8	0.37
Fayette	6.3	0.47	<0.01	12.7	11.7	31.2	0.92	6.7	1.01
Miami	6.6	0.35	<0.01	11.5	16.4	34.8	1.42	6.2	1.57

Table 3. Chemical properties of the studied soils.

Steady state infiltration rate (I_s) and steady state soil loss (SL_s) data are presented in Table 4. All of the soils except the Chelsea fine sand developed surface seals, but to a highly variable extent. This was reflected in an exponential decrease of the initial infiltration rate to a lower steady state rate (Figures 1-4). Three of the soils exhibited I_s of less than 7 mm/hr when untreated. Treatment affects on I_s and SL_s (Table 4) were considerable and highly variable.

Since Chelsea did not have a sealing problem none of the treatments had a significant affect on SL_s or I_s. The surface treatment of PAM had no significant difference from the control on I_s, but did significantly lower SL_s for Catlin and Fayette. Surface application of FBCBA significantly increased I_s on all soils but Chelsea. It also decreased SL_s for the soils except Miami. Since FBCBA releases electrolytes which flocculates clays, this process must be important in surface sealing. The PAM+FBCBA(P+F) treatment behaved similar to FBCBA alone.

Soil	Treatment							
	Control	PAM	FBCBA	P+F	Control	PAM	FBCBA	P+F
	Infiltration (mm/hr)				Soil Loss (g/sq m/s)			
Blount	6.17b	6.83b	17.8a	20.6a	0.113a	0.095ab	0.065ab	0.030b
Catlin	15.8b	24.4b	46.0a	39.1a	0.140a	0.023b	0.008b	0.013b
Chelsea	62.6a	36.8b	66.6a	60.0a	0.008a	0.045a	0.000a	0.020a
Fayette	2.63c	3.42c	14.9a	10.2b	0.213a	0.075b	0.080b	0.078b
Miami	3.42b	3.02b	14.2a	3.60b	0.090a	0.148a	0.158a	0.140a

Means with like letters are not significantly different at the P=0.05 level between treatments according to Duncans multiple range test.

Table 4: Steady state infiltration rate and soil loss.

Figures 1-4 show the precipitation, infiltration runoff and soil loss rates as a function of time during the run for Catlin and Miami. These two soils are very similar, in parent material and genesis as well as texture and clay mineralogy. The main difference between these two soils is the state of aggregation (Table 2) and the soil chemical properties mainly the Mg/Ca ratio and organic matter (Table 3). The Catlin soil has larger and more stable aggregates than Miami and a much lower sealing index. It also has very low Mg/Ca ratio.

It seems very likely that the high aggregate stability of Catlin relative to Miami is partially due to the low content of Mg (Keren, 1991). This high stability is reflected in the low soil loss and high infiltration rate as seen in Figure 1. Compared to Miami in Figure 3, Catlin has much greater infiltration and less soil loss as a result of this greater stability. Although Catlin is much less susceptible to surface sealing than Miami, it still has a positive effect of adding PAM+FBCBA (Figure 2) indicating this treatment is a viable erosion control treatment even for stable soils. This conclusion is also valid for the Blount and Fayette soils (Table 4).

The Miami soil on the other hand is not stable and is highly susceptible to surface sealing. Unlike the other soils that surface sealed, this soil did not have a positive effect of surface treatment with any of the treatments including PAM+FBCBA (Figure 4). The only explanation for its completely different behavior is the high Mg content causing dispersion and low infiltration rate even in the presence of an electrolyte source (FBCBA) and a binding agent (PAM). Although FBCBA is also a source of Ca, it is unlikely that during the short duration of rainfall that the Mg/Ca ratio would change.

4 Conclusions

Surface sealing was a problem for all soils studied except the sandy Chelsea soil. The addition of soil amendments produced variable results. Generally, the FBCBA treatment performed better than PAM alone, but when applied together worked better. For soil with a high Ca status, the

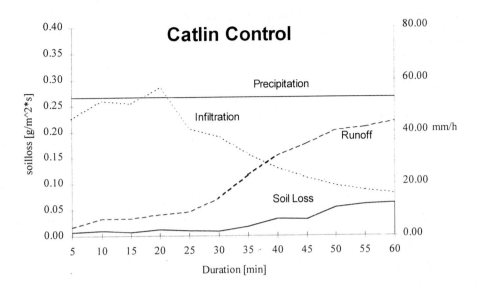

Figure 1: Soil loss, infiltration and runoff rates for Catlin with no amendments.

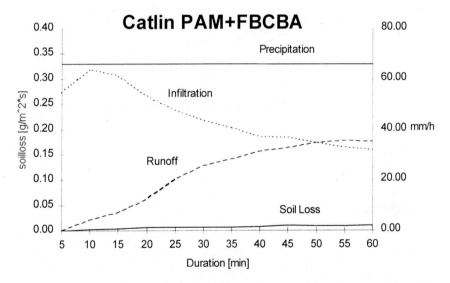

Figure 2: Soil loss, infiltration and runoff rates for the PAM plus FBCBA treatment for Catlin.

electrolyte affect was considerable and soil erosion was almost completely eliminated under the conditions of the experiment. However, for a similar soil with high Mg content, erosion control using PAM+FBCBA was not effective because of the dispersive nature of Mg. The effect of the low Mg/Ca ratio of the Catlin soil relative to the high Mg/Ca ratio of the Miami soil resulted in a final infiltration rate of more than seven times under the same experimental conditions even though the soils were similar in most other respects. As a result, the electrolyte effect was not observed for the Miami soil. More research is needed to separate the specific effect of the Mg/Ca ratio on the exchange complex, the electrolyte effect and type of charged surfaces in soils before large scale application of these control methodologies are applied to farm fields.

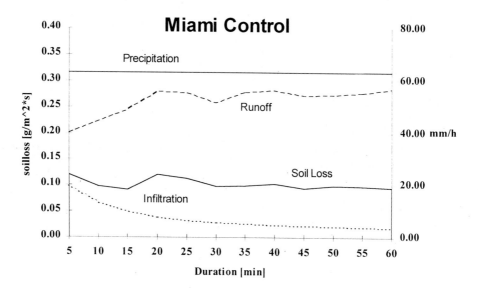

Figure 3: Soil loss, infiltration and runoff rates for Miami with no amendment.

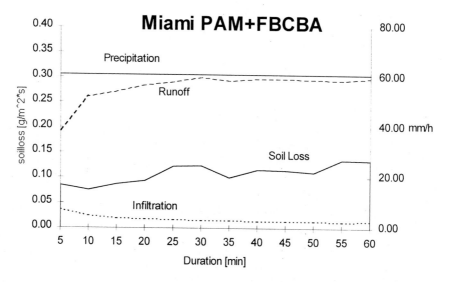

Figure 4: Soil loss, infiltration and runoff rates for the PAM plus FBCBA treatment for Miami.

References

Hairsine, P. and McTainsh, G. (1986): The Griffith tube: A simple settling tube for the measurement of settling velocity of aggregates. Australian Environmental Studies Working Paper 3/86.

Keren, R. (1991): Specific Effect of Magnesium on Soil Erosion and Water Infiltration. Soil Sci Soc. Am. J. **55**, 783-87.

Lachapelle, G., Cannon, M.E., Gehue, H., Goddard, T.W. and Penny, D.C. (1994): GPS System for Integration and Field Approaches to Precision Farming. Navigation **41(3)**, 323-35.

Norton, L.D., Shainberg. I. and King, K.W. (1993): Utilization of gypsiferous amendments to reduce surface sealing in some humid soils in the eastern USA. CATENA SUPPLEMENT **24**, 77-82.

Norton, L.D. (1995): Mineralogy of high calcium/sulfur-containing coal combustion by-products and their effect on soil surface sealing. In: Agricultural Utilization of Urban and Industrial By-Products (D.L. Karlen, R.J. Wright and W.O. Kemper, eds.), Am. Soc. of Agron. Special Publication Number 58, Am. Soc. Agron., Madison, WI, 87-106.

Reichert, J.M. and Norton, L.D. (1994): Fluidized bed bottom-ash effects on infiltration and erosion of swelling soils. Soil Sci. Soc. Am. J. **58**, 1483-1488.

Reichert, J.M. and Norton, L.D. (1996): Fluidized bed combustion bottom-ash effects on infiltration and erosion of variable-charge soils. Soil Sci. Soc. Am. J. **60**, 275-282.

Singer, M.J., Jenitzky, P. and Blackard, J. (1982): The influence of exchangeable sodium percentage on soil erodibility. Soil Sci. Soc. Am. J. **46**, 117-121.

Addresses of authors:
Darrell Norton
Katerina Dontsova
USDA-ARS
National Soil Erosion Research Laboratory
1196 Soil Building
Purdue University
West Lafayette, Indiana 47907-1196
USA

Roughness and Sealing Effect on Soil Loss and Infiltration on a Low Slope

K. Helming, M.J.M. Römkens & S.N. Prasad

Summary

Rills are the major source of soil loss from sloping fields. It is generally accepted that soil surface roughness reduces runoff and soil loss, yet, little quantiative information is available. This study investigates the effect of soil surface roughness on infiltration and soil loss during the stages of seal development and runoff generation, steady-state flow conditions, and rilling.

Flume studies using a rainfall simulator consisted of applying (i) a sequence of four rainstorms of constant rainfall amount with decreasing intensity, and (ii) two 30 min overland flow regimes to a loess soil of 2 % slope. Treatments consisted of rough, medium-rough and smooth soil surface conditions.

During the initial stage of seal development and runoff generation, roughness affected the time of incipient runoff and the reduction in the seal hydraulic conductance. Furthermore, spatial runoff distribution patterns led to high soil losses for rough surface conditions. The second stage was characterized by near steady-state values of infiltration and soil loss rates during the low intensity rainstorms. The effect of surface roughness on infiltration and soil loss was minor. The third stage represented high overland flow rates that induced flow concentration. Whereas soil loss increased on the smooth surface due to rill formation that was characterized by the development of headcuts, no rills developed on the rough surfaces, though preferred pathways for flow were evident. The results suggest that surface roughness may be a factor in reducing the risk of rill development and soil erosion on low slopes.

Keywords: Roughness, sealing, infiltration, soil loss, rilling.

1 Introduction

Soil erosion is a complex phenomenon involving many components and interacting processes. Runoff and soil erosion only occurs when rainfall intensity exceeds the infiltration rate. During the early stages of a rainstorm, tilled soils usually have a high water infiltration capacity, such that runoff will be delayed. However, raindrop impact and slaking of clods and aggregates and subsequent rearrangement of solid particles leads to the formation of surface seals. These surface seals appreciably reduce the effective hydraulic conductivity (McIntyre, 1958), so that surface sealing is an important factor in the development of runoff and soil erosion. Surface roughness substantially affects sealing and thus the partition of rainfall into infiltration and runoff. Various roughness parameters can be used to describe the effect of surface roughness on erosion processes: clod size distribution which affects the resistance to soil detachment by raindrop impact (Römkens & Wang, 1987), the specific surface area which affects the amount of rainfall energy impacting per unit area (Helming et al., 1993), the depressional storage which affects soil detachment by raindrop impact due to the presence of ponded water (Hairsine et al., 1992), as well as incipient runoff after infiltration rate decreases below rainfall intensity (Huang & Bradford, 1990). Thus, roughness is an important condition that reduces sealing and delays runoff. Roughness effects are especially evident during the early stages of the erosion process, when seal development and runoff generation are the major factors of interest. Little information is available about the effectiveness of roughness in reducing soil erosion risks during successive rainstorms when runoff tends to concentrate and flow pathways are established. Since this process is highly affected by slope steepness (Huang, 1995), methods to reduce the risk of rill development and growth should be easily obtained on low slopes. We hypothesized that a rough soil surface created by usual tillage practices might be a sufficient measure to substantially reduce the risk of flow concentration and rilling on low slopes. This practice has two advantages: it can easily be obtained by secondary tillage practices, and it does not adversely affect the established management of intensive crop production. The objective of this study was to evaluate the effect of roughness on infiltration, runoff, and soil loss during a series of simulated rainstorms and overland flow regimes on a low slope.

2 Experimental approach

2.1 Equipment

Simulated rainfall studies with different surface roughness conditions were conducted on a flume of 3.7 m x 0.61 m x 0.23 m (length, width, depth). A detailed description of the experimental setup is given elsewhere (Römkens et al., 1996). A multiple-intensity rainfall simulator consisting of three oscillating VeeJet nozzles placed 1.64 m apart was mounted above the flume. The simulator is in concept and design similar to the single nozzle simulator described by Meyer and Harmon (1979). For overland flow studies, an inlet tank was attached to the upper end of the flume. Water was admitted as uniform flow to the soil bed over a level, baffled edge on the downstream side of the tank. A removable laser microreliefmeter (Römkens et al. 1988) was mounted on top of the flume. Its automatic movement into the longitudinal and transverse directions allowed the digitization of the soil surface microrelief within the flume with a height resolution of 0.25 mm.

2.2 Soil and soil bed preparation

The Ap material of a Glossic Fragiudalf with a particle size distribution of 18 % clay, 79 % silt, 2 % sand, and with 1.1 % organic carbon was used in this study. The flume soil bed was prepared in three stages. In the first stage, perforated drainage pipes were embedded into a 30 mm thick sand layer on the bottom of the flume. In the second stage, a 130 mm thick layer of soil, sieved to pass a 4 mm screen, was packed on top of the sand layer. Packing was conducted in a careful and reproducible way to achieve a controlled, uniform density. In the third stage, the upper 70 mm of the soil bed was packed with air dry soil, sieved through screens to obtain the desired degree of surface roughness. We used sieve sizes of 2 mm, 27 mm, and 56 mm to prepare a smooth, medium rough, and rough surface condition, respectively. With this procedure surface roughness was determined by the diameter of the biggest clods of the surface material. The rough and medium rough surface conditions represented different seedbed roughnesses commonly observed in the field. The smooth surface represented an artificial condition usually assumed in computational models describing overland flow.

2.3 Experimental procedure

Experiments were performed at 2 % slope steepness. The soil was air-dried before use in each experiment. Soil bulk density ranged between 1.3 and 1.5 Mg m^{-3}. Each experiment consisted of applying a series of four 45 mm rainstorms and two overland flow regimes. The rainstorm intensity sequence started with 60 mm h^{-1} for the first rainstorm, followed by 45 mm h^{-1}, 30 mm h^{-1}, and 15 mm h^{-1} for the second, third, and fourth rainstorm, respectively. The two overland flow regimes of 380 l h^{-1} and 1034 l h^{-1} lasted 30 min each and simulated high flow rates. During the rainstorms and overland flow tests samples were intermittently taken at intervals of 4 to 5 minutes for gravimetric determination of runoff rate and sediment concentration. Before and after each test, the topography of the soil surface was determined using a laser microreliefmeter. Grid distance was 3 mm resulting in a digital elevation map with 111 875 data points per m². The microrelief data were used to characterize surface roughness and to visualize runoff flow paths and the spatial distribution of runoff.

3 Results and discussion

3.1 Rainstorm tests

The infiltration and soil loss relationships on a sealing soil usually follow a distinct pattern (Moore & Singer, 1990). At the beginning of the rainstorm, the infiltration rate is similar to the rainfall intensity. The development of the seal reduces the surface hydraulic conductance and the infiltration rate decreases sharply followed by a rapid increase in the soil loss rate until a maximum is reached. Subsequently, the infiltration rate and soil loss rate attain steady-state condition indicating that seal development is completed. This distinct pattern was reflected in the infiltration and soil loss rates obtained during the rainstorm tests presented in Fig. 1 and Fig. 2. The duration of the preponding situation was 6 to 8 minutes longer for the rough and medium rough surface than for the smooth surface indicating the runoff delay effect of rough surfaces. Seal development was reduced on the rough surface conditions with big stable aggregates of up to 56 mm diameter, which have the effect of increasing the duration of high infiltraton rates. The decrease in the infiltration rate was sharpest for the smooth surface. This rapid change reflected the effect of a reduced surface hydraulic conductance due to seal development. However, the near equilibrium infiltration rate for the smooth surface was slightly higher than that for medium rough surface condition. As may be inferred from the visualisations of the surfaces obtained from laser microreliefmeter measurements (Fig. 3a), runoff occured as spatially uniform flow covering the entire surface area with low flow depth. Low flow velocity leads to a relatively long detention time of water on the surface with thicker waterfilms, reduced raindrop energy impact, lessened surface sealing, and therefore enhanced infiltration. Soil detachment was low, and the material that was detached could not readily be transported by runoff thus resulting in low values of soil loss rates for the smooth surface. On the rough and medium rough surfaces, soil loss increased later but more rapidly and yielded a higher degree of steady-state soil loss rate. In this case, surface flow was distributed between and around the clods routing the runoff into several flow paths (Fig. 3b). Within these flow paths runoff depth and flow velocity was high increasing both the detachment of soil material and the transport capacity of the flow.

The subsequent rainstorms were characterized by equilibrium infiltration and soil loss rates (Fig. 1 and Fig. 2). The process of seal development was more or less completed during the first storm, and infiltration rates as well as soil loss rates gained steady-state conditions. The differences in infiltration rates between the three different roughness conditions diminished continuously with rainfall, and also soil loss rates approached similar low values. Apparently, the differences in the initial roughness had almost disappeared during the first rainstorm. Microrelief measurements confirm this observation. The specific surface area, which is an index sensitive to microrelief variations (Helming et al., 1993), decreased to values of about 1.1 indicating that roughness elements had been levelled off to a high degree during the first storm (Fig. 4).

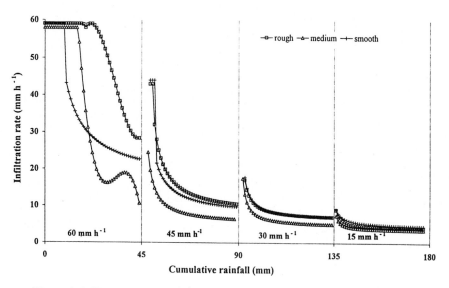

Figure 1: Infiltration rate (mm h^{-1}) for the four rainstorms and three surface conditions.

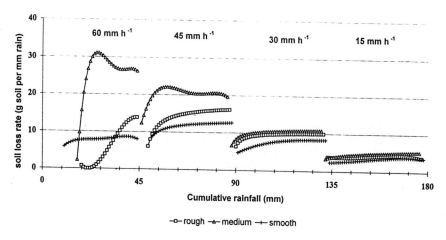

Figure 2: Soil loss rate (g soil per mm rain) for the four rainstorms and three surface conditions.

In short, the results from rainfall tests suggest that surface roughness had an appreciable effect on soil loss and infiltration rate during the seal development stage, but that the effect on soil loss was minor during the subsequent stage of steady-state sheet flow.

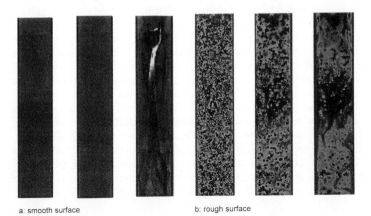

a: smooth surface   b: rough surface

Figure 3: Black and white maps of soil surface topography derived from laser microreliefmeter measurements for the initial situation (left), after the first rainstorm (middle), and at the end of the experiment (right). Pixel size was 3 mm x 3 mm, the area of each picture is 0.6 m x 2.8 m.

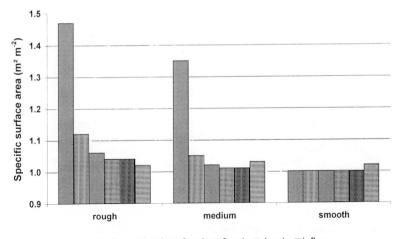

□ before ▨ 1. rain ▨ 2. rain ▨ 3. rain ■ 4. rain ▨ inflow

Figure 4: Parameter specific surface area describing surface roughness changes for the three surface conditions.

3.2 Overland flow test

Two overland flow tests were conducted to study the effect of different initial roughness conditions on flow concentration, rill development, and soil loss for high flow rates. The tests were carried out on the wet, sealed surfaces after the rainfall tests. The infiltration rates were about 3 mm h^{-1} for both overland flow tests and this value was not affected by the initial roughness conditions. Yet, soil loss was appreciably affected by the initial roughness conditions (Fig. 5).

Whereas soil losses were small for all roughness conditions during the first test with 380 l h^{-1} flow intensity (equivalent to 173 mm h^{-1}), soil loss increased substantially for the initially smooth surface during the second, high intensity test with 1034 l h^{-1} (470 mm h^{-1}). As can be observed in Fig. 3a, flow concentrated on this surface thereby altering the flow regime from an initially, unifomly distributed shallow overland flow to concentrated flow with one main flow path. The flow concentration led to a sharp increase in flow depth and flow velocity, which enhanced soil detachment and flow transport capacity. The flow regime on the rough surfaces did not undergo these changes. Runoff was already spatially distributed during the very first rainstorm within several flow paths between and around the clods. The flow paths had stabilized during successive rainstorms of lesser intensity, and rilling as characterized by headcut development did not occur even during the high flow rates on the initially rough surfaces (Fig. 3b). Thus, the initial differences in the roughness conditions, although diminished during the rainfall tests, substantially affected the erosion process during high intensity flow rates. Based on these findings, we may postulate that rilling and soil loss is appreciably affected by the initial roughness condition. The risk of rilling seems to be higher on smooth surfaces than on rough surfaces.

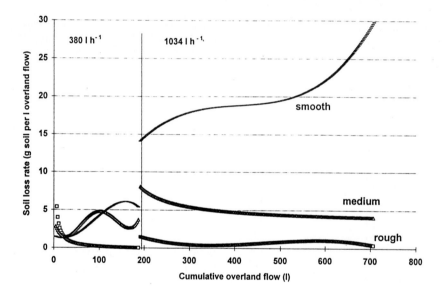

Figure 5: Soil loss rate (g soil per l overland flow) for the two overland flow tests and three surface conditions.

4 Conclusions

A flume study with simulated rainfall and high intensity overland flow was conducted to determine the effect of surface roughness on infiltration, soil loss, and rilling. Results suggest that the roughness effect may be grouped into three stages of the soil erosion process. During the first stage of seal development and runoff generation, the time for incipient runoff and the reduction of the seal hydraulic conductance was highly influenced by surface roughness. Roughness dependent spatial flow distribution affected soil loss during this stage. Near equilibrium rates of infiltration and soil loss occured in the second stage, and no appreciable effect of roughness could be measured. The development of channelized flow due to high flow rates in the third stage were substantially affected by roughness. Whereas the concentration of flow within one main flow path led to a pronounced increase in soil loss on the smooth surface, no changes in the flow regime occurred on the rough surfaces and soil loss rates remained at a low level. Runoff volume was not affected by roughness conditions.

It may be concluded that for the conditions of a low slope, surface roughness may be a factor in reducing the risk of rill development and soil loss. Roughness can easily be varied with secondary tillage practices and it does not affect the economics of crop production.

References

Hairsine, P.B., Moran, C.J. and Rose, C.W. (1992): Recent developments regarding the influence of soil surface characteristics on overland flow and erosion, Aust. J. Soil Res. **30**, 249-264

Helming, K., Roth, Ch.H., Wolf, R. and Diestel, H. (1993): Characterization of rainfall - microrelief interactions with runoff using parameters derived from digital elevation models (DEMs), Soil Technology **6**, 273-286

Huang, C. and Bradford, J.M. (1990): Depressional storage for markov-gaussian surfaces, Water Resources Research **26**, 2235-2242

Huang, C. (1995): Empirical Analysis of slope and runoff for sediment delivery from interrill area, Soil Sci. Soc. Am. J. **59**, 982-990

McIntyre, D.S. (1958): Permeability measurements of soil crusts formed by raindrop impact, Soil Science **85**, 185-189

Meyer, L.D. and Harmon, W.C. (1979): A multiple intensity rainfall simulator for erosion research on row sideslopes, Transaction of the ASAE **22**, 100-103

Moore, D.C. and Singer, M.J. (1990): Crust formation effects on soil erosion processes, Soil Sci Soc. Am. J. **54**, 1117-1123

Römkens, M.J.M. and Wang, J.Y. (1987): Soil roughness changes from rainfall, Transactions of the ASAE **31**, 408-413

Römkens, M.J.M., Wang, J.Y. and Darden, R.W. (1988): A laser microreliefmeter, Transaction of the ASAE **31**, 408-413

Römkens, M.J.M, Prasad, S.N. and Helming, K. (1996): Sediment concentration in relation to surface and subsurface hydrologic soil conditions, Proceedings of the 6th Federal Interagency Sedimentation Conference, Las Vegas, Nevada, USA, Vol. 2, IX9-IX16

Addresses of authors:
Katharina Helming
Center for Agricultural Landscape and Land Use Research (ZALF)
Eberswalder Straße 84
D-15374 Müncheberg, Germany
M.J.M. Römkens
USDA-ARS
National Sedimentation Laboratory
Oxford, MS 38655, USA
S.N. Prasad
University of Mississippi
University, MS 38677, USA

Effect of Different Land Preparation Practices on Crop Production and Soil Compaction

N.M. El-Mowelhi, M.S.M. Abo Soliman, S.A. Abd El-Hafez & S.A. Hassanin

Summary

Two field trials were conducted at the Sakha Agricultural Research Station Farm at the North Delta, Egypt, during the 1994 growing season. In the first field trial, four different land preparation practices and their impact on rice crop and soil compaction were evaluated. Medium tillage treatment achieved the highest values of grain yield of rice and water utilization efficiency, increased the values of soil bulk density and altered soil infiltration characteristics. The recommended tillage improved the soil conditions (decreased soil bulk density and raised soil infiltration values) as compared to other treatments: traditional, minimum and medium tillages.

In the second field trial, three different land-levelling practices and two levels of tillage were applied before planting the cotton crop. The treated ground surface slope resulted in the highest seed cotton yield, and recorded the highest values of water utilization efficiency compared to other treatments (flat and traditional land levelling). The deep plowing treatment received more irrigation water than the surface plowing, while the differences in seed cotton yield was not significant. At the same time deep plowing reduced soil compaction. It is also clear that the precision land-levelling treatments had an unfavourable effect on both bulk density and infiltration rate characteristics of the soil. This trend could be attributed to compaction as a result of the passing of heavy equipment used in finishing such work.

Keywords: Soil compaction, tillage, land-levelling, rice, cotton

1 Introduction

The compaction process is defined as the change in volume for a given mass of soil. This simple process causes progressively serious problems in agricultural top soils which are often affected by machinery operating in the field. The resulting difficulties involve excess soil hardness reduced soil permeability to water and air flow and a resulting loss of crop yield (Ragavan et al., 1976). Therefore, optimum tillage operations should encourage root development and penetration and provide an optimum air-water balance, as well as maximum water storage capacity and moderate gas exchange for the root growth (Suliman et al., 1993).

Soil bulk density provides direct information on the level of soil compaction even through air permeability and saturated hydraulic conductivity. Moreover, infiltration rate is the most important characteristic directly affected by soil properties. Thus, it is influenced by tillage processes (Abo Soliman, 1984; El-Sayed and Ismail, 1994).

Rice has an unique place in Egyptian agriculture as a crop of great economic significance. Land preparation, method of planting and proper water management are among the most important factors which could significantly contribute increasing rice productivity (El-Gibali and Mahrous 1970; Abd El-Hafez 1982; Abo-Soliman et al., 1990; Hegazy et al., 1992; El-Serafy et al., 1993).

Cotton is one of the most important crops in Egypt. Eid et al. (1988) and Saied (1992), reported that cotton yield and irrigation efficiency were increased as a result of land-levelling (dead level and 0.1% ground surface slope). The slope treatment led to an obvious impact on bulk density and soil infiltration values. At the same time, the plough is the most used farm implement for alternating the conditions of the soil to a suitable physical environment for the cotton crops. El-Sayed and Ismail (1993) reported that the minimum tillage technique consumed less energy. However, it did not give enough yield to justify its use, while the improved tillage produced the greatest cotton yield.

The objective of the present investigation is to study the effect of different tillage techniques on some soil characteristics which affect the performance of surface irrigation and yield of some summer crops (rice and cotton) at North Delta in Egypt.

2 Materials and methods

Two separated field trials were conducted at the Sakha Agricultural Research Station farm in 1994. The soil of the experimental sites are clayey, non saline, non-alkaline soil [EC_e 3.8 dS m^{-1}, ESP = 11.2% and the hydraulic characteristics, f.c = 42% and W.P 21.6% for surface layer (0-30 cm)]. The total area was about 11 *feddans* (one feddan = 4200 m^2) provided with an adequate field subsurface drainage system.

2.1 The first field trial

The objective of this experiment was to evaluate different tillage techniques of land preparation with methods of planting on some soil properties and rice yield. Rice (Giza 175 variety) was planted on May 7 and transplanted on June 8, 1994. The experimental design consisted of Complete Randomized Blocks with four replicates. The plot area was 1500 m^2 (100 m length × 15 m width). The four treatments were as follows:

A. Traditional tillage:
Chiseling twice to a working depth of 15 cm followed by puddling process (wet-levelling) with a broadcasting planting method (pregerminated seed were broadcasted by hand).

B. Minimum tillage:
Disk harrowing (twice) followed by a puddling process with dibbling planting method (pregerminated seeds were sown on a hill structure, 4-6 seeds/hill).

C. Medium tillage:
Mold board and disk harrowing followed by puddling process, with transplanting method (32 days-old seedlings were transplanted in puddled soil at 20 × 20 cm spacing using 4-5 seedlings per hill).

D. Recommended tillage:
Accordingly to the Rice Research and Training Center Mold board and disk harrowing followed by a levelling scraper without puddling, with a seed drill method (pregerminated seeds were sown with 12 rows drum seeder with 20 cm spacing between rows).

2.2 The second field trial

To study the effect of land levelling and tillage practices on cotton yield, aspects of water management, and some soil physical properties. Cotton (Giza 77/85) was planted on March 23, 1994 and received six irrigations. A split plot design with four replicates was used. The plot size was 900 m^2 (120 m length × 7.5 m width). The main plots were subjected to the land-levelling treatments and the tillage practices occupied the sub plots.

Treatments:
L: Land levelling
L_1: 0.1% ground surface slope
L_2: Basin level or dead level (0.0%)
L_3: Traditional land-levelling practiced by the farmers
T: Tillage
T_1: Surface plowing (0-20 cm): Chiseling twice (two different ways) and disk harrowing to a working depth of 20 cm followed by a levelling scraper.
T_2: Deep plowing (0-50 cm): Subsoiling plowing twice, chiseling twice and disk harrowing followed by a levelling scraper.

In the field trials, all experimental treatments experienced the same agricultural practices as usual at North Delta.

All the data were statistically analyzed according to Snedecor and Cochran (1971).

Characters studied:
1- Advance and recession of irrigation water (EWUP, 1984).
2- Rice and seed cotton yields.
3- Water consumptive use (Israelsen and Hansen, 1962).
4- Total amounts of irrigation water applied to the different treatments using a cut-throat flume 30 × 90 cm.
5- Water use, water utilization and water consumptive use efficiencies (Israelsen and Hansen, 1962).
6- Bulk density (Vomocil, 1957).
7- Infiltration rate and cumulative infiltration (Garcia, 1978).

3 Results and discussion

3.1 The first field trial

3.1.1 Rice yield and its components:

Data (Table 1) indicate that there were significant differences in the grain yield among the tillage methods. Medium tillage with transplanting gave the highest rice grain and straw yields (4.08 and 4.01 tons/feddan, respectively). The other tillage treatment gave comparable lower grain yield. Minimum tillage treatment gave the lowest grain and straw yield (1.78 and 2.84 tons/feddan, respectively). With regard to plant height, it was significantly affected by different tillage treatments. The tallest plants occurred under medium tillage treatments, while the shortest plants were obtained by minimum and recommended tillage treatments. Regarding panicle length it was insignificantly affected by the tillage methods.

Treatments	Grain yield Tons/fed	Straw yield Tons/fed	Plant height cm	Panicle length cm
Traditional tillage	2.82	3.03	72.14	18.92
Minimum tillage	1.78	2.84	62.83	19.69
Medium tillage	4.08	4.01	76.97	18.50
Recommended tillage	2.04	2.88	63.42	19.00
F. test	**	**	**	n.s.
L.S.D. 0.05	0.099	0.221	7.15	-
L.S.D. 0.01	0.137	0.302	9.73	-

** = Significant at 1% level
n.s. = Not significant

Table 1: Rice yield and its components as affected by different method of tillage.

In general, weed control and difficulty in insuring an adequate stand might have contributed to lower yields under traditional, minimum and recommended tillage treatments as compared to medium tillage treatment. It should be mentioned here that similar results were obtained by Abd El-Hafez (1982), Abo Soliman et al. (1990), Hegazy et al. (1992) and El-Serafy et al. (1993). They reported that the transplanting method gave adequate weed control and achieved high rice production compared to the other methods.

3.1.2 Water measurements

a) Amount of irrigation water delivered:
The average amount of water delivered for each treatment is presented in Fig. 1. The water requirements for rice under such conditions at North Delta were 8148.5, 8059.9, 7149.7 and 9167.7 m^3/feddan for treatments of tillage, traditional, minimum, medium and recommended, respectively. One possible explanation to the aforementioned results may be due to the medium tillage with transplanting method which had been established under an excess puddling process. This decreased the percolation losses more than the other treatments. A similar trend was reported by Abo Soliman et al. (1990), Hegazy et al. (1992) and El-Mowelhi et al. (1995).

b) Actual water consumption:
Values of actual water consumption of rice under different methods of rice land preparation were determined and calculated from the data obtained from the drum cultures technique (Fig. 1) The values of water consumption were low at the early stage of growth, increased through July and August, and then declined due to prevailing agronomic and climatic conditions in this area (Doorenbos and Kassam, 1979). The actual seasonal consumptive use of rice crop was found to be 790 mm for medium tillage treatment and approximately 1000 mm for the other treatments.

c) Water losses or percolation rates:
Percolation rates were higher at the beginning of the season, then reached an approximately constant value, then matched the basic infiltration rates and reached its minimum rate at maturity stage. The total amount of water losses through the rice growing season was approximately equal to the actual evapotranspiration. The recommended tillage treatment recorded the highest values of water losses or percolation, followed by traditional and minimum tillage treatments, while medium tillage treatment achieved the lowest values. Generally, medium tillage with transplanting method could be applied at North Delta to save on irrigation and water losses. Similar findings were reported by Abd El-Hafez (1982), Abo Soliman et al. (1990) and El-Mowelhi (1995).

Land preparation practices

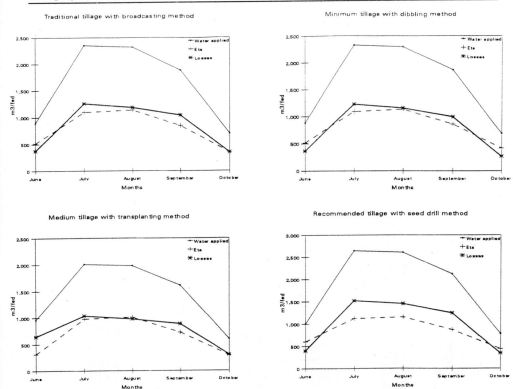

Fig. 1: *Water applied, actual water consumption (Eta), and water losses (deep percolation) m^3/feddan under different land preparation and planting methods of rice.*

Soil depth in cm	Before Planting	After planting rice			
		Traditional tillage	Minimum tillage	Medium tillage	Recommended tillage
0-10	1.01	1.20	1.25	1.30	1.14
10-20	1.12	1.24	1.29	1.33	1.20
20-30	1.21	1.30	1.34	1.37	1.25
30-40	1.29	1.36	1.38	1.44	1.33

Table 2: *Average values of bulk density (g/cm^2) as affected by different methods of rice land preparation*

3.1.3 Bulk density

The bulk density values obtained by the core method show the dense nature of the soils (Table 2.). Values of bulk density were higher after harvesting than before planting for all treatments of rice land preparation. The medium tillage with transplanting method increased the bulk density in the field of study as compared to the other methods, while the best treatment in that respect was recommended tillage with a seed drill method. The relative high values of bulk density after harvesting rice are directly or indirectly related to the effect of heavy equipment used in the tillage operation, and hoof pressure of cattle breaking the macro-aggregates into micro-aggregates (Abd El-Hafez, 1982; Abo Soliman, 1984).

3.1.4 Basic infiltration rate (IR)

The infiltration rate is a function of the magnitude of the forces causing water movement through soil pores which can be controlled by tillage and soil management. On the other hand, the rate at which a soil absorbs water decreases rather rapidly with time. After several hours, it usually becomes nearly constant, turned the basic infiltration rate (Garcia, 1978). Values of basic infiltration rate (mm/hr) were determined before and after planting rice for the different treatments (Fig. 2.)

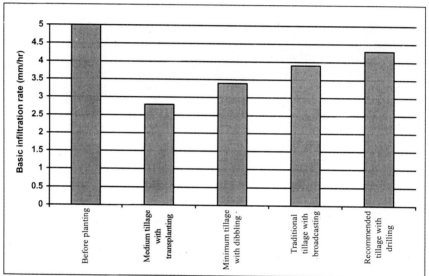

Fig. 2: Basic infiltration rate (mm/hr) as affected by methods of land preparation and planting of rice.

Values of (IR) decreased after planting compared to the values before planting. More reduction in the corresponding values were accompanied by the medium tillage with transplanting method. The best treatment was recommended tillage with seed drill method. These findings may be attributed to decreasing permeability as a result of the puddling practice which reduced the non-capillary pore space. Similar results were obtained by Abd El-Hafez (1982), Abo Soliman (1990) and El-Serafy et al. (1993).

3.2 The second field trial

3.2.1 Seed cotton yield

Table 3 indicates that there were highly significant differences in seed cotton yields among land-levelling treatments. The 0.1% ground surface slope treatment resulted in the highest seed cotton yield (1074.8 kg/feddan), followed by the dead level treatment (1016.9 kg/feddan), while the traditional land levelling treatment recorded the lowest yield (734.3 kg/feddan).

On the other hand, a slight difference was found (not significant) between the two tillage practices. Average seed cotton yield was 1027.8 and 1016.0 kg/feddan for T_1 and T_2 treatments respectively. Moreover, the interaction between the different treatments had a significant effect on seed cotton yield. These findings are in agreement with Semaika and Rady (1987), Eid et al. (1988), and Saied (1992).

Treatments	Seed cotton (kg/fed)
Land levelling (L):	
Traditional (R)	734
Dead level (L_1)	1017
0.1% slope (L_2)	1075
F test	
L.S.D. 0.05	62.80
L.S.D. 0.01	96.56
Tillage practices (T):	
(T_1) (Surface plowing)	1028
(T_2) (Deep plowing)	1016
Interaction:	
L X T	Significant at 5% level

Table 3: The mean values of seed cotton yield as affected by the different treatments (kg/feddan).

Treatments	Consumption (cm)	Total amounts of water applied (cm)	*Crop water use efficiency (kg/cm)	**Water utilization efficiency (kg/cm)	***Water consumption efficiency (%)
Land levelling:					
Traditional (R)	71.18	96.11	10.31	7.64	74.06
Dead level (L_1)	73.64	86.24	13.81	11.79	85.38
0.1% slope (L_2)	75.77	82.48	14.18	13.04	91.86
Tillage:					
Surface (T_1)	73.83	88.35	13.92	11.63	83.56
Deep (T_2)	75.03	95.88	13.54	10.59	78.25

* Crop water use efficiency (W.U.E) $= \dfrac{\text{Kilogram of seed cotton yield}}{\text{Consumptive use (cm)}}$

**Water utilization efficiency (W.Ut.E) $= \dfrac{\text{Kilogram of seed cotton yield}}{\text{Total amounts of water applied (cm)}}$

***Water consumptive use efficiency (W.C.U.E.) $= \dfrac{\text{Consumptive use (cm)}}{\text{Total amount of water applied}} \times 100$

Table 4: Average values of water consumption, total amounts of water applied, crop water use efficiency, water utilization efficiency and water consumption efficiency.

3.2.2 Water measurements

Table 4 provides data on water consumption for the various treatments.

a) Crop water consumption:

The ground slope (0.1%) treatment for cotton recorded the highest values of consumption (75.77 cm) followed by the dead level treatment (73.64 cm). The lowest value was obtained from the traditional land levelling treatment (71.18 cm). However, the deep plowing treatment consumed more water in comparison with the surface plowing treatment.

b) Total amounts of irrigation water:

Traditional land-levelling treatment received the highest amounts of irrigation water among the other land levelling treatments. The deep plowing treatment, however, received more irrigation water than the surface treatment.

c) Water use, water utilization and water consumption efficiencies

The 0.1% ground surface slope treatment achieved the highest values of W.U.E, W.Ut.E. and W.C.U.E. followed by dead level treatment, while the lowest ones were obtained by the traditional treatment. On the other hand, the surface plowing treatment utilized and consumed water more efficiently compared to the deep plowing treatment.

The highest values of W.U.E, W.Ut.E and W.C.U.E. are associated with the highest values of seed cotton yield, less water consumed and the lowest amount of irrigation water delivered. These results are in agreement with Semiaka and Rady (1987), Eid et al. (1988) and Saied (1992).

d) Advance and recession:

During the first irrigation (planting irrigation), advance and recession times of irrigation water were recorded (Fig.3) for three land-levelling treatments.

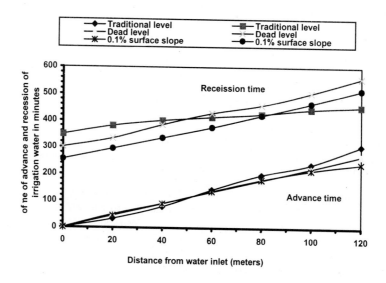

Fig. 3: Advance, recession and opportunity time of infiltration for the different land levelling treatment

It is obvious that the arrival times of the applied irrigation water to the ends of the irrigation runs were 240, 270 and 306 minutes for 0.1% slope, dead level and traditional land-levelling treatments respectively. The differences between time of advance and recession (infiltration opportunity time) were approximately the same at the different stations along the irrigation runs of dead level and 0.1% slope treatments, while they differ too much with the traditional treatment The infiltration opportunity time was high on the upper part of the irrigation run and decreases along the irrigation run.

3.2.3 Soil bulk density

Average values of soil bulk density (Table 5) are relatively high under the conditions of precision land-levelling and this trend may be true due to the compaction resulted from passing of heavy equipment used in finishing such work. On the other hand deep ploughing reduces soil compaction and thus slightly decreases alittle bit the soil bulk density compared to surface ploughing.

Treatments	Bulk density (g/cm^3) for different depths (cm)			
	0-15	15-30	30-45	45-60
Before Exp.	1.10	1.18	1.25	1.32
Land levelling				
Traditional	1.17	1.21	1.29	1.38
Dead level	1.23	1.34	1.42	1.53
0.1% ground slope	1.26	1.41	1.45	1.56
Tillage:				
Surface	1.20	1.27	1.35	1.42
Deep	1.22	1.26	1.32	1.38

Table 5: Soil bulk density (g/cm^3) as influenced by the different treatments.

3.2.4 Infiltration rate and cumulative infiltration:

It is clear that the precision land-levelling treatments had a negative impact on the infiltration characterization of the soil. Both the dead level and 0.1% ground surface slope treatments decreased the infiltration rate and altered the cumulative infiltration depth in comparison to the traditional treatment.

The basic infiltration rates were found to be 15.0, 6.0 and 5.0 mm/hr for traditional, dead level and 0.1% slope treatments respectively. No clear effect was noticed due to the different tillage treatments on the soil infiltration characteristics. The same trend was obtained by Saied (1992) and El-Mowelhi (1995).

4 Conclusions

Based on the results of this study the following conclusions can be drawn:
1- Medium tillage with transplanting method achieved the highest grain yield of rice and water utilization efficiency. On the other hand, recommended tillage with seed drill rice planting resulted in improved soil conditions.
2- Laser land-levelling is considered to be a proper technique to save water and increase cotton yield, as compared to unlevelled land. At the same time, surface plowing represents the optimum treatment for water management.
3- In light of the results, there is a need for more elaborate studies and consideration of out the most suitable farm machinery for different crop patterns and optimum use of available land in Egypt.

References

Abd El-Hafez, S.A. (1982): Effect of irrigation and fertilization on rice yield and soil properties. Ph.D. Thesis, Fac. of Agric. Mansoura Univ.

Abo Soliman, M.S.M. (1984): Studies on some physical and chemcial properties of some soils of Middle Delta in Egypt. Ph.D. Thesis, Fac. Agric., Mansoura Univ.

Abo Soliman, M.S.M., Hegazy, M.H., Hammouda, F.M. and Abd El-Hafez, S.A. (1990): Effect of preceding crop, algalization and nitrogen fertilization on some soil chemical properties and rice yield in salt affected soil. Proc. 4th Conf. Agron. Cairo, 15-16 Sept. 1990. Vol. (1).

Doorenbos, J. and Kassam, A.H. (1979): Yield response to water. FAO Irrigation and Drainage Paper, No. 33, FAO, Rome, 1974.

Egypt Water Use and Management Project (EWUP) (1984): Improving Egypt's irrigation system in the old lands. EWUP Final Report, International Press.

Eid, M., El-Taweel, M., Ibrahim, M.A.A., Ainer, N.G., Sherif, M.A., Wahba, M.M., Abd El-Mallak, K.K., El-Khader, E.A. and Gad El-Rab, G.M. (1988): Controlled irrigation for field crops production within the context of improved farming system at Minya. Agric. Res. Center. Soil & Water Res. Inst. Field Irrigation and Agroclimatology. Conf. 20-23 June, 1988. Giza, Egypt.

El-Gibali, A.A. and Mahrous, F.N. (1970): Water requirements for rice crop in Northern Delta. First Agric. Rice Res. Conf. Cairo, Jan. 24-26.

El-Mowelhi, N.M., Hegazy, M.H., Mahrous, F.N. and Benjamen, I. (1995): Evaluating some important agricultural practices of rice to maximize water utilization efficiency in Northern Delta soil. Proc. of Second Conf. of On-Farm Irrigation and Agroclimatology. June 1995, P. 196-205.

El-Sayed, A.S. and Ismail, E.S. (1994): Effect of different tillage techniques on some soil properties and cotton yield. Misr. J. Eng. **11 (4)**: 929-941.

El-Serafy, A.M., Sayed, K.M. and Abo Soliman, M.S.M. (1993): Evaluation of different rice planting methods under different salinity levels. J. Agric. Sci. Mansoura Univ. **18 (9)**: 3767-3774.

Garcia, J. (1978). Soil-Water Engineering Laboratory Manual. Department of Agricultural and Chemical Engineering, Colorado State University, Fort Collins, Colorado, U.S.A.

Hegazy, M.H., Abo Soliman, M.S.M., Abou El-Soud, M.A. and Abd El-Hafez, S.A. (1992): Evaluating different planting methods of rice grown under salt affected soil. Proc. 5th Conf. Agron. Zagazig, Vol. (1): 64-70, 1992.

Israelsen, O.W. and Hansen, V.E. (1962): Irrigation principles and practices. 3rd Edit. John Wiley and Sons., New York.

Raghavan, G.S.V., Mukyes, E., Amir, I., Chasse, M. and Broughton, R.S. (1976): Prediction of soil compaction due to Off-Road Vehicle traffic. ASAE. Vol. 19 No. 4, P: 610-613, 1976.

Saied, M.M.M. (1982): Effect of land levelling and irrigation discharge on cotton yield and irrigation efficiency. Ph.D. Thesis, Fac, Agric., El-Mansoura Univ., Egypt.

Semaika, M.R. and Rady, A.H. (1987): Land levelling as an important water management operation and its impact on water resources in Egypt. International Commission on Irrigation and Drainage. Egyptian National Committee Proceedings. Vol. 11, 1987.

Snedecor, G.W. and Cochran, W.G. (1971): Statistical Methods. Sixth Edition. Iowa State Univ. Press Ames. Iowa.

Suliman, A.E., Nasr, G.E.M. and Adawy, W.M.T. (1993): A study on the effect of different systems of tillage on physical properties of the soil. Misr. J. Ag. Eng. **10(2)**: 169-189.

Vomocil, J.A. (1957): Measurements of soil bulk density and penetrability. A review of methods. Adv. Agric. Press, New York, London

Addresses of authors:
N.M. El-Mowelhi
M.S.M. Abo Soliman
S.A. Abdul El-Hafez
Somaya A. Hassanin
Soil, Water and Environmental Research Institute, Agricultural Research Center
9 Cairo University Street
Giza, Egypt

Monitoring, Control and Remediation of Soil Degradation by Agrochemicals, Sewage Sludge and Composted Municipial Wastes

S.E.A.T.M. van der Zee & F.A.M. de Haan

Summary

To assess sustainability of soil use it is necessary to take into account standards of soil quality as well as standards for other environmental compartments, such as ground and surface water. We develop a conceptual approach for the assessment of sustainability that can be easily extended to incorporate additional factors. We illustrate our approach using examples of fertilizer, sludge and compost use and show that anticipation of adverse effects is important in view of the difficulty in reversing contamination by heavy metals.

Keywords: heavy metals; balance calculations; sustainability; bioavailability; leaching.

1 Introduction

For many years soils that are in agricultural use have been affected by different treatments to improve their properties. These properties include the capacity to provide nutrient elements for crops or vegetation as well as the capacity to provide water and air to plant roots. In addition, pesticides are often applied to limit the hazards and reductions in crop yield. Examples of applied constituents include commercial fertilizer, manure (and other organic fertilizer), sewage sludge, and compost.

Although the beneficial aspects of various applied constituents for attaining or maintaining sufficient and good quality yields are beyond question, adverse effects have been recognized during the past decades. These adverse effects differ with regard to their location and nature. Thus, pesticides or contaminating agents may adversely affect soil ecology, surface water ecology, and the quality of ground water. Fertilizers have been recognized as an environmental problem in view of eutrophication. Both an inbalance in nutrient element availability to crops and livestock and excessive nutrient availability, for example surface waters, may cause problems. Furthermore, concerns have risen regarding the emission of greenhouse gases to the atmosphere.

For sustainable use of soil, the gradual or sudden degradation of important soil functions is unacceptable. However, to assess whether or not soil use (including the application of different constituents) may be considered sustainable is not simple, as adverse effects may require years or decades before they become apparent. Hence, warnings are often not heeded and may be too late by the time that effects are observed. For this reason, predictions and warnings need to be convincing and based on reliable data. Unfortunately, soil quality standards (which should serve as a guideline for assessing sustainability) are seldomly developed with the aim of ascertaining sustainability. This is due in part to historic reasons: standards were needed quickly to be able to discriminate between what was clearly acceptable and what was not, and had to be simple and reproducible. This poses the problem that, in many countries, we often still deal with a lack of concepts for assessing sustainability, and that only some tools for the practical application of such concepts have been developed.

In this paper we consider sustainability of soil use for a rather restricted type of problem: voluntary and involuntary applications and inputs of constituents to the soil system causing adverse effects. We develop an approach that may be more adequate for assessing sustainability as well as some concepts that may be of use for the practical implementation of this approach, and provide some examples.

2 Choice of standards

In view of soil contamination problems, many countries have developed standards that should be indicative of whether or not contamination is hazardous. Often, these standards define total element or constituent contents that distinguish contaminant levels that are perceived differently by the authorities, e.g., hazardous vs. non-hazardous. These standards may still be of use when they reflect e.g. *in situ* biological availability or toxicity for organisms (Marinussen et al., 1996). This is the aim of the Dutch ecotoxicologically-based soil intervention values (or functions). These functions, which we denote generically by G_E here, vary not only as a function of the constituent but also as a function of soil composition (clay or organic matter content, pH, etc.). This indicates that perhaps not the total content in soil, but rather a biologically available fraction (e.g. dissolved fraction) is of environmental interest. This has of course been recognized in conventional soil fertility research in the past century, that has been dedicated to a large part to identifying the bio-availability of nutrient elements with different chemical fractionation techniques (Houba et al., 1996). In fact, it is likely that availability for plants is a complicated nonlinear function of various soil and plant parameters. In restricted cases (e.g. for similar soils), bio-availability can be adequately assessed (approximately) by single extractants and subsequently used for correlative fertilization recommendations (Houba et al., 1996).

Besides *in situ* effects, the application of different compounds may have effects elsewhere. Due to a limited retention capacity, compounds may be emitted to the atmosphere, leached to ground water or be released to surface waters through runoff or erosion. In addition, compounds may be intercepted by biota (e.g. crops) and accumulate in the produce. Hence, sustainable soil use implies that neither soil quality standards nor other standards are violated. Because leaching and runoff (and bio-availability for some organisms) are related to the concentration of a compound in the soil solution, the assessment of sustainability involves standards with regard to both total contents and concentrations in solution. An obvious approach of such an assessment for an environmental compartment in the context of applications is a balance approach as is common in systems analysis. This approach is illustrated below.

3 Balance approach

Soil, or part of soil such as the ploughed layer, may be regarded as a grey or black box that can be characterized by input rates, output rates, and changes in the state variables of the box. Sustainable treatment of soil implies that it is not adversely affected by changes in the state variables (protected by soil quality standards, G_E) and that output rates do not violate standards for other compartments. In agreement with the common approach of systems analysis, we develop a balance equation for soil as was done earlier by Boekhold and Van der Zee (1991). Omitting details that may be more relevant for 'fine-tuning' in a later stage, we assume a linear relationship between accumulated compound in soil and its concentration in the soil solution. Then, we obtain for a ploughed layer of thickness L

$$\frac{dG}{dt} = A - BG - CG \qquad (1)$$

where symbols are defined in the appendix. In (1), the first right hand side term identifies the constant input rate, the second one the outputs via harvested products and the last one the outputs via involuntary losses to ground and surface water and emission to the atmosphere.

Although the coefficients A, B, and C may vary as a function of time, their long term average may be considered to be constant for our purpose (i.e., to evaluate the current situation). Hence, (1) may be integrated to yield

$$G(t) = \frac{A(1 - e^{-(B+C)t})}{(B+C)} \qquad (2)$$

for the particular case of initially negligible contents in soil ($G(t=0)=0$). However, the mentioned constants depend on compound, environmental and soil properties. In particular, if we assume that plant uptake occurs via the soil solution and is therefore controlled by the concentration in solution, c, as is also the case for the leaching rate, we obtain (in view of $c=G/\theta R$)

$$B.G = \frac{k_p}{\theta}\frac{G}{R}, \quad C.G = \frac{v}{L}\frac{G}{R} \qquad (3)$$

where we recognize R as the retardation factor. For e.g. heavy metals, R is known to depend on organic matter content, pH, and other soil solution properties (Chardon, 1984, Temminghoff et al., 1994, 1995). In particular, the complicated dependency of R on system and compound properties makes a generalization of soil standards that are effect-related a major problem. Before continuing to some practical concepts, it may be worthwhile to point out that the combination of (2) and (3) results in a decay of the exponential of (2) as a function of t/R: this implies a rescaling of time, i.e., effects become apparent only for time periods that are R times as large as in case of absence of chemical interactions of compound and soil matrix.

4 Practical concepts

In view of the complicated relationships that are implicit to (2) and (3), the practical assessment of sustainability with these equations is difficult. Fortunately, the balance equation (1) suggests an alternative that takes standards of soil quality as well as other relevant standards into account.. This alternative is based on the concept of the **discrepancy factor** (F_d), which was developed by De Haan and Van der Zee (1993) using the approach of Ferdinandus et al. (1989). First, we observe that sustainability requires that no violations of standard either now or in future occur and that the total content G at infinite time is found from (2) to equal

$$G(t = \infty) = \frac{A}{B+C} \qquad (4)$$

At this steady state controlled by the current treatment of soil in terms of application rates and immissions, we know that the input rate (A) equals the sum of the rates of removal by harvest ($r_U=B.G(\infty)$) and leaching ($r_L=C.G(\infty)$). If we define an acceptable removal rate (r_U^a) by harvest as being equal to the harvested crop yield times the tissue concentrations that are in agreement with good crop quality (with standards defined nationally or by the EC) and acceptable leaching rate (r_L^a)

according to ground water recharge times ground water quality standards (or the stand-still principle of ground water concentrations advocated by Van der Zee et al. (1990)), we may define a discrepancy factor according to

$$F_d = \frac{A}{r_U^a + r_L^a}. \tag{5}$$

The power of the concept of a characteristic number such as the discrepancy factor is, that as the denominator follows directly from simply measurable properties (yield, mean recharge of ground water) and existing, independent standards for produce and ground water quality, this factor is accessible without having detailed knowledge of soil and environmental properties. If F_d is larger than unity, either crop or ground water quality standards will in due time be violated according to current input rates, A, and such input rates should not be considered sustainable.

By comparing discrepancy factors of different compounds, it becomes easy to assess which compound is applied most abundantly and will therefore lead to the largest violation of existing standards. The concept of discrepancy factor has already been used in The Netherlands. Thus, manure regulations have been formulated in terms of phosphate equivalents, because P is more abundant in manure with regard to crop requirements than nitrogen or potassium (Van der Zee, 1988).

Before considering other examples, we note that the balance approach enables the assessment of other characteristic numbers that are of use in practical soil protection. If F_d exceeds unity, either crop quality or ground water quality or both will, in the long term, become unacceptable. Comparison of discrepancy factors for different compounds may identify the most "abundant" compound, but does not identify where problems will be greatest. However, by calculating the **critical sustainability factor**, given by

$$F_c = \text{MAX}(\frac{1}{G_E}, \frac{B}{r_U^a}, \frac{C}{r_L^a}) \tag{6}$$

we can identify whether soil ecology, crop quality, or ground water quality is most threatened. For example, soil ecology is most threatened if F_c is equal to $1/G_E$. Furthermore, it is easy to assess for the linear balance that the most threatened quality standard is also the first one to be exceeded, which indicates the most efficient direction for protective measures. In summary, when the most hazardous compound has been identified by its F_d, the most threatened quality standard follows from (6).

5 Extension of the balance

In the balance equation (1), we have only accounted for a certain input rate, and removal rates due to leaching and harvesting. For pesticides, also degradation and volatilization may occur. In case of linear partitioning, both processes are linear with regard to G, and can therefore be easily introduced into (1). In fact, also tarra losses (removal of compounds associated with soil that adheres to harvested plant material) and losses via erosion imply the removal of soil with a particular compound content G, and are therefore linear with regard to G. Therefore, the incorporation of all these processes into the balance would simply lead to

$$\frac{dG}{dt} = A - k.G \tag{7}$$

where k is found by summation of B, C (both of eq.1) and similar rate coefficients for degradation, volatilization, tarra, and erosion losses. Integrated, we obtain (2) where the factor $(B+C)$ is replaced by k.

6 Illustrations

The inputs of heavy metals via different fertilization strategies are likely to lead to positive balances in The Netherlands (Van der Zee et al., 1990).

	Cd	Cu	Pb	Zn
Input				
Agricultural crops				
- commercial crops (cf)	0.55	1.7	4.7	15.5
- sewage sludge + cf	0.84	91.0	55.0	250
- cattle manure + cf	0.38	6.0	6.2	21.0
- pigs manure + cf	0.21	30.0	5.0	59.5
- poultry manure + cf	0.15	15.5	4.6	60.5
Horticultural crops (cf)	1.35-1.6	3.6-4.3	5.5-7.5	27.5-330
Precipitation	0.2	3.2	13.0	20.0
Removal				
Agricultural crops for				
- human consumption	0.30-0.25	3.6-11.0	0.1-0.5	36.0-110
- animal fodder	0.15-0.75	3.6-13.5	7.0-31.0	36.0-135
Horticultural crops	0.43-0.85	3.7-7.5	1.7-4.5	37.0-75.0
Groundwater recharge	0.1	1.3.	1.3	7.5

Table 1: Calculated input and removal rates (in mg/m^2 year). Removal rates are in agreement with current Dutch standards for crop quality, and the stand-still principle for ground water quality. (After: Ferdinandus et al.1989)

To construct Table 1 we considered acceptable removal rates via harvested crop according to existing Dutch standards for different crops, leaching losses were based on a recharge of ground water of 0.25 m/y and the stand-still principle of mean Dutch ground water quality. The inputs followed from documented compositions of various fertilizers: the indicated fertilizer was applied until the required N, P, or K was met, and for the remaining nutrient elements commercial fertilizer (cf) was used. Precipitation was calculated from documented atmospheric deposition (Ferdinandus et al., 1989). Clearly, immision rates are larger than removal rates that are in agreement with crop and ground water standards assumed in this paper. In particular, Cd in fertilizer (cf), Cu in pigs and chicken manure, and lead from atmospheric deposition give rise to an insustainable situation.

Taking the acceptable removal and atmospheric deposition of Table 1 as a point of departure, we may also consider the amendment of soil fertility with (relatively clean) compost (De Haan and Van der Zee, 1993): Table 2.

Three scenarios with a maximum tolerated application rate of 1.5 (grassland; scenario 1), 3 (arable land; scenario 2), and 6 metric tons of compost per hectare per year (other use; scenario 3) were considered in view of the 1991 Dutch General Administrative Order on the use of other organic fertilizers. The remaining nutrient requirement was accounted for by additional fertilization. Again, we find positive balances and we obtain discrepancy factors that differ for Cd (6), Cu (2.7), Pb (21), and Zn (1.3). Table 2 reveals that the largest violations may be expected due to excessive involuntary inputs of lead. The considered scenario of Table 2 is not such that realistic changes in e.g. atmospheric deposition would alter this conclusion. A more detailed study for different farming systems and particular soils (rather than the mean Dutch situation) is in preparation and will also address the critical sustainability factor and time scales.

metal	Cd	Cu	Pb	Zn
removal	0.13-0.95	4.9-14.8	1.4-32.3	43.5-143
Atm. deposition	0.2	3.2	13	20
compost/fert. 1	0.59	10.0	16	36
compost/fert. 2	0.66	18.8	33	65
compost/fert. 3	0.88	36.6	62	124

Table 2. Acceptable removal (by leaching and uptake) and input by atmospheric deposition and compost/fertilizer additions (mg/m2yr).

The persistency of metal accumulation in soil may be large, which has implications for natural remediation of soil quality. In the city of Utrecht, soil metal contents have been assessed in inner and outer city greens. Compared with reference values for rural soils (Lexmond et al., 1986), the inner city values appeared to be larger (and correlated with organic matter) whereas the outer city values were not (and were correlated with clay content). Using the metal contents for clay, silt and sand fractions provided by Lexmond et al. (1986), the metal content arising from anthropogenic applications (denoted G^a) can be estimated and related with the cumulative application $A^c = \rho.L.G^a$ where ρ is the soil bulk density and L the thickness of sampled soil. With an empirical relationship between organic matter content and bulk density, ρ can be calculated. If A^c were constant, a correlation between G^a and organic matter content is expected in view of the effect of the latter on density. However, both a good correlation (95-99% confidence level) was observed for all metals (Cd, Cu, Ni, Pb, Zn) in the inner city and A^c-values varied with normal frequency distribution. This suggests that the cumulative application is correlated with organic matter content (and not with clay content), which appeared to be the case for the inner city but not for the outer city greens. These correlations, the significant accumulation in inner city and limited accumulation in outer city, and the correction that was made for bulk density effects of organic matter, suggest that heavy metals have been applied together with organic matter. Indeed, until about 1980 sewage sludge had been used in the inner city greens for fertilization. This practise was discontinued because of the measured heavy metal contents in the sewage sludge.

Depending on different properties, such as R, remediation of adverse effects by natural causes may involve long time periods. For this reason, we consider that active remediation of diffuse contamination in agricultural soils is unlikely to be realistically attainable. After all, retardation factors for e.g. heavy metals are often in the range exceeding 10 or 100 and for a nutrient such as phosphate it was found to exceed approximately 40 for Dutch sandy soils (Van der Zee, 1988, Van der Zee and Bolt, 1990).

An illustration of the effect of R on the time periods involved and on the necessity for early recognition of hazards is the leaching of 1,2-dichloropropane. This compound is a by-product of the pesticide 1,3-dichloropropene which was forbidden as a contaminant in the pesticide formulation by 1980. However, at that time its application for about one decade to soil had already created a slug of contaminants that continued to leach to ground water. When it was measured by monitoring networks in ground water the province Drenthe in the early 1980s (Figure 1), modeling of the well-known geohydrology indicated that EC ground water quality standards would be exceeded by a factor of 500-1000 by the year 2030 as leaching would continue to build up concentrations. Predictions were made that it would take until the year 2100 before ground water quality would become acceptable again, whereas remediation was recognized to be difficult (Beugelink, 1987).

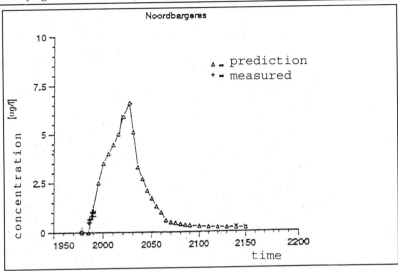

Figure 1: Concentration development of 1,2 dichloropropane in pumping water at Noordbargeres well field (Drenthe) as a function of time. Predictions and measured concentrations (in µg/l).

7 Conclusions

The use of various materials for amending soil is associated with simultaneously occurring risks of soil quality deterioration due to the involuntary application of contaminating compounds. To assess whether such use may be regarded as being sustainable requires the further development of soil quality standards. This development should be aimed at acquiring standards that better reflect biological availability of contaminants (and therefore are better correlated with ecotoxicological risks) as well as being compatible with standards that have been developed for other environmental compartments such as ground and surface water, or economic produce. In this paper, we described a balance approach that combines these requirements. Additionally, we discussed some concepts for sustainability assessment, which make use of characteristic numbers such as the discrepancy factor and the critical sustainability factor. These numbers may be more practical instruments for assessing sustainability than considering contaminant balances as such. Both balances and characteristic numbers can be easily adapted to take other aspects (e.g. erosion, biodegradation) into account.

With balance considerations for heavy metals, we were able to show that currently accepted fertilization practice and atmospheric deposition rates may not be sustainable for The Netherlands on average. Thus, use of sewage sludge as a fertilizer should be discouraged and use of compost requires this material to contain less heavy metals in future. For pesticide use, an example was given to illustrate that the time scales may be large. This has two main consequences: anticipation of potential problems may be too late, and remediation may require a lenghty time period. Both consequences are a major motivation to improve our instruments for predicting whether or not anthropogenic influences on soil quality may cause environmental problems.

References

Beugelink, G.P. (1987): Future concentrations of dichloropropane in raw water of Noordbargeres (Dr.) pumping station, (in Dutch), RIVM report 728618001, Bilthoven, The Netherlands.

Boekhold, A.E., and S.E.A.T.M. van der Zee (1991): Long term effects of soil heterogeneity on cadmium behaviour in soil, J. Contam. Hydrol., **7**, 371-390, 1991

Chardon, W.J. (1984): Mobility of cadmium in soil, PhD thesis, Wageningen Agric. Univ. (in Dutch), 200 pp.

De Haan, F.A.M., and S.E.A.T.M. van der Zee (1993): Compost regulations in The Netherlands in view of sustainable soil use, in: Science and Engineering of Composting: Design, Environmental, Microbiological and Utilization Aspects, (Hoitink, H.A.J., and Keener, H.M., (eds.), Renaissance Publ., Worthington, OH, USA, 507-522.

Ferdinandus, H.N.M., Th.M. Lexmond, and F.A.M. de Haan. (1989): Heavy metal balance sheets as criteria for the sustainability of current agricultural practices (in Dutch), Milieu **4**, 48-54.

Houba, V.J.G., Th.M. Lexmond, I. Novozamsky, and J.J. van der Lee (1996): State of the art and future developments in soil analysis for bioavailability assessment, Sci. Total Environ. **178**, 21-28.

Lexmond, Th.M., Th. Edelman, and W. van Driel (1986): Preliminary reference values and current background contents for a number of heavy metals and arsenic in the topsoil of nature reserve soils and arable soils (in Dutch), in: Bijlagen behorende bij het advies Bodemkwaliteit, VTCB, Leidschendam.

Marinussen, M.P.J.C., S.E.A.T.M. van der Zee, and F.A.M. de Haan (1996): Cu accumulation by Lumbricus rubellus as affected by total amount of Cu in soil, soil moisture and soil heterogeneity, Soil Biology and Biochemistry, in press.

Temminghoff, E.J.M., S.E.A.T.M. van der Zee, and M.G. Keizer (1994): The influence of pH on the desorption and speciation of copper in a sandy soil, Soil Sci., **158(6)**, 398-408.

Temminghoff, E.J.M., S.E.A.T.M. van der Zee, and F.A.M. de Haan (1995): Speciation and calcium competition effects on cadmium sorption by sandy soils at various pH levels, Eur. J. Soil Sci., **46**, 649-655.

Van der Zee, S.E.A.T.M. (1988): Transport of reactive contaminants in heterogeneous soil systems, PhD thesis Wageningen Agricultural University, 283 pp.

Van der Zee, S.E.A.T.M., H.N.M. Ferdinandus, A.E. Boekhold, and F.A.M. de Haan (1990): Long term effects of fertilization and diffuse deposition of heavy metals on soil and crop quality, Plant Nutrition - Physiology and Applications, Kluwer Ac. Publ., 323-326.

Van der Zee, S.E.A.T.M., and G.H. Bolt (1990): Deterministic and stochastic modelling of reactive solute transport, J. Contam. Hydrol., **7**, 75-93.

Van Luit, B. (1984): Cadmium uptake by crops, (in Dutch), Landbouwkundig Tijdschrift, **96**, 19-20.

Appendix: definition of symbols

The total content (dissolved and adsorbed) of a compound in soil which is denoted by G consists of dissolved (θc, with θ the volumetric moisture fraction and c the concentration in solution) and adsorbed (ρs, with ρ the dry bulk density and s the adsorbed quantity on mass basis) quantities. Defining the retardation factor $R = 1 + \rho s/\theta c$, we have $G = \theta.R.c$. The change of content, G, as a function of time, t, depends on the difference between input rate (A) and output rates ($B.G + C.G$, where B is the rate parameter due to loss of compound in harvested plant material and depends linearly on the k_p which is the ratio of content in plant over content in soil, and C is the leaching rate parameter, which depends on flow velocity v and thickness of the ploughed layer L). To derive the balance for the plough layer (eq. 1), it is assumed that this layer is perfectly mixed. The balance (1) is derived per volume of plough layer and has to be converted (Ferdinandus et al., 1989) to assess changes per hectare surface area.

Address of authors:
S.E.A.T.M. van der Zee
F.A.M. de Haan
Department of Soil Science and Plant Nutrition
Wageningen Agricultural University
P.O. Box 8005
6700 EC Wageningen, The Netherlands

The Influence of Organic Manure on *Striga hermonthica* (Del.) Benth. Infestation in Northern Ghana

B. Kranz, W.D. Fugger, J. Kroschel & J. Sauerborn

Summary

In the savannah regions of Africa intensified land use brought an increasing infestation of the parasite *Striga hermonthica* Del. Benth. in cereal crops. Therefore witchweed (*S. hermonthica*) is often described as an indicator plant for decreasing soil productivity. In Northern Ghana the interrelation of soil fertility and *Striga* infestation was investigated in low-infested compounds fields (0.1% affected planting holes) and highly-infested bush fields (45.5% affected planting holes). Manure application was one of the main differences between the two field types: In the compound fields organic manure was applied in 9 out of 10 years on average. The bush fields received organic manure or mineral fertiliser in 4 out of 10 years. Due to a long-term application of organic manure the contents of phosphorous, nitrogen and organic carbon were significantly higher in the compound fields. The availability of nutrients (NH_4, NO_3, $P_{Bray\ I}$, K, Ca, Mg), the microbial biomass and the microbial activity benefited significantly by manure treatment compared to the bush fields. Fallow periods of two years were too short to maintain the soil productivity in the bush fields. Since the availability of organic manure is limited, alternative measures such as green manure application or crop rotation with legumes are required to maintain or increase the productivity of the bush fields.

Keywords: *Striga hermonthica*, organic manure, compound fields, Northern Ghana

1 Introduction

Striga hermonthica Del. Benth., a parasitic angiosperm, is a major threat to cereal crops in the savannah regions of Africa. Yield losses averaged 24 % (10-31 %) but in areas of heavy infestation, losses reached 90-100 % in some years (Hess et al., 1996). Traditional methods of controlling weeds such as hand-weeding are ineffective since a significant crop damage caused by *Striga* occurs before it emerges from the soil. The increasing *Striga* infestation in the past years is often related to decreasing soil productivity (Ogborn, 1984; Bebawi, 1987). In Northern Ghana organic manure application is common in compound fields. In bush fields traditional fallow periods are reduced as a consequence of increasing land pressure. In addition a lower infestation is observed in the compound fields compared to the bush fields. A field study was undertaken to investigate whether the organic manure application had a significant impact on soil fertility and whether this could explain a lower *Striga* infestation on the compound fields.

2 Materials and methods

In 15 villages of the Tolon/Kumbungu District of Northern Ghana, 17 farms were selected with one compound and one bush field, respectively. Compound fields are situated close to the compound (220 m on the average) while the locationof the bush fields is more remoted (1650 m on the average). Regardless of the host crop type, infested bush fields were chosen. The maize crops of the compound field were rarely infested by *Striga*. At the end of the cropping season 1994, the farming practice of the past 10 years was ascertained and the *Striga* infestation rate recorded (random sample survey). Soil samples were taken for standard analysis (one mixed sample/field from 0-30 cm) at the beginning of the 1995 rainy season.

3 Results

In the region, host plants of *Striga hermonthica* (e.g. maize (*Zea mays* L.), sorghum (*Sorghum bicolor* (L.) Moench), millet (*Pennisetum americanum* (L.) Leeke)) form the staple food. In the compound fields maize was continuously monocropped. In the bush fields mixed cropping of hosts was regular (e.g. maize/sorghum, maize/groundnut) while sole non-host crops such as cassava (*Manihot esculenta* Crantz) and trap crops (e.g. cotton (*Gossypium* ssp.), groundnut (*Arachis hypogea* L.) or cowpea (*Vigna unguiculata* (L.) Walp.) were cultivated for less than two years of the past ten years. For land preparation the vegetation and crop residues were burned down. In both field types the burning practice hardly differs, therefore a similar impact on mineralisation was assumed. In the bush fields the fallow periods lasted two years on average while in the compound fields the introduction of fallow was linked to customs and rather was exceptional. The manure management showed pronounced differences: In the compound fields organic manure was applied for over 9 years on average (2-10 years) of the past ten years. The bush fields received organic manure or mineral fertiliser in 3-4 years only (Table 1).

Frequency of crops, burning, fallowing, and manure application in the past 10 years	Compound fields Mean (Years)	Stdev.[a]	Bush fields Mean (Years)	Stdev.[a]
Host crops	9.1	1.4	7.2	2.4
Nonhost crops	0.1	0.2	0.6	1.1
Trap crops	0.2	0.7	0.7	1.0
Burning for land preparation	2.8	4.0	3.6	4.1
Fallow periods	0.7	1.1	2.0	2.1
Organic manure application	8.1	2.4	0.7	2.0
Mineral fertiliser application	0.7	2.2	3.0	3.3

[a] Stdev.: Standard deviation

Table 1: Cultivation practices in compound and bush fields in Northern Ghana, results of the field reports for the past 10 years

Thirty % of the compound fields contained *Striga* seeds but an emergence of the parasite was rare. In the bush fields, emerged *Striga* plants were observed at 45.5 % (10-81 %) of the planting holes. On average, the infection rate and the seed bank was moderate (Table 2). In the surveyed farms effective control measures were not established. For weed control the fields are weeded three times but as the last weeding is done before the emergence of *Striga* the parasite is not effected.

	Compound fields		Bush fields	
	Mean	Stdev.[a]	Mean	Stdev.[a]
Affected planting holes (%)	0.1	0.2	45.5	25.6
Emerged *Striga* plants / infected host plant	0.2	0.7	3.3	2.6
Striga seed content (seeds 100 g soil^{-1})	0.4	0.7	10.7	14.1

Stdev: Standard deviation

Table 2: Occurrence of Striga hermonthica in compound and bush fields in Northern Ghana

The soils were sandy loams (chromic luvisoles) with a gravel content of 20 %. The different cultivation practice had no impact on the soil texture since the particle size distribution was not affected. The soil pH was slightly acid in both field types. The total nitrogen content (N_t) was moderate while the total phosphorous content (P_t) was low in both field types. Except for a good potassium content in the compound fields, the exchangeable bases (K, Ca, Mg) had low or moderate concentrations in both fields.

The application of organic manure promoted the soil organic matter content (2.54 % in the compound fields vs. 1.42 % in the bush fields). In the compound fields the analyses revealed a significant increase of the nutrient contents and exchangeable nutrients of about 50 % on average. For P_{org} and base saturation the differences were not significant. The cation exchange capacity (CEC) and the exchangeable bases are important parameters of the productivity and the nutrient supply of soils. Both factors were supported by the application of organic manure. The increase of the soil microbial biomass by 76 % reflected the improvement of the soil conditions in the compound fields as well. The dehydrogenase activity represents the extent of microbial turnover. Corresponding to the microbial biomass the enhancement of the microbial activity reached 70 % compared to the bush fields (Table 3).

4 Discussion

In the former traditional systems of Northern Ghana a fallow of 10-15 years alternate with a cultivation period of 4-6 years in the bush fields (Runge-Metzger, 1993). In the course of an increasing land pressure the fallow was reduced to 2 years. This period is ineffective to maintain the productivity of the soils as the analysed nutrient contents, their availability and the humus contents were significantly lower in the bush fields compared to the compound fields. Consequently the microbial biomass and the microbial activity decreased as well (Table 3). The difference between the two field types was mainly attributed to the long-term application of organic manure which was the most prominent difference regarding soil management (Table 1).

The dispersal of *Striga* seeds happens mainly through contaminated crop seeds and to a minor extent through cattle dung or wind (Berner et al., 1994). It was assumed that the probability of infection was equal in both fields. In addition host crops were more frequently grown in the compound fields compared to the bush fields which would favour a build-up of a *Striga* seed bank. In fact the *Striga* infestation was higher in the latter (Table 1 and 2). Improved soil management including organic manure and crop residue management was described as an effective method to control *Striga* (Ransom and Odhiambo, 1994; Chidley and Dennan, 1987; Watt, 1936). Therefore the different *Striga* infestation level on the two field types was linked to soil productivity. But since most of the nutrient factors correlated with each other it was impossible to differentiate the impact of a single soil factor on *Striga hermonthica* under farming conditions.

Soil characteristics	Compound fields mean	Stdev.[b]	Bush fields mean	Stdev.[b]	Sign. level[a]
Soil Physical Characteristics:					
Particle size distribution (%)					
Clay (<2 µm)	5.9	1.9	6.1	1.2	n.s.
Silt (2-50 µm)	26.5	7.4	26.0	7.4	n.s.
Sand (>50 µm < 2 mm)	67.5	7.3	67.9	8.1	n.s.
Gravel (>2mm)	23.8	16.9	21.3	14.8	n.s.
Soil Chemical Characteristics:					
C_{org} (%)	1.27	0.63	0.71	0.17	**
PH ($CaCl_2$)	6.52	0.64	6.13	0.41	+
N_{total} (%)	0.13	0.06	0.09	0.01	*
$NH_{4\,min}$ (mg kg^{-1})	4.4	1.6	2.3	0.9	**
$NO_{3\,min}$ (mg kg^{-1})	18.8	19.9	6.1	3.7	*
P_{total} (mg kg^{-1})	370	464	145	37.1	+
P_{org} (mg kg^{-1})	154	259	69.8	36.7	n.s.
$P_{(Bray\,I)}$ (mg kg^{-1})	28.3	38.5	4.2	2.8	*
K_{exch} (mg kg^{-1})	500	411	146	101	**
Ca_{exch} (mg kg^{-1})	1743	1183	755	493	+
Mg_{exch} (mg kg^{-1})	245	331	92.4	36.1	+
CEC (cmol$_c$ kg^{-1})	13.13	12.8	5.42	2.58	*
Exch. Bases (cmol$_c$ kg^{-1})	12.46	13.1	4.98	2.63	*
Base Saturation (%)	89.53	8.5	90.76	4.99	n.s.
Soil Microbial Characteristics:					
Biomass C (µg C g ds^{-1} ^{c})	198.2	158.4	47.2	23.7	**
Dehydrogenase (µg INTF g ds^{-1} ^{c})	111.2	87.4	32.1	18.9	**

[a] Sign. level: Significance level P: +<0.1; *<0.05; **<0.01; ***<0.001; n.s.: not significant;
[b] stdev.: standard deviation
[c] ds:dried soil;

Tab. 3: Soil characteristics of Striga hermonthica in compound and bush fields in Northern Ghana

The reduced *Striga* infestation was related to indirect effects such as an increased resistance of the well-supplied host plants against pathogens (e.g. phytoalexines). An enhanced microbial activity might diminish the *Striga* seed bank and suppress the germination of the parasite as direct effects of the manure treatment.

To maintain the soil productivity, organic manure application could be an alternative for fallows in intensified land use. Since the availability of farm yard manure or other organic materials is limited (Runge-Metzger, 1993), alternative measures such as green manure application, cover crops or crop rotation with legumes are recommended. For the interrelation of soil productivity and *Striga* infestation, these measures should be integrated in *Striga* control programs.

Acknowledgements

The facilities for the field work were provided by the Savannah Agriculture Research Institute (SARI), Nyankpala, Northern Ghana. The standard analyses were done by the laboratory of soil chemistry. The work was funded by the Agency for Technical Cooperation (GTZ), Eschborn, Germany.

References

Bebawi, F.F. (1987): Cultural Practices in Witchweed Management, in: Parasitic Weeds in Agriculture, Vol.I. Musselman L.J. (eds.), *Striga* CRC Press, Boca Raton, Florida, USA, 159-171.

Berner, D.K., Cardwell, F.K., Faturoti, B.O., Ikie, F.O. and Williams, O.A. (1994): Relative Roles of Wind, Crop Seeds and Cattle in Dispersal of *Striga*, Plant Disease **78 (4)**, 402-406

Chidley, V.L. and Drennan, D.S.H. (1987): Effect of Sorghum Root Residue on *Striga asiatica* (L.) Kuntze Infection. In: H.C. Weber & W. Forstreuter (eds.), Parasitic Flowering plants Proceedings of the 4th ISPFP, Marburg, 819-828

Hess, D.E., Obilana, A.B. and Grard, P. (1996): *Striga* Reaserch at ICRISAT, in Advances in Parasitic Plant Research; Proceedings of the 6th International Parasitic Weed Symposium; Cordoba, Spain, 827-834

Ogborn, J.E.A. (1984): *Striga* Research Priorities with specific-reference to Agronomy. In: Ayensu E.S., H. Doggett, R.D. Keynes, J. Marton-Lefèvre, L.J. Musselman, C. Parker and A. Pickering (eds.), *Striga*, biology and control. International Council of Scientific Union Press, Paris 195-212

Ransom, J.K. and Odhiambo, G.D. (1994): Long-term Effects of Fertility and Hand-weeding on *Striga* in Maize, Biology and Management of *Orobanche*, Proceedings of the 3rd International Workshop on *Orobanche* and related *Striga* research, Pieterse, A.H. J.A.C. Verkleij, and S.J. Borg (eds.), Amsterdam, The Netherlands, Royal Tropical Institute, 513-519

Runge-Metzger, A. (1993): The Economics of *Striga* Control in different Farming Systems in Northern Ghana, Report for the over regional GTZ-project: Ecology and Management of Parasitic Weeds, PN 87.3450.9-01.200 Eschborn, Germany

Watt, W.L. (1936): Control of *Striga* in Nyanza Province, Kenya; The East African Agriculture Journal, 320-322

Addresses of authors:
Brigitte Kranz
University of Hohenheim (380)
D-70593 Stuttgart, Germany
W.D. Fugger
University of Göttingen
Institute of Agronomy and Animal Health in the Tropics
Griesebachstraße 6
D-37077 Göttingen, Germany
J. Kroschel
Deutsche Gesellschaft für Technische Zusammenarbeit (GTZ) GmbH
University of Hohenheim (380)
D-70593 Stuttgart, Germany
J. Sauerborn
University of Giessen
Tropical Crop Science
Schottstraße 2
D-35390 Giessen, Germany

Influence of Long-Term Heavy Application of Pig Slurry on Soil and Water Quality in Latvia

T.K. Haraldsen, V. Jansons, A. Spricis, R. Sudars & N. Vagstad

Summary

Soil investigations and monitoring of the quality of surface water, drainage water and ground water have been carried out at three large Latvian pig farms in 1995 and 1996. Heavy applications of pig slurry caused very high leaching and runoff of plant nutrients, and increased nitrate concentrations in ground water wells. The problems were mainly related to areas receiving much more slurry than the fertilizer demand for the crops.

Keywords: Leaching, manure, nitrogen, pig farms, phosphorus, slurry

1 Introduction

Many large and specialized collective farms were established in different parts of the former Soviet Union during the 1970's and 1980's. The Baltic states (Estonia, Latvia and Lithuania) specialized in animal production and pig farms were built with a capacity up to 70 000 slaughter pigs per year. Due to the former Soviet regulations, acreage for slurry application was calculated according to the N-balance, allowing a maximum application of 200 kg N/ha. Consequently, amounts of manure applied surpassed the fertilizer requirements of the crops, leading to increased risk of phosphorus accumulation in soils and runoff and leaching of plant nutrients to rivers and lakes. Pollution problems due to heavy applications of pig manure or slurry have also been reported from many European countries and USA (Gerritse 1981; Unwin 1981, 1986; Sharpley et al. 1991).

Independence from the Soviet Union initiated a dramatic change of agricultural structure in all Baltic states: most state and collective farms were privatized. In Latvia the transition to private farms has been the following:
1. State farms have been divided into smaller specialized share companies. The buildings and equipment are used as previously, but available land for cultivation and utilization of manure may be reduced due to different owners.
2. State farms have been divided into small orginal farms, according to the ownership rights in 1940. Most of the farmers have no farming experience, and low input agriculture for self-consumption is most common. Only few successful farmers have sufficient knowledge to produce for market. These farms use increasing amounts of pesticides and fertilizers.
3. Share companies owned by the workers of the former state or collective farms have been established. This caused no great change in agricultural practice, but less use of fertilizer and pesticides have been common due to the market situation and the economy.

The impact on the environment of major Latvian pig farms has been investigated in a joint project between Latvian and Norwegian research institutions. In spring 1995, the Department of

Environmental Engineering and Management, Latvia University of Agriculture, Centre for Environmental Studies, University of Latvia, and Centre for Soil and Environmental Research (Norway) started monitoring the surface water, drainage water and ground water, and investigated the soil at four Latvian pig farms. This report discusses the results from three of the farms: Bauska, Ulbroka and Vecauce, representing the three groups of transition to private farms.

2 Material and methods

2.1 Pig farms

The pig farm "Strauti" at Uzvara, near Bauska, is located in the upper part of the Lielupe river catchment in the central part of Latvia. This farm was established in 1970 and reached full production in 1976 (12,000 fattening pigs per year, 55,000 m^3 pig slurry per year). Until 1987 tractor-moved tankers were used for slurry application. In 1987 a slurry irrigation system was installed (226 ha). Today the pig production is 10,000 fattening pigs, but the slurry utilization area is only 50 ha, due to changes in land use after the privatization. The field is drained with tile drainage at 1.1-1.3 m. Some properties of the imperfectly drained, calcareous, silt loam soil at Bauska are shown in Table 1.

Location	Horizon	Depth cm	Sand %	Silt %	Clay %	pH (H$_2$O)	Org. C. g/100 g	CEC cmol/kg
Bauska	Ap	0-30	32	61	7	7.2	2.3	19.1
	Eg	30-42	29	60	11	7.9	0.5	9.2
	Bwg	42-58	43	48	9	8.0	0.1	7.3
Ulbroka	Ap1	0-25	63	28	9	6.3	2.8	12.5
	Ap2	25-40	59	32	9	6.1	2.2	13.4
	EB	40-52	54	35	11	5.8	1.5	8.4
	Bw	52-62	86	8	6	5.7	0.7	5.7
Vecauce	Ap	0-35	56	29	15	7.3	2.4	17.6
	Bw	35-56	47	32	21	7.5	0.5	12.5
	Bg	56-80	49	29	22	7.8	0.2	10.8

Table 1: Soil properties in the monitoring sites

Water sampling points at Bauska are shown in Table 2. Water samples were collected each month. At B 3, B 4, and B 5 no samples could be taken in dry periods, and these sampling points were covered by snow in the winter period.

The pig farm at Ulbroka is located just outside the border of Riga. This farm was one of the greatest pig farms in Latvia in the Soviet period with a production of 50,000 fattening pigs per year (24,000 pig places). The farm was built in 1974-75, and reached full production in 1982-84. The last three years the pig farm has been under private ownership, but the land of the previous collective farm is now property of small private farms in the area. The annual pig production is now 2500 fattening pigs and 8000 piglets. The present slurry production is 50-60 m^3 per day, while it was 300-350 m^3 per day in the period 1982-1990. About 400 ha of land was used for slurry utilization and dumping in the Soviet period. Today slurry is offered to private farmers in the area and spread by heavy tankers. Some properties of the moderately well to well-drained sandy soils at Ulbroka are shown in Table 1.

Sampling point	Description	Acreage ha	Crops	Farming system
B 1	Small catchment	800	Different crop rotations	Intensive farming, dumping site and pig farm
B 2	Small catchment, part of B 1	750	Different crop rotations	Intensive farming, former slurry utilization area
B 3	Small catchment, upper part of B 2	50	Pastures 30 ha, forest 20 ha	Slurry utilization 1984-1992, extensive farming since 1993
B 4	Drainage channel, part of B 1	50	Grassland 1992-1996	Slurry dumping site
B 5	Outlet of tile drain, part of B 4	14.7	Grassland 1992-1996	Slurry dumping site

Table 2: Description of sampling points at Bauska

Water quality at Ulbroka has been checked in three small channels and streams (Table 3). Samples from ground water wells at farms in the area have also been analysed.

Sampling point	Description	Acreage ha	Crops	Farming system
U 1	River catchment, agricultural land and forest	2020	Different crop rotations	Mixed farming, slurry utilization since 1975
U 2	Channel, agricultural land	120	Mixed crops, 1995-1996	Intensive farming, slurry utilization since 1975
U 3	Small drainage channel, part of U 2	30	Cereals, 1995-1996	Intensive farming, slurry utilization since 1975

Table 3: Description of sampling points at Ulbroka

The Vecauce state research farm is located in western Latvia in the upper part of the Lielupe river catchment. This farm is typical of the continuation of collective farming systems. The pig farm was established in 1987 for production of 6000 fattening pigs per year. In 1991 a slurry separation system was installed. Liquid manure was stored in a lagoon and used for irrigation at an area of 32 ha. Since 1991 up to 250 m^3 pig slurry per ha/year has been applied. The solid manure was stored in heaps and was partly composted before use at farm land outside the investigated site. Some properties of the imperfectly to poorly-drained, calcareous sandy loam soil at Vecauce are shown in Table 1. The field is systematically drained with tile drains at 1.2 m.

Water samples have been collected at four locations (Table 4).

Sampling point	Description	Acreage ha	Crops	Farming system
V 1	Outlet of tile drain, part of V 3	3.5	Ley 1992-1995, cereals 1996	Extensive farming up to autumn 1995
V 2	Outlet of tile drain, part of V 3	3.5	Cereals 1995-1996	Intensive farming, slurry irrigation
V 3	Small catchment, channel	60	Ley and cereals, 1995-1996	Slurry irrigation 30 ha, without irrigation 30 ha
V 4	Outlet of tile drain	16.2	Cereals	Intensive farming without irrigation

Table 4: Description of sampling points at Vecauce

2.2 Laboratory methods

Water analyses were carried out in Latvia according to standard methods. The Latvian laboratory has shown satisfactory precision in intercalibration tests against Nordic accredited laboratories. Soil phosphorus was analysed in Norway according to the AL-method (Egnér et al. 1960).

3 Results and discussion

Slurry irrigation at Bauska caused very high concentrations of nitrogen and phosphorus in runoff and drainage water (Figs. 1 and 2). Loss of phosphorus through drainage system after application of pig slurry has also been found in Norwegian experiments (Oskarsen et al. 1996). The phosphorus content of the topsoil at Bauska was high (P-AL 11-19 mg/100 g). The dumping area (50 ha) significantly influenced the water quality downstream. The concentrations of total-P in the brook (Fig. 2, B 1) were similar to the level in runoff from the dumping area B 4. For nitrogen the concentrations in the B 1 catchment were diluted compared with the slurry dumping area B 4 (Fig. 1). This dilution indicates that other sources for P-pollution also exist. Runoff from solid manure storage at the pig farm is a possible explanation. In water from the catchment B 2 where no slurry has been applied recently, there were very low concentrations of nitrogen. Some periods of increased phosphorus concentrations in this catchment may be due to runoff from cow manure storage at a dairy farm in the catchment.

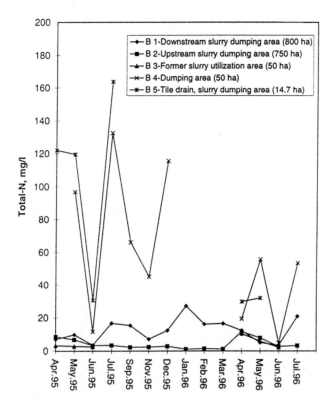

Fig. 1: Nitrogen concentrations in catchments at a large pig farm in Bauska, Latvia.

The sandy soils at Ulbroka have good absorption capacity of phosphorus, and very high levels of P-AL in the topsoil (19-69 mg/100 g). This might explain the very low concentrations of total-P in the runoff from this area (average (U 1 - U 3) 0.04 mg P/l - figures not shown). Most of the excess water feeds the groundwater, and the drainage channels monitored had short periods of discharge in spring and autumn. Accordingly, the concentrations of nitrate in the ground water wells at farms in the area have been analysed. Almost all wells close to the area, which are influenced by slurry utilization and lagoons, have more than 10 mg/l NO_3-N, which is the limit for acceptable drinking water quality in Latvia. The common level of NO_3-N in these ground water wells is 10-60 mg/l, but samples with more than 100 mg/l NO_3-N have been found. In areas where the amounts of slurry applied have been in accordance with fertilizer requirements for plants, good water quality was found. In such areas nitrate-N concentrations less than 5 mg/l have been typical.

At Vecauce the slurry utilization has not been found to influence the loss of phosphorus through drainage water (average (V 1, V 2, V 4) 0.02 mg P/l - figures not shown). The phosphorus values in the topsoil are moderately high to high at Vecauce (P-AL 5-12 mg/100 g), and no effect of the slurry irrigation during five years has been found. Ploughing of large areas of old leys in autumn 1995 caused a large release of nitrogen in the snowmelt period in 1996 (Fig. 3, V 1 and V 3). Gustafson (1987) observed similar patterns in Swedish experiments. Increased release of

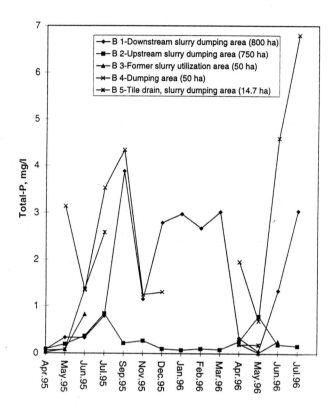

Fig. 2: Phosphorus concentrations in catchments at a large pig farm in Bauska, Latvia.

nitrogen in spring was not observed at the area with cereals the previous year (V 2). The water samples from Vecauce show differences in N-leaching between ley farming V 1 and intensive arable farming V 4 (Fig. 3). It has not been possible to detect differences in N-leaching between pig slurry irrigated fields and intensive arable farming with commercial fertilizer at Vecauce.

4 Conclusions

Environmental problems related to pig farming in Latvia are mainly located in areas receiving much more slurry than fertilizer demand for the crops. Considerable losses of nitrogen and phosphorus have been found from slurry dumping areas. Long term heavy applications of pig slurry have caused increased nitrate concentrations in ground water wells used for drinking water. The water quality in many wells may be harmful to human health. Reduction of livestock density and increased acreage for application of manure will reduce the losses of nitrogen and phosphorus, and improve the water quality. Cooperation between farms will be important in solving the problems.

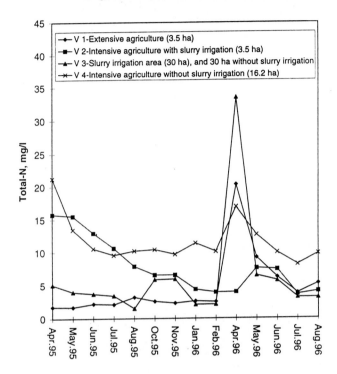

Fig. 3: Nitrogen concentrations in small catchments at Vecauce, Latvia.

References

Egnér, H., Riehm, H. und Domingo, W.R. (1960): Untersuchungen über die chemische Bodenanalyse als Grundlage für der Beurteilung des Nährstoffzustandes der Böden. II. Chemische Extraktionsmethoden zur Phosphor und Kaliumbestimmung. Statens Jordbruksforsök, särtrykk och småskrifter no. 133, Uppsala.

Gerritse, R. G. (1981): Mobility of phosphorus from pig slurry in soils. In: T.W.G. Hucker and G. Catroux (eds.), Phosphorus in Sewage Sludge and Animal Waste Slurries, London, D. Reidel Publishing Company, pp. 347-366.

Gustafson, A. (1987): Nitrate leaching from arable land in Sweden under four cropping systems. Swedish J. Agric. Res. 17, 169-177.

Oskarsen, H., Haraldsen, T.K., Aastveit, A.H. and Myhr, K. (1996): The Kvithamar field lysimeter. II. Pipe drainage, surface runoff and nutrient leaching. Norw. J. Agric. Sci. 10, 211-228.

Unwin, R.J. (1981): Phosphorus accumulation and mobility from large applications of slurry. In: T.W.G. Hucker and G. Catroux (eds.), Phosphorus in Sewage Sludge and Animal Waste Slurries, London, D. Reidel Publishing Company, pp. 333-343.

Unwin, R.J. (1986): Leaching of nitrate after application of organic manures. Lysimeter studies. In: A.Dam Kofoed, J.H. Williams and P.L'Hermite (eds.), Efficient Land Use of Sludge and Manure, London, Elsevier, pp. 158-167.

Sharpley, A.N., Carter, B.J., Wagner, B.J., Smith, S.J., Cole, E.L. and Sample, G.A. (1991): Impact of long-term swine and poultry manure application on soil and water resources in Eastern Oklahoma. Tech. Bull. T-169, Okla. State Univ., Stillwater, OK, 50 pp.

Addresses of authors:
Trond Knapp Haraldsen
Nils Vagstad
Jordforsk
Centre for Soil and Environmental Research
N-1432 Ås, Norway
Viesturs Jansons
Ritvars Sudars
Latvia University of Agriculture
Faculty of Rural Engineers
Department of Environmental Engineering and Management
Akademijas iela 19
LV-3001 Jelgava, Latvia
Andris Spricis
University of Latvia
Faculty of Chemistry
Centre for Environmental Studies
Kr. Valdemara iela 48
LV-1013 Riga, Latvia

Protection of Soils from Contamination in Swiss Legislation

M. Hämmann, S.K. Gupta & J. Zihler

Summary
The comprehensive Swiss protection policy for contaminated soils is based on preventive, predictive and curative instruments which are legally implemented. Preventive instruments are emission control measures at the point of origin of pollutants and soil input control. Soil monitoring and risk assessment are the instruments of predictive soil protection. Restrictive and clean up measures to avert hazards from contaminated soil belong to the curative soil protection. Guide, trigger and clean-up values have been established in order to assess the hazardous impact of pollutants on long-term soil fertility, man and animal health. Exceeded guide values can shed light on long-term soil fertility. When a trigger value is exceeded, a hazard to man or animal health is possible. A case-specific investigation confirms or refutes the assumed hazard. Exceeded clean-up values indicate an existing hazard. Curative measures include land use restriction, land use ban, gentle or hard clean-up of the soil.

Keywords: Soil protection; legal instruments; guide value; trigger value; clean-up value

1 Introduction

The contamination of soils in Switzerland continually increased over the last 50 years due to industrialisation and intensive agricultural land use (Meyer 1991). The on-going increase of emissions and their impact on man, air quality, water and soils led to the first Swiss Environmental Protection Law (USG 1983) which is fundamental to the preventive soil protection policy. In addition to other important goals (e.g. air quality control, water protection, environmental impact assessment), soil contamination should be minimized.

Up to now, the soil was considered as a sink for pollutants, but not as a source posing a hazard to other environmental compartments. In the revised Environmental Protection Law (USGrev 1995), which came into force in 1997, the legal basis for wider soil protection was achieved. Curative measures to avert hazards caused by contaminated soil as well as the physical protection of soil (e.g. soil erosion, soil compression) was legalised. The soil is considered as a source of pollutants (e.g. heavy metals, organic pollutants) which could be hazardous to man, animals or plants. The revised Environmental Protection Law, the existing Development Planning Law (RPG 1979), and the corresponding ordinances will create a sound basis for sustainable land use in Switzerland.

2 Preventive soil protection and related legal instruments

Preventive soil protection focuses on the fate of pollutants in the environment from their source to the input into soil. Therefore preventive soil protection does not begin in the soil, but at the point of emission. Figure 1 gives an overview of different pollutant sources and their pathway to the soil and shows the necessary instruments to avoid soil contamination.

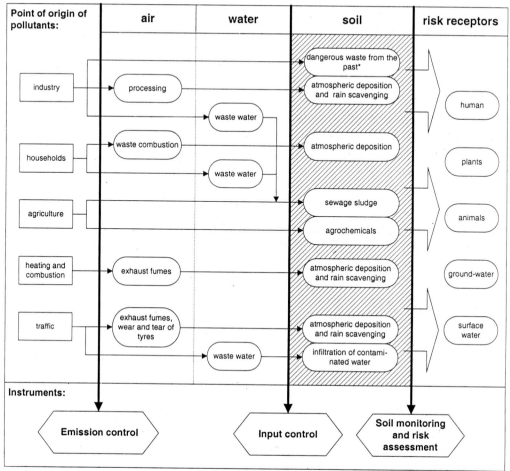

Fig. 1: *Important sources of soil pollutants, their pathways and limitation measures.*

* Dangerous waste from the past is legally distinguished from soil contamination. However, the trigger and clean up values are applied for the assessment of such contaminated sites.

Preventive soil protection aims on one hand to avoid and limit the emission of pollutants, and on the other hand to control and reduce the input of pollutants into other environmental compartments. The realisation of these aims is achieved by different ordinances which are based on the Environmental Protection Law.

The main instruments of the Clean Air Ordinance (LRV 1986) relating to emission control are emission standards for industrial facilities, combustion and heating facilities, declaration and measurement of emission rates of industrial facilities, technical surveillance and fuel quality. Soil input control is achieved by air quality monitoring. Air quality standards are stipulated not only for classic air pollutants (e.g. nitrogen oxides, sulphur oxides), but also for typical soil pollutants (e.g. lead, cadmium). Standards exist for pollutants in floating dust and deposition of pollutants as well. When air quality standards are not met, measures in the field of emission limitation must be taken (e.g. tightening of emission standards).

The Ordinance on Hazardous Substances (StoV 1986) regulates the handling and assessment of the environmental impact of substances and products. The following regulations are stipulated: Limitation and prohibition of substances (e.g. polychlorinated biphenyles) and products containing hazardous substances (e.g. Cadmium-containing items), take-back guarantee from producers (e.g. batteries), product control by the producer with respect to environmental impact, declaration of new substances introduced in the market, quantity and quality standards for substances (e.g. sewage sludge, compost, agrochemicals, fertilisers) and optimum information for the purchaser.

The Technical Ordinance on Waste (TVA 1990) regulates the limitation of the environmental impact through wastes. It stipulates requirements for facilities of household and industrial waste incineration and therefore reduces airborne soil contamination. Additionally it regulates the quality control process for waste (i.e. compost, cinders).

The drafted Agriculture Law (LWGrev 1996) suggests direct financing of farmers as an incentive instrument. The main requirement for direct financing is sustainable agricultural land use. Suitable soil protection is part of this and is mentioned in the drafted law. Detailed regulations on Integrated Production systems (IP) will follow. Other legal regulations include also soil protection measures.

3 Predictive soil protection: soil monitoring and risk assessment

Soil monitoring is necessary to assess the effectiveness of the above-mentioned preventive regulations and to elucidate environmental risks. The Ordinance relating to Pollutants in Soil (VSBo 1986), which currently is revised (VBBo 1997), regulates soil monitoring and the measures to be taken, if long-term soil fertility is no longer assured as well as the risk assessment of severely contaminated soils (Fig. 2).

Soil is considered to be fertile if (i) it is characterised by a diverse and biological active community, possessing a typical structure and undisturbed capacity for decomposition, (ii) it makes possible undisturbed growth and development of plants, plant communities (both natural and influenced by

	Pollutant concentrations in the soil [1]	
Inorganic pollutants	Total content	Mobile content
Cadmium (Cd)	0.8	0.02
Chromium (Cr)	50	–
Cobalt (Co)	–	–
Copper (Cu)	40	0.7
Fluorine (F)	700	20
Lead (Pb)	50	–
Mercury (Hg)	0.5	–
Molybdenum (Mo)	5	–
Nickel (Ni)	50	0.2
Thallium (Tl)	–	–
Zinc (Zn)	150	0.5
Organic pollutants	Total content	
Benzo(a)pyrene (BaP)	0.2	
Polycyclic aromatic hydrocarbons (PAH) [2]	1	
Dioxins and furans (PCDD/F)	5 [3]	

1) in mg/kg dry soil matter (extraction methods according to VBBo (1997))
2) sum of 16 PAH according to U.S. EPA
3) in ng I-TEF/kg (International toxic equivalency factors)

Table 1: Guide values for soil pollutant concentrations (VBBo 1997).

Erratum

page 632: Fig. 1 has to be replaced by Fig. 2 as shown below

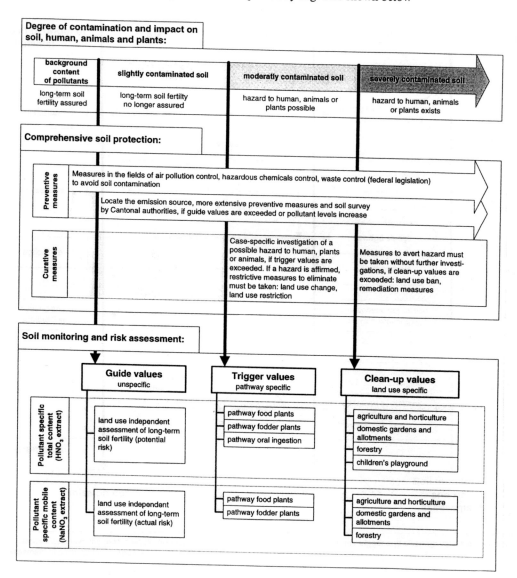

Fig. 2: *Impacts of contamination, comprehensive soil protection, soil monitoring and risk assessment.*

man), (iii) it assures that plant products are of good quality and wholesome for man and animals and (iv) if the direct exposure does not affect human and animal health (VSBo 1986, VBBo 1997). The regulations are guide values for inorganic and organic pollutants to assess long-term soil fertility (Table 1), the National Soil Monitoring Network NABO (BUWAL 1993; Desaules 1993; Desaules et al. 1996) and survey of already-contaminated soils by Cantonal authorities (e.g. AGW 1994; Baudepartement des Kantons Aargau 1994). Long-term soil fertility may also be threatened by an appreciable increase in pollutant levels even if a guide value has not been exceeded. Where guide values do not exist, case-specific assessments must be carried out (Gysi et al. 1991) to determine whether or not soil fertility is assured. Guide values for inorganic pollutants have been set for the HNO_3-extractable total content to assess the potential hazard and in some cases for the $NaNO_3$-extractable mobile content to assess the actual hazard to soil fertility (Gupta et al. 1996). A guide value is deemed to have been exceeded if at least one of the two values (HNO_3 extract or $NaNO_3$ extract) is exceeded.

If pollutant levels increase or the guide values are exceeded, the Cantons (Swiss states) shall investigate the source of the pollutants. If measures in the fields of air quality control, hazardous substance control or waste control prove insufficient to prevent further increases in pollutant levels in the affected area, the Cantons are required to adopt more extensive measures (Fig. 2). Decisions have to be made on a case by case basis as to further measures to be taken to preserve soil fertility (e.g. stipulation of renovation measures for industrial facilities).

Fundamental to the revision of the Environmental Protection Law (USGrev 1995), the risk assessment has been introduced in the revised Ordinance relating to Pollutants in Soil (VBBo 1997). The implementation is based on the Risk Assessment Plan for Contaminated Soils in Switzerland (Vollmer and Gupta 1995). For this purpose trigger values and clean-up values were proposed to permit an assessment of the associated hazard to man, animals and plants (Fig. 2).

The clean-up values, which are specific to certain land uses, indicate a level of contamination at which the current form of land use ceases to be possible without hazard to man, animals or plants. Corrective measures must be taken immediately to eliminate the hazard. Three land-use categories were proposed: agriculture and horticulture, domestic gardens and allotments and children's playgrounds for each of which pollutant-specific clean-up values are established (Fig. 2, Table 3).

Pollutant concentrations below the clean-up values are assessed differentially with the help of trigger values demonstrating a possible hazard to man, animals or plants. They trigger detailed, case-specific studies to assess the presumed hazard to man, animals or plants. If the hazard is confirmed, eliminating measures are initiated. Trigger values are specified for direct and indirect exposure pathways from the soil to man, animals and plants (Fig. 2, Table 2).

Trigger and clean up values were established for the most important inorganic pollutants (Cadmium, Copper, Lead, Zinc) as well as for organic pollutants (polycyclic aromatic hydrocarbons PAH, polychlorinated biphenyles PCB, dioxins and furans PCDD/F). For the groundwater pathway, intervention values were established for inorganic and organic pollutants. They are not part of the soil protection legislation but in legislation concerning dangerous waste from the past (AL+LV 1997). They are determined from soil samples with a leaching test.

	Pollutant concentrations in the soil for different exposure pathways		
	Total content (mobile content) [1]		
Inorganic pollutants	Food plants	Fodder plants	Direct soil uptake
Cadmium (Cd)	2 (0.02)	2 (0.02)	10
Copper (Cu)	–	150 (0.7)	–
Lead (Pb)	200	200	300
Organic pollutants	Food plants	Fodder plants	Direct soil uptake
Benzo(a)pyrene (BaP)	2	2	2
Polycyclic aromatic hydrocarbons (PAH) [2]	10	10	10
Dioxins and furans (PCDD/F)	5 [3]	20 [3]	20 [3]
Polychlorinated biphenyles (PCB)	0.2	0.2	0.1

1) in mg/kg dry soil matter (extraction methods according to VBBo (1997))
2) sum of 16 PAH according to U.S. EPA
3) in ng I-TEF/kg (International toxic equivalency factors)

Table 2: Trigger values for soil pollutant concentrations (VBBo 1997).

	Pollutant concentrations in the soil for different land use categories		
	Total content (mobile content) [1]		
Inorganic pollutants	Agriculture and horticulture	Domestic gardens and allotments	Children's playgrounds
Cadmium (Cd)	30 (0.1)	20 (0.1)	20
Copper (Cu)	1'000 (4)	1'000 (4)	–
Lead (Pb)	2'000	1'000	1'000
Zinc (Zn)	2'000 (5)	2'000 (5)	–
Organic pollutants	Agriculture and horticulture	Domestic gardens and allotments	Children's playgrounds
Benzo(a)pyrene (BaP)	20	20	20
Polycyclic aromatic hydrocarbons (PAH) [2]	100	100	100
Dioxins and furans (PCDD/F)	1000 [3]	100 [3]	100 [3]
Polychlorinated biphenyles (PCB)	3	1	1

1) in mg/kg dry soil matter (extraction methods according to VBBo (1997))
2) sum of 16 PAH according to U.S. EPA
3) in ng I-TEF/kg (International toxic equivalency factors)

Table 3: Clean-up values for soil pollutant concentrations (VBBo 1997).

The trigger and clean-up values for inorganic pollutants for food plants and fodder plants are determined by means of soil-plant relations (Hämmann and Gupta 1997). In the case of food plants, the quality objectives used are the permissible maximum concentrations of metals in foods as laid down in the Swiss Ordinance for Food Quality (FIV 1995), supplemented by the Federal German guide values (ZEBS values) for foods (BGA 1993). Fodder plants are assessed in accordance with the maximum levels set in the Swiss Ordinance for Feedingstuffs Quality (FMBV 1995) or, should these not exist for a particular pollutant, with the handicap threshold values (Kessler 1993). Those values give a threshold of daily pollutant intake which could affect the animal's metabolism. The additional effects of oral soil ingestion during grazing and crop contamination due to clinging soil particles have been taken into consideration for fodder plant trigger values. The derivation of the trigger values for oral soil ingestion is based on epidemiological studies and the Provisional Tolerable Weekly Intake (PTWI) for Cadmium and on blood level criteria for Lead (Hämmann and Gupta 1997). As for the guide values, trigger and clean up values for inorganic pollutants are set for the total and mobile content. They are deemed to be exceeded if one of the two contents is exceeded. Assessment of the oral soil ingestion exposure pathway is based only on total content, because it is assumed that this extractant at least simulates the quantity of pollutants that can be dissolved and made available by the acid medium of the gastrointestinal tract.

The derivation of trigger and clean-up values for the organic pollutants is contained in BUWAL (1997) and in Häberli et al. (1997).

4 Curative soil protection to avert hazards to man, animals and plants

Curative measures can be classified into two groups, (i) restrictive measures with an administrative character, and (ii) technical remediation measures. Technical remediation measures can be harsh or gentle. Gentle remediation measures preserve or improve soil fertility. A set of curative measures is shown in Table 2.

Restrictive measures must be taken by Cantonal authorities, if trigger values for food and fodder plants are exceeded and the case specific investigation reveals a hazard to man, animals or plants. The necessary limitation could be either a land use restriction (e.g. prohibition of cultivation of accumulative plants) or land use change (e.g. change from food plant to ornamental plant cultivation).

Measures to avert the hazard must be taken immediately once the applicable clean up value is exceeded. The Cantons are required to initiate measures whereby the contamination of or damage to the soil is to be reduced at least sufficiently to permit cultivation without hazard. Where the clean up value for children's playground is exceeded, the Canton shall stipulate a land use ban. If the playground is, however, to be used, harsh remediation measures are necessary.

		Remediation measures (technical)		Restrictive measures (administrative)
	Stabilisation of pollutants	*Soil decontamination*	*Soil removal*	
Harsh	⇒ Seal and close the site ⇒ Vitrification ⇒ Dilution of contaminated soils	⇒ *Ex situ* soil washing ⇒ Thermal soil treatment ⇒ Electromigration	⇒ Incineration ⇒ Strip off contaminated layers ⇒ Dispose of	⇒ Land use ban (e.g. forbid children to play) ⇒ Land use change (e.g. change from food to fodder plants)
Gentle	⇒ *In situ* immobilisation by metal binding agents or liming	⇒ Microbial decomposition of organic pollutants ⇒ *In situ* soil leaching ⇒ Controlled mobilisation of metals and phytoextraction		⇒ Land use restriction (e.g. cultivation of less accumulative food plants)

Table 4: Curative measures for contaminated soils to avert hazards to man, animals or plants.

5 Conclusions

A comprehensive soil protection policy is needed for sustainable land use and for the preservation of human health. The presented Swiss approach is a powerful instrument to avoid further soil contamination and to assess and manage already existing soil contamination. The basic ideas of soil protection are exemplary and may be useful to other nations which want to establish their own soil protection policy. The described instruments are then to be implemented in their respective legal system, taking into consideration their geographical, economical and social environment.

References

AGW (1994): Der Pilotlauf zum Kantonalen Bodenbeobachtungsnetz (KABO) Zürich - Kurzfassung und Synthese, Zürich: Amt für Gewässerschutz und Wasserbau (AGW).
AL+LV (1997): Verordnung über die Sanierung von durch Abfällen belasteteten Standorten (Altlastenverordnung), Vernehmlassungsentwurf Bern: Eidg. Departement des Innern.
Baudepartement des Kantons Aargau (1994): Bodenbeobachtung im Kanton Aargau - Belastungszustand der Böden 1991/1992, Aarau: Baudepartement des Kantons Aargau - Abteilung Umweltschutz.
BGA (1993): Richtwerte für Schadstoffe in Lebensmitteln (Bekanntmachung des BGA), Bundesgesundheitsblatt **5**, 210-211.
BUWAL (1997): Dioxine und Furane -Standortbestimmung (Schriftenreihe Umwelt Nr. 290 - Umweltgefährdende Stoffe), Bern: Bundesamt für Umwelt, Wald und Landschaft.
BUWAL (1993): NABO Nationales Bodenbeobachtungsnetz - Messresultate 1985-1991 (Schriftenreihe Umwelt Nr. 200 - Boden), Bern: Bundesamt für Umwelt, Wald und Landschaft.
Desaules A. (1993): Soil monitoring in Switzerland by the NABO-Network: Objectives, Experiences and Problems, in Schulin R., A. Desaules, R. Webster, B. von Steiger, eds., Soil monitoring - Early detection and surveying of soil contamination and degradation, Basel: Birkhäuser Verlag.

Desaules A., K. Studer, S. Geering, E. Meier and R. Dahinden (1996): Die Nationale Bodenbeobachtung in der Schweiz - Konzept, Stand und Perspektiven, in Rosenkranz D., G. Einsele, H.-M. Harress, eds., Bodenschutz: Ergänzbares Handbuch der Massnahmen und Empfehlungen für Schutz, Pflege und Sanierung von Böden, Landschaft und Grundwasser, Berlin: Erich Schmidt Verlag.

FIV (1995): Verordnung über Fremd- und Inhaltsstoffe in Lebensmitteln (Fremd- und Inhaltsstoffverordnung), SR 817.021.23, Bern: Eidgenössische Drucksachen- und Materialzentrale.

FMBV (1995): Futtermittel (Futtermittelverordnung, Futtermittelbuch-Verordnung, Anhänge), SR 916.307/ 916.307.1, Bern: Eidgenössische Drucksachen- und Materialzentrale.

Gupta S.K., M.K. Vollmer and R. Krebs (1996): The importance of mobile, mobilisable and pseudo total heavy metal fractions in soil for three level risk assessment and risk management. Sci.Total.Envir. **178**, 11-20.

Gysi C., S.K. Gupta, W. Jäggi and J.-A. Neyroud (1991): Wegleitung zur Beurteilung der Bodenfruchtbarkeit, Liebefeld-Bern: Eidg. Forschungsanstalt für Agrikulturchemie und Umwelthygiene.

Häberli K., M. Hämmann and S.K. Gupta (1998): Richt-, Prüf- und Sanierungswerte für organische Schadstoffe im Boden - Fallbeispiel Polyzyklische Aromatische Kohlenwasserstoffe PAK (Umweltmaterialien - Boden), Bern: Bundesamt für Umwelt, Wald und Landschaft (in press).

Hämmann M. and S.K. Gupta (1997): Derivation of trigger and clean-up values for inorganic pollutants in soil (Environmental materials No. 83 - Soil), Bern: Federal Office of Environment, Forests and Landscape (in press).

Kessler J. (1993): Schwermetalle in der Tierproduktion, Landw.Schweiz **6**, 273-277.

LRV (1986): Luftreinhalteverordnung, SR 814.318.142.1, Bern: Eidgenössische Drucksachen- und Materialzentrale.

LWGrev (1996): Botschaft zur Reform der Agrarpolitik: Zweite Etappe (Agrarpolitik 2002), in Bundesblatt 40 (Band IV), Bern: Eidgenössische Drucksachen- und Materialzentrale.

Meyer K. (1991): Bodenverschmutzung in der Schweiz, Liebefeld-Bern: Nationales Forschungsprogramm Boden.

RPG (1979): Bundesgesetz über die Raumplanung (Raumplanungsgesetz), SR 700, Bern: Eidgenössische Drucksachen- und Materialzentrale.

StoV (1986): Verordnung über umweltgefährdende Stoffe (Stoffverordnung), SR 814.013, Bern: Eidgenössische Drucksachen- und Materialzentrale.

TVA (1990): Technische Verordnung über Abfälle, SR 814.015, Bern: Eidgenössische Drucksachen- und Materialzentrale.

USG (1983): Bundesgesetz über den Umweltschutz (Umweltschutzgesetz), SR 814.01, Bern: Eidgenössische Drucksachen- und Materialzentrale.

USGrev (1995): Umweltschutzgesetz - Änderung vom 21. Dezember 1995, Bern: Eidgenössische Drucksachen- und Materialzentrale.

VBBo (1997): Verordnung über Belastungen des Bodens, Vernehmlassungsentwurf, Bern: Bundesamt für Umwelt, Wald und Landschaft.

Vollmer M.K. and S.K. Gupta (1995): Risk assessment plan for contaminated soils in Switzerland - General procedure and case study for Cadmium, Liebefeld-Bern: Swiss Federal Research Station for Agricultural Chemistry and Hygiene of Environment.

VSBo (1986): Verordnung über Schadstoffe im Boden, SR 814.12, Bern: Eidgenössische Drucksachen- und Materialzentrale.

Addresses of authors:
M. Hämmann
S.K. Gupta
Swiss Federal Research Station for Agroecology and Agriculture
Institute of Environmental Protection and Agriculture (IUL) Liebefeld
CH-3003 Bern, Schweiz
J. Zihler
Swiss Federal Office for Environment, Forests and Landscape (BUWAL)
CH-3003 Bern, Schweiz

Fertilizer P Transformations in Loamy and Clayey Oxisols of Central Brazil

A. Freibauer, J. Lilienfein, M. Ayarza, J.E. da Silva & W. Zech

Summary

Fertilized arable lands on loamy and clayey Oxisols of Central Brazil were compared with native Cerrado savanna with respect to the P-status. Soil samples were partitioned into four inorganic (P_i) and three organic (P_o) P fractions by a modified Hedley-fractionation. Phosphorus fertilization increased short and medium term bioavailable P in arable lands in comparison with native Cerrado. In the loamy soil, ten years of P_i application resulted in an enrichment of inorganic P only. In the clayey soil, both inorganic and organic P increased and stable P_o was formed after fertilization. In loamy soils, the accumulation of inorganic fertilizer P in P_i fractions only indicates rapidly degrading systems, whereas clayey soils may be considered more stable.

1 Introduction

Phosphorus deficiency and high P sorbing capacities are the major limitations to intensive cropping of the Cerrado-Oxisols of Central Brazil (Goedert, 1983). In general, cultivation without fertilization results in P loss, especially of P_o (Beck and Sanchez, 1994). Phosphorus fertilization increases the percentage of inorganic P (P_i) forms, especially of labile P_i and may compensate for losses of P_o (O'Halloran et al., 1987; Beck and Sanchez, 1994). The transformation of inorganic P into organic binding mainly depends on biological activity. Inorganic P fertilization may result in an accumulation of P_o, if adequate C and N supplies are available. The accumulation of P_i, only, is typical for rapidly degrading systems as this may be interpreted as an indicator of low biological activity (Stewart and Tiessen, 1987).

The sequential P extraction procedure, developed by Hedley et al. (1982) allows the partition of soil P according to bioavailability and determination of the fate of P fertilizer by a conceptual model of P pools and their transformations (Tiessen and Moir, 1993). Data from sequential P fractionations for Oxisols are limited to a few studies (Cross and Schlesinger, 1995).

The objectives of this study are to answer the following questions: 1) Which P fractions are enriched under fertilized cropping systems? 2) How does texture influence the transformation of inorganic fertilizer?

2 Materials and Methods

2.1 Study area and soil sampling

The study was carried out in the region of Uberlândia (Minas Gerais), Brazil. The treatments are ten year old arable lands (AR; 75 - 80 kg ha^{-1} yr^{-1} inorganic P fertilizer) and native Cerrado (CE).

The soils are classified as a coarse-loamy isohyperthermic Typic Haplustox and a very fine isohyperthermic Anionic Acrustox (Soil Survey Staff, 1994). Topsoil properties are given in Table 1.

For each system of treatment, three homogeneous 10x10 m² subplots were chosen. From each subplot one representative sample (0 - 12 cm), consisting of several subsamples, was taken at the beginning of the rainy season.

2.2 Phosphorus fractionation

A modified Hedley fractionation (Hedley et al., 1982), following Tiessen and Moir (1993), was used to sequentially partition soil P into organic (P_o) and inorganic (P_i) P fractions in air-dry (40° C) samples sieved to < 2 mm. The fractionation scheme and a characterization of the obtained fractions are given in Figure 1.

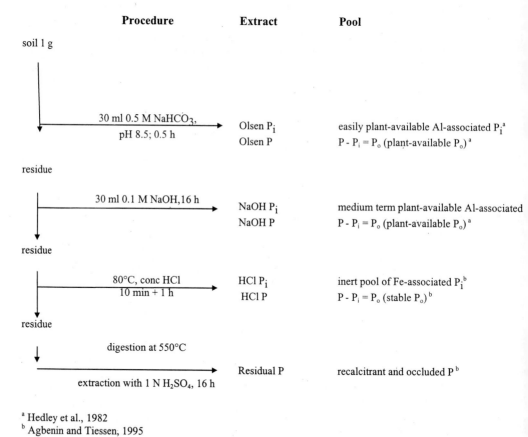

[a] Hedley et al., 1982
[b] Agbenin and Tiessen, 1995

Figure 1: *Fractionation scheme and characterization of the obtained P fractions.*

The procedure of Tiessen and Moir (1993) was shortened by the resin and dilute HCl steps as these fractions will contain only traces of P in well drained weathered soils (Agbenin and Tiessen,

1995). In addition, Bicarb-P (16 h shaken) used by Tiessen and Moir (1993), was replaced by Olsen-P (0.5 h shaken, Page et al., 1982) as this allows comparison with the literature data on plant-available P.

Inorganic P was measured colorimetrically (molybdene blue method, Page et al., 1982) with a PERKIN-ELMER 550 SE UVIVIS spectralphotometer. Total P was measured colorimetrically after oxidizing the organic matter (Bowman 1989) and with inductively coupled plasma - atomic emission spectroscopy (ICP-AES) (GBC Integra XMP). Results of the two methods agreed well. Organic P (P_o) was calculated as the difference of P and P_i.

2.6 Statistical analyses:

Main and interactive affect means were tested by using Tukey's honestly significant difference (HSD) mean separation test with a 5 % significance level. Variance analyses were performed by using STATISTICA for Windows 5.1.

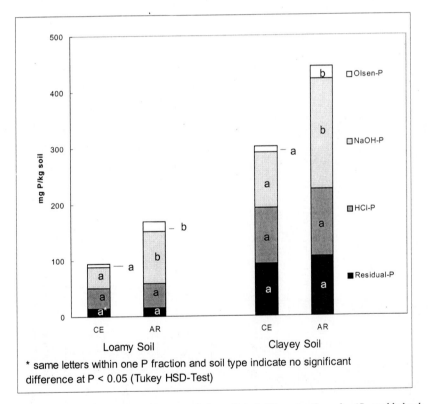

Figure 2: Partitioning of total P in the loamy and clayey Oxisol. CE: native Cerrado, AR: arable lands.

Table 1: Surface soil (0 - 12 cm) properties of the loamy and clayey Oxisols under study (data partly from Neufeldt, 1996):

	C	N	S	P	Fe_o^a	Fe_d^b	Al_o^a	Al_d^b	Ca	Mg	K	Na	Al	Fe	ECEC	BS^c	pH	pH	texture^d [%]		
	[g kg^{-1}]		[mg kg^{-1}]		[g kg^{-1}]				cmol(+) kg^{-1} soil							[%]	(H_2O)	(KCl)	clay	silt	sand
loamy soil																					
Cerrado	10.3	0.7	132	95	0.64	19	0.46	2.1	0.15	0.07	0.10	0.01	0.57	0.4	0.93	36	4.7	3.7	18	0	82
arable lands	7.8	0.6	89	170	0.50	14.3	0.32	3.3	1.67	0.29	0.08	0.01	0.00	0.2	2.06	100	5.9	5.6	17	0	83
clayey soil																					
Cerrado	23.3	1.4	178	303	1.85	44	1.66	7.0	0.09	0.07	0.12	0.01	1.01	0.8	1.34	22	4.6	4.0	70	7	23
arable lands	21.5	1.39	197	444	0.13	48	0.11	1.1	3.22	0.47	0.11	0.01	0.00	0.2	3.82	100	5.6	5.9	66	13	21

^{a} oxalate soluble ^{b} dithionate citrate soluble ^{c} base saturation, calculated for ECEC
^{d} clay: < 0.002 mm; silt: 0.002 - 0.05 mm; sand: 0.05 - 2 mm

Table 2: Phosphorus fractions of the loamy and clayey Oxisols

	Olsen		NaOH		HCl		Residual-P^a	ΣP_i	ΣP_o	ΣP
	P_i	P_o	P_i	P_o	P_i	P_o				
	[mg P/kg soil]									
Loamy soil										
native Cerrado	2.5 a'	3.3 a	14 a	24 a	36 b	n.d.^{b}	15 a	67 a	27 a	95 a
arable lands	15 b	3.1 a	53 c	40 a	43 b	n.d.	16 b	127 b	44 a	170 b
Clayey soil										
native Cerrado	4.3 a	6.2 a	33 a	66 a	100 a	n.d.	93 a	230 a	73 a	303 a
arable lands	14 b	8.6 b	95 b	100 b	94 b	33	106 a	278 a	142 a	444 b

^{a} Residual-P is assumed to be completely inorganic ^{b} calculated P_o in the HCl extract (HCl-P - HCl-P_i) equals zero
Values followed by the same letter within one soil are not significantly different at $P < 0.05$ according to Tukey's HSD mean separation

3 Results and discussion

3.1 Partitioning of P fractions

The most important P fraction in all systems is NaOH-P. Generally, P concentrations decrease in the line NaOH-P > HCl-P > Residual-P > Olsen-P. In clayey Cerrado, the first three fractions are similar. (Figure 2, Table 1). Total P is three times higher in the clayey soil than in the loamy one but plant available P (Olsen-P and NaOH-P) is only doubled. Phosphorus fixation is higher in the clayey soil because of a higher Fe content (Table 1) and maybe because of a higher C content (Tiessen et al., 1984).

In native Cerrado of both soils, inorganic Olsen-P, which is used as a reference value for plant supply, is below the critical range for soybeans in Cerrado-Oxisols (6 - 8 mg P kg^{-1}, Smyth and Sanchez, 1982). As expected, the total P concentration is higher in arable lands than in native Cerrado due to fertilization. The high application rates of fertilizer P in arable lands allow the growth of soy beans (Table 1).

Repeated annual P fertilization results in a significant increase of plant available P fractions (Olsen-P, NaOH-P). This agrees with the findings of O'Halloran et al. (1987), and Beck and Sanchez (1994). Recalcitrant P (HCl- P_i, Residual-P) is not affected even after ten years of fertilization.

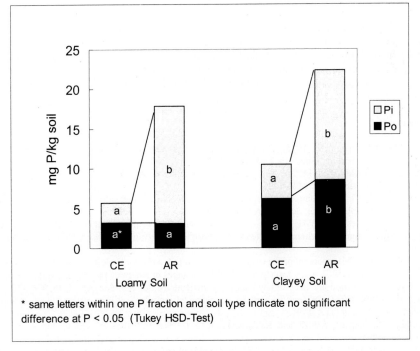

Figure 3: Inorganic and organic Olsen-P in loamy and clayey Oxisols; CE: native Cerrado, AR: arable lands

3.2 P_i/P_o ratios in the extracts:

In the loamy soil, fertilization results in an increase of the plant available inorganic P fractions but not of P_o (Table 1, Figure 3).

In the clayey arable lands, however, both inorganic and organic plant available P forms are enriched compared with native Cerrado and stable HCl-P_o was formed (Table 1). Nevertheless, the P_i/P_o ratio increases in both soils, since the enrichment of the inorganic P forms is more apparent than that of the organic ones.

Microbial biomass is higher in the clayey than in the loamy soil (Renz, personal communication). In agreement with the findings of O'Halloran et al. (1987), the higher effective CEC and carbon and nutrient supplies in the clayey soils (Table 2) may result in higher biological activity in the clayey than in the loamy soils. This may explain why fertilizer P is partly transformed into organic P in the clayey, but not in the loamy soil. Thus, according to Stewart and Tiessen (1987), arable lands on the loamy soils seem to be rapidly degrading systems, whereas cropping on the clayey soil seems to be more sustainable.

4 Conclusions

The percentage of plant-available P of the total P concentration is smaller in the clayey than in the loamy soil due to a higher percentage of the more recalcitrant P fractions.
Fertilizer-P accumulates in labile and medium labile forms. In loamy soils, inorganic P fractions are enriched and fertilization compensates for loss of P_o. In clayey soils, P fertilization results in an accumulation of inorganic and organic P fractions. According to Stewart and Tiessen (1987) the accumulation of fertilizer P in the inorganic P fractions only, indicates rapidly degrading systems. This is the case in the loamy soil. The increase of inorganic and organic P fractions after P fertilization of the clayey soil indicates a more stable system with higher microbial activity.

References

Agbenin, J.O. and Tiessen, H. (1995): Phosphorus forms in particle-size fractions of a toposequence from northeast Brazil. Soil Sci. Soc. Am. J. **59**, 687-1639.
Beck, M.A. and Sanchez, P.A. (1994): Soil phosphorus dynamics during 18 years of cultivation on a Typic Paleudult. Soil Science. **34**, 1424-1431.
Bowman, R.A. (1989): A sequential extraction procedure with concentrated sulfuric acid and dilute base for soil organic phosphorus. Soil Sci. Soc. Am. J. **53**, 362-366.
Cross, A.F. and Schlesinger, W.H. (1995): A literature review and evaluation of the Hedley fractionation: Applications to the biogeochemical cycle of soil phosphorus in natural ecosystems. Geoderma. **64**, 197-214.
Goedert, W.J. (1983): Management of the Cerrado soils of Brazil: a review. J. Soil Sci. **34**, 405-428.
Hedley, M.J., Stewart, J.W.B. and Chauhan, B.S. (1982): Changes in inorganic and organic soil phosphorous fractions by cultivation practices and by laboratory incubations. Soil Sci. Soc. Am. J. **46**, 970-976.
Neufeldt, H. (1996): Effects of land use systems on chemical and physical properties of Cerrado-Oxisols, Brazil. Jahresbericht GTZ, 1995.
O'Halloran, I.P., Stewart, J.W.B. and Kachanoski, R.G. (1987): Influence of texture and management practices on the forms and distribution of soil phosphorus. Can. J. Soil Sci. **67**, 147-163.
Page, A.L., Miller, R.H. and Keeney, D.R. (1982): Methods of soil analysis, Part 2. Chemical and microbiological properties. Agronomy Monograph no. 9 (2nd edition), ASA-SSSA, Madison, Wi. USA, 1159 p.
Smyth, T.H. and Sanchez, P.A. (1982): Phosphate Rock and superphosphate combinations for soybeans in a Cerrado Oxisol. Agron. J. **74**, 730-735.
Soil Survey Staff (1994): Keys to Soil Taxonomy. SMSS Technical Monograph No. 19, 6th edition.
Stewart, J.W.B. and Tiessen, H. (1987): Dynamics of organic phosphorus. Biogeochemistry **4**, 41-60.

Tiessen, H. and Moir, J.O. (1993): Characterization of available P by sequential extraction. p. 75-86. In M.R. Carter (ed.). Soil sampling and methods of analysis. CRC Press, Boca Raton, FL.

Tiessen, H., Stewart, J.W.B. and Cole, C.V. (1984): Pathways of phosphorus transformations in soils of different pedogenesis. Soil Sci. Soc. Am. J. **48**, 853-858.

Addresses of authors:
A. Freibauer
J. Lilienfein
W. Zech
Department of Soil Science and Soil Geography
University of Bayreuth
D-95440 Bayreuth
Germany
M. Ayarza
J.E. da Silva
EMBRAPA-CPAC/CIAT
Bl 20 km 18, C.P. 08.223
70.301-970 Planaltina (DF)
Brazil

Nitrogen Losses from Soils in the Indian Semi-arid Tropics

A.K. Patra

Summary
To evaluate ammonia volatilization, nitrate leaching, and denitrification under tropical conditions, experiments were undertaken in the laboratory and in the field for several soil types with different crops and moisture contents. Studies with ^{15}N labelled urea on three different Vertisols (pH >7.7) indicated that losses from surface applications were significantly controlled by pH and moisture, being 29.5, 33.5 and 33% under wet and 37, 42, and 40.5% under moist-dry conditions with pH values of 7.7, 8.2, and 9.3, respectively. The N leaching experiment for Alfisols and Vertisols showed that continuous heavy rainfall in August, a characteristic feature of the Indian monsoon, causes NO_3^- migration over 50% of the applied N to a depth below 60 cm. Denitrification under submerged conditions may exceed more than 60% of the applied N when KNO_3 was used in the presence of easily decomposable organic-C.

Keywords: Nitrogen, semi-arid tropics, ammonia volatilization, leaching, denitrification)

1 Introduction

Liberalization of the Indian economy by the Indian Government has gradually led to a withdrawal of many subsidies. Nitrogen fertilizers are still supported because, like other parts of the developing world, it has been given top priority by Indian farmers. A sudden escalation in price of N fertilizers may reduce the amount of applied N, and result in a large reduction in agricultural production. The decontrol of prices for N fertilizers, however, may be expected sooner or later.

The impact of this decision on P and K fertilizers is already evident. The average ratio of applied NPK in India changed from 5.9:2.1:1 in 1991-92 to 9.8:3:1 in 1993-94 (Awasthi, 1995). In India, 70% of the cultivated area (136.2 million ha) is under rainfed agriculture. Production here is unstable and suffers from poor soil fertility and resource constraints.

Because of poor soil N availability, adequate fertilizer N application is necessary to achieve higher crop yields. Urea is the dominant N fertilizer, i.e. more than 80% of applied total N. However, the efficiency of crops to added N fluctuates due to variable rainfall, low available N to the plant, and due to losses of available N. Because of the importance of agronomic yields sources for soil N loss, e.g. ammonia volatilization, leaching and denitrification need to be estimated preferably by direct methods, in order to determine actual losses and the relationship to environmental variables. This paper deals with experiments on the mechanisms of N losses for soils of the Indian semi-arid tropics (SAT).

2 Materials and methods

To assess the magnitude of NH_3 volatilization in Vertisols, a study was conducted using ^{15}N labelled urea N. Urea granules were placed on the top of 15 cm soil columns (Vertisols) collected from three sites varying in pH value, electrical conductivity and cation exchange capacity. Two contrasting moisture treatments were selected: one near field capacity (wet) and another with intermittent wetting of the soil surface before allowing the columns to dry (moist dry). The study was conducted for 7 days; soil sampling was made immediately after applying the granules and after 7 days.

To study leaching of N and other nutrients, such as P, K, Ca and S, an experiment was conducted on a medium black soil for 2 years during 1994 and 1995, using field lysimeters (1 m depth, 1.3 x 1.3 m^2 area), with fodder maize intercropped with fodder cowpea (*Vigna unguiculata* L. Wasp.) in the wet season (as rainfed) and berseem (*Trifolium alexandrinum*) in winter (with irrigation) (Patra et al. 1998). In another experiment movement of NO_3^- in an Alfisol and a Vertisol was measured in the field using bromide (Br$^-$) as a tracer for NO_3^- during June-September 1992. Bromide as NaBr was applied at a rate of 200 kg ha^{-1} and its movement was estimated weekly up to a depth of 60 cm in Alfisol and 100 cm in Vertisol (Patra and Rego, 1996).

For denitrification, simulating flooded conditions, a green house experiment was conducted on a specially fabricated cylindrical PVC column (100 cm length x 7.5 cm outer diameter) having a closed chamber on the top of it, with known volume, to tap the emanating N_2O and N_2 from soil and prevent mixing with the atmosphere (Patra, 1989). These pots were filled with 2.5 kg surface (0-20 cm) soil (*Typic ustochrept*). On the flooded soil aqueous solutions of fertilizer N was applied though urea, ammonium sulphate and KNO_3 (10% atom excess) with or without organic C (0.5%). The level of N was at 0 and 100 kg^{-1} soil. Gas samplings were taken from the 2nd day after fertilizer application at zero time and at an interval of 24 h. This was continued up to 7th day after fertilizer application, and thereafter on day 12, 17, 22, 27 and 32 in each case at zero time and after 24 h. Gas samples were analysed by Mass Spectrometer for $N_2O + N_2$. After 32 days the columns were dissembled, and soil samples were analysed for total N and residual fertilizer N for mass-balance studies.

3 Results and discussions

3.1 Ammonia (NH$_3$) volatilization

The results (Fig. 1) indicate that the recovery of added N after 7 days was markedly reduced in all soils. In the wet treatments, 70.5, 66.5 and 67.1% ^{15}N urea N was recovered from Vertisol 1, 2 and 3, respectively. Thus N losses from these soils were substantial. In the moist dry treatment the extent of N losses was still higher, at 37, 42 and 40% from Vertisol 1, 2 and 3, respectively. This indicates that a moist dry soil after the application of urea allows greater N losses than a wet soil. Although in this study measurement of NH$_3$ losses was not made directly, nevertheless, it can be assumed that the loss was mainly due to NH$_3$ volatilization because there was no leaching and the moisture level was not high enough to create anaerobic conditions for denitrification.

Earlier studies have indicated that urea hydrolysis is rapid in these soils. Urea was almost hydrolysed within 24 h after application at a soil moisture content near field capacity and soil temperatures at 27-37°C (Sahrawat, 1984). Due to rapid hydrolysis of urea, the concentration of NH$_4^+$ increased on the soil surface and the alkaline pH of the soil facilitated the volatilization of NH$_3$. The NH$_3$ volatilization is large in soils of the Indian SAT, especially if urea fertilizer is applied by spreading granules on the soil surface. Several soil and climatic factors influence NH$_3$ volatilization from soil (Freney et al., 1983). NH$_3$ volatilization of the surface applied urea may be particularly large following a light rainfall (<15 mm) that is sufficient to moisten the soil but not enough to leach the urea to any substantial depth, most likely because of an increase in the rate of urea hydrolysis (Craig and Wollum, 1982).

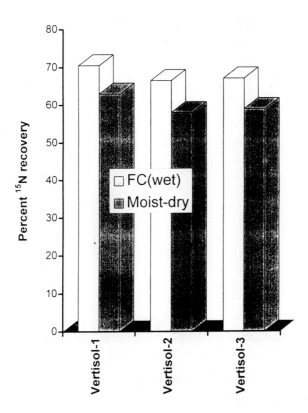

Fig. 1: Percent recovery of surface applied ^{15}N labelled urea-N from Vertisols 7 days after application. The pH values of Vertisol 1, 2, and 3 were 7.7, 8.2 and 9.3, respectively

3.2 Leaching

The results show (Fig. 2) that of the nutrients, N is most susceptible for leaching below the 1 m depth - as high as 140 kg ha^{-1} in 1995 - and a large portion (60-70%) of this amount occurred during 27-34 weeks. This study also revealed that a significant amount of N leaching occurred during early monsoon when the mineralization is expected to be high with the initial rain just after summer.

In contrast to N, the leaching of K was about 7 kg ha^{-1} yr^{-1} and Ca was up to 63 kg^{-1} ha^{-1} which were almost similar in both years. However, in this study, there was no measurable P and SO$^{2-}_4$ in the leachate (Fig. 2).

Studies with Br$^-$ as a tracer for NO$_3^-$ indicated that NO$_3^-$ leaching in the profiles of an Alfisol and a Vertisol may be rapid (Table 1).

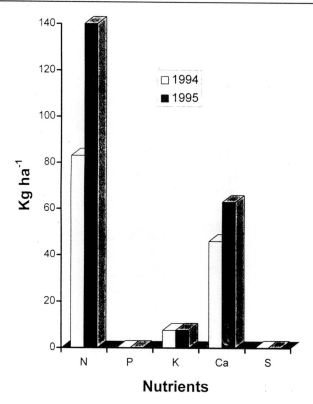

Fig. 2: Leaching of nutrients from Lysimeters during July-September in 1994 and 1995 in a medium black soil (Vertisols), at Jhansi, India

Sampling Dates 1992	Total precipitation after preceding sampling (mm)	Total evaporation after preceding sampling (mm)	Cumulative precipitation (mm)	Cumulative evaporation (mm)	Percentage of applied Br⁻ recovered from the Alfisol profile (0-60 cm)	Percentage of applied Br⁻ recovered from the Vertisol profile (0-100 cm)
15 June	-	-	-	-	98.5	101.4
22 June	61.4	61.7	61.4	61.7	93.7	90.1
30 June	31.8	47.6	93.2	109.3	95.8	91.1
06 July	84.6	33.8	177.8	143.1	58.1	70.1
13 July	27.2	48.4	205.0	191.5	93.5	94.2
20 July	3.8	43.7	208.8	235.2	85.8	90.0
28 July	23.4	45.6	232.2	280.8	78.2	89.6
05 Aug	53	40.3	285.2	321.1	80.5	72.0
17 Aug	164.8	45.2	450.0	366.3	4.2	36.0
25 Aug	17.4	34.3	467.4	400.6	3.8	54.3
04 Sept	49.5	35.7	516.9	436.3	2.3	45.2
11 Sept	8	25.7	524.9	462.0	4.0	58.7

Tab. 1: Precipitation, evaporation, and percentage of applied bromide recovery from the profile of an Alfisol and a Vertisol on different dates after NaBr application, at ICRISAT centre, Hyderabad, India.

The rainfall in SAT is erratic, it occurs often in large storms punctuated by discrete dry spells (Huda et al., 1988). During excessive rainy periods large part of the soil-N, especially NO_3^-, may be leached beyond the root zone by the excess water percolating down the soil profile. The magnitude of N use efficiency or losses in semi-arid regions of India is strongly influenced by climate and many soil factors (such as soil depth, native fertility) and their interactions. For example, intensive mineralization and leaching of NO_3^- coincided at the beginning of the rainy season in this region: the first rain causes a flush of N mineralization known as the 'Birch effect' (Birch, 1958) resulting in rapid accumulation of NO_3^- in the top soil. The next heavy rain may leach the NO_3^- leading to substantial losses of N that are neutral (urea) or anionic (NO_3^-, NO_2^-) in form. In semi-arid tropical region, losses of N by leaching depend upon the coincidence of high concentrations of soluble N in soil and occurrence of high rainfall events. The analysis of 30-year rainfall data of the Jhansi (India) region (Patra and Singh, 1996) revealed that the probability of two consecutive wet weeks with rainfall over 20 to 50 mm is greatest between 30 July - 5 August (Fig. 3) and this is just before the crop growing season (Patra and Singh, 1996). Consequently, NO_3^- leaching and N deficiency for the crop can be expected, as these periods are critical growth periods for other food and fodder crops such as maize (*Zea mays*).

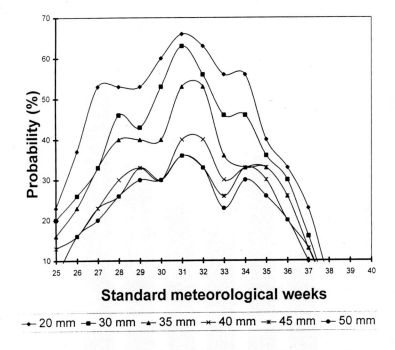

Fig. 3: Probabilities for occurrence of two consecutive wet weeks at Jhansi region, India

Thus, both the above experiments on leaching revealed that early August is a critical period for NO_3^- leaching. Obviously there is a need for strategies to increase N use efficiency and avoid the contamination of ground water in this region. The fertilizer should be applied when plants have their highest demand.

DAF	Urea	Urea + OC	Ammonium Sulphate (AS)	AS + OC	KNO_3	KNO_3 + OC
2	0.82	1.79	0.46	2.17	6.06	12.64
3	1.03	4.64	1.92	5.46	10.85	24.62
4	7.15	11.11	6.72	11.19	19.75	13.58
5	3.06	6.18	3.18	5.74	7.62	8.58
6	9.86	18.71	4.18	15.18	4.22	3.65
7	12.64	5.49	10.98	4.17	3.20	2.92
12	2.85	2.85	3.75	4.54	3.16	1.75
17	2.74	6.18	0.89	7.55	2.65	0.98
22	3.75	2.64	2.87	1.17	1.94	3.46
27	1.29	3.75	4.17	0.64	2.16	1.54
32	1.04	3.45	2.64	2.14	0.48	1.02

Tab. 2: Denitrification (N_2O+N_2) rates (mg N m^{-2} h^{-1}) as influenced by different N sources, in a loam (Typic ustochrept) soil, at New Delhi, India. DAF = days after fertilization, OC = organic C.

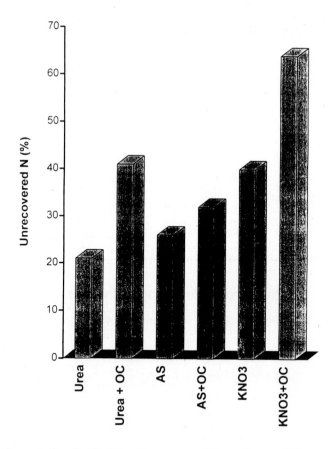

Fig. 4: Percent of applied fertilizer-N unrecovered form a loam soil (Typic ustochrept), at New Delhi, India

3.3 Denitrification

Results clearly showed the occurrence of N losses through denitrification, being greatest in the first 12 days after fertilizer application (Table 2). Of the N sources, KNO_3 showed maximum fluxes of $N_2O + N_2$ in the first week after fertilizer application. Application of organic-C along with fertilizer N showed tremendous increases in flux particularly during the first week of fertilizer application. The lowest losses of N were observed for $(NH_4)_2SO_4$. The mass balance studies at the end of 32 days showed that 61% was unaccounted for KNO_3 when applied with organic-C (Fig. 4). This was perhaps due to no limitation of NO_3^- and C particularly during the early days after fertilizer application. The KNO_3 being a source of NO_3^--N it was subjected to maximum loss under submerged conditions, which was perhaps through denitrification as there was no chance of leaching and the possibility of NH_3 volatilization was remote. Burford and Bremner (1975) demonstrated the capacity of surface soils for denitrification when submerged and it was highly significantly correlated ($r = 0.99**$) with their content of soluble or readily decomposable organic-C.

4 Conclusions

- Losses through NH_3 volatilization were high when urea was applied on the surface of a Vertisol with pH > 7.7 at the early monsoon, when rainfalls were generally light and the crop growing season begins.
- N leaching was maximum during the first half of August.
- Denitrification was high when NO_3^- source (e.g. KNO_3) was applied together with organic-C and soil remained flooded even for one week.

However, more direct measurements are required using ^{15}N tracer with other soils and agroclimatic regions of the Indian SAT. Suitable technology needs to be adopted to minimize such losses and reduce the danger of soil fertility decline of the region. More field data are needed for the development of N-cycling models under such environments.

Acknowledgements

AKP is grateful to Prof. R.J. Haggar, Technical Coordinator, and Dr. Bhag Mal, Field Coordinator of the Indo-UK project on 'Forage Production and Utilization', and Dr. S.C. Jarvis, Group Leader (Soils), IGER, North Wyke, Devon, UK, for their kind help and encouragement for participating in this conference. The financial assistance provided by the British Council on behalf of the ODA, UK is also gratefully acknowledged.

References

Awasthi, U.S. (1995): Chairman's Speech. Fert. News **40(1)**, 24-28.

Birch, H.F. (1958): The effect of soil drying on humus decomposition and nitrogen availability, Plant Soil, **10**, 9 – 31

Burford, J.R. and Bremner, J.M. (1968): Relationship between denitrification capacities of soils and total water soluble and readily decomposable soil organic matter. Soil Biol. Biochem. **7**, 389-394

Craig, J.R. and Wollum, W.G. (1982): Ammonia volatilization and soil nitrogen changes after urea and ammonium nitrate fertilization of *Pinus taeda* L. Soil Sci. Soc. Am. J. **46**, 409

Freney, J.R., Simpson, J.R. and Denmead, O.T. (1983): Volatilization of ammonia. In: Freney, J.R. & Simpson, J.R. (Eds). Gaseous loss of nitrogen from plant-soil systems. Martinus Nijhoff / Dr.W.Junk. The Hague, 32.

Huda, A.K.S., Pathak, P., Rego, T.J. and Virmani, S.M. (1988): Agroclimatic considerations for improved soil and water management and efficient use in semi-arid India. Fert News **33**, 51 - 57.

Patra, A.K. (1989): Studies on Denitrification in the Rice Rhizosphere, Ph.D. Thesis. Division of Soil Sci. & Agril. Chemistry, IARI, New Delhi 110 012, India.

Patra, A.K. and Singh, J.B. (1996): Predicting critical periods of nitrate leaching in a semi-arid region of India. Fert. News **41(5)**, 51 - 53.

Patra, A.K., Pahwa, M.R. and Pradeep Behari (1998): Nitrogen mineralization and leaching losses in a soil under forage crops. J. Range Management & Agroforestry **18(2)**, (in press)

Patra, A.K. and Rego, T.J. (1996): Movement of bromide as a tracer for nitrate in a semi-arid tropical Alfisol. Fert Res. **45**, 111-116

Sahrawat, K.L. (1984): Effects of temperature and moisture on urease activity in semi-arid tropical soils. Plant and Soil **78**, 401 - 408.

Address of author:
Ashok K. Patra
Indian Grassland and Fodder Research Institute
Jhansi 284 003
Uttar Pradesh, India

Changes in Plant Available Phosphorus and Its Soil Indices Following the Introduction of Moderate Fertilization

K. Voplakal

Summary

The effects of various combinations of high application rates of organic manure and reduced application rates of commercial fertilizers (in comparison with application rates used in the preceding stage) were investigated in a number of soil representatives in microparcel trials as exerted on the development of available phosphorus and indices of the phosphate regime. The results of the trials confirm that the soil phosphate regime is beneficially influenced particularly by experimental treatments with organic manuring, where fixation is restricted or the production of Fe-phosphate fraction is suppressed. Different increases in the content of available phosphorus after fertilizing depend on a number of pedological properties - in addition to the soil texture, especially on the soil pH/KCl and on the soil buffering capacity.

Keywords: Stagno-gleyic Luvisol, spodic Cambisol, solifluctional loess loam, indices of soil phosphate regime, phosphorus mobilization

1 Introduction

Prior 1990, the price of fertilizers and of other agrochemicals were subsidized in the State and Collective farms of the former Czechoslovakia. This often resulted in the application of increasing amounts to the soil without regard to the environment. However, the situation changed in the early 1990s, when subsidies were abolished and ecological reasons forced a considerable reduction in the total agrochemical inputs into the soil. We thus took the opportunity to investigate the effect of fertilization and liming on soil properties, especially on the behaviour of phosphorus. Starting in 1989 the system of intensive fertilization and liming was abandoned and, after a two year interruption when no fertilizers were applied, a new schedule of fertilizing treatments was introduced in 1991 that differed from the preceding one by a large reduction in commercial fertilizer application rates and by an emphasis on organic manure applications. This permitted us to study the development of pedological properties (the behavior of soil phosphorus) under new management conditions. In some experimental plots with organic manuring the mineral fertilizers were not applied at all (alternative management methods), in order to study the influence of these conditions on the time development of plant nutrient levels, particularly phosphorus concentration, in different soils.

2 Materials and methods

The small-scale field experiments consisting of 12 different fertilizing treatments were established on soil representative of different pedoclimatic conditions. During the period 1982 - 1988, heavy, gradated doses of industrial fertilizers and lime were applied. Then, after two years without any fertilization, a new schedule of fertilizer application, typical for an alternative or ecological method of farming, was introduced, in which mostly organic fertilizers were applied, supplemented by strongly reduced doses of industrial fertilizers. The schedule of experimental treatments and its modification are contained in Table 1. The set of soils on which microparcel trials were established can be divided into two groups: the first soil group comprises heavier texture soils (average content of clay particles is 31%) with acid reaction (average pH/KCl 5.5), little resistance to acidification and, with respect to the phosphate regime, less convenient (unfavourable fractional composition of the soil phosphate). Only a small portion of phosphorus applied in fertilizers assumes forms available to plants (Mano and Hague, 1991). These soils include: stagno-gleyic Luvisols and albic Luvisols on loess-loam and albic gleyic Luvisols on solifluctional loess loam. The soils of the second group provide better conditions for phosphorus sorption and transformations; their reaction approaches neutral (pH/KCl 6.1) and their buffering capacity to balance acidification is higher. Orthic Luvisols on loess and, with respect to the finer texture (and consequently lower sorption), Cambisols on clay slate and spodic Cambisols on orthogneiss (in which the average clay particles content amounts cca 22%) can be included in this group. Treatments $(NPK)_1$ - $(NPK)_3$ used in 1982 - 1988 consisted of gradated nutrient rates (as converted per hectare = 0.01 km^2): 300 - 600 - 900 kg NPK, 2 and 5 tons CaO (treatments Ca_1, Ca_3) and compost application rates 10 and 50 tons (treatments org_1 and org_3). An uniform application rate of fertilizer has been used since 1991: total element amount 200 kg per hectare (20 tons per km^2); 35 kg P, 55 kg K, 20 kg Mg, 90 kg N (treatment N_1) or 45 kg N in Spring and 45 kg N in Autumn (treatment N_2). The plant available phosphorus content was determined by the method of Mehlich-II (1978). The phosphate regime indices (or factors) were determined as follows: the capacity (FQ) factor according to Amer et al. (1955), the intensity (FI) factor according to Aslying (1954) and the pH/KCl according to the well-tried method of Agrochemical Soil Testing.

Variant No.	1982-1988	since 1991			
1	0	N_2	PK	Mg	org
2	Ca_1	N_1	PK	---	org
3	Ca_3		PK	---	org
4	org_1	N_1	PK	Mg	---
5	org_3		PK	Mg	---
6	$(NPK)_1$	N_1	PK	Mg	org
7	$(NPK)_2$		PK	Mg	org
8	$(NPK)_3$			Mg	org
9	$(NPK.Ca)_3$				org
10	$(NPK.org)_3$			Mg	---
11	$(NPK.Ca.org)_1$	N_2	PK	Mg	---
12	$(NPK.Ca.org)_3$	0			

Table 1: Survey of the experimental treatments of the microparcel plots, 1982-88 and since 1991.

3 Results and discussion

The change of the fertilization schedule made it possible to evaluate the residual effect of the previously applied organic matter on the increase of the efficiency of the current application of phosphorus fertilizers. In other trials, it was possible to demonstrate the influence of the current applied organic matter on mobilization of the potentially available soil phosphate reserves, accumulated in the soil as a consequence of high rates of mineral fartilizers applied in the course of 1980s. It has been possible to prove that a parallel application of both the organic matter and reduced doses of phosphate fertilizer promoted a transformation of the phosphate into plant available forms. In addition, we wished to determine, if the organic fertilization can affect the soil phosphate regime not only in the plough-layer. For the time development of the plant available phosphorus level and its soil regime indices description, the linear trend equations ($y = a + b.t$) were used ($t =$ years). Statistical analyses are contained in table 2 for pH/KCl, plant available P content and both soil phosphate regime indices (FI and FQ) values.

					1st soil group:				
treat.	pH/KCl		P		FQ		FI		
	σ_x	$\bar{x}$	σ_x	$\bar{x}$	σ_x	$\bar{x}$	σ_x	$\bar{x}$	
1	0.17	4.86	1.46	30.5	0.75	11.0	0.005	0.103	
3	0.24	7.17	3.20	52.5	2.08	44.3	0.036	0.539	
5	0.31	5.95	4.36	78.1	2.85	53.8	0.040	0.845	
8	0.34	4.81	5.78	128.4	4.55	75.9	0.066	1.029	
9	0.27	6.80	6.08	119.3	5.35	83.6	0.061	1.280	
12	0.27	6.96	5.84	143.1	4.50	92.3	0.071	1.694	
					2nd soil group:				
1	0.28	5.67	4.29	68.1	1.07	21.0	0.013	0.161	
3	0.42	7.23	5.77	82.4	1.05	26.8	0.018	0.272	
5	0.37	6.19	5.55	118.1	3.89	90.6	0.055	0.920	
8	0.24	5.43	6.98	178.9	7.64	104.7	0.053	1.078	
9	0.29	7.12	10.52	164.1	5.41	98.4	0.062	1.186	
12	0.41	7.10	10.47	190.4	6.95	115.8	0.075	1.834	

Table 2: Standard deviation σ_x and the average values $\bar{x}$ of P, FQ, FI and pH/KCl in selected experimental treatments.

	pH/KCl		P - available (mg.kg^{-1})	
treat.	1. soil group	2. soil group	1. soil group	2. soil group
1	5.08 - 0.11.t	5.59 + 0.09.t	14.7 + 7.9.t	31.9 + 18.1.t
2	6.49 - 0.17.t	6.91 - 0.23.t	22.7 + 5.5.t	37.1 + 17.3.t
3	7.61 - 0.22.t	7.73 - 0.25.t	41.9 + 5.3.t	46.0 + 18.2.t
4	5.15 - 0.02.t	5.79 + 0.08.t	30.9 + 1.9.t	56.5 + 11.1.t
5	5.99 - 0.02.t	6.23 - 0.02.t	68.5 + 4.8.t	89.5 + 14.3.t
6	4.69 + 0.07.t	5.18 + 0.17.t	47.9 + 4.3.t	81.6 + 13.6.t
7	4.51 + 0.14.t	5.21 + 0.11.t	74.3 + 8.1.t	90.0 + 22.8.t
8	4.55 + 0.13.t	5.19 + 0.03.t	123.6 + 2.4.t	147.9 + 15.5.t
9	7.21 - 0.20.t	7.54 - 0.21.t	107.1 + 6.1.t	133.1 + 15.5.t
10	5.14 + 0.24.t	5.99 + 0.03.t	161.6 + 3.0.t	174.3 + 11.5.t
11	6.48 - 0.15.t	6.65 - 0.11.t	68.2 + 3.8.t	80.3 + 15.5.t
12	7.36 - 0.20.t	7.54 - 0.22.t	136.1 + 3.5.t	182.4 + 8.4.t

Table 3: Linear trends of the average pH/KCl values and of the plant-available phosphorus level (since 1991) of both soil groups in the individual experimental treatments.

It can be seen in Table 3 that these increases are incomparably higher in the 2nd group of soils. The higher effectiveness of fertilizing is a result of the more favourable pH/KCl level and of soil resistance to acidification. In the 1st group of soils with unsaturated sorption complex, an overwhelming portion of fertilizer phosphorus is bound to less available forms, particularly to the fraction of Fe-phosphates, so the effectiveness of phosphoric fertilizing (the proportion of available phosphorus formed from the total P application rate supplied by fertilizers) is relatively low (treatment no. 1). A gradual decrease in the high values of pH/KCl is observed in treatments Nr. 2 and 3 under the new fertilizing scheme that originally received intensive liming (Table 1). Previous liming was superfluous especially in treatment no. 3 for soils of the 2nd group, and resulted in partial phosphorus retrogradation through the formation of tertiary Ca-phosphates. The effect of undesirable excessive liming is fading away in the present experimental stage, accompanied by a gradual decrease in pH/KCl to reach the neutral reaction (Fig. 1 and 2). The conditions for transformations of phosphorus fractional composition have become more favorable in this way also for the soils of 1st group, which is reflected in a greater increase in the proportion of its available forms (Rab and Guaggio, 1990) and values of phosphate regime indices (Tabs. 3 and 4, treatments 2 and 3). Naturally, an increase in the level of available phosphorus is faster in the 2nd group of soils in all experimental treatments, both after mineral fertilizing (treatments 4 and 5) and compost applications (treatments 8 and 9), as well as when combinations of both fertilizing practices were used (treatments 1 - 3 and 6 - 7) if compared with the 1st group of soils (see Table 3).

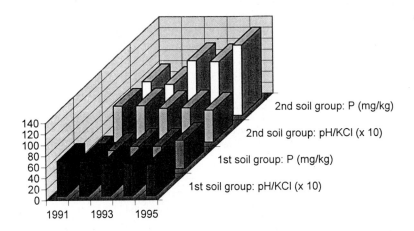

Fig. 1: Development of plant available phosphorus level (mg P/kg) and of pH/KCl values in the initially limed plots (low doses).

Differences between the time development of available phosphorus content and the values of phosphate regime indices were recorded in soils of the 2nd group in those experimental treatments where neither phosphoric fertilizers nor organic manure (treatments nos. 10 and 12) have been applied since 1991. While there has been no decrease in the content of available phosphorus until now (Tab. 3, Fig. 3) due to the ongoing mobilization of phosphate reserves that accumulated as a result of organic and mineral fertilizing in the preceding experimental stage, a deficient balance (Table 4) is instantly obvious from the phosphate regime indices (Fig. 3). They apparently react to changes in the conditions of the studied soils more readily than do extraction methods used to determine plant available phosphorus level.

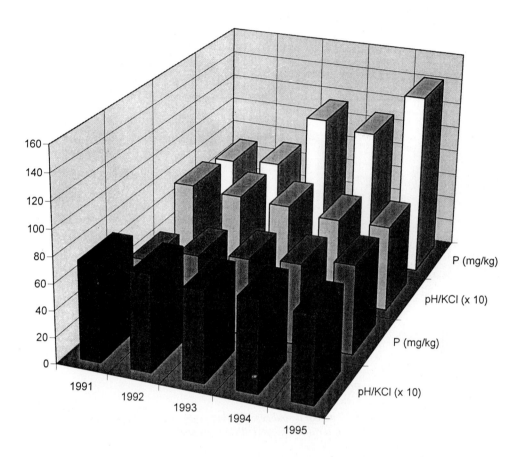

Fig. 2: Development of plant available phosphorus level (mg P/kg) and of pH/KCl values in the initially limed plots (high doses).

The data also reveal that organic manuring benefits the soil phosphate regime: trends of the time increase in available phosphorus are steeper in organically-manured treatments compared to treatments where phosphoric fertilizers have been applied. Organic manuring significantly supports the production of available phosphorus forms by decreasing the proportion of Fe-phosphate fraction (Voplakal and Damaška, 1981) in favor of forms that are more easily available to plants. Treatments 4 and 5, which originally received different doses of compost, currently receive the same doses of phosphoric fertilizer, but increases in available phosphorus are different (they are higher in treatment no. 5, in which higher doses of compost were applied previously). Hence it is likely that the size of the residual effect of organic fertilizer (applied previously) on effective transformations of phosphoric fertilizers (applied currently) is proportionate to the amount of organic manure applied (Tabs. 3 and 4).

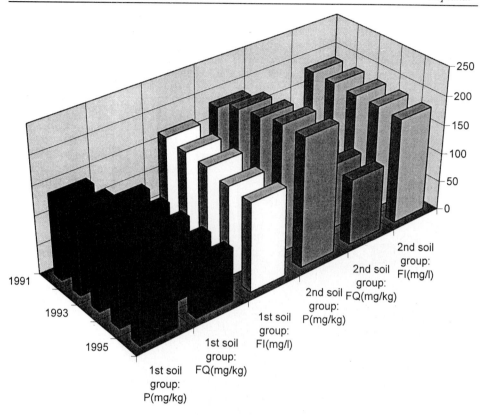

Fig. 3: Development of soil P and of its indices: FQ (mg P/kg) and of 1000 × Fl (mg P/l) in the non-fertilized plots (both soil groups).

Treat.	FQ (mg P . kg^{-1})		FI (mg P . l^{-1})	
	1st soil group	2nd soil group	1st soil group	2nd soil group
1	10.6 + 0.67.t	14.3 + 3.35.t	0.095 + 0.004.t	0.143 + 0.009.t
2	33.9 + 2.1.t	20.9 + 3.94.t	0.344 + 0.009.t	0.434 + 0.017.t
3	40.4 + 1.94.t	18.2 + 4.3.t	0.475 + 0.032.t	0.228 + 0.022.t
4	31.9 + 1.32.t	53.5 + 3.16.t	0.308 + 0.017.t	0.550 + 0.025.t
5	48.3 + 2.74.t	82.9 + 3.87.t	0.745 + 0.050.t	0.898 + 0.022.t
6	18.3 + 1.18.t	38.6 + 3.10.t	0.237 + 0.006.t	0.245 + 0.017.t
7	38.6 + 3.11.t	60.5 + 4.86.t	0.511 + 0.041.t	0.635 + 0.035.t
8	70.0 + 2.95.t	96.8 + 3.94.t	0.947 + 0.041.t	1.048 + 0.015.t
9	80.0 + 1.79.t	86.5 + 5.97.t	1.236 + 0.022.t	1.028 + 0.079.t
10	82.6 - 0.96.t	134.9 + 1.51.t	1.835 - 0.071.t	1.832 - 0.006.t
11	44.7 + 2.83.t	64.8 + 3.17.t	0.550 + 0.027.t	0.675 + 0.026.t
12	92.9 - 0.30.t	117.5 - 0.87.t	1.739 - 0.045.t	1.846 - 0.006.t

Table 4: Time development trends of phosphate capacity - FQ (mg P.kg^{-1}) and intensity - FI (mg P.l^{-1}) indices values in groups (plough layer).

The fact that organic manuring supports the mobilization (making them available) of potentially available soil phosphorus reserves was demonstrated in our incubation trials with glucose, peptone and other easily mineralizable materials that do not contain phosphorus (Voplakal and Sirový, 1986). In experimental treatments 8 and 9, which were formerly fertilized with high application rates of phosphoric fertilizer and which currently receive organic manuring, relatively large time increases in available phosphorus are recorded (2.5 - 15.9 mg P yearly). Treatment no. 7 shows even higher increases (8.1 - 22.8 mg P), not only the mobilization effect of the currently applied organic matter is obviously reflected here, but also the above-mentioned support of phosphorus bond from phosphoric fertilizer (that was applied in combination with compost).

treat.	fertilization	1st soil group	2nd soil group
1	N$_2$PKMg.org	13.1 + 1.38.t	11.5 + 1.43.t
2	N$_1$PK - org	16.9 + 2.70.t	10.2 + 1.72.t
3	PK - org	20.0 + 3.10.t	11.3 + 1.86.t
4	N$_1$PKMg -	28.1 - 1.06.t	29.4 - 2.08.t
5	PKMg -	37.3 - 1.67.t	44.3 - 3.89.t
6	N$_1$PKMg.org	14.2 + 0.79.t	11.4 + 1.67.t
7	PKMg.org	17.1 + 2.55.t	18.9 + 2.71.t
8	Mg.org	23.5 + 0.69.t	20.7 + 3.04.t
9	org	21.9 + 2.71.t	21.1 + 3.00.t
10	Mg -	49.4 - 2.81.t	41.4 - 1.89.t
11	N$_2$PKMg -	32.5 - 1.14.t	23.2 - 0.71.t
12	0	51.1 - 2.15.t	42.1 - 3.10.t

Table 5: *Time development trends of the phosphate capacity index - FQ values (mg P.kg^{-1}) in the experimental treatments in the subsoil.*

Table 5 highlights that the beneficial effects of organic fertilizing on the phosphate regime are not bound to the topsoil horizon only but a certain increase in the values of the regime capacity index can also be seen in the subsoil horizons, such as in experimental treatments with organic manuring. As a rule, it is very difficult to increase the content of available phosphorus in the subsurface soil layers by common applications of phosphoric fertilizers; an overwhelming portion of phosphorus regularly remains fixed at the place of application. Therefore, organic manuring is one way to enrich the sub-surface soil layer with available phosphorus and to decrease inbalance in the content of available phosphorus between the topsoil and the plow pan.

References

Amer, F., Bouldin, D. and Black, C.A. (1955): Characterization of soil phosphorus by anion exchange resin absorption and ^{32}P equilibration, Plant and Soil, **6**, 391-408.

Aslying, H.C. (1954): The lime and phosphate potential of soils, the solubility and availability of phosphates, Royal Veter. Agric. Yearb., 1-50.

Mano, T. and Hague, I. (1991): Forms and distribution of inorganic phosphates and their relation to available phosphorus, Tropical Agriculture, **68**, 2-8.

Mehlich, A. (1978): New extractant for soil test evaluation of phosphorus, potassium, magnesium and calcium, Comm. Soil Sci. Plant Anal., **9**, 477-485.

Rab, D. and van Guaggio, J.A. (1990): Extractable phosphorus availability indexes as affected by liming, Comm. Soil Sci. Plant Anal., **21**, 1267-1276.

Voplakal, K. and Damaška, J. (1981): Ukazatelé fosforeèného re zimu pùd v závislosti na intenzitì hnojení, Rostl. výroba, **27**, 503 - 509.

Voplakal, K. and Sirový, V. (1986): Úèinek krátkodobé inkubace pùdy s nìkterými organickými látkami na mobilitu pùdního fosforu, Vìdecké práce VÚMOP Praha, **4**, 93 - 97.

Address of author:
Karel Voplakal
Research Institute for Soil and Water Conservation
Žabovřeská 250
156 27 Prague, Czech Republic

Acidic Deposition on Forest Soils:
A Threat to the Goal of Sustainable Forestry

E. Matzner & N. Dise

Summary

The input of free- (H^+ ions) and potential- (NH_4^+ ions) acids by atmospheric deposition can cause long-term detrimental changes to soil chemical properties which are not easily reversed. Forests in industrialized regions are especially vulnerable to these changes, which include soil acidification and base cation depletion, mobilization of aluminum and other toxic metal ions from the soil solid phase into soil solution, damage to plant roots by aluminum toxicity, and damage to groundwater or runoff through acidification or enhanced nitrate levels. Chronic exposure to acid deposition can threaten the health of trees, the stability of forest ecosystems and the regional ground water quality.

Management decisions to bring about sustainable use of a forest must take into account the long-term history, current effects and future prognosis of acid deposition on the forest. Sulfate deposition across much of Europe has been greatly reduced in the last decade, whereas nitrate and ammonium deposition continue to increase. In Asia, sulfate deposition is expected to at least double within the next 15 years due to greatly increased industrialization. The resulting damage will undoubtedly augment other stresses put on forest ecosystems in Asia by the rapidly-growing population. Limiting this damage will only be possible with a thorough knowledge of the potential effects of acid deposition on different forest ecosystems.

Reversing the damage caused by acid deposition through decreasing sulfur or nitrogen deposition may not be rapid or simple. Depending on the soil type, the duration of the acid input, the stand management history and other factors, the recovery of acidified soils and waters may be delayed by years or decades by the remobilization of accumulated sulfate from the soil, and by a decreased supply of calcium and magnesium due to reduced emission of particulates. Due to differing retention mechanisms, the response of forest ecosystems to decreased sulfate deposition is in general slower than the response to decreased nitrate deposition or decreased ammonium deposition.

Over the long term, to reach the goal of sustainable forestry in industrialized countries, the emission of acidifying substances and of nitrogen compounds needs to be adjusted to the carrying capacity of different soils and ecosystems. The "critical load" concept, which estimates the highest tolerable deposition load for different soils and ecosystems, is a useful approach to quantify target emission rates. The emission of acidifying substances today far exceeds the critical load in many areas of Europe. Over the short term, the effects of acidification can be temporarily mitigated by liming, an (albeit expensive) option now undertaken on a large scale over many areas of Europe.

Keywords: Acid deposition, forest, soil acidification, soil chemistry, sustainability

1 Introduction

Due to the emission of acid forming gases (SO_2 and NO_x), large amounts of acidity are deposited by wet and dry deposition on soils and vegetation in industrialized regions. Since most of the pollutants deposited on vegetation are washed to the soil, the soil has to bear the main deposition load. The acids formed in the atmosphere are partly neutralized by dust (mainly Ca-carbonates and -silicates) and by NH_3 to form Ca^{2+} and NH_4^+ ions in precipitation. Because forest canopies have a higher filtering efficiency for air pollutants than agricultural soils, and many agriculture soils in sensitive areas are limed periodically, the problems associated with soil acidification and acidic deposition are mostly relevant to soils under non- agricultural use - especially forest soils.

The natural processes involved in the acidification of soils and their modification by acid deposition have been extensively studied and reviewed in the past (van Breemen, 1983; Reuss and Johnson, 1986; Ulrich, 1991; Ulrich and Sumner, 1991; Schwertmann et al., 1986). In general it has been shown that chronically high acidic deposition increases the rate of: (1) soil acidification (defined as loss of acid neutralizing capacity), (2) loss of base cations from soil exchange sites and (3) release of Al ions into the soil solution.

By thus changing the soil conditions in a way detrimental to plant growth, acid deposition is a threat to the goal of sustainable forestry. Here we define sustainable forestry as the sustainable use of forest ecosystem products, mainly woody biomass and drinking water. Sustainable forestry requires the sustainability of forested areas, forest growth, chosen species composition, soil nutrient status and seepage water quality.

Although important throughout the world, sustainable forestry is perhaps most critical in Asia. Currently accounting for more than 55% of the total world population, the population of Asian countries is expected to reach about 3.6 billion by the year 2010 (Bhatti et al., 1992). With an economy already comparable to that of America or Western Europe, the rate of economic growth on the Asian continent is far higher than for any other world region. Such a combination of population growth and industrialization has already put immense pressure on natural resources which will only increase in the future.

We will concentrate here on:
- development of air pollutant deposition in Central Europe and Asia,
- qualitative differences between anthropogenic and natural acidification of forest soils,
- recent findings on soil changes in acid forest soils as induced by air pollution deposition,
- the reversibility of soil acidification by reduced deposition,
- the fate of deposited N in forest soils, and
- possible countermeasures.

2 Deposition of acids and acid precursor ions in Central Europe

One of the longest records of deposition of air pollutants in forest ecosystems is the Solling data set from Germany which began in 1969 and continues to the present day. Recent evaluations by Ulrich (1994), Matzner and Meiwes (1994), Manderscheid et al. (1995) and Wesselink et al. (1995) indicate that the chemical climate of Central European forests has changed dramatically over this time:
- SO_4^{2-} and H^+ deposition and throughfall fluxes decreased by about 50% from 1980 to 1990
- NH_4^+ and NO_3^- deposition remained constant at a high level from 1970 to 1990, but throughfall fluxes increased slightly
- Ca^{2+} and Mg^{2+} deposition and throughfall fluxes decreased by about 50% from 1970 to 1990
- Both Na^+ and Cl^- deposition and throughfall fluxes showed no trend from 1969 to 1990.

These observations correspond to trends in SO_2^-, dust- and N-emissions in West Germany during the past 20 years, and the trends in emission and deposition seems to be representative for Central

Europe and Northern America (Christophersen et al., 1990; Driscoll et al., 1989; Probst et al., 1995; Hedin et al., 1994). Due to economic changes and recent air pollution legislation a major decrease in S deposition can also be expected in Eastern Europe in the near future. In contrast to this, N emission and deposition in Central Europe will presumably remain at about the present level during the next decade. This is because the major sources of NO_x emission (automobiles) and NH_3 emission (agriculture) remain relatively stable.

In most areas of Central Europe inputs of H^+, N, S, and Ca^{2+} are high in relation to the internal cycling of these elements by plant uptake and mineralization of litter. Changes in the deposition of these elements might thus not only influence soil conditions but might also strongly impact the element budget of the whole ecosystem and the physiology of the trees (Matzner and Murach, 1995).

The H^+ load of the soil buffer system resulting from deposition consists of H^+ deposition and of H^+ formed during the turnover of deposited NH_4^+ in the soil. Whereas the input of free H^+ has decreased due to decreasing SO_2, the proportion of H^+ formed in the soil during the turnover of deposited NH_4^+ has increased. The H^+ load resulting from N turnover depends mainly on the amount of NH_4^+ deposited, the degree of nitrification, and on the rate of NO_3^- loss with seepage, and varies substantially amoung different ecosystems. The uptake of NH_4^+ by plants and microorganisms is connected with an equivalent release of H^+ into the soil solution while the nitrification of NH_4^+ results in the production of 2 mol H^+ per mol NH_4^+. That is why agricultural soils which receive high fertilization of NH_4^+ have a very high soil acidification rate.

Although the H^+ load of many Central European forest soils has decreased in total, the remaining load from deposition and internal production today still exceeds by far the rate of H^+ consumption by silicate weathering in soils derived from fairly resistant bedrock. As an example, the total H^+ load of the soil from deposition in the Solling spruce site in 1980 was 4.8 kmol $ha^{-1}yr^{-1}$, with 3.8 from H^+ deposition and 1.0 from N turnover. In 1991 the total load was 2.9 kmol $ha^{-1}yr^{-1}$, with 1.3 from H^+ deposition and 1.6 from N turnover (Manderscheid et al., 1995). The rate of proton consumption in the soil by weathering of primary silicates and the associated release of base cations was estimated as only 0.4 kmol $ha^{-1}yr^{-1}$ (Matzner, 1989).

3 Deposition of acids and acid precursor ions in Asia

An evaluation of the development and extent of acid deposition in Asia is hampered by a lack of data. Long-term studies of emission and deposition are nonexistent and short-term studies have only become available in recent years. In addition, detailed studies on the effects of pollutant emission on different habitats in Asia are rare. The Asian continent contains every major biome on Earth, from alpine tundra to wet tropical forests. Some of the temperate forest biomes are very similar in soil chemistry and tree species to their counterparts in North America and Europe. Studies on the effects of acid deposition on these forest types in the West may in some cases be extrapolated to their Asian counterparts. Other ecosystems are greatly different from European or North American forests and will require major new research initiatives to understand their sensitivity.

The nature of acid deposition in Asia today is in many ways similar to that of North America or Western Europe some 40--50 years ago: dominated by sulfur and characterized by high spatial variability with very high SO_2 concentration around urban areas and much lower levels in rural areas. Most acid precursors are emitted low to the ground from households and small industries. The pollutant emission is primarily from the burning of high-sulfur coal and secondarily from fuel oil, which together account for more than 90% of the continent's energy supply. China contributes nearly 70% of Asia's total SO_2 emission (Bhatti et al., 1992).

Wet sulfate deposition estimated from a full-year study of 5 sites in China ranged from 5 to 84 kg SO_4-S $ha^{-1}yr^{-1}$, with an average of 30 kg SO_4-S $ha^{-1}yr^{-1}$ (Galloway et al., 1987). Ammonium deposition was also comparable to that of many areas of Europe (average of 6 kg NH_4-N $ha^{-1}yr^{-1}$;

range 3-10), although lower than that of highly impacted areas such as the Netherlands and northern Germany. High ammonium deposition is generally linked to areas of intensive agriculture and areas of dense human/animal population. In contrast, the wet deposition of nitrate measured in the study was fairly low: some 2--3 kg NO_3-N $ha^{-1}yr^{-1}$. The primary source of NO_x emission in the West is from transportation, and there continue to be relatively low numbers of motor vehicles in China.

Balancing the high deposition of sulfate and ammonium in China found in the study was relatively high deposition of buffering dusts such as calcium-carbonate or -oxide. The major reasons for this are (1) the essential lack of any particulate removal technology on emission sources, and (2) a large area of calcareous soils in the north. Thus the high deposition of sulfate is not associated by markedly low precipitation pH (4-5 in southern areas and about 6.5 in the north).

There is no doubt that the emission of acidifiying substances will greatly increase in Asia in the future due to rapid population growth and industrialization. Already between 1979 and 1992 emission of SO_2 in China increased by 40% (Bhatti et al., 1992). Energy plans in many Asian countries, especially China, India, Indonesia, Thailand and the Koreas, call for a very large increase in coal combustion in the future. Even the most optimistic projections, such as improved energy efficiency and installation of scrubbers on all new power plants in Asia, project a doubling in SO_2 emission over 1986 levels by 2010 (Foell and Green, 1990). A more likely estimate is closer to a tripling. China will be responsible for the majority of these emissions, and by 2010 will emit more SO_2 than all of western and eastern Europe do today (Bhatti et al., 1992). NO_x emission and nitrate deposition in Asia is also expected to double or triple from 1986 levels. Emission and deposition of ammonium is closely tied to agricultural use of land, which will become even more intensive as population increases.

As local air pollution issues increase in severity, Asian countries will no doubt go the route of their European and North American counterparts some 40 years ago and construct larger coal-fired power plants and factories with taller smokestacks. As in the West this will have two main effects: local acute pollution will decrease and regional long-distance pollution will increase. In other words, the high spatial variability in pollution concentration will greatly decline. This will mark the start of true "Western-style" long-range acid deposition in Asia and will probably occur within the next 25 years.

The effect of a dramatic increase in the emission of acidifying precursors may be exacerbated by special conditions existing in Asia. For instance, the Asian land surface receives more intense solar radiation over more of the year than most of Europe or North America, and average air temperatures are higher. This, together with large areas of stable air masses over much of the year and overall higher rainfall increases the efficiency of atmospheric chemical reactions. Therefore, a greater proportion of SO_2 and NO_x will be converted to sulfuric and nitric acid.

The major sources of this acid deposition are expected to be from northeast China, and the large cities of Southeast Asia and India. Areas receiving the heaviest load of acid deposition will vary seasonally, since both wind and deposition patterns differ dramatically between summer and winter in many areas of Asia. Wet deposition of acid sulfates and nitrates will dominate in the summer monsoon season and substances will be transported over longer distances. Dry deposition with shorter transport distances will characterize the winter pattern (Bhatti et al., 1992).

4 Major effects of acid deposition on forest ecosystems

Direct foliar damage due to acid deposition is only rarely documented unambiguously. In contrast, detrimental changes by acid deposition to soils, soil biota, groundwater and surface water are well known. Here we summarize the major recorded effects.

4.1 Anthropogenic vs. natural sources of soil acidity

Evaluation of the effects of acidic deposition on forest soils requires an understanding of natural acidification processes (van Breemen et al., 1983; Bredemeier et al., 1990; Matzner, 1989) and a criterion for distinguishing between natural and human-induced soil acidification. Natural soil acidification is driven by the production of weak acids such as carbonic and organic acids and by the accumulation of base cations in biomass, which leaves an equivalent amount of H^+ in the soil. In addition, nitric acid formed during nitrification (bacterial conversion of NH_4^+ to NO_3^-) may result in soil acidification if the nitrate is leached from the ecosystem (accompanied by base cations).

The degree of soil acidification can be measured by various soil parameters, commonly soil pH, acid neutralization capacity, base saturation of cation exchange capacity (CEC), exchangeable acidity and soil solution chemistry. Soil pH and base saturation are easily measurable and thus widely used to describe the intensity of soil acidification. Whereas base saturation and pH of A and upper B horizons might be low even in soils not affected by acidic deposition (e.g. Ugolini et al., 1987), low base saturation of deeper B and C horizons seems to be typical for systems impacted by acidic deposition (Ulrich, 1995, Johnson et al., 1991). Karltun (1994) reported increasing acidification (deceased pH and decreased base saturation) of deeper soil layers from north to south in Sweden corresponding to a gradient of increasing acid deposition. The degree of acidification (pH, base saturation) of deeper soil layers in forest soils has been the subject of several studies in different forest soils of Germany (overview in Matzner and Davis, 1997). Deeper soil layers were often found to be highly acidic with low base saturation again attributed to the effect of acidic deposition.

Many field and laboratory studies have demonstrated that acidic deposition increases the rate of base cation leaching. However, in a study of depth gradients of pH and base saturation in pristine temperate forests of New Zealand (not subjected to acid deposition), Matzner and Davis (1997) found that under conditions of high (> 1000 mm) rainfall, even deeper soil layers can show low base saturation of CEC. Thus reduced base saturation cannot exclusively be attributed to the impact of acidic deposition and land use but may also be caused by natural leaching of carbonic acid.

The soil solution chemistry is vitally important to soil organisms and roots. Acidity and nutrients are also transferred to groundwater via soil solution, and characteristic soil solution changes occur under chronic deposition of acidity. Soil solution chemistry may be a more specific indicator of the influence of acid deposition on forest soils than base saturation of CEC or soil pH (Schecher and Driscoll, 1988; Reuss and Johnson, 1986). The input of H^+ together with "mobile anions" (Seip, 1980) like SO_4^{2-} and NO_3^- into the soil solution leaches basic cations (Ca, Mg, Na, K) from the soil until their available pools are depleted, where upon the soil enters the aluminum buffer range (Ulrich, 1991). This is marked by a drop in soil pH to about 4.2 and an increase in the soil solution concentration of Al^{n+}. Because of this, in European forest ecosystems the concentrations of SO_4^{2-}, NO_3^- and Al^{n+} in soil solution are highly correlated (Dise and Wright, 1995). We consider high Al^{n+} in the soil solution the major differentiating effect of acidic deposition vs. natural processes of soil acidification in temperate forests. Soil solutions under natural conditions have very low Al^{n+} concentrations (e.g. Davis, 1990).

4.2 Al-release

As stated above, the release of Al^{n+} ions into the soil solution during H^+ buffering is a major effect of acid deposition in acid forest soils. Because of the potential toxicity of Al^{n+} ions to plant roots, soil organisms, and fish, and because of the acidifying potential of Al^{n+} ions in deeper, less acidified soil layers and groundwater, the chemistry of Al in soils is a key to understanding the detrimental effects of acidic deposition in terrestrial and aquatic ecosystems. The Al chemistry of soils and soil solutions and its regulation by solid phase properties has thus been addressed in several recent papers (Berggren and Mulder, 1995; Mulder and Stein, 1994; Franken et al., 1995; Dahlgren and Walker, 1993; Matzner and Prenzel, 1992).

The aqueous geochemistry of aluminum is complex but important to understand because some species of aluminum in soil solution are far more toxic to biota and tree roots than others. Aluminum is most toxic in its reactive ionic forms: $Al(OH)_2^+$, $Al(OH)^{2+}$ and Al^{3+}. In general, the higher the charge, the more toxic the species. Monomeric aluminum, Al^{3+}, dominates in low pH soils containing little organic matter. Although upper soil horizons may have high concentrations of total aluminum, this is often complexed with organic matter so the (toxic) reactive aluminum concentration is low.

Ulrich (1995) and Matzner and Murach (1995) concluded that high levels of aluminum ions (as well as increased N availability) reduce the fine root biomass of trees in relation to the above-ground biomass. In addition, the roots are often concentrated in the upper organic layer (where aluminum is complexed), which may be only a few centimeters thick. Such a shallow root distribution can lead to drought stress and greatly increased susceptibility to wind throw. The concentration of Al^{3+} at which damage occurs depends on the tree species and the concentration of counterions like Ca^{2+}. Rost-Siebert (1985) first suggested that the Ca/Al molar ratio in soil solution could be a useful ecological indicator of the risk of Al-toxicity. This was recently corroborated by Cronan and Grigal (1995) in a review of approximately 300 studies.

4.3 Sulfate desorption and reversibility of soil acidification

Adsorption of sulfate on soil solids is accompanied by the release of an equivalent amount of OH-ions and desorption with an equivalent OH-consumption. The latter causes acidification of the soil solution. The behaviour of soil sulfate pools is thus of critical interest for predicting the effects of decreasing sulfate deposition on soil chemistry and the reversibility of soil acidification.

The concentration of SO_4^{2-} in soil solution and runoff is decreasing in many areas of the industrialized world as a result of decreasing SO_4^{2-} deposition (Newell, 1993; Christophersen et al., 1990; Wright et al., 1993). However, the proportional decrease in soil solution SO_4^{2-} is often less than in deposition (Wesselink et al., 1995; Alewell et al., 1997). Forest soils have accumulated SO_4^{2-} in the past, with the amount depending on soil conditions and SO_4^{2-} deposition. The amount of SO_4^{2-} in forest soils is highly variable and may be in the range of 30 to 1000 kg. ha^{-1} (Erkenberg et al., 1996). Predicting the effects of decreasing SO_4^{2-} inputs on soil solution chemistry requires knowledge of the behaviour of the previously stored soil SO_4^{2-}, namely of the proportion of the reversibly bound (water-soluble) fraction and the shape of the desorption isotherm.

Charlet et al. (1993) showed in a laboratory study using both soils of the German Harz mountains and oxide-coated reference minerals that the largest part of total soil SO_4^{2-} is reversibly bound. Alewell (1995) supported this conclusion in a laboratory study using different soils from Germany and concluded that most previously-adsorbed soil sulfate will eventually be mobilized if SO_4^{2-} deposition decreases. This suggests that the acidification of soil solutions is to a large extent reversible, but that the recovery of acidified soil solution will be delayed by SO_4^{2-} desorption. The length of the time lag depends on the amount of SO_4^{2-} stored and the shape of the desorption isotherm. Given current SO_4^{2-} concentrations in most forest soils observed so far, the desorption isotherms are relatively flat, only small amounts being released during the first stages of decreasing SO_4^{2-} inputs. Only if the SO_4^{2-} inputs (and soil solution concentrations) drop to far lower values than present will desorption increase. In soils with relatively high SO_4^{2-} contents, it will take several decades until the SO_4^{2-} output with seepage will equal a greatly reduced SO_4^{2-} input (Alewell et al., 1995).

In contrast to this, decreasing N deposition will immediately lead to decreasing NO_3^- concentrations in the soil solution (Alewell et al., 1997, Tietema et al., 1995) because of the high biological turnover of N and low inorganic soil pools.

4.4 Base cations in soil solutions

In acidified forest soils with low base saturation, the deposition of Ca^{2+} and Mg^{2+} adds significantly to the turnover or these elements within the ecosystem. For example, the average (1973--1985) annual Ca^{2+} deposition in the Solling spruce stand, at about 18 kg ha^{-1}yr^{-1} (Matzner, 1989) amounts to about 20% of the total exchangeable Ca^{2+} pool of the mineral soil (about 100 kg ha^{-1}). Thus a decrease of the Ca^{2+} input from deposition rapidly affects internal Ca^{2+} cycling (Manderscheid et al., 1995; Wesselink et al., 1995; Driscoll et al., 1989; Christophersen et al., 1990) with the consequence of a decreased Ca^{2+} in soil solution, and increased risk of Al-toxicity to tree roots, if other factors remain constant. The same relationship between deposition and soil solution concentration was found for Mg^{2+} at the Solling sites. Thus, it can be expected that with the reduction of dust emissions -- as the major source of deposited Ca^{2+} and Mg^{2+} -- the detrimental effect of acid depositions on soil solution chemistry will increase.

4.5 Fate of deposited nitrogen in soils

The true N input to many coniferous forests of Central Europe exceeds 30 kg ha^{-1}yr^{-1} if canopy uptake of N deposition is included. The input is much higher than the annual N demand of trees for increment and reaches about 50% of the annual internal N flux in coniferous forests. Most forest ecosystems still accumulate N from deposition to a substantial degree, since leaching losses are much smaller than deposition rates. Denitrification seems to be of less importance to the N budget (Brumme and Beese, 1992).

Increasing N content of leaves, needles and litter (Sauter, 1991; Matzner, 1988) and increased growth (Kauppi et al., 1992; Klädtke, 1995) is one sink of nitrogen. The other is N accumulation in the soil. Increased N deposition appears to increase the rate of net N-mineralization (Mc-Nulty et al., 1991; Gundersen et al., in press), but in the long run enhanced N availability may reduce the rate of decomposition because of the inhibition of lignin-decomposing fungi by N (Fog, 1988; Berg et al., 1993; Berg and Matzner, 1996). This is confirmed by long-term studies of humus accumulation in the German Solling forest and in Scotland (Billet et al., 1990; Matzner, 1989). Additionally, decreasing C/N ratios of the forest floor in Germany and elsewhere with increasing deposition (von Zezschwitz, 1985; Hildebrand, 1994; Mc Nulty et al., 1991) indicate that the soil organic N-pool is a major sink of deposited N. The accumulation of NH_4^+ in exchangeable form or fixed in interlayers of clay minerals is of minor importance as a long-term sink of deposited N in forest soils (Matschonat and Matzner, 1995).

4.6 Soil biota, ground- and surface water quality

It is beyond this paper to fully review the effects of acid deposition on water quality and soil biota, tree roots and tree nutrition. Here we comment only on some major effects.

Acidification changes the species composition of the forest floor and soil (Wittig et al., 1985). Earthworms are restricted to higher pH and certain mycorrhiza fungi are also highly sensitive to acidification and N excess (Kottke and Oberwinkler, 1992). Their disappearence affects the ability of trees to absorb water and nutrients.

The acidification of surface waters and damage to aquatic biota resulting from the transfer of acidity with seepage water to groundwater and springwater was one of the first detrimental effects of acidic deposition observed (e.g. Seip, 1980). The transfer of acidity is caused by the deposition of the "mobile" anions NO_3^- and SO_4^{2-} from the atmosphere (Reuss, 1991) and was first observed in the sensitive areas of Scandinavia and North America with shallow soils of low buffering capacity.

Groundwater acidification is today also occurring in parts of Central Europe with deeper soils receiving acidic deposition (Bayerisches Landesamt für Wasserwirtschaft, 1995). This will raise problems for drinking water supply because of the acidity and metal content of these waters.

5 Sensitive forest ecosystems and their distribution

Around the world, forest ecosystems most at risk from high sulfate deposition are those with acid-sensitive soils derived from base-poor bedrock with a low buffering capacity. Forests underlain by soils with low sulfate adsorption capacity are also susceptible, as are forests on soils with a high rate of removal of soil base cations, for example, areas of intensive forestry or agricultural use (without a return of cations via liming and fertilization).

Among these soil types, the most susceptible are forests located in areas of moderately high rainfall (ca 500–1000 mm/yr). Mountainous regions are especially sensitive because they are often the first drop-off point for acid- and moisture-loaded air masses, and because of the high concentration of pollutants in fog. Coniferous forests are more susceptable than broadleaf trees because of their higher filtering efficiency for atmospheric pollutants and resulting higher deposition load. The sensitive areas of Europe have been mapped recently (Posch et al., 1995) and their distribution is according to the above mentioned criteria. Those areas at highest risk from increasing acid deposition in Asia are shown in Table 1.

Area	Vegetation type
Siberia, northern Mongolia, extreme northeast China	Taiga - coniferous forests (pine, spruce) with some deciduous species (birch, aspen, larch)
Northeast China, mountain regions of southern China, Korea, and Japan, lower Himalayan slopes in India and Nepal, upper mountain region of southern India	Temperate - broad leaf and mixed forests of spruce, pine, fir, birch, aspen, oak, and cedar
Himalayan foothills in India, montane regions of southeastern Asia, southern China, and southern India	Subtropical montane - evergreen forests of conifers (fir, pine, cedar, spruce) and deciduous species
Upper mountain regions of India, China and Nepal	Alpine - coniferous forest (spruce and fir) with birch

Table 1: Areas in Asia with forest vegetation at highest risk from acid deposition (modified from Bhatti et al., 1992)

6 Countermeasures

The observed changes in soils, vegetation and groundwater quality caused by acidic deposition indicate that the present load of air pollutant deposition on forests is too high in many areas in Europe and Asia. In Europe, sulfate deposition will continue to decrease and nitrogen deposition will increase or remain constant in the foreseeable future. In Asia, both sulfate and nitrate deposition will increase, sulfate dramatically so, within the next 1-2 decades.

To identify a threshold value for the pollution of soils and ecosystems by acidic deposition, the concept of a "critical load" for the deposition of acidity, sulfur and nitrogen has been developed. The "critical load" is defined as the "highest tolerable deposition load that would not lead to detrimental changes in the functioning of soils and ecosystems" (Posch et al., 1995). The background for such calculations are assumptions about buffer processes and capacities of soils by silicate weathering and the sensitivity of different ecosystems to N deposition. Although the estimates for critical loads are still often empirical and highly approximate, they indicate that today large areas of the industrialized world receive a total load of acidifying substances exceeding the capacity of ecosystems to buffer it. Matching emission and deposition rates to calculated critical loads is a necessary step to a sustainable forest ecosystem use.

The liming of forest soils is restricted to the above-ground application of lime, since plowing soils is effectively impossible in forests. The positive effects of liming are thus mostly restricted to the upper soil horizons, where increase of pH, higher nutrient cation status and improvement of soil solution chemistry (increased Ca^{2+} and Mg^{2+} and decreased Al^{3+}) have been reported. Liming is today a common practice in many forested areas of Europe. However, detrimental effects are also reported, including increased rates of nitrification (which can trigger the leaching of NO_3^-) and the restriction of nutrient availability to upper soil profiles, which can concentrate root biomass to these layers (Kreutzer, 1995). Thus liming is a useful short-term tool to mitigate the effects of soil acidification, but must be accompanied with the reduction of the emissions.

7 Summary and conclusions

As a consequence of acidic precipitation the chemical status and nutrient availability of many forest soils are changing rapidly. For Central Europe, the recent changes in acid forest soils chemistry as a result of atmospheric deposition can be summarized as follows:

- The H^+ load of the soil buffer systems due to atmospheric deposition is still high and the proportions have shifted from deposition of H^+ to H^+ generated during the turnover of deposited NH_4^+.
- The mitigating effects of decreasing SO_4^{2-} and H^+ depositions on the chemical status of soil solutions are delayed by desorption of soil SO_4^{2-} and decreasing deposition of Ca^{2+}.
- Due to Al complexing with organic matter in the surface soil horizons, Al stress to tree roots may be highest in B horizons.
- N deposition causes increased plant N availability and increased NO_3^- output with seepage. The latter also impedes the recovery of soil solution chemistry brought about by decreased SO_4^{2-} deposition.
- High N deposition changes the internal ecosystem cycling of nitrogen. In particular, needle and soil N pools increase and organic layer C/N ratios decrease. The long-term implications of this are not known, but could result in decreasing degradability of humus.

The major environmental risks of acidification, namely (1) depletion of base cations from the soil and increasing soil solution aluminum, (2) acid stress to plant roots and soil organisms, and (3) ground- and springwater acidification, will continue to be high in sensitive areas of Central Europe. In addition, ecosystem storage of nitrogen will increase. In general, regional-scale acid deposition does today not exist in Asia as it does in the West. Pollution is still characterized by highly localized patches of intense sulfur emission and deposition. However, this pattern will change in the near future: greatly expanded industrialization will increase regional pollution to levels similar to Central Europe, and even remote forest ecosystems may receive greatly increased deposition of sulfate and nitrate.

These developments are not compatible with the concept of sustainable land use. Both short- and long-term countermeasures must be taken. In impacted areas of central Europe, liming is a short-term, expensive option to mitigate past damage. The only effective long-term solution to avoid further damage to forests, soils, groundwater and surface water is to reduce the emission of SO_2, NH_3 and NO_x. The "critical load" concept is an important step in this direction. In many cases recovery will not be immediate -- this is especially true for reductions in sulfur deposition, where soils require time to desorb sulfate. The knowledge and experience gained through several decades of acid deposition in Europe should be used to set mitigation policies for Asia, where increased acid rain is inevitable in the next two decades.

Acknowledgements

This review was supported by the German Ministry of Education, Science, Research and Technology (BMBF) Grant No. BEO-51-0339476A.

References

Alewell, C. (1995): Sulfat-Dynamik in sauren Waldböden: Sorptionsverhalten und Prognose bei nachlassenden Depositionen, Bayreuther Forum Ökologie, University of Bayreuth,. Band **19**, 1-185.

Alewell, C., Matzner, E. and Bredemeier, M. (1995): Reversibility of sulfate sorption in acid forest soils: A transfer of laboratory studies to the field with MAGIC, Informationsberichte des Bayerischen Landesamtes für Wasserwirtschaft **3**, 171-174.

Alewell, C., Bredemeier, M., Matzner, E. and Blanck, K. (1997): Effects of experimentally reduced acid deposition on soil solution chemistry in a forest ecosystem, J. Environ. Qual. 26/3, in press

Bayerisches Landesamt für Wasserwirtschaft (1995): Informationsberichte des Bayerischen Landesamtes für Wasserwirtschaft, Heft 3/95.

Berg, B., McClaugherty, C., De Santo, A.F., Johansson, M.B. and Ekbohm, G. (1993): Decomposition of litter and soil organic matter - can we distinguish a mechanism for soil organic matter buildup?, Mitt. d. Österr. Bodenkundlichen Gesellschaft **47**, 11-41.

Berg, B. and Matzner, E. (1996): Effect of N deposition on decomposition of plant litter and soil organic matter in forest systems, Environmental Reviews, in press.

Berggren, D. and Mulder, J. (1995): The role of organic matter in controlling aluminium solubility in acidic mineral soil horizons, Geoch. Cosmoch. Acta **59**, 4167-4180.

Bhatti, N., Streets, D.G. and Foell, W.K. (1992): Acid rain in Asia, Environmental Management **16**, 541-562.

Billett, M.F., Fitzpatrick, E.A. and Cresser, M.S. (1990): Changes in the carbon and nitrogen status of forest soil organic horizons between 1949/50 and 1987, Environmental Pollution **66**, 67-79.

Bredemeier, M., Matzner, E. and Ulrich, B. (1990): Internal and external proton load to forest soils in Northern Germany, J. Environ. Qual. **19**, 469-477.

Brumme, R. and Beese, F. (1992): Effects of liming and nitrogen fertilization on emissions of CO_2 and N_2O from a temporate forest, Journal of Geophysical Research **97**, 851-858.

Charlet, L., Dise, N. and Stumm, W. (1993): Sulfate adsorption on a variable charge soil and on reference minerals, Agriculture, Ecosystems and Environment **47**, 87-102.

Christophersen, N., Robson, A., Neal, C., Whitehead, P.G., Vigerust, B. and Henriksen, A. (1990): Evidence for long-term deterioration of streamwater chemistry and soil acidification at the birkenes catchment, southern Norway, J. Hydrol. **116**, 63-76.

Cronan, C.S. and Grigal, D.F. (1995): Use of Calcium/Aluminium ratios as indicators of stress in forest ecosystems, J. Environ. Qual. **24**, 209-226.

Dahlgren, R.A. and Walker, W.J. (1993): Aluminium dissolution rates from selected spodosol B Horizons: II. Effect of pH and solid phase aluminium pools, Geoch. Cosmoch. Acta **57**, 57-66.

Davis, M.R. (1990): Chemical composition of soil solutions extracted from New Zealand beech forests and West German beech and spruce forests, Plant and Soil, 237-246.

Dise, N.B. and Wright, R.F. (1995): Nitrogen leaching from European forests in relation to nitrogen deposition, Forest Ecology and Management **71**, 153-161.

Driscoll, C.T., Likens, G.E., Hedin, L.O., Eaton, J.S. and Borman, F.H. (1989): Changes in the chemistry of surface waters, Environ. Sci. Technol. **23**, 137-143.

Erkenberg, A., Prietzel, J. and Rehfuess, K.E. (1996): Schwefelausstattung ausgewählter europäischer Waldböden in Abhängigkeit vom atmogenen S-Eintrag, Z. Pflanzenernähr. Bodenk. **159**, 101-109.

Foell, W.K. and C.W. Green (1990): Proceedings of the second annual workshop on acid rain in Asia, 19-22 November 1990, Asian Institute of Technology, Bankok, Thailand.

Fog, K. (1988): The effect of added nitrogen on the rate of decomposition of organic matter, Biological Reviews **63**, 433-462.

Franken, G., Pijpers, M. and Matzner, E. (1995): Al chemistry of the soil solution in an acid forest soil as influenced by percolation rate and soil structure, Europ. J. of Soil Science **46**, 613-619.

Galloway, J.N., Zhao, D., Xiong, J. and Likens, G.E. (1987): Acid rain: China, United States, and a remote area, Science **236**, 1559-1562.

Gundersen, P., Emmett, B.A., Kjonaas, O.J., Koopmans, C.J. and Tietema, A. (1997): Impact of nitrogen deposition on nitrogen cycling in forests: a synthesis of NITREX data, Forest Ecology and Management, in press.

Hedin, L.O., Granat, L., Likens, G.E., Buishand, T.A., Galloway, J.N., Butler, T.J. and Rohde, H. (1994): Steep declines in atmospheric base cations in regions of Europe and North America, Nature **367**, 351-354.

Hildebrand, E.E. (1994): Der Waldboden- ein konstanter Produktionsfaktor?, Allg. Forst. Zeitschrift **49/2**, 99-104.

Johnson, D.W., Cresser, M.S., Nilsson, I.S., Turner, J., Ulrich, B., Binkley, D. and Cole, D.W. (1991): Soil changes in forest ecosystems: Evidence for and possible causes, Proc. Royal Soc. of Edinburgh 97B, 81-1161.

Karltun, E. (1994): Principal geographic variation in the acidification of Swedish forest soil, Water, Air and Soil Pollution **76**, 353-362.

Kauppi, P.E., Mielikäinen, K. and Kuusela, K. (1992): Biomass and carbon budget of european forests, 1971-1990, Science **256**, 70-74.

Klädtke, J. (1995): Untersuchungen zum Wachstum der Wälder in Europa - Ein Überblick, UBA-Texte **28**, 120-130.

Kottke, I. and Oberwinkler, F. (1992): Air Pollution and Interactions between Organisms in Forest Ecosystems. Tesche, M. and Feiler, S. (eds.) Tharandt/Dresden. 242-244

Kreutzer, K. (1995): Effects of forest liming on soil processes, Plant and Soil **168-169**, 447-470.

Manderscheid, B., Matzner, E., Meiwes, K.-J. and Xu, Y. (1995): Long-term development of element budgets in a Norway Spruce (*Picea abies* (L.) Karst.) forest of German Solling area, Water, Air and Soil Pollution **79**, 3-18.

Matschonat, G. and Matzner, E. (1995): Quantification of ammonium sorption in acid forest soils by sorption isotherms, Plant and Soil **168-169**, 95-101.

Matzner, E. (1988): Der Stoffumsatz zweier Waldökosysteme im Solling, Berichte des Forschungszentrums Waldökosysteme d. Univ. Göttingen, Reihe A, Band 40, 1-217.

Matzner, E. (1989): Acidic Precipitation: Case Study: Solling, West Germany. In: D.C. Adriano and M. Havas (eds.), Advances in Environmental Science: Acid Precipitation, New York: Springer Verlag, 39-83.

Matzner, E. and Prenzel, J. (1992): Acid deposition in the German Solling area: Effects on soil solution chemistry and Al-mobilization, Water, Air and Soil Pollution **61**, 221-234.

Matzner, E. and Meiwes, K.-J. (1994): Long-term development of element fluxes with bulk precipitation and throughfall in two forested ecosystems of the German Solling area, J. Environ. Qual. **23**, 162-166.

Matzner, E. and Murach, D. (1995): Soil changes induced by air pollutant deposition and their implication for forests in Central Europe, Water, Air and Soil Pollution **85**, 63-73.

Matzner, E. and Davis, M. (1997): Chemical soil conditions in pristine Nothofagus forests of New Zealand as compared to German forest, Plant and Soil **186**, 285-291.

Mc-Nulty, S.G., Aber, J.D. and Boone, R.D. (1991): Spatial changes in forest floor and foliar chemistry of spruce-fir forests across New England, Biogeochemistry **14**, 13-29.

Mulder J. and Stein, A. (1994): The solubility of aluminium in acidic forest soils: Long-term changes due to acid deposition, Geochimica et Cosmochimica Acta **58**, 85-94.

Newell, A.D. (1993): Inter-regional comparison of patterns and trends in surface water acidification across the United States, Water, Air and Soil Pollution **67**, 257-280.

Posch, M., de Smet, P.A.M., Hettelingh, J.-P- and Downing, R.J. (1995): Calculation and mapping of critical thresholds in Europe. Coordination Center for Effects, RIVM, Bilthoven, RIVM Report No. 259101004.

Probst, A., Fritz, B. and Viville, D. (1995): Mid-term trends in acid precipitation, streamwater chemistry and element budgets in the Strengbach catchments (Vosges mountains, France), Water, Air and Soil Pollution **79**, 39-59.

Reuss, J.O. and Johnson, D.W. (1986): Acid Deposition and the acidification of soils and waters, Springer Verlag.

Reuss, J.O. (1991): The transfer of acidiy from soils to surface waters. In: B. Ulrich and M.E. Sumner (eds.), Soil Acidity, Springer Verlag, 203-218.

Rost-Siebert, K. (1985): Untersuchungen zur H^+- und Al-Ionen-Toxität an Keimpflanzen von Fichte (*Picea abies* Karst.) und Buche (*Fagus sylvatica* L.) in Lösungskultur, Berichte des Forschungszentrum Waldökosysteme d. Univ. Göttingen, Band 12.

Sauter, U. (1991): Zeitliche Variationen des Ernährungszustand nordbayerischer Kiefernbestände, Forstwissenschaftliches Centralblatt **100**, 13-33.

Schecher, W.D. and Driscoll, T.C. (1988): An evaluation of the equilibrium calculations within acidification models: The effect of uncertainty in measured chemical components, Water Rec. Res. **24**, 533-450

Schwertmann, U., Süsser, P. and Nätscher, L. (1986): Protonenpuffersubstanzen in Böden, Z. Pflanzenernähr. Bodenk. **150**, 174-178.

Seip, H.M. (1980): Acidification of fresh waters - sources and mechanisms. In: D. Drablos and A. Tollan (eds.), Ecological Impacts of Acid Precipitation. SNSF-Project, Oslo, 358-366.

Tietema, A., Wright, R.F., Blanck, K., Boxman, A.W., Bredemeier, M., Emmett, B.A., Gundersen, P., Hultberg, P., Kjonaas, O.J., Moldan, F., Roelofs, J.G.M., Schleppi, P., Stuanes, A.O. and van Breemen, N. (1995): NITREX: The timing of response of coniferous forest ecosystems to experimentally-changed nitrogen deposition, Water, Air and Soil Pollution **85**, 1623-1628.

Ugolini, F.O., Stoner, M.G. and Marrett, D.J. (1987): Arctic pedogenesis: 1. Evidence for contemporary podzolization, Soil Science **144**, 90-100.

Ulrich, B. (1991): An ecosystem approach to soil acidification. In: B. Ulrich and M.E. Sumner (eds.), Soil Acidity, Springer Verlag, Berlin, 28-79.

Ulrich, B. and Sumner, M.E. (1991): Soil Acidity, Springer Verlag, Berlin.

Ulrich, B. (1994): Nutrient and acid-base budget of central european forest ecosystem. In: D.L. Godbold and A. Hüttermann (eds.), Effects of Acid rain on Forest Processes, Wiley-Liss, Inc., New York, 1-50.

Ulrich, B. (1995): Der ökologische Bodenzustand - seine Veränderung in der Nacheiszeit, Ansprüche der Baumarten, Forstarchiv **66**, 117-127.

van Breemen, N., Mulder, J. and Driscoll, C.T. (1983): Acidification and alkalinization of soils, Plant and Soil **75**, 283-308.

von Zezschwitz, E. (1985): Qualitätsänderungen des Waldhumus, Forstwissenschaftliches Centralblatt **104**, 205-220.

Wesselink, L.G., Meiwes, K.J., Matzner, E. and Stein, A. (1995): Long-term changes in water and soil chemistry in a Spruce and Beech forest, Solling, Germany, Env. Sci. & Techn. **29**, 51-58.

Wittig, R., Werner, W. and Neite, H. (1985): Der Vergleich alter und neuer pflanzensoziologischer Aufnahmen: Eine geeignete Methode zum Erkennen von Bodenversauerung?, VDI-Berichte 560, Kolloquium Goslar, 18.-20. Juni 1985

Wright, R.F., Lotse, E. and Semb, A. (1993): RAIN project: results after 8 years of experimentally reduced acid deposition to a whole catchment, Can. J. Fish Aquat. Sci. **50**, 258-268.

Addresses of authors:
E.Matzner
Department of Soil Ecology
University of Bayreuth
D-95440 Bayreuth, Germany
N. Dise
Department of Earth Sciences
The Open University
Milton Keynes, MK7 6AA, U.K.

Release of Cd, Cr, Ni, Pb and Zn to the Soil Solution as a Consequence of Soil Contamination and Acidification

J. Csillag, A. Lukács, K. Bujtás & T. Németh

Summary

Batch experiments were carried out on samples of the "A" horizon of a light textured, slightly acidic brown forest soil in order to study the impact of metal loading rate, soil water content, adsorption time and soil acidification on the release of Cd, Cr, Ni, Pb and Zn into the soil solution. The metals were applied either as multicomponent nitrate salt solution or as metal-enriched sewage sludge. The soil loading rates were in the following order: Cd<Ni<Cr=Pb<Zn. The liquid phases of unloaded, loaded and loaded+dried+acidified soil samples were obtained by centrifugation using a rotor speed corresponding to -1500 kPa (conventional wilting point of plants).

Cadmium, Ni and Zn entered the liquid phase in relatively higher amounts than the less mobile Cr and Pb as related to their previous loading rates. Much higher adsorption was found in the sludge than in the soil with the exception of Ni. Chromium and Pb were adsorbed by the sludge to such a high extent that their adsorption hardly increased when the metal-enriched sludge was added to the soil. While element concentrations in the soil solution after the acidic treatment were generally proportional to the previous loading rates, with the exception of the strongly adsorbed Cr (Cd<Cr<Ni<Pb<Zn), the recoveries expressed as percentage of the applied amounts reflected the order of the element mobilities (Cr<Pb<Ni≅Cd≅Zn). The results indicate that considerable release of metals into the soil liquid phase occurs only under extreme conditions (excessive metal contamination and strong acid pollution).

Keywords: Soil contamination, heavy metals, sewage sludge, soil acidification, soil solution.

1 Introduction

One form of chemical degradation of soils (Logan, 1990) is their contamination with potentially toxic heavy metals present as typical pollutants in industrial and communal sewage sludges. Application of such waste materials on agricultural fields necessitates comprehensive risk assessment, taking into account various scenarios and exposure pathways (Chang et al., 1992). Heavy metals represent especially serious environmental risks when they are released into the liquid phase of the soil. The concentration of heavy metals in the soil solution is regarded as an indicator of the mobile pool of metals in soils (Kabata-Pendias and Adriano, 1995). Data concerning heavy metal concentrations in the soil solution are available in the literature (Behel et al., 1983; Bergkvist, 1987; Campbell and Beckett, 1988; cit. in Kabata-Pendias and Pendias, 1992), but to make a comparison is very difficult because of the wide range of different techniques applied for soil solution sampling (Sposito, 1989; Lorenz et al., 1994; Wolt, 1994; Keller, 1995; Castilho et al., 1996).

The storage capacity of the soil for the heavy metals and the concentrations of the elements in the soil solution are influenced by environmental changes such as soil acidification. A serious ecological consequence of acid pollution may be an increase in the solubility, mobility and bioavailability of the stored chemicals, e.g. toxic heavy metals (Herms and Brümmer, 1984; Løbersli et al., 1991; Evans et al., 1995; Reddy et al., 1995).

The aim of our research was to follow the changes in soil solution concentrations of Cd, Cr, Ni, Pb and Zn, resulting from excessive contamination levels and from changes of soil water content and duration of adsorption time, or caused by the soil acidification process.

2 Materials and methods

The experiments were carried out on samples of the "A" horizon of a light textured, slightly acidic brown forest soil. Its main characteristics were: $pH(H_2O)$ 6.3; $pH(KCl)$ 5.3; organic matter content 1.2%; cation exchange capacity 8.5 me/100g soil; clay+silt fractions (<0.02mm) 18.9%; clay fraction (<0.002mm) 12.8%.

Cadmium, Cr, Ni, Pb and Zn nitrates as multicomponent solutions were added to air-dried soil samples. The order of the loading rates of the elements corresponded to their maximum permitted amounts in sewage sludges for use on agricultural lands as specified by the Hungarian Technical Directive (1990), ($1L$ in Table 1). For modelling provocative overloadings, metals were added to the soil in amounts equivalent to 10- and 100-times the permitted levels ($10L$ and $100L$, resp.).

The $100L$ metal loading rate was applied not only in soil, but also in sludge and in soil+sludge samples. The main characteristics of the compressed communal sewage sludge used in the experiments are listed in Table 2. Original metal contents in this sludge were comparable to or less than the limit values shown in Table 1.

elements	in sludge [mg/kg dry matter]	resulting soil loadings ($1L$) [mg/kg soil]*
Cd	15	0.125
Cr	1000	8.33
Ni	200	1.67
Pb	1000	8.33
Zn	3000	25.0

* calculated by assuming an average sludge application practice (500 t/ha sewage sludge with 5 % dry matter content incorporated into the 20-cm surface soil layer)

Table 1: *Maximum permissible amounts of heavy metals in sewage sludges for application onto agricultural lands (Hungarian Technical Directive, 1990)*

dry matter content [%]	inorganic	concentrations in the sludge [mg/kg dry matter]				
		Cd	Cr	Ni	Pb	Zn
20.6	48.2	12.3	217	109	210	3026

Table 2: *Some characteristics of the compressed communal sewage sludge*

Water contents (Θ) of the soil samples, originating from the additions of the metal salt solution or metal-spiked sludge to the air-dried soil were within the field water content range, between -250 and -0.1 kPa. The wet, contaminated soil samples were homogenized three times during a one week equilibration period, then the soil solutions were separated by centrifugation. For the $10L$ and $100L$ loading rates soil suspensions with soil : metal salt solution ratios of 1:1, 1:2, 1:2.5, 1:5 and 1:10 were also prepared. Suspensions were shaken for 1 hour then the liquid phases (extracts) were separated with centrifugation or paper filtering immediately or after one week.

The impact of soil acidification on the release of these elements to the soil solution was studied on samples of the same soil from the metal+sludge contaminated (100L) upper layer of a column experiment (for a detailed description of the column experiment, see Bujtás et al., 1995). The air-dried samples were rewetted to maximum water capacity (MC= -0.1 kPa) with distilled water, or 0.001, 0.1 and 1.5 mol/L HNO$_3$ solutions, and equilibrated for one week.

Soil solutions from samples of the unloaded, loaded and loaded + dried + acidified soils were obtained by centrifugation using a rotor speed corresponding to -1500 kPa, i.e. conventional wilting point of plants. By this method soil solution fractions of physically well-defined energy status may be separated. As they are retained in the soil with suctions less than the wilting point, they are considered to be directly available for plant uptake (Csillag et al., 1995).

Cadmium, Cr, Ni, Pb and Zn concentrations of all solution samples (applied metal salt solutions, soil solutions and extracts) were determined by inductively coupled plasma (ICP) spectrometry.

The recovery of the heavy metals in the acidification experiment was calculated from concentrations measured in the soil solutions (c_s) and in the applied metal salt solution (c_m), as

$$\% = 100 \, c_s / c_m. \qquad [1]$$

3 Results and discussion

3.1 Effects of loading rate, water content and adsorption time

The heavy metal concentrations in the liquid phase of the uncontaminated soil samples were negligible or were below the detection limit of the ICP measurement (see the concentrations e.g. at field capacity in Table 3). When the maximum permitted amounts of the elements (1L) were applied, their transfer to the soil solution was very small. At provocative overloadings (10L and mainly 100L) metal concentrations in the soil solution significantly increased. In accordance with literature data (Kabata-Pendias and Adriano, 1995) Cd, Ni and Zn entered the liquid phase in relatively higher amounts than the less mobile Cr and Pb as compared to their previous loading rates (Tables 1 and 3).

loading rates	soil water potential	Cd	Cr	Ni	Pb	Zn
[L]	[kPa]		[mg/L]			
0 L	-14	<d.l.	0.03	0.02	0.24	0.41
1 L	-18	0.01	0.11	0.08	0.05	1.2
10 L	-20	0.46	0.46	13	0.62	189
100 L	-20	75	937	970	1392	15420

0 L: without loading; 1 L: see Table 1.; <d.l.: below detection limit

Table 3: Heavy metal concentrations in the liquid phase of soil samples contaminated to different levels, after 1 week adsorption time

For all elements, loading rates and adsorption times, the metal concentrations in the liquid phase (c) increased at lower soil moisture contents over the whole water content range, i.e. in soil solutions and soil extracts. Table 4 shows - on the example of 10L loaded soil samples at 1 week adsorption time - that this effect was more pronounced in the field moisture range (in the soil solutions above -0.1 kPa water potential) than in soil extracts used in routine laboratory analyses.

Also, element concentrations in the liquid phase were higher at field capacity than at maximum water capacity.

However, it is important to emphasize that concentrations of the applied metal salt solutions varied in the different soil water content treatments: it was necessary to apply solutions of lower metal concentrations to the soil samples of higher final water contents to ensure the same total metal loadings at each value of Θ. Nevertheless, the $c - \Theta$ relationships make it possible to predict the effect of soil water content on the release of the elements to the soil solution at constant contamination levels.

The concentrations of heavy metals in the soil extracts at 10L and 100L were found to decrease somewhat when the adsorption time increased from 1 hour to 1 week. This is shown at 10L for all elements (Table 4). The explanation of the relatively great decrease of Cr concentration in the soil extracts at the longer adsorption time at 10L and especially at 100L needs further studies.

soil : water ratio	kPa	1 hour adsorption time					1 week adsorption time				
		Cd	Cr	Ni [mg/L]	Pb	Zn	Cd	Cr	Ni [mg/L]	Pb	Zn
1 : 0.10[a]	-63						0.70	0.44	15.3	0.88	307
1 : 0.13 (FC)[a]	-20						0.46	0.46	13.0	0.62	189
1 : 0.21 (~MC)[a]	-0.3						0.22	0.23	5.1	0.13	130
1 : 1[b]		0.072	0.50	1.61	0.083	32.3	0.052	0.042	1.1	0.052	33.2
1 : 2[b]		0.036	0.30	0.77	0.071	18.0	0.024	0.033	0.57	<d.l.	16.0
1 : 2[c]		0.038	0.13	0.78	0.062	17.0	0.028	0.026	0.51	<d.l.	10.8
1 : 2.5[c]		0.032	0.18	0.69	0.075	13.3	0.022	0.031	0.41	<d.l.	8.4
1 : 5[c]		0.016	0.15	0.37	<d.l.	6.5	0.012	0.044	0.27	<d.l.	6.8
1 : 10[c]		<d.l.	0.14	0.15	0.054	4.1	<d.l.	0.042	0.09	<d.l.	1.7

L: loading rate; kPa: soil water potential; FC and MC: field and maximum water capacities; the liquid phase was separated
[a] with centrifugation using a rotor speed corresponding to -1500 kPa,
[b] with centrifugation using a rotor speed corresponding to -500 kPa,
[c] with paper filtering (see also Csillag et al., 1995)

<d.l.: below detection limit

Table 4: Heavy metal concentrations in the liquid phase of the soil at 10L loading rate

3.2 Effect of sludge application

When the soil was contaminated to 100L with metal-spiked communal sewage sludge containing also substantial amounts of adsorptive inorganic and organic materials, metal concentrations in the soil solution were many times lower (Table 5c) than after application of the pure multicomponent metal salt solution (Table 5a). Generally higher adsorption was found in the sludge itself (Table 5b) than in the soil (Table 5a), with the exception of Ni. Chromium and Pb were adsorbed by the sludge to such an extent (Table 5b) that their adsorption hardly increased any more when soil and sludge were present together (Table 5c).

Environmental behaviour of metals may differ when they are applied as metal salt solutions, as sludges enriched artifically with metals, or as sludges having high metal contents (Logan, 1990; Chang et al., 1992). In the present experiment considerable differences were measured in metal concentrations of soil solutions from samples in which the heavy metals might be in different chemical forms, due to the different treatments (absence or presence of sludge).

Treatment	Cd [mg/L]	[%]	Cr [mg/L]	[%]	Ni [mg/L]	[%]	Pb [mg/L]	[%]	Zn [mg/L]	[%]
a. soil	75	33	937	87	970	33	1392	82	15420	26
b. sludge	5.3	95.2	3.7	99.95	1023	31	17	99.80	10590	52
c. soil + sludge	3.8	96.6	0.8	99.99	417	72	5.6	99.92	4598	79

% = 100 $(c_m - c_s)/c_m$, where c_s = metal concentrations in the liquid phases of the soil, sludge, and soil+sludge, and c_m = metal concentrations in the appied metal salt solutions; L: loading rate

Table 5: Heavy metal concentrations [mg/L] in the liquid phases and adsorption [%] in the metal-contaminated (100 L) soil, sludge, and soil+sludge at field water capacity

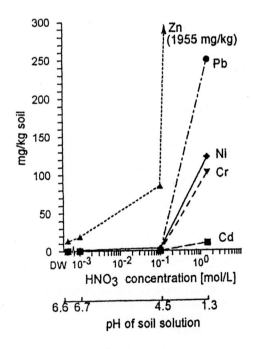

Figure 1: Heavy metal concentrations in the liquid phase of the contaminated soil (100L) treated with HNO_3 solutions of different pH values (soil water content: maximum water capacity; L: loading rate; DW: distilled water)

3.3 Effect of soil acidification

Acidic treatment of the previously contaminated soil promoted the release of the elements into the soil solution. The element concentrations in the liquid phase of air-dried, then acidified samples from the metal+sludge contaminated (100L) upper soil layer of the column experiment (Bujtás et al., 1995) were proportional to the loading rates (Cd<Ni<Cr=Pb<Zn, see Table 1) with the exception of the strongly adsorbed Cr (Cd<Cr<Ni<Pb<Zn, Fig. 1). As expected, the element concentrations increased at higher contamination levels and when the acidity of the extractant was higher (Figs. 1 and 2). However, only extremely strong acidification (1.5 mol/L HNO_3) increased substantially the element concentrations in the soil solution, with the exception of Zn which at 100L loading was released also by the less strong acidic treatments (the decrease of pH in the soil solution at increasing acid loads is also shown on Fig. 1).

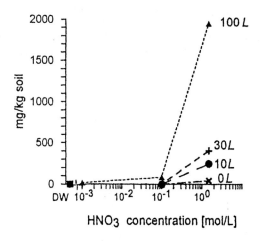

Figure 2: Zn concentrations in the liquid phase of variously contaminated soil samples treated with HNO_3 solutions of different pH values (soil water content: maximum water capacity; L: loading rate; DW: distilled water)

While element concentrations in the soil solution after soil acidification reflected mostly the loading rates, recoveries (see Eq. [1]) were more closely related to the mobilities of the elements, being higher for the more mobile Cd, Ni and Zn than for Cr and Pb (Fig. 3). Recovery increased relatively little even for Cd, Ni and Zn, when the soil was treated with 0.1 mol/L acid. A likely explanation may be the retention effect of adsorptive inorganic and organic materials present in the sludge-treated soil. Chromium and Pb were even more strongly adsorbed in the sludge-treated soil than Cd, Ni or Zn. Desorption of Cr and Pb was only 15% and 30%, resp., even in the strong (1.5 mol/L) acidic treatment, while 80-90% of Cd, Ni and Zn present in the sludge-treated soil were released by the same treatment (Fig. 3).

Our results are in accordance with literature data on influence of pH on solution concentration, and on mobilization of heavy metals due to acidification (Herms and Brümmer, 1984; Kiekens and Cottenie, 1985; Kabata-Pendias and Pendias, 1992; Hornburg and Brümmer, 1993).

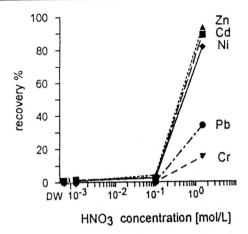

Figure 3: Recovery (%) of heavy metals after acidic treatment of the contaminated soil (100L) (%: see Eq. [1]; soil water content: maximum water capacity; L: loading rate; DW: distilled water)

4 Conclusions

In the light textured, slightly acidic soil with low cation exchange capacity and organic matter content, which was selected for our studies, only a negligible transfer of Cd, Cr, Ni, Pb and Zn to the soil solution occurred when metal salts were applied at concentrations corresponding to the official limits. Thus sludges, if applied in accordance with the regulations, may not pose an immediate environmental hazard: even such a soil with relatively low buffering capacity may retain the potentially toxic heavy metals. Nevertheless, excessive loading by waste materials containing high amounts of toxic elements, or an extreme decrease in soil pH, might substantially increase heavy metal concentrations in the soil solution.

In agreement with literature data (Chang et al., 1992), metals applied in the form of sludges were generally less mobile than soluble metal salts in our experiments. In certain cases application of sludges as soil amendments may decrease the harmful effects of soil acidification by decreasing mobilization of the heavy metals.

Acknowledgements

This research was supported under Grant No. DHR-5600-G-00-1056-00, Program in Science and Technology Cooperation, Office of the Science Advisor, U.S. A.I.D., and by the Hungarian National Scientific Research Fund under grant No. T023221.

References

Behel, D., Nelson, D.W. and Sommers, L.E. (1983): Assessment of heavy metal equilibria in sewage sludge-treated soil, J. Environ. Qual. **12**, 181-186.
Bergkvist, B. (1987): Soil solution chemistry and metal budgets of spruce forest ecosystems in S. Sweden, Water Air Soil Pollut. **33**, 131-154.
Bujtás, K., Csillag, J., Pártay, G. and Lukács, A. (1995): Distribution of selected metals in a soil-plant experimental system after application of metal-spiked sewage sludge. In: M.H. Gerzabek (ed.), Proc.

XXVth Annual Meeting of ESNA, Castelnuovo Fogliani (Piacenza/ Italy), Seibersdorf: Österreichisches Forschungszentrum Ges. m.b.H. Austria, 99-105.

Campbell, D.J. and Beckett, P.H.T. (1988): The soil solution in a soil treated with digested sewage sludge, J. Soil Sci. **39**, 283-298.

Castilho, P. del, Bril, J., Chardon, W.J., Römkens, P.F.A.M. and Oenema, O. (1996): Effect of changes in soil factors on soil solution cadmium, Hrvat. Vode **4**, 125-134.

Chang, A.C., Granato, T.C. and Page, A.L. (1992): A methodology for establishing phytotoxicity criteria for chromium, copper, nickel, and zinc in agricultural land application of municipal sewage sludges, J. Environ. Qual. **21**, 521-536.

Csillag, J., Tóth, T. and Rédly, M. (1995): Relationships between soil solution composition and soil water content of Hungarian salt-affected soils, Arid Soil Res. and Rehabilitation **9**, 245-260.

Evans, L. J., Spiers, G.A. and Zhao, G. (1995): Chemical aspects of heavy metal solubility with reference to sewage sludge amended soils, Int. J. Env. Anal. Chem. **59**, 291-302.

Herms, U. and Brümmer, G. (1984): Einflußgrößen der Schwermetallöslichkeit und -bindung in Böden, Z. Pflanzenernähr. Bodenk. **147**, 400-424.

Hornburg, V. and Brümmer, G.W. (1993): Verhalten von Schwermetallen in Böden 1. Untersuchungen zur Schwermetallmobilität, Z. Pflanzenernähr. Bodenk. **156**, 467-477.

Hungarian Technical Directive (1990): Land- and forest applications of waste waters and sewage sludges, MI-08-1735-1990. (S 02), (In Hung.)

Kabata-Pendias, A. and Pendias, H. (1992): Trace Elements in Soils and Plants, 2nd edition, CRC Press, Boca Raton.

Kabata-Pendias, A. and Adriano, D.C. (1995): Trace metals. In: J.E. Rechcigl (ed.), Soil Amendments and Environmental Quality, Boca Raton: CRC Lewis Publishers, 139-167.

Keller, C. (1995): Application of centrifuging to heavy metal studies in soil solutions, Commun. Soil Sci. Plant Anal. **26**, 1621-1636.

Kiekens, L. and Cottenie, A. (1985): Principles of investigations on the mobility and plant uptake of heavy metals. In: R. Leschber, R.D. Davis and P. L'Hermite (eds.), Chemical Methods for Assessing Bio-available Metals in Sludges and Soils, London: Elsevier Applied Science Publishers, 32-41.

Løbersli, E., Gjengedal, E. and Steinnes, E. (1991): Impact of soil acidification on the mobility of metals in the soil-plant system. In: J.P. Vernet (ed.), Heavy Metals in the Environment. Trace Metals in the Environment. 1., Amsterdam: Elsevier, 37-53.

Logan, T.J. (1990): Chemical degradation of soil. In: R. Lal and B.A. Stewart (eds.), Soil Degradation. Advances in Soil Science, Vol. 11, New-York: Springer-Verlag, 187-221.

Lorenz, S.E., Hamon, R.E. and McGrath, S.P. (1994): Differences between soil solutions obtained from rhizosphere and non-rhizosphere soils by water displacement and soil centrifugation, European J. Soil Sci. **45**, 431-438.

Reddy, K.J., Wang, L. and Gloss, S.P. (1995): Solubility and mobility of copper, zinc and lead in acidic environments, Plant and Soil **171**, 53-58.

Sposito, G. (1989): The Chemistry of Soils, Oxford University Press, New York.

Wolt, J.D. (1994): Soil Solution Chemistry. Applications to Environmental Science and Agriculture, Wiley & Sons, New-York.

Address of authors:
J. Csillag
A. Lukács
K. Bujtás
T. Németh
Research Institute for Soil Science and Agricultural Chemistry of the
Hungarian Academy of Sciences
Herman Ottó u. 15
H-1022 Budapest, Hungary

An Assessment of Rehabilitation of Strongly Acidic Sandy Soils

A. Badora & T. Filipek

Summary

An evaluation of the influence of lime and farmyard manure on the degree of aluminium detoxication was carried out in eastern Poland. A pot experiment and a field trial on a strongly acidified sandy soil demonstrated the usefulness of lime and farmyard manure (FYM) in the rehabilitation of soils where aluminium toxicity was a problem. With applications of 30 t/ha of FYM or $CaCO_3$ acc. to 1 hydrolytic acidity there was a measurable decrease of the toxic aluminium ions in the soil and an increase in the amount of exchangeable calcium and available phosphorous. Manure appeared to be the most efficient agent to increase the content of available magnesium in soil. The implications for rehabilitation of strongly acid soils are discussed.

Keywords: Acidic soils, aluminium toxicity, magnesium toxicity, assessment of rehabilitation, available magnesium, lime and organic fertilizers

1 Introduction

Highly acid soils are widespread in agricultural land in east Europe and elsewhere. Soil acidification is a consequence of multiple physical, chemical and biological processes occurring in the soil (Parker et al., 1989; Furrer, 1991). Highly acidic soils easily undergo chemical degradation. Rehabilitation of strongly acidic soils is difficult.

One consequence of soil acidification is the increasing solubility and mobility of aluminium. In mineral soils Al^{3+} and Al-hydroxy cations are the dominant acidic interchange cations (McBridge, 1994). High levels of Al are considered to be toxic to plants. Where soluble aluminium in the soil is more than 70 mg/kg of soil, it is enough to cause symptoms of toxicity. Where the concentration was 90 mg/kg of soil, all plants died (Badora, 1994; Filipek and Badora, 1994).

Soil humus contains a variety of acidic functional groups, some of which have high specificity for Al complexation or chelation (McBridge, 1994). The acid buffering capacity and the cation exchange capacity are other important functions of soil organic matter (Schnitzer and Kahn, 1978; Stevenson,1982). Binding of micronutrients, trace metals, pesticides and other toxic compounds to organic matter, together with binding of clay minerals and hydrogen oxides control the chemostate of soil systems and the bio-availability and mobility of these compounds (Schnitzer and Kahn, 1978; Aiken et al., 1985). This has considerable relevance for the rehabilitation of strongly acidic soils.

In agreement with the well-known Schultz-Hardy rule, trivalent cations are more effective in coagulating humic matter than divalent cations, which in turn are more effective than monovalent ones (Stevenson, 1982). For a variety of tri-, di-, and monovalent cations, flocculation of humic substances follows the order:

$$Fe^{3+} \sim Al^{3+} > H^+ > Ca^{2+} > Sr^{2+} > Mg^{2+} > K^+ > Na^+ > Li^+$$

As long as the soil is rich in calcium, soil pH does not drop below 7.0. Anywhere Ca is at low concentrations in soils, or absent, we deal with the system H_2O-C_2O. For the partial pressure of CO_2 between 0.32 mbar and 50 mbar, the calculated range of pH is from 5.7 to 4.9. For this pH range, no measurable activity of strong acids was observed (Schwertmann,1987). If the pH decreases below 4.2, protons are being taken by polymeric forms of aluminum hydroxy complexes. That is the way that aluminium changes to soluble forms in the soil solution:

$$Al_6(OH)_{15}^{3+} + 15H^+ + 21H_2O = 6Al(H_2O)_6^{3+}$$

This paper presents the results of a series of pot and field experiments designed to assess the influence of organic and lime-based fertilizers on the degree of Al ion detoxication. The purpose of the work was to evaluate treatment options for rehabilitation of strongly acidic soils which are widespread in agricultural lands.

2 Materials and methods

2.1 Pot trial

A split-plot design was used to arrange 32 pots (eight treatments and four replications) to assess the effectiveness of various rehabilitation agents for remediation of strongly acidic soils. Each pot contained 5 kg (dry weight) of soil which was collected from a field that had grown yellow lupin. Fertilizer (NPK was added as (NH_4NO_3)at a rate of 0.15 g N/kg; P as $CaHPO_4 2H_2O$ at 0.07 g/kg of soil; K (as KCl) at 0.15 g/kg of soil. The four treatments (with four replications) were:

Without added Al
(Soil only @ 5 kg per pot)
1 Soil only + NPK
2 Soil + NPK+ Peat@10 g/kg of soil
3 Soil + NPK + calcium oxide
   acc. To 1 hydrolytic acidity (Hh)
4 Soil +NPK+ oxide Ca-Mg acc. To 1 Hh

With added Al
($AlCl_3 6H_2O$ @100 mg/kg of soil)
1' Soil + NPK+ Al
2' Soil +NPK+Al
3' Soil + CaO + Al
4' Soil + NPK+ Ca-Mg + Al

2.2 Field experiment

A field of 2500 m^2 was divided into 5 blocks of 500 m^2. A randomised block design was used without replication. The soil was similar to that used in the pot trial but the field had just grown a crop of buckwheat. Each plot received NPK fertilizer at the following rates: N 60 kg/ha; P 20k/ha and K 60 kg/ha.

Five treatments were applied:
1. Control (no additives)
2. $CaCO_3$ acc. to 1 hydrolytic acidity (Hh)
3. $CaCO_3$ x Mg CO_3 acc. to 1 Hh
4. transient peat @ 30 t/ha
5. farmyard manure (FYM) @ 30 t/ha

2.3 Soil analyses

Soil samples were taken from each of the nine blocks. After drying at room temperature, the soil samples were passed through a 0.5 mm sieve. Determinations were made for pH in 1M KCl and in 0.01M $CaCl_2$; hydrolytic acidity after Kappen; exchangeable acidity and mobile aluminium after Solokov; amount of exchangeable cations extracted with CH_3COONH_4 (Na, Ca, K by flame photometry, Mg by ASS); and available nutrients (P and K after Egner and Mg after Schachtschabel) Average values for the four blocks are shown in Table 1 with the calculated percentage of exchangeable cations (Ca, Mg, K. and Na) and acidic cations (H, Al).

Agents	pH 1MKCl	pH 0.01MCaCl$_2$	hydrolytic acidity	exchangeable acidity	soluble aluminium
				[mmol(+)/kg soil]	
			Pot experiment		
0					
1	3.8	4.0	36.38	9.80	8.75
1`	3.6	3.8	39.00	18.46	17.46
peat					
2	4.0	4.3	38.63	8.40	7.99
2`	3.8	4.0	39.38	12.51	11.51
calcium oxide					
3	5.9	6.2	13.13	0.69	0.57
3`	5.7	6.0	12.00	1.51	1.31
oxide Ca-Mg					
4	5.8	6.1	14.63	0.69	0.57
4`	5.7	6.0	14.63	0.80	0.70
Agents			Field experiment		
0					
	4.0	4.1	31.92	12.22	10.68
calcium carbonate					
	5.3	5.5	16.33	1.58	1.41
carbonate Ca-Mg					
	4.3	4.5	26.84	6.52	5.57
peat					
	4.0	4.2	26.79	9.85	8.71
FYM					
	4.5	4.7	22.54	5.85	4.95

FYM - farmyard manure
1 2 3 4 - combinations with the addition of agents
1` 2` 3` 4` - combinations with the addition of 100 mg/kg soil of Al ($AlCl_3 6H_2O$) before the addition of agents

Table 1: State of soil acidification and content of soluble aluminium

2.4 Statistical analyses

For the pot trial an analysis of variance was performed. In the unreplicated field trial correlation and regression coefficients were calculated.

Agents	Ca	Mg	K	Na	SC	Ca	Mg	K	Na	H	Al
					mmol(+)/kg .soil						
Pot experiment											
0											
1	5.36	0.21	2.43	0.20	44.58	12.02	0.47	5.45	0.45	61.98	19.63
1'	6.86		4.35	0.17	50.38	13.62		8.63	0.34	42.76	34.65
peat											
2	7.49	0.14	2.11	0.17	56.53	13.25	0.25	3.73	0.30	68.34	14.13
2'	14.35	0.62	4.67	0.30	59.32	24.19	1.05	7.87	0.51	46.98	19.40
calcium oxide											
3	34.31	0.62	2.49	0.47	51.02	67.25	1.22	4.88	0.92	24.62	1.11
3'	51.77	0.21	4.54	0.69	69.21	74.80	0.30	6.56	1.00	15.45	1.89
oxide Ca-Mg											
4	17.47	13.37	1.79	0.25	47.51	36.77	28.14	3.77	0.53	29.59	1.20
4'	22.46	20.98	2.94	0.35	61.36	36.60	34.19	4.79	0.57	22.70	1.15
Field experiment											
0	1.74	0.30	1.78	0.08	46.50	3.75	0.65	3.83	0.17	68.64	22.96
calcium carbonate	13.67	0.37	1.15	0.12	34.72	39.37	1.07	3.31	0.35	47.03	8.87
carbonate Ca-Mg	3.94	1.35	1.30	0.07	39.07	10.08	3.46	3.33	0.18	68.70	14.25
peat	2.33	0.23	1.23	0.13	39.42	5.91	0.58	3.12	0.33	67.96	22.10
FYM	8.19	0.62	1.66	0.10	38.06	21.52	1.63	4.36	0.26	59.22	13.01

FYM - farmyard manure
1 2 3 4 - combinations with agents
1' 2' 3' 4' - combinations with the addition of 100 mg/kg soil of Al ($AlCl_3$ $6H_2O$) before the addition of agents
SC - sum of cations

Table 2: Content of exchangeable cations in soil and their percentage

3 Results and discussion

The use of either lime or farmyard manure had a greater effect in reducing Aluminium toxicity than reducing soil acidity. This was clear from both the pot experiment and the field trial (Table 2). Not all of the ameloriation treatments were equally effective. Peat had little or no effect in reducing soluble Al whereas FYM was the most effective in either pot or field situation and soluble Al was reduced to below the level considered toxic to plants (Badora 1994; Filipek and Badora, 1994).

There was an different interaction between the effects of peat and FYM on pH and on soluble aluminium in the soil. This might be explained by the chelating effect of FYM on the Al ions. Peat was unable to eliminate Al toxicity although it had been reported by Haug (1984) that adding peat to soil could decrease the values of mobile aluminium in soil even where pH was low. Some trends in this direction were observed in the present work (Table 2).

Motowicka-Teralak (1988) stated that the degree of binding of ions by the organic matter depended on the kind of organic matter. McBridge (1994) in the Organic Matter Complexation Model, explained that the pH is an order of preference of functional groups for Al^{3+}. At pH 4.5 the preference of Al^{3+} for phenolic groups is higher than for carboxylate groups. The phenolic groups have a higher affinity for Al^{3+} and bond with aluminium into stronger complexes and the solubility of Al decreases. The strong complexation of Al^{3+} causes drastic fall out when the pH is less than 4.5. In this case carboxylate groups of organic matter are responsible for the bond of aluminium ions. These groups dissociate readily, compared to phenolic groups. Because of this, the complexation is not strong and Al^{3+} can easily go into solution.

Agents		P	K	Mg
			m/kg soil	
			Pot experiment	
Nil	1	76.51	87.76	5.26
	1`	85.54	154.6	6.53
peat	2	70.02	77.42	6.02
	2`	78.40	159.8	13.02
calcium oxide	3	102.0	82.54	5.23
	3`	98.55	149.5	9.26
oxide Ca-Mg	4	80.34	57.00	44.60
	4`	84.85	87.70	46.23
Agents		Field experiment		
Nil		29.09	70.67	11.92
calcium carbonate		32.98	40.11	12.33
carbonate Ca-Mg		32.02	67.83	27.12
peat		29.14	30.00	12.14
FYM		32.63	53.00	18.38

FYM - farmyard manure
1 2 3 4 - combinations with agents
1` 2` 3` 4` - combinations with the addition of 100 mg/kg soil of Al
($AlCl_3 6H_2O$) before the addition of agents

Table 3. Content of available nutrients in soil

The extent of dissociation of acidic groups in humus is much higher when Al ions are present in the solution, as opposed to situations when only base cations are in solution. Decreasing the toxic fraction of aluminium in soil is an important element of soil recultivation, because high concentrations of this element in soil will determine, to a great degree, its chemical degradation

(Filipek and Badora, 1994). The addition of lime acc. to 1 Hh to soil (both experiments) didn't bring the pH to 7 but reduced the soluble aluminium to a nontoxic level for the particular crop to be grown (Badora, 1994; McBridge, 1994; Motowicka-Terelak, 1988; Parker et al., 1989). The lime oxide decreased the aluminium solubility, but this compound is not suitable for use on light, sandy soils.

Independent variable		Dependent variable	Soluble Al	Exchangeable Ca	Na	Available P	Mg
pH	KCl	rxy	-0.82	0.96	0.78	n.s.	n.s.
		byx	-42.62	172.77	1.27	n.s.	n.s.
pH	CaCl$_2$	rxy	-0.88	0.96	0.76	n.s.	n.s.
		byx	-42.09	157.98	1.14	n.s.	n.s.
Soluble Al		rxy	n.s.	-0.85	-0.63	-0.38	-0.29
		byx	n.s.	-2.95	-0.05	-0.05	0.06

n.s. - not significant
rxy; byx - correlation and regression coefficients

Table 4: Correlations in soil for some key elements

In both experiments the use of any lime fertilizer and, in the field experiment, the manure greatly activated (in most cases) the increase of the exchangeable forms of calcium and magnesium and decreased the amount of exchangeable potassium in soil (Table 2). The reason of those differences was probably the easier sorption of magnesium and calcium cations by the sorption complex of soil during its low acidity and ion antagonism between Mg^{2+} and Al^{3+} and Ca^{2+} and Al^{3+} (Table 4).

An increased concentration of Ca in the nutrient solution from 1 to 5 mM counteracted the damaging effect of Al (Keltjens and Kezheng, 1993). Figure 1 shows that the percentage of soluble Al in the sum of cations in soil from the pot experiment decrease about 90% compared to control (combinations without agents) after the use of either oxide. In the field experiment (Figure 2) the best result came from calcium carbonate and FYM. In the first case the decrease of soluble Al compare to control (0 agents) was about 60% and in the second case about 40%. The use of lime and organic fertilizers in soil increased the amounts of available phosphorus (Table 3). In the field experiment there was an increase of phosphorus after the addition of calcium carbonate, carbonate calcium-magnesium and farmyard manure.

Adding manure to soil most likely caused the humic-forming effects, followed most probably by the decrease of the chemical sorption between aluminium and phosphorus ions. Because of that phosphorus ions became available for plants (Pokojska, 1979).

4 Conclusions

Rehabilitation of strongly acid soils is difficult, especially where induced aluminium toxicity is causing yield reductions. However, where it is available, applications of FYM at rates of 30 t/ha or lime can ameliorate the situation by reducing soluble Al and/or lowering pH to a point where Al does not go so readily into solution. The present work demonstrates that application of FYM is the most efficient method to increase the content of available magnesium in the soil and thus counter the effect of soluble aluminium.

Further research is required to explore the complex interactions between soil type, pH, crop type or cultivar and aluminium toxicity in strongly acidic sandy soils.

Assessment of rehabilitation of strongly acidic sandy soils 687

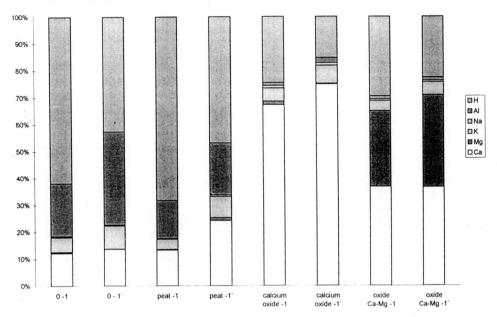

Figure 1: The part of cations [%] in the sum of cations in soil from pot experiment with the addition of different rehabilitation agents.

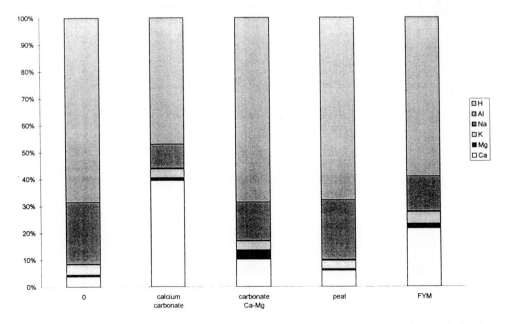

Figure 2: The part of cations [%] in the sum of cations in soil from field experiment with the addition of different rehabilitation agents.

References

Aiken, G. R., McKnight, D.M., Wershaw, R.L. and MacCarthy, P. (eds.) (1985): Humic Substances in Soil, Sediment, and Water; Wiley Interscience: New York

Badora, A. (1994): Unfavourable influence of mobile aluminium on cereal plants. Zeszyty Problemowe Postepow Nauk Rolniczych **413**, 9-16

Bartlett, R. I. and Riego, D.C. (1972): Toxicity of hydroxy aluminum in relation to pH and phosphorus. Soil Science **114(3)**, 194-200

Filipek, T. and Badora, A. (1994): Skutki fizjologiczne silnego zakwaszenia gleb dla roslin zbozowych. Fragmenta Agronomica **1(41)**, 53-60

Furrer, G. (1991): Theorie der Bodenversauerung: Das Zusamenspiel verschiedener Ursachen. Bulletin der bodenkundlichen Gesellschaft **15**, 5-18.

Haug, A. (1984): Molecular aspects of aluminum toxicity. CRC Critical Review Plant Science **1**, 345-373

James, B. R. and Riha, S.J. (1984): Soluble aluminium in acidified organic horizons of forest soils. Canadian Journal of Soil Science **64(4)**, 637-646

Keltjens, W. G. and Kezheng, T. (1993): Interactions between aluminium, magnesium and calcium with different monocotyledonous and dicotyledonous plant species. Plant and Soil **155/156**, 485-488

McBridge, M. B. (1994): Environmental chemistry of soils. Oxford University Press, New York, pp. 169-206

Motowicka-Terelak, T. (1988): Szkodliwosc dla roslin aktywnych form Al i Mn oraz sposoby ich neutralizacji w glebach kwasnych. Materialy z sympozjum, Olsztyn

Parker, D. R., Zelazny, L.W. and Kinraide, T.B. (1989): Chemical speciation and plant toxicity of aqueous aluminium. In: T. E. Lewis (ed.), Environmental chemistry and toxicology of aluminium, printing 1990. Lewis Publishers Inc., Chelsea, Michigan, 117-145.

Pokojska, U. (1979): Geochemical studies on podzolization. Part I. Podzolization in the light of the profile distribution of various forms of iron and aluminium. Roczniki Gleboznawcze **30(1)**, 207-209

Schitzer, M. and Khan, S.U. (1978): Soil Organic Matter, Elsevier: Amsterdam

Stevenson, F. J. (1982): Humus Chemistry: Genesis Composition, Reactions; Wiley Interscience: New York

Address of authors:
Aleksandra Badora
Tadeusz Filipek
Department of Agricultural Chemistry
University of Agriculture
Akademicka 15 str., P.O. Box 158
20-050 Lublin, Poland

Small-scale Distribution of Airborne Heavy Metals and Polycyclic Aromatic Hydrocarbons in a Contaminated Slovak Soil Toposequence

W. Wilcke, J. Kobza & W. Zech

Summary

In this study soil "pools" (horizons, aggregate and particle size separates) were identified in which airborne heavy metals and polycyclic aromatic hydrocarbons (PAHs) accumulate. Along a deposition gradient caused by an Al smelter in Central Slovakia we sampled organic layers and A horizons of a soil toposequence under beech (*Fagus sylvatica* L.) consisting of three Dystrochrepts. We analyzed Cd, Cu, Ni, Zn and 20 PAH in the organic layers, in bulk soil, aggregate surface and core fractions, and PAH only in particle size separates of the A horizons.

The concentrations of Cd, Cu, Ni, Zn, and PAHs in the organic layers were markedly higher than those in the mineral soil (up to 1.3 mg kg^{-1} Cd, 53 mg kg^{-1} Cu, 32 mg kg^{-1} Ni, 148 mg kg^{-1} Zn, 80 mg kg^{-1} PAHs). As the concentrations increased while approaching the Al smelter, this pointed to the smelter as the source. The heavy metal concentrations generally were lower in the aggregate surface than in the core fractions. This probably was caused by preferential acidification of the aggregate surfaces resulting in preferential weathering and leaching and by a higher plant uptake from the surfaces than from the cores of aggregates. The depletion decreased as distance to the Al smelter decreased. Cu and Zn were even enriched at the aggregate surfaces of the site nearest to the smelter. Thus, the depletion of lithogenic heavy metals was compensated by the preferential sorption of deposited heavy metals. Like the heavy metals, PAHs were more enriched at aggregate surfaces nearer to the smelter. In addition, the percentage of total PAHs bound to the sand fraction increased as distance to the smelter decreased. Thus, airborne pollutants are sorbed preferentially to aggregate surfaces and to the sand fraction. This may affect the bioavailability of the pollutants.

Keywords: Heavy metals, polycyclic aromatic hydrocarbons, small-scale heterogeneity, Al smelter, Slovakia.

1 Introduction

All ecosystems are affected by ubiquitous atmospheric deposition of inorganic and organic pollutants such as heavy metals and PAHs (Führer et al., 1988; Wild and Jones, 1995). As these substances are only little mobile in soils, they tend to accumulate (Brümmer et al., 1986; Jones et al., 1989). The ongoing accumulation of airborne pollutants in soil endangers plant growth, soil biological activity, and crop and groundwater quality.

Possible negative effects depend on the bioavailability of a pollutant. Generally, the full concentration of a substance is not available for biotics or leaching into the groundwater, but only a compound and soil specific fraction, the active "pools" (Brümmer et al., 1986). Thus, to assess the impact of airborne pollutants on ecosystems it is necessary to know in which pools depositional pollutant inputs accumulate.

The objective of this study was to identify soil pools in which airborne heavy metals and PAH emitted by an Al smelter in Central Slovakia accumulate preferentially.

2 Materials and Methods

2.1 Study sites and soils

Three sites along a slope with beech (*Fagus sylvatica* L.) vegetation were sampled. The study sites are situated at a distance of 1320 m (site 1), 1780 m (site 2) and 2180 m (site 3) from the centre of an Al smelter in the valley of the Hron near Ziar in Central Slovakia (Fig. 1). The elevations are 244-251 m (Al smelter), 275 m (site 1), 420 m (site 2) and 480 m a.s.l. (site 3). The organic layers and A horizons were sampled from two Typic Dystrochrepts (sites 1 and 3) and an Aquic Dystrochrept (site 2). Tab. 1 summarizes physical and chemical properties of the A horizons.

2.2 Aggregate fractionation

Aggregates of approximately 2 to 20 mm in diameter were selected manually from all samples. The aggregates were mechanically separated into a surface and a core fraction by the method of Kayser et al. (1994). The method was adjusted to peel off a surface fraction of approximately one third of the aggregate's mass.

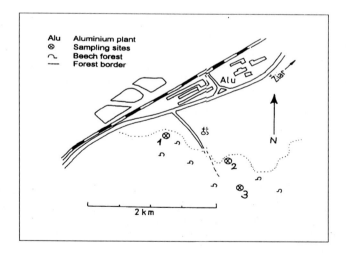

Fig. 1: Location of the study sites.

2.3 Texture class fractionation

Bulk soil was fractionated into the particle size classes < 2µm (clay), 2-20 µm (silt), and 20-2000µm (sand) according to Christensen (1992).

2.4 Extraction and analysis

Cd, Cu, Ni, and Zn were extracted by digestion with concentrated $HNO_3/HClO_4$ according to Zeien and Brümmer (1989). Metals were analyzed by atomic absorption spectrometry with Varian SpectrAA 400 and 400Z devices. Twenty PAHs were determined according to Hartmann (1996). A Hewlett Packard 5890 Series II GC and a Hewlett Packard 5971 A mass selective detector were used for PAH determination.

Using common standard procedures, the following bulk soil properties were determined: effective cation exchange capacity (ECEC) with 1 M NH_4-acetate, pH 7 (Page et al., 1982), base saturation (BS), total C with the C/N-Analyzer (Elementar vario EL). Fe oxides were extracted according to Schwertmann (1964, poorly crystalline Fe oxides, oxalate buffer = Fe_o) and according to Shuman and Hargrove, (1985, crystalline Fe oxides, ascorbic acid in oxalate buffer = Fe_a).

	Depth [cm]	pH_{KCl}	C_{org}	Sand	Silt	Clay	ECEC [$mmol_c kg^{-1}$]	BS [%]	Fe_o	Fe_a
			----------[g kg^{-1}]----------						[g kg^{-1}]	
Site 1	0-2	4.14	24.6	595.6	244.0	160.4	54.8	55.7	2.1	3.4
Site 2	0-2	3.85	16.7	536.2	293.6	170.2	65.8	38.2	1.5	2.1
Site 3	0-8	3.87	20.8	678.2	179.8	142.0	65.0	54.0	0.6	0.8

Tab. 1: Physical and chemical properties of the A horizons.

	Cd	Cu	Ni	Zn	PAH
	--[mg kg^{-1}]--				
Site 1	0.16	15.4	5.2	26.3	2.5
Site 2	0.24	6.4	3.8	44.1	1.7
Site 3	0.17	10.7	1.8	29.7	1.2

Tab. 2: Heavy metal concentration and sum concentration of 20 PAH (PAH) of the A horizons.

3 Results and discussion

The Cd, Cu, Ni, Zn, and PAH concentrations of the organic layers are markedly higher than in the A horizons, indicating a high recent input of these pollutants (Tab. 2, Filipinski and Grupe, 1990). The increase in concentrations towards the Al smelter identifies the smelter as the point source (Fig. 2).

The low heavy metal concentrations in the A horizons may be due to the high mobility of heavy metals in these acid soils (Brümmer et al., 1986). The mobility may even be enhanced by high fly ash depositions which result in an increase of the ionic strength in the soil solution causing heavy metal desorption or preventing their adsorption (Deschauer, 1995). Additionally, it contains F which is used for Al production and may form easily soluble fluoro-metal-complexes (Polomski et al., 1982).

The preferential leaching of aggregate surfaces generally results in lower heavy metal and PAH concentrations in the exterior than in the interior of soil aggregates (Wilcke, 1996). The preferential sorption of airborne heavy metals and PAH at aggregate surfaces increasingly compensates for the leaching losses as the pollutant burden of the soil increases (Figs. 3, 4).

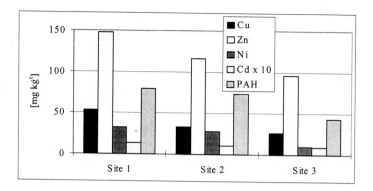

Fig. 2: Heavy metal concentrations and sum of the concentrations of 20 PAH in the organic layers.

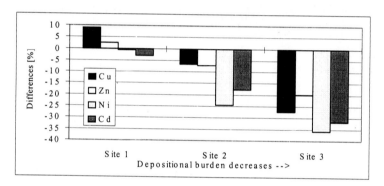

Fig. 3: Differences in heavy metal concentrations between aggregate surface and core fractions in % of the core's concentration (negative: depletion; positive: enrichment at the surface).

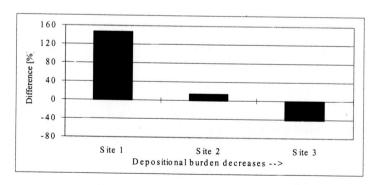

Fig. 4: Differences in the sum of PAH concentrations between aggregate surface and core fractions in % of the core's concentration (negative: depletion/positive enrichment at the surface).

The PAH concentrations in the A horizons decrease as the particle size increases. Highest PAH concentrations of sites 2 and 3 are found in the clay fractions (Tab. 3). However, when normalized to the organic matter concentrations, the highest PAH concentrations (µg PAH/g C_{org}) at all sites are found in the silt fraction which may be explained by the highest aromaticity of the silt-bound organic matter (Gauthier et al., 1987; Guggenberger et al., 1995).

	Site 1			Site 2			Site 3		
	C_{org}^a	PAH	PAH_{OM}	C_{org}^a	PAH	PAH_{OM}	C_{org}^a	PAH	PAH_{OM}
[µm]	[g kg^{-1}]	[µg kg^{-1}]	[µg PAH (g C)$^{-1}$]	[g kg^{-1}]	[µg kg^{-1}]	[µg PAH (g C)$^{-1}$]	[g kg^{-1}]	[µg kg^{-1}]	[µg PAH (g C)$^{-1}$]
< 2	42.0	3164	75.3	43.5	2749	63.2	75.2	3275	43.6
2-20	21.5	3766	175.2	9.7	2020	208.3	12.6	1606	127.5
> 20	16.8	1272	75.7	10.2	723	70.9	8.2	346	42.2

^{a}The sand fraction includes particulate organic matter (POM).

Tab. 2: C_{org} concentrations, sum concentrations of 20 PAH normalized to the dry weight of the sample (PAH) and normalized to the organic matter concentrations (PAH_{OM}) of the three particle size fractions.

The percentage of the total PAH concentrations bound to silt and sand increases as the depositional burden, i.e. the concentrations in the organic layer, increases, whereas the percentage bound to clay decreases (Fig. 5). Recently deposited PAH preferentially bind to the organic matter of coarser particles. The percentage of the PAH sum concentration at site 1 is more than double of that at site 3. In the course of humification the PAH may later be incorporated into finer particles (Guggenberger et al., 1994).

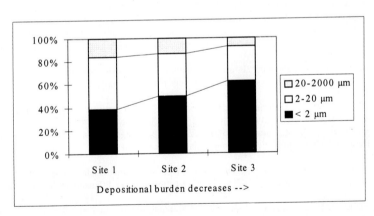

Fig. 5: Distribution of the sum of the concentration of 20 PAHs in particle size classes.

4 Conclusions

The emissions of an Al smelter have resulted in a marked accumulation of heavy metals and PAHs in soils near the smelter through atmospheric deposition. Airborne heavy metals and PAHs preferentially bind to aggregate surfaces. Furthermore, PAHs are sorbed preferentially to the organic matter of sand and silt-sized particles. This may have impacts on the reactivity and bioavailability of pollutants. Thus, to assess environmental hazards, both the bulk soil concentration of pollutants and the small-scale distribution have to be considered.

Acknowledgements

We thank Dr. V. Linkeš, Ilse Thaufelder, Stefan Reuschel, Jutta Mosbach, Ralf Berghofer, and the coworkers of the Soil Fertility Research Institute Banská Bystrica for their help in sampling, sample preparation, and performing the analyses. We are indebted to the Bavarian Staatskanzlei for funding this project.

References

Brümmer, G.W., Gerth, J. and Herms, U. (1986): Heavy metal species, mobility and availability in soils. Z. Pflanzenernähr. Bodenk. **149**, 382-398.
Christensen, B.T. (1992): Physical fractionation of soil and organic matter in primary particle size and density fractions. Adv. Soil Sci. **20**, 1-90.
Deschauer, H. (1995): Eignung von Bioabfallkompost als Dünger im Wald (Untersuchungen zu den Auswirkungen einer Bioabfalldüngung auf den Elementumsatz, die Schwermetalldynamik und das Verhalten von polyzyklischen aromatischen Kohlenwasserstoffen in einem humus- und nährstoffarmen Kiefernbestand). Bayreuther Bodenkundl. Ber. **43**, 169p.
Filipinski, M. and Gruppe, M. (1990): Verteilungsmuster lithogener, pedogener und anthropogener Schwermetalle in Böden. Z. Pflanzenernähr. Bodenk. **153**, 69-73.
Führer, H.-W., Brechtel, H.M., Ernstberger, H. and Erpenbeck, C. (1988): Ergebnisse von neuen Depositionsmessungen in der Bundesrepublik Deutschland und im benachbarten Ausland. DVWK-Mitteilungen **14**, 1-122.
Gauthier, T.D., Seitz, W.R. and Grant, C.L. (1987): Effects of structural and compositional variations of dissolved humic materials on pyrene K_{OC} values. Environ. Sci. Technol. **21**, 243-248.
Guggenberger, G., Christensen, B.T. and Zech, W. (1994): Land-use effects on the composition of organic matter in particle-size fractions of soil: I. Lignin and carbohydrate signature. Eur. J. Soil. Sci. **45**, 449-458.
Guggenberger, G., Zech, W., Haumaier, L. and Christensen, B.T. (1995): Land-use effects on the composition of organic matter in particle-size fractions of soil: II CPMAS and solution ^{13}C NMR analysis. Eur. J. Soil Sci. **46**, 147-158.
Hartmann, R. (1996): Polycyclic aromatic hydrocarbons (PAHs) in forest soils: critical evaluation of a new analytical procedure. Intern. J. Environ. Chem. **62**, 161-173.
Jones, K.C., Grimmer, G., Jacob, J. and Johnston, A.E. (1989): Changes in the polynuclear aromatic hydrocarbon content of wheat grain and pasture grassland over the last century from one site in the UK Sci. Tot. Environ. **78**, 117-130.
Kayser, A., Wilcke, W., Kaupenjohann, M. and Joslin, J.D. (1994): Small-scale heterogeneity of soil chemical properties. I. A rapid technique for aggregate fractionation. Z. Pflanzenernähr. Bodenk. **157**, 453-458.
Page, A.L., Miller, R.H. and Keeney, D.R. (1982): Methods of soil analysis. Part 2., 2nd ed., Agron. Monogr., 9. ASA and SSSA, Madison, WI, USA, 1159 p.
Polomski, J, Flüheler, H. and Blaser, P. (1982): Fluoride-induced mobilisation and leaching of organic matter, iron and aluminium. J. Environ. Qual. **11**, 452-456
Schwertmann, U. (1964): Differenzierung der Eisenoxide des Bodens durch Extraktion mit Ammoniumoxalat-Lösung. Z. Pflanzenernähr. Düng. Bodenk. **105**, 194-202.
Shuman, L.M. and Hargrove, W.L. (1985): Effect of tillage on the distribution of Mn, Cu, Fe, and Zn in soil fractions. Soil Sci. Soc. Am. J. **49**, 1117-1121.

Wild, S.R. and Jones, K.C. (1995): Polynuclear aromatic hydrocarbons in the United Kingdom environment: a preliminary source inventory and budget. Environ. Pollut. **88**, 91-108.

Wilcke, W. (1996): Kleinräumige chemische Heterogenität in Böden: Verteilung von Aluminium, Schwermetallen und polyzyklischen aromatischen Kohlenwasserstoffen in Aggregaten. Bayreuther Bodenkundl. Ber. **48**, 136 p.

Zeien, H. and Brümmer, G.W. (1989): Chemische Extraktion zur Bestimmung von Schwermetallbindungsformen in Böden. Mitteilgn. Dtsch. Bodenkundl. Gesellsch. **59**, 505-510.

Addresses of authors:
Wolfgang Wilcke
Wolfgang Zech
Institute of Soil Science and Soil Geography
University of Bayreuth
D-95440 Bayreuth, Germany
Jozef Kobža
Soil Fertility Research Institute
Mládežnícka 36
SK-97400 Banská Bystrica, Slovakia

Heavy Metal Pollution of Irrigated Soils in Ningxia, China

A. Kollender-Szych, J. Breburda, P. Felix-Henningsen & H. Trott

Summary

The recent accelaration of industrial management causes a distinct heavy metal pollution of water and soils in the irrigation area of Ningxia Region, China. The contents of Cr and Ni in the soils exceed the tolerable contents of these metals defined for European soils. Cu and Zn contents are 2-3 times higher than their contents in the geochemical background of European soils and in loessial soils from the Ningxia Upland. For soils in the neighbourhood of industrial centers an ascending tendency of heavy metal contents was noted. It can be concluded that industrial pollution is responsible for the high heavy metal contents of irrigated soils in the whole region.

Keywords: Ningxia, arable soils, heavy metals

1 Introduction

The Ningxia Hui Autonomous Region is located in the northern central part of China where the climate is semi-arid to arid. Hot summers and cold winters, strong winds, low annual average precipitation of 200-300 mm decreasing from southeast to northwest, strong radiation and a high annual potential evaporation of 1,600-2,000 mm preclude agricultural management except where irrigation occurs. In such cases, the approach is for surface irrigation with complete flooding of the soil surface. (Bishop et al., 1967; Plath, 1991; Soil Atlas of China, 1986).

The Yinbei irrigation plain of Ningxia is situated in a transitional area between the middle and the upper course of the Yellow River. It covers an area of arable land of 192,800 ha (Xie, 1995). The cultivation history of this area by irrigation is more than 2,000 years old and led to severe salinization of the soils (Di, 1993; Wang, 1986; Xie, 1993, 1995). Climatic conditions, high content of soluble salts in the soil, irrigation management and the relatively high level of groundwater, all helped to increase salinization. Accordingly, over 40 % of the arable soils show a decrease of 50 % or more of the total productivity due to salinization (Vomocil, 1987; Xie, 1995).

Anthropogenic activities not only induced salinization problems. During the last century an acceleration of industrial development took place throughout the region. In recent years many factories were built and the industrial sewage and waste rich in heavy metals were often led directly into the river. By irrigation the polluted load was conveyed on the irrigated soils.

In Ningxia the major source of irrigation water is the Huang He River, also known as the Yellow River because of its silt-laden water (Breburda, 1983). The sediment concentrations of the river water are amongst the highest of the world. About 2.5 - 3.5 % of the total sediment load were diverted from the Huang He by irrigation water, which was appraised to be between 63 to 84 million tons per year (Fullen et al., 1993; Chen, 1993). The silt content of the Huang He water is highly variable with an average of 3wt.-%. However, in the overflooding period it can reach up to 58wt.-% (Du et al., 1990). This variable sediment load is applicated by irrigation management on the soils, resulting in elevation of their surfaces. The main part of the sediment load originates from spurs of the Loessial Upland in the south of the region. While transported by water, the eroded loessial material divides into particle fractions of various diameters. The particle size composition is related to the variable flow-speed of the suspension (the irrigation water), before it is sedimented on the soil surface during irrigation.

According to their higher sorption capacity on internal and external surfaces in relation to coarse silt and sand particles (Fiedler & Rössler, 1988; Lichtfuss, 1985), the clay (< 0.002 mm) and fine silt (0.0063 - 0.002 mm) of the soil ought to show higher contents of heavy metals. In order to determine the negative influence of contamination by irrigation due to industrialization, sites with arable soils at the irrigation area of the Yinbei Plain were investigated.

Figure 1: Geographical localization of the soil sites

2 Materials and methods

Five agriculturally relevant "irrigating warped soils", assumed to be characteristic for the irrigated area, were studied. Their geographic location along the western river side of the Yellow River is shown in Fig. 1. The soils were classified as Cumulic Anthrosols (FAO - UNESCO, 1989); according to the World Reference Base - Irragric Anthrosols (Spaargaren, 1994). They were mainly built up by more than 2,000 years of continued irrigation management and sedimentation of the suspended particles on the soil surface. Particle size composition and chemical analyses were done according to Schlichting et al. (1995). Total contents of heavy metals were determined by X-Ray-Fluorescence-Spectrometry. Soluble fractions were extracted with 1M ammonium acetate and 0.02 M EDTA-solution (Anderson, 1975; Lakanen.& Erviö, 1972), and analyzed by AAS.

3 Results and discussion

The soils differ in particle size composition and chemical properties according to applied irrigation and fertilization methods (Tab.1). The soil profile from the area of Zhongwei, next to the Loessial Upland, shows the highest content of clay, fine silt, $CaCO_3$, organic matter and the highest cation exchange capacity (CEC). Somewhat lower results can be noticed for the profile from the Yinchuan area and profile 4 from the Pingluo area. The lowest values were obtained for Qintongxia (profile 2) and profile 5 from the Pingluo area.

profiles	Horizons	cm	percentage of particles with dia in mm:					pH 0,01m CaCl2	$CaCO_3$ %	OM %	CEC cmol$_c$/100 g
			2 - -0,063	0,063 - -0,02	0,02 - -0,0063	0,0063 - -0,002	< 0,002				
1. Zongwei	Ap	0 - 18	7,98	17,31	21,31	19,91	33,49	7,70	13,56	2,13	22,02
	Ap/B	18 - 34	8,05	19,19	2,58	24,15	46,04	7,69	14,92	1,43	21,93
	B	34 - 56	8,66	24,12	6,55	20,60	40,08	7,70	14,67	1,22	21,01
	C	< 56	0,95	19,23	3,79	23,66	52,37	7,69	13,84	0,74	22,03
2. QinTongx	Ap	0 - 26	8,34	45 21	17,47	9,41	19,57	7,62	11,54	1,71	15,97
	B	26 - 54	3,10	34,79	19,21	14,88	28,01	7,74	13,52	0,91	18,13
	B/C	54 - 71	2,38	38,82	26,72	10,99	21,09	7,86	13,52	0,79	16,85
	C	71 - 81	2,58	46,89	31,25	7,56	11,78	7,78	13,83	0,80	14,69
3. YinChuan	Ap	0 - 20	7,63	25,11	18,62	14,29	34,35	7,83	13,64	1,75	20,28
	Ap/B	20 - 40	13,92	25,60	24,19	11,37	24,92	8,00	12,17	1,24	18,63
	B	40 - 110	2,66	14,68	26,89	18,38	37,38	7,90	13,58	1,01	22,75
	C	<110	14,74	17,76	14,81	13,32	39,38	7,79	11,60	0,43	20,97
4. PingLuo	Ap	0 - 17	9,73	29,96	21,39	13,66	25,26	7,79	12,24	2,10	18,99
	Ap/B	17 - 27	10,98	32,30	17,40	15,15	24,17	7,78	12,55	1,53	18,70
	B	27 - 39	12,45	36,38	8,19	15,28	27,70	7,81	11,94	1,24	18,44
	B	39 - 63	10,94	31,39	10,75	14,92	32,00	7,82	12,19	1,06	20,29
	B/C	63 - 75	2,30	14,37	13,76	23,19	46,38	7,85	16,21	0,84	25,03
	C	75 - 100	1,35	46,03	33,07	7,01	12,54	7,70	11,83	0,71	14,09
5. Pingluo	Ap	0 - 19	14,48	49,90	15,62	5,90	14,11	7,63	10,55	1,75	13,66
	Ap/B	19 - 36	12,28	50,30	15,58	6,96	14,87	7,76	11,00	1,33	13,13
	B	36 - 53	8,43	43,86	17,99	9,66	20,05	7,98	11,63	1,30	16,22
	B	53 - 75	9,33	37,54	19,75	11,49	21,90	7,95	12,04	1,69	18,04
	C	75 - 100	5,12	38,30	26,14	10,31	20,19	7,89	12,51	1,24	16,86

Tab. 1: Particle size composition and some properties of the analyzed soils.

Particle size composition and chemical properties associated with the smaller fractions are the result of continually and regularly applied irrigation with muddy water of the Huang He River. The contents of silt and clay in the river water are variable, with higher values in longer overflooding periods when the current of the river is slower. Accordingly, the current of the river is slowest in the area of Zhongwei and Yinchuan and fastest in the Qintongxia area near the water reservoir and in the area of Pingluo 5.

Profiles	horizon	Cr Total	soluble	Zn total	soluble	Cu total	soluble	Ni total	soluble	Pb total	soluble
1. Zhong-wei	Ap	134	0,48	88	2,22	44	7,57	86	0,51	23	6,14
	Ap/B	133	0,25	85	1,62	47	6,95	86	0,67	22	5,33
	B	134	0	79	1,41	43	5,67	81	0,55	22	4,14
	C	133	0	94	0,76	40	3,21	83	0,23	22	3,08
2. Qin-Tongx	Ap	126	0,37	66	1,63	42	4,94	78	0,28	20	4,24
	B	127	0,28	74	0,96	56	3,83	84	0,16	21	3,9
	B/C	129	0,67	65	1,08	38	3,56	78	0,01	20	3,82
	C	124	0,66	58	0,89	34	2,65	78	0,03	19	3,28
3. Yin-Chuan	Ap	124	0,4	82	0,87	57	3,35	87	0,08	22	4,14
	Ap/B	127	0,12	72	0,77	45	2,94	82	0,12	23	3,91
	B	133	0,96	96	0,87	48	3,02	87	0,18	25	4,78
	C	132	0,54	79	1,04	40	3,27	78	0,02	28	5,43
4. Pingluo	Ap	133	0,1	82	1,45	41	3,34	80	0,27	21	4
	Ap/B	150	0,3	102	1,39	31	3,66	82	0,18	39	3,96
	B	125	0,23	70	1,19	39	3,76	82	0,07	22	3,56
	B	140	0,35	77	0,98	36	3,56	88	0,03	35	3,5
	B/C	135	0,46	97	0,84	51	4,16	88	0,02	23	4,01
	C	122	0,24	69	0,55	44	2,61	76	0	19	3,33
5. Pingluo	Ap	121	0	64	1,85	40	2,36	78	0,34	23	3,41
	Ap/B	117	0,74	67	1,14	28	2,77	76	0,2	17	3,29
	B	125	0,37	77	0,83	56	2,54	79	0,19	25	3,35
	B	125	0,57	78	0,87	43	2,58	80	0,11	24	3,82
	C	124	0,07	76	0,63	34	3,59	78	0,05	20	3,63

Tab. 2: Total and soluble (in ammonium acetate/EDTA) contents of heavy metals in the analyzed soils (in mg/kg).

The total contents of heavy metals do not differ strongly between the individual profiles. (Tab. 2). However, in relation to the above-mentioned model they are similarly arranged. Therefore, the profiles from the Qintongxia area and Pingluo 5 show the lowest average contents of heavy metals, the profile from the Zhongwei area the highest for Zn and Ni, the profile from Yinchuan the highest average content of Cu, and at Pingluo 4 the highest average content of Pb.

It must be emphasized that the contents of Cr and Ni in all profiles from the top horizon to a depth of 100 cm and more exceed the tolerable contents of these metals as defined for German soils (Cr = 100 mg/kg, Ni = 50 mg/kg; (AbfKlärV, 1992). The Cu and Zn contents are 2-3 times higher than determined in parent rocks and unpolluted soils of Europe (Lichtfuß, 1985; Czarnowska, 1995) and in loessial soils from Ningxia Loessial Upland (Kollender-Szych & Battenfeld, 1993). Only the total contents for Pb are within the tolerable limits.

Comparing the obtained results for the soil profile from theYinchuan area with results from this area as obtained earlier (Kollender-Szych, 1993), an increase of total contents of Cr, Cu and Ni was noticed. Within 3 years the average content of Cr in irrigated soil profiles from the same area increased about 70 % from 76 mg/kg to 129 mg/kg. Within the same time the average content of Cu increased from 33 mg/kg to 48 mg/kg (45 %), and Ni from 54 mg/kg to 84 mg/kg (56 %). Only the Zn contents remained at the same level. The average contents of Pb were lower.

Yinchuan is the capital of the Ningxia Autonomous Region and most of the industrial centres of the region are located in its suburbs. It can be assumed that the rapid growth of industry results in an increase of water pollution. Together with increasing water consumption by irrigation of about 50 % within the last years (Kollender-Szych, 1993), soil pollution with heavy metals also increased, caused by polluted water of the Huang He River.

The soluble and easily extractable fraction of heavy metals mainly bound to organic matter and potentially available for plants show distinct differences between the profiles and also according to depth, within a single profile (Tab. 2). The highest contents of soluble Zn, Cu, Ni and Pb were determined for soil samples from the Zhongwei area, decreasing with increasing depth. The depth function is similar for the profile from Qintongxia, although the soluble contents are minor. In the profiles from the Pingluo area only the contents of soluble Zn decrease with depth.

The contents of heavy metals in soils depend not only on the chemical composition of the soils, the content and type of clay minerals and organic matter but also on pH, the redox potential and the composition of the soil moisture (Blume, 1992; Brümmer, 1992; Herms & Brümmer, 1984; Schimming, 1990). Their mobility depends on the soil properties and the soil processes determining also the form of bonds between the metal and the specific soil component - the types of clay minerals or oxides of manganese, iron or aluminium and humus (Brümmer, 1992; Fiedler & Rössler, 1988; Kuntze, 1988; Lichtfuss & Brümmer, 1981; Schimming, 1990; Schulthess & Huang, 1990; Sposito et al., 1982). To evaluate the relationship between total and soluble contents of the heavy metals, organic matter, $CaCO_3$, sorption capacity and particle size composition, a linear correlation analysis was undertaken (Tab.3).

In the analyzed soils the total contents of Cr, Zn and Ni are significantly related to the fine silt and clay content and sorption capacity which depends on these fractions. Lichtfuss (1985) showed that in a soil developed of a morainic marl with about 25 wt.-% of clay and fine silt, more than 70 % of total contents of Zn and Cu are bound to these fractions. In the analyzed soils from the irrigation area of the Yinbei plain the average contents of the <0.0063 mm fraction were between 65 wt.-% and 27 wt.-% Although we do not have results about the total contents of heavy metals in the singular particle size fractions, the significant relation between the total contents of Cr, Zn and Ni and clay as well as fine silt confirm the observations of Lichtfuss (1985).

Looking at the amount of heavy metals per unit of fine silt plus clay, the quotient for different horizons within one profile is almost constant for the non-polluted glacial deposits investigated by Lichtfuss (1985). Similar results were obtained by Kollender-Szych & Battenfeld (1993) for non-irrigated and non-polluted soils from the Loessial Upland in the southern part of Ningxia. The same quotient calculated for the "irrigating warped soils" of the Ningxia region shows that in horizons with a higher percentage of the finer fractions, the content in mg of a specific metal bound to 10 mg of the <0.0063mm diameter fraction decreases (Fig.2).

In horizons with minor percentages of fine silt and clay, e.g. in the profiles from Qintongxia and Pingluo, relatively more heavy metals were found. This lead to the assumption that a large part of heavy metals is not bound to the finer fractions. They can be included in coarser particles, bound to their surface, etc. This may be due to the high amount of heavy metals deposited by irrigation water exceeding the binding capacity of fine silt and clay.

The obtained results of linear correlation analysis also show that the total content of Cr is significantly correlated to the total contents of Zn, Ni and Pb (Tab. 3). Regarding organic matter, only the content of soluble Zn shows a significant correlation with it. It confirms the great affinity of this metal in alkaline soils to organic matter and hydroxides (Schachtschabel et al., 1992).

	Cr total	Cr Soluble	Zn total	Zn soluble	Cu total	Cu soluble	Ni total	Ni soluble	Pb total	Pb soluble
O.M.	0,0029 n.s.	-0,1947 n.s.	0,0792 n.s.	0,6502 ***	0,0454 n.s.	0,3247 n.s.	0,1025 n.s.	0,4839 *	0,0094 n.s.	0,2404 n.s.
CaCO3	0,3963 n.s.	0,0634 n.s.	0,4881 *	-0,0551 n.s.	0,3548 n.s.	0,4992 *	0,6207 **	0,3373 n.s.	-0,0762 n.s.	0,2862 n.s.
CEC	0,6428 ***	0,0204 n.s.	0,7904 ***	0,0547 n.s.	0,3755 n.s.	0,4926 *	0,8171 ***	0,3187 n.s.	0,3217 n.s.	0,5597 **
< 0,002 mm	0,5266 **	-0,1131 n.s.	0,7110 ***	0,0393 n.s.	0,3381 n.s.	0,4728 *	0,7086 ***	0,3981 n.s.	0,2076 n.s.	0,4474 *
0,0063 - 0,002mm	0,6177 **	-0,1211 n.s.	0,7598 ***	0,1622 n.s.	0,2998 n.s.	0,6245 **	0,7617 ***	0,5252 *	0,2075 n.s.	0,4652 *
0,02 - 0,0063mm	-0,3031 n.s.	0,3806 n.s.	-0,3181 n.s.	-0,3066 n.s.	-0,0337 n.s.	-0,3907 n.s.	-0,3603 n.s.	-0,5246 n.s.	-0,2254 n.s.	-0,1010 n.s.
0,063 - 0,02mm	-0,5899 n.s.	-0,0287 n.s.	-0,7540 n.s.	-0,1283 n.s.	-0,3577 n.s.	-0,5477 **	-0,7299 ***	-0,3567 n.s.	0,2479 n.s.	-0,6801 ***
Cr total										
Cr soluble	-0,0687 n.s.									
Zn total	0,7368 ***	0,0008 n.s.								
Zn soluble	0,2173 n.s.	-0,1234 n.s.	0,0303 n.s.							
Cu total	-0,0932 n.s.	-0,0332 n.s.	0,2187 n.s.	-0,1309 n.s.						
Cu soluble	0,3690 n.s.	-0,1196 n.s.	0,2890 n.s.	0,6536 ***	0,1278 n.s.					
Ni total	0,5340 **	0,0665 n.s.	0,6506 ***	0,0796 n.s.	0,4638 *	0,4204 n.s.				
Ni soluble	0,2170 n.s.	-0,2850 n.s.	0,2837 n.s.	0,7234 ***	0,1629 n.s.	0,7566 ***	0,2639 n.s.			
Pb total	0,7897 ***	-0,0075 n.s.	0,5189 *	0,0759 n.s.	-0,1411 n.s.	-0,0155 n.s.	0,3731 n.s.	-0,0909 n.s.		
Pb soluble	0,3471 n.s.	0,2401 n.s.	0,3796 n.s.	0,5391 **	0,2323 n.s.	0,7232 ***	0,3882 n.s.	0,4907 *	0,1353 n.s.	

Number of samples = 23
n.s. = not significant, *: p = 0,05, **: p = 0,01, ***: p = 0,001
O.M. = organic matter
CEC = cation exchange capacity

Tab. 3: Relationships (r = correlation coefficient) between the contents (in mg/kg) of heavy metals (total and soluble) and the contents (in %) of organic matter, CaCO3, the CEC (in cmol_c/100 g soil), the contents (in %) of clay, the fine, middle and coarse silt fractions and the relationships between the both forms of the heavy metals to each other.

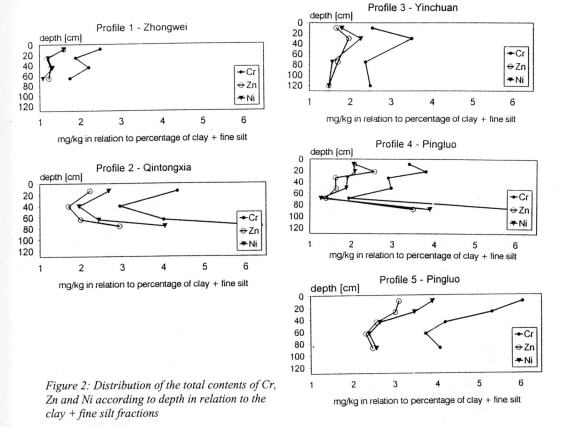

Figure 2: Distribution of the total contents of Cr, Zn and Ni according to depth in relation to the clay + fine silt fractions

Summarizing, the results show a high heavy metal load of the arable soils of the Ningxia region due to irrigation and industrialization next to the Huang He River. The increasing pollution of the soils of the Ningxia Region due to the rapid growth of industrial production was also pointed out by Xie (1995). The increase of production cannot be stopped; its positive influence on the development of the region is indisputable. But the signs of soil pollution are more than obvious and demand appropriate management.

References

AbfKlärV (1992): Klärschlammverordnung, Bundesgesetzblatt, T.I, 912-932
Anderson, A. (1975): Relative efficiency of nine different soil extractants, Swedish J.Agr.Res.5, 125-135.
Bishop, A. A., Jensen, M. E., Hall, W. A. (1967): Surface irrigation Systems, Irrigation of agricultural lands 11, Agronomy, 865-884
Blume, H.-P. (1992): Handbuch des Bodenschutzes. Bodenökologie und -belastung, ecomed, Landsberg/Lech.
Breburda, J. (1983): Bodenerosion-Bodenerhaltung, DLG, Frankfurt (Main).
Brümmer, G. (1992): Nährstoffe, 276-300. In: Schachtschabel et al., Lehrbuch der Bodenkunde, Ferdinand Enke, Stuttgart
Chen, W. (1993): The characteristics of arid soil degradation in Ningxia and its controlling methods, Proc. Intern. Worksh. on Classification and Management of arid-desert soils, China Sc. a. Techn.Press, Beijing, 136-13

Czarnowska, K. (1995): Ogólna zawartość metali ciężkich w skałach osadowych okruchowych jako tło geochemiczne gleb. OBREM, Łódź, Poland

Di, X. (1993): Desertification processes in desert steppe regions, a case study of Ningxia, Giess. Abhdlg zur Agrar- u. Wirtschaftsforschg. d. europ. Ostens **185**, 73-79

Du, G., Lei, W., Zhou, M., Zhang, J. (1990): Fluvo-aquic soils and calcic concretion black soils, in: Soils of China - Inst. Soil Sc, Academia Sinica, Beijing.

FAO - UNESCO (1989): Soil map of the world, Revised legend, Rome.

Fiedler, H.J., Rössler, H.J. (1988): Spurenelemente in der Umwelt, Ferdinand Enke-Verlag, Stuttgart.

Fullen, M.A., Fearnehough, W., Mitchell, D.J., Trueman, I.C. (1993): An investigation in desert reclamation in Ningxia Autonomous Region, Proc. Intern. Worksh. on Classification and Management of arid-desert soils, China Sc. a.Technol. Press, Beijing, 75-80

Herms, U., Brümmer, G. (1984): Einfluss der Schwermetalllöslichkeit und -bindung in Böden, Z. Pflanzenern. Bodenk, **147**, 400-424.

Kollender-Szych, A. (1993): The soils of the Yinchuan flood plain with special emphasis on the salt-affected soils of Tong Dong, Giess.Ahdlgn. zur Agrar- und Wirtschaftsforschg. des europ. Ostens, **185**, 125 - 141

Kollender-Szych, A., Battenfeld, S. (1993): Distribution of heavy metals in some soils of deserts and waste land of Ningxia, Proc.Intern. Worksh. on Classification and Management of arid-desert soils, China Sc. a. Technol. Press, Beijing, 455 - 460

Kuntze, H. (1988): Bodenkunde, Ulmer-Verlag, Stuttgart.

Lakanen, E., Erviö, R. (1972): A comparison of eight extractants for the determination of plant available micronutrients in soils, Acta Agr. Fenniae, **123**, 223-232.

Lichtfuß, R. (1985): Korngrössenfraktionierung und Bestimmung der minerogenen Spurenelementgehalte in Böden aus jungpleistozänem Geschiebemergel, Mittlg. Dtsch. Bodenkdl. Ges. **43**, 409 - 414.

Lichtfuß, R., Brümmer, G. (1981): Natürlicher Gehalt und anthropogene Anreicherungen von Schwermetallen in den Sedimenten von Elbe, Eider, Trave und Schwentine, CATENA **8**, 251-264.

Plath, B. (1991): Ländliche Umweltprobleme und Umweltpolitik in der Voksrepublik China, Studien zur integrierten ländlichen Entwicklung 36, Weltarchiv GmbH, Hamburg

Schachtschabel et al. (1992).:, Lehrbuch der Bodenkunde, Ferdinand-Enke-Verlag, Stuttgart, 13. Auflage.

Schimming, C.G. (1990): Metalle, 258-304. In: Blume, H.P., Handbuch des Bodenschutzes: Bodenökologie und Bodenbelastung, vorbeugende und abwehrende Schutzmaßnahmen, ecomed, Landsberg/Lech.

Schlichting, E., Blume, H.-P., Stahr, K.(1995): Bodenkundliches Praktikum, Blackwell Wissenschaftsverlag, Wien.

Schulthess, C.P., Huang, C.P. (1990): Adsorption of heavy metals by silicon and aluminium oxide surfaces on clay minerals", Soil Sc. Soc. Amer. **54**, 679-694.

Soil Atlas of China (1986): Cartographic Publishing House, Beijing, China

Spaargaren, O.C. (ed.) (1994): World Reference Base for Soil Resources, Draft ISSS-AISS-IBG, ISRIC, FAO, Wageningen/Rome.

Sposito, G., Lund, L.J., Chang, A. C. (1982): Trace metal chemistry in arid-zone field soils amended with sewage sludge. I. Fractionation of Ni, Zn, Cd and Pb in solid phases, Soil Sc. Soc. Amer. **46**, 260-264.

Vomocil, J.A. (1987): Report on a trip to Ningxia-People's Republic of China to study food system resources - pers. inform.

Wang, J.-Z. (1986): Main soils in Ningxia Autonomous Region and Alashanzuoqi Inner Mongolia, Soil Sc. Soc. of China, Current Progr. in Soil Res., 562 - 573

Xie, G. (1993): Soil salinization and ist control on the Ningxia flood plain. Giess. Abhdlg. zur Agrar- u. Wirtschaftsforschg. d. europ. Ostens, **185**, 81-86.

Xie, G. (1995): Über die Versalzung und Alkalisierung der Böden der Yinbei Ebene, Diss. Universität Giessen

Address of author:
Anna Kollender-Szych
J. Breburda
P. Felix-Henningsen
H. Trott
Institute of Soil Science and Soil Conservation, Wiesenstraße 3-5
D-35390 Giessen, Germany

Forms of Selected Heavy Metals and Their Transformation in Soils Polluted by the Emissions from Copper Smelters

A. Karczewska, L. Szerszeń & C. Kabała

Summary
The aim of the paper was to determine mobile and bioavailable forms of heavy metals (Cu, Pb and Zn) in soils polluted by emissions from copper smelters, and to assess their changes under different pH and moisture conditions. 6 soil samples differing in their texture and sorption properties, were collected from the vicinity of copper smelters in the area of LGOM (Poland). Forms of Cu, Pb and Zn were determined by the means of single and sequential extractions. The results showed high contributions of soluble and easily mobilizable forms (9.7-61 % Cu, 8.0-50 % Pb), which confirmed the enhanced risk of leaching the metals from soils, especially from sandy ones. The effects of different soil pH and moisture on metals mobility were examined in a two-year pot experiment carried out with two LGOM soils and with an artificially polluted soil. Spring wheat was used as a testing plant. Concentrations of Cu, Pb and Zn were determined in leachates, wheat tissues and in soils. Both soil pH and watering schemes affected the mobility of metals, especially in sandy soils. The lowest bioavailability and extractability occurred under treatments of high pH and high watering. Neutral salts proved to be much more suitable than 1M HCl or EDTA for determining the bioavailable forms of heavy metals in polluted soils.

Keywords: Soil, copper smelter, heavy metals, pH, moisture, mobility, bioavailability, extraction

1 Introduction. Copper metallurgy in Poland

The paper focuses on the properties of soils polluted with heavy metals in the area of *LGOM (Copper Industry Region of Legnica and Głogów,* Poland), the biggest European centre of copper mining and metallurgy (Figure 1). The centre was established in 1959 and now consists of 4 mines, 3 ore-dressing plants, 3 smelters and a rolling mill, and produces 400 thousand tons of copper yearly.

Metallurgical activity has always been an environmental hazard, but in the 1960's and 1970's little attention was paid to environmental protection, and as a consequence serious pollution problems developed. Particularly dangerous effects were caused by smelters *(Legnica, Głogów I* and *Głogów II)*, the emissions from which contained toxic gases (especially SO_2) and dusts rich in heavy metals Cu, Pb, Zn and Cd (Table 1). Beginning in the 1980's, emissions were reduced (Table 2), but the already existing soil pollution around the smelters remained and is still of great environmental concern.

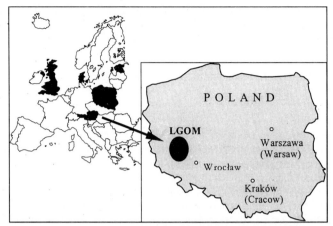

Figure 1. Location map of the Copper Industry Region of Legnica and Głogów (LGOM), Poland

Dust:		Heavy metals content in dust, %						
		Cu	Pb	Zn	Cd	As	Mn	Fe
emitted from the smelters	min	0.1	1.7	0.4	0.00	0.03	*	*
(different sources),	max	22	53	9.8	0.62	0.07	*	*
(Rutkowski, 1993)	mean	16	6.9	1.2	0.01	0.03	*	*
collected from the snow	min	2.6	0.4	0.04	*	0.04	0.2	2.1
and soil surface	max	11	7.3	0.2	*	0.09	0.6	4.3
(Weber, 1995)								

Table 1: Chemical composition of dust emitted from the smelters (* no data).

Emissions	Yearly emissions pollutants and dust in years:			
	1980	1985	1990	1994
SO_2, 10^3 tons	151.9	77.6	47.5	30.3
CO, 10^3 tons	314.9	192.6	120.4	5.2
Total dust, 10^3 tons	28.9	8.8	8.1	1.4
Copper, tons	2857	297	163	78
Lead, tons	3107	453	118	63

Table 2: Emissions from copper smelters in 1980-1994 (Dobrzański and Byrdziak, 1995).

2 Soil pollution and remediation in the vicinity of copper smelters.

Strong accumulation of heavy metals in soils occurred in an area covering more than 100 km². Metal concentrations in soils increased dramatically, primarily in soil surface horizons (Szerszeń et al., 1991; 1993; Karczewska, 1995). At a distance of up to 1 km from the smelters, the concentrations of heavy metals were found to be as high as: 250-10,000 mg/kg Cu, 90-18,000 mg/kg Pb, 0.3-10.9 mg/kg Cd and 55-4000 mg/kg Zn (Roszyk and Szerszeń, 1988). In general, the smaller the distance from the emission source the higher heavy metal concentration in the soils. Not only soils were affected by smelter emissions. Plants were also contaminated both as a result of airborne depositions and through metal uptake from polluted soils. The areas of extremely strong metal accumulation have been purchased by the smelters and excluded from typical agricultural usage. Sanitary protection zones were formed around the smelters (2840 ha close to the smelters *Głogów* and 1128 ha close to *Legnica*), where intensive reclamation was carried out involving removal and dumping of the surface soil layers from the closest smelter vicinity and afforesting other, less contaminated areas of former arable lands and greenlands. Strong efforts were made to immobilise metals accumulated in soils and to reduce their leaching and plant uptake. As the mobility of heavy metals depends on soil sorption properties (CEC) and pH (Mc Bride, 1989; Hornburg and Brümmer, 1993; Kabata-Pendias and Pendias, 1993), soils were limed in the whole area of sanitary protection zones, and in selected fields some other treatments were undertaken to improve soil properties, such as the addition of brown coal mixed with lime, introducing synthetic zeolites or fertilising with phosphorus in order to form insoluble metal compounds. Those measures, although efficient in reducing metal mobility, did not ensure permanent metal immobilisation. Smelter environmental services reported the incidental increase of Cu and Cd concentrations in underground waters and there are still many unanswered questions concerning the forms and fate of metals accumulated in soils. Some current research projects deal with this problem, focusing on metals forms and their transformation caused by the changes of soil conditions (Karczewska, 1995). The effects of both soil pH and redox potential on heavy metal forms in polluted soils were examined in some papers (Mc Bride, 1989; Kabata-Pendias and Pendias, 1993; Karczewska, 1995), however, only preliminary results were published referring to soil moisture as the modifying factors.

The aim of this paper is to describe transformations of metal forms in polluted soils under different pH and moisture conditions.

3 Methods for determining heavy metals forms in soils - an overview.

Although the content of heavy metals in soils is an important indicator of soil contamination, it is not sufficient to characterise this environmental hazard as it depends on the forms of metals rather than on their total concentration. The method most frequently used for describing bioavailable forms of metals is that of chemical extraction with an agent which was previously checked and showed a high correlation between the amounts of metals extracted and real plant uptake tested by certain cultivated plants with various soil properties. The extractants commonly used are neutral salts (potassium or calcium chlorides, ammonium nitrate or acetate) at various concentrations, as well as complexing salts EDTA (ethylenediaminetetraacetic acid) and DTPA (diethylenetriaminepentaacetic acid) (Queuauviller et al., 1993). In Poland, 1M HCl has been chosen as the standard for bioavailable metal forms. Though it is a good indicator for metals bioavailability in unpolluted soils, it does not seem to be appropriate for soils polluted with metallurgical dusts. This problem will be recalled later on in the paper.

During the last ten years, sequential extraction procedures have been developed to describe the forms of heavy metals in soils. They separate certain operationally defined metal fractions believed to be bound to different soil components (Hirner, 1994). Two of the most commonly used procedures are those developed by Tessier et al. (1979) and by Zeien and Brümmer (1991). The latter distinguishes the following fractions: 1) mobile, 2) exchangeable, 3) occluded in MnOx, 4) organically bound, 5) occluded in amorphous FeOx, 6) occluded in crystalline FeOx and 7) residual, and seems to be suitable for polluted soils.

It should be added that there have been some other attempts to determine the forms of metals in soils in a direct way, e.g. using scanning electron microscopy (Weber, 1995). However, in this manner the most dispersed forms cannot be detected while they are actually those of the highest mobility and greatest environmental importance.

4 Analytical methods.

Soil samples described here, were taken from the humus horizons in the area of sanitary protection zones. Six samples differing strongly in their properties were chosen for close examination. In order to determine the forms of heavy metals in polluted soils, both single and sequential extractions (according to Zeien and Brümmer, 1991) were applied. Apart from the standard 1M HCl solution, other reagents were also used for "bioavailable" metal forms, i.e. the neutral salts 1M KCl, 0.1M $CaCl_2$, 1M NH_4NO_3, a complexing agent 0.0025M EDTA and diluted 0.1M HCl. Different extraction times (1-24 hours) and various solid phase to solution ratios (1:2.5-1:25) were used to examine desorption equilibria.

No.	Distance from the smelter km	Texture percent of fractions, mm				pH 1M KCl	C org %	CEC mmol /100g	Total content of metals, mg/kg			
		>0.1	0.1–0.02	<0.02	<0.002				Cu	Pb	Zn	Cd
1	0.3 L*	10	48	42	4	4.8	1.10	11.3	360	170	103	0.5
2	0.5 L	6	54	40	8	7.1	1.16	7.40	540	185	190	1.0
3	0.2 G**	55	30	15	3	5.9	0.78	5.17	1050	243	250	1.1
4	3.0 G	92	4	4	1	7.4	0,25	3.86	270	106	36	0.6
5	3.5 G	37	23	40	15	6.7	1.35	19.6	290	127	80	0.6
6	4.5 G	31	15	54	28	6.9	2.34	38.2	250	99	94	0.7
7	***	78	14	8	3	5.1	0.40	2.26	1600	500	27	6.2

* L - Legnica, ** G - Głogów, *** - artificially polluted

Table 3: Selected properties of examined soils.

In order to determine the effects of soil pH and moisture on the transformations of metal forms in soils, a two-year pot experiment was carried out with two of the previously examined *LGOM* soils (No.2,3), differing in their texture and sorption properties, and with an artificially polluted soil (No.7), to which insoluble forms of metals (CuO, $PbSO_4$ and CdO) were added. The most relevant data on soil properties are given in Table 3. At the beginning of the experiment, soil reaction was modified and adjusted to 3 different pH levels: 4.5-5.0, 5.5-6.0 and 6.7-7.0, using either a mixture of diluted nitric and sulphuric acids or calcium carbonate. The second factor to be tested in the experiment was soil moisture. Thus, three different watering schemes were applied and induced three kinds of moisture conditions: mild (optimum watering: 100 %), wet (150 %) and extremely wet (200 %). All experimental variants were examined in three replicates and statistical calculations were made using a T-test with a reliability level of 0.95. Spring wheat var. "Henika" was chosen as a testing plant. Heavy metals concentrations were determined in the leachates, in wheat tissues (straw and grains) and finally in soils, taken from a depth of 5-15 cm. Both total concentrations and extractable amounts of metals in soil were determined.

5 Results I. Forms of heavy metals in soils.

Results of sequential extraction showed an unusually high percentage of easily soluble and mobilizable metals forms (fractions 1 and 2), which was observed not only for the metals known as relatively mobile (Cd, Cu and Zn), but also for usually immobile Pb. Cadmium fractionation did not give satisfactory results (low recovery ratio due to analytical problems), so it is not reported on here.

The data on Cu and Pb fractionation are illustrated in Fig. 2. The sum of fractions 1 and 2, expressed in the percentage of total metals content, varied from 9.7 % Cu and 8.0 % Pb in heavy loam to 61 % Cu and 50 % Pb in sandy soil, and showed a negative correlation with soil CEC. Fraction 1 was much lower in case of Pb (0.0-2.6 %) than Cu (1.8-16.5 %) and Zn (1.2-28 %). These results differed greatly from those of unpolluted soils of similar origin where mobile and exchangeable fractions of Cu and Pb together are less than 10 % (Tessier et al., 1979; Zeien and Brümmer, 1991). Unfortunately, at present any satisfactory explanation of such unusually high percentages of mobile metal fractions in the examined soils cannot be given and further research is required.

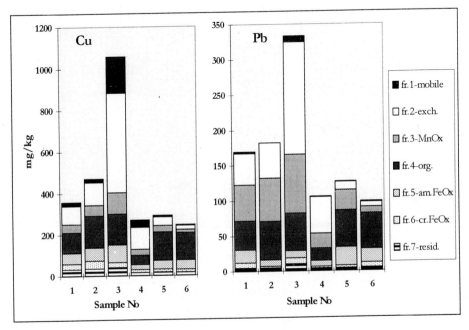

Figure 2: Cu and Pb fractions in soils according to the method of Zeien and Brümmer (1991).

Relatively high amounts of both Cu and Pb were also determined as organically bound forms (fraction 4) whereas Zn occurred mostly in fractions 7 and 6, particularly in heavy soils. The data obtained from the single extractions confirmed the relationships described above. Both Cu and Pb were easily and almost totally soluble not only in 1M HCl but also in its more diluted 0.1M solution (Fig. 3). The amounts of Cu and Pb dissolved in 1M HCl were as high as 90-99 % of total concentrations and depended neither on soil CEC nor on its reaction pH. On the other hand, the bioavailability of metals will apparently be conditioned by those properties, which gives the evidence for 1M HCl being unsuitable for determining the bioavailable metals forms in polluted soils. A similar conclusion may be drawn for 0.0025 M EDTA, the extracting capacity of which was also very high and varied from 76 % Cu and 60 % Pb to 100 % Cu and 97 % Pb. Zinc, representing not only metallurgical but also a geogenic origin in the soils examined, was neither as highly soluble in acids as Cu and Pb, nor as easily extractable by complexing salt EDTA.

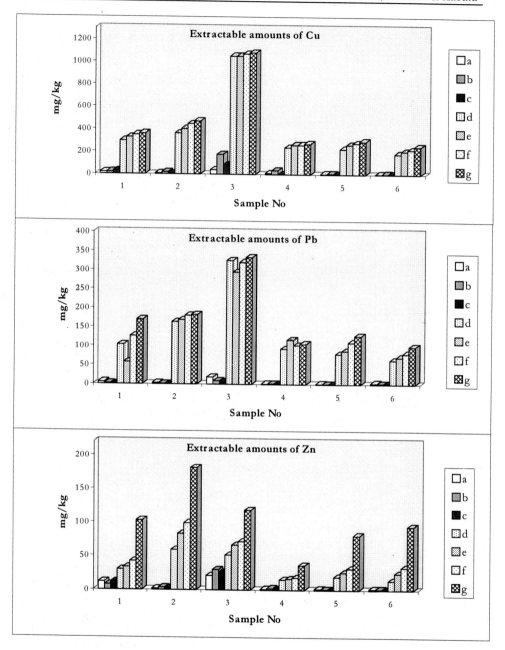

Figure 3: Results of single extractions of Cu, Pb and Zn from soils:
a) 1 M KCl b) 1 M NH_4NO_3 c) 0.1 M $CaCl_2$ d) 0.0025 M EDTA-Na
e) 0.1 M HCl f) 1 M HCl g) total

Soil No.	Watering	pH initial/final	Cu leachate mg/dm³	Cu wheat straw mg/kg	Cu wheat grain mg/kg	Cu extract CaCl₂ mg/kg	Pb leachate mg/dm³	Pb wheat straw mg/kg	Pb wheat grain mg/kg	Pb extract CaCl₂ mg/kg	Zn leachate mg/dm³	Zn wheat straw mg/kg	Zn wheat grain mg/kg	Zn extract CaCl₂ mg/kg
2	mild	5.0/5.7	0.38	14.1	10.7	7.0	0.15	2.8	0.0	4.2	0.33	41	132	7.9
	wet	5.0/6.3	0.35	10.9	9.4	4.2	0.17	2.2	0.0	7.5	0.21	22	107	4.8
	mild	6.5/6.3	0.30	14.0	10.9	2.5	0.19	2.6	0.3	7.8	0.18	29	104	4.6
	wet	6.5/6.4	0.29	11.3	8.1	2.5	0.22	2.8	0.3	7.5	0.10	24	61	2.0
3	mild	5.0/5.7	10.0	58	*	90	0.31	5.7	*	5.8	18.3	34	*	32
	wet	5.0/6.0	5.8	**	**	62	0.22	**	**	5.0	14.6	**	**	22
	mild	6.3/6.0	0.6	61.5	*	60	0.16	7.7	*	6.0	3.6	41	*	16
	wet	6.3/6.2	0.3	33.5	*	42	0.12	4.0	*	5.2	2.1	30	*	11
7	mild	5.2/5.0	3.7	260	*	120	0.33	21.5	*	64	7.2	32	*	4.3
	wet	5.2/5.3	1.4	252	*	68	0.17	14.8	*	34	8.3	12.8	*	2.8
	mild	6.4/5.5	1.4	14.1	10.2	42	0.20	3.4	0.2	28	5.8	19.3	44	1.7
	wet	6.4/5.9	0.36	18.9	10.1	22	0.14	2.5	0.3	16	0.5	13	33	0.9

*   Wheat did not bear grains
** Plant material unsuitable for analysis (poor growth)

Tab. 4: *Effects of soil pH and watering scheme on forms of Cu, Pb and Zn in soils (pot experiment).*

Extraction procedures with neutral salts (1M KCl, 1M NH₄NO₃, 0.1M CaCl₂) gave much more satisfactory results, reflecting the differences in soil properties, i.e. CEC and pH. Equilibria were obtained after 6-24 hours depending on soil texture. The results showed also the dependence of metals extractability on solid phase to solution ratio, and from this a rough conclusion can be drawn about the importance of soil moisture which apparently determines the desorption of heavy metals from polluted soils and their further diffusion in soil solution.

As all the neutral salts showed similar suitability for extracting the mobile fractions of Cu, Pb (and presumably other metals), each of them could be chosen for the next part of the experiment. The 0.1M CaCl₂ was used because of its relatively high similarity to natural composition of soil solution.

6 Results II. Transformation of metals forms under various pH and moisture conditions

In the pot experiment containing polluted soils, the soluble forms of heavy metals differed significantly between the treatments under different watering. Selected results for Cu, Pb and Zn are given in Table 4. The data are mean values of three replicates. For each of three examined soils four treatments are presented, i.e. combinations of low and high pH and mild and wet soil moisture. Well-marked initial pH differences between the soil groups partly disappeared during the experiment and, at the end, the pH did not differ very much. Final pH values of all wet combinations were higher than those of respective mild ones. The differences were not very high (0.1-0.6 pH units) but significant at P=0.95. Concentrations of all metals (Cu, Pb, Zn and Cd) in leachates (Table 4) tended to be lower in the wet combinations than in the mild ones, the differences being more significant in the case of sandy soils (No. 3 and 7) than the loamy one (No. 2). In the latter, Pb concentrations in leachates from wet treatments were even insignificantly higher than from the mild ones. Exactly the same tendencies were observed for the amounts of metals in wheat tissues, both in the straw and grains. In some pots with sandy soils, the plant growth was very poor and wheat did not bear grains.

Metals bioavailability was well reflected by the extraction with 0.1M CaCl₂ and significant correlations were found between amounts of metals extracted and the concentrations in the wheat tissues. The only exception being Pb in the loamy soil for which such a correlation was not found.

7 Conclusions

1. All soils examined contain high amounts of Cu and Pb.
2. The contributions of soluble and easily mobilizable forms of metals in soils polluted by emissions from copper smelters are high, which confirms the enhanced risk of metals leaching from the soils, especially from sandy ones.
3. Neutral salts, and in particular $0.1M$ $CaCl_2$, are much more suitable than $1M$ HCl or EDTA for determining the mobile / bioavailable metals forms in polluted soils.
4. Both soil pH and watering schemes affected the changes of metals forms, especially in the sandy soils. The lowest metals bioavailability and extractability occurred under treatments of high pH and high watering.

References

Dobrzański, J. and Byrdziak, H. (1995): Impact of Polish copper industry on the environment. Z. Probl. Post. N. Roln. **418**, 399-405 (in Polish).

Hirner, A.V. (1992): Trace element speciation in soils and sediments using sequential chemical extraction methods. Int. J. Environ. Anal. Chem. **46**, 1992, 77-85.

Hornburg, V. and Brümmer, G.W. (1993): Verhalten von Schwermetallen in Böden. Untersuchungen zur Schwermetallmobilität. Z. Pflanz. Bodenk. **156**, 467-477.

Kabata - Pendias, A. and Pendias, H. (1993): Biogeochemia pierwiastków śladowych. PWN, Warszawa (in Polish).

Karczewska, A. (1996): Metal species distribution in top- and sub-soil in the area affected by copper smelter emissions. Appl. Geoch. **11**, 35-42.

Karczewska, A. (1995): Changes in metal speciation in polluted soils from copper smelter sanitary protection zones as resulted from soil flooding. Zesz. P. Post. N. Roln **418**, 487-493, (in Polish).

Mc Bride, M.B. (1989): Reactions controlling heavy metal solubility in soils. in: Advances in Soil Science, vol.10, Springer Verlag, New York, 1-56.

Queuauviller, P.H., Rauret, G. and Griepink, B. (1993): Single and sequential extraction in sediments and soils. In: Rauret, G. (ed.), Workshop on Sequential Extraction in Sediments and Soils. Intern. J. Env. Anal. Chem. **51**, 231-235.

Roszyk, E. and Szerszeń, L. (1988): Accumulation of heavy metals in arable layer of soils in the protection zones of copper smelters, Rocz. Glebozn. **39**, 135-156 (in Polish).

Rutkowski, J. (1993): The sources of air pollution. Wyd. PWr. (in Polish).

Szerszeń L., Chodak, T. and Karczewska, A. (1993): Areal, profile and time differentiation of heavy metal content in soils in the vicinity of copper smelters in LGOM, Poland, in: Integrated Soil and Sediment Research. A Basis for Proper Protection. Eijsackers and Hamers eds., Kluwer, Dordrecht, 279-281.

Szerszeń, L., Karczewska, A., Roszyk, E. and Chodak, T. (1991): Distribution of Cu, Pb and Zn in profiles of soils adjoining copper metallurgic plants. Rocz. Glebozn. **42**, 3/4,199-206 (in Polish).

Tessier, A., Campbell, P.G. and Bisson, M. (1979): Sequential extraction procedure for the speciation of particulate trace metals. Anal. Chem. **51**, 844-851

Weber, J. (1995): Submicromorphological characteristics of soil environment changed by copper smelters emissions, ZN AR Wrocław, (in Polish).

Zeien, H. and Brümmer, G.W. (1991): Ermittlung der Mobilität und Bindungsformen von Schwermetallen in Böden mittels sequientieller Extraktionen. Mitt. Dtsch. Bodenkundl. Gesellsch. **66 / I**, 439-442.

Address of authors:
Anna Karczewska
Leszek Szerszeń
Cezary Kabała
Institute of Soil Science and Agricultural Environment Protection, Agricultural University of Wrocław
Ul. Grunwaldzka 53
50-357 Wrocvław, Poland

Elution of Zn, Pb and Cd in Soil
Under Field and Laboratory Conditions

M. Gebski, S. Mercik & K. Sommer

Summary

Waste lime from a zinc factory, containing high amounts of Zn, Pb, and Cd, were applied to soils in the Experimental Field of Warsaw Agricultural University in 1985, 1988 and 1991. In the third year following each liming, the contents of Zn, Pb and Cd soluble in 1M HCl were determined for particular genetic soil horizons. In the laboratory, leaching of Zn, Pb and Cd from this soil was studied. The samples were treated with different calcium compounds ($CaCl_2$, $CaSO_4$ and $Ca(OH)_2$) in quantities equivalent to 1, 1.5 and 3 times of soil CEC, pH 8.1. The results of the field experiment indicated that zinc was predominantly translocated down and lead was least mobile. In the laboratory, the greatest amounts of all heavy metals were leached from soil samples where $CaCl_2$ was applied and the least after the addition of $Ca(OH)_2$. With the addition of water (equivalent to 600 mm precipitation), as much as 6% of Zn and 46% of Cd on $CaCl_2$ treatment from total content of these elements was leached. The results obtained under field and laboratory experiments indicate that to restrict the phytoavailability of heavy metals in contaminated soils, it is necessary to use fertilization technology which leaches away heavy metals from soil layers with roots or to maintain soil pH at a suitable level.

Keywords: Waste lime, heavy metals, calcium compounds, leaching

1 Introduction

As a result of the use of industrial and municipal wastes for agricultural purposes, an increased level of heavy metals in soil can be detected. In such soils, a higher absorption of heavy metals occurs in cultivated plants, which in turn may pose a risk for people (Sauerbeck and Styperek 1988). The use of the above-mentioned waste as a fertilizer may also lead to the leaching of certain metals (in particular, cadmium and zinc) to field run-off water (Dudka and Chlopecka 1990). An increased movement of these elements was observed in light soil and soil with lower pH values (Korte et al., 1976; Tackett et al., 1986).

Brümmer et al. (1986) found that the application of lime to fields contaminated with heavy metals is the most commonly proposed procedure to reduce the mobility and bioabsorption of heavy metals in soil. The effect of this procedure is short-term and does not lead to permanent decontamination of the soil. Permanent decontamination can be accomplished by biological methods such as heavy-metal harvesting (Kuntze et al., 1984), physical and chemical methods (Körbitzer and Müller 1990; Weilandt 1990) and also through leaching heavy metals beyond the root system (Bieniek and Fischer 1991). However, the biological methods are time-intensive, while the physical and chemical methods are expensive.

The most effective way to remove heavy metals from the upper soil layer may be to speed up their drainage to deeper soil layers, beyond the reach of the root system, if the drainage water is not to used for drinking. The principal purpose of this study was to specify the intensity of leaching of zinc, cadmium, and lead from soil in a laboratory, and movement of these elements down through soil layers under field conditions.

2 Materials and methods

A laboratory experiment was carried out at the Institute of Agricultural Chemistry of the University of Bonn. Experiments were conducted in cylindrical glass tubes, each with a diameter of 2 cm, height of 67 cm, and capacity of 0.37 kg of soil. Soil for the columns was collected from the ploughing layer (0-25 cm) of the field experiment conducted at the Experimental Field of Warsaw Agricultural University (WAU). In that field, high dosages of post-factory waste lime (8.2 t ha^{-1} lime), which contained high amounts of heavy metals, was applied in Autumn 1985 and 1988. Total input of heavy metals to the soil were: 279 kg Zn, 40 kg Pb and 0.75 kg Cd ha^{-1}. The following additional elements were added to the soil one year before using the soil samples for the laboratory experiment, in the specified amounts; ZnO, Pb(CH$_3$COO)$_2$ and Cd(NO$_3$)$_2$. The soil was contaminated with these metals and contained 310 ppm Zn, 93 ppm Pb, and 3.2 ppm Cd.

To the columns with this soil, three kinds of calcium compounds, such as CaCl$_2$, CaSO$_4$ or Ca(OH)$_2$ were added. The calcium compounds were introduced in three applications in quantities equivalent to 1.0, 1.5, 3.0 times that of the soil exchange capacity, measured by the Mehlich method, excluding Ca in CEC. Soil exchange capacity according to the Mehlich method amounted to 213 mmolJE 1000 g^{-1}. Leaching of heavy metals was conducted under hermetic conditions. For rinsing soil in the columns, redistilled water was used in an amount corresponding to 600 mm of rainfall. The water was introduced into the soil columns in three consecutive applications equivalent to 200 mm rainfall. The applications were spaced at two-month intervals.

The field experiment was conducted at the WAU Experimental Field. Waste lime was applied from the zinc factory which contained 15% Ca, 8% Mg and high amounts of zinc (average 17000 ppm), lead (average 2800 ppm) and cadmium (average 46 ppm). Lime was applied to the field three times in 1985, 1988, and 1991, or twice in 1985 and 1988 in amounts of 4.1 (1Ca) or 8.2 (2Ca) t ha^{-1} lime. Total input of heavy metals were 70 or 140 kg Zn, 12 or 24 kg Pb and 0.19 or 0.38 kg ha^{-1} year^{-1} Cd, with 4.1 or 8.2 t ha^{-1} waste lime respectively.

Limes were applied in the following 5 treatments:
1) 3x1Ca = 4.1 t ha^{-1} lime without heavy metals, in 1985, 1988, 1991
2) 3x1Cah = 4.1 t ha^{-1} lime with heavy metals (h), in 1985, 1988, 1991
3) 2x1Cah = 4.1 t ha^{-1} lime with heavy metals (h), in 1985, 1988
4) 3x2Cah = 8.2 t ha^{-1} lime with heavy metals (h), in 1985, 1988, 1991
5) 2x2Cah = 8.2 t ha^{-1} lime with heavy metals (h), in 1985, 1988

The experiment was carried out on soil containing the following amounts of silt and clay (particles < 0.02 mm) in particular soil layers: 15-17% in Ah (0-25 cm), 10-12% in Al (25-45 cm), and 25% in Bt/C and the C layer (below 45 cm). The humus content was 1.3%. In the third year after each application of lime (1988, 1991, and 1994), the amounts of Zn, Pb, and Cd in the above-noted soil layers were determined by the Rinkis method (1M HCl soil extract). On this basis, conclusions were drawn on the intensity of heavy metal leaching through the soil profile. Measurement of the amount of heavy metals in soil extracts from laboratory and field experiments were carried out using the ASA method.

Heavy metal	Portion Of Leachates[a]	0	CaCl$_2$			CaSO$_4$			Ca(OH)$_2$		
			1	1.5	3	1	1.5	3	1	1.5	3
Zn	I	207	5380	8670	100000	115	373	1020	87	95	174
	II	308	2309	4856	41720	265	564	701	368	56	115
	III	376	154	556	2736	245	600	975	214	200	227
Cd	I	1	2648	3695	6365	4	4	3	4	2	1
	II	2	1664	1393	4480	9	5	2	2	2	2
	III	1	138	179	134	1	1	3	1	1	2
Pb	I	44	153	163	873	53	40	56	60	59	41
	II	19	23	17	60	14	12	12	16	28	20
	III	18	10	10	27	5	6	9	11	9	14

Doses and forms of Ca calculated on the base 1, 1.5 and 3 CEC pH 8.1

[a]Portion of leachates = equivalent to 200 mm precipitation.

Table 1: Content of heavy metals in leachates ($\mu g \cdot dm^{-3}$)

3 Results and discussion

The amount of heavy metals in the leachate in the laboratory experiment varied within a broad range: 56-10,000 µg Zn, 1-6 365 µg Cd and 5-873 µg·dm^{-3} Pb (Table 1). The intensity of leaching of Zn, Cd, and Pb depended on the type of calcium compound applied to the soil, the amount of calcium, as well as the duration of water percolation in the soil column. The highest degree of leaching of Zn, Cd, and Pb occurred after applying CaCl$_2$. CaSO$_4$ and Ca(OH)$_2$ usually did not increase leaching of these metals in comparison with the control column lacking calcium. Only through the application of medium (1.5) and high (3.0) doses of CaSO$_4$ was the leaching of zinc higher than in soil without calcium. Leaching of Zn, Cd and Pb increased with dosages of CaCl$_2$ and, in the case of Zn, also with higher dosages of CaSO$_4$. As indicated by Herms and Brümmer (1984), an increased concentration of easily diluting salts in soil solution is accompanied by strong desorption of heavy metals in the soil. Presumably, this might explain an increased leaching of Zn and Cd which was noted in our experiment in relation to the application of CaCl$_2$. Bingham et al. (1984) note that a higher amount of Cl$^-$ ions in the soil solution is accompanied by an increased mobility of Cd in the soil. The cause of this phenomena is the creation of easily-dissolving chlorine bonds with cadmium such as CdCl$^+$, CdCl$_2$, CdCl$_3^-$, CdCl$_4^{2-}$.

The greatest leaching of all heavy metals in combination with CaCl$_2$ occurred with the first portion of 200 mm rainfall (Table 1). In the case of lead, also after the application of CaSO$_4$ and Ca(OH)$_2$, the greatest amount of Pb was in the first leachates. It is worth highlighting, however, that the zinc was leached most in the third rain portion from soils to which small amounts of CaSO$_4$ and Ca(OH)$_2$ had been applied.

The lowest proportion of leached heavy metals in the soil was lead (0.14%) after application of CaCl$_2$ in relation to their total amount in the soil (relative leaching). A significantly higher relative leaching in the soil was shown by zinc (6.5%). Cadmium had the highest mobility; leaching of this element was 46.5% (Fig. 1). Herms and Brümmer (1984) noticed a similar relation in mobility of heavy metals in soil: Cd>Zn>Pb.

In the field experiments, a marked increase was noted in the amount of zinc in the soil (dissolved in 1M HCl) in association with the amount of applied waste lime (Fig. 2). This increase is less noticeable in deeper layers of soil. In the layer below 45 cm the differences under all combinations are minimal. Czarnowska and Gworek (1991) note that in Poland, the "natural" amount of zinc in such a soil is typically stated to be up to 30 ppm, "elevated" zinc in soil is stated between 30 - 100 ppm, and "contaminated" soil exists when the amount of zinc in the soil exceeds 100 ppm. In our experiments,

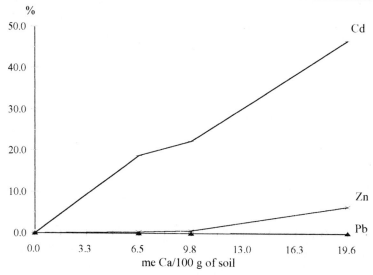

Fig. 1: Relative leaching of heavy metals, as a % of their total content in soil, using $CaCl_2$.

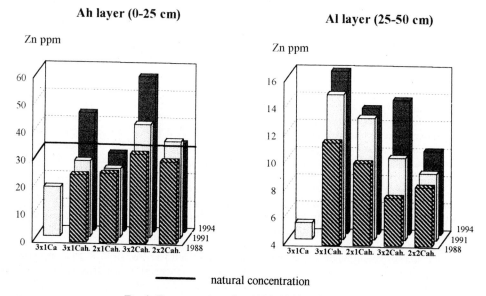

Fig. 2: Zn content in soil in 1988, 1991 and 1994.

an elevated amount of zinc occurred with three times application of waste lime in smaller amounts (4.1 t ha^{-1}) and with three and two times of lime in amounts of 8.2 t ha^{-1} were applied. In 1994 less zinc in the soil was noted in combinations when waste lime was twice applied than with three applications. It is worth emphasizing that even with a triple application of waste lime, the soil remained in the "elevated" zinc amount class, and not in the "contaminated" range.

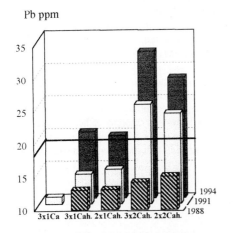
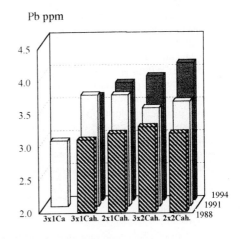

Fig. 3: Pb content in soil in 1988, 1991 and 1994.

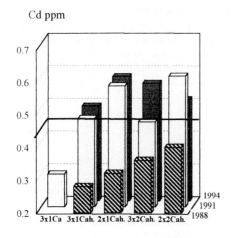
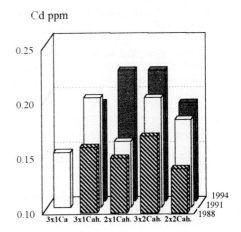

Fig. 4: Cd content in soil in 1988, 1991 and 1994.

A noticeable increase of lead amounts in the soil after applying waste lime was observed only in the 0 - 25 cm soil layer, and to a minor degree in the lower soil layer (Fig. 3). Only through a triple application of low-dosage waste lime and with a double and triple application of higher dosages could the soil be considered as containing an elevated amount of lead. A concentration of 17 - 43 ppm Pb is considered to constitute an elevated amount of lead in the soil (Czarnowska and Gworek 1991).

During the 4 years after the last waste lime application (1991), and even 7 years after following the last application, the amount of lead in the soil did not decrease.

A marked increase in the amount of cadmium in the soil after the application waste lime was observed in the 0-25 cm layer and a lesser increase was obtained in the 25-45 cm layer (Fig. 4). However, double and triple applications of waste lime increased the amount of cadmium beyond the amount considered as natural only in the 0-25 cm layer. For this type of soil, the natural level is up to 0.45 ppm Cd (Czarnowska and Gworek 1991).

The field experiment results show that even a triple application of high dosages of waste lime containing several times more zinc, lead and cadmium than the norm did not elevate the amounts of heavy metals soluble in 1M HCl for the soil to be considered as contaminated. As indicated in guidelines (Kabata-Pendias et al., 1993), soil containing "elevated" amounts of heavy metals can be utilised to grow all field production crops, with the exception of specified vegetables. The increased amounts of the studied heavy metals in deeper soil layers shows that the speed of movement of these metals occurs in the following order: Zn>Cd>Pb.

4 Conclusions

Even a triple application of high doses of waste lime containing several times more Zn, Pb, and Cd than foreseen by allowable norms did not increase the amount of the heavy metals in the soil such that, the soil could be considered as contaminated. Lime application to the soil effectively limited the dissolution of the heavy metals in the soil and the elements almost exclusively remained in the 0 - 25 cm soil layer. Among the examined heavy metals, the highest intensity of leaching to the infiltrating water was shown by Cd, followed by zinc, with the lowest noted in the case of lead in the laboratory experiment. The field experiments showed the following leaching sequence: zinc> cadmium> lead. Among the calcium compounds applied to the soil ($CaCl_2$, $CaSO_4$ and $Ca(OH)_2$), calcium chloride most effectively speeded up leaching of the heavy metals from the soil. Through the use of calcium chloride in the soil, 46.5% of cadmium was leached from the soil, noticeable less zinc (6.5%), and very little lead (0.14%) in comparison with the total content of these elements in the soil. It indicates, that calcium chloride may be successful use to removing of cadmium and zinc from the soil, and the same to it permanent decontamination.

References

Bieniek D., Fischer K. (1991): Dekontamination schwermetallbelasteter Böden mit organischen Komplexbildern. Erkundung und Sanierung von Altlasten. Berichte vom 7. Bochumer Altlasten-Seminar. Brookfield, Rotterdam, 37-44.

Bingham F.T., Sposito G., Strong J.E. (1984: The effect of chloride on the availability of cadmium. J Environ Qual. **13**: 71-74.

Brümmer G.W., Gerth J., Herms U. (1986): Heavy metal species, mobility and availability in soils. Z Pflanzenernähr Bodenkd. **149**: 382-389.

Czarnowska K., Gworek B. (1991): Syntetyczne przedstawienie tla geochemicznego pierwiastków sladowych w glebach uprawnych wytworzonych ze skal osadowych. Materialy Sympozjum "Mikroelementy w rolnictwie". Wroclaw, 37-41.

Dudka S., Chlopccka A. (1990): Effect of solid-phase speciation on metal mobility and phytoavailability in sludge-amended soil. Water, Air, and Soil Pollution **51**: 153-160.

Herms U., Brümmer G. (1984): Einflussgrössen der Schwermetalllöslichkeit und -bindung in Böden. Z Pflanzenernähr Bodenkd. **147 (3)**: 400-424.

Kabata-Pendias A., Motowicka-Terelak T., Piotrowska M., Terelak H., Witek T. (1993): Ocena stopnia zanieczyszczenia gleb i roslin metalami ciezkimi i siarka. IUNG, Pulawy.

Korte N.E., Skopp J., Fuller W.H., Niebla E.E., Alesii B.A. (1976): Trace element movement in soils: Influence of soil physical and chemical properties. Soil Science **122**: 350-359.

Körbitzer B., Müller K.J. (1990): Sanierung Hg-kontaminierter Böden durch chemische Laugung und elekrolytische Hg-Abscheidung. Vortrag bei der DECHEMA-Jahrestagung am 01.06.1990.

Kuntze H., Pluquet E., Stark J.H., Coppola S. (1984): Current techniques for the evaluation of metal problems due to sludge. Processing and use of sewage sludge. Proceedings of an Symposium held in Brighton 27-30.09.1983, L'Hermite P. & H. Ott (eds.), Dordrecht, 396-403.

Sauerbeck D., Styperek P. (1988): Schwermetallakkumulation durch Klärschlammanwendung - Ergebnisse von 25 langjährigen Feldversuchen. VDLUFA-Schriftenreihe. **23**: 489-506.

Tackett S.L., Winters E.R., Puz M.J. (1986): Leaching of heavy metals from composted sewagesludge as a function of pH. Can J Soil Sci. **66 (4):** 763-765.

Weilandt E. (1990): Thermische Behandlung von schwermetallkontaminierten Böden im Drehrohrofen. IWS-Schriftenreihe **10**: 249-263.

Addresses of authors:
Marek Gebski
S. Mercik
Department of Agricultural Chemistry
Warsaw Agricultural University-SGGW
Rakowiecka 26/30
02-528 Warszawa, Poland
Karl Sommer
Agriculturchemisches Institut der Rheinischen
Friedrich-Wilhelms-Universität Bonn
Meckenheimer Allee 176
D-53115 Bonn, Germany

Heavy Metals in Soils of Costa Rican Coffee Plantations

S. Kretzschmar, M. Bundt, G. Saborió, W. Wilcke & W. Zech

Summary
The use of fungicides, fertilizers, and atmospheric depositions result in heavy metal inputs into the soils of Costa Rican coffee plantations, which may endanger plant growth and ground water quality. In this study we examined the Al and heavy metal status of soils in Costa Rican coffee plantations. Samples were collected from A horizons of 16 soils in coffee plantations and of 2 soils in natural forest used as reference (mainly Oxisols and Andisols). Aggregates were selected manually from the A horizons and mechanically separated into core and surface fractions. We used a seven step sequential extraction procedure to analyze Al, Cd, Cu, Fe, Mn, Pb, and Zn in bulk soil and aggregate fractions. Copper concentrations were found to be high (generally between 100 to 330 mg kg^{-1}). The Al and Fe concentrations (45-134 and 21-115 g kg^{-1}, respectively) were higher than those in soils of temperate regions. The concentrations of the other metals were comparable to background concentrations in soils of temperate climates (Cd: 0.03-0.33, Mn: 156-4965, Pb: 4-57, Zn: 31-141 mg kg^{-1}). Metal partitioning in the seven fractions was similar to that of soils in temperate regions, however, the studied soils, especially the Oxisols, contained lower percentages of plant available metal concentrations, because of strong leaching. A principal component analysis revealed three metal groups: the mainly lithogenic Al, Cu, and Fe, the anthropogenically affected Cd, Mn, and Zn, and the predominantely atmospheric deposited Pb. This interpretation was supported by the depth distribution. The concentrations of lithogenic metals increased as depth increased, those of anthropogenic metals decreased as depth increased. Additional support was given by the small-scale distribution on aggregate level. The Pb from atmospheric deposition was enriched at aggregate surfaces compared with aggregate cores.

Keywords: Al, heavy metals, Costa Rica, coffee plantations, anthropogenic influence.

1 Introduction

Heavy metal inputs into soils of Costa Rican coffee plantations are associated with fungicides (Cu, Mn, Fassbender and Bornemisza, 1987), P fertilizers (Cd, Zn, Boysen, 1992) and probably atmospheric deposition (Pb). As heavy metals accumulate in soil (Brümmer et al., 1986) this may have impacts on coffee yield and quality.

Whereas there are many studies on the heavy metal status of soils in temperate regions (e.g. Shuman, 1985; Wilcke and Döhler, 1995; Zeien, 1995), such studies are scarce for tropical soils.

The objective of this study was to examine the heavy metal status of soils in Costa Rican coffee plantations.

2 Materials and methods

2.1 Study sites and soils

We sampled A, B, and C horizons of 16 sites under coffee and 2 sites under natural forest as reference (mainly Andisols and Oxisols[1]) in the main coffee cultivation areas of Costa Rica (Fig. 1). The parent material in most cases are andesitic pyroclastica. Table 1 summarizes the range of physical and chemical properties of the A horizons.

Fig. 1: Location of the study sites (Anonymous, 1989, modified).

2.2 Aggregate fractionation

Aggregates of approximately 2 to 20 mm in diameter were selected manually from all samples and mechanically separated into a surface and a core fraction with the method of Kayser et al. (1994).

2.3 Extraction and Analysis

Aluminium and heavy metals were extracted with a seven step sequential extraction procedure (Zeien and Brümmer, 1989). This procedure comprises the following fractions: (1) readily soluble and exchangeable, (2) specifically adsorbed and other weakly bound species, (3) bound to Mn oxides (4), bound to organic substance (5), bound to amorphous Fe oxides (6), bound to crystalline Fe oxides, and (7) residuum (mainly bound within silicates). Aluminium and Fe were included in the study as reference metals whose concentrations are only little affected by atmospheric deposition.

---

[1] Oxisols were classified mainly on the base of the low CEC without counting the number of weatherable minerals. As the studied soils received several fresh volcanic ash inputs it is possible that the number of weatherable minerals is too high to classify an Oxisol. We nevertheless keep our classification to underline the dominance of kaolinite and gibbsite resulting in a very low CEC.

Using common standard procedures, the following bulk soil properties were determined: cation exchange capacity (CEC) with 1 M NH$_4$ acetate, pH 7 (Page et al., 1982), base saturation (BS) as the percentage of Ca+Mg+K+Na of the CEC$_{eff}$, total carbon with the C/N-Analyzer (Elementar vario EL). Poorly crystalline Fe oxides were extracted with the oxalate buffer method (= Fe$_o$, Schwertmann, 1964), crystalline Fe oxides with the dithionite citrate buffer (= Fe$_d$, Mehra and Jackson, 1960). Aluminium, Ca, Cd, Cu, Fe, K, Mg, Mn, Na, Pb, and Zn were analyzed by atomic absorption spectrometry using Varian SpectrAA 400 and 400 Z devices.

Clay [g kg^{-1}]	pH(KCl)	CEC [mmol$_c$ kg^{-1}]	BS [%]	C$_{org}$ ----------[g kg^{-1}]----------	Al$_o$	Fe$_o$/Fe$_d$
210-880	3.7-6.6	61-203	19-99	22.2-102.9	2.4-42.5	0.03-0.53

Tab. 1: Range of physical and chemical properties of the A horizons.

2.4 Statistical analysis

Statistical analysis was performed with the software package STATISTICA for Windows 5.1 (Loll&Nielsen, Hamburg, Germany). All data sets were tested for normality with the Kolmogorov-Smirnov test. Principal component analysis was conducted using factor extraction with an eigenvalue > 1 after varimax rotation. The data set was log transformed prior to principal component and cluster analysis, to reduce the dominance of high values.

3 Results and Discussion

3.1 Total metal concentrations

The concentrations of Al, Cu, and Fe are higher than in soils of temperate regions, those of Cd, Mn, Pb, and Zn are comparable (Wilcke and Döhler, 1995, Tab. 1). The Cu concentrations in most of the soils are extraordinarily high. They exceed, in almost all studied soils, the threshold value of the German AbfKlärV (Anonymous, 1992) of 60 mg kg^{-1}.

Element	Range of concentrations Costa Rica [mg kg^{-1}]	"Normal" concentrations in German soils (Scheffer/Schachtschabel, 1989; Wilcke and Döhler, 1995) [mg kg^{-1}]
Al	45,000-134,000	average earth crust: 70,000
Cd	0.03-0.33	0.05-1
Cu	30-330	1-40
Fe	21,000-115,000	2,000-50,000 average earth crust: 42,000
Mn	156-4,965	20-800
Pb	4-57	0.1-60
Zn	31-141	3-100

Tab. 2: Range of metal concentrations.

3.2 Metal fractions

Generally, the studied soils contain higher percentages of strongly bound metals (associated with Fe oxides and silicates, fractions 5, 6, and 7 of the Zeien and Brümmer (1989) procedure) than soils of temperate regions (Fig. 2). This may be explained by the more intense leaching.

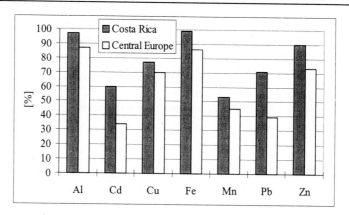

Fig. 2: Percentages of strongly bound metals (associated with Fe oxides and silicates) in Costa Rica and Central Europe (Wilcke, 1996).

3.3 Anthropogenic metal input

Three metal groups may be distinguished based on a principal component analysis (Fig. 3). Copper, Al, and Fe are mainly lithogenic, Cd, Mn, Zn, and especially Pb are affected by anthropogenic inputs.

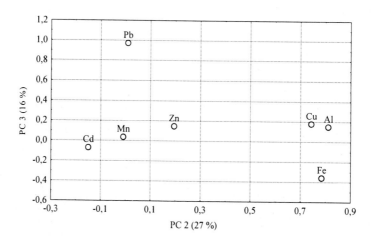

Fig. 3: Illustration of 18 soil samples from Costa Rican coffee cultivation areas in the coordinate system of two principal components (PC) after varimax rotation according to the heavy metal pattern.

In Fig. 3 we chose principal components 2 and 3 as this graph illustrates best the differentiation into three groups.

The interpretation of the principal component analysis is supported by the depth distribution of the metal concentrations and by the small-scale distribution of the metals on aggregate level. The average concentrations of mainly lithogenic metals increase as depth increases (Al, Cu, and Fe), those of mainly anthropogenic metals decrease as depth increases (Cd, Pb, Fig. 4).

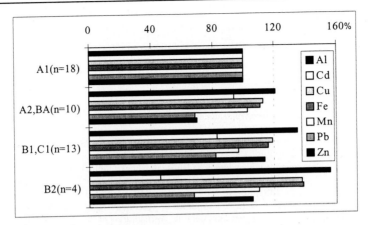

Fig. 4: Average depth distribution of Al and heavy metals. (The concentrations of the subsoil horizons are given in % of the uppermost A horizon, for Pb two sites which had low mainly lithogenic concentrations substantially increasing with depth were eliminated).

Generally, preferential leaching mainly of easily mobilizable metals from aggregate surfaces results in lower metal concentrations of the aggregate surface fraction than of the aggregate core fraction. Higher metal concentrations in the surface than in the core fraction of soil aggregates indicate metal input (Wilcke, 1996). Thus, the small scale distribution of Pb confirms Pb inputs into most of the studied soils (Fig.5).

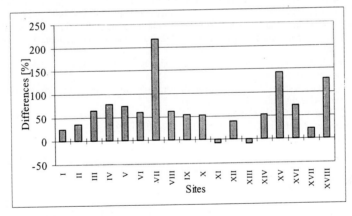

Fig. 5: Differences in Pb concentrations between aggregate surface and core fractions in % of the core's concentration (negative depletion/positive enrichment at the surface).

4 Conclusions

Soils of Costa Rican coffee plantations are affected by anthropogenic Cd, Cu, Mn, Pb, and Zn input. This results in heavy metal accumulation especially of Pb in the topsoils. As a consequence of the high lithogenic Cu concentrations the Cu input via fungicides does not result in a marked change of the depth distribution. Thus, Cu shows the same distribution pattern as the mainly lithogenic Al and Fe.

Acknowledgements

We are indebted to the DFG for funding this project (Ze 154/28-1).

References

Anonymous (1989): Manual de recomendaciones para el cultivo del café. Programa Cooperativo ICAFE-MAG, 140 p.

Anonymous (1992): Klärschlammverordnung (AbfKlärV) vom 15. April 1992. Bundesgesetzblatt Z 5702 A, Teil I, 912-927.

Boysen, P. (1992): Schwermetalle und andere Schadstoffe in Düngemitteln. UBA Fachberichte 92, 54 S.

Brümmer, G.W., Gerth, J. and Herms, U. (1986): Heavy metal species, mobility and availability in soils. Z. Pflanzenernähr. Bodenkd. **149**, 382-398.

Faßbender, H.W. and Bornemisza, E. (1987): Química de suelos con énfasis en suelos de América Latina, Instituto Interamericano de Cooperación para la Agricultura (IICA), Serie de libros y materiales educativos **24**, San José, Costa Rica, 398 p.

Kayser, A., Wilcke, W., Kaupenjohann, M. and Joslin, J.D. (1994): Small-scale heterogeneity of soil chemical properties. I. A rapid technique for aggregate fractionation. Z. Pflanzenernähr. Bodenkd. **157**, 453-458.

Mehra, O.P. and Jackson, M.L. (1960): Iron oxide removal from soils and clays by a dithionite-citrate system buffered with sodium bicarbonate. Clays Clay Min. **7**, 317-327.

Page, A.L., Miller, R.H. and Keeney, D.R. (1982): Methods of soil analysis. Part 2., 2nd ed., Agron. Monogr. **9**. ASA and SSSA, Madison, WI, USA, 1159 p.

Scheffer/Schachtschabel (1989): Lehrbuch der Bodenkunde, 12th ed., Enke, Stuttgart, Germany, 491 p.

Schwertmann, U. (1964): Differenzierung der Eisenoxide des Bodens durch Extraktion mit Ammoniumoxalat-Lösung. Z. Pflanzenernähr. Düng. Bodenkd. **105**, 194-202.

Shuman, L.M. (1985): Fractionation method for soil microelements. Soil Sci. **140**, 11-22.

Wilcke, W. (1996): Kleinräumige chemische Heterogenität in Böden: Verteilung von Aluminium, Schwermetallen und polyzyklischen aromatischen Kohlenwasserstoffen in Aggregaten. Bayreuther Bodenkundl. Ber. **48**, 136 p.

Wilcke, W. and Döhler, H. (1995): Schwermetalle in der Landwirtschaft - Quellen, Flüsse, Verbleib - KTBL-Arbeitspapier 217, 98 p.

Zeien, H. (1995): Chemische Extraktionen zur Bestimmung der Bindungsformen von Schwermetallen in Böden. Bonner Bodenkundl. Abhandl. **17**, 284 p.

Zeien, H. and Brümmer, G. (1989): Chemische Extraktion zur Bestimmung von Schwermetallbindungsformen in Böden. Mitteilgn. Dtsch. Bodenkundl. Gesellsch. **59**, 505-510.

Addresses of authors:
Sigrid Kretzschmar
Maya Bundt
Guillermo Saborió
Wolfgang Wilcke
Wolfgang Zech
Institute of Soil Science and Soil Geography
University of Bayreuth
D-95440 Bayreuth, Germany

Action Values for Mobile (NH$_4$NO$_3$-Extractable) Trace Elements in Soils Based on the German National Standard DIN 19730

A. Prüeß

Summary
This brief report discusses the derivation of background values (German: Hintergrundwerte or Referenzwerte), action values (Prüfwerte) and threshold values (Grenzwerte or Belastungswerte or Maßnahmenwerte) for mobile trace elements in soils. The method chosen to determine mobile amounts of trace elements in soils is an ammonium nitrate extraction procedure: 20 g soil to 50 ml of 1 molar ammonium nitrate solution, shaken for 2 hours. This procedure became the German national standard DIN 19730 in 1997. The background, action and threshold values have been integrated into regulations for soil protection in Baden-Württemberg, which was the first such regulation in Germany. In combination with the determination of the so-called total contents of elements in soils, the DIN 19730 can be used as a basis for risk assessment, for limiting agricultural use and as a sensitive instrument to detect soils with high amounts of potentially leachable elements.

Keywords: Trace elements, mobile, ammonium nitrate extraction, standard

1 Introduction

Mobile trace elements in soils can adversely affect the quality of food and fodder plants, the growth of plants, the activity of soil microorganisms and the quality of the groundwater. The objective of this work was to assess background values above which the mobility of elements in soils should be monitored and to derive action values above which plants respectively soil solution should be watched and threshold values above which agricultural use of soils should be limited (Figure 1).

2 Materials and methods

The assessment of background values and the derivation of action and threshold values are based on trace element analyses of 400 paired soil/plant samples and 300 soil samples from field surveys in SW-Germany (Figure 2). The analyses were done by electrothermal atomic absorption and mass spectrometer with inductive coppled plasma (ICP-MS).

After a review of relevant literature, an ammonium nitrate solution was chosen for the estimation of amounts of mobile trace elements in soils (e.g. Symeonides & McRae 1977, Zeien & Brümmer 1988, Sauerbeck & Lübben 1991). The following extraction parameters proved to be the most "effective" for routine analysis of mineral soils (Prüeß et al. 1991) and was finally accepted for the DIN 19730 as the standard method: Place 20 g air dry soil in a shaking bottle (100-150 ml), add 50 ml ammonium nitrate solution (1 mol/l) and shake for 2 hours at 20 revolutions per min. at room temperature. Then allow the solid particles to settle for 15 min. Decant the supernatant solution and filter (0.45 µm). Dispose the first 5 ml of the filtrate. Collect the remaining solution in a 50 ml bottle.

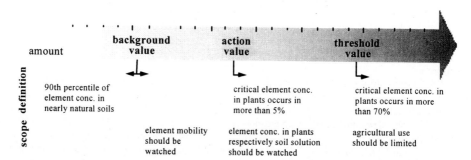

Figure 1: Scope and definition of background, action and threshold values

The DIN 19730 (1997) also gives information on the analytical procedures (e.g. dilution factors) and lists minimum concentrations which should be quantified accurately for good results in the field of soil protection (Table 1). These concentrations represent typical concentrations in pH-neutral to alkaline soils (see section 3.1 below).

		1 mol/l ammonium nitrate solution [µg/l]
Ag	Silver	< 0.4
As	Arsenic	10
Be	Beryllium	1
Bi	Bismuth	< 0.4
Cd	Cadmium	2
Co	Cobalt	20
Cr	Chromium	4
Cu	Copper	100
Hg	Mercury	< 0.4
Mn	Manganese	2000
Mo	Molybdenum	10
Ni	Nickel	100
Pb	Lead	< 8
Sb	Antimony	10
Tl	Thallium	4
U	Uranium	1
V	Vanadium	10
Zn	Zinc	100

Table 1: Minimum concentrations which have to be quantified accurately with the ammonium nitrate extraction method (DIN 19730) for good results in the field of soil protection

Mobile (NH₄NO₃-extractalbe) trace elements in soils

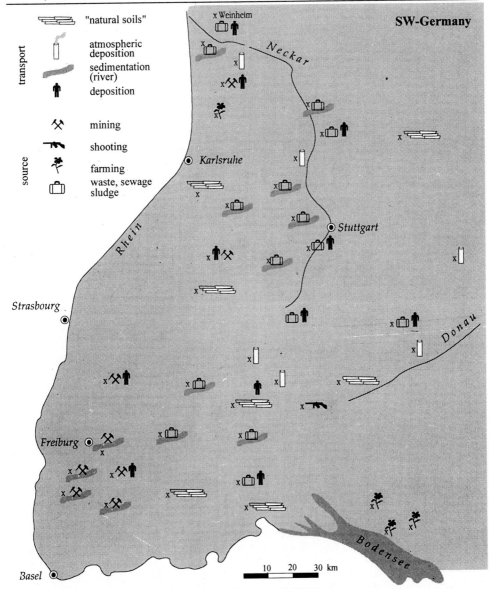

Figure 2: Sampling locations in SW-Germany

3 Results
3.1 Background values

The soil samples were classified according to their pH-values. Figure 3 shows examples of frequency distributions of ammonium nitrate extractable trace element amounts in nearly natural soils. The 90th percentiles of the respective groups were defined as background values (Table 2).

The background values for mobile elements demonstrate the important role of pH on the mobility of trace elements in soils. However these background values should only be used in soil protection in association with background values for the total element concentrations.

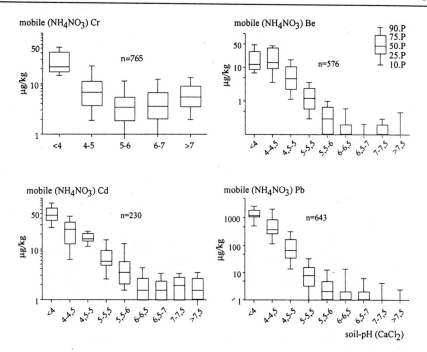

Figure 3: NH$_4$NO$_3$-extractable amount of trace elements in "natural soils" (P = percentile)

		[μg/kg] ammonium nitrate extractable amount (DIN 19730) soils were classified according to the soil-pH (CaCl$_2$)								
		< 4.0	4.0-4.5	4.5-5.0	5.0-5.5	5.5-6.0	6.0-6.5	6.5-7.0	7.0-7.5	> 7.5
Ag	Silver	1.5	1.5	1.5	1.5	1.5	1.5	1.5	1.5	1.5
As	Arsenic	60	50	40	40	40	40	40	45	50
Be	Beryllium	60	40	20	5	1	0.6	0.4	0.4	0.4
Bi	Bismuth	1	1	1	1	1	1	1	1	3
Cd	Cadmium	80	50	20	15	10	5	3	3	3
Co	Cobalt	500	500	200	70	30	25	20	20	20
Cr	Chromium	50	40	15	12	10	10	12	15	15
Cu	Copper	300	280	250	250	250	250	300	350	400
Hg	Mercury	1	1	1	1	1	1	1	1	1
Mn	Manganese	30 000	28 000	25 000	20 000	15 000	10 000	5 000	4 000	3 000
Mo	Molybdenum	10	10	10	25	30	50	60	70	110
Ni	Nickel	1 000	1 000	600	300	250	200	200	200	200
Pb	Lead	3 000	2 000	150	30	15	10	6	4	3
Sb	Antimony	5	5	5	5	7	10	20	30	40
Tl	Thallium	50	30	20	15	12	10	12	15	15
U	Uranium	5	4	3	3	3	3	3	4	5
V	Vanadium	40	30	20	15	15	15	15	20	30
Zn	Zinc	5 000	4 000	3 000	1 000	300	200	170	130	100

Table 2: Background values (90th percentile) for mobile trace elements in soils

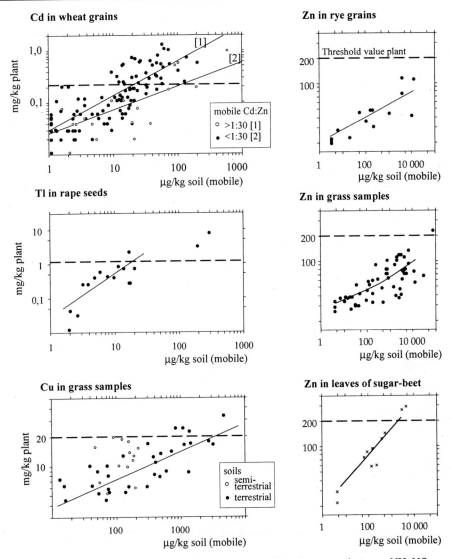

Figure 4: Trace element concentrations in food or fodder plants in relation to NH4NO3-extractable amounts in the respective topsoil samples

3.2 Action values

Figure 4 shows examples of relationships of trace element concentrations in edible plants with ammonium nitrate extractable amounts of these elements in the corresponding topsoil samples. Provided that the plant species in question are able to take up the respective element via the roots and to transport it into its edible plant parts in significant amounts (such as Cd in wheat grain, Tl in rape seed, Zn in sugar beet leaf), the correlations were close whenever the range of element concentrations in the soil samples collective was large.

The action values were derived either from regression or contingency analyses (e.g. Cd in wheat grain) or from field observation of growth inhibition (e.g. As, Ni). The action values were set to meet the following condition: if the amount of mobile elements in the soil is below the action value, critical element concentrations in plants (as taken from German food and fodder threshold values; Bundesgesundheitsamt 1990, Anonymous 1988, Sauerbeck 1982 and other literature) were not exceeded in 95% of the examined cases.

In this way, action values for soils related to plant quality could be calibrated for As, Cd, Cu, Pb, Tl, Zn and those related to plant growth for As, Cu and Zn (Table 3, see details in Prüeß 1992). For Cr, Ni and Co the action values were assessed from observations at only one site with critical contents in the plants and large extractable concentrations in the soils. For Ag, Be, Bi, Mo, Sb, U and V no critical contents in plants were found, even on soils with large extractable concentrations of these elements.

Action values for the activity of soil microorganisms were derived from information in the literature. Action values for water quality in topsoils were set identical with the background values for soils with pH < 4.0 (mostly forest soils) for Be, Cd, Co, Ni, Pb and Zn.

Element		plant food	plant fodder	plant growth	soil micro-organisms	waterquality in topsoil	waterquality in subsoil
		[µg/kg] ammonium nitrate extractable amount (DIN 19730)					
Ag	Silver	•	•	•	5	5	1.5
As	Arsenic	100	100	600	•	100	50
Be	Beryllium	•	•	•	•	60	15
Bi	Bismuth	•	•	•	•	•	•
Cd	Cadmium	20	20	•	•	80	25
Co	Cobalt	•	150	500	•	500	150
Cr	Chromium	•	•	50	100	100	15
Cu	Copper	•	800	2000	1000	1000	400
Hg	Mercury	•	•	•	5	5	1
Mn	Manganese	•	•	30000	•	•	•
Mo	Molybdenun	•	500	•	•	•	•
Ni	Nickel	•	•	1000	•	1000	600
Pb	Lead	50	50	•	•	3000	200
Sb	Antimony	•	•	•	•	1000	40
Tl	Thallium	30	30	•	•	•	30
U	Uranium	•	•	•	•	25	5
V	Vanadium	•	•	•	100	100	30
Zn	Zinc	•	5000	10000	•	5000	1500

•: no adverse effects expected under field conditions below or in the range of the mentioned action values

Table 3: Action values for mobile trace elements in soils

3.3 Threshold values for soils

The treshold values for soils were set as low as possible and to meet the following condition: if the amount of a mobile element in the soil is above the respective threshold value, critical element concentrations in plants (above the German threshold values for food and fodder plants) would be expected to be exceeded in more than 70% of the examined cases.

As an example, threshold values for the amount of mobile cadmium in soils with regard to limiting cultivation of food and fodder plants are given in Table 4. The respective food and fodder plants were grouped according to their ability to accumulate cadmium.

element		[μg/kg]	food plants	fodder plants
Cd	Cadmium	40	endive, broccoli, oats, carrot, lettuce, cress, leek, chard, parsley, radish, chives, black salsify, celeriac, sun flower, spinach, wheat	leaves of sugar beet, sunflower
		100	berries, cauliflower, potato, kohlrabi, radish, red cabbage, tomato, savoy, onion	grass, wheat straw

Table 4: Threshold values for mobile cadmium in soils for limiting agricultural use

4 Conclusions and comments

The ammonium nitrate extraction method is a standard method for the estimation of mobile trace elements in soils (e. g. Liebe et al. 1995). To develop a model for predicting bioavailability of trace elements in soils, the ammonium nitrate extractable amount can be used as the basic input parameter. To further develop this model other factors controlling plant availability (e. g. antagonism of trace elements) and leachability must be quantified.

The ammonium nitrate extractable amount is used along with the so-called total element concentration as supplementary information for risk assessment (BADEN-WÜRTTEMBERG 1993). The method is a useful instrument for limiting agricultural use and is a sensitive instrument to detect soils with large amounts of potentially leachable elements. But the ammonium nitrate extractable amount should not be misinterpretated as being equivalent to the plant available or leachable amount.

The ammonium nitrate extraction method poses high demands on analytical equipment. If laboratories can meet this requirement, the method is suitable for routine analyses and is applicable for a wide range of soils and elements.

Acknowledgements

Credit is due to many people who were involved with this project, especially V. Schweikle, G. Turian and T. Nöltner, and to the Ministry of Environment in Baden-Württemberg for founding the research.

References

Anonymous (1988): Das geltende Futtermittelgesetz vom 29.06.1988. Bundesgesetzblatt, 1/28: 869.
Baden-Württemberg (1993): Dritte Verwaltungsvorschrift des Umweltministeriums zum Bodenschutzgesetz über die Ermittlung und Einstufung von Gehalten anorganischer Schadstoffe im Boden (VwV Anorganische Schadstoffe). Gemeinsames Amtsblatt des Landes Baden-Württemberg (GABl), Heft **30**: 1029-1036; Stuttgart.
Bundesgesundheitsamt (1990): Richtwerte für Schadstoffe in Lebensmitteln. Bekanntmachungen des BGA **12**: 224-226.
DIN [Deutsches Institut für Normung Hrsg.] (1997): Bodenbeschaffenheit, Extraktion von Spurenelementen mit Ammoniumnitratlösung. Beuth Verlag, DIN 19730; Berlin.
Liebe F., G.W. Brümmer & W. König (1995): Ableitung von Prüfwerten für die mobile Fraktion potentiell toxischer Elemente in Böden Nordrhein-Westfalens. Mitt.Dtsch.Bodenkdl.Ges.**76**: 345-348; Oldenburg.

Prüeß A., Turian G. & Schweikle V. (1991): Ableitung kritischer Gehalte an NH4NO3-extrahierbaren ökotoxikologisch relevanten Spurenelementen in Böden SW-Deutschlands. Mitt. Dtsch. Bodenkdl. Ges. **66/1**: 385-388, Oldenburg.

Prüeß A. (1992): Vorsorgewerte und Prüfwerte für mobile und mobilisierbare, potentiell ökotoxische Spurenelemente in Böden. Verlag Ulrich E. Grauer, Wendlingen, Stuttgart, 145 S.

Sauerbeck D. (1982): Welche Schwermetallgehalte in Pflanzen dürfen nicht überschritten werden, um Wachstumsbeeinträchtigungen zu vermeiden? Landw.Forsch., Sh. **39**: 108-129.

Sauerbeck D. & S. Lübben (1991): Auswirkungen von Siedlungsabfällen auf Böden, Bodenorganismen und Pflanzen. Forschungszentrum Jülich, Berichte aus der ökologischen Forschung, Band **6**: 417 S.; Jülich.

Symeonides C. & McRae S.G. (1977): The assessment of plant available cadmium in soils. J.Environ.Qual. **6**: 120-123.

Zeien H. & Brümmer G.W. (1988): Analysenvorschrift zur Extraktion der mobilen und leicht nachlieferbaren Schwermetalle. Methodenvorschrift des Instituts für Bodenkunde der Universität Bonn. unpublished.

Address of author:
Andreas Prüeß
UMEG Gesellschaft für Umweltmessungen und Umwelterhebungen GmbH
Daimlerstraße 5b
D-76185 Karlsruhe, Germany

Criteria for Soil Contamination of Organic and Inorganic Pollutants in the Czech Republic

E. Podlešáková & J. Němeček

Summary

Contamination limits were set up for hazardous trace elements and for principal groups of persistent organic pollutants. They correspond to the upper boundary of variability of natural and anthropogenic background concentrations. Pollution limits for hazardous trace elements, aimed at the food chain and crop protection, are also being developed. Preliminary studies of trace elements mobility and uptake by crops support the necessity for developing soil-differentiated criteria for mobile elements which are pH-dependent. Extremely high contents of Cu, As, Pb and Cr (V) due to either geogenic (non-soluble) or fluvial anthropogenic (slighty soluble) loads, do not cause crop contamination.

Keywords: Soil contamination; trace elements; mobility; transfer into plants; persistent organic xenobiotics.

1 Introduction

During the 1980's it was generally held that the soils in the Czech Republic were highly contaminated. This view was based on indirect indices derived mostly from severe atmospheric loads with SO_2, N oxides and particulate matter and from excessive use of agrochemicals in agriculture. Actual data on immission fall-out of hazardous trace elements (HTEs) and persistent organic pollutants (POPs) were scarce.

At the beginning of the 1990's Kloke's limiting values differentiated between two groups of soils, namely the coarse-textured and other soils, and were used as universal criteria for maximum permissible contents of HTEs. Kloke's values were modified in the Czech Republic for use with results of analyses using 2M HNO_3. Concerning POPs, the original Dutch A values (MHSPE 1988) have been adapted in the Czech Republic as critical limits for agricultural soils. Pollution criteria for forest and organic soils have not yet been established. The original Dutch limits (A, B and C values for HTEs and POPs) have been accepted for assessments in industrial areas (MHSPE 1988).

However, it was found that apart from the old-loads sites in urban and industrial areas, the contamination of soils is much lower than it was presupposed (Podlešáková and Němeček 1995). Limiting values derived from Kloke's values turned out to be too high to indicate the contamination of agricultural land, as was shown by so-called retrospective monitoring, which used the stored soil samples taken during the systematic soil surveys (1960-1972) (Podlešáková and Němeček 1995). Further information was obtained from our inventories of HTEs and POPs in ecologically critical areas of the Czech Republic and from the recently completed inventory of HTEs, made by the Central Institute for Supervising and Testing in Agriculture (Penk 1994). A nation-wide testing of contamination of fodder plants also supports this conclusion (SVA-CAFI 1996).

This established the necessity for a new comprehensive system of reference values of soil loads, compatible with analogous standards used abroad. The first level of soil load - characterized by the so-called soil contamination limits - refers to the natural and/or anthropogenic diffuse background. The second level - indicating a district soil pollution - corresponds to a significant loss of soil multifunctionality and represents critical concentrations of harmful substances in the soil, which if exceeded threatens the food chain or may cause considerable inputs into plants. The third level, i.e. the extreme concentrations of harmful substances in the soil, indicates an immediate hazard for human health and a need for remediation. The purpose of our research was to derive criteria characterising the first and second levels of soil loads.

2 Materials and Methods

The geometrical means of background concentrations of 12 (As, Be, Cd, Co, Cr, Cu, Hg, Mn, Ni, Pb, V, Zn) trace elements (total contents in HNO_3 + HF + $HClO_4$, and contents in 2M HNO_3 at room temperature) were set up for 13 soil - lithological groups and their upper ranges of variation ($GM.GD^2$; GM=geometric mean, GD=geometric deviation; 90% percentiles) were calculated for the most typical groups. Similar values were calculated for main groups of persistent organic pollutants (mono-, polyaromatic, chlorinated and petroleum hydrocarbons) in immission - affected and immission-free regions. The system of soil-differentiated contamination limits for HTEs and non-differentiated limits for POPs was developed.

The investigation of the mobility of HTEs (0.025M Na_2EDTA - pH 4.7, 2 g in 50 ml, shaking 20 min.; 0.01 $CaCl_2$ - 5 g in 50 ml, shaking 120 min.; 1M NH_4NO_3 - 20 g in 50 ml, shaking 120 min.) were conducted in parallel with research in the transfer of HTEs into plants. The latter included: a) pot experiments on representative soils with a load artificially added, b) pot experiments on soils loaded naturally due to different pollution processes (testing plants: radish, spinach, barley), c) field investigations (analysing/fodder crops and soils on which they were grown).

Three different paths of soil contamination were studied: the airborne and fluvial anthropogenic contamination and the geogenic loads.

Soil – lithological groups	Hg	As	Cd	Be	Pb	Zn	Cu	V	Mn	Co	Cr	Ni	Mo
Sands													
Sandstones		25	0.40	3.5	70	120	50	150	1100	25	100	50	
Loesslike sediments – argiluviation													
Loesses, loesslike sediments				4.0			60					60	2
Polygenetic loams					80	150		180	1300	30	130		
Sedimentary shales	0.5	30	0.5										
Clayes, marls							70					70	
Acid granites				5.0	90		50	150	1100		100	50	
Gneisses, micaceous schists					160		70	190	1400	40	180	70	3
Ultramafic rocks				4.0	80	>200	>100	>300	>2000	>50	>200	>80	2.5

Table 1: Contamination limits of soils of different lithological groups (total content, $mg.kg^{-1}$ d.m.)

3 Results and discussion

3.1 Soil contamination

For simplification, the contamination limits for HTEs (Podlešáková et al. 1994, 1996) are only given here for the nine most important soil-lithological groups differentiated according to their textural (granulometry, argilluviation of sedimentary parent materials) and mineralogical properties (Table 1). These values do not take into account the geogenic (mineralogical) extremes and the anthropogenic fluvial loads. The latter are characterised not only by increased contents but also increased solubilities of some HTEs, especially Cr, As, Zn, Pb, Cu, Mn and Ni in 2M HNO_3 and Na_2EDTA. In soils derived from transported products of weathering of mafic and ultramafic rocks, geogenic contents of Co, Cr, Ni, V are very high and therefore only minimum values are indicated in Table 1. These values exceed the commonly-accepted maximum permissible contents. Other geogenic extremes refer to As, Pb, Cu, (Be, Cd) in acid magmatic and metamorphic rocks. A combination of airborne anthropogenic loads and of geogenic loads in large impacted areas of North and North-West Bohemia makes their differentiation necessary for the evaluation of loads at

Substances	Proposal		German upper limit (Tebaay et al.1993)	Dutch target value (MHSPE 1994)
	Czech geometric mean	upper limit		
Monocyclic Aromatic Hydrocarbons (MAH)				
Benzene (B)	5	30	-	50
Toluene (T)	5	30	-	50
Xylene (X)	5	30	-	50
Ethylbenzene (Eb)	5	40	-	50
Polycyclic Aromatic Hydrocarbons (PAH)				Σ PAH
Fluoranthene (Fl)	60	300	300	
Pyrene (P)	50	200	-	
Benzo(b)fluoranthene (B(b)F)	15	100	200	
Phenanthrene (Ph)	20	150	-	
Indeno(cd)pyrene (I(cd)P)	5	50	-	1000
Anthracene (A)	10	50	-	
Benzo(a)pyrene (B(a)P)	10	75	100	
Benzo(a)anthracene (B(a)A)	10	100	200	
Benzo(k)fluoranthene (B(k)F)	10	50	100	
Benzo(ghi)perylene (B(ghi)P)	10	50	-	
Chrysene (Ch)	10	50	-	
Naphtalene (N)	5	50	-	
Chlorinated Hydrocarbons (ClH)				
PCB	2	20	-	20
HCB	2	20	-	2.5
DDT	1	15	-	2.5
DDE, DDD	1	10	-	2.5
HCH lindan	1	10	-	1.0
Others				
Styrene (S)	3	50	-	-
Petroleum Hydrocarbons ($mg.kg^{-1}$)	30	100	-	50

Table 2: Soil contamination limits ($\mu g.kg^{-1}$)

both the contamination and pollution level. The geogenic loads were estimated on the basis of elements distribution in soil profiles, their low mobilities, and the occurrence in non-common rocks and metallogenic zones.

Later we used a similar method to derive the non-differentiated contamination limits for POPs included in the Dutch list (Němeček et al. 1996) (Table 2). We compared our results with German and Dutch reference values (MHSPE 1994) and found them particulary similar to the data of Tebaay et al. (1993).

An overview of the percentages of soil samples showing district anthropogenic and geogenic loads in ecologically critical regions is presented in Table 3. The predominating geogenic loads with As, Be, Cu, Pb, Zn in acid rocks with Cr, Ni, V in ultramafic rocks and with Co and Mn in all mentioned rocks disprove previous conclusions about a generally high degree of anthropogenic soil contamination. Anthropogenic loads of both HTEs and POPs (Tables 3 and 4) show the following quantitative descending sequence with the following qualitative features:

- 1) inundated Fluvisols downstream of industrial cities (Cd, Zn, Hg, Pb, Cu, Cr; all groups of POPs, specific load with B(a)A, A, N; influence of waste waters),
- 2) metropolitan area of Prague (Cd, Pb, Hg, Cu, Zn; PAHs, specific loads with B(a)P, B(k)F, B(ghi)P; influence of traffic and industrial emissions),
- 3) North-Moravian region (Cd, Zn, Pb, Hg - also Cr, Mn; PAHs, specific loads with Ch, B(a)A, B(k)F, B(ghi)P; combustion of black coal, metallurgic industry),
- 4) North and North-West Bohemian regions (As, Be, Cd, PAHs, monocyclic hydrocarbons, HCB; combustion of brown coal, chemical industry).

The field investigations of the loads of fodder crops with HTEs and POPs showed low levels of contamination even in regions severely affected by immission fall-outs.

3.2 Soil pollution

At the very beginning of the transfer-to-plants studies, soil samples were loaded artificially. The results of repeated experiments then showed high, slowly-diminishing transfer due to uncompleted diffusion equilibria (Hornburg and Brümmer 1993). A comparison of soil-specific plant uptakes of different elements (Cd, Zn, Ni, Cu), allowing for their different mobilities, led to a soil vulnerability classification (accomplished hitherto for 6 soil - lithological groups).

The data of the pot experiments with soils contaminated in the field were processed by factor analysis (Table 5). A preliminary evaluation revealed two district groups of trace elements concerning their mobilities and transfers. For Mn, Cd, Zn, Co and Ni the first factor loads correlate with their mobile fractions (including Be) and plant uptakes (excluding Be) and the second factor loads with their total contents and the contents of potentially mobilizable (EDTA) and mobile forms. For Cu, Pb, As and Cr the first factor loads correlate with their total contents and the contents of some of their mobile and mobilizable forms, but not with their plant uptakes.

A comparison of the above finding with the data from field investigations indicates a higher correlation between the element mobilities and the plant uptake in pot experiments. Less mobile elements are mostly not being taken up by plants under field conditions. This statement is confirmed by data presented in Table 6, indicating that even very high slightly soluble geogenic or more soluble anthropogenic fluvial loads of Cu, Cr, As and Pb do not cause surpassing of the critical plant loads, except in a few extremes.

A high mobility of the soil Cd manifests itself in pot experiments through exceedance of the critical plant loads. Due to soil liming, the Cd contents in crops were reduced markedly below the critical level. The liming also has a strong effect on the reduction of mobility of Zn and Ni. However only very high contents of these elements in very acid soils causes an exceedance of the critical plant loads (Table 7).

Soil contamination criteria, Czech Republic

District, site	Ecological risks	No. of samples	As A	As G	Be A	Be G	Cd A	Cd G	Co A	Co G	Cr A	Cr G	Cu A	Cu G	Hg A	Hg G	Mn A	Mn G	Mo A	Mo G	Ni A	Ni G	Pb A	Pb G	V A	V G	Zn A	Zn G
Č. Krumlov	South Bohemian immission-free area	30	0	0	0	20	0	0	0	0	0	3	0	0	0	0	0	3	0	0	0	3	0	0	0	3	0	3
Prachatice		48	0	6	0	12	0	2	0	0	0	8	0	0	0	0	0	0	0	0	0	8	0	2	0	0	0	4
Pelhřimov		77	0	12	0	4	0	0	0	3	0	0	0	0	0	0	0	1	0	0	0	3	0	0	0	0	0	0
Chomutov	North Bohemian and North West Bohemian immission - fall out areas: mining (brown coals), power plants, chemical industry	115	3	11	18	11	1	0	16	0	12	0	12	0	0	0	6	0	0	0	6	0	10	0	17	0	0	6
Most		63	25	19	22	5	13	0	5	0	14	0	11	0	1	0	3	0	1	0	19	0	21	0	6	0	0	8
Teplice		75	18	23	26	20	17	4	15	0	21	0	11	0	0	0	11	0	11	0	17	0	27	0	9	0	0	13
Ústí n. Labem		78	26	6	13	7	29	0	34	0	26	0	13	0	4	0	18	0	0	0	20	0	12	0	19	0	0	4
Litoměřice		55	4	0	0	0	2	0	13	0	18	0	4	0	5	0	0	0	5	0	9	0	5	0	4	0	0	4
Děčín		28	4	3	0	0	0	0	0	0	11	0	0	0	0	0	3	0	0	0	11	0	2	0	0	0	0	3
Cheb		47	4	42	6	11	8	0	2	0	0	0	0	0	6	0	0	6	19	0	0	0	6	0	17	0	0	2
Sokolov		54	4	54	5	33	24	0	2	0	7	0	5	0	2	0	0	7	2	0	0	0	7	0	31	0	0	22
Karlovy Vary		59	12	12	10	17	14	0	15	0	15	0	14	0	0	0	29	0	15	0	6	0	7	0	36	0	0	10
Ostrava-Karviná	North Moravian immission-fallout area:mining (black coal), metalurgy	33	6	0	0	0	30	0	0	0	6	0	3	0	3	0	15	0	3	0	0	0	6	0	0	0	36	0
Frýdek-Místek		50	4	0	0	0	48	0	0	0	4	0	4	0	0	0	12	0	6	0	0	0	14	0	2	0	36	0
Prague	traffic immission	39	0	0	0	0	36	0	0	0	0	0	33	0	38	0	0	0	-	-	0	0	33	0	0	0	28	0
Fluvisols	flooding-waste waters	33	7	0	29	0	93	0	7	0	31	0	56	0	88	0	33	0	-	-	10	0	53	0	0	0	88	0

A ... anthropogenic contamination
G ... geogenic contamination

Table 3: Percentage of soil samples exceeding the background values of trace elements in different districts/sites of the Czech Republic

	District (site)	Sites number	PAH												MAH				CIH	
			Fl	P	Ph	B(b)F	B(a)A	A	B(a)P	I(cd)P	B(k)F	B(ghi)P	Ch	N	% sites	B	T	X	Eb	PCB	HCB
emmission free regions	Č. Krumlov	22	9	4	9	4	9	14	1	0	0	0	0	14	0	15	0	0	0	0	0	3
	Prachatice	32	9	12	3	3	6	6	6	3	3	3	3	3	9	16	0	0	3	3	3	0
North-Bohemian region	Chomutov	34	9	3	0	9	6	3	12	6	9	6	0	0	12	24	0	3	3	3	0	3
	Most	39	23	8	13	13	5	5	8	8	8	10	13	13	13	32	8	18	15	15	2	0
	Teplice	48	23	19	8	4	4	6	6	8	6	10	8	10	10	32	2	4	10	15	0	0
	Ústí n. Labem	33	15	9	9	9	9	12	9	6	9	6	3	3	15	24	0	0	3	0	6	3
	Litoměřice	29	14	0	0	3	17	0	3	0	3	0	0	0	7	17	0	0	0	0	3	14
	Děčín	27	7	4	4	7	7	4	7	4	7	11	11	11	11	22	0	0	0	0	18	0
West-Bohemian region	Cheb	32	9	6	34	6	22	3	6	0	0	0	0	19	12	42	22	0	0	0	31	0
	Sokolov	33	24	21	36	12	45	12	9	9	6	12	39	51	51	69	0	0	0	0	21	6
	Karlovy Vary	36	19	17	19	14	36	11	8	8	5	14	22	11	11	32	0	0	0	0	3	0
North-Moravian districts	Ostrava+Karviná	25	20	24	36	20	36	12	16	8	20	12	36	0	0	48	4	0	0	0	4	0
	Frýdek-Místek	20	15	0	40	10	35	10	5	0	5	5	20	10	10	60	0	0	0	0	0	5
urban area	Prague	63	67	10	10	62	40	33	60	13	62	52	41	0	0	78	0	0	0	0	43	0
Inundations	Fluvisols	17	100	41	47	94	94	47	94	53	94	88	6	100	100	100	0	0	12	23	82	35
Reference values	μg.kg^{-1}		300	200	150	100	100	50	75	50	50	50	50	50	50	-	30	30	30	40	20	20

PAH   Polyaromatic Hydrocarbons
MAH   Monoaromatic Hydrocarbons
CIH   Chlorinated Hydrocarbons
Fl   Fluoranthene
P   Pyrene

Ph   Phenanthrene
B(b)F   Benzo(b)fluoranthene
B(a)A   Benzo(a)anthracene
A   Anthracene
B(a)P   Benzo(a)pyrene

I(cd)P   Indeno(cd)pyrene
B(k)F   Benzo(k)fluoranthene
B(ghi)P   Benzo(ghi)perylene
Ch   Chrysene
N   Naphtalene

B   Benzene
T   Toluene
X   Xylene
Eb   Ethylebenzene

Table 4: Anthropogenic loads assessment of POPs (% of samples exceeding the background of POPs)

Characteristics * n = 162 abbreviation	\multicolumn{10}{c	}{TRACE ELEMENTS FACTOR LOADS** - F1, F2}

Characteristics * n = 162 / abbreviation	Mn	Cd	Zn	Co	Ni	Be	Cr	Cu	As	Pb
pH/KCl	-0.871	-0.836	-0.846	-0.748	-0.530	-0.790				
total content - TO	0.848	0.916	0.847	0.811	0.778		0.460	0.787	0.911	0.705
mobilizable fraction (Na_2EDTA) - ED	0.907	0.887	0.914	0.857	0.817		0.875	0.883	0.875	0.809
mobile fraction ($CaCl_2$) - MC	0.833	0.894	0.851	0.786	0.856	0.892		0.857		
mobile fraction (NH_4NO_3) - MN	0.891	0.907	0.868	0.787	0.884	0.931	0.561	0.933	0.852	
potential mobility - ED/TO			0.649		0.571	0.886	0.947	0.812		0.632
mobility - MC/TO	0.919	0.916	0.807	0.861	0.682	0.802			0.582	0.974
mobility - MN/TO	0.919	0.874	0.885	0.909	0.896	0.934	0.669	0.877	0.772	
mobility - MC/ED	0.849	0.811	0.807	0.726		0.739			0.727	0.965
mobility - MN/ED	0.847	0.769	0.775	0.825	0.735	0.858	0.664	0.662		
radish (TE in DM) - R for F1	0.865	0.605	0.520	0.483	0.506					
radish (TE in DM) - R for F2										
% of F1 LOAD	51.3	47.1	44.4	44.8	35.3	40.9	26.5	37.0	34.1	31.7
% of F2 LOAD	20.8	20.4	21.3	22.9	27.0	16.6	23.8	18.0	18.7	24.8

* Cox, clay omitted,
** : F1 > 0.5
     F2 > 0.5
DM = dry matter

Table 5: Factor analysis of trace element mobilities and transfers (pot experiments with radish)

Element	Plant			max.per. contents°	Soil					load
	Critical load	Content	TF		Total content	pH	Na₂EDTA % total	NH₄NO₃ % total	NH₄NO₃ content	
As	4 *	0.44	0.0004	30	1200	5.7	4	0.05	0.55	G
	2 ***	0.10	0.0003		359	5.3	14	0.12	0.44	A
Pb	40 *	2.54	0.008	140	306	5.3	29	0.08	0.26	G
	45 ***	11.90	0.003		3660	5.3	70	0.08	2.90	A
Cu	20 +	5.20	0.026	100	200	4.0	4	0.07	0.13	G
		10.80	0.027		394	6.4	71	0.62	2.44	A
Cr	3 **	1.20	0.001	200	1230	6.2	0.1	<0.001	<0.01	G
		1.00	0.0004		238	6.5	16	<0.01	<0.01	A
Ni	5 **	14.90	0.01	80	1500	6.2	10	0.10	1.47	G
		5.80	0.02		325	6.4	53	1.27	4.12	A
Zn	200 +	164	0.68	200	240	4.6	22	4.58	11	G
		50	0.08		619	6.4	68	0.49	2.7	A
Cd	1 *	0.55	0.67	1	0.8	3.3	92	0	0.16	G
	1.14 **	1.20	0.02		52	5.3	11	76	1.44	A

G: Geogenic loads of soils;
A: Anthropogenic loads of soils;
TF: transfer factors

*   Czech standard values for fodder plants, regulation 194/1996 Ministry of Agriculture
**   Czech standard values for fodder plants, regulation 264/1993, Ministry of Agriculture.
*** Bundesgesetzblatt Futtermittelverordnung, 1/28, 1988.
+   Magnicol, R.D., Beckett, P.H. 1985: Plant and Soil, 85, p. 107 - 129.
°   Czech regulation 13/1994 Ministry of Environmental Protection

Table 6: Example of geogenic and anthropogenic soil loads (extremes) and fodder contamination in field conditions (concentrations in mg.kg⁻¹, d.m.)

	element	Cd			Ni			Zn		
	variant	A	B	%	A	B	%	A	B	%
	pH	4.83	6.55		4.83	6.55		4.83	6.55	
	total	0.71	0.67		20.8	20.6		168.8	157.6	
	Na₂EDTA	0.32	0.25	22	4.2	3.4	19	21.,1	19.1	9
Content Mg.kg⁻¹	CaCl₂	0.14	0.02	86	0.25	0.11	56	2.,9	0.3	89
	NH₄NO₃	0.14	0.03	79	0.69	0.02	97	4.7	0.5	89
	spinach	20.1	13.9	31	1.8	0.9	50	355.6	175.5	51
	radish	2.1	1.5	29	1.8	1.4	22	71.9	36.6	49
	Na₂EDTA	0.46	0.38	17	0.14	0.13	7	0.13	0.12	8
Mobility	CaCl₂	0.21	0.04	81	0.011	0.005	55	0.018	0.002	89
	NH₄NO₃	0.21	0.05	76	0.021	0.001	95	0.028	0.004	86
Transfer Factors	spinach	15.3	20.6		0.02	0.04		0.88	0.96	
	radish	3.1	2.,5		0.07	0.07		0.45	0.25	

A ... nonlimed soils; B ... limed soils

Table 7: Influence of liming on Cd, Ni and Zn behaviour in Cambisols and plants

Acknowledgements

This study was performed with the financial support of the Grant Agency of the Czech Republic (contract 502/94/0138) and Ministry of Agriculture (RE 096 000 6153).

References

Bundesgesetzblatt Futtermittelverordnung (1988): 1/28.
Hornburg, V. & Brümmer, G.W. (1993): Verhalten von Schwermetallen in Böden: 1. Untersuchungen zur Schwermetallmobilität, Z. Pflanzenernähr. Bodenk., **156**, 467-477.
Magnicol, R. D. & Beckett, P.H. (1985): Critical tissue concentrations of potentially toxic elements, Plant and Soil **85**, 107-129.
Ministry of Agriculture (1993): Czech standard values of fodder plants, regulation 264/1993.
Ministry of Agriculture (1996): Czech standard values for fodder plants, regulation 194/1996.
Ministry of Environmental Protection (1994): Czech regulation 13/1994.
MHSPE (Ministry of Housing, Spatial Planning and Environment) (1988): Soil Clean-up guidelines, The Hague, 28 p.
MHSPE (Ministry of Housing, Spatial Planning and Environment) (1994): Circular on intervention values for soil remediation, The Hague, 19 p.
Němeček, J., E. Podlešáková, M. Pastuszková (1996): Proposal of soil contamination limits with persistent organic xenobiotic substances in the Czech Republic (in Czech), Rostlinná Výroba **42** (2), 49-53.
Penk, J. (1994): Content of hazardous trace elements in agricultural soils of the Czech Republic (I.), Central Institute for Supervising and Testing in Agriculture (in Czech), Brno, 62 p.
Podlešáková, E., N me ek, J. & Hálová, G. (1994), "Background values of potentially hazardous elements in the soils of the Czech Republic (total contents)" (in Czech), Rostlinná Výroba **40** (12), 1095-1105.
Podlešáková, E., J. Němeček (1995): Retrospective Monitoring and Inventory of Soil Contamination in Relation to Systematic Monitoring, Environmental Monitoring and Assessment, Kluwer Academic Publishers, Netherlands **34**, 121-125.
Podlešáková, E., J. Němeček, Hálová, G. (1996): Proposal of soil contamination limits with hazardous trace elements for the Czech Republic (in Czech), Rostlinná Výroba **42** (3), 119-125.
SVA-CAFI (State Veterinary Administration and Czech Agricultural and Foot Inspection) (1996): Contamination of food chains by xenobiotics in the Czech Republic in 1995. Information Bulletin (in Czech), 277 p.
Tebaay, R. H., Welp, G. & Brümmer, G.W. (1993): Gehalte an polycyklischen aromatischen Kohlenwasserstoffen (PAK) und deren Verteilungsmuster in unterschiedlich belasteten Böden, Z. Pflanzenernähr. Bodenk. **156**, 1-10 p.

Addresses of authors:
Eliška Podlešáková
Research Institute for Soil and Water Conservation
Žabovřeská 250
15627 Prague, Czech Republic
J. Němeček
Czech University of Agriculture
Dept. Soil Science and Geology
Kamýcká 127
16521 Prague, Czech Republic

Possible Approaches for In Situ Restoration of Soils Contaminated by Zinc

K. Wenger, T. Hari, S.K. Gupta, R. Krebs, R. Rammelt & C.D. Leumann

Summary

Gentle remediation techniques must be developed that are both ecologically and economically viable, in order to achieve an effective curative concept in soil protection. Two *in situ* techniques - mobilization (enhanced accessibility for "remediation" plant uptake) and immobilization (stabilization) of heavy metals in soil - are discussed.

The "soluble" heavy metal fraction in soil can be decreased by increasing pH (e.g. liming, alkaline fertilization) or by increasing cation binding capacity (e.g. addition of clay minerals or gravel sludge) (Krebs 1996, Krebs and Gupta 1994). Pot experiments showed that gravel sludge greatly reduced the "soluble" heavy metal fraction in soil. It appears that additional agricultural land use is possible with decreased risk of the entry of heavy metals into the food chain.

Our research group is presently investigating the scope of gentle decontamination of soil in collaboration with the Swiss Federal Institute of Technology in Zürich and private organizations. One strategy involves the mobilization of heavy metals in soil to increase their exportation by plants. Batch experiments showed that citric acid is a good mobilizing organic acid for zinc. Organic acids such as citric acid are found in natural soil systems and are degradable by soil microorganisms.

Keywords: Heavy metal, zinc, gentle remediation, stabilization, mobilization, organic acids

1 Introduction

There are several thousand hectares of soil in Switzerland which are contaminated by heavy metals (i.e. Zn, Cd, Cu, Pb, etc.). These are not abandoned sites according to the legal definition (USG_{rev} 1995, USG 1986), but areas with a diffuse or point source of pollutants. This includes areas around older waste incinerators, industrial sites, soil along busy roads, or areas with intensive use of waste-based fertilizers. In most areas the risk to man, plants, and animals is presently not acute, but nevertheless, soil fertility and groundwater are affected. Through metal loaded cultivated plants grown on certain polluted sites, there is even the hazard of entry of pollutants into the food. Apart from reducing the entry of pollutants into soils, the development of remediation techniques for these contaminated sites is an important task. The main goals of any remediation technique are to restore soil functions, avert hazards to man and animals, protect groundwater, and provide an undisturbed plant growth (USG_{rev} 1993). Today, few gentle decontamination techniques are available which are readily applicable to field conditions. Other than land use changes or land use bans, there are only harsh remediation techniques to reduce the hazards to man, plants, and animals. Harsh remediation techniques often have the consequence of (at least) partly damaging soil fertility. Moreover, these methods often require an *ex situ* treatment which is very expensive and thus are not applicable for large areas. Consequently gentle remediation techniques must be developed that consider environmental and economical concepts.

1.1 Possible solutions

Two techniques - mobilization (enhanced accessibility for "remediation" plant uptake) and immobilization (stabilization of heavy metals in soils) - will be discussed. Figure 1 shows the concept of these two techniques.

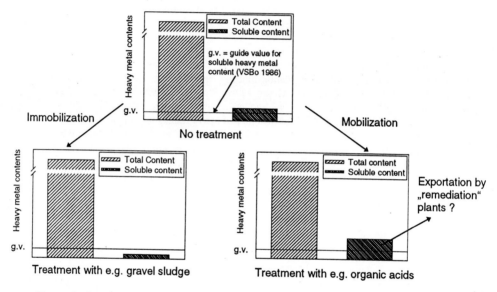

Figure 1: Concept of immobilization and controlled mobilization of heavy metals in soil

1.1.1 Mobilization of heavy metals in soil to enhance their accessibility to "remediation" plants

High metal uptake by plants is intended to eliminate risk and restore soil fertility. The feasibility of this technique depends on plants which are able to sufficiently extract heavy metals from soil. There are some plants, so called hyper-accumulator plants, that have an unusually high uptake of heavy metals from contaminated sites. Biomass production of these hyper-accumulator plants is, however, usually very small. Thus, calculated decontamination time varies from 100 to several hundreds of years. One possibility to shorten this time is to grow high biomass plants that have been selected for a high rate of heavy metal uptake. Another possibility is to enhance the plant available heavy metal fraction in soil, because the metal uptake by plants also depends on their bioavailability.

The greatest part of heavy metals is adsorbed on soil particles and is not available for plants. Therefore, a controlled increase of the plant available heavy metal fraction is proposed. Heavy metals can be mobilized by decreasing soil pH (Gupta 1992, Herms and Brümmer 1980) or by augmentation of complexing agents. By decreasing soil pH, the uptake of heavy metals by plants can be increased by a factor of two to three (Hasselbach and Boguslawski 1991). The addition of inorganic acids leads to an accumulation of the corresponding anions, nitrate, chloride, and sulfate, in soil. Therefore, it is preferable to use organic agents that are degradable by microorganisms. Wallace et al. (1974) described the mobilization by synthetic chelates, e.g. EDTA (ethylenedinitrilotetraacetic acid) or NTA (nitrilotriacetic acid). Heavy metal uptake by plants could be increased by the addition of EDTA or NTA to soil (Balmer and Kulli 1994, Jorgensen 1993, Wallace et al. 1974). These synthetic ligands, especially EDTA, are barely degradable by microorganisms. For this reason, it is preferable to use soil-borne agents. Depending on their degradation rate, it may be necessary to add such agents several times to maintain an optimal concentration of "soluble" heavy metal during the vegetation period.

1.1.2 Immobilization of heavy metals

The main goal of immobilization is to reduce the risk of an uncontrolled heavy metal transfer in groundwater and biosphere. To attain this, the plant available heavy metal fraction in soil must be reduced. This can be achieved by increasing soil pH, with liming, or by increasing metal binding capacity, with the addition of clay minerals, iron oxides, or waste products such as gravel sludge. On average, gravel sludge contains about 45% clay minerals. The addition of these substances should not (or only slightly) affect the availability of nutrient ions for plants.

2 Methods and materials

2.1 Soil used for experiments

The soil samples used for pot experiments (immobilization of heavy metals) and for laboratory studies (mobilization and incubation experiments) were collected from the topsoil (0 to 20 cm) of arable soil in Rafz (Switzerland) treated with sewage sludge. The soil had a sandy loam texture with 1.9% Corg and a pH (H_2O) of 6.7. Total heavy metal concentrations determined by nitric acid extraction according to the Swiss ordinance relating to pollutants in soil (VSBo 1986) amounted to 860 mg / kg dry soil for zinc.

2.2 Mobilization of zinc with different agents

Soil samples were dried (40 °C) and sieved (< 2mm). Aliquots of 30 g were treated with various doses of NTA (2, 5, 25 mmol / kg dry soil), citric, oxalic, phtalic, salicylic, and nitric acid (5, 25, 100 mmol / kg dry soil), respectively. The samples were analyzed for $NaNO_3$-extractable zinc.

2.3 Incubation experiment: effects and degradation dynamic of organic acids in soil

Aliquots of 500 g of dry soil (40 °C) were mixed with various concentrations (5, 25 mmol / kg dry soil) of citric, oxalic, phtalic acid and NTA. The samples were rehydrated with 120 ml deionized water and kept 4/5 covered with an opaque sheet (NTA with a glass sheet) in a climate chamber by 20 °C (± 2 °C). The soil was moistened twice weekly. Thus, water content was almost about field capacity. Samples were taken after 2, 17 and 35 days and analyzed for $NaNO_3$-extractable zinc and pH (H_2O).

2.4 Pot experiments: Immobilization of zinc

Soil samples were dried at 40 °C and sieved < 1cm. Aliquots of 400 g of dry soil were weighted in plastic pots and rehydrated to 50% field capacity. Each pot experiment was fertilized with 16 mg Mg as $MgSO_4$ and 52 mg N as NH_4NO_3. Various doses (1.25, 2.5, 5, 10, 20, 80 g / kg dry soil) of Na-montmorillonite or gravel sludge were added in powder form to each pot. Gravel sludge, as a waste product of the gravel industry, normally contains about 45% clay minerals. Table 1 shows the heavy metal contents of gravel sludge used for experiments. The concentrations of heavy metals in the gravel sludge were much lower than concentrations tolerated in sewage sludge according to the Swiss Ordinance on Substances (StoV 1986). Relative to the heavy metal concentration already present in soils, metal input due to the experimental application of the gravel sludge is insignificant.

After one week, 200 mg of red clover (Trifolium pratense) was sown and grown in a growth chamber under controlled conditions: 16 h artificial light, 25 °C; 8 h darkness, 15 °C.

Parameter [units]	Component	Content of Gravel sludge (Krebs 1996)	Tolerance value for sewage sludge (StoV 1986)	Guide value for total metal content in soil (VSBo 1986)
Mineralogical composition [weight percent]	Clay minerals Carbonate	~ 42 ~ 31		
CEC effective $[cmol_c/kg][cmol_c/kg]$		156		
Ranges of heavy metal contents extractable with 2 M HNO_3 [mg/kg dry soil]	Cd Cu Zn Pb	0.13 15 53.5 10.6	5 600 2000 500	0.8 50 200 50

Table 1: Heavy metal content of gravel sludge, Tolerance value for sewage sludge (StoV 1986), and guide value for total metal content in soil (VSBo 1986)

3 Results

3.1 Laboratory studies: Mobilization of zinc with different agents and effects and degradation dynamics of organic acids in soil

Figure 2 shows that the synthetic agent NTA mobilized about 15 times more zinc than the other tested agents (NTA was analysed after an incubation time of 2 days). Among natural organic agents, citric acid mobilized the most; about three times more zinc than other natural organic agents and nitric acid.

Figure 3 shows the degradation dynamics of different organic acids and their effects on $NaNO_3$-extractable zinc concentrations. With the addition of 5 mmol / kg dry soil of organic acids, the mobilizing effect of the natural organic agents almost disappeared after two days. With NTA, a decreasing effect occurred after 20 days. By adding 25 mmol / kg dry soil of natural organic acid, the mobilizing effect of the agents decreased after 17 days. After 17 days, there was even less "soluble" zinc in the soil than without the addition of agents. With NTA, no decreasing effect of

In situ restoration of soils contaminated by zinc

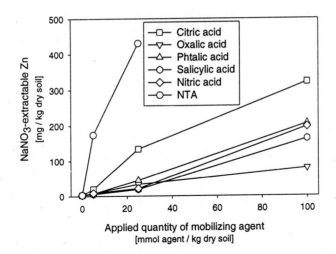

Figure 2: Mobilizing effect on NaNO$_3$-extractable zinc by several concentrations of synthetic and natural ligands and nitric acid

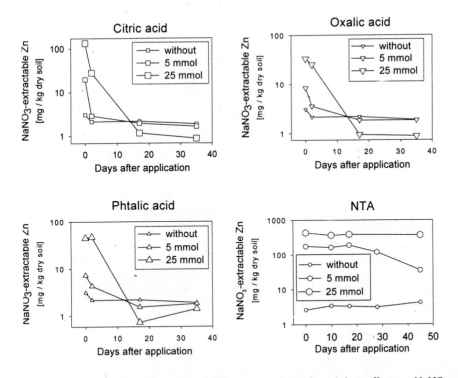

Figure 3: Degradation dynamics of different organic acids and their effects on NaNO$_3$-extractable zinc concentrations by various doses of mobilizing agents (0, 5, and 25 mmol agent / kg dry soil)

"soluble" zinc-concentration was observed after one month. Further batch experiments showed that the mobilizing effect of NTA disappeared after 63 days. The organic acids decreased soil pH up to 0.5 units by the addition of 5 mmol / kg dry soil and up to 2 units by the addition of 25 mmol / kg dry soil. In the case of 5 mmol / kg acid application, soil pH returned to the original pH after 2 days. In the case of 25 mmol / kg acid application, soil pH remained low for several days but returned to the original soil pH after 17 days. Nitric acid decreased the soil pH for the entire experimental time.

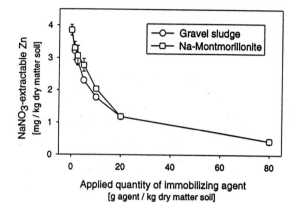

Figure 4.1: Decrease of $NaNO_3$-extractable zinc after addition of binding agents

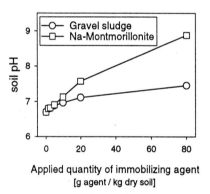

Figure 4.2: Effect of binding agents on soil pH (H_2O)

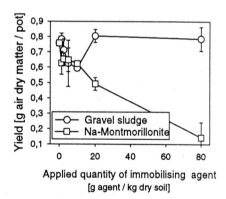

Figure 4.3: Effect of binding agents on biomass yield of red clover (Trifolium pratense)

3.2 Pot experiments: Immobilization of zinc

The effectiveness of Na-montmorillonite and gravel sludge as binding additives is shown in Figure 4.1. Both of them reduced the soluble heavy metal fraction in soils by about a factor of eight. In contrast to Na-montmorillonite, gravel sludge only slightly affected soil pH (Fig. 4.2), but did not affect the plant yield of red clover (Trifolium pratense) (Fig. 4.3).

4 Discussion

4.1 Mobilization of heavy metals in soils

The laboratory studies showed that the mobilizing effect of the synthetic ligand NTA is unattainable by the tested natural organic agents. The effects of soil treatment with NTA as well as with citric acid on soil pH were similar, but the metal complex equilibrium constants of NTA are higher than those of the tested natural organic ligands (Martell and Smith 1989). Therefore, it is assumed that the greater mobilizing effect of NTA is based mainly on its greater complexing capacity. Among the tested natural organic acids, citric acid was the most effective in mobilizing heavy metals. Citric acid has been identified as a plant exudate and was also found in the xylem sap, where it is assumed to translocate iron and zinc (White et al. 1981). Senden et al. (1995) showed that after pre-incubation with citric acid, total tomato plant uptake of cadmium increased twofold. His speciation calculations showed that Cd in xylem may be, at least in part, complexed in citric acid. Thus, it is assumed that at least some plants are able to take up this complex.

By treatment with 25 mmol / kg dry soil of NTA no decreasing effect of "soluble" zinc concentration was observed after one month. On the one hand, NTA could have mobilized so much zinc that there was a zinc-toxic effect on microorganisms. On the other hand, NTA may itself have had toxic effects on microorganisms.

By treatment with 25 mmol / kg dry soil of natural organic acids, the mobilizing effect of the agents decreased after 17 days. A hysteresis effect was observed. The sudden enormous supply of soil-borne organic agents could have led to a greater reproduction of microorganisms that live by degrading these organic substances. When food and other required materials accumulate before a population starts growing, it can result in an "overshoot" in population growth (Nicholson 1954, cited in: Odum 1983). This could be evidence of a fast degradation of soil-borne organic acids by microorganisms.

The mobilization of heavy metals in soil will increase the hazard of plant toxicity, as well as the risk of metal leaching to deeper soil layers or to groundwater. To minimize this risk, the agent dose and the frequency of agent application must be adapted to the heavy metal uptake capacity of the particular plant (Fig. 5). The hazard of metal leaching will also be minimized by fast degradation of natural organic acids in soil. The treatment with mobilizing agents should not affect "remediation" plant growth.

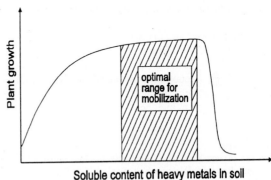

Figure 5: Optimal range for mobilization with no growth restriction for „remediation" plant

4.2 Immobilization of heavy metals

Experiments showed that treatment with gravel sludge greatly reduced the $NaNO_3$-extractable zinc in soil. The heavy metal uptake by plants will also be reduced (Krebs 1996). Thus, agricultural land use remains possible. Immobilization is a helpful instrument for soil use management.

Furthermore, gravel sludge as a waste product of the gravel industry is available in very large quantities at a low price. However, since heavy metals still remain in the soil, it is not a long-term solution. Some questions relating to leaching of pollutants by particulate transport or the long-term stability of heavy metal binding still need to be answered.

5 Conclusions

5.1 Mobilization of heavy metals in soil to enhance their accessibility by plants

Treatment with organic acids led to an increase of $NaNO_3$-extractable zinc in soil. The synthetic ligand NTA mobilized the most zinc. Among the tested natural organic acids, citric acid mobilized the most zinc. NTA was slowly degraded in soil. In contrast, natural organic acids degraded very fast.

5.2 Immobilization of heavy metals

Immobilization techniques reduce the risk of an uncontrolled heavy metal transfer to groundwater and the biosphere. The addition of gravel sludge reduced the $NaNO_3$-extractable heavy metal fraction in soil. No adverse effects could be observed by using gravel sludge. Thus, agricultural land use remains possible.

6 Outlook

The main task is to find further appropriate soil-borne substances for mobilizing the heavy metals in different soils. Substances that are exuded by plants or identified in xylem or phloem sap of plants are preferable. It must be tested whether mobilization of heavy metals with these organic substances leads to an enhancement of metal uptake by plants.

6.1 Fertilization to achieve maximum biomass

Another approach to enhance the soluble heavy metal fraction in soil is to decrease soil pH by nitrogen fertilizing. There will be a release of protons by nitrification of these fertilizers. Nitrogen fertilizing can enhance the proton quantity as well as the biomass. The effect of this proton release on the soluble fraction of heavy metals in soil, as well as possible adverse side effects of excessive fertilizing must be investigated. Another question is whether high biomass production induced by fertilization leads to high heavy metal uptake by plants.

References

Balmer M. and Kulli, B. (1994): Der Einfluss von NTA auf die Zink- und Kupferaufnahme durch Lattich und Raygras, Diplomarbeit, Fachbereich Bodenschutz am Institut für terrestrische Oekologie, ETH-Zürich, Zürich.

Gupta S. K. (1992): Mobilizable metal in anthropogenic contaminated soils and its ecological significance. In: J. P. Vernet (ed.), Impact of heavy metals on the environment, Elsevier, Amsterdam.

Hasselbach G. und von Boguslawski, E. (1991): Bodenspezifische Einflüsse auf die Schwermetallaufnahme der Pflanzen. In: D. Sauerbeck und S. Lübben (eds.), Auswirkungen von Siedlungsabfällen auf Böden, Bodenorganismen und Pflanzen, Berichte aus der ökologischen Forschung, Band 6.

Herms U. und Brümmer, G. (1980): Einfluss der Bodenreaktion auf Löslichkeit und tolerierbare Gesamtgehalte an Nickel, Kupfer, Zink, Cadmium und Blei in Böden und kompostierten Siedungsabfällen, Landwirtsch. Forschung **33**, 4, 408-423.

Jorgensen S. E. (1993): Removal of heavy metals from compost and soil by ecotechnological methods, Ecological Engineering **2**, 89-100.

Krebs R. (1996): In Situ Immobilization of Heavy Metals in Polluted Agricultural Soil - an Approach to Gentle Soil Remediation, Dissertation ETH Zürich, Nr. 11838, p.110.

Krebs R. and Gupta, S.K. (1994): Sanfte Sanierung schwermetallbelasteter Böden, Agrarforschung **1**, 8, 349-352.

Martell A. E. and Smith, R.M. (1989): Critical stability constants, 2nd edition, Plenum Press New York

Odum E. P. (1983): Grundlagen der Oekologie, Band 2: Standorte und Anwendung, 2nd edition Stuttgart, New York.

Senden M. H. M. N., van der Meer, A.J.G.M., Verburg, T.G. and Wolterbeek, H.Th. (1995): Citric acid in tomato plant roots and its effect on cadmium uptake and distribution, Plant and Soil **171**, 333-339.

StoV (1986): Verordnung über umweltgefährdende Stoffe (StoV), Verordnung des Schweiz. Bundesrates, Eidg. Drucksachen- und Materialzentrale, Bern, SR 814.015.

$USG_{rev.}$ (1995): Umweltschutzgesetz - Aenderung vom 21. Dezember 1995, Bern.

USGrev (1993): Botschaft zu einer Änderung des Bundesgesetzes über den Umweltschutz, SR 93.053, Bern.

USG (1986): Bundesgesetz über den Umweltschutz (Umweltschutzgesetz), Eidg. Drucksachen- und Materialzentrale, Bern, SR 814.01.

VSBo (1986): Swiss Ordinance on Pollutants in Soil, Verordnung des Schweiz. Bundesrates, Eidg. Drucksachen- und Materialzentrale, Bern, SR 814.12.

Wallace A., Mueller, R.T. and Alexander, G. (1974): Effects of high levels of NTA on metal uptake of plants grown in soils, Agron. J. **66**, 707-708.

White M. C., Decker, A.M. and Chaney, R.L. (1981): Metal complexation in xylem fluid. I. Chemical composition of tomato and soybean stem exudate, Plant Physiol. **67**, 292-300.

Addresses of authors:
K. Wenger
T. Hari
S.K. Gupta
R. Rammelt
C.D. Leumann
Swiss Federal Research Station for Agroecology and Agriculture FAL
Institute of Environmental Protection and Agriculture IUL Liebefeld
CH-3003 Bern, Switzerland
R. Krebs
Institute of Terrestrial Ecology
ETH Zürich
Grabenstraße 3
CH-8952 Schlieren, Switzerland

Soil Degradation Caused by Industrialization and Urbanization

W.E.H. Blum

Summary

Soil degradation through industrial and urban development is described as sealing and physical, chemical and biological degradation processes, caused by traffic and transport, urban and industrial activities. Degradation is defined as a measurable loss or reduction of soil functions or soil uses.

Approaches for dealing with these problems are discussed on the basis of reversibility versus irreversibility, concluding that these forms of degradation considerably reduce the multifunctioality of soils, thus harming future generations.

Keywords: Soil functions, urban and industrial development, soil degradation (sealing, physical, chemical and biological processes), sustainable land management.

1 Introduction

Soil degradation arising from industrial and urban development has been observed for centuries. However, since the 1950's, this development has reached unprecedented dimensions. In this paper, different forms of soil degradation will be analyzed on the basis of a comprehensive concept. The various forms of degradation caused by this development will be explained.

Soil degradation can be defined as the measurable loss or reduction of soil functions or soil uses. Therefore degradation should be defined on the basis of specific soil functions or soil uses.

2 The main functions and uses of soil and land and the competition between them

Six main functions and uses of soil and land can be distinguished, three ecological ones and three others, linked to socio-economic, technical and industrial uses (Blum, 1988).

The three ecological functions and uses are:
- production of biomass, ensuring food, fodder, renewable energy and raw materials, a function which is well known;
- filtering, buffering and transformation between the atmosphere, the groundwater and the plant cover, thus protecting the environment, including human beings, especially against the contamination of the ground water and of the food chain. This function becomes increasingly important, because of the many solid, liquid or gaseous, inorganic or organic depositions on which soils react through mechanical filtration (in its pore space), physico-chemical adsorption and precipitation at its inner surfaces (mainly pore walls) or microbiological and biochemical mineralization or metabolization of organic compounds (see Fig. 1). The latter may also

contribute to global changes through the emission of gases from the soil into the atmosphere (see dotted line in Fig. 1). As long as these processes can be maintained by the soil, there is no danger for the groundwater or the food chain. However, this capacity of soil is limited and varies according to specific soil conditions.
- Soil is a biological habitat and gene reserve, containing more species in number and quantity than all other biota together. Soil biota become increasingly important for many technical processes, especially biochemical, biotechnological and bioengineering ones.

Moreover, three technical, industrial and socio-economic functions and uses can be defined:
- The use of soils as a spatial base for technical, industrial and socio-economic structures and their development, e.g. industrial premises, housing, transport, sports, recreation, dumping of refuse, etc.;
- Soils are also used as a source of raw materials (clay, sand, gravel, minerals in general, and others), as well as a source of energy and water;
- Finally, soils are an important geogenic and cultural heritage, forming an essential part of the landscape in which we live and concealing paleontological and archeological treasures of high importance for the understanding of the history of mankind.

Therefore, a definition of land use should include all these main functions which were often used concomitantly in the same area. The problem is the competition between the different uses (Blum 1994a).

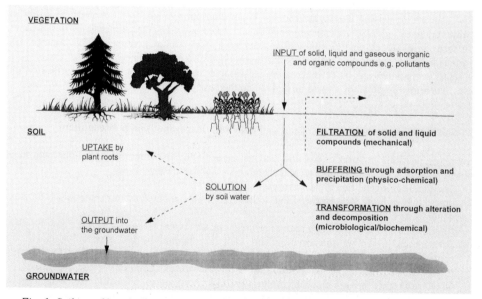

Fig. 1: Soil as a filter, buffer and transformation system between atmosphere, biosphere and hydrosphere.

At least three different categories of competition between the six main uses of soil and land can be distinguished:
- exclusive competition between the use of land for infrastructure, as a source of raw materials

and as geogenic and cultural heritage on one hand, and for agricultural and forest production, filtering, buffering and transformation activities, and as gene reserve on the other hand;
- intensive interactions between infrastructural land use and its development on one hand and agriculture and forestry, filtering, buffering and transformation as well as biodiversity on the other hand;
- Moreover, intensive competition also exists between the three ecological soil and land uses themselves.

In the context of soil degradation by urbanization and industrialization, especially the first two categories must be considered.

3 Urban and industrial development and sealing of soils

The exclusive competition between the different land uses becomes evident by the sealing of soils through urban and industrial development, e.g. the construction of roads, of industrial premises, of houses, of sporting facilities, and the use of soils e.g. for the dumping of refuse, all this being known as the process of urbanization.

The growth of urban population on a world-wide level as a measure of urbanization is shown in Table 1. Urban population increased from 1970-1990 on all continents; in 1990 South America had a higher degree of urbanization than Europe. Looking at the growth of the 35 largest cities between 1970 and 1990 (Blum, 1994b), urbanization is clearly an exponential process. Ninety per cent of the global population increase occurs in developing countries, and 2/3 of it is concentrated in urban agglomerations. Estimates of the United Nations indicate that, while in 1984 34 metropolitan areas existed with populations greater than 5 million, in the year 2025 about 93 metropolitan areas of the same size are expected.

AREA	1970 %	1990 %
Europe	67	73
South America	60	76
North America	58	71
Africa	23	34
Asia	24	29
World	37	43

Table 1: Increase of urban populations from 1970-1990 (United Nations Environmental Data Report, 1991/1992).

Data from Europe for the 1970's show that, for example, in Austria about 35 ha were sealed per day. The data for Western Germany at the same time indicate about 140 ha per day. Even though these figures may vary to a certain extent, due to differing statistical approaches by the respective national institutions, it is clear that during the main development of industrialization and transport infrastructure in central Europe in the 1970's, enormous surface areas were irreversibly sealed. One exception was Switzerland, which was already well aware that its agricultural land reserves were mainly concentrated in the western part of the country, where at the same time the bulk of industrial and urban development was taking place (Bodenkundliche Gesellschaft der Schweiz, 1985). Taking another example, Egypt, it becomes evident that all big urban agglomerations in this country can be found on the 3.8% of fertile land, in the southern Nile River valley and in its delta between Damietta and the Rosetta branch, near the Mediterranean Sea. It is also known that all these agglomerations are still expanding on this fertile land, which underlines the problem of sealing as an irreversible loss of soil multifunctionality.

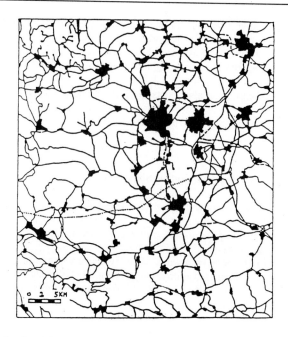

Fig. 2: Sealing of soils and landscapes by settlements and roads (Schröder and Blum, 1992).

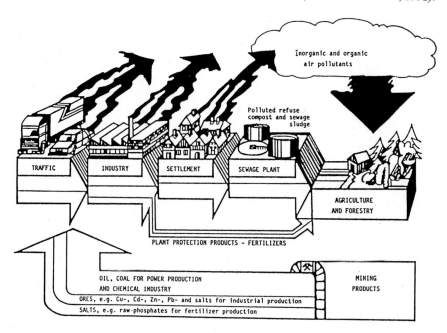

Fig. 3: Soil pollution by heavy metals through excessive use of fossil energy and raw materials (Blum, 1988).

4 Loads from technical infrastructures onto agricultural and forest land surfaces

Such a development can be seen in detail by the spatial distribution of roads, villages and cities in Fig. 2, illustrating the sealing of soils and landscapes by settlements and roads at a very large scale in the southern part of Germany. From all these linear and point sources, loads are put into the adjacent agricultural and forest areas (see also Sauerbeck, 1982), which leads to a second form of competition. This can be seen in Fig. 3, indicating that mining products are taken from inert positions in the inner part of the earth and put on to the land surface by different processes, such as traffic, industrialization and urbanization on three different pathways: the atmospheric pathway, the waterway and by terrestrial transport (Blum 1996). The main question in this context is the capacity of the biota, including humans, in the loaded terrestrial and aquatic ecosystems to withstand these loads which are still increasing every day (Blum 1994 c). In this context, it seems also necessary to point out that soils are the next to last sink for many of these inorganic and organic depositions, the last one being the bottom of the ocean.

	Lead	Cadmium	Nickel	Zinc
Rural areas	110-290	3-16	20-110	180-1800
Urban areas	365-1100	4-37	37-290	365-1100
(maximum values)	(14600)	(1100)	(4380)	(328500)
Forest areas	400-580	20-30		

Table 2: Indicative values for heavy metal deposition in the Federal Republic of Germany in $g \times ha^{-1} \times year^{-1}$ (Bundesminister des Inneren, FRG, 1985)

The worldwide mobilization factors for different elements are shown by Blum (1996). Regarding organic compounds it is known that worldwide about 100,000 different chemical compounds are produced with an annual increase of about 1,000 new substances, most of them organics. All these compounds are recycled to the soil through atmospheric transport (exhausts, e.g. gases, dust, aerosols etc.), transport in water (open lakes, rivers, ground water, irrigation water etc.) and terrestrial transport (refuse deposits, sewage sludges, use of chemicals, e.g. fertilizers, soil conditioners, insecticides, herbicides, fungicides, pharmaceutics, and others), see also Adriano (1986). An example is given in Table 2, by indicative values for heavy metal deposition in the Federal Republic of Germany in $g \times ha^{-1} \times y^{-1}$ (Bundesminister des Inneren, FRG 1985), showing that rural areas receive less heavy metal depositions in relation to forest areas and urban agglomerations. Forests filter solid, gaseous and aerosol compounds from the atmosphere. The absolute values may have changed in the meantime (Schulte and Blum, 1997), but the relative distribution of airborne pollutants in different areas with different land uses has certainly remained the same. When considering soils and measuring impacts through soil analysis, a certain precaution has to be taken, because the distribution of these heavy metals is quite different with regard to mechanical soil treatments. Whereas ploughing dilutes these elements and brings them down to ploughing depths, under undisturbed conditions, e.g. under forests and grasslands, distinctive accumulations of these elements in the uppermost soil horizons occur, as shown in Table 3.

In the following, the main forms of soil degradation caused by traffic and transport, by urban activities and by industrial processes will be presented in detail.

forest soil depth (cm)	pH	Pb	Cd	Zn	garden soil depth (cm)	pH	Pb	Cd	Zn
0-2	4,7	1020	3,37	1020	0-2	7,3	187	0,97	488
2-5	4,6	240	0,84	268	2-5	7,5	193	0,92	467
5-10	4,5	45	0,37	116	5-10	7,5	190	0,85	437
10-20	4,3	23	0,25	79	10-20	7,5	177	0,81	381
20-40	4,3	21	0,22	68	20-40	7,5	151	0,67	317
40-60	4,2	21	0,21	65	40-60	7,7	31	0,15	76
60-80	4,4	16	0,20	63	60-80	7,9	18	0,08	40

Table 3: Lead, cadmium and zinc content in $mg \times kg^{-1}$ in a vicinal forest and garden soil, under similar air pollution conditions (Häni & Klötzli, 1984)

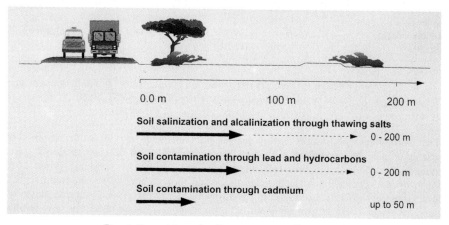

Fig. 4: Deposition of pollutants near traffic routes.

5 Main forms of soil degradation through traffic and transport

The main forms of soil degradation through traffic and transport are based on physical and physico-chemical processes through the application of thawing salts in winter (e.g. NaCl: up to 20-50 $kg \times m^{-1}$ road $\times y^{-1}$), mainly in the northern hemisphere, and soil erosion through concentration of surface water runoff from the sealed surfaces, especially in tropical and subtropical environments, but not only there. Moreover, chemical and biological soil degradation are caused by contamination with inorganic compounds, e.g. heavy metals and other elements, as well as by organic compounds, e.g. polycyclic aromatic hydrocarbons (PAHs) and others. Fig. 4 shows the deposition of pollutants near traffic routes. The distribution of different PAHs in soils near a highway in the northern part of the city of Bonn, Germany, are described by Teebay et al., 1993.

6 Main forms of soil degradation through urban activities

Soil degradation through urban activities is mainly due to the enormous consumption of air, water and other goods within towns, which can be analyzed by the flow of materials through urban agglomerations (e.g. in tons/day or tons/year). An example for the city of Vienna with 1.6 million inhabitants is given in Fig. 5. In this figure it can be seen that the town consumes every day 560,000 tons of water, 100,000 tons of air, and undefined amounts of energy, of construction material, disposals and durable goods, emitting 127,000 tons of gas and producing 550,000 tons of sewage, which does not include about 8,000 tons of solid wastes and undefined quantities of consumer goods (Daxbeck et al., 1994).

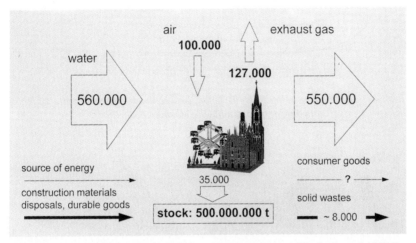

Fig. 5: Flow of goods through the city of Vienna in tons/day (Daxbeck et al., 1994).

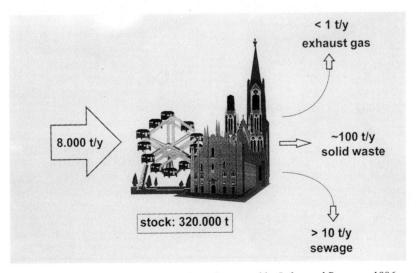

Fig. 6: Flow of copper through a city in tons/year (estimated by Lohm and Brunner, 1996 on the basis of Bergbäck et al., 1995).

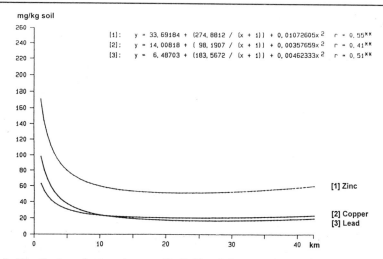

Fig. 7: Distribution of copper in top soils (0-20 cm), between metropolitan Vienna and the eastern state border, in a distance of 40 km (Köchl, 1987).

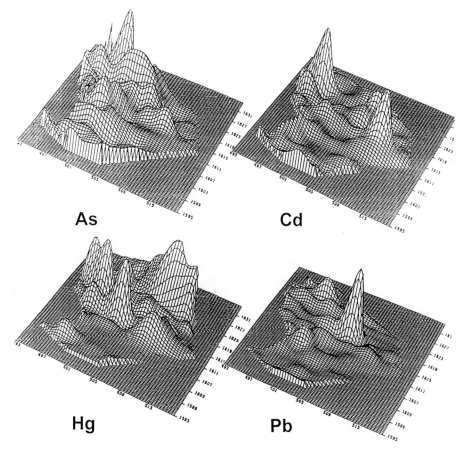

Fig. 8: Distribution of As, Cd, Hg, and Pb (mg/kg) in 373 top soils (0-5 cm) of metropolitan Manila (Manila, Caloocan, Quezon, and Pasay)/Philippines (Pfeiffer et al., 1988).

Looking at the flow of copper in the same city in tons x y^{-1} (see Fig. 6), estimated by Lohm and Brunner (1996), on the basis of data from Stockholm, Sweden (Bergbäck et al., 1995), it can be seen that about 8,000 tons of copper per year are used in the city. Only 100 tons per year leave the city as solid waste and slightly more than 10 tons per year as sewage and less than one ton as gas exhaust. The rest is added to the already existing stock of 320,000 tons. The distribution of the copper emissions can be found in the soils around the city, by depositions up to a distance of 40 km. The copper content in this area diminishes from 100 mg x kg^{-1} top soil in the city center to about 30 mg x kg^{-1} top soil at a distance of about 40 km (see Fig. 7).

The distribution of other airborn heavy metals around Vienna reveals that the deposition of lead occurs closer to the city because of the exhaustion by motor-vehicles close to the soil surface, whereas nickel, emitted by industrial processes through high chimneys, is deposited at a further distance (Köchl, 1987). These examples are not only valid for Europe but for many other parts of the world, as can be seen from Fig. 8 (Pfeiffer et al., 1988), showing the distribution of As, Cd, Hg and Pb in the metropolitan area of Manila/Philippines. Sakagami et al. (1982) analyzed the distribution of heavy metal content in surface soils of the metropolitan area of Tokyo/Japan.

Thus, soil degradation through urban activities can be subdivided into physical degradation, through sealing and erosion, by the concentration of surface water runoff and compaction, e.g. through heavy machinery on and near construction sites. Physico-chemical and biological degradation is caused on the atmospheric pathway, e.g. through the deposition of acid rain, inorganic compounds, e.g. heavy metals from combustion and corrosion of goods and organic compounds, mainly from combustion and other processes, e.g. PAHs. On the waterway, soils are contaminated through irrigation with polluted surface water and groundwater, deriving from uncontrolled outlets, insufficient sewage systems and others. Both forms of degradation are highly diffusive. Moreover, through direct sewage water application, salinization and alcalinization occurs. The main contamination, mostly in a concentrated form, is caused by controlled and uncontrolled deposition of city rubble, household refuse, ashes, slacks and other forms of waste (Burghardt, 1994; Meuser, 1993), put onto the soil through terrestrial transport, thus forming the parent material for new soil formation (Blume, 1989).

7 Main forms of soil degradation through industrial activities

Soil degradation through industrial activities occurs mainly in a concentrated form and can be subdivided into physical degradation, e.g. through the sealing of large surfaces for technical production, transport, and storage, including parking areas. Physico-chemical and biological degradation occurs similar to urban activities, but generally more concentrated and more intensive. Large-scale industrial accidents prompting public concern, show that chemical escape and hazardous waste were the cause for Seveso (1976), Bhopal (1984) and Basel (1986). Release of radioactive material occurred in Three-Mile-Island (1974) and Chernobyl (1986). Such spectacular accidents indicate only the tip of the iceberg, whereas minor contamination and pollution problems occur each day around industrial sites through the processing of pulp and paper, organic petro-chemicals, fertilizers, petroleum refining, foundries, metal works and many other industrial processes, emitting heavy metals and other compounds at a near and medium distance from the industrial production sites. An example is the industrial area in the south-eastern part of Hamburg/Germany (see Fig.9), showing the distribution of Cd, Sn, As, Cu, Pb in mg x kg^{-1} soil at two different distances from the above-mentioned industrial area. Such contaminations are even manifested in river sediments which are contaminated by sewage water (see Hintze,1982). A general picture of soil degradation in Europe by chemical deterioration is given by Van Lynden, 1995, based on dates of the GLASOD (Global Assessment of Soil Degradation) project. There are numerous further data now available regarding soil degradation by industrial activities.

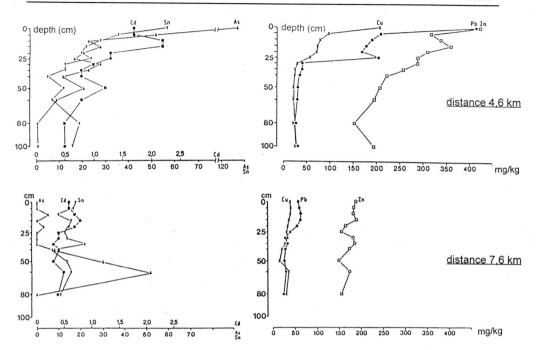

Fig. 9: Content of Cd, Sn, As, Cu, Pb and Zn (mg/kg) in soils at two distances from an industrial area in the south-eastern part of the city of Hamburg/Germany (Lux, 1982, mod.).

8 Evaluation of the physical, physico-chemical and biological degradation processes

Many of the above-cited forms of soil degradation through industrialization and urbanization, such as sealing, severe acidification, pollution by heavy metals and other elements, pollution by xenobiotic organic compounds, deposition of non-soil material, e.g. city rubble, ashes, slacks, are more or less irreversible, because soils act as a sink. "Irreversible" is defined as the non-reversibility by natural forces or technical remediation measures within 100 years, corresponding to about four human generations. Few processes of degradation, such as compaction, contamination by biodegradable organics or by small amounts of heavy metals are reversible through different techniques or soil remediation through natural forces, e.g. bioturbation and bio-accumulation.

In conclusion, soil degradation by industrialization and urbanization can be defined as an unbalanced and excessive use of one or several functions of soils at the cost and risk of others. But which are the consequences, especially in the case of irreversible degradation? How can these soils be used by future generations? Which approaches are needed to deal with those problems?

One approach was shown by Kloke, 1988 (see Fig10), indicating that with increasing concentrations of harmful elements or compounds in the soil, three different levels of land use should be distinguished. At very low levels of contamination, the multifunctionality can be maintained. At the next level, only restricted use of these soils is possible, e.g. only growth of specific food or even non-food vegetation. At the third level, remediation has to be considered. But is remediation technically feasible? Is it politically acceptable? Those questions are primarily not scientific ones, but socio-economic and political ones. Scientists can contribute by providing scenarios in order to develop sound concepts for politicians and decision-makers.

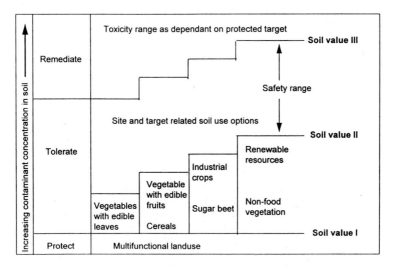

Fig. 10: A three-step system for the use of polluted soils (Kloke, 1988)

9 Conclusions

Urban and industrial activities clearly, measurably and in many cases irreversibly reduce the multifunctionality of soils.

Through these activities, the capacity of soils to produce biomass, to act as a filter, buffer and transformation medium (clean food, clean water, biodiversity), to produce and to protect genes and to be a geogenic and cultural heritage, concealing archeological and biological treasures, is constantly and increasingly and in most of the cases irreversibly reduced.

Therefore, it seems necessary to enlarge our traditional concept of soil degradation (mainly aiming at physical soil deterioration, e.g. through erosion by water and wind of agricultural and forest land), to include physical, chemical and biological forms of degradation through industrialization and urbanization.

In view of the exponential growth of urbanization and industrialization, especially in Africa, Asia and Latin America, such an enlargement of the definition of soil degradation is urgently needed.

References

Adriano, D.C. (1986): Trace elements in the terrestrial environment. Springer, N.Y.
Bergbäck,B., Hallin, P.O., Hedbrandt, J., Johansson, K. and Lohm, U. (1995): Metals in the Urban Environment. In: Metals in the Urban and Forst Environment-Ecocycles and Critical Loads, Report 4435, Swedish Environmental Protection Agency, Stockholm.
Blum, W.E.H. (1988): Problems of Soil Conservation.- Nature and Environment Series **39**, Council of Europe, Strasbourg.
Blum, W.E.H. (1994 a): Sustainable Land Management with Regard to Socioeconomic and Environmental Soil Functions - a Holistic Approach. In: Wood, R.C. and Dumanski, J. (Eds.): Proceedings of the International Workshop on Sustainable Land Management for the 21st Century. Volume 2: Plenary Papers, 115-124.- The Organizing Committee. International Workshop on Sustainable Land Management. Agricultural Institute of Canada, Ottawa, 1994.
BLUM, W.E.H. (1994 b): Sustainable Land Use for Sustainable Food Production in Africa. - Proceedings of a Seminar organised by COSTED-IBN in Accra/Ghana, 5. - 6. April 1994, pp. 92 - 107, COSTED-IBN, Madras/India, 1994.

Blum, W.E.H. (1994 c): Soil Resilience - General Approaches and Definition. - Transactions 15th World Congress of Soil Science, Vol. 2a, 233-237, Acapulco, Mexico,1994.

Blum, W.E.H. (1996): Soil Pollution by Heavy Metals-Causes, Processes, Impacts and Need for Future Actions.- Mittlg. Österr. Bodenk.Gesellsch. **54**, 53-78, 1996.

Blume, H.-P. (1989): Classification of soils in urban agglomerations. Catena 16, 269-275.

Bodenkundl. Ges. der Schweiz (1985): Boden- bedrohte Lebensgrundlage. Sauerländer Verlag Aarau, Frankfurt, Salzburg.

Bundesminister des Inneren (Ed.) (1985): Bodenschutzkonzeption der Bundesregierung.- Kohlhammer Stuttgart.

Burghardt, W. (1994): Soils in urban and industrial environments. Z. Pflanzenernähr. Bodenk. **157**, 205-214.

Daxbeck,H., Merl, A., Obernosterer, R. and Brunner, P.H. (1994): Die Stoffflußanalyse als Instrument für eine nachhaltige urbane Entwicklung.- Studie zur Wiener Internationalen Zukunftskonferenz - WIZK 94 UTEC Absorga, Wien.

Häni,H.and Klötzli, F. (1984): Schwermetalle in Klärschlamm und Müllkompost. In: Merian, E. (Ed.): Metalle in der Umwelt, 153-162.- Verlag Chemie Weinheim.

Hintze, B. (1982): Erste Ergebnisse von Untersuchungen zur Geochemie von Schwermetallen in Böden und Sedimenten des Elbtals. Mittlg. Dtsch. Bodenkundl. Ges. **33**, 95-104.

Kloke, A. (1988): Das "Drei-Bereiche-System" für die Bewertung von Böden und Schadstoffbelastung. VDLUFA-Schriftenreihe 28/2, Kongreßband 1988, 1117-1127.

Köchl, A. (1987): Die Belastung der Böden des Marchfeldes mit Schadstoffen. Österr.Ges.für Natur-u. Umweltschutz, Wien.

Lantzy, R.J. and Mackenzie, F.T. (1979): Atmospheric Trace Metals: Global Cycles and Assessment of Man's Impact.- Geochim. Cosmochim. Acta 43, 511-526.

Lux. W. (1982): Schwermetallverteilung in Böden im Südosten Hamburgs. Mittlg. Dtsch. Bodenkundl. Ges. **33**, 81-89.

Lohm U. and Brunner, P. (1996): personal communication

Möller, D. (1985): Der globale biogeochemische Schwefelzyklus.- In: Komission Umweltschutz beim Präsidium der Kammer der Technik (Hrsg): Umweltschutz in der Land- und Forstwirtschaft.-Technik und Umweltschutz, Luft- Wasser-Boden-Lärm, **31**, 35-65, VEB Deutscher Verlag für Grundstoffindustrie, Leipzig.

Meuser, H.(1993): Technogene Substrate in Stadtböden des Ruhrgebietes. Z. Pflanzenernähr. Bodenk. **156**, 137-142.

Pfeiffer, E.-M., Freytag, J., Scharpeenseel, H.-W., Miehlich, G. and Vicente, V. (1988): Trace elements and heavy metals in soils and plants of the southeast Asian Metropolis Metro Manila and of some rice cultivation provinces in Luzon, Philippines. Hamburger Bodenkundl. Arbeiten, Bd.**11**.

Sakagami, K., Hamada, R. and Kurobe, T. (1982): Heavy metal contents in dust fall and soil of the National Park for Nature Study in Tokyo. Mittlg. Dtsch. Bodenkundl. Ges. 33, 59-66.

Sauerbeck, D.(1982): Probleme der Bodenfruchtbarkeit in Ballungsräumen. Mittlg. Dtsch. Bodenkundl. Ges. **33**, 179-193.

Schroeder,D. and W.E.H. Blum (1992): Bodenkunde in Stichworten. 5th Ed. Hirt-Bornträger, Berlin, Stuttgart.

Schulte,A.and Blum, W.E.H. (1997): Schwermetalle in Waldökosystemen.- In: Matschullat J., H.J. Tobschall und H.-J. Voigt (Hrsg.): "Geochemie und Umwelt", 53-74, Springer Verlag, Berlin, Heidelberg, 1997.

Teebay, R.H., Welp, G. and Brümmer, G.W. (1993): Gehalte von polycyclischen aromatischen Kohlenwasserstoffen (PAK) und deren Verteilungsmuster in unterschiedlich belasteten Böden. Z. Pflanzenernähr. Bodenk. **156**, 1-10.

United Nations Environmental Data Report (3rd Ed.) (1991/92).

Van Lynden, G.W.J.(1995): European Soil Resources-Current status of soil degradation, causes, impacts and need for action. Nature and Environment **71**, Council of Europe, Strasbourg.

Address of author:
Winfried E.H. Blum
University of Agricultural Sciences, Institute of Soil Science, Gregor-Mendel-Str. 33, 1180 Wien, Austria

Promoting Better Land Husbandry in the Reclamation of Surface Coal-mined Land

M.J. Haigh

Summary

The Better Land Husbandry (BLH) philosophy suggests that soil conservation is best served by promoting local land management strategies that benefit both the land user and the soil. BLH strategies work with the soil's regenerative capacities to improve the vitality of the total soil ecosystem. BLH emphasises creating favourable soil structure and managing soil organic matter rather than merely preventing physical soil loss. This paper offers a case study of a BLH approach to the regeneration of opencast coal-lands in Wales. Here, many officially 'reclaimed' lands are in poor condition. Attempts to reclaim this land by masking the spoil with a thin layer of alien topsoil have not proved universally successful. Trafficking and the natural accelerated weathering of minestones have raised soil densities to levels which inhibit plant growth and foster accelerated runoff. A different approach involving soil self-creation is being developed. Countering soil compaction may be the key. While grassed spoils have a lower density and a greater capacity to absorb runoff than unvegetated, the effect is too superficial. At Blaenavon, flood peaks on reclaimed grasslands remain more than twice those on adjacent moorland. However, trees affect soil density, hydrological processes and erosion rates more dramatically than grass. Thus, long term forest fallowing may prove the best route to the regeneration of self-sustaining soils. In Wales, the problem remains to demonstrate to local communities that forestation is a valid option for the restoration of waste lands.

Keywords: Land reclamation, better land husbandry, soil quality, sustainability, forest fallowing

1 Introduction: The better land husbandry approach.

Better Land Husbandry (BLH) is a new wave in soil conservation thinking (Shaxson 1996). It is based on the understanding that accelerated soil erosion is caused by soil degradation and that this is the consequence of inappropriate land use. It suggests that preventing soil erosion is less important to agricultural sustainability than preserving soil quality. What matters is not the degree to which soil may be kept in one particular place, but the quality and health of the whole of the soil system available for agricultural production.

In respect of this, BLH strategists frame solutions that have four aspects.
1) They aim to improve the biophysical properties of soil/vegetation systems to the point where they are best capable of resisting degradation and hence do not display its symptom, accelerated erosion.
2) They aim to modify land management strategies to favour both improved soil quality and the needs of local communities.

3) They aim to foster and propagate the innovations needed to match the means of the land user with the needs of the soil
4) They intend to achieve this by working for, or under the guidance of, the land user's community.

BLH solutions contrast with those offered by traditional soil conservation in several ways. First, while traditional soil conservation stresses the prevention of physical soil loss, BLH is more concerned with overall changes in soil quality, especially the loss of voids in the soil and the vitality of the soil organic system (Shaxson 1992, Doran 1996). BLH proponents argue that it is the space in the soil where the most important soil processes occur: water movement, gas exchange, and biological activity and hence that soil conservationists would be better advised to consider the conservation of soil spaces more than their physical matrix.

Second, while traditional soil conservation seeks to establish itself as a separate, specialist enterprise, BLH prefers that its ambitions for the preservation of soil quality are seen as an integral part of local land husbandry systems. BLH proponents argue that as long as land users conceive soil conservation as something apart from the routines of agricultural production, they will be tempted to neglect it.

Third, while traditional soil conservation functions by dispensing advice through experts, who are often outside experts appointed to support policies devised by Governments or aid agencies, BLH strategists act only as facilitators. BLH proponents recognise that, when it comes to the character and capabilities of a farm's soils and the limitations in the local agricultural enterprise, the individual with the greatest understanding and expertise is likely to be the land user (Hudson and Cheatle 1994). Rather than allowing the land users to remain the passive recipients of advice and assistance, BLH strategists aim to empower local communities to develop their own existing expertise, and to gather expertise from elsewhere where required, to create a self-sustainable local economy.

BLH is best known from projects in village communities in the developing world. In practice, the social and economic aspects of BLH are the least explored and the most difficult to apply and so a large proportion of BLH literature is devoted to these topics (*vide* ABLH 1994-1996, Hudson 1994, Shaxson et al. 1989). By contrast, this paper focuses on the biophysical thinking that underpins BLH. It offers a case study of a situation where many problems are associated with the physical attributes of the soil. Broadly speaking, the biophysical argument runs as follows. 1) Accelerated runoff is encouraged by poor soil structure. 2) Accelerated erosion is encouraged by accelerated runoff, poor surface cover and unstable soil aggregates. 3) Poor surface cover and unstable soil aggregates are symptoms of low quality soils. 4) Low quality soils are encouraged by an ineffective soil organic system or inadequate organic matter management by land users. 5) Accelerated erosion and runoff is permitted by an ineffective organic system, which is often the consequence of inappropriate land husbandry.

2 BLH Approach to land reclamation

This application of BLH involves the reclamation of surface coal-mine disturbed, 'opencast', coal lands near Blaenavon in South Wales (Haigh 1995). Recent years have seen much critical debate about the quality of lands reclaimed from surface coal mines (Bridges 1992, Walley 1994). Some suggest that the future of the industry depends upon its capacity to decommission surface mines successfully (Haigh 1993). From the viewpoint of an affected community, the greatest environmental impact of mining is the production of reclaimed land. The impacts of active mining are transitory but reclaimed lands remain. Increasingly, surface mine operators realise that their acceptability to a community depends on their track record in land restoration (cf Cragg et al. 1995, Haigh 1995). Sadly, the record is far from perfect. In Wales and England over 78,000 ha are affected by mineral extraction and wastes. There are many under-reclaimed, contaminated and degrading coal-mine-disturbed lands.

Reclamation is supposed to transform an initially sterile and sometimes toxic spoil into a productive and self-sustaining soil. Present practice assumes that this can be achieved by masking the problem beneath a thin layer of alien topsoil. Frequently, this only provides a temporary cosmetic mask.

In nature, soils are complex living systems evolving over many centuries as a partnership between substrate, local climate, drainage, topography and land usage. Land reclamation subjects them to rough treatment. New topsoils, perhaps soils stored from the pre-mine environment, are carpeted over a subsoil from which that soil has become entirely divorced. Mine spoils are new materials, quite different to those of the pre-mine environment. They have new, often unfavourable, physical, chemical, and hydrological properties (Kilmartin 1989).

Clearly, such rough treatment would stress any living system. The immediate consequence is a fragile system struggling to adapt to its new situation. Sometimes, the quality of the substrate and the vitality of the soil system is such that it survives and adapts. Equally, sometimes, unfavourable influences from the subsurface, surface soil erosion, inadequate surface vegetation cover and/or inappropriate land usage, combine to overwhelm the soil system (Haigh 1992). Since the architecture of soils is created and preserved by organic processes, the consequence of soil ecosystem failure is soil structural collapse.

Compaction is a general problem of land reclamation sites, especially in black coal mining areas (Haigh 1995). It may be caused, initially, by poor land reclamation practice, and latterly by poor land management (cf Bragg 1983, Sweigard and Escobar 1989). Trafficking by ordinary farm machinery can cause compaction to depths of 45 cm, and heavy machinery to depths of 90 cm (Barnhisel 1988). However, even soils which do not suffer initial compaction suffer secondary compaction due to settlement and properties inherent to minespoils. These include chemical toxicity, which abets the dieback of the soil biota (reducing soil aggregation and the formation of macropores) and accelerated weathering.

Minestones, recently broken from compressed rock horizons, are fractured by pressure release and the process of excavation. When first exposed, surface weathering, freeze/thaw, wetting/drying, exploits these weaknesses. Poorly consolidated shales and mudstones quickly disintegrate releasing their constituent clays (Taylor 1974). Wet sieving newly-exposed spoils can find the percentage passing a 2.0 mm sieve soaring from 2 - 99%. Thus, initially coarse minespoil deposits quickly develop a compact clayey matrix.

These clays provide much of the sediment removed by erosion while that not removed accumulates in the soil as a compacted layer. On the site of the former opencast coal-mine at Varteg Hill (reclaimed: 1963), near Blaenavon, the proportion of clay in the <2mm fraction is greater in the subsoil than at the surface (>70% vs <60%). The clay-enriched layer lies at 20-40 cm depth. Below lie unweathered open textured minespoils (Haigh et al 1994).

Site, Age, Texture (Source)	Depth (cm)	Bulk Density (g/cm^3)
Pwll Du, Blaenavon, S. Wales / 45 years / mudstones, shales and sandstones (Haigh et al. 1994). Two tests: grass cover < 5% / grass cover > 30%. Sample size: 5 - 18.	00 - 05	1.66 / 1.55
	05 - 30	1.77 / 1.79
	10 - 30	1.76 / 1.77
	30 - 50	1.67 / 1.79
Blaenant, Blaenavon, S. Wales / 10-15 years / mudstones, shales and sandstones. Grass cover: >90%. Sample size: 14	00 - 05	1.39
	05 - 10	1.83 / 1.86
	30 - 50	1.87

Table 1: Bulk Density of Surface Coal Mine Disturbed Lands at Blaenavon, Wales (cf. Haigh et al. 1994)

Ramsay argues that soil bulk density is the best single indicator of land reclamation success (Ramsay 1986, p.31). Natural clay through silt-loam soils have densities of 1.00 - 1.60 g/cm^3 and sands to sandy loams: 1.20 - 1.80 g/cm^3. However, densities of 1.8 g/cm^3 and upwards drastically inhibit plant growth (Barnhisel 1988). Density records from surface coal-mined lands lie close to and often exceed this 1.8 g/cm^3 threshold. Results from the reclaimed sites of former opencast coal-mines near Blaenavon are typical (cf Table 1).

3 Hydrological impacts

In combination, clay accumulation and compaction have a major impact on runoff. Hydrological monitoring on the site of the former opencast coal-mine Blaenant (reclaimed: 1978-1986), Blaenavon, confirms that reclaimed land drains more rapidly than the land it replaces. At Blaenant, the duration of stream flow is far shorter than that from neighbouring undisturbed moorland. In around 50% of 60 storm events examined, peak flows from the reclaimed site were more than double those from the undisturbed catchment. In addition, even though flow patterns involved complex interactions between pipes and/or fissure systems, a larger proportion of the stream flow was generated by waters moving close to the soil surface (Kilmartin 1995, cf Addis et al. 1984).

Similar processes may be demonstrated at plot scale. At Blaenavon, erosion and runoff were studied by means of a standard device, the *ORSTOM Delta Lab Rainfall Simulator*. Designed to replicate convectional storm downpours on metre-square runoff plots, this has been employed for the evaluation of runoff and sediment yield in European Vineyards and in tropical agricultural fields (Gril et al. 1989). It consists of a motorised, computer controlled, swinging nozzle - mounted upon a 4 metre high frame which ensures that droplets accelerate to natural terminal velocity.

These tests compared the hydrological behaviour of unvegetated spoils, naturally revegetated spoils, and grassed, topsoil-mantled, spoils. Test plots were established on Pwll Du Opencast (reclaimed: 1947 - 1948) and the adjacent Blaenant Opencast (reclaimed: 1978 - 1986) ((Haigh et al. 1994, Kilmartin 1995). The Pwll Du tests simulated rainfall on two, 15° slope, plots - one unvegetated, the other with a thin grass cover (50%) developed on a thin humus layer of 3-5cm. The Blaenant test plot was set in pasture which had a dense surface layer of grass and moss over 10cm of applied topsoil (Table 1). Artificial rainfall was delivered at an intensity of 248 mm/hour for 5 minutes, equivalent to a 19 mm layer of water, and a summer cloudburst such as might be received once in several years.

Pwll Du reclaimed surface coal mine (1948)	Unvegetated plot on 15° slope	50% grass covered plot on 15° slope
Time to surface runoff initiation in seconds	42	46
Total depth of runoff (mm)	9.5	9.0
Volume of eroded sediment (ml)	230	60
Infiltration of rainwater (mm)	9.5	10

Table 2: Field results from Rainfall Simulation Tests at Pwll Du 1993 (Runoff from a standard storm depositing 19 mm at 248 mm/hour through 5 minutes).

Erosion losses from the unvegetated plot were about a quarter of those on the vegetated plot - a result broadly in line with long term records from the long term monitoring of erosion pins on the same site (Haigh et al 1994). Around 50% of the rainfall on the unvegetated site was converted to runoff while a little less, 47%, was returned as runoff on the thinly vegetated slope. By contrast, the topsoil layer on the densely vegetated Blaenant soils soaked up all of the artificial rainfall with no surface ponding and no surface runoff. Rainwater percolated through the topsoil and ran along at the junction between topsoil and subsoil-spoil layers. Losses of topsoil and vegetation associated with seepage scars, which develop where throughflow re-emerges at the surface, are common features on site.

4 Remedial treatment

Many researchers seek a mechanical remedy for poor drainage on reclaimed lands. However, deep ripping and subsoiling causes further disruption of the fragile minestones, the release of still more clays into the soil and greater soil compaction. A different approach is needed. Scullion (1992) inoculates surface mined lands with earthworms. Sown in strips by the thousand, their tunnelling reduces compaction while their secretions help bind newly liberated clays into stable soil crumbs.

However, the traditional way of restoring damaged soils is by long rotation forest fallowing. Trees are very efficient biomass generators adding more organic material to the soil, both above and below ground, than other plants. They are associated with relatively large arrays of soil organisms, including earthworms (Haigh et al. 1994). Their deep roots involve a greater depth of the raw minestones than grass and, with a little encouragement, penetrate to the less compacted spoil layers beneath the "cap" of trapped clays.

Preliminary results from tree planting trials at Blaenavon confirm that tree growth on minespoils is more affected by soil compaction than by variations in soil fertility. Early trials of two officially recommended methods of tree planting, pit planting and notch planting, found that, after the second year, pit planted trees grew more rapidly. Folk wisdom held that trees planted in soil pits would become 'pot bound', their roots would not penetrate the surrounding spoils. In fact, excavation of two *Fraxinus excelsior*, five years after planting as 2-year-old bare-root whips, showed that both pit-planted and notch-planted trees had extended their roots into the adjacent spoils and that the pit-planted tree had developed a more substantial root system (Flege 1995).

For forestation to be effective, a dense tree canopy must be established quickly. It is necessary that the trees close their canopy, out-compete perennial grasses, take over the soil system, and foster enough organic input to the soil to bind together newly formed fines into water stable soil aggregates. Sadly, trees grow very slowly on compacted soils. Dense planting, adding one stem per square metre or less, helps offset this problem and accelerates canopy closure (cf Gentcheva-Kostadinova and Haigh 1988). Ground preparation that allows the trees roots to get established also helps. The greatest success at Blaenavon has come from planting into slit trenches back-filled with gravel and organic mulch. This approach seems preferable to the 'quick fix' being promoted by official agencies in Wales. Spraying sewage sludge to fertilise minespoils achieves rapid tree growth (cf Weavers 1992). However, since both minespoils and sewage sludge are famously loaded with metals, the practice risks creating lands which are permanently contaminated.

Recultivating mine spoils with trees may help reduce the tendency for compaction. If these new soils drain more easily, less water remains at the soil surface and the possibility of soil erosion is reduced. If they hold water more effectively and provide a better environment for life, then the soil ecosystem will thrive, creating a self-catalysing cycle of positive environmental change.

5 Implementation of Better Land Husbandry Practices

The Better Land Husbandry (BLH) philosophy argues that soil conservation is best served by working with land users to promote locally acceptable land management strategies that benefit both the land user and the soil. In Wales, the incidence of local communities taking ownership of the issue of land reclamation is increasing. If forest recultivation is to be adopted as a appropriate practice, it must also be shown to be the best economic option for the restoration of the waste lands. At present, this view is not entirely accepted. However, the idea is being promoted by: 1) involving opinion formers, in this case environmentalists from the ngo 'Earthwatch', in the research process; 2) involving local people and school groups in tree planting; 3) enlisting support from the media; and 4) supporting those who link improved environmental quality to local economic regeneration.

Like growing trees, propagating the forest fallowing idea is a long slow process. To date, twelve *Earthwatch* Field Camps have been organised at Blaenavon involving a total of 350 environmentalist volunteers in tree planting and measurement. A first school group visited the site in March 1995. Two more followed in March 1997. Media involvement has helped publicise the approach through television, radio and newspaper reports. Inevitably, this has attracted both supporters and opponents. Opposition has come from local officials, who feel that their former work is being criticised, and from sheep farmers who oppose any loss of pasture, even low grade pasture, to forest. However, increasingly, elected officials are persuaded that environmental improvement is a route to economic regeneration in an area that remains one of the most impoverished in Europe. In 1991, Blaenau Gwent Council planned to create 3550 new jobs in 5 years. Its intention was to attract new firms to the area. To achieve this, it needed to offer potential industrialists, among other things, 'a pleasant local environment' (Smith 1991). This was prejudiced by the degraded remains of former coal mines.

Acknowledgements.

A version of this paper was presented as part of the WASWC contribution to the First European Conference and Trade Exposition of the International Erosion Control Association, Spain, June 1996, and appears in the conference 'Lecture Book' 1. The author thanks Mr Benedict Sansom and Mr Richard Pearson for their work on the rainfall simulation and the volunteers of Earthwatch for their sponsorship of the field work in Wales.

References

ABLH (1994-1996): Enable: Association for Better Land Husbandry, Newsletter 1-5. Oxford, Oxford Brookes University (SSL).
Addis, M.C, Simmons, I.G. and Smart, P.L. (1984): The environmental impact of an opencast operation in the forest of Dean. Journal of Environmental Management **19**, 79-95.
Barnhisel, R.I. (1988): Correction of physical limitations to reclamation: In: L.R. Hossner (ed.), Reclamation of Surface-Mined Lands I. Boca Raton, Fla., CRC, 191-211
Bragg, N.C. (1983): The Study of Soil Development on Restored Opencast Coal Sites. ADAS Land and Water Service Report RD/FE/9, 4pp.
Bridges, E.M. (ed) (1992): Quality of land restoration. Land Degradation and Rehabilitation **3(3)**, 153-180.
Cragg, W., Pearson, D. and Cooney, J. (1995): Ethics, surface mining and the environment. In: R.J. Singhal, A. Mehotra, J. Hadjigeorgiou and R. Poulin (eds.), Mine Planning and Equipment Selection '95. Rotterdam, Balkema, 645-650.
Doran, J.W. (1996): Soil health and sustainability. Advances in Agronomy **16**, 1-54.
Flege, A. (1995): Preliminary findings for roots excavated at Bryn Llamarch. Cincinnati, University of Cincinnati, unpublished report. 3pp.
Gentcheva-Kostadinova, Sv. and Haigh, M.J. (1988): Land reclamation and afforestation research on the coal-mine disturbed lands of Bulgaria. Land Use Policy **5(1)**, 94-102.
Gril, J.J., Canler, J.P. and Carsoulle, J. (1989): Benefit of permanent grass and mulching for limiting runoff and erosion in vineyards. Experimentation using rainfall simulation in the Beaujolais. Soil Technology Series **1**, 157-166.
Haigh, M.J. (1992): Problems in the reclamation of coal-mine disturbed lands in Wales. International Journal of Surface Mining and Reclamation 6, 31-37.
Haigh, M.J. (1993): Surface mining and the environment in Europe. International Journal of Surface Mining and Reclamation **7(3)**, 91-104.

Haigh, M.J. (1995): Surface mining in the South Wales environment. In: R.K. Singhal, A. Mehotra, J. Hadjigeorgiou and R. Poulin (eds.), Mine Planning and Equipment Selection '95. Rotterdam, Balkema, 675-682.

Haigh, M.J., Gentcheva-Kostadinova, Sv. and Zheleva, E. (1994): Evaluation of forestation for the control of accelerated runoff and erosion on reclaimed coal-spoils in South Wales and Bulgaria. In: Environmental Restoration Opportunities Conference (Munich), Proceedings, Arlington, Va., American Defence Preparedness Association, 127-149

Hudson, N.W. (1994): Land Husbandry. London, Batsford.

Hudson, N.W. and Cheatle, R.J. (eds.) (1994): Working with Farmers for Better Land Husbandry. London, Intermediate Technology Publications.

Kilmartin, M.P. (1995): Modelling rainfall/runoff on reclaimed land based on an adjacent natural catchment: a study in S. Wales, U.K. In: R.K. Singhal, A. Mehotra, J. Hadjigeorgiou and R. Poulin eds., Mine Planning and Equipment Selection '95. Rotterdam, Balkema, 717-722

Kilmartin, M.P. (1989): Hydrology of reclaimed surface coal-mined land: a review. International Journal of Surface Mining 3, 71-83.

Ramsay W.J.H. (1986): Bulk soil handling for quarry restoration. Soil Use and Management 2, 30-39.

Scullion, J. (1992): Reestablishing life in restored topsoils. Land Degradation and Rehabilitation 3, 161-169.

Shaxson, T.F. (1996): Principles of good land husbandry. Enable 5, 4-13.

Shaxson, T.F. (1993): Organic materials and soil fertility. Enable 1, 2-3.

Shaxson, T.F. (1992): Erosion, soil architecture and crop yields. Journal of Soil and Water Conservation 47, 433.

Shaxson, T.F., Hudson, N.W., Sanders, D.W., Roose, E. and Moldenhauer, W.C. (1989): Land Husbandry: A Framework for Soil and Water Conservation. Ankeny, Iowa: WASWC/SWCS.

Smith, R. (1991): Application by British Coal for Opencast Coal Extraction at Pwll Du etc. Gwent County Council, Proof of Evidence. Cwmbran, GCC.

Sweigard, R.J. and E. Escobar (1989): A field investigation into the effectiveness of equipment alternatives in reducing subsoil compaction. Mining Science and Technology 8, 313-320.

Taylor, R.K. (1974): Colliery spoil heap materials - time dependent changes. Ground Engineering, July 1974, 24-27.

Walley, C. (1994): Carving out a future? Rural Wales, Summer 1994, 22-24.

Weavers, P. (1992): Sewage sludge as an agent in reclamation to forestry. UNECE, Symposium on Opencast Coal Mining and the Environment, ENERGY/WP.1/SEM.2/R.41, 1.

Address of author:
Martin J. Haigh
World Association of Soil and Water Conservation (Europe)
C/o Department of Geography (SSC)
Oxford Brookes University
Oxford OX3 0BP, England

Towards Soil Quality Standards for Reclaimed Surface Coal-mined Lands

M.J. Haigh

Summary

Soil quality standards designed to focus land reclamation agencies on potential problems, supported by sustained post-project inspection and enforcement, may be the best route to securing the long term productivity of reclaimed lands. This paper opens discussion on the determination of appropriate measures of soil quality by discussing two aspects: soil architecture and soil pH. It suggests that 'successfully reclaimed' land should have soil bulk densities not exceeding 1.6 g/cm^3 within 50 cm, and 1.8 g/cm^3, within 100 cm of the surface and a soil pH not exceeding the bounds pH (3.0) 3.5 - 8.5 (9.0) within 150 cm of the soil surface. Since, many problems associated with soil decline develop slowly, these conditions should obtain in tests conducted at reclamation, and 10 and 20 years subsequently. It is emphasised that the environmental standards proposed are arbitrary and that the inter-relationships between bulk density, soil acidity and other aspects of the soil system are complex.

Keywords: Soil quality standards, bulk density, pH, heavy metals, post project inspection

1 Introduction

The environmental impacts of most surface coal mine activities are transitory but reclaimed land remains forever. Land reclamation after surface mining involves complete landscape reconstruction. New landforms, drained by newly created watercourses, are veneered with artificial topsoils which are enriched with chemical and organic additives. Frequently, the restoration process creates problems in the soil that lead to land degradation (cf Haigh 1992).

Although some surface coal-mine disturbed land may be reclaimed to urban or new industrial uses, much is returned to biological production. This biological productivity is a function of soil quality (cf: Blum 1987, Shaxson 1993). Soil quality on post-industrial land is, thus, a key issue for environmental regulation. Advising the creation of such legislation should be a major concern of the soil science profession. This paper starts this process by considering two fundamental issues: soil density and acidity.

2 Towards regulation

In the USA, legal action may be brought against any activity that damages the environment (Haley and O'Keefe 1994). Soon, it may be possible for land owners to secure compensation from industries that reduce soil vitality on their land. Where harm or damage is done to a third party, those responsible may have to pay compensation. Legislation already exists in many nations.

The Council of Europe's "Convention on Civil Liability" (December 1991) provides that the industrial operator/land owner shall be responsible for all damage caused in the environment, will continue to be liable even after closure, that several operators shall be jointly and severally liable and that, unless proven that damages result from activities prior to its interest, the last operator shall be liable. EU discussions include the concern that this wide definition of responsibility could inhibit the future development of lands where potential costs of damages and for clean-up may be high. Already, many insurance companies limit their liabilities by restricting cover to single catastrophic events and excluding gradual and cumulative problems (Haley and O'Keefe 1994).

Soil degradation is widespread on post-industrial lands. It is a problem more easily prevented than cured. Establishing soil quality standards which focus land reclamation agencies on potential problems, may be the best route to underwriting long-term soil quality on reconstructed lands.

3 Setting standards: Problems

Setting appropriate, meaningful, and applicable environmental standards for soils on reclaimed surface coal mine disturbed lands is a thankless task. There is a chronic shortage of research data and massive variability in the character of sites. Academic researchers, recognising this complexity, tend to create complex and impractical indicators. Industrialists consider all regulation a nuisance but prefer rules which are clear, objective, simple, attainable and which provides an appropriate factor of safety for their work. Environmentalists may appreciate the complexity of environmental processes but, being worried about enforcement, prefer measures which are both meaningful and easily, if necessary clandestinely, verified. It may be impossible to make any proposal which will satisfy all three lobbies. Whatever is proposed is certain to be regarded as crude to the point of absurdity by the scientific lobby, fussy and impractical by the industrial lobby, and lax to the point of license by the environmentalists.

The problem is compounded because there is no single, widely - accepted, index of soil quality (cf Doran 1996). The relationship between soils and bioproductivity is complex. It varies with climate, soil type, vegetation type, moisture regime, soil structure, soil texture, soil chemistry, soil ecology and land management (Boone 1986). Quality must be evaluated in terms of a number of interlinked "favourable" soil properties. Van Breemen (1993, p.186) defines a "favourable soil property" as one which allows an increases in net primary production from higher plants on a plot with uniform vegetation. Many soil properties favour biological vitality but only some are readily measured.

Some argue that soil quality may be assessed indirectly, by means of easily measured indicators of a soil's capacity to provide a favourable environment for life (Shaxson 1992, 1993, Doran 1996). In this respect, spaces in the soil (where the movements of water and gases, the growth of roots and soil (micro)organisms all occur) seem centrally important.

4 Bulk Density

Soil bulk density is called the best single indicator of land reclamation success (Ramsey 1986, p.31). It is easily measured (Baize 1992). High quality soils, such as those found under forest and old pasture, have low density. Their favourable architecture is preserved by soil aggregates of high (water-)stability which are regenerated by organic activity (Lynch and Bragg 1985). By contrast, reclaimed surface coal-mine sites, especially in black coal mining areas, often have soils of high density and low stability. Soil becomes compacted by trafficking of machinery during construction (cf Sweigard and Escobar 1989). Further compaction occurs through settling, the accelerated weathering of mudstones and shales newly exposed at the surface, and/or by the dieback of the soil biota, perhaps aided by chemical toxicity (Haigh 1992, 1995).

Increased soil density affects the soil strength and the resistance to root penetration. It reduces the number of macropores and the number of pores large enough for root hairs to enter. Associated changes include increases in the soil's capacity to hold water to its particles, decreases in its permeability to air and water, and a corresponding increase in runoff and erosion (Kilmartin 1989). High soil density encourages anaerobic soil conditions and unfavourable microbiological processes such as denitrification (Sheptukhov et al. 1982). As a result, at higher densities, soils tend to be less favourable habitats for life.

In nature, clay, clay loam, and silt loam soils may have densities of 1.00 - 1.60 g/cm^3 and sands and sandy loams from 1.20 - 1.80 g/cm^3 (Ramsey 1986). The upper ranges of these natural values are less favourable for plant growth. Agriculturalists normally set optimum soil bulk density levels below 1.3 g/cm^3 (Zhengqi Hu et al. 1992, p.131, Medvedev 1990, p.66). The 'critical bulk density', where root growth ceases, varies with soil water content, soil structure, soil texture (soil clay and silt plus clay percentages), and plant species (Jones 1983, Barnhisel 1988). Roots do not easily penetrate clays of densities above 1.46 g/cm or sandy soils of densities above 1.75 g/cm (Veihmeyer and Hendrickson 1948). In Britain, foresters advised that soil densities should not exceed 1.5 g/cm^3 to 50 cm depth and 1.7 g/cm^3 to 100 cm on disturbed lands where trees will be grown (Moffat and Bending 1992).

In sum, the upper limit for organic penetration of a soil layer is around 1.8 g/cm^3 while problems become serious above 1.6 g/cm^3 (cf Verpraskas 1988, Lal et al. 1989). So, a first nod toward a soil density standard might be that soil bulk densities on reclaimed should not exceed 1.6 g/cm^3 in the 0 - 50 cm depth layer and 1.8 g/cm^3, in any 10 cm zone of the 0 - 100 cm layer. Further, because many reclaimed soils autocompact, these conditions should obtain in checks conducted at reclamation, then 10 and 20 years subsequently.

5 Acidity and soil pH.

Soil pH is another good, easily measured, indicator of soil quality. Key plant nutrients become less available at extreme pH values whilst others become available in toxic amounts. Low pH is a common feature of industrial sites. Coal spoils often develop very low pH (1.9-4.0) through the weathering of ferrous sulphide by chemical and biological agencies (eg. Bacillus ferro-oxidans). In Britain, coal lands often begin with a pH of around 5.0 on reclamation and end up with a pH 3.0 or below after a few years (cf Doubleday 1969). The release of acid mine drainage waters is a major concern in North America and Europe.

Low pH has negative soil quality impacts. However, different plants have different tolerances to soil acidity. Nevertheless, soil bacteria and actinomycetes do not thrive, while the oxidation / fixation of nitrogen is curtailed, in mineral soils of pH 5.5 and lower, while at pH 5.0 and below, soil phosphates tend to become unavailable to plants.

The Referentiel Pedologique characterises levels below pH 3.5 as hyperacid and those between pH 3.5 and 4.2 as very acid (Baize 1992). Brady (1984, p.728) calls soils with pH levels between 5.0 and 4.5 very strongly acid, those below pH 4.5 extremely acid, and those with a pH above 8.7 very basic. It is suggested that pH 3.0 is the limit for forest recultivation of industrial wastes (Strzyszcz 1992). The British prefer the bounds pH 3.5 - 8.5 (Moffat and Bending 1992). The suggestion, here, is that reclaimed lands should not be hyperacid. A pH of 3.0 (pH 3.5 minus 0.5 - the maximum range of normal seasonal variation) might be offered as the minimum acceptable whilst, perhaps pH 9 (pH 8.5 plus 0.5) should be taken as the maximum in any 10 cm layer between 0 and 150 cm of the soil surface. This condition should obtain at reclamation and after both 10 and 20 years.

It is sometimes recommended that extreme pH materials should be buried, 1.5 metres or more below the soil surface (Moffat and Bending 1992). However, Moffat (1995) points out that trees planted on disturbed sites in Britain require minimum soil depths up to 2 and 3 metres to obtain adequate soil moisture reserves. In fact, several factors may cause buried contaminants to affect the soil surface. These include: construction, mixing through cultivation, rising in ground waters during floods, bioturbation as by worms, uptake by plants or capillary rise of moisture in dry conditions, vapour phase transfer, and erosion, especially gully incision. Thus, the best policy for dealing with such waste is neutralisation rather than burial (cf Doyle 1976).

6 Conclusion

"Reclaimed" land is the permanent end product of surface coal mining. Much of the land reclaimed in earlier decades is now in poor and deteriorating condition (Haigh 1992). Since land degradation is more easily prevented than cured, it is suggested that these problems are tackled by encouraging improvements in the quality of soil restoration. These improvements must be guided by recognised and enforced environmental standards. To be effective, such standards must be consistent, meaningful, objective, easily checked and enforced. Regrettably, because of the complexity of the soil system, these standards must also be arbitrary. Nevertheless, two simple indices are described by way of illustration.

Changes in bulk density do not necessarily correlate with plant productivity. Nevertheless, in the absence of better indicators, bulk density, which is widely recognised and easily measured, is recommended for the evaluation of reclamation success on surface-coal-mine disturbed lands (cf Ramsey 1986). Certainly, no-one argues that soils which develop higher bulk density, reduced porosity or greater strength also become more adverse environments for plant life. Much the same may be argued for soils with very low (or very high) pH.

In other contexts, it has been suggested that industry should take the lead in developing soil quality standards, because self-imposed standards are less likely to be problematic than those imposed by an outside agency (Haigh 1995, Davy 1979). However, industry might benefit from more help and advice from those with specialist skills in soil remediation. Those with a concern for the future of the environment need to take a more proactive role in determining what that environment will be like. There is little point complaining about damages done in ignorance by industrial operations if this situation is a consequence of the profession's own failure to provide guidance.

Acknowledgements

This paper was originally presented to the United Nations Economic Commission for Europe's Working Party on Coal. An extended version, aimed at the surface mining industry, was published in the International Journal of Surface Mining, Reclamation and Environment 9, 1995.

References

Barnhisel R.I. (1988), Correction of physical limitations to reclamation. In: L. R. Hossner (ed.), Reclamation of Surface-Mined Lands I, Boca Raton, Florida, CRC Press, 191-211
Baize, D. ed. (1992): Referentiel Pedologique. Paris, AFES/INRA.
Blum, W. (1987): Soil Conservation Problems in Europe. Strasbourg, Council of Europe.
Boone, F.R. (1986): Towards soil compaction limits for crop growth. Netherlands Journal of Agricultural Science **34**, 349-360.

Brady, N.C. (1984): The Nature and Properties of Soils, 9th ed., New York, Macmillan.
Davy, T.G. (1979): An industry view of the problems of dealing with land reclamation legislation. Canadian Land Reclamation Association, Proceedings **4**, 277-283.
Doubleday, G.P. (1969): The assessment of colliery spoil as a soil forming material. North of England Soils Discussion Group, Proceedings **6**, 5-13.
Doran, J.W. (1996): Soil health and sustainability. Advances in Agronomy **16**, 1-54.
Doyle, W.S. (1976): Deep coal mining waste disposal technology. New Jersey, Noyes Data Corporation Environmental Technology Review **29**, 1-392.
Haigh, M.J. (1995): Soil quality standards for reclaimed coal-mine disturbed lands: a discussion paper. International Journal of Surface Mining, Reclamation and Environment **9(4)**, 187-202.
Haigh, M.J. (1992): Problems in the reclamation of coal-mine disturbed lands in Wales. International Journal of Surface Mining and Reclamation **6**, 31-37.
Haley, G. and O'Keefe, J. (1994): Contaminated land - what implications for the City? London and Brussels, S.J. Berwin and Co.
Jones, C.A. (1983): Effect of soil texture on critical bulk densities for root growth. Soil Science Society of America, Journal **47**, 1208-1214.
Kilmartin, M.P. (1989): Hydrology of reclaimed surface coal-mined land: a review. International Journal of Surface Mining **3(2)**, 71-83.
Lal, R. Hall, G.F. and Miller, F.P. (1989): Soil degradation I. Land Degradation and Rehabilitation **1(1)**, 51-69.
Lynch. J.M. and Bragg, E. (1985): Microorganisms and soil aggregate stability. Advances in Soil Science **2**, 133-171.
Medvedev, V.V. (1990): Variability of the optimal soil density and its causes. Pochvovdeniye 1990(5), 20-30. (Eurasian Soil Science **21**, 1991, 65-75).
Moffat, A. (1995): Minimum soil depths for the establishment of woodland on disturbed ground. Arboricultural Journal **19(1)**, 19-27.
Moffat, A. and Bending, N. (1992): Physical site evaluation for community woodland establishment. Forestry Commission (U.K.) Research Information Note **216**, 1-3.
Ramsey, W.J.H. (1986): Bulk soil handling for quarry restoration. Soil Use and Management **2(1)**, 30-39.
Shaxson, T.F. (1993): Organic materials and soil fertility. Enable **1**, 2-3.
Shaxson, T.F. (1992): Erosion, soil architecture and crop yields. Journal of Soil and Water Conservation **47(6)**, 433.
Sheptukhov, V.N., Voronin, A.I. and Shipliov, M.A. (1982): Bulk density of the soil and its productivity. Agrokhimiya 1982(8), 91-100.
Strzyszcz, Z. (1992): Biological reclamation of phytotoxic Tertiary formations composing overburden of the brown coal mine in Zar region. UNECE, Working Party on Coal, Symposium on Opencast Coal Mining and the Environment, ENERGY/WP.1/SEM.2/ R.6, 1-2.
Sweigard, R.J. and Escobar, E. (1989): A field investigation into the effectiveness of equipment alternatives in reducing subsoil compaction. Mining Science and Technology **8**, 313-320.
Van Breemen, N. (1993): Soils as biotic constructs favouring net primary productivity. Geoderma 57, 183-211.
Veihmeyer, F.J. and A.H. Hendrickson (1948): Soil density and root penetration. Soil Science **65**, 487-493.
Verpraskas, M.J. (1988): Bulk density values diagnostic of restricted root growth in coarse textured soils. Soil Science Society of America, Journal 52, 1117-1121.
Zhengqi Hu, Caudle, R.D. and Chong, S.K. (1992): Evaluation of farmland reclamation effectiveness based on reclaimed mine soil properties. International Journal of Surface Mining and Reclamation **6**, 129-135.

Address of author:
Martin J. Haigh
World Association of Soil and Water Conservation (Europe)
C/o Department of Geography (SSC)
Oxford Brookes University
Oxford OX3 0BP, England

Sorption of Organic Chemicals in Urbic Anthrosols

Qinglan Wu, H.-P. Blume, L. Rexilius, S. Abend & U. Schleuß

Summary

The behaviour of organic chemicals entering the soil environment strongly depends on their sorption on soil particles. In soils from urban wasteland areas (e.g. sewage sludge, brick and mortar debris, municipal wastes, coal-mine deposits, flyash), the composition of organic matter and clay differs very much from that in natural soils. Therefore, different sorption behaviour of organic chemicals must be expected. This study investigated the sorption properties of different Urbic Anthrosols in order to assess soil affinity. The composition of organic matter in various Urbic Anthrosols was analysed, and the adsorption isotherms of three reference chemicals (atrazine, 2,4-D, pentachlorophenol (PCP)) were determined. For most of the soils studied, a linear positive correlation (r^2: 0.7 - 0.9) between $\lg(K_f)$ and $\lg(C_{org})$ could be found for each of the three chemicals. Soil samples with low pH or high coal contents deviated from this correlation. Our results suggest that apart from the contents of organic carbon and clay, the coal content and soil pH are also important parameters for the assessment of soil affinity in Urbic Anthrosols.

Keywords: Atrazine, 2,4-D, PCP, Sorption, Urbic Anthrosols

1 Introduction

Studies on the sorption of different contaminants in urban wasteland areas have received increasing attention due to the ongoing industrialisation and urbanisation in the world. The behaviour of organic chemicals in soils, such as retention, biodegradation, evaporation, transport and uptake by plants, strongly depends on their sorption in soils. Therefore, the adsorption constant is always an important parameter used in different models for predicting the fate of organic chemicals entering the soil environment (e.g. Wagenet and Rao 1990; Yong et al. 1992; Blume and Ahlsdorf 1993).

This paper focuses on the study of the sorption properties of organic chemicals in Urbic Anthrosols, a knowledge of which is still rather limited with respect to published data. This knowledge may provide the basis for the estimation of soil affinity by means of simple soil parameters. Similar investigations have already been carried out on natural soils (e.g. Kukowski 1989; Green and Karickhoff 1990; Blume and Ahlsdorf 1993).

Urbic Anthrosols are soils having, to a depth of more than 50 cm, an accumulation of wastes from mines, town refuse, fills from urban developments etc. (FAO-UNESCO 1989). As the composition of their organic matter and clay minerals differ widely from that of natural soils, divergent soil sorption properties are expected.

The sorption of three different reference organic compounds on Urbic Anthrosols was studied. The role of soil organic matter, soil minerals, soil pH and especially the role of coal on the sorption is discussed in detail.

2 Materials and methods

2.1 Soils and sampling

Soil samples of very different Urbic Anthrosols (cf. Table 1) were collected in the urban wasteland areas of the cities of Kiel, Halle, Rostock and Stuttgart, all in Germany. The samples were air-dried and sieved to 2 mm.

Horizon	Depth (cm)	pH (CaCl$_2$)	Clay (%)	C_{org} (%)	CEC (mmol$_c$/kg)	C/N
1. Urbic Anthrosol (Ockerreduktosol*) from sewage sludge						
Ah	0--20	7.4	25.9	9.2	525	11
C_1	20--38	7.4	35.5	9.9	671	12
C_2	38--55	7.8	29.6	12.0	547	14
2. Urbic Anthrosol (Norm-Pararendzina*) from brick and mortar debris						
Ah	0--20	7.3	14.4	2.1	143	17
C_1	20--45	7.6	8.3	1.0	70	24
C_2	45--80	7.5	7.2	0.6	58	34
3. Urbic Anthrosol (Pararendzina-Hortisol*) from municipal wastes						
Ah_1	0--12	6.8	8.1	7.5	209	27
Ah_2	12--30	7.1	6.3	5.1	131	39
C_1	30--53	7.3	6.6	5.2	95	58
C_3	90--106	7.4	4.6	11.1	124	61
C_4	106--150	7.4	4.1	0.6	52	63
4. Urbic Anthrosol (Humuspararendzina*) from flyash						
Ah	0--12	7.7	5.8	14.5	264	89
C_1	12--30	7.8	3.1	23.5	179	121
C_2	30--60	7.8	7.0	17.5	132	113
5. Urbic Anthrosol (Humusregosol*) from coal-mine deposits						
Ah	0-5	3.5	26.9	7.0	265	59
C_1	5-15	3.1	18.8	5.3	234	58
C_2	15-140	2.9	13.6	4.7	169	67

*German classification (AG Bodenkunde 1995)

Tab.1: Selected properties of some Urbic Anthrosols (<2 mm)

In Table 1 selected characteristics of some of the investigated Urbic Anthrosols are presented. Most of these soils have pH-values ranging from neutral to slightly alkaline. Only the soil samples from coal-mine deposits show rather low pH-values of about 3. The measured contents of organic carbon include the carbon originating from vegetation and animal sources, but also from sewage sludge, wood and different types of coal. The large difference in the C/N ratio in the C-horizon in profile 3 and in all horizons in profile 4 may be attributed to the coal-containing parent material such as burnt wood in municipal waste and the coal residue in flyash.

2.2 Reference chemicals

The adsorbates atrazine (6-chloro-N-ethyl-N'-(1-methylethyl)-1,3,5-triazine-2,4-diamine), 2,4-D (2,4-dichlorophenoxyacetic acid) and PCP (pentachlorophenol) were chosen as representatives of s-triazines, halogenated phenoxycarboxylic acids, and polyhalogenated phenols, respectively. They differ widely regarding their chemical properties and sorption mechanisms (Fig. 1).

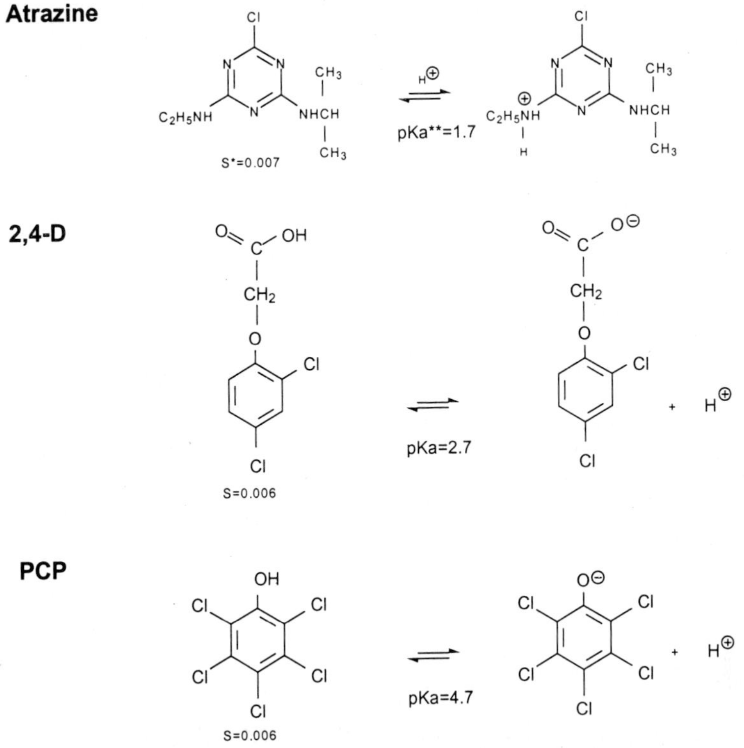

*Solubility in distilled water (g/100 g; 20°C), ** Acidity constant

Fig. 1: Acid-base equilibria of atrazine, 2,4-D and PCP and their water solubilities (Rippen 1984)

It should be considered that all of the three reference chemicals are ionizable organic compounds (IOC), and that they can exist either as protonated or deprotonated species in the aqueous phase. At low pH ($pH<pK_a$), the weakly basic atrazine can be protonated to its cationic form, while the weak-acidic 2,4-D and PCP remain in their undissociated neutral forms. As the pH increases ($pH>pK_a$), atrazine changes from its cationic to the neutral form, and 2,4-D and PCP change from their neutral to the dissociated anionic forms. Not only the charge of the molecule, but also the solubility of IOCs can differ significantly with pH. Consequently the sorptivity of IOCs may change with pH as different sorption mechanisms may be involved.

2.3 Characterization of soil organic matter

The non-humified materials in soil organic matter such as lipids, hemicelluloses, cellulose, lignin, and proteins were separately determined with wet-chemical methods (Beyer et al. 1993; Schlichting et al. 1995).

For the determination of the coal content, soil samples were first extracted with the solvent mixture toluene/ethanol 1:2.3 (v/v) to remove the lipid fraction (Wu et al. 1995). Lipid-free soil samples were then extracted with 0.5 M NaOH in N_2-atmosphere to remove the fulvic and humic acids and further treated with H_2O_2 (30 %) to oxidize the residual organic matter. The organic carbon remaining in the sample after these three treatments was defined as coal. According to the results of ^{13}C-NMR-spectroscopic analyses this fraction mainly consists of polyaromatic hydrocarbons (Wu et al., 1997).

2.4 Measurement of Adsorption Isotherms

Adsorption isotherms of atrazine (Abend et al. 1995), 2,4-D and PCP were determined using the batch-equilibrium method with 0.01 M $CaCl_2$ at 25 °C. The Freundlich adsorption constant K_f (with the concentration in solid phase in mg/kg and in liquid phase in mg/L) was used as a measure of the soil affinity for each sorbate.

4.0 g of air-dried soil (< 2 mm) were mixed with 10 ml of 0.01 M $CaCl_2$. After 16 h end-over-end shaking, the suspensions were mixed with 10 ml of 0.01 M $CaCl_2$ containing either atrazine or 2,4-D or PCP in different concentrations. After another 16 h shaking, the suspensions were centrifuged at 10,000 rpm (Cryofuge, Heraeus) for 20 min, and 10 ml of the clear supernatant were used for the determination of the sorbate concentration.

Atrazine in the supernatant was extracted into 8 ml methanol by using C_{18}-solid-phase extraction (SPE) (LC-18, Supelco) and determined by HPLC/UVD (Qiao 1992).

2,4-D and PCP in the supernatant were extracted into 10 ml of dichloromethane/acetone 9:1 (v/v) with liquid-liquid partition and determined by GLC/ECD as their methylated products after derivatization with diazomethane (Rexilius and Wu, 1995 (unpublished)) following DFG (1979).

Adsorption isotherms were also measured on soil samples after treatment with 30 % H_2O_2, i.e. after removal of the organic matter.

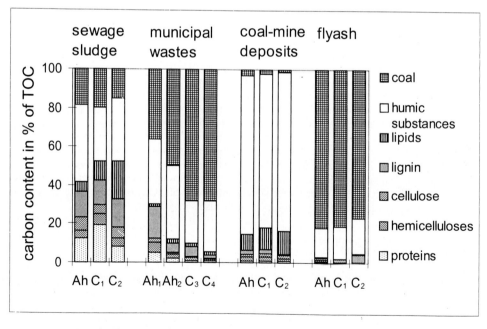

Fig. 2: *The composition of organic matter in different Urbic Anthrosols*

3 Results and discussion

3.1 Organic matter in Urbic Anthrosols

According to our knowledge of natural soils, the organic matter plays an important role in the sorption of organic chemicals. Thus, the composition of soil organic matter was further analysed. The difference in the composition of organic matter in various Urbic Anthrosols can be clearly demonstrated in Fig. 2.

High coal contents were observed in the samples of municipal waste and flyash. It is remarkable that the coal contents in soil samples from the brown-coal mine deposits are very low; instead the contents of humic substances are relatively high. This result shows that the composition of coal of different origin (e.g. charcoal, brown coal, flyash) varies in a wide range; this may lead to different soil sorption properties.

3.2 Sorption properties of Urbic Anthrosols

Measurements of the adsorption isotherms of atrazine, 2,4-D and PCP were carried out on bulk soils and H_2O_2-treated soils in order to investigate the role of soil organic matter in the sorption of organic chemicals.

Substrate (Horizon)	C_{org} %	Clay %	Atrazine K_f	K_{fm}	K_f/K_{fm}*	2,4-D K_f	K_{fm}	K_f/K_{fm}	PCP K_f	K_{fm}	K_f/K_{fm}
Sewage sludge (C_1)	9.2	35.5	7	1.4	4.1	6	4	1.6	791	232	3.4
Brick/mortar debris (Ah)	2.1	14.4	2	0.3	4.8	2	1	2.1	15	4	3.6
Municipal waste (C_3)	11.1	4.6	26	8.0	3.2	9	3	3.1	84	27	3.1
Coal-mine deposit (Ah)	7.0	26.9	261	59.3	4.4	72	6	12.9	1,874	273	6.9

* For details see text.

Tab. 2: The Freundlich sorption constants of atrazine, 2,4-D and PCP on bulk soils (K_f) and on H_2O_2-treated soils (K_{fm}) from different horizons in various Urbic Anthrosols

On the example of atrazine, the K_f-values in various bulk soils in Tab. 2 vary in a wide range from 2 (brick and mortar debris) to 261 (coal-mine deposit). The respective Freundlich sorption constants of the H_2O_2-treated soils (K_{fm}-values) are much lower than those of bulk soils. It is noticeable that the K_f/K_{fm} ratios for all tested substrates range from 3 to 5. This indicates that the humic substances are the dominant sorbents for atrazine. The same conclusion can be drawn from the results obtained for 2,4-D and PCP, respectively. However, for soils with very low organic carbon contents the role of clay minerals can become more important.

Fig. 3 shows the influence of organic matter on the sorption of atrazine, 2,4-D and PCP. For the majority of the soil samples tested, a linear positive correlation between $lg(K_f)$ and $lg(C_{org})$ could be found for each of the three compounds. The outliers, which exhibit a very high soil affinity, are the samples with either low pH-values or high coal contents.

Similar relationships were also found in natural soils with pH-values ranging from 5 to 7 (Kukowski 1989). Comparing the two correlation curves for the sorption of atrazine in natural soils and in Urbic Anthrosols, two observations should be pointed out: a) The slopes of both correlation curves lie at about one, i.e. the K_f-value increases almost proportionally with the increasing organic carbon content. This result justifies the use of K_{OC}-values ($K_{OC}=100 \cdot K_f/C_{org}$) for the assessment of the sorption properties of most Urbic Anthrosols, and b) The intercepts of the correlation curves of natural soils and Urbic Anthrosols are very similar for each of the three reference chemicals. This implies that the organic matter in different soils shows similar affinities to each of the three sorbates. This fact appears somehow surprising in view of the big differences in the composition of their soil organic matter (cf. Fig. 2). It seems that the hydrophobic interaction is the dominant bonding force for the sorption of these three chemicals and that the sorption is not site-specific.

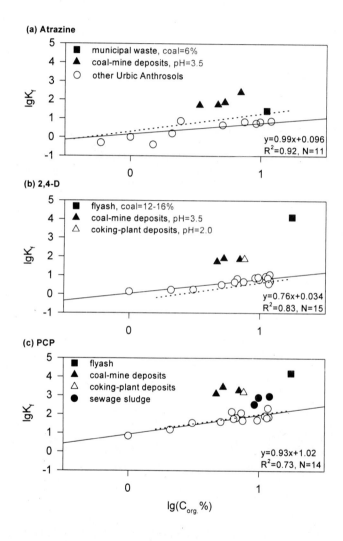

Fig. 3 : Correlation between the organic carbon content and the Freundlich sorption constant (K_f with the concentration in solid phase in mg/kg and in aqueous phase in mg/L) for (a) atrazine, (b) 2,4-D and (c) PCP on Urbic Anthrosols ___ and the corresponding correlation found in natural soils by Kukowski (1989) ... .

The deviation of the results for acid soils from deposits of coal-mines and coking plants can be attributed to the lower soil-pH. pH-dependent charge development of the soil components and protonation of the chemicals may be involved. The effect of pH on the sorption in natural soils has already been described in the literature (e.g. Kukowski 1989; Green and Karickhoff 1990; Stapleton et al. 1994).

Atrazine is a weakly basic chemical with a pK_a value of 1.7. Due to the enhanced protonation on the normally negative-charged surface of soil particles, considerable amounts of atrazine can be protonated at a pH-value two units above the pK_a (Green and Karickhoff 1990). There is abundant evidence for cation-exchange effects as the predominant bonding mechanism for atrazine in acid soil (e.g. Senesi and Testini 1980). On the other hand, it is also known that atrazine is not protonated at soil pH-values greater than two units above its pK_a and that other weak interactions, such as H-bonding and hydrophobic attractions, become more important (Bouchard and Lavy 1985; Laird et al. 1994; Martin-Neto et al. 1994).

Weakly acidic organic chemicals, such as 2,4-D (pK_a: 2.7) and PCP (pK_a: 4.7) may exist either as undissociated molecules or as their corresponding anions. At lower pH, adsorption occurs through hydrophobic bonding of the undissociated molecules, and the sorption is not site-specific. Adsorption increases with a decrease of pH, which is caused by a decrease in the percentage of the dissociated anion (Kukowski 1989; Cheng 1990; Stapleton et al. 1994).

The extremely high affinity of flyash to organic substances may be caused by its high coal content, which interacts in a characteristically different manner with organic chemicals. Furthermore, the porosity and specific surface area which can vary from coal type to coal type, markedly affect their sorption properties. Thus, the role of the different coal-types must be thoroughly considered as a special problem in the study of the sorption properties of Urbic Anthrosols.

4 Conclusions

The present study has shown the similarity and specificity of the sorption properties between Urbic Anthrosols and natural soils. From the results elaborated in this study, the following conclusions can be drawn:
- Apart from the organic carbon content, the clay content and the soil pH, the **coal content** is an important parameter which can strongly affect the sorption properties of Urbic Anthrosols.
- Linear correlation between $lg(K_f)$ and $lg(C_{org})$ was found for the sorption of atrazine, 2,4-D, and PCP by Urbic Anthrosols with neutral pH and low coal contents.

According to the close relationship between $lg(K_f)$ and $lg(C_{org})$, soil affinity to organic chemicals similar to the three tested chemicals can be estimated from the organic carbon content, under the condition that the Urbic Anthrosols have neutral soil pH and not too high coal contents. In case of higher clay contents, the role of clay minerals has to be adequately considered.

The estimation of soil affinity based on the organic matter ($C_{org}\%$) has already been successfully used for natural soils (Karickhoff 1981; Mingelgrin and Gerstl 1983; Green and Karickhoff 1990). The suitability of this method is restricted to non-ionic or weakly basic (e.g. atrazine) or weakly acidic chemicals (e.g. 2,4-D and PCP), whose sorption is dominantly based on hydrophobic interactions and is not site-specific.

In the case of Urbic Anthrosols, more research is still needed to develop an effective concept for estimating soil affinity to different chemicals. The reliability of the elaborated correlation between $lg(K_f)$ and $lgC_{org}\%$ should be verified on a number of different soils and with other reference chemicals. Furthermore, the influence of soil pH and coal content should be investigated in greater detail.

Acknowledgments

The authors sincerely appreciate the excellent technical assistance of Mrs. D. Rexilius who carried out the analyses for the determination of the adsorption isotherms of 2,4-D and PCP and of soil organic matter. This research project is part of a cooperation between the universities of Berlin (TU), Halle, Kiel, Rostock and Hohenheim, and was financially supported by the Federal Ministry of Education, Science, Research and Technology (BMBF), Bonn (Project No. 033195A3).

Symbols

C_{org}: Content of organic carbon in soil (expressed in %)
K_a: Acid dissociation constant
K_f: Freundlich adsorption constant (concentration in solid phase: mg/kg; concentration in aqueous phase: mg/L)
K_{fm}: Freundlich adsorption constant of soils treated with 30 % H_2O_2
K_{OC}: The ratio $100 \times K_f / C_{org}$.

References

Abend, S., Wu, Q. and Blume, H.-P. (1995): Atrazin-Bindung in Böden anthropogener Substrate, Mitteil. Dtsch. Bodenkdl. Ges. **76**, 199-202

AG Bodenkunde der Geologischen Landesämter (1995): Bodenkundliche Kartieranleitung, 4. Auflage Schweizerbart, Stuttgart, 392 pp.

Beyer, L., Wachendorf, C. and Köbbemann, C. (1993): A simple wet-chemical extraction procedure to characterize soil organic matter (SOM), I. Application and recovery rate, Commun. Soil Sci. Plant Anal. **24**, 1645-1663

Blume, H.-P. and Ahlsdorf, B. (1993): Prediction of pesticide behaviour in soil by means of simple field tests, Ecotox. Environm. Safety **24**, 313-332

Bouchard, C.D. and Lavy, T.L. (1985): Hexazinone adsorption-desorption studies with soil and organic adsorbents, J. Environ. Qual. **14**, 181-186

Cheng, H. H. (1990): Organic residues in soils: Mechanisms of retention and extractability, Int. J. Environ. Anal. Chem. **39**, 165-171

DFG (Deutsche Forschungsgemeinschaft) (1979): Rückstandsanalytik von Pflanzenschutzmitteln; Mitteil. VI der Senatskommission für Pflanzenschutz-, Pflanzenbehandlungs- und Vorratsschutzmittel, Methodensammlung der Arbeitsgruppe „Analytik", 5. Lieferung [Methode 43-(480)-(8)], VCH-Verlagsges. mbH, Weinheim (Germany)

FAO-UNESCO (1989): Soil Map of the World - Revised Legend; ISRIC Technical Pap. 20, Wageningen (The Netherlands) 138 pp.

Green, R.E. and Karickhoff, S.W. (1990): Sorption estimates for Modeling. in: H.H. Cheng (ed.): Pesticides in the Soil Environment: Processes, Impacts, and Modeling. SSSA Book Series No. 2, Madison, WI (USA), 79-101

Karickhoff, S. W. (1981): Semi-empirical estimation of sorption of hydrophobic pollutants on natural sediments and soils, Chemosphere **10**, 833-846

Kukowski, H. (1989): Untersuchungen zur Ad- und Desorption ausgewählter Chemikalien in Böden, SchrR. Institut für Pflanzenernährung und Bodenkunde, Nr. 7, Universität Kiel, 190 pp.

Laird, D.A., Yen, P.Y., Koskinen, W.C., Steinheimer, T.R. and Dowdy, R.H. (1994): Sorption of atrazine on soil clay components, Environ. Sci. Technol., **28**, 1054-1061

Martin-Neto, L., Vieira, E.M. and Sposito, G. (1994): Mechanism of atrazine sorption by humic acid: a spectroscopic study, Environ. Sci. Technol., **28**, 1867-1873

Mingelgrin, U. and Gerstl, Z. (1983): Reevaluation of partitioning as a mechanism of non-ionic chemicals' adsorption in soils, J. Environ. Qual. **12**, 1-11

Qiao, X. (1992): Das Verhalten von Atrazin und Isoproturon in ausgewählten Böden aus Mittelhessen: Abnahme, Abbau und Verlagerung. Dissertation, Universität Gießen, 130 pp.

Rexilius, L. and Wu, Q. (1995): Gaschromatograph Determination of 2,4-D and PCP as their methylated products in Urbic Anthrosols. Pflanzenschutzamt des Landes Schleswig-Holstein, Kiel (Unpublished).
Rippen, G. (1984): Handbuch der Umweltchemikalien. Ecomed, Landsberg
Schlichting, E., Blume, H.-P. and Stahr, K. (1995): Bodenkundliches Praktikum, 2nd edition, Blackwell, Berlin (Germany), 280 pp.
Senesi, N. and Testini, C. (1980): Adsorption of some nitrogenated herbicides by soil humic acid, Soil Sci. **130**, 314-320
Stapleton, M.G., Sparks, D.L. and Dentel, S.K. (1994): Sorption of pentachlorophenol to HDTMA-Clay as a function of ionic strength and pH, Environ. Sci. Technol. **28**, 2330-2335
Wagenet, R.J. and Rao, P.S.C. (1990): Modeling Pesticide Fate in Soils. in H. H. Cheng (ed.): Pesticides in the Soil Environment: Processes, Impacts and Modeling, SSSA Book Series No. 2, Madison, WI (USA), 52-77
Wu, Q., Schleuß, U. and Blume, H.-P. (1995): Investigation on soil lipid extraction with different organic solvents, Z. Pflanzenernähr. Bodenk. **158**, 347-350
Wu, Q., blume, H.-P. and Schleuß, U. (1997): Humuskörper. In: H.-P. Blume and U. Schleuß (eds.), Bewertung anthropogener Stadtböden. SchrR. Inst. Pflanzenernähr. Bodenk. Universität Kiel **39, 22-31, 346**
Yong, R.N., Mohamed, A.M.O. and Warkentin, B.P. (eds.) (1992): Principles of Contaminant Transport in Soils, Developments in Geotechnical Engineering **73**, Elsevier, Amsterdam, 327 pp.

Addresses of authors:
Qinglan Wu
Hans-Peter Blume
Institute of Plant Nutrition and Soil Science
University of Kiel
Olshausenstraße 40-60
D-24118 Kiel, Germany
Lutz Rexilius
Amt für ländliche Räume Kiel
Abteilung Pflanzenschutz
Westring 383
D-24118 Kiel, Germany
Sven Abend
Institute of Inorganic Chemistry
University of Kiel
Olshausenstraße 40-60
D-24118 Kiel, Germany
Uwe Schleuß
Ecological Research Centre
University of Kiel
Schauenburgstraße 112
D-24118 Kiel, Gemany

Monitoring of Soil Degradation Caused by Oil Contamination

J. Ammosova & M. Golev

Summary
The influence of oil pollution on surface reflectivity of grey-brown (Luvisols) and podzolic soils (Podzoluvisols) as well as on reflectivity of samples of the same soils in the laboratory were studied. The results show that strongly oil contaminated-soils can be ascertained by remote sensing in cases where unpolluted soils are also present in the area.

Keywords: Soil pollution, oil contamination, remote sensing

1 Introduction

The problem of soil pollution caused by petroleum and oil products is of great importance in environmental protection.
Methods for determining oil content in soil are based on extracting soil samples using different organic solvents (Orlov & Ammosova, 1994).
Remote sensing is another tool for obtaining high resolution information very new quickly. The possibility of using remote sensing to assess soil oil pollution using soil reflectivity, in conjunction with other analytical methods, is very important for monitoring ongoing soil degradation (Orlov et al., 1993).

2 Materials and methods

We studied the influence of soil pollution on surface reflectivity of grey-brown (Luvisols) and podzolic soils (Podzoluvisols) as well as the influence of oil pollution on the reflectivity of gray-brown soils. More than 100 soil samples were obtained in various regions of the Apsheron peninsula. These samples had varying levels of oil pollution in loamy sand, sandy clay loam, and sandy loam mechanical composition (Orlov et al., 1989, 1990, 1993, 1994).
We also studied the influence of oil pollution on the reflectivity of podzolic soils (Podzoluvisols) in a model-ling experiment. Table 1 contains some characteristics of the podzolic soils.
Reflection spectra (from 400 to 750 nm) were obtained using a spectrophotometer "SF-14". Hydrocarbon content was determined by extraction with hexane using a "Soxtec" apparatus to determine if any correlation existed between soil reflectivity and content of oil hydrocarbons.
An estimation of oil pollution according the quantity of C_{org} was determined by dry combustion of samples in oxygen flow on a C automatic analyser (system AN-7529, Russia).

3 Results and discussion

Non-contaminated gray-brown and podzolic soils had an increase to the red zone reflec-tion lines of spectras, without visible bands (Fig. 1). The reflection coefficients were 13-15% in the blue-violet zone of the spectra and increased to 32-35% in the red zone. The integral reflection is formed at 27-30%.

The reflection ability decreases for soil under oil pollution. The lines of the reflection spectra are nearly horizontal under high soil pollution and sometimes a decline to the red zone of the spectra occurs (Fig. 1). The reflection coefficient fell in the blue-violet zone spectra to 9-11%, and in the red zone spectra to 8-13%. The integral reflection of highly polluted soil was 10-12%.

A reflection spectra of oil polluted soil samples after hexane extraction was similar to that of unpolluted soils (Fig. 1). The levels of integral reflection decreased to 25-30%, but such data was lower than that for background soils (Tab. 2).

An inverse relation was found between level of soil oil pollution (calculated by hexane extract) and reflective ability. The correlation coefficient between integral reflection and content of hydrocarbon of oil was -0.85 to -0.89 at a level of probability P=0.95.

The reflectivity of soil decreased under oil pollution. Soils contaminated by oil after the procedure of extraction by hexane had similar spectra to non-polluted soils. The graphs used data of integral reflection and content of oil hydrocarbons. This dependence is described by an exponential equation:

$$\rho_\Sigma = \rho_{\Sigma(min)} + Ae^{-k*(\%HC)}.$$

where ρ_Σ - integral coefficient of reflection, HC - content of hydrocarbons of oil, $\rho_{\Sigma(min)}$ - minimum reflection of strongly polluted soil, A and K - coefficients.

The level of soil pollution by oil can be determined tentatively by graphs of integral reflection dependence.

On the basis of experimental data the equation parameters were calculated for gray-brown soils of various mechanical compositions. For podzolic soils we obtained very similar results in the model experiment (Table 1).

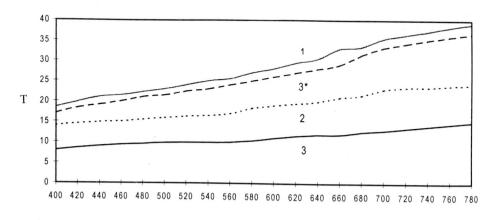

1 - unpolluted soils; 2 - mildly polluted soils; 3 - strongly polluted soils; 3* - after hexane extraction

Figure 1: The reflection spectra of gray-brown soils versus various levels of oil pollution.

Location	Horizon	Depth, cm	Texture	C_{org}, %	A	$\rho_{\Sigma(min)}$	k
Tzarevo, Pushkin district, Moscow region	Ah	0-12	Sandy loam	2.06	13.4	10.5	0.19
	Al	13-25	Loamy sand	0.38	21.5	10.6	0.41
	AlBt	26-51	Loamy sand	0.25	28.0	10.9	0.35
	Bt	52-90	Sandy loam	0.15	25.4	10.0	0.33
Tzarevo, Pushkin district, Moscow region	Ah	0-11	Sandy loam	1.97	13.6	10.7	0.19
	Al	12-23	Loamy sand	0.31	21.4	10.6	0.41
	AlBt	24-54	Loamy sand	0.25	27.8	10.9	0.33
	Bt	55-107	Sandy loam	0.15	25.2	10.1	0.32
Chashnikovo, Zelenograd district	Ah	0-13	Loam	2.58	15.9	11.5	0.21
	Al	14-31	Loamy sand	0.34	18.0	11.1	0.37
	AlBt	32-63	Sandy loam	0.14	18.8	10.4	0.34
	Bt	64-98	Loam	0.12	16.5	11.4	0.28
Chashnikovo, Zelenograd district	Ah	0-10	Loam	2.27	16.0	11.5	0.21
	Al	11-29	Loamy sand	0.33	18.0	11.1	0.37
	AlBt	29-42	Sandy loam	0.14	19.1	11.0	0.34
	Bt	43-88	Loam	0.12	19.6	11.1	0.28
Chashnikovo, Zelenograd district	Ap	0-25	Clay loam	3.31	12.9	14.5	0.10
	Al	26-37	Clay loam	0.78	18.5	14.4	0.21
	Bt	38-58	Clay	0.19	22.4	14.4	0.13
	Bw	59-100	Clay loam	0.11	24.0	13.9	0.11
	C	100-...	Clay loam	0.07	26.8	13.5	0.10
Chashnikovo, Zelenograd district	Ap	0-25	Clay loam	2.98	13.2	14.5	0.27
	Al	26-37	Clay loam	1.04	18.9	14.3	0.18
	Bt	38-58	Clay	0.30	22.5	14.4	0.14
	Bw	59-100	Clay loam	0.14	24.0	13.7	0.11
	C	100-...	Clay loam	0.05	26.8	13.7	0.10

Table 1: Characteristics of podzolic soils (Podzoluvisols) from the Moscow region (A, ρ_Σ (min), see chapter 3)

Depth	Coefficient of reflection ρ_Σ, %		
	before extraction	after extraction	
cm		hexane	chloroform
Point 1 - 40 years after pollution			
0-10	9.70	25.6	-
10-20	13.0	27.0	27.4
20-60	14.9	33.0	27.6
60-80	15.6	30.4	29.1
Point 2 - 100 years after pollution			
0-7	15.3	25.9	26.1
7-42	27.7	28.6	28.1
42-55	18.5	23.6	25.8
55-69	21.9	46.9	27.7
69-100	33.9	35.2	31.3

Table 2: The reflection ability of gray-brown oil polluted soils before and after extraction of oil components.

An estimate of soil oil pollution can be determined by graphs of integral reflection dependence (Fig. 2).

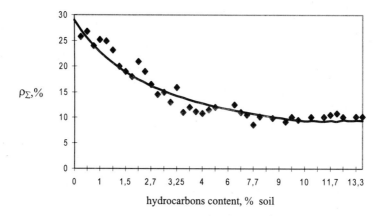

Figure 2: The dependence of reflection from oil hydrocarbons content. Line is loam sandy gray-brown soil. $\rho\Sigma$ is the integral coefficient of reflection.

We have distinguished the following levels of oil pollution of gray-brown soils, which corresponds to levels of integral reflection:

	ρ_Σ
unpolluted	more than 27
mildly polluted	27-22
moderately polluted	22-14
strongly polluted	less than 14.

The change in reflection ability can permit a quick evaluation of soil conditions in regions of oil-field and oil-pipelines. This method is easy and economical for obtaining regional characteristics of soils.

Remote sensing of soils requires much experimental data on the reflection ability of oil and oil-products polluted soils.

A method to determine oil pollution in soils by their reflectence does not require the extraction of pollutants from samples. The method can be used for satellite and other types of monitoring of soils, but the quantity of oil pollution can be determined by this method only for light coloured soils (podzolic, brown, nut-brown, gray-brown, serozem, etc.)

The total quantity of carbon-containing pollutants in soils, silts and rocks can be determined using the difference between C_{org}-contents in polluted and non-polluted sections with the background content of C_{org}.

We used a standard laboratory method for determining C_{org}-content by dry combustion in oxygen or air flow (Betelev, 1994). The quantity of C_{org} is determined from the quantity of CO_2 produced due to combustion of the organic matter sample.

Experiments have shown that oil pollution of natural objects can be measured by this method of analysis. The oil contains 84-88% (on average 86%) of carbon. Thus, the content of oil can be calculated using C_{org}-content and a conversion factor of 1.16.

Similar calculations can be performed for any kind of carbon-containing pollutant if its chemical composition is known.

The analysis by the proposed method was carried out using gas-analysers, CHN-analysers and instruments for micro-analysis in which C_{org}-content is determined by dry combustion. For our analysis we used the gas-analyser AH-7529 made in Russia.

The oil pollution is measured using the difference between C_{org}-contents in polluted and non-polluted sections of samples. But in actual situations, this C_{org} in non-polluted soils is very difficult to determine exactly.

It was ascertained in our model experiment that an inverse relation between integral reflectivity and content of total carbon is exponential for podzolic soils treated by various doses of oil. Thus the value of total carbon can be used instead of content of hydrocarbons extracted by hexane to determine pollution level in soils, silt and deposits. It is probably possible to estimate this level comparing values of total carbon content in contaminated and non-polluted soils, but only in conjunction with the previously-described method of spectral reflectivity.

4 Conclusions

Methods of chromatography, mass-spectrometry, infra-red spectrometry and luminescence are traditionally used for investigating organic pollution. All these methods demand extraction of pollutants from samples through the action of organic solvents. Such extraction is not always complete. Orlov and Ammosovà (1994) extracted oil components using organic solvent hexane C_6H_{14} from three types of soils which were artificially polluted by oil. The initial concentration of oil in polluted soils varied from 4 to 15%; the period of interaction between soil and polluting oil before extraction varied from 0 to 5 months. The average per cents of extraction of oil components by hexane were: in turfy-podzol soil 66%, in chernozem soil 61%, and in grey-brown soil 63%. The percent extraction had a tendency to decrease with increasing period of pollution, especially for old pollutants in which pollutants had been transformed into forms difficult for extraction as a result of oxidation and microbiological processes. Thus, in many cases traditional methods of analysis connected with extraction can give incorrect estimates of pollution. In addition, the methods of chromatography, mass-spectrometry and infrared spectrometry permit the investigation of only some components of pollutants and do not determine the general quantity of pollution. Lumines-cence determines the quantity of pollutants only at low concentrations (no more than 0.1-0.5%). All these methods are not good for monitoring soil degradation caused by oil contamination.

Dehumified and polluted soils over vast areas can be attested by means of remote soil monitoring to reveal quantitative dependence between the coefficients of spectral reflectivity and physical and chemical soil properties in natural and industrial landscapes.

References

Betelev N.P. (1994): A new method for determination of C_{org} in soils and rocks to estimate pollutant levels. Proc. of 1-st Int. Cong. on Environment Geotechnics. Edmonton, 133-136 (in English).

Orlov, D.S., Ammosova, Ja.M. & Bocharnikova, E.A. (1989): The influence of oil pollutants on reflection spectra of gray-brown soils of Apsheron. Biol.Sci. **5**, 92-95 (in Russian).

Orlov, D.S., Ammosova, Ja.M., Bocharnikova, E.A. & Lapuchina, O.V. (1990): Use of reflection capacity of oil polluted soil in aerospace monitoring. Proc.All Union. Meeting on "Aerospace methods in Soil Science and their use in agriculture", Moscow, Nauka, 161-166 (in Russian).

Orlov, D.S., Ammosova, Ja.M. & Bocharnikova, E.A. (1993): Distance monitoring of oil polluted soils by method of soil reflection spectra. Proc. Inter. SPJE Optical monitoring of the environment. Proceedings Series. V.2107, 307-314 (in English).

Orlov, D.S. and Ammosova, Ja.M. (1994): Methods for control of soils polluted by oil and oil products. Soil-ecological monitoring and protection of soils. D.S.Orlov and V.D.Vasilievskaya (eds.), Publishing House of Moscow State University: 219-231 (in Russian).

Address of authors:
Iana Ammosova
Michael Golev
Soil Science Department
Moscow State University
Vorob´evy Gori
119899 Moscow, Russia

Soil Physical Processes in Sealing Systems of Waste Dumps Affecting the Environment

Th. Baumgartl, B. Kirsch & M. Short

Summary

Mineral substrates are often used as a natural sealing material to avoid pollution or for reconstruction of landscapes. However, any decision about the utilization of a certain material needs prior knowledge about its properties and dynamic behaviour in response to environmental conditions. An important factor is the hydraulic setting of the constructed system and its reaction to changes. The main factors influencing the hydraulic properties are climate, vegetation and soil water balance.

Material rich in clay shows a great capacity for swelling and shrinkage and hence crack formation. This process increases mechanical stability, which is necessary in the longterm for a self-sustainable system. Such an increase in the degree of structure often goes along with an increase in the hydraulic permeability, which in the given case shall be avoided. Substrates rich in salt show a more pronounced interrelation of these processes. In dry conditions soil can be even more stabilised, yet shows high hydraulic permeabilities due to the intensive flocculation capacity. With wetting the structure collapses and can cause severe problems resulting in erosion or slumping of slopes.

Therefore the properties of mineral substrates must be investigated and interpreted very thoroughly to understand their behaviour under possible existing environmental conditions in order to avoid any negative outcome of their usage.

Keywords: Hydraulic permeability, soil mechanics, crack formation, swelling and shrinkage, erosion

1 Introduction

The increasing number of waste disposal sites and increasing awareness of the problems caused by waste dumps makes it necessary to minimize or prevent environmentally negative impacts. This can be achieved by creating a safe and self-sustainable system. One system is the encapsulation of waste dumps by mineral substrates.

However encapsulation is exposed to several possible stresses including: mechanical stress, hydraulic stress, hydro-mechanical stress and physico-chemical stress.

Such stresses influence deformation, consolidation, structure change by hydraulic stress, tensile stress, structure change by aging, hydrodynamic stress and seepage water. Furthermore heat production by metabolic processes in the waste body and climatic conditions influence the properties of the sealing. Chemical reactions can be important in sealings where oxygen flux is not prevented by the sealing and can result in the cause of acid mine drainage.

Therefore a system has to be created which:
- maintains the hydraulic permeabilities within the allowed limits,
- reduces the effect of swelling and shrinkage and hence crack formation,
- is resistant to mechanical stresses (shear resistance),
- reduces infiltration without erosion,
- includes a high capacity for tensile strength.

Some of the characteristics are in contradiction to the properties of the material. Low hydraulic conductivity is not only achieved by a high degree of compaction of the material but also by using clay rich material as the main texture. However clays are very capable for water storage and show a high capacity of swelling and shrinkage. This process of volume change and particle orientation by water bound forces very often results in the development of cracks with each process of wetting and drying, and eventually in an increase of the hydraulic permeability. In addition mechanical strength (shear strength, tensile strength) is affected not only by the existence of water, but also by the material properties of the soil material in respect to the behaviour in a water saturated and unsaturated system.

Consequences of the processes which take place in the encapsulation which is exposed to the atmosphere are different from those of the bottom of the landfill. The top part of the encapsulation is exposed to climatic and biotic (vegetation) influences and almost without overburden. The bottom sealing is partly carrying high loads, i.e. mechanical stability is required. Processes occur which are difficult to determine precisely such as heat production, and hence water movement can cause crack formation.

Two examples, one focusing on the top sealing and one on the bottom sealing, will show factors and processes which can lead to a change of the system to an environmentally more negative state.

2 Top sealing

A recultivated site (age 10 years) of a coal mine in Queensland, Australia, shows remarkable signs of gully (max. depth: 2 m) and subsoil erosion, respectively. This site was covered with spoil (non-developed 'soil' substrate), whereas a site next to it was covered with soil. This site did not show obvious signs of erosion. The orientation of both slopes was the same, as well as the slope length and inclination.

2.1 Methodology

On both slopes 'soil' and 'spoil' a plot of 100 m x 200 m was chosen for detailed investigations to understand and clarify the processes which caused the erosion. The following aspects were investigated:
- vegetation cover: the percentage of the soil surface covered by vegetation was estimated by dividing the plot in 48 subplots (1 x 1 m^2), which were distributed evenly over the plot area.
- biomass production: determination of the dry matter production of the vegetation (mainly grass: buffel grass, Rhodes grass) by harvesting the biomass at different levels above surface (0-25 cm, 25-50 cm, 50-75 cm, 75-100 cm, >100 cm) in one randomly chosen subplot of each plot
- bulk density of soil samples of one randomly chosen subplot of each plot in the depths 0-5 cm, 25-30 cm, 50-55 cm
- acidity (average of total depth 0-50 cm)

Soil physical processes in sealing systems of waste dumps 799

- mechanical parameters: determination of cohesion and angle of internal friction in the laboratory. Sampled soil material of two depths of each plot was refilled in soil cylinders with the corresponding bulk density of that depth. The tests were carried out after precompaction of the soil samples with different loads (0, 10, 20, 30, 50, 70, 100 kPa).

2.2 Results

The site 'spoil' was not evenly covered by vegetation (Fig. 1). The vegetation cover of the plot varied from 5% to 100 %, the average percentage of the spoil covered by vegetation was < 50%. The site 'soil' was nearly fully covered by vegetation. The plant species were the same but the distribution of the species was not determined.

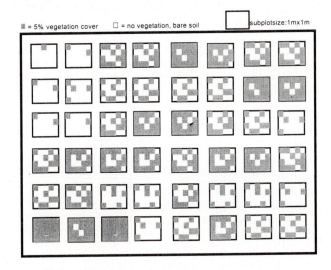

Fig. 1: Distribution of vegetation cover of site 'spoil'

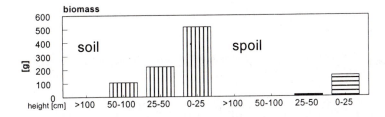

Fig. 2: Biomass production of site 'soil' and 'spoil' in 4 levels (dry weight)

The results of the dry matter production (Fig. 2) show much higher values on the site 'soil' than on the site 'spoil'. The higher above ground biomass production is always correlated with higher root production, which stabilises the soil. Furthermore it can be assumed that in a rain event, the interception is much higher on the site 'soil'. The amount of water and the velocity of the runoff as well as the energy input of the rain on the soil surface would be then reduced to a much greater extent on the test site 'soil'.

The bulk density (Fig. 3) showed higher values of the dry soil mass/volume on the site 'spoil' for the investigated depths. The bulk density of the top layer 'soil' is much higher as compared to the top layer 'spoil'.

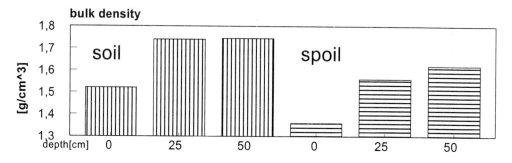

Fig. 3: Bulk density of site 'soil' and 'spoil' in 3 depths

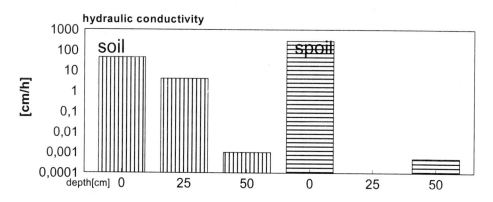

Fig. 4: Hydraulic conductivity of site 'soil' and 'spoil' in 3 depths

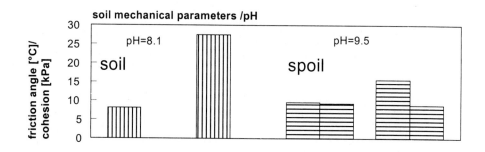

Fig. 5: Cohesion and angle of internal friction of site 'soil' and site 'spoil' (2 depths); pH-values of 'soil' and 'spoil'

The hydraulic conductivity (Fig. 4) decreases in both test sites with depth. The hydraulic conductivity is highest in the site 'spoil' in the top layer, but also lowest at 50 cm. Rain can easily infiltrate in the `spoil´ material, however the water flux is more reduced with depth and will more likely cause saturated conditions in the `spoil´ than in the `soil´ site. The lack of a correlation between bulk density and hydraulic conductivity is an important hint towards the importance of the degree of structure development in respect to water flow, whereas bulk density does not correspond with hydraulic permeability and is therefore not a definite measure.

The soil mechanical parameters cohesion and angle of internal friction showed similar values for the cohesion (Fig. 5), but a much higher value in the angle of internal friction in the site 'soil'. The site 'spoil' does not show great differences in the compared depths 0-25 cm and 25-50 cm. From the soil mechanical point of view the 'spoil' is less stabile than the 'soil' at the same unsaturated hydraulic conditions.

2.3 Discussion

The hydraulic permeability, which is a main factor for the distribution of runoff water into the soil, decreases with depth at both sites. The hydraulic conductivities are not much different at the same depths for each site. Considering the fact of increase in bulk density with depth and decreasing hydraulic conductivity, it could be concluded that for heavy rainfall events the infiltration is high in the top layers, but water flow is then reduced with depth, which causes subsoil interflow with destabilisation. However the bulk density is lower in the site 'spoil', i.e. a higher pore volume exists. Thus more water should be transported, which would then reduce the risk of a water saturated and less stabile soil. Combining these results with those of soil mechanical tests, it becomes obvious that the site 'spoil' is much more susceptible to hydrostatic stress because of the reduced shear stability. Mechanical stability is diminished under wet conditions and even more reduced with the existence of sodium (high pH-values, electrical conductivity) (Benito et al., 1993). The degree of dispersion increases, the interparticle bondings are reduced as well as the bulk density because of swelling effects. As a consequence, the hydraulic permeability becomes smaller due to reduced pore continuities, although the pore volume is higher. The instability of the 'spoil' leading to gully erosion is therefore due mainly to a low stability of the material combined with drastically reduced hydraulic permeabilities with depth. Cracking and the formation of gullies is often correlated (Trzcinka et al 1993). However, the existence of cracks in the site 'soil', which could not be found in the site 'spoil', did not increase erosion, i.e., structure formation of the less-dispersive soil material led to a stabilisation and increase of stability by aggregation (Baumgartl and Horn, 1991).

In general, the appearance of gully erosion is not bound to certain climatological conditions (Mäckel, 1976), but interacts with soil physical, soil mechanical and soil chemical properties. Furthermore, bulk density alone is not a factor describing the phenomena of gully erosion. Only in combination with soil physical, soil mechanical and soil chemical studies can the process be explained more completely.

3 Bottom sealing

A sealing consisting out of a mineral substrate has to fulfill certain criteria. Most important is a low hydraulic permeability, which is best achieved by homogeneous soil substrate. In practice clay rich material often forms stabile clods, which are supposed to be homogenized after overwheeling at an optimum water content with heavy rollers. Furthermore this sealing underlies the risk of drying and wetting which can change the pore size distribution and the pore structure.

3.1 Investigation

The risk of crack development of highly compacted clay under different loads (constant loading with 20, 50, 100 kPa) was determined by visualisation and measurement of hydraulic properties of a Kaolinite-clay (80% Kaolinite; Proctor density), which was wetted to saturation and dried by heating wires (temperature: 40 Celsius) in 3 cycles. The tests were carried out in cylinders (diameter: 10cm, height: 4cm), the samples were scanned in the Computer Tomograph at a height of 2cm after the 3rd cycle.

3.2 Results

Even intensiv compaction of preaggregated clay material is not able to destroy aggregates as results of the computer tomography scanning showed. In each layer the aggregated soil is still existent and the aggregates are less destroyed the greater is the distance from the compactive force. The hydraulic conductivity is low only because of thin highly compacted layers, where the aggregates have been destroyed. Once the homogeneity is disturbed by crack formation, the hydraulic conductivity will drastically increase.

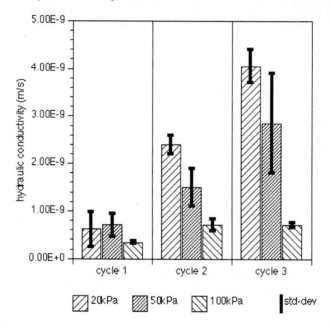

Fig. 6: Hydraulic conductivity under constant load after 3 wetting and drying cycles

The question whether crack formation is possible even under load by a change of the hydrological situation (water input/output, heat flow) was determined in a laboratory experiment. The measurement of the hydraulic conductivity (Fig. 6) showed a decrease in the permeability with increasing load. After each drying event, however, the hydraulic conductivity increased under load conditions and was therefore highest after the 3rd drying cycle. The computer tomograph scanning of the middle layer showed a load dependent crack development after the 3rd cycle. Cracks mainly developed under the load 20 kPa (Fig. 7).

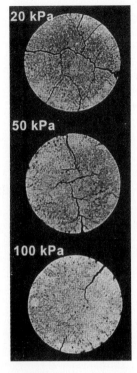

3.3 Discussion

Crack formation of a high compacted soil material is possible including loaded conditions. It seems that the increase in the hydraulic conductivity is, on the one hand, regulated by the number and length of the cracks. On the other hand, crack formation causes much higher pore continuities. In the saturated situation the cracks are closed due to swelling, but still present. Thus the hydraulic conductivity is higher in the sample with the most intensive crack development. The risk of an increase of the hydraulic permeability increases with drying of the sealing (Nagel et al., 1996). The process of aggregation due to matric forces, which are often much higher than any possible mechanical load, changes the particle orientation and pore size distribution. It results in higher continuities and a higher hydraulic conductivity. Furthermore the use of prestructured material, which is not completely homogenized during the construction of a sealing, can increase the risk of loosing the hydraulic properties when its possible negative properties are hidden by a very heterogenous construction of layers.

Fig. 7: Crack visualisation of differently loaded samples by computer tomography after 3 wetting and drying cycles

References

Baumgartl, T. & Horn, R. (1991): Effect of aggregate stability on soil compaction. Soil and Tillage Research. **19**, 203-213.

Benito, G., Gutierrez, M. & Sancho, C. (1993): The influence of physico-chemical properties on erosion processes in badland areas, Ebro basin, NE-Spain. Z. Geomorph. N. F. **37 (2)**: 199-214.

Mäckel, R. (1976): Ist die Röhrenbildung (Piping) klima- und substratbedingt? Z. Geomorph. N. F. **20 (4)**: 476-483.

Nagel, A., Baumgartl, T. & Horn, R. (1996): Änderung der Durchlässigkeit von mineralischen Basisabdichtungen als Folge von Entwässerungszyklen. Z. f. Kulturtechnik und Landentwicklung. **37 (1)**: 9-13.

Trzcinka, B.W., Loughran, R.J. & Campbell, B.L. (1993): The geomorphology of a field tunnel system: a case study from Australia. Z. Geomorph. N. F. **37 (2)**: 237-247.

Addresses of authors:
Th. Baumgartl
Inst. Plant Nutrition and Soil Science
University of Kiel
Olshausenstraße 40
D-24118 Kiel, Germany
B. Kirsch
Norwich Park Mine
Dysart 4745, Qld, Australia
M. Short
CMLR
The University of Queensland
St. Lucia, 4072 Qld., Australia

Soil Erosion on the Yamal Peninsula (Russian Arctic) Due to Gas Field Exploitation

A. Sidorchuk & V. Grigorév

Summary

Arctic cryogenic soils in the western part of the Yamal peninsula are poor and unstable because of the high ice content. The combination of high natural erosion potential and anthropogenic influence causes extremely intensive rates of erosion. The mean rate of gully length growth is 20-30 m per year on loam deposits and 150-200 m per year on sands. Intensity of sheet erosion on the bare slopes is up to 30-50 t ha^{-1} per year on sands and up to 7-8 t ha^{-1} per year on loam.

Complicated processes of thermoerosion and thermokarst, which take place under permafrost conditions, prevent the use of hydrotechnical methods for soil conservation. The main measures to prevent erosion are 1) to avoid unstable geomorphic units for construction; and 2) to improve the quality of vegetation cover in the area of gas exploitation activities.

Keywords: Gully erosion and thermoerosion, Yamal peninsula, gas fields exploitation, soil conservation under permafrost conditions

1 Introduction

Natural processes of erosion are widespread in the arctic tundra of Russia. The main natural factors of erosion in this region are 1) high density of rivers combined with the relief amplitude of up to 40-45 meters; 2) sufficient precipitation (350-400 mm year^{-1}); 3) low soil permeability due to the permafrost and therefore high runoff coefficients (up to 0.9-1.0); 4) high erodibility of frozen and thawing soils due to the combined mechanical and thermal action of flowing water (the so-called process of thermoerosion).

This natural high erosion hazard has been greatly increased recently by human impact. In areas of gas production and transportation facilities the erosion potential increases due to: 1) deterioration of the vegetation cover due to industrial development; 2) increased snow storage on the slopes due to excessive snow accumulation near buildings and roads; 3) an increase of the runoff coefficient on impermeable surfaces of the urbanized territories and roads; 4) local industrial and urban sources of warm water; 5) exploitation of sand-pits, gas and oil fields, and construction of pipe lines and ditches. The combination of high natural erosion potential and human interference causes extremely intensive gully, rill and sheet erosion. The old stable drainage lines are subjected to erosion, and new gullies cut into previously gently sloping elongate depressions. This gully growth leads to the formation of badlands and failures of engineering constructions. Rill erosion causes serious damage to the bare slopes of the road embankments and within the exploitation camps.

The main methods for soil and water conservation has been designed for the Temperate Zone, and there is no experience of their application in the conditions of continuous permafrost. It is necessary to devise special methods to reduce erosion and thermoerosion within an already disturbed field and to prevent erosion in this extremely unstable landscape.

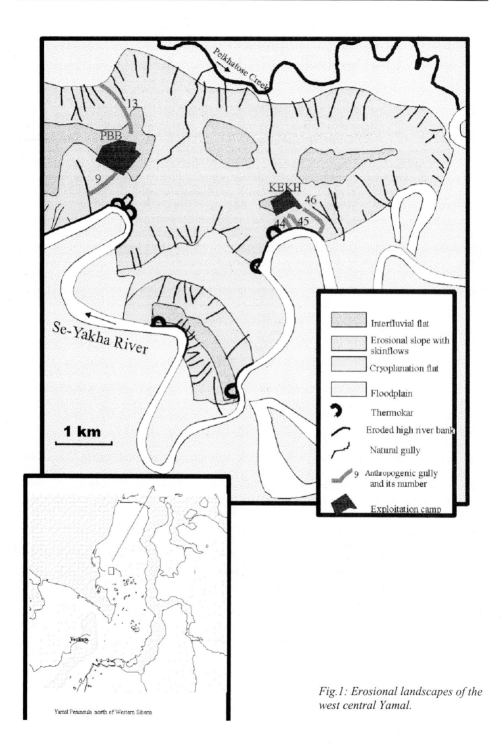

Fig.1: Erosional landscapes of the west central Yamal.

2 Climate of the west-central Yamal

The Yamal peninsula is situated at the northern part of the West Siberian plain and has an area of about 122,000 km^2. At the west-central part of the Yamal peninsula the mean annual air temperature is -8.3° C. In January (the coldest month of the year) the mean temperature is -21.8° C (minimum of -52° C). The mean temperature of August (the warmest month) is 6.7° C (maximum of 28° C). The air temperature is below 0° C for 223 days per year. Snow covers the territory from October till June. The thickness of snow cover at the beginning of the snow melt period does not exceed 0.3-0.4 m on gentle slopes and flat interfluves, but it can exceed 3-5 m in gullies and creek valleys and near steep river and lake banks due to wind transport. The depth of runoff for the period of snowmelt is 220-250 mm. Its main source being snow packs in erosion landforms.

Rainfall occurs generally in June – September; its total mean duration is 470 hours with a maximum up to 900 hours within this season. The mean rainfall for this period is 140 mm with a maximum of 357 mm and a minimum of 25 mm. The mean daily rainfall maximum is 12 mm (absolute maximum is 40 mm) with mean maximal 30 minutes' intensity of 0.8 mm min^{-1} and up to 12 mm min^{-1}.

Low winter temperatures cause the formation of deep permafrost layer with very low (almost zero) permeability. The summer thaw layer thickness reaches its maximum (0.6-1.2 m) in August-September, but the soil is often highly saturated with water because of melting of ground ice. Under such conditions the runoff coefficient is up to 0.9-1.0 for the period of snow melting in June. It decreases to 0.3-0.4 in August-September. Accordingly, even low precipitation produces relatively high surface water flow.

3 Natural destructive processes

The main natural destructive processes are 1) river channels and cryogenic lake migrations; 2) massive ground ice melting, cryoplanation; 3) thaw slumps and active layer detachment failure (skinflow); 4) gully erosion (Fig.1).

The main river channels on the Yamal peninsula have meandering patterns. The rate of bank erosion is usually 0.3-0.5 m per year and up to 2-3 m per year. River erosion causes formation of steep bare slopes at the concave sides of meanders, accompanied by various destructive processes.

Cryoplanation is the process of massive ground ice melting, mainly in the outcrops in the steep banks of the rivers. It leads to formation of thermokars up to 250 m wide and 300-500 m long. Cryoplanation is generally accompanied by gully erosion. These processes together can form extensive low terraces because the sheets of massive ground ice are about 50 m thick and have areas of several square kilometers.

Skinflows often occur during a wet summer, if it follows 2 or 3 warm summers. Numerous laminations of ice appear near the lower boundary of the thaw layer and the bulk cohesion of the loamy slope deposits decreases to 50-100 Pa. In this situation a heavy block of water-saturated thaw deposits can rapidly slide over the ice-rich surface to a distance up to several hundred meters even on slopes with an inclination of 1-2° or less. Newly exposed soil is subjected to intensive erosion, and a new gully can be formed the following year after the skinflow. The most favorable conditions for such an event took place in 1989-90. Skinflows occurred on 2.0% of the area on the slopes of loamy high terrace, and the vegetation cover was destroyed. The mean area of individual slumps was 7300 m^2, and the mean depth was 1.2 m. The numerous remnants of the old skinflows cover all the surface of these slopes.

The natural gully erosion occurs mainly on terraces with height 20-45 m above sea level. The high (30-45 m) terrace consists of loam and clays with a massive cryogenic structure. Steep riverside slopes of this terrace are dissected by numerous natural bank gullies, usually 50-70 m long. Some of these gullies are up to 1-2 km long in the areas where the ice content in eroded deposits is high. There is a net of long gentle troughs on the flat surface of the terrace. Their density is 2.3 km km^{-2}. Natural and man-induced gullies usually follow these troughs where the vegetation cover is deteriorated or the volume of runoff increases. The lower terrace (20-25 m above sea level) is composed of fine silty sand with ice wedges in the upper layer and massive ground ice in the lower layer. The natural gullies are more widespread on this surface, and the main part of the drainage network is eroded due to the high erodibility of fine sands.

4 Erosion in the west-central Yamal

Investigations conducted in 1990-1995 concentrated on the territories of two exploitation camps (PBB and KEKH), where the vegetation cover was significantly destroyed by heavy tractors, and the thaw water supply was increased due to snow accumulation near buildings and road embankments.

A dendritic system of rills develops rapidly on the bare slopes with irregular microrelief. The rill depth varies from 0.1 m to 0.5-0.6 m, and the width varies from 0.3 to 1.5 m. Their density is about 0.1 m m^{-2} at the upper part of the slopes and 0.05 m m^{-2} in the middle part of the slopes. At the lower part of the slopes the rills become gullies. On a slope with inclination 3.5° and area 1 ha, situated in the basin of gully 44, the depth of runoff was 171 mm and the erosion rate was 40 t ha^{-1} during the snow melt period of 6-27 June 1991. At the basin of gully 45, on the adjacent slope with an inclination of 2.5° and an area of 2.8 ha, the depth of runoff was 132 mm and the erosion rate was 6 t ha^{-1}. During the rainfall event of 8-9 August 1990, when the maximum rainfall intensity for 30 minutes reached 0.44 mm min^{-1} (20 year frequency), the depth of runoff was 11 mm for both slopes. The rate of erosion for this event at the first slope was 4.3 t ha^{-1}, and for the second slope it was 5.6 t ha^{-1}. Mainly different levels of human activity explain this variance on these slopes at different times.

The process of the water and sediment transport in the gullies is even more complicated. The water and sediments from the basin collect at the gully head, but at the beginning of the thaw period the gully is full of snow. The rate of snowmelt in gullies depends strongly on snow porosity and the quantity and temperature of melt water supplied from the upper part of the watershed. For example, at the first week of snow melt (6-10 June 1991) in gully 9, where the area of slopes above the gully head was 7-10 ha, and in gully 46, where this area was 10 ha, the inside snow packs (2-3 m thick) were completely cut by meandering channels with vertical snow banks. The flows were continuous with only several snow bridges remaining in the lower parts of the gullies. The sediments from the catchments and gully beds were delivered to the gully mouths. Gully 45, where the area of the slopes above the gully head was 3 ha, and gully 44, where this area was 1 ha, were completely filled by snow. Up to 83% of sediments, supplied to the gully head, was withheld by snow within the gully at the beginning of the snowmelt period. Only at the end of the thaw period did active erosion begin of the deposited sediments and gully bottom. The average rate of erosion and thermoerosion in the gullies during the snowmelt period of 1991 was 0.8-1.2 m for the month, and the depth of runoff was 150 mm.

Summer and autumn rains have duration of 74-171 hours per month on the west central Yamal. Only 2-4 events per month are characterized by rainfall depths more than 1 mm day^{-1}. Some of these rains cause intensive gully erosion. During 8-9 August 1990 the average rate of erosion and thermoerosion in the gullies at the KEKH camp was 0.5-0.6 m day^{-1}. The rate of erosion in the middle section of the gully 9 during the rainfall period of August 1993 was twice that of the period of snow melt in June, despite the fact that the runoff in June was six times higher.

Narrow rectangular trenches with depths of 0.6-1.4 m (and up to 2.5 m) and width 0.4-0.6 m are usually formed by erosion and thermoerosion at the gully heads or bottoms during the snowmelt. Such trenches with vertical walls are stable only in frozen deposits. When the thaw layer reaches a depth more than 0.5-0.8 m, the sides of the gully become unstable. Shallow landslides quickly transform the gully cross-section shape to triangular or trapezoidal, and this process continues from days to weeks.

The alternating processes of quick intensive incision and rapid sidewall slumping results in gully deepening and growth in length and volume. Repeated instrumental leveling of gully longitudinal profiles showed relatively low mean rates of deepening. For the period 1991-1995 the increase of the mean depth of gullies 9 and 45 was 0.6 m, and that of gully 44 was 0.9 m. At the upper section of gully 46, the depth increase was 1.3 m, but in its lower section deposition occurred with a thickness of sediments 0.7 m. The mean rate of gully deepening is about 10 times less than the rate of incision at the periods of snow melt and rainfall runoff. Analysis of the airphotos reveals that gully length growth is also a complicated process. The main head of gully 44 was stable since 1970, but a long and deep rill was formed at the middle part of the catchment, and the second active head was developed at the convex upper part of the slope. The length of gully 45 was 165 m in 1988, 190 m in 1989, 210 m in 1990, 230 m in 1991, and 280 m in 1995. The rate of the main gully head growth decreased in time, but as in the case of gully 44, the second head was formed at a distance of 400 m from the gully mouth in 1991-95. The same trend is obvious for gully 46. The rate of gully length growth was 40 m per year between 1988-1991, and 10 m per year between 1991-1995. In 1989-1991 a gas flame was operated at the bank of the gully; that was the period of its intensive deepening. Gully 9 did not exist in 1986; only 240 m long shallow elongated depression existed. After PBB camp construction in 1986-87, erosion and thermoerosion were initiated due to the increase in water supply. In 1988 the gully length was 450 m, in 1989 it was 740 m, and in 1990 the length was 940 m. The gully head reached the buildings of PBB camp, and here its growth was stopped by continuous filling of the head by heavy loam from the banks by bulldozers. Nevertheless in 1995 the gully was 25 m longer, than in 1991.

5 Soil conservation under conditions of continuous permafrost.

Several methods to stop gully growth were used on the territory of Bovanenkovskoye gas field of the west central Yamal peninsula. All these methods were designed for the temperate zone, and all of them proved to be useless under conditions of permafrost. The check dam was constructed at the head of gully 9, but a new gully head had passed around the check dam in 1995. The erosion cut was filled with sediments from gully sides by bulldozer, but every year it was renewed by gully erosion. Several wall cuts in gully 9 were covered by technical textiles. Cuts with small subcatchments were stabilized, but in most of them the cover was destroyed by erosion that took place around the covers.

These cases highlight, that human developmental activities in the arctic tundra, accompanied by deterioration of the vegetation and an increase of runoff causes intensive erosion. This is due to low permafrost permeability, high runoff, high erodibility of bare soils with high ice content, and low slope stability. For existing gullied basins, it is very difficult to stop erosion and thermoerosion. To minimize it several methods can be tried: mechanical snow removal from gully catchment areas; vertical drainage of industrial and rainfall waters; covering of disturbed slopes with a peat layer; filling of the gullies with heavy loam and a peat cover. Two main general measures are recommended: 1) to remove all buildings and construction from unstable interfluvial surfaces and erosion slopes with skinflows to more stable floodplains; 2) to recultivate vegetation cover near roads and pipe lines, especially where they cross unstable areas.

The first measure has been applied on the Bovanenkovskoye gas field. The location of engineering facilities on the floodplain causes problems with riverbank erosion and flooding (Sidorchuk and Matveev, 1994). The second approach has been intensively studied for arctic soil conditions, and hydraulic thresholds of erosion stability of tundra soil with different vegetation cover qualities have to be taken into account.

The biomass in typical tundra of the Yamal peninsula varies from 500 to 2000 t m^{-2}, and 10-25% of this amount is represented by plant roots in the soil (Vasilevskaya et al., 1986). Thin (less that 1 mm in diameter) living and dead roots penetrate into the soil aggregates, gather together and increase soil cohesion. Field and laboratory experiments show that the bulk soil cohesion C_h (in Pa $10^{5)}$ increase rapidly with the content of thin roots R (kg m^{-3}):

$$C_h = C_0 \exp(0.05R) \qquad (1)$$

Here C_0 is cohesion of the same soil, but without vegetation roots.

The critical shear stress and velocity of erosion initiation in cohesive soil are mainly controlled by the forces of friction and by cohesion. Critical velocity U_{cr} (m s^{-1}) can be calculated with the formula (Grigor'ev et al.,1992):

$$U_{cr} = 2.25\sqrt{d + 0.18 C_h^{2.25}} \qquad (2)$$

Here d - mean diameter of soil aggregates (m). Empirical coefficients include the influence of flow turbulence, soil porosity, and the variance of measured values.

When the flow velocity U is less than the critical velocity of erosion initiation, U_{cr}, both erosion and thermoerosion rates are lower than the rate of soil formation. The critical velocity for a given soil type can be increased by improving the vegetation cover quality (that is, increase of soil top layer cohesion). Formulas (1) and (2) were used with the models of sheet erosion (Grigor'ev and Sidorchuk, 1996) and gully erosion (Sidorchuk, 1996) to determine the threshold qualities of vegetation cover for recultivating the area, which will be crossed by pipe lines on the Bovanenkovskoye gas field. This critical quality varies from one basin to another due to different catchment morphology and is changed if water supply from the catchment is increased by human activity. For example (Fig.2), the basin of gully 46 in natural conditions will be stable in the case U_{cr} = 0.48 m s^{-1}, and the vegetation cover quality R not less than 20 for clays, 31 for loam and 41 kg m^{-3}

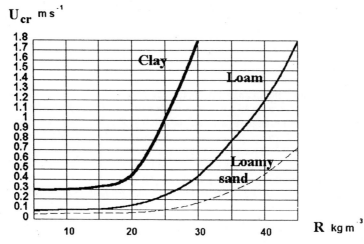

Fig.2: *Diagram for evaluation of the vegetation cover quality R for a given critical velocity U_{cr} and soil type.*

for loamy sands. Under conditions of snow storage increase at the KEKH camp, this basin will be stable in the case of $U_{cr}=0.65$ m s^{-1}, and vegetation cover quality not less than 22 for clays, 33 for loam and 43 kg m^{-3} for loamy sands.

6 Conclusion

1) The heavy and local constructions are not suited for soil conservation under permafrost conditions with complicated thermoerosion and thermokarst processes. Once initiated, thermo-gully growth can not be stopped by methods normally used in a temperate zone.
2) The methods to reduce erosion on already disturbed fields are: mechanical snow removal from gully catchment areas; vertical drainage of industrial and rainfall waters; covering of disturbed slopes with a peat layer; filling of the gullies with heavy loam and a peat cover.
3) The main steps to prevent erosion are:
   - avoiding unstable geomorphic units for construction;
   - improvements of vegetation cover quality at gas exploitation sites.
4) The method was developed to evaluate the quality of vegetation cover required to protect soils of different types from erosion and thermoerosion.

References

Grigor'ev, V., Krasnov, S., Kuznetsov, M., Larionov, G. and Litvin L. (1992): Methods of Prediction and Prevention of Irrigation Erosion, Moscow, Moscow Univ. Publ. House. (in Russian)

Grigor'ev, V.Ya. and Sidorchuk, A.Yu. (1996): Forecast of Rain Erosion of Tundra soils on the Yamal Peninsula, Eurasian Soil Sci. **28**, p.351-357.

Sidorchuk, A. (1996): Gully Thermoerosion on the Yamal Peninsula. In: O.Slaymaker (ed.), Geomorphic Hazards, Chichester, Wiley & Sons.

Sidorchuk, A. and Matveev, B. (1994): Channel Processes and Erosion Rates in the Rivers of the Yamal Peninsula in Western Siberia, IAHS Publ. **224**, p.197-202.

Vasilevskaya, V.D., Ivanov, V.V. and Bogatyryev, L.G. (1986): The Soils of the Northern Western Siberia, Moscow, Moscow Univ. Publ. House. (in Russian)

Address of authors:
Aleksey Sidorchuk
Victor Grigor´ev
Geographical Faculty
Moscow State University
119899 Moscow, Russia

Degradation of Silicate Minerals by *Bacillus mucilaginosus* Using *Bacillus intermedius* RNase

F.G. Kupriyanova-Ashina, G.A. Krinari, A.L. Kolpakov & I.B. Leschinskaya

Summary

The effect of Bacillus intermedius RNase on spores of *Bacilus mucilaginosus* was investigated. It was shown that the enzyme stimulated germination of spores at a concentration of 0.001 mkg/ml. The spores at stages of activation and initiation appear to be most susceptible to the enzyme action. The effect of RNase on *B. mucilaginosus* multiplication correlates with intensification of bacterial leaching of bauxites. The fibrils of outer surfaces and interfibrillar substance were decomposed by the bacteria intensified by RNase.

Keywords: Ribonuclease, stimulation, leaching

1 Introduction

Alumosilicates are the foundation of the mineral part of soil. Studies of their degradation are important for understanding soil formation. The bacterial destruction of silicates proceeds rather slowly. Consequently the intensification of this process by acceleration of bacteria reproduction is an urgent problem. Alumosilicates are almost inert to various inorganic solvents but it is known that they are degraded by *Bacillus mucilaginosus* (Malinovskaya et al., 1990; Mishustin et al., 1981).

It is well known that exogenous RNases stimulate reproduction of microorganisms (Egorov et al., 1994; Kupriyanova-Ashina et al., 1996) and, as a consequence, the intensive synthesis of exopolysaccharides and acid metabolites can take place. It is be assumed that a stimulation of *Bacillus mucilaginosus* growth by RNase and increase in silicate minerals destruction are the associated processes. That is why we selected the system - exogenous RNase - *Bacillus mucilaginosus* bacteria - silicates for study in order to evaluate the destruction of silicate in the presence of RNase microdoses.

2 Material and methods

The following silicates were studied: kaolinite-$Al_4[Si_4O_{10}](OH)_8$; microcline (potassium alumosilicate)- $KAlSi_3O_8$. Bauxite comprising: gibbsite, kaolinite, quarz, microcline, FeO(OH). The minerals were added to a cultivation medium A27 (1 g/l). The cells were cultivated in a liquid and agar nutrient medium A27 at 30° C (Andreev and Kirikilitsa, 1996). To stimulate bacterial growth, extracellular alkaline RNase of *Bacillus intermedius* with a specific activity $1·10^6$ U/mg of protein was used.

The concentration of the bacteria or spore material was made 20 · 10^6 spores/ml of medium.

The physiological stage of spores germination was determined as sensitivity to Ultra-Violet irradiation (UV) (Agre and Kalakutski, 1977).

The quantity of SiO_2 was determined spectrophotometrically (555 nm) through the appearance of a blue Si-Mo complex (Kupriyanova-Ashina et al., 1995). The quantity of Al_2O_3 was determined according to Tihonov (1965).

The identification of inherent mineral crystalline phases was carried out by X-ray scanning the reciprocal space of axial textures, which provided separate registration of reflexes, belonging to different crystallographic zones (Krinary and Halitov, 1991).

3 Results and discussion

Introduction of the RNase in concentrations of 0.01 and 0.001 μg/ml into the medium with microcline accelerated germination of the *B. mucilaginosus* spores by 40% - 50% in comparison with the control (Fig.1). The effect of the RNase depended on the quantity and the incubation time and can be attributed to the physiological state of organisms.

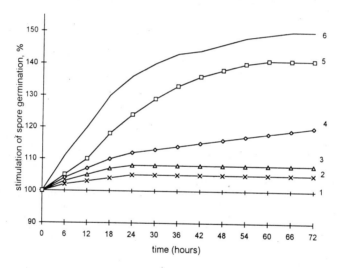

Fig. 1: *Influence of RNase on germination of B. mucilaginosus spores dependent on enzyme concentration and incubation time. Line 1 - the medium without enzyme; the medium containing 10 (line 2), 1 (line 3), 0.1 (line 4), 0.01 (line 5), 0.001 (line 6) μg/ml of enzyme.*

It was shown that *B. mucilaginosus* spores were in the activation stage during 6 h of incubation in the medium with microcline. After 8 hr of incubation the stage of germination was started upon the increse of sensitivity of the spores to UV. After 24 hr of incubation the stage of spores growing was commenced.

The best target for the maximum stimulating effect of RNase (0.001 μg/ml) were the spores on the activation (4 hr) and initiation (8 hr) stages of their germination (Table 1). The stimulating effect of the RNase was evolved during 6 hr and reached its maximum after 48 hr incubation with enzyme (Fig.1).

Stage of germination, hr	Number of cells 10^6 Experiment	Number of cells 10^6 control	Effect of stimulation, %
Activation (4)	96.0 ±7.1	40.2±3.4	140
Initiation (8)	100.3±8.3	43.1±3.4	132
Growing (24)	140.2±9.8	125.4±9.9	111

Table 1: RNase influence on spore germination for different physiological states

Stage of spore germination in the medium, hr	Al_2O_3 concentration (mg/ml) in the medium after 96 h incubation control	experiment	%	SiO_2 concentration (mg/ml) in the medium after 96 h incubation control	experiment	%
Activation (4)	3.03±0.05	7.01±0.09	131	13.02±0.01	48.04±0.09	268
Initiation (8)	3.05±0.08	6.03±0.07	98	18.07±0.08	49.01±0.05	171
Growing (24)	1.51±0.05	2.02±0.02	34	24.12±0.08	25.08±0.04	4

Table 2: Al_2O_3 and SiO_2 leaching from potassium alumosilicates (microcline) under the action of RNase (0.001 µg/ml) on spores at different stages of generation

During incubation of the B. mucilaginosus spores in the presence of the stimulating dose of RNase (0.001 µg/ml), the maximum intensity of the microcline leaching coincided with the stages of activation and initiation of spores germination that were the most sensitive to the exogenous RNase. The issue of the element of aluminium and of silica into the solution grew up to 1.36-1.19 times - 2.5-2 times, respectively (Table 2). Thus, the leaching of silica was more intensive than that of aluminium.

The intensity of the bacterial decomposition of silicate minerals in the presence of RNase was shown for kaolinite too. At inoculation the B. mucilaginosus spores at the activation stage to the kaolinite containing medium RNase at a dose of 0.001 µg/ml, stimulated the issue of alumina and silica into the medium up to 42% and 18%, respectivity after 5 days. The predominance of silica over alumina was accounted for by the localization of these elements in the kaolinite lattice structure (Table 3).

Experiment Conditions	Al_2O_3 content, mg	Al_2O_3 extraction, %	SiO_2 content, mg	SiO_2 extraction, %
Original kaolinite	11.24	100	14.35	100
After bacterial leaching during 5 days				
Control		5.87		16.6
Experiment		8.36		19.6
Stimulation, %		142.4		118.1
After bacterial leaching during 10 days				
Control		4.98		10.5
Experiment		6.22		24.1
Stimulation, %		124.9		229.5

Table 3: RNase`s effect on the bacterial decomposition of kaolinite

The treatment of the bauxite by *B. mucilagenosus* spores led to an increase of the alumina and silica quantity in the liquid phase culture up to 52.4% and 6.4% in comparison with the control after 10 days of incubation (Table 4). The predominance of alumina is likely to be connected with the rapid decomposition of gibbsite and X-ray amorphous phases of free alumina to be found in ore.

Experiment Conditions	Al_2O_3 content, mg	extraction, %	SiO_2 content, mg	extraction, %
Original bauxite	13.47	100	4.86	100
After bacterial leaching during 5 days				
Control		8.4		23.6
Experiment		13.1		27.7
Stimulation, %		155.9		117.4
After bacterial leaching during 10 days				
Control		8.4		46.3
Experiment		12.8		49.3
Stimulation, %		152.4		106.5

Table 4: RNase's effect on the bacterial decomposition of bauxite

The mechanism of decomposition of silicates and alumosilicates by soil bacteria under natural conditions is still unclear. According to some scientists, decomposition of silicates is caused by such bacteria low-molecular metabolites as amino- and organic acids (Egorov et al., 1994). Other studies suggested that the leading part in silicates' destruction belongs to mucoids of capsule bacteria, i.e., polysaccharides, containing uronic acids (Mishustin et al.,1981; Avakyan, 1985; Malinovskaya et al., 1990). We assume the contradiction can be overcome by including mineralogical data along with biochemical data.

It is known that chemical composition and crystal structure are not the only characteristics that determine the speed of bacterial decomposition as well. Thus the reactive ability of various kaolinites in such a process ranges widely and depends greatly on the geometry of interlayer and intercrystallic space (Yakhontova et al., 1991). The explanation can be found in the presence of a certain critical size of a reaction zone or molecules of biochemical reagents, beyond the limits of which the access to a soluble surface of a mineral becomes impossible. This hypothesis was confirmed by a direct experiment on bacterial decomposition of a natural aggregate of chrysotile-asbestos, crystalline fibres of which, i.e., fibrils, have a form of tubes.

Figure 2 represents diffractive spectra of the same sample before (curve 1) and after bacterial treatment (curve 2). The two spectra were standardized according to the amplitude to the reflection of 200 clinochrysotile (11). Besides the reflection of clinochrysotile 20-1 and 201 (maxima are 12,13) which were weak because they belonged to a different crystallographic zone, the reflex of 020 parachrysotile is present. Reflections of various semi-crystalline phases of silicon earth (silica) were also recorded (Brown, 1980). The first of them corresponds to the reflection of the 111th phase of cubic cristobalite -(0), reflections 3-6 to the phase of tetragonal cristobalite, reflections 7-9 to the phase of low tridymite.

As shown in Figure 2, the intensity of the reflection of the 111th cubic cristobalite remained practically constant after the bacterial treatment of the sample by the *B. mucilaginosus* spores in the presence of RNase (0.01μg/ml). The concentration of parachrysotile was slight, the reflections of all the other phases of free silicon earth practically disappeared. It should be noted that the cubic cristobalite did not present other X-ray reflections while scanning any other crystallographic zones, except 111. In other words, we are dealing with a one - dimensionally arranged cristobalite-like fragments, stretched along the fibril axis. Consequently this phase together with H_2O molecules occupies intrafibrillar space. As for other crystalline and semi-crystalline phases of silicon earth (silica), they are not textured, i.e they are localized in interfibrillar space.

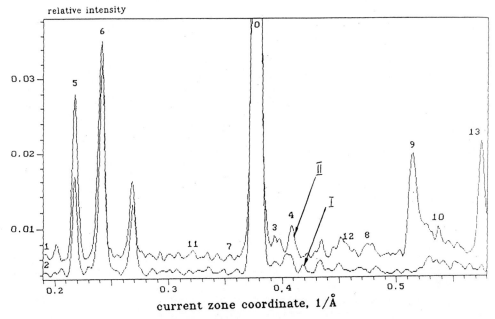

Fig.2 Diffractive spectrum of scanning the HOO zone of natural chrysotile-asbestos aggregate. I - Initial (original) sample; II - After bacterial treatment.

Thus the fibrils outer surfaces and interfibrillar substance were decomposed by the bacteria intensified by RNase. According to our data, interfiblrillar space has a diameter of 30-40 A (Krinary and Halitov, 1995), which is sufficient for the penetration of low-molecular bacteria metabolites, in contrast to macromolecules of protein and polysaccharide. The preservation of free silicon (silica) in intercapillary space of parachrysolite indicated the direct contact of a polysaccharide capsule of silicate bacteria with the mineral's surface. It provided lattice's decomposition. During the growth of a microbial population under natural conditions, RNases were secreted from lysated cells. These microdoses of RNases are able to increase the number of microorganisms and as a consequence the quantity of low-molecular metabolites and exopoly-saccharides in the environment. As a result of the reaction between the above - mentioned compounds and silicates, the degradation of the lattice and dissolving of minerals could take place.

References

Agre, N.S. and Kalakutski, L.V. (1977): Development of actinomycetes, Nauka, Moskva.
Avakyan, Z.A. (1985): Silicon compounds in a solution by quartz bacterial degradation, Microbiologiya **54**, 2, 301-306.
Andreev, P.I. and Kirikilitsa, S.I. (1996): Microbial concentration of bauxite, Naukova Dumka, Kiev.
Brown, G. (1980):Silica minerals, Chapter 6. Associated Minerals, London: Mineralogical Society, 378-380.

Egorov, S.Yu., Zakharova, N.G., Naumova, E.S. and Kupriyanova-Ashina, F.G. (1994): The influence of RNase on survival rate and symbiotic properties of *Rhizobium meliloti* in soil, Microbiology **63,3**, 484-488.

Krinary, G.A and Halitov, Z.Y. (1991): Method of scanning of reciprocal space of axial textured and its applications to structural investigation, Materials Science Forum, Switzerland, 79-82 , 191-196.

Krinary, G.A and Halitov, Z.Y. (1995): Physico-technical problem of mineral development, Rus. Sci. Acad. Siberian Branch **4**, 89-95.

Kupriyanova-Ashina, F.G, Lutskaya, A.Yu., Kolpakov, A.I, Kepecheva, R.M. and Leschinskaya, I.B. (1996): Effect of RNase from *Bacillus intermedius* on growth of *Saccharomyces cerevisiae*, Prikl. biochim.i microbiol. **32, 2**, 260-264.

Malinovskaya, I.M., Kosenko, L.V., Vocelko, S.K. and Podgorsky, V.S. (1990): Role of *Bacillus muciloginosus* polysaccharide in degradation of silicate minerals, Mikrobiologiya **59, 1**, 70-78.

Mishustin, E.N., Smirnova, G.N. and Lohmacheva, R.A. (1981): Destruction of silicates by the microorganism and using siliceon bacterias as bacterial fertilizable preparations, Izv.AN SSSR ser.biol. **5**, 698-708.

Tihonov, V.N. (1965): Photometric determination of aluminium with xylenol orange, Jurnal analiticheskoi himii **20, 9**, 941.

Yakhontova, L.K., Grudyev, A.P., Krinary, G.A., Morozov, V.P. and Sidyakyina, G.G. (1991): X-ray and intercalation characteristics of kaolinite as a criterion of its stability, USSR Acad.of Sci. Report **330, 6**, 13.

Address of authors:
F.G. Kupriyanova-Ashina
G.A. Krinari
A.I. Kolpakov
I.B. Leschinskaya
Microbiology Department
Kazan State University
420008 Kazan, Russia

GeoEcology paperback

L.M. Reid & T. Dunne

Rapid Evaluation of Sediment Budgets

164 pages / numerous figures, tables
ISBN 3-923381-39-5
list price: DM 69,-/US $ 49,-

Abstract

Many land-management decisions would be aided by an understanding of the current sediment production and transport regime in a watershed, and of the likely effects of planned land use on that regime. Sediment budgeting can provide this information quickly and at low cost if reconnaissance techniques are used to evaluate the budget. Efficient budget construction incorporates seven steps: **careful definition of the problem to be addressed; collection of background information; subdivision of the project area into uniform sub-areas; interpretation of aerial photographs; fieldwork; analysis; and checking of results.**
Methods used in fieldwork and analysis must be selected according to the types of hillslope and channel processes active and the level of precision required. Methods for evaluating erosion and sediment transport rates are described, and four examples are given to demonstrate budget applications and construction.
Retrieval terms: watershed, cumulative impact, erosion, sediment transport, sediment budget, land-use planning, watershed-analysis.

CATENA VERLAG GMBH   *GeoScience Publisher*

Advances in GeoEcology 27

Rorke B. Bryan (Editor)

Soil Erosion, Land Degradation and Social Transition:
Geoecological Analysis of a Semi-arid Tropical Region, Kenya

hardcover / 256 pages / 12 different articles
numerous figures, photos and tables
ISBN 3-923381-36-0
list price: DM 189,-/US $ 126,-
standing order price Advances in GeoEcology: DM 132,30/US $ 88.20
(follow-up series of CATENA SUPPLEMENTS)

Contents

Preface
R.B. Bryan
Land degradation and the development of land use policies in an transitional semi-arid region
S. Schnabel
Using botanical evidence for the determination of erosion rates in semi-arid tropical areas
J.L. Kiyiapi
Structure and characteristics of *Acacia tortilis* woodland on the Njemps Flats
R.B. Bryan
Microcatchment hydrological response and sediment transport under simulated rainfall on semi-arid hillslopes
D.J. Snelder
Productivity of eroded rangelands on the Njemps Flats
D.J. Oostwoud Wijdenes & J. Gerits
Runoff and sediment transport on intensively gullied, low-angled slopes in Baringo District

J.J.P. Gerits
The potential for irrigated agriculture on the Njemps Flats
M.J. Kamar
Natural use of stone and organic mulches for water conservation and enhancement of crop yield in semi-arid areas
A.C. Yobterik & V.R. Timmer
Nitrogen mineralization of agroforestry tree mulches under saline soil conditions
E.K. Kireger & T.J. Blake
Genetic variation in dry matter production, water use efficiency and survival under drought in four *Acacia* species studied in Baringo, Kenya
D.J. Oostwoud Wijdenes & R.B. Bryan
Gully headcuts as sediment sources on the Njemps Flats and initial low-cost gully control measures
R.B. Bryan & S. Schnabel
Estimation of sedimentation rates in the Chemeron reservoir

CATENA VERLAG   GeoScience Publisher

GeoEcology textbook

M. Kutilek & D. Nielsen

Soil Hydrology

Textbook for students of soil science, agriculture, forestry, geoecology, hydrology, geomorphology and other related disciplines

370 pages / numerous figures, tables
ISBN 3-923381-26-3
list price: DM 59,-/US $ 39,-

We have intended to present an introduction to the physical interpretation of phenomena which govern hydrological events related to soil or the upper most mantle of the earth's crust. The text is based upon our teaching and research experience. The book can serve either as the first reading for future specialists in soil physics or soil hydrology. Or, it can be a source of basic information on soil hydrology for specialists in other branches, e.g. in agronomy, ecology, environmental and water management. (from Authors)

Contents (shortened)

1	Soils in Hydrology	6	Elementary Soil Hydrologic Processes
1.1	Soils	6.1	Principles of Solutions
1.2	Concepts of Soil Hydrology	6.2	Infiltration
2	Soil Porous System	6.3	Soil Water Redistribution and Drainage after Infiltration
2.1	Soil Porosity	6.4	Evaporation from a Bare Soil
2.2	Classification of Pores	6.5	Evapotranspiration
2.3	Methods of Porosity Measurement	7	Estimating Soil Hydraulic Functions
2.4	Soil Porous Systems	7.1	Laboratory Methods
2.5	Soil Specific Surface	7.2	Field Methods
3	Soil Water	8	Field Soil Heterogeneity
3.1	Soil Water Content	8.1	Variability of Soil Physical Properties
3.2	Measurement of Soil Water Content	8.2	Concept of Soil Heterogeneity
4	Soil Water Hydrostatics	8.3	Spatial Variability and Geostatistics
4.1	Interface Phenomena	8.4	Scaling
4.2	Soil Water Potential	8.5	State-space Equations for Multiple Locations
4.3	Soil Water Retention Curve	9	Transport of Solutes in Soils
5	Hydrodynamics of Soil Water	9.1	Solute Interactions
5.1	Basic Concepts	9.2	Miscible Displacement in a Capillary
5.2	Saturated Flow	9.3	Miscible Displacement in Surrogate Porous Media
5.3	Unsaturated Flow in Rigid Soils	9.4	One-dimensional Laboratory Observations
5.4	Two Phase Flow	9.5	Theoretical Descriptions
5.5	Flow in Non-rigid (Swelling) Soils	9.6	Implications for Water and Solute Management
5.6	Non-Isothermal Flow		

CATENA VERLAG GMBH   *GeoScience Publisher*

Please, send your orders to:
CATENA VERLAG GMBH, Ärmelgasse 11, D-35447 Reiskirchen, Germany, phone (49)6408-64978, fax (49)6408-64978
USA/Canada: CATENA VERLAG GMBH, P.O. Box 1897, Lawrence KS 66044-8897, USA, phone (913)843-1221, fax (913)843-1274

Advances in GeoEcology Formerly **CATENA SUPPLEMENTS**

Dan H. Yaalon & S. Berkowicz (Editors)

History of Soil Science
- International Perspectives -

Advances in GeoEcology 29
(follow-up series of CATENA SUPPLEMENTS)
438 pp   1997   DM 264,00 / US$ 176.-
ISBM 3-923381-40-9

The book presents a wideranging perspective on the history of soil science comprising a collection of 22 papers. Following an overview on the main paradigms, developments of the concepts of humus, horizons, classification of soil types and soil series usage are treated in specific chapters. Some selected topics in the history of soil chemistry and soil physics are treated in detail. A number of articles deal with regional aspects of soil science and the contribution of some outstanding personalities from the 18th to the 20th centuries. This is the first original history in soil science in the English language.

H.-P. Blume & S.M. Berkowicz (Editors)

Arid Ecosystems

Advances in GeoEcology 28
(follow-up series of CATENA SUPPLEMENTS)
229 pp   1995   DM 189,00 / US$ 126.-
ISBN 3-923381-37-9

K. Auerswald, H. Stanjek & J.M. Bigham (Editors)

Soils and Environment
Soil Processes from Mineral to Landscape Scale

Advances in GeoEcology 30
(follow-up series of CATENA SUPPLEMENTS)
422 pp,   1997   DM 189,00 / US$ 126.-
ISBN 3-923381-41-7

CATENA VERLAG GMBH   *GeoScience Publisher*